Proceedings of the Sixth International Conference on Cognitive Modeling

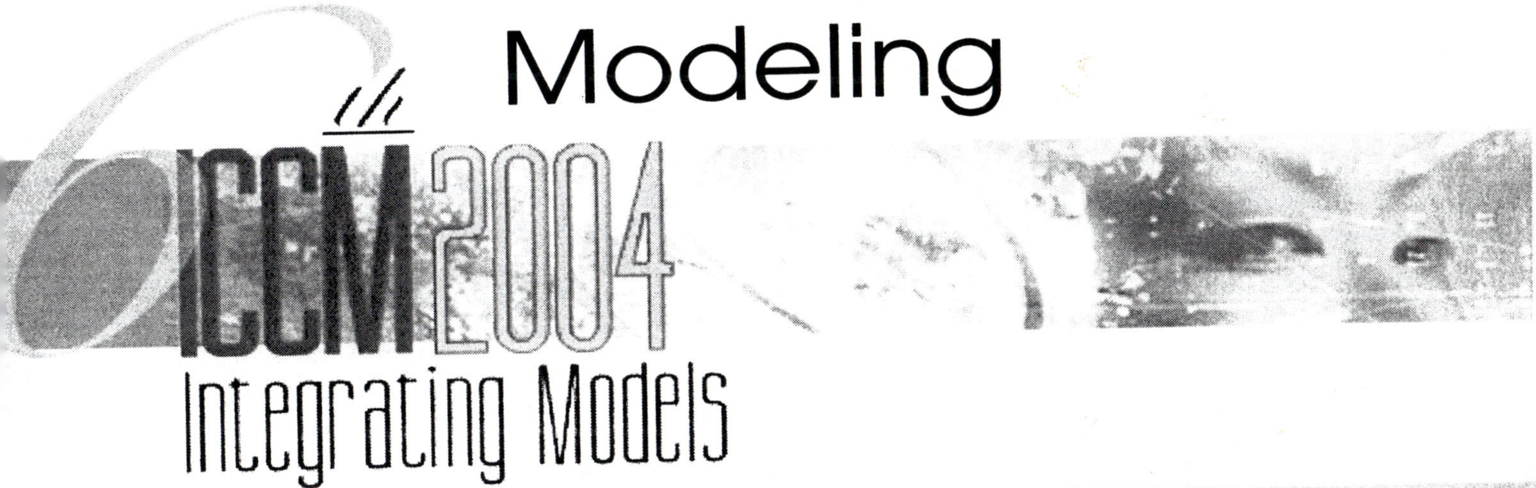

July 30–August 1, 2004

Carnegie Mellon University
University of Pittsburgh
Pittsburgh, Pennsylvania, USA

Edited by: Marsha Lovett, Christian Schunn
Christian Lebiere & Paul Munro

Sponsored by

Air Force Office of Scientific Research
Air Force Research Laboratory
Army Research Laboratory
CMU Psychology
MicroAnalysis and Design
NASA
National Science Foundation
Office of Naval Research
University of Pittsburgh, LRDC

LEA LAWRENCE ERLBAUM ASSOCIATES PUBLISHERS
Mahwah, New Jersey London
2004

Distributed by

Lawrence Erlbaum Associates, Inc., Publishers
10 Industrial Avenue
Mahwah, New Jersey 07430

ISBN 0-8058-5426-6

Books published by Lawrence Erlbaum Associates are printed on acid-free paper, and their bindings are chosen for strength and durability.

Printed in the United States of America

Poster Abstracts

Doctoral Consortium Posters

Preface

ICCM-2004 — Sixth International Conference of Cognitive Modeling

ICCM brings researchers together who develop computational models that explain/predict cognitive data. The core theme of ICCM2004 was **Integrating Computational Models**: integration of diverse data through models of coherent phenomena; integration across modeling approaches; and integration of teaching and modeling.

ICCM2004 sought to grow the discipline of computational cognitive modeling by providing

- a sophisticated modeling audience for cutting-edge researchers
- a forum for integrating insights across alternative modeling approaches (including connectionism, symbolic modeling, dynamical systems, Bayesian modeling, and cognitive architectures) in both basic research and applied settings, across a wide variety of domains, ranging from low-level perception and attention to higher-level problem-solving and learning
- a venue for planning the future growth of the discipline

More concretely, the conference included 1) a careful peer-review process of 6-page paper submissions, shepherded by a large set of knowledgeable conference program committee members, 2) poster-abstracts to include late-breaking work in the area, 3) prizes for the best student papers (Newell Award) and the best applied modeling paper (Siegel-Wolf Award), 4) a doctorial consortium to build bridges between the newest crop of cognitive modeling PhDs, and 5) competitive modeling symposia that compare and contrast differing approaches to the same phenomena. The first four features were carried over from previous years. The fifth feature was a novel feature of ICCM2004, although it extended a similar general intent from previous meetings of ICCM.

ICCM Committee

Conference Chairs
Marsha Lovett, Carnegie Mellon University
Christian Schunn, University of Pittsburgh
Christian Lebiere, Micro Analysis and Design, Inc.
Paul Munro, University of Pittsburgh

Local Arrangements
Carmela Rizzo, University of Pittsburgh
Stacey Becker, Carnegie Mellon University

Proceedings Editors
Christian Schunn, University of Pittsburgh
Christian Lebiere, Micro Analysis and Design, Inc.
Marsha Lovett, Carnegie Mellon University
Paul Munro, University of Pittsburgh

Doctoral Consortium
David Touretzky, Carnegie Mellon University
Wayne Gray, Rensselaer Polytechnic Institute

Tutorials Committee
Chair, Frank Ritter, The Pennsylvania State University
Glenn Gunzelman, Air Force Research Laboratory
Randy Jones, Soar Technologies, Inc, and Colby College
Josef Nerb, Universität Freiburg

Web Support
Anthony Harrison, University of Pittsburgh

Publications Support
Jody Getsy, University of Pittsburgh
Judith Brooks, Carnegie Mellon University

Program Committee

John Anderson
Carnegie Mellon University

Axel Cleeremans
Université Libre de Bruxelles

Gary Cottrell
University of California, San Diego

Dietrich Dorner
Universität Bamberg

Bob French
Université de Liège

Wayne Gray
Rensselaer Polytechnic Institute

Andrew Howes
Cardiff University

Rick Lewis
University of Michigan

Stephan Lewandowsky
The University of Western Australia

Art Markman
University of Texas, Austin

Mike Mozer
University of Colorado, Boulder

David Plaut
Carnegie Mellon University

Erik Reichle
University of Pittsburgh

Frank Ritter
The Pennsylvania State University

Gerhard Strube
Universität Freiburg

Ron Sun
Rensselaer Polytechnic Institute

Niels Taatgen
Carnegie Mellon University

Kurt VanLehn
University of Pittsburgh

Alonso Vera
NASA Ames Research Center

Janet Wiles
The University of Queensland

Richard Young
University of Hertfordshire

Sponsors

Air Force Office of Scientific Research

Air Force Research Laboratory

Army Research Laboratory

Carnegie Mellon University Department of Psychology

Micro Analysis and Design, Inc.

NASA Ames Research Center

National Science Foundation

Office of Naval Research

University of Pittsburgh Learning Research and Development Center

Invited Addresses

Cognitive Simulation in the Large

Kenneth D. Forbus (forbus@northwestern.edu)
Computer Science and Education
Northwestern University
1890 Maple Avenue, Evanston, IL 60201, USA

One of the most exciting directions in cognitive simulation is exploring larger-scale phenomena. Now it is possible to tackle phenomena that would have been impossible even a few years ago, due to the confluence of three factors: progress in modeling basic cognitive processes, dramatic increases in computational power, and the availability of large-scale representational resources. This talk will discuss some methodological issues raised by larger-scale simulations, and some examples of larger-scale simulations that our group has created. Work in progress on Companion Cognitive Systems, a new architecture which use structure-mapping centrally in reasoning and learning, exploiting a million-fact subset of the Cyc knowledge base and cluster computing, will also be discussed.

How Practice Makes Perfect

Michael C. Mozer (mozer@colorado.edu)
Computer Science and Institute of Cognitive Science
University of Colorado

Although human learning is typically viewed as a phenomenon that occurs on a time scale of days or weeks, experimental evidence suggests that our second-to-second experiences have significant, long lasting effects on cognition and neocortex. For example, in behavioral studies of _repetition priming_, processing of a stimulus becomes more efficient as a result of prior presentation of the stimulus; neuroimaging studies robustly demonstrate _repetition suppression_, a stimulus-specific decrease in neural activity resulting from mere exposure; and neurophysiological studies have observed rapid changes to neural representations that underlie repetition suppression and that indicate improved discriminability in the face of noise. We interpret these ubiquitous adaptation phenomena via models at two different levels of abstraction. In a neural model, we propose an unsupervised learning mechanism that achieves an item-specific gain of neural responses. In a dynamical probabilistic model, we characterize changes to the time course of information transmission with experience. Through these complementary models, we obtain an understanding of how practice leads to the refinement of cognitive and motor skills.

This work is a joint collaboration with Michael Colagrosso, David Huber, Todd Mytkowicz, and Richard Zemel.

Allen Newell Award for Best Student Paper

Deciding When to Switch Tasks in Time-Critical Multitasking

Yelena Kushleyeva (yk45@drexel.edu)
Dario D. Salvucci (salvucci@cs.drexel.edu)
Frank J. Lee (fjl@cs.drexel.edu)
Department of Computer Science, Drexel University
3141 Chestnut St., Philadelphia, PA 19104

Abstract

While cognitive modeling has begun to make good progress in accounting for human multitasking behavior, current models typically focus on externally-driven task switching in laboratory-task settings. In contrast, many real-world complex tasks, particularly time-critical tasks, involve internally-driven multitasking in which the person her/himself decides when to switch between tasks. In this paper we propose an adaptation of the ACT-R cognitive architecture that incorporates a notion of elapsed time for the current goal and uses time to determine when to switch away from the current task. We demonstrate the usefulness of this mechanism in an application to a dynamic, time-critical dual search task, showing how an ACT-R model can account for various aspects of human subjects' switching behavior.

Introduction

In the push to generalize to increasingly complex real-world tasks, cognitive modeling and cognitive architectures have begun to address several important challenges, including direct interaction with realistic environments and complex integration of lower-level performance with higher-level planning and decision making. One of the most important challenges today involves making the leap from single-task laboratory experiments to real-world, time-critical tasks in which a person performs tasks together — in other words, addressing the fundamental yet ill-understood skill of human multitasking in time-critical tasks.

Recent modeling work on multitasking has focused primarily on situations where external cues are provided as a signal initiating switching between tasks. Such situations provide an excellent framework for analyzing the dynamics of the reaction time, and thus the distribution of attentional resources, at the time when switching between tasks takes place — for instance, the analysis of switching-time costs (STC) and the psychological refractory period (PRP) has been done for various types of tasks, discrete and continuous, successive and concurrent (Kieras et al., 2000). However, it has been argued (e.g., Burgess et al, 2000) that relying on external signals for switching between tasks is less common in real-life multitasking situations. Instead, many real-life scenarios, particularly in time-critical environments, involve *internally-driven* multitasking in which the person decides when to switch between tasks (e.g., the tasks of driving and dialing a cell phone). While the environment may indirectly define certain parameters of task switching, reliance on direct environmental feedback as a signal to switch would imply processing that is too slow for time-critical situations (Rasmussen, 1983). In addition, environmental feedback is not even always available in real-life situations; it may also be significantly delayed and so only useful for adjusting future, but not current, behaviors. For this reason, the ultimate distribution of task switch points under time-critical conditions must be signaled by processes internal to human cognition.

The modeling of internally-driven multitasking proves to be an interesting challenge for current cognitive architectures such as ACT-R (Anderson et al., in press), EPIC (Meyer & Kieras, 1997), and Soar (Newell, 1990). One critical component of internally-driven multitasking in time-critical contexts is the ability to implicitly reason about time and the temporal aspects of each task. Intuitively this can be understood in terms of the increasing "pressure" over time that a person working on one task may feel to switch to another task. Possibly, this psychological phenomenon is what allows people to effectively balance cognitive resources across tasks. However, the current major architectures do not incorporate time-sensitive mechanisms that affect behavior on the scale of single-task execution.

To address this issue, we turn to the ACT-R cognitive architecture (Anderson et al., in press) and show that relatively minor changes to mechanisms already built into ACT-R allow us to implicitly reason about time and capture task-switching behavior in time-critical tasks. We evaluate the proposed architectural changes in the context of a simple experimental environment whose task demands are based on our analysis of the available neuropsychological tests (see Burgess, 2000). In developing an ACT-R model that incorporates the proposed changes, we found that the model nicely captured several aspects of people's temporal sensitivity in the task and their resulting behavior of when to switch tasks.

Deciding When to Switch Tasks in the ACT-R Architecture

As a step toward modeling internally-driven task switching in time-critical environments, we first examine some of the neuropsychological bases of real-world multitasking and explore the mapping of these concepts onto the ACT-R architecture. Given this insight, we then propose a small but powerful change to the current architecture that incorporates time into the task-switching decision and thus account for empirical phenomena such as those described in the subsequent sections of the paper.

ACT-R and Neuropsychological Bases

Analysis of the demands that real-life multitasking situations place on the individual allows us to identify three major skill sets whose involvement is crucial for satisfactory multitasking performance (Burgess et al., 2000):

- The ability to create and schedule future intentions.
- The facility to remember/maintain those intentions, as well as prioritize them.
- The ability to switch from carrying out one intention to another when the appropriate moment in time is finally reached.

In identifying the brain structures that are expected to be most significantly involved in producing these types of skills, we note two important points related to modeling and multitasking. First, both creation and maintenance of intentions, or goals, through prospective cognition rely heavily on the activity of the Dorsolateral Prefrontal Cortex (DLPFC), which immediately points to the possibility that this aspect of multitasking behavior should be taken care of by the existing ACT-R "goal buffer," since the activity of this buffer is thought to account for neural processing in DLPFC (Anderson et al., in press). In this paper, we focus on the problem of when to switch away from the current goal in the buffer; in a complementary paper, we focus on the problem of choosing what goal to execute next (Salvucci et al., 2004).

Second, the ability to initiate switching between tasks and to do so at appropriate points in time is closely related to the notion of interval timing and prospective memory, and thus, anatomically, to basal ganglia (Nenadic et al, 2003; Diedrichsen et al, 2003). This connection to basal ganglia leads us to suggest that, in ACT-R, execution of prospective goals should be managed on the level the production system, which is thought to account for the activity of the basal ganglia (Anderson et al., in press). More specifically, we claim that such management centers on the production-level conflict resolution in ACT-R, which chooses the next production instantiation based on "expected gain" (described below). It is this last aspect of human multitasking and its possible ACT-R implementation in terms of conflict resolution that is our focus in this paper.

Procedural Memory and Conflict Resolution

The contents of procedural memory in ACT-R are composed of condition-action production rules. The "condition" part of each such rule is utilized in selection of appropriate productions to be activated; this process is accomplished through matching condition parts of the rules to the current states of the system buffers, including the goal buffer, declarative memory, and the buffers of perceptual modules. However, any given combination of buffer states may potentially match the conditions of multiple production rules; yet, the serial nature of ACT-R's cognitive processing allows only one production to be activated within one cognitive cycle. For this reason, the conflict-resolution mechanism in ACT-R selects a single production rule out of the pool of rules whose condition parts match the current state of the system.

Conflict resolution attempts to select a production rule that maximizes the *expected gain* of the rule — namely, the gain that model expects to receive from executing the next production. Expected gain is defined as the value PG-C. Here, P is the probability that selecting a particular production will eventually lead to successful completion of the goal objective, G is the value of the goal, and C is the expected cost of completing the goal if a given production is chosen on this particular step (Anderson et al., in press). While the conflict resolution process attempts to select a production with the highest expected utility, ACT-R also includes logistic noise on expected gain to introduce some level of randomness to the rule selection process.

Analysis of conflict resolution within the ACT-R procedural memory shows that all parameters involved in conflict resolution are encoded through temporally static variables. The parameter G, representing the subjective value of successfully accomplishing a given goal and expressed as the amount of time the model (or person) is willing to spend on that particular goal, is currently kept static at 20 seconds for all models and all goals within those models. Parameters C (the cost of accomplishing the current goal given that a particular production is chosen) and P (expected probability of successfully achieving the objective of the goal if the given production is chosen) can be modified through the ACT-R utility learning mechanisms (Anderson et al., in press). However, such modifications take place only at such points that are explicitly marked as clear successes or failures in execution of the current goal, and thus such dynamics of the parameters C and P take place at a higher timescale than what is needed to make conflict resolution reflective of the passage of time within a single instance of a goal execution. Thus, the system of conflict resolution in ACT-R can currently be regarded as temporally static if looked at on the level of the execution of a single goal.

This aspect of the ACT-R architecture can be questioned in terms of psychologically plausibility, given that people are clearly capable of adjusting their behavioral strategies with constantly varying levels of time pressure (Maule et al, 2000). Thus, we propose slight modifications to the existing formulation of the conflict resolution parameters that allow us to resolve this issue and also enable us to rely on the same essential conflict resolution mechanism for deciding when to switch away from the current task, particularly in the context of time-critical multitasking.

Proposed Architectural Modifications

We propose to redefine the parameter G as a representation of the subjective value of continuing to execute the current goal. Since utilities throughout the ACT-R architecture are expressed in terms of time, we attempt to maintain consistency with this idea and define G more specifically as the amount of time the model is willing to spend on continuing attempts to achieve the current goal.

To implement this idea, we propose that the initial value of parameter G be set at goal creation — that is, when a production rule creates and sets the new declarative goal chunk in the goal buffer. The initial value of G represents the duration of time that the model is willing to spend on achieving the goal in its entirety, from the start and to the moment the goal is abandoned in favor of another one. From the moment of goal creation, the value of parameter G linearly decreases with the passage of time, ultimately decreasing to 0 as the desired time expires. Such a redefinition of G removes the previously static temporal nature of conflict resolution: because production utilities are calculated as PG-C, as the time goes and the value of G decreases, the system will go from favoring productions that have high probability P of leading to a successful state, to favoring productions that promise low future execution costs C.

The resulting dynamics of the conflict resolution process are very interesting even simply because they allow models to change their strategies for performing tasks with the passage of time. This may affect models in both single and multiple task situations. But more importantly for us here, this new mechanism enables us to rely on conflict resolution for the initiation of switching between tasks. We can now introduce one extra production rule for every goal within a model, such that this new production, when activated, causes the goal to give up control. We further assign this production very low values of parameters P and C — which is intuitively reasonable, since switching away from a goal cannot lead to a successful state *within* that goal (thus low value of P), and similarly, switching away from a goal means that the cost of executing that goal immediately goes down to nearly zero (thus low value of C).

Low values of P and C will ensure that this newly introduced production rule will be activated at or near the end of the time period that the model intended to spend on execution of the corresponding goal: because P is low, this production will not likely be activated at the beginning of goal execution, since the relatively high initial value of G will cause the conflict resolution system to favor productions with high P (and thus high product PG); However, as the value of G goes down with time, conflict resolution becomes increasingly more biased towards selecting low-cost production rules, and as G approaches zero, a production with nearly-zero cost will likely to be selected even if its P value is nearly zero as well.

In addition to ensuring that switch times be near their intended values, reliance on conflict resolution mechanism allows us to utilize the built-in logistic distribution of the expected gain noise (Anderson et al., in press) to guide our predictions about the level of precision we may expect in the distribution of switch times.

Design of the Validation Task

The two major factors guiding us in designing the experimental task were the literature on neuropsychological tests available for evaluation of human multitasking skills and the need for such level of task simplicity that would facilitate integration of the task with computational models.

Guiding Principles

There is a strong understanding in psychological, particularly the neuropsychological, literature that skills involved in human multitasking compose a subset quite distinct from the rest of the skills enabling executive functioning. For instance, it has long been shown that certain types of brain damage nearly exclusively affect the ability to carry out previously planned intentions, thus greatly impairing the person's ability to effectively function in the everyday life, while having very limited impact on performance on traditional standardized tests of general intelligence, and even on the tests of executive functioning (Shallice & Burgess, 1995).

This notion has led to development of specialized tests that evaluate the ability to execute prior intentions, such that the level of performance on those tests would reliably correlate with the person's ability to carry out basic everyday tasks involving multitasking, such as shopping, cooking, etc. Some of the tests thus developed were the Six Element Test (Shallice & Burgess, 1991) and the Greenwich Test (Burgess, 2000). Because both tests involve performing a range of fine motor movements that are difficult to model, we were not able to employ these tests directly; instead, we analyzed the procedures of the two tests to derive guiding principles for designing our own experimental procedure:

- Encourage subtask scheduling on the scale of seconds to minutes, rather than a sub-second scale, as this appears to be optimal for evaluation of the prospective memory aspect of human multitasking, and generally better represents the demands of real-life multitasking environments.
- Provide such rules and instructions that would create a continuing conflict between the need to stay with the current task for prolonged intervals, in order to maximize the given performance measure, and the need to switch between tasks, to prevent negative consequences on performance.
- Provide feedback on progress both along the time dimension and in terms of accuracy in order to allow people to tune their performance by possibly modifying future planning strategies; however, ensure that feedback information is only available after the corresponding action and only upon specific request by the subject, and thus cannot be utilized as a performance-guiding forward cue.
- In general, exclude any environmental cues that could be interpreted as external signals for initiation and/or termination of subtasks, thus ensuring that execution of previously scheduled subtasks is governed exclusively by internal signaling.
- Encourage task prioritization; enable the experimenter to manipulate task priorities throughout the course of the experiment.

Experimental Task

These guiding principles led us to the an experimental task and procedure which can describe briefly as follows. In the experiment, the subject was simultaneously presented with two similar tasks, of which only one could be active at any given point in time. The instructions were given to work on both tasks within the allocated time. Each task required the subject to repeatedly perform visual search on a sequence of

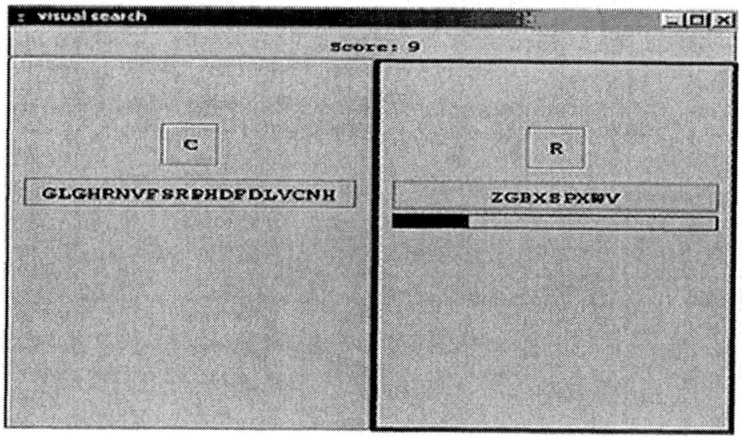

Figure 1: Screenshot of the experimental window.

letters with the goal of determining whether or not a specified target letter is present in the sequence, as shown in Figure 1. The subject was asked to respond to each search query by pressing one of the two designated keys, one key ('y') meaning that the target is present, and the other key ('n') meaning that target is absent from the sequence. In order to motivate the subject to answer accurately, a correctness score was calculated and displayed to the subject. Each correct response earned 1 point in the scoring, while an incorrect response deducted 10 points. Subjects were encouraged to answer as fast as they could, since that would allow them to reach a higher score within the limited time interval.

In order to ensure that the subject regularly switches between the two presented tasks and does not commit to working on only one of them, each task had a timer associated with it. Each timer was reset whenever the subject answered search queries of the task corresponding to this particular timer. However, whenever a task was abandoned in favor of the other task, the timer of the first task was able to run uninterrupted at constant speed up until the point where it reached its endpoint, where the timer would "expire," thus causing a very significant decrease (25 points) in the subject's performance score. The subject was instructed to treat the timers with high priority and to be sure to switch between tasks with a frequency high enough to avoid either of the timers ever reaching their expiration points.

The tasks were further complicated by the fact that the timer of the currently inactive task was hidden until that task was activated again. Thus, the subject had to work in the situation where he or she had to switch from one task to another without being able to rely on any external cues to guide in determining the appropriate time to do so.

In order to prevent the subject from adopting the strategy of switching between tasks after every single query, we manipulated the difficulty of the two tasks: the length of the letter sequence in one task was kept at 20, and in the other task it was either 5 or 9. The subjects were explicitly encouraged to spend as much time as possible on the easier task, with the notion that such strategy would allow them to reach a higher performance score within the time allocated for the experiment.

Participants and Procedure

In the experiment, which included 10 university students (2 women, 8 men), each participant was first instructed as to the basic structure of the task. Then, the participant performed two experimental blocks in each of two conditions: one in which the timer expired in 30 seconds, and another in which the timer expired in 15 seconds. In each block, the participant was allowed to practice the task for 5 minutes, and then performed 15 minutes of actual experimental trials. For simplicity, in this paper we present only the results for the 30-second timer condition; the results for the 15-second timer condition are reasonably similar and not discussed herein.

ACT-R Model

Given the experimental task described above, we set out to develop an ACT-R model that incorporates the proposed architectural mechanism, thus validating that the mechanism can indeed generate interesting emergent predictions in task-switching behavior.

Description

We first developed an isolated model that adequately performed the required search-and-answer procedure in a single-task context. The model adhered to the following action pattern:

- encode the target letter;
- sequentially search the letter sequence (the mechanism of simultaneous encoding and attention shifts was used to achieve a search speed high enough to match the one observed in people);
- if the target is encountered during search, respond by pressing the 'y' key;
- if end of the letter sequence is reached, respond by pressing 'n'.

We do realize that this model oversimplifies the real human interaction with the experimental environment. For instance, it does not incorporate the action redundancy that was observed in the general task procedure and especially in the search strategy (evidence for this comes in part from the fact that people were able to perform search significantly faster under increased time pressure, and this speedup was only

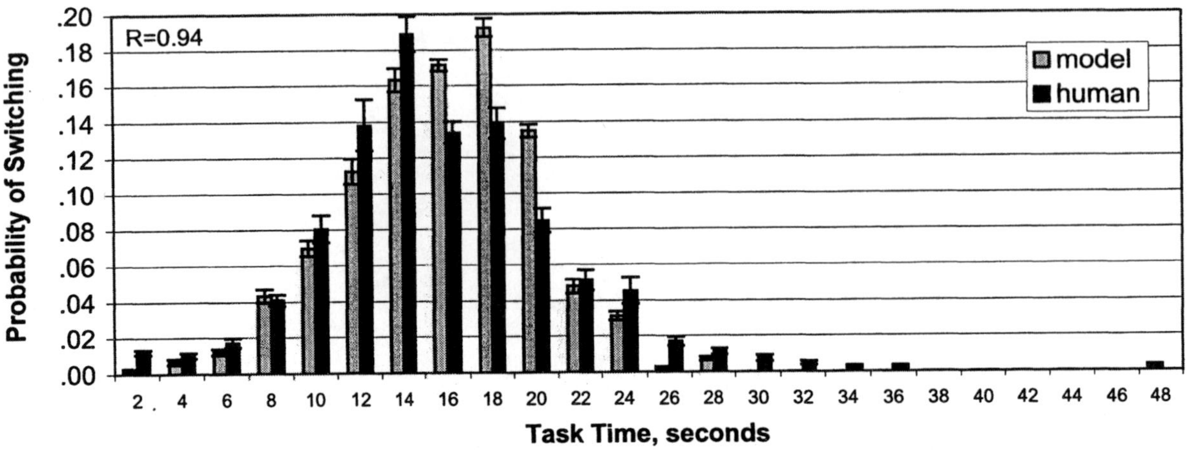

Figure 2: Comparison of experimental results and model predictions. Experimental data were obtained using ten subjects, each performing the experimental task for 15 minutes; to match these conditions, the model was also run for ten 15-minute intervals.

poorly correlated with the increase in error rate). However, its problems notwithstanding, this model provides a close enough approximation of human performance to allow us to focus our study on the initiation of task switches.

Enabling the model to perform two independent tasks within a given time interval by regularly switching between the tasks required only those changes described earlier: we had to introduce one extra production into each task with the intention that this new production, when activated, would cause the goal to cede control to the other task goal.

To fit this model to the behavior of the experiment participants, we varied one parameter, the value of ACT-R's expected gain noise, which was set to a value of 2.0 (instead of the default value of 0.5). We explain this by the fact that the conflict resolution system in our model was relying on perfect system time, while in people the corresponding processing system is likely to receive temporal information that is already noisy. And most importantly, the very idea of having the value of the current objective as a dynamic construct may by itself call for reevaluation of expected gain noise parameters. In addition, we preset the C values of the production rules that create a new goal to include the cost of this new goal; this additional cost was derived from the average time needed by the model to complete these goals. The final model results were compiled from 10 simulations of 15 minutes each (the same amount of data as for human subjects).

Model Predictions

The most important result that we obtained from experiments with the model was that the conflict resolution system relying on the redefined dynamic goal value parameter G is capable of producing a probability distribution for the times of switching between tasks that is very close to the one obtained for human subjects, as shown in Figure 2, with R=.94 (calculated only from 2-30 s). The shape of the distribution is an emergent prediction of the time-sensitive conflict resolution; the exact size of this distribution is a function of the expected gain noise.

In addition to achieving the desired shape of the probability distribution, we were also able to account for the fact that the majority of our human subjects were more likely to switch away from a task whenever continuing with that task required them to perform a discrete time-consuming subtask — that is, both the model and human subjects were more likely to switch before a 5-letter search than before a 9-letter search, as shown in Figure 3.

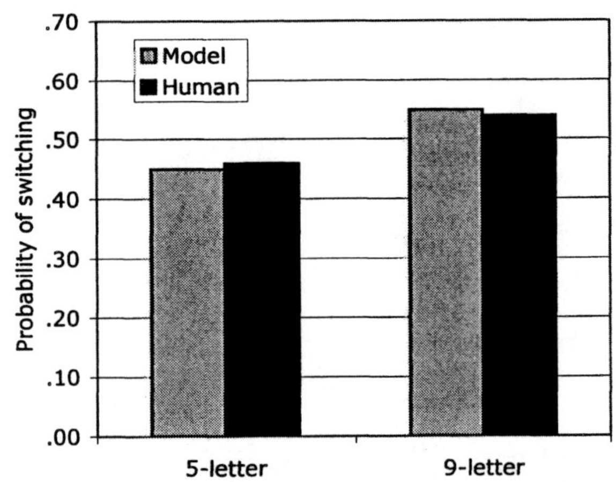

Figure 3: Probability of switching given that the next trial is a 5-letter or 9-letter search.

The model predicts this effect because the C values for subgoal-launching productions include the cost of the subgoal itself, thereby making the 9-letter search more costly and thus more likely to lead to a switch. Interestingly, although we are not modeling learning in the current model, we believe that such definition of the C parameters suggests a simple and intuitive way of learning conflict resolution parameters and nicely accounts for the tendency to avoid time-consuming tasks, for production rules.

Discussion

Overall the new mechanism of a time-sensitive conflict resolution seems to account nicely for the decision of when to switch tasks, particularly during time-critical performance. One aspect of multitasking not accounted for in this work, however, is *what* to do next — that is, once you switch away from the current task, deciding what task (or goal) should follow. In a companion paper (Salvucci et al., 2004), we propose a mechanism that generalizes the goal buffer into a goal set, scheduling goals according to "urgency" and thus help to decide what to do next. Also, Altmann and Trafton (2002) have proposed an activation-based model for retrieving the next goal from declarative memory. Such work on deciding what to do next is complementary to our work here and could eventually be integrated to form a more cohesive account of task scheduling and multitasking.

Another interesting aspect of the new mechanism emerged from the treatment of time as associated with individual goals. The current mechanism focuses on the time elapsed for an individual goal, but time sensitivity does not currently cross goal boundaries. Another approach would allow the G value of the current goal to influence the G value for subgoals. The older ACT-R goal stack (Anderson & Lebiere, 1998) discounted subgoals' G values when pushed onto the stack; however, this (now defunct) mechanism had some aspects of this propagation but was somewhat rigid in implementation. While we believe that our mechanism has the potential to generalize to such phenomena, more in-depth studies could clarify how the propagation of time sensitivity across goals might be incorporated into the architecture.

Acknowledgments

This work was supported by ONR grant N00014-03-1-0036 to the second author. We thank Alex Chavez for helpful ideas and suggestions.

References

Altmann, E. M., & Trafton, J. G. (2002). Memory for goals: An activation-based model. Cognitive Science, 26, 39-83.

Anderson, J. R., Bothell, D., Byrne, M. D., Douglass, S., Lebiere, C., & Qin, Y. (in press). An integrated theory of the mind. Psychological Review.

Anderson, J.R., & Lebiere, C. (1998). *The atomic components of thought*. Hillsdale, NJ: Erlbaum.

Burgess, P.W. (2000). Strategy application disorder: the role of the frontal lobes in human multitasking. *Psychological Research*, *63*, 279-288.

Burgess, P.W., Veitch, E., De Lacy Costello, A., Shallice, T. (2000). The cognitive and neuroanatomical correlates of human multitasking. *Neuropsychologia*, *38*, 848-863.

Diedrichsen, J., Ivry, R.B & Pressing J. (2003). Cerebellar and basal ganglia contributions to interval timing. In Meck, W.H. (Ed.), *Functional and Neural Mechanisms of Interval Timing*. 457-483. CRC Press.

Kieras, D.E., Meyer, D.E., Ballas, J.A., & Lauber, E.J. (2000). Modern computational perspectives on executive mental processes and cognitive control: Where to from here? In S. Monsell & J. Driver (Eds.), *Control of Cognitive Processes: Attention and Performance XVIII* 681-712. Cambridge, MA: MIT Press

Maule, A., Hockey, G. R. J., Bdzola L (2000). Effects of time pressure on decision making under uncertainty: changes in affective state and information processing strategy. *Acta Psychologica*, *104*, 283 – 301.

Nenadic, I., Gaser, C., Volz, H., Rammsayer, T., Hager, F., Sauer, H. (2003). Processing of temporal information and the basal ganglia: new evidence from fMRI. Experimental Brain Research, 148, 238-246.

Newell, A. (1990). *Unified theories of cognition*. Cambridge, MA: Harvard University Press.

Rasmussen, J. (1983) Skills, rules and knowledge; signals, signs and symbols, and other distinctions in human performance models. *IEEE Transactions: Systems, Man and Cybernetics*, *SMC-13*, 257-267.

Salvucci, D. D., Kushleyeva, Y., & Lee, F. J. (2004). Toward an ACT-R general executive for human multitasking. To appear in Proceedings of the Sixth International Conference on Cognitive Modeling.

Shallice T., Burgess P.W. (1995). Supervisory control of action and thought selection. In: Baddeley A.D., Weiskrantz L. (Eds.) *Attention: selection, awareness and control: a tribute to Donald Broadbent*. Oxford: Clarendon Press.

Shallice T., Burgess P.W. (1991). Deficits in strategy application following frontal lobe damage in man. *Brain*, *114*, 727-741.

Surface and Phonological Dyslexia: Modes of Processing Instead of Components of Processing

Daragh E. Sibley (dsibley@gmu.edu)
Department of Psychology, George Mason University
Fairfax, VA 22030-4444 USA

Christopher T. Kello (ckello@gmu.edu)
Department of Psychology, George Mason University
Fairfax, VA 22030-4444 USA

Abstract

Two connectionist models are reported that mimicked the defining features of the double dissociation between phonological and surface dyslexia in word reading. One model was a feed-forward, three-layer perceptron, and the other included recurrent connections. Neither model contained an architectural separation of sublexical and lexical processes, nor of phonological and semantic processes. Analyses showed that the double dissociation was simulated because the control parameter *input gain* shifted the models between conjunctive and componential modes of processing. The dissociation was not simulated by any kind of damage to separate system components. The models are discussed in the context of current accounts of surface and phonological dyslexia.

Introduction

In most current models of word reading, two routes of processing are proposed to compute the sound of a word given its printed form (Coltheart et al., 1993, 2001; Plaut et al., 1996; Harm & Seidenberg, 1999). One route is *sublexical* in that it extracts and utilizes regularities in the mapping from components of orthography (e.g., letters) to components of phonology (e.g., phonemes). The other route is *lexical* in that it uses either localist or semantic representations of words. Localist and semantic representations are both considered lexical in this context because they bear little or no systematic relationship to the components of orthography and phonology.

The evidence for sublexical and lexical routes of processing in word naming has come from a number of sources (for a review, see Plaut et al., 1996), but here we focus on evidence from selective reading impairments that occur as a result of brain damage. One class of impairment termed *phonological dyslexia* is characterized by poor reading aloud of nonwords (e.g., SHONG), with relatively intact word reading (Behrmann & Bub, 1992). Conversely, *surface dyslexia* is characterized by poor reading aloud of words with irregular spelling-sound correspondences (e.g., PINT), with relatively intact regular word and nonword reading (Funnell, 1983).

The complementary impairments of phonological and surface dyslexia have a straightforward explanation in terms of separate sublexical and lexical routes of processing. The impairment in nonword reading that characterizes phonological dyslexia is hypothesized to occur as a result of damage to the sublexical route. Damage should have this effect because performance on nonword stimuli presumably relies on the componential regularities between spelling and sound (e.g., the sounds corresponding to SH, O, and NG can be pieced together to form a plausible nonword pronunciation). By contrast, the impairment in irregular word reading that characterizes surface dyslexia is hypothesized to occur as a result of damage to the lexical route. Damage should have this effect because performance on irregular word stimuli presumably relies on word-specific knowledge (e.g., one must know the word PINT in order to pronounce it correctly).

There are a number of reasons why sublexical and lexical routes have been proposed to account for surface and phonological dyslexia, but perhaps the most important one is the logic of *double dissociations*. If two components of a cognitive system operate independently, then they will make independent contributions to the behavior of that system. Thus, a change in one component will have no effect on the behavioral contributions of the other. Such complementary effects on behavior constitute a double dissociation, and they are often interpreted as strong evidence that independent system components underlie the dissociated behaviors.

Surface and phonological dyslexia constitute a double dissociation, and many researchers have interpreted this double dissociation as evidence for sublexical and lexical routes. Some researchers have alternatively proposed that semantic and phonological components may underlie this dissociation (Patterson & Ralph, 1999), but this alternative also assumes that two system components underlie the dissociation, as in the dual-route hypothesis.

Despite appearances, even the purest cases of phonological and surface dyslexia do not necessitate the existence of two processing routes or two system components of any kind (Plaut, 1995; Van Orden, Pennington, & Stone, 2001). A single-component explanation is always a logical possibility, but without a specific single-component account, it is just a logical possibility. With a single-component account, dissociations would continue to stand as evidence for the existence of two system components in word reading.

Recently, Kello and Plaut (2003) reported on a model of word reading that offers a true single-component explanation of the double dissociation between phonological and surface dyslexia. The model was inspired by the basic question of how reading acquisition builds upon the prior learning that occurs during spoken language acquisition (also see Plaut & Kello, 1999). The acquisition of spoken language requires the mapping from sound to meaning (comprehension) and meaning to sound (production). In the context of a connectionist approach to lexical processing, a single, distributed level of representation can be learned to mediate the bi-directional mapping between the phonological and semantic attributes of words. The question raised by this approach is, how do printed word forms make contact with the bi-directional mapping learned during spoken language acquisition?

The answer offered by Kello and Plaut (2003) is that orthography maps into the level of representation that mediates semantics and phonology (see Figure 1, inset), rather than mapping into semantics and phonology themselves. This architecture effectively "kills two birds with one stone": the mediating representations provide access to both the semantic and phonological forms, so only one route is necessary. While there is a logical appeal to this single-route architecture, it is unclear how it could account for the dissociation between surface and phonological dyslexia because there is no architectural separation of sublexical and lexical processes.

In a subsequent analysis of this single-route model, Kello (2003) showed that an impairment in the processing of nonwords could, in fact, be dissociated from an impairment in the processing of irregular words. The method used to simulate this dissociation was to shift the model between sublexical and lexical "modes" of processing by means of a control parameter termed *input gain* (see Model Architecture section). The simulation results showing the dissociation are graphed in Fig 1. The use of input gain was inspired by simulations of other behavioral data (see Kello & Plaut, 2003), but the function of input gain such that it caused this dissociation was never elucidated.

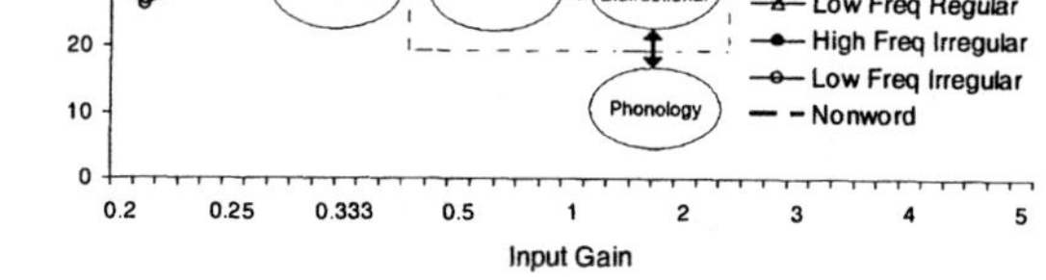

Figure 1: Results from the single-route model of word reading reported in Kello (2003)

Current Work

The goal of the current work was to use computational analyses to determine how input gain had its dissociating effect in the simulation of word reading reported by Kello (2003). To facilitate these analyses, the essential principles and mechanisms of the single-route model of word reading were distilled into two very simple connectionist models: one model was a feed-forward, 3-layer perceptron, and the other included recurrent connections from the output layer back to the hidden layer. These models did not simulate word reading or dyslexia, although their results are compared with data from experiments with phonological and surface dyslexics. Rather, the main purpose of the models was to replicate the dissociating effect of input gain in a more controlled and analytically tractable context. This tractability enabled us to determine that input gain can shift a connectionist model between *componential* and *conjunctive* modes of processing, and that it is this function of input gain that was responsible for the simulation of phonological and surface dyslexia reported by Kello (2003).

Simulation Methods

Input and Output Representations were constructed from a 12 dimensional binary space. Out of $2^{12} = 4096$ possible input patterns, one fourth (1024) were chosen at random to constitute the corpus of known items. Each chosen input pattern was associated with one output pattern. Output patterns were created in three steps. First, each input pattern was copied to its corresponding output pattern (i.e., the identity mapping). Second, frequencies were assigned to each item according to Zipf's law, $f = r^{-0.5}$, where r was an arbitrarily assigned rank from 1 to 1024. Third, the bit value of each dimension, for each output pattern, was flipped with a probability governed by Zipf's law, $p = 0.82r^{-0.5}$. The result of this formula was that the more frequent items were more likely to be irregular, and more likely to be more irregular (i.e., have more flipped values), compared with the less frequent items. The multiplicative constant of 0.82 was set such that there was a 5% probability on average of flipping each target value across the set of known items. There were 580 fully regular items (no flipped bits), and 444 irregular items with at least one flipped bit per item. The 3072 remaining patterns served to test the generalization of learning to novel items.

Each of the 12 input dimensions were coded by two input units, one coding on-bits as 1 and off-bits as 0, the other coding the reverse of the first. This $x|1-x$ coding scheme was used because the models were trained via backpropagation. In backpropagation, no learning will occur on a unit's sending weights when the activation value of that unit is zero. Therefore, the $x|1-x$ coding scheme ensured that weight derivatives were generated for every input dimension, on every training episode. The $x|1-x$ coding was not necessary for the output units, so there were only 12 output units, each one corresponding directly to one of the 12 output dimensions.

The input and output representations captured the essential properties of quasi-regularity as it is implemented in most connectionist models of word reading. Specifically, each input unit had a systematic relationship with one output unit, much like the way that each orthographic unit would have a systematic relationship with at least one phonological unit (e.g., a unit for the letter P in the initial position would have a systematic relationship with a unit for the phoneme /p/ in the first position). Moreover, these relationships were never entirely systematic, much like real quasi-regularity in spelling-sound correspondences.

Model Architecture. In both the feed-forward and recurrent models, the input units were fully connected to 200 hidden units, and the hidden units were fully connected to the output units. In the recurrent model, the output layer was also fully connected back to the hidden layer. The number of hidden units was determined through pilot testing to be about 50 units more than the minimum needed to learn the mapping. However, results were very similar over a range of hidden unit numbers. Unit activations were calculated with the hyperbolic tangent function,

$$a_j^{[t]} = \tanh\left[\gamma\left(I_j^{[t-1]} + \Delta t\left(I_j^{[t]} - I_j^{[t-1]}\right)\right)\right],$$

where γ was input gain, Δt was an integration constant fixed at 0.166, and $I_j^{[t]}$ was the net input at time t. For the feed-forward network, there was no time course of processing, so there was no integration constant, and activations were simply a function of the instantaneous net input. Input gain was fixed at 1 during training, and varied during testing (see next section). The net input to each unit was calculated as the dot product between the activation vector over its sending units, and the weight vector over its incoming connections. The hyperbolic tangent is analogous to the logistic, except it has asymptotes at +1 and −1 instead of 0 and 1. Low levels of input gain flatten the activation function, making it more linear. High levels sharpen it into a step function, making it more nonlinear.

Forward connection weights were initialized to random values in the range ±0.1, and recurrent weights (for the recurrent model) were initialized in the range ±0.5. A larger range was used for recurrent weights to ensure that they had a substantial impact on processing. Weights were learned by gradient descent,

$$\Delta w_{ij} = \eta\left(\partial E/\partial w_{ij}\right),$$

where w_{ij} was the connection weight from unit j to i, η was the learning rate (fixed at 0.001), and E was cross-entropy error (Rumelhart et al., 1995), which was scaled by each item's frequency. Prior to each weight update, each of the 1024 items in the corpus were presented once to the network. Weight derivatives were calculated for each item as follows: input units were set to a given item's input pattern, activations were propagated forward through the network, and an error signal was calculated from the difference between actual and target outputs. In the feed-forward model the error signal was then backpropagated to generate the weight derivatives. In the recurrent model, activations were propagated forward for 18 ticks, error was injected on the last 12 ticks, and then error was backpropagated in time.

Weight updates were repeated until every output unit was within 0.1 of its target for every item in the training corpus. This criterion was reached in the feed-forward model after 62000 passes through the corpus, and in the recurrent model after 56000 passes.

Testing Procedure. The models were tested by setting the input units to a given input pattern, and recording the output activations. In the recurrent network, activations were recorded on the first tick for which all 12 output nodes were within 0.1 of an asymptote. If a node did not reach this criterion after 18 ticks, the output was judged as incorrect. The criterion for correct performance was having the activations of all 12 output units on the target side of zero. Targets for items in the corpus were set according to each item's output pattern. Targets for the 3072 novel items were set according to each item's input pattern, i.e., the identity mapping.

To dissociate item-based and regularity-based processing, input gain was varied as a single control parameter over the hidden units. The reported levels of input gain were between 0.33 and 3. This range was chosen to show asymptotic performance at the lower and upper ends. Performance did not change substantially beyond this range.

Simulation Results

To simplify the presentation of results, known items were divided into categories of high frequency and low frequency. The high frequency category consisted of the 256 most frequent trained items (upper quartile), and the low frequency category consisted of the 256 least frequent trained items (lower quartile). Mean accuracies for the feed-forward model are graphed in Figure 2 as a function of input gain and item type (high and low frequency regular, high and low frequency irregular, and novel). The same are graphed for the recurrent model in Figure 3.

Figures 2 and 3 show that both models exhibited a clear dissociation in performance on irregular items compared with novel items. At low levels of input gain, generalization of the identity mapping to novel inputs was essentially perfect, as was performance on regular items. By contrast, performance on irregular items dropped to 0%, at which point all inputs were computed as the identity mapping. For irregular items, application of the identity mapping can be considered as a *regularization* error because, for the quasi-regular domain constructed here, the identity mapping is the regular mapping.

At high levels of input gain, performance on known items was better than performance on novel items. At their maximum difference, mean accuracies in the feed-forward model were 97% for known items, and 46% for novel items. The same comparison in the recurrent model was 94% and 35%, respectively. The biggest difference between the models was that performance on known items was near ceiling at all high levels of input gain in the feed-forward model, whereas performance on known items dropped off at sufficiently high levels of input gain in the recurrent model. This difference is explained by the fact that high levels of input gain amplify the effect of nonlinearities in the activation function on processing (see below). Recurrence caused this amplification to increase to the point of distorting the course of processing. Given that the dissociation was maintained despite this distorting effect, we did not see it as informative with respect to the dissociating effect of input gain.

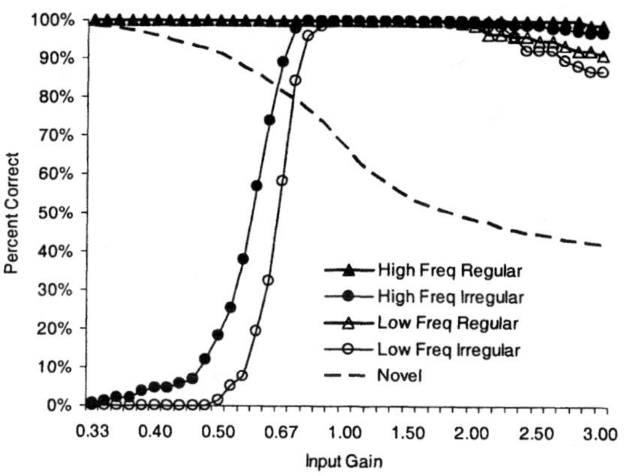

Figure 2: Mean accuracies for the feed-forward model

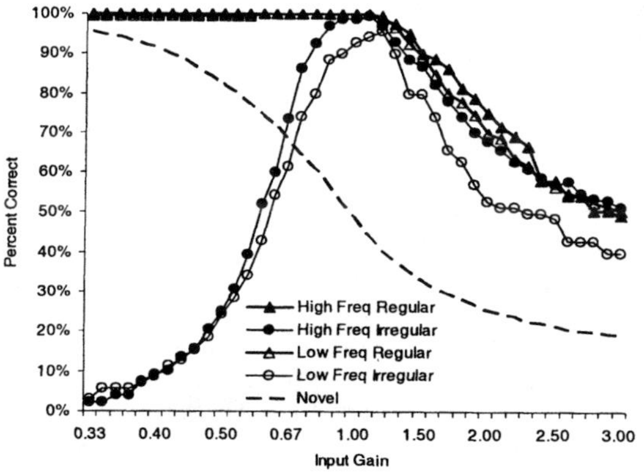

Figure 3: Mean accuracies for the recurrent model

The models' behavior was reminiscent of the double dissociation seen in surface and phonological dyslexia. Table 1 compares the word naming performance of two surface dyslexic patients (MP, Behrmans & Bub, 1992; KT, McCarthy & Warrington, 1986) with each model's performance at low levels of input gain. These levels were chosen to best match the patients' patterns of performance. Table 2 similarly compares the word naming performance of a phonological dyslexic patient (WB, Funnell 1983) with each model's performance. The purpose of these tables was only to draw a relation between the current models and dyslexia; as stated earlier, the models were not intended to simulate dyslexia.

Table 1. Surface dyslexic MP and KT, compared with feed-forward (FF) and recurrent (Re) models at different levels of input gain, on high frequency (HF), low frequency (LF), regular (Reg), and irregular (Irr) items.

	MP	*FF.66*	*Re.71*	***KT***	*FF.55*	*Re.55*
HFReg	**95%**	100%	100%	**100%**	100%	100%
LFReg	**98%**	100%	100%	**89%**	100%	99%
HFIrr	**93%**	89%	86%	**47%**	57%	25%
LFIrr	**73%**	58%	74%	**26%**	19%	24%
Novel	**96%**	82%	65%	**100%**	86%	82%

Table 2. Phonological dyslexic WB, compared with feed-forward (FF) and recurrent (Re) models at different levels of input gain.

	WB	*FF 3*	*Re 3*
Words/Known	**89%**	93%	44%
Novel	**0%**	42%	19%

The results reported here show that the manipulation of input gain as a single control parameter, over a single level of representation, caused a double dissociation in both models. Thus, we can conclude that the dynamics produced by recurrent connectivity is not required to give input gain its dissociating effect. While this result is informative, it does not fully elucidate the computational principles by which input gain has its dissociating effect. The following analyses were designed to explicitly show that input gain affects the extent to which processing is componential versus conjunctive, and that it is this property of input gain that is responsible for its dissociating effect.

Componential Versus Conjunctive Processing. In the context of the current models, componential processing occurs when each input dimension is used independently of all other input dimensions in computing an output pattern. That is, each input dimension is treated as an independent component of the input pattern. By contrast, conjunctive processing occurs when conjunctions of input dimensions are used to compute outputs.

We measured the degree of componential versus conjuctive processing in the feed-forward and distributed models by manipulating the value of a single input dimension, while holding the remaining dimensions constant at a neutral value of 0.5. In this way, the manipulated dimension was responsible for any changes in output activations. As a given input dimension was manipulated, activation values were monitored for the 11 output units that did *not* correspond to the manipulated input dimension. The polarity of each activation value (i.e., how close it was to an asymptote) was then calculated, and these polarities were averaged across all output dimension, for manipulations of all 12 input dimensions. This calculation was a measure of the conjunctivity of processing because higher values indicated that a change in one input dimension was affecting multiple output dimensions that it did not correspond to, and lower values indicated that a change in one input dimension only affected its corresponding output dimension.

This measure of conjunctivity is plotted in Figure 4 as a function of input gain and model type. The figure shows that, in both models, the conjunctivity of processing increased with higher levels of input gain. In the next two sections, we explain why componential processing at low levels of input gain could not support the processing of irregular items, and why conjunctive processing at the high levels of input gain could not support the processing of novel items.

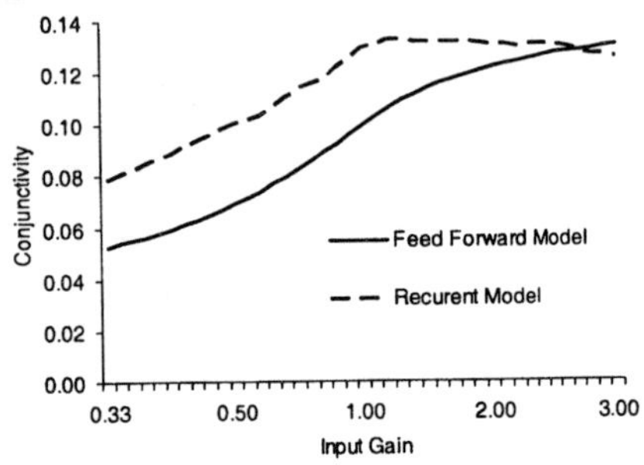

Figure 4: Conjunctivity as a function of input gain

Low Levels of Input Gain. Low levels of input gain caused the sigmoidal shape of the hyperbolic tangent function to flatten and become more linear. Linear hidden units can only support a linearly separable mapping between the inputs and outputs. It is also the case that input dimensions are processed as independent components in a linearly separable mapping. In the quasi-regular mapping created for these models, the identity mapping was linearly separable, whereas the exceptions to the identity mapping were not.

Moreover, hidden units operated in their linear range early in training because the positive and negative random initial weights tended to cancel each other out on any given input; therefore, net inputs to the hidden units tended to be small early in training, as if input gain was low (even though input gain remained fixed at 1 throughout training). The consequence is that the linearly-separable identity mapping was learned in the linear range of the hidden units early in training, and exceptions to the identity mapping were learned only after hidden unit activations moved closer to their asymptotes. An analysis confirmed this statement: after only 30 epochs of learning, the model applied the identity mapping to all 4096 possible input patterns, and hidden unit activations were 0.35 away from zero on average, i.e., they were mostly operating in their linear range. Therefore, one can hypothesize that low levels of input gain invoked the componential, linearly-separable identity mapping that was learned early in training (further analyses are needed to test this hypothesis).

High Levels of Input Gain. High levels of input gain cause the sigmoidal shape of the activation function to sharpen and more closely mimic a step function. The consequence of this sharpening is that hidden unit activations are more likely to be near their asymptotes. This point is important because it is the asymptotic behavior of hidden units that enables them to use conjunctions of input values in mapping their inputs onto their outputs (O'Reilly, 2001). The XOR function is the quintessential example in which the conjunction of two input values must be considered in order to produce the correct output, and at least one nonlinear hidden unit is necessary to compute this function in a feed-forward neural network. In the current quasi-regular mapping, it was the irregularities that engaged the asymptotic behavior of the hidden units in order to process conjunctions of input values. On this logic, one can say that high levels of input gain placed a greater emphasis on conjunctive processing of input values.

Given this functional effect of input gain, why would an emphasis on conjunctive processing cause a selective impairment in the processing of novel input patterns? The answer begins with the fact that the conjunctive processing was used to handle irregularity in the quasi-regular mapping, whereas it is the regularity in this mapping that provides the basis of correct performance on novel inputs. Therefore, the conjunctions learned for irregular inputs will tend to be incorrectly applied to novel inputs at high levels of input gain.

This logic leads one to ask, why was performance on regular known items intact at high levels of input gain? If conjunctions were learned only to process irregularities, then one would have to conclude that the emphasis on conjunctions at high levels of input gain should interfere with the identity mapping for all input patterns, both novel and known. What is missing here is that conjunctions were learned not only to process the irregularities for known items, but regularities as well. This consequence of using conjunctions necessarily followed from the use of distributed representations over the hidden units. When the asymptotic behavior of hidden units was engaged through learning, it affected the processing of all known inputs, both irregular and regular, because every hidden unit contributed to the processing of every input dimension, for every input pattern. Therefore, conjunctions had to be learned for the regularities in the known items, and these conjunctions supported correct performance on all known items at high levels of input gain.

Conclusions

The current simulations demonstrated how a double dissociation can occur in a non-modular system via the manipulation of a control parameter. Analyses showed that input gain shifted the current models between conjunctive and componential modes of processing. This shift was reminiscent of the word naming impairments that characterize phonological and surface dyslexia. Conjuctive modes impaired the processing of novel items, whereas componential modes impaired the processing of irregular items.

The current simulations and the single-route model of word reading (Kello & Plaut, 2003; Kello, 2003) comprise the beginnings of a single-route alternative to dual-route theories of word reading. They address one of the biggest challenges to any single-route theory of word reading, namely, the neuropsychological evidence for separable sublexical and lexical processing routes. However, many challenges remain. How would a single-route alternative be consistent with evidence for separable brain regions correlated with sublexical and lexical processing, to the extent that such evidence exists? What would be the neural mechanisms that underlie the hypothesized function of input gain? Could a large-scale single-route model account for any of the more detailed findings in the vast literature on word reading? These questions await further research.

Acknowledgments

This work was funded in part by NIH Grant MH55628, and NSF Grant 0239595. The computational simulations were run using the Lens network simulator (version 2.6), written by Doug Rohde (http://tedlab.mit.edu/~dr/Lens).

References

Behrmann, M., & Bub, D. (1992). Surface dyslexia and dysgraphia: Dual routes, single lexicon. *Cognitive Neuropsychology, 9*, 209-251.

Coltheart, M., Curtis, B., Atkins, P., & Haller, M. (1993). Models of reading aloud: Dual-route and parallel-distributed-processing approaches. *Psychological Review, 100*, 589-608.

Coltheart, M., Rastle, K., Perry, C., Langdon, R. & Ziegler, J. (2001). DRC: A dual route cascaded model of visual word recognition and reading aloud. *Psychological Review, 108*, 204-256.

Funnell, E. (1983). Phonological processes in reading: New evidence from acquired dyslexia. *British Journal of Psychology, 74*, 159-180.

Harm, M. W., & Seidenberg, M. S. (1999). Phonology, reading, and dyslexia: Insights from connectionist models. *Psychological Review, 163*, 491-528.

Joanisse, M. F., & Seidenberg, M. S. (1999). Impairments in verb morphology following brain injury: a connectionist model. *Proceedings of the National Academy of Sciences of the United States of America, 96*, 7592-7597.

Kello, C. T. (2003). The emergence of a double dissociation in the modulation of a single control parameter in a nonlinear dynamical system. *Cortex, 39*, 132-134.

Kello, C. T., Plaut, D.C. (2003). Strategic control over rate of processing in word reading: A computational investigation of the temp-naming task. *Journal of Memory and Language*, 48, 207-232.

McCarthy, R., & Warrington, E.K. (1986). Phonological reading: Phenomena and paradoxes. Cortex, 22, 359-380.

McClelland, J. L., & Patterson, K. (2002). Rules or connections in past-tense inflections: What does the evidence rule out? *Trends in Cognitive Sciences, 6*, 465-472.

Minsky, M., & Papert, S. (1969). Perceptrons: An introduction to computational geometry. Cambridge, MA: MIT Press.

O'Reilly, R. C. (2001). Generalization in interactive networks: The benefits of inhibitory competition and Hebbian learning. *Neural Computation, 13*, 1199-1241.

Patterson, K. & Ralph, M. (1999). Selective disorders of reading? *Current Opinion in Neurobiology, 9*, 235-239.

Plaut, D. C. (1995). Double dissociation without modularity: Evidence from connectionist neuropsychology. *Journal of Clinical and Experimental Neuropsychology, 17*, 291-321.

Plaut, D. C. & Gonnerman, L. M. (2000). Are non-semantic morphological effects incompatible with a distributed connectionist approach to lexical processing? *Language and Cognitive Processes, 15*, 445-485.

Plaut, D. C., McClelland, J. L., Seidenberg, M. S., & Patterson, K. (1996). Understanding normal and impaired word reading: Computational principles in quasi-regular domains. *Psychological Review, 103*, 56-115.

Plaut, D. C., & Kello, C.T. (1999). The emergence of phonology and from the interplay of speech comprehension and production: A distributed connectionist approach. In B. MacWhinney (Ed.), *The Emercenge of Language* (381-415). Maweh, NJ: Erlbaum.

Pinker, S. (1999). *Words and Rules: The Ingredients of Language*. New York: Basic Books.

Pinker, S., & Ullman, M. T. (2002). The past and future of the past tense. *Trends in Cognitive Sciences, 6*, 456-463.

Rumelhart, D. E., Durbin, R., Golden, R., & Chauvin, Y. (1995). Backpropagation: The basic theory. In C. Yves & D. E. Rumelhart (Eds.), *Backpropagation: Theory, architectures, and applications* (pp 1-34). Hillsdale, NJ: Lawrence Erlbaum.

Rumelhart, D. E., Hinton, G. E., & Williams, R. J. (1986). Learning representations by back-propagating errors. *Nature, 323*, 533-536.

Rumelhart, D. E., McClelland, J. L. (1986). On learning the past tenses of English verbs. In D. E. Rumelhart, McClelland, J. L., and The PDP Research Group (Eds.), *Parallel distributed processing: Explorations in the microstructure of cognition* (pp. 216-271). Cambridge, MA: MIT Press.

Ullman, M. T., Corkin, S., Coppola, M., Hickok, G., & et al. (1997). A neural dissociation within language: Evidence that the mental dictionary is part of declarative memory, and that grammatical rules are processed by the procedural system. *Journal of Cognitive Neuroscience, 9*, 266-276.

Van Orden, G.C., Pennington, B.F., & Stone, G.O. (2001). What do double dissociations prove? *Cognitive Science, 25*, 111-172.

Siegel-Wolf Award for Best Applied Paper

Automated GOMS–to–ACT-R Model Generation

Robert St. Amant (stamant@cs.ncsu.edu)
Department of Computer Science, North Carolina State University
Raleigh, NC 27695 USA

Frank E. Ritter (ritter@ist.psu.edu)
School of Information Sciences and Technology, The Pennsylvania State University
University Park, PA 16802 USA

Abstract

We describe a system, G2A, that produces ACT-R models from GOMS models containing hierarchical methods, visual and memory stores, and control constructs. Because GOMS is a more abstract formalism than ACT-R, a single GOMS operator might be plausibly translated in different ways into ACT-R productions (e.g., a GOMS Look-for operator might be carried out by different visual search strategies in ACT-R). Given a GOMS model, G2A generates and evaluates alternative ACT-R models by systematically varying the mapping of GOMS operators to ACT-R productions. In experiments with a text editing task, G2A produces ACT-R models with predictions that are within 5% of GOMS model predictions. In the same domain, G2A also generates ACT-R models that give good predictions of overall task duration for actual users (within 2% error), though the models are much less accurate at a detailed level.

Introduction

A recent trend has appeared in cognitive modeling for human-computer interaction: researchers are making the capabilities of cognitive architectures much more accessible to developers and designers who lack modeling expertise (Ritter et al., 2003). Salvucci and Lee's ACT-Simple (2003) automatically generates ACT-R models (Anderson et al., in press) from a language similar to KLM-GOMS. Techniques developed by John et al. (2004) automatically build ACT-R models (via KLM and ACT-Simple) from actions demonstrated by interface designers. A system by St. Amant, Horton, and Ritter (2004), in the domain of cell phone modeling, partially automates the generation of ACT-R models from the specification of a keypad and menu hierarchy.

The inspiration for the work described in this paper is ACT-Simple and to a lesser extent TAQL (Yost, 1993). ACT-Simple demonstrates the feasibility of automatic translation of a high-level, GOMS-like specification language into detailed ACT-R models—an exciting and significant achievement. That said, the ACT-Simple translation process and its results have limitations. Each operator in the source model is transformed into one or two ACT-R productions in a static translation process. These productions are chained together in a mostly linear fashion. The productions make almost no use of the environment (e.g., all visual processing is represented as shifts in attention between two fixed locations.) Essentially, the generated ACT-R models have relatively little that is "cognitive" about them: there is no input of information from the environment, there are no memory retrievals for information processing, and there is no decision-making. This is not to slight ACT-Simple—these constraints are mainly dictated by KLM-GOMS, which is intended to provide a simple, largely external description of performance.

Our research explores the same basic problem addressed by ACT-Simple, that of translating high-level specifications into more detailed cognitive models, but shifts the focus from simplicity to representational power. We have built a system, G2A, that translates models from GOMSL into ACT-R. GOMSL is an abstract but rich modeling language within the GOMS family (Kieras, 1999). Going beyond the capabilities of KLM-GOMS, GOMSL allows representation of mental objects, working memory storage, primitive internal and external operators, composite methods, and various flow-of-control constructs. G2A supports all of these capabilities in its translation process. Our goal is to allow the expression of models as simple as KLM-GOMS models, while letting modelers add more complexity if it is needed.

We chose ACT-R as a target language (e.g., rather than EPIC, Kieras & Meyer, 1997) to leverage existing work on model translation. We chose GOMSL as a source language because of a detailed manual documenting its structure and use (Kieras, 1999). G2A has been tested using 34 examples taken from the GOMSL manual (mostly verbatim, but with a few syntax corrections). The most complex GOMSL model given takes up about four pages—190 lines of code containing 15 mental object definitions, 11 methods, and two sets of selection rules. To match this model, the translation process generates an ACT-R model of five chunk definitions, 23 chunks, 79 productions, and various auxiliary constructs (over 1500 lines of formatted model code). The ACT-R model's predictions closely match the predictions of the source GOMSL model. Using the same GOMSL model as a starting point, G2A can also automatically generate different models that give reasonable predictions of actual user performance, based on different assumptions to match different evaluation criteria.

The goal of developing G2A was not simply to translate GOMS models to ACT-R models, which turned out to be demanding in programming terms but conceptually relatively simple.[1] Our work provides insight into some

[1] G2A is available at www4.ncsu.edu/~stamant/G2A.

of the more interesting problems faced by the cognitive modeling community today. How does one go about extending simple models into more detailed ones? More generally, what is the space of models appropriate for a given task and performance data? (With G2A, the "degrees of freedom" are related to the flexible mapping of actions to productions, visual processing, and environment specification.) How can different, potentially complex models be integrated? (We argue that integrating high-level specifications is often easier than integrating detailed models themselves.) How can ACT-R modeling concepts be made accessible to novice modelers already familiar with GOMS formalisms? (G2A provides an explicit computational account of how some common ACT-R idioms are used to represent specific behaviors.) Finally, even if the feasibility of generating ACT-R models from GOMSL models is not an unexpected result, it is worthwhile to see such an expectation verified.

GOMSL translation

GOMSL syntax is comparable to that of a procedural programming language, as shown in Figure 1. For reasons of space, in our discussion we will assume basic familiarity with GOMS and ACT-R concepts, and we will not explicitly treat architectural differences; some details are also elided. G2A begins by parsing a GOMSL model into an intermediate representation[2] appropriate for further translation as described below.

Methods and flow of control. GOMSL methods are sequences of steps, where each step can contain one or more operators. For example, in Figure 1 the first step in the method Edit Document is the Store operator. G2A follows ACT-Simple in its representation of control: every production contains a condition test of the ?state slot of a Goal chunk and an action to update that slot. Sequences of operators are translated into productions that are ordered by appropriately sequenced ?state values. Step execution is not always sequential within a method; Decide statements support branching based on predicate tests Goto allows arbitrary transfer of control to labeled steps. These forms are translated to appropriately set ?state values in the relevant productions.

Each GOMSL method satisfies a goal; instead of calling operators, a method can "invoke" another method with an Accomplish-goal (AG) statement. This flow of control translates to the creation of a subgoal that is pushed on the goal stack.[3] When a method completes, it ends with a Return-with-goal-accomplished (RGA) statement; this becomes a pop action in ACT-R. Selection rules, which govern the choice of methods when multiple methods can accomplish a given goal, are translated similarly to Decide statements.

[2]For this we used an off-the-shelf LALR parser, written by Mark Johnson, from the online AI Repository at CMU.

[3]The models produced by G2A include a small number of ACT-R 4 constructs (Anderson & Lebiere, 1998) that are deprecated in ACT-R 5 (Anderson et al., in press); we are working to make the system ACT-R 5/6 compliant.

```
LTM_item: Cut_Command
  Name is Cut.
  Containing_Menu is Edit.
  Menu_Item_Label is Cut.
Method_for_goal: Edit Document
  Step. Store First under <current_task_name>.
  Step Check_for_done.
  Decide: If <current_task_name> is None, Then
    Delete <current_task>; Delete <current_task_name>;
    Return_with_goal_accomplished.
  Step. Get_task_item_whose Name is <current_task_name>
    and_store_under <current_task>.
  Step. Accomplish_goal: Perform Unit_task.
  Step. Store Next of <current_task>
    under <current_task_name>;
    Goto Check_for_done.
```

Figure 1: GOMSL object and method (partial).

Objects and working memory. GOMSL supports declarative object representations, including objects in long term memory (LTM-items), objects available through visual processing (Visual-items), and task descriptions (Task-items). Each object consists of a named collection of property–value pairs. GOMSL objects are translated directly into ACT-R chunks. Each generated chunk is given a unique identifier for reference.

Objects in GOMSL are brought into working memory as named tags, such as <current_task_name> in Figure 1. Properties of an object stored in a tag are also immediately available for processing. G2A captures this functionality via a slot in the Goal chunk for each tag in a GOMSL program. A change to the contents of a tag by a GOMSL operator translates to an update of the corresponding slot in a Goal chunk.

For convenience, GOMSL allows method arguments (pseudo-parameters) to be defined. These are implemented as implicit Store operators, which record values in global tags for the information being passed between methods in AG forms. When a production creates a subgoal, it copies tag values from the current goal to the subgoal before pushing it on the stack.

Operators. GOMSL defines a number of primitive operators that carry the basic load of modeling performance. Aside from the control-flow forms mentioned above, G2A handles Keystroke, Type-in, Click, Double-click, Hold-down, Release, Point-to, Home-to, Speak, Look-for-object-whose, Get-task-item-whose, Store, Delete, Recall-LTM-item-whose, Verify, and Think-of. These encompass all GOMSL operators except Wait-for-object-whose and Wait-for-auditory-object-whose, which we expect to add soon.

For most of these operators, G2A contains one G translation and one or more A translations:

- A-type translations use converted AC-R idioms to generate productions that perform activities equivalent to the GOMSL operator. For example, Point-to expands to a Hand-to-mouse production followed by a Move-cursor production, imposing appropriate conditions on the manual state.

- *G*-type translations use GOMSL-specified duration values (e.g., a Keystroke requires 280 ms). In most cases they generate the same productions as *A*-type constructs.

Think-of is one exception to this scheme; it only has a *G* translation, with the same fixed duration of other mental and visual operators, 1,200 ms. The creation of an ACT-R model using G2A requires choosing a specific *A* or *G* translation for each of the operators in a source GOMSL model. The more complex *A* translations are as follows.

Double-click/A, Hold-down/A, and Release/A are translated into ACT-R Mouse-click actions—ACT-R lacks a direct implementation of these low-level mouse operations as atomic actions. A Double-click/single-action, with duration comparable to a *G*-type translation, provides a partial workaround.

Point-to/A, as mentioned above, expands to two productions, taking a target as an argument. This target can be either a literal symbol, an object stored in a tag, or a literal or object stored in the property of a tag. The default G2A translation process generates a random location associated with the object or literal, which is stored for later pointing actions to the same target. This location is used in the Move-cursor production. (Numerical screen coordinates can be used if provided in the GOMSL model, but this requires specialized task-specific code.)

Recall-LTM-item-whose/A and Get-task-item-whose/A both take a list of predicates and a store tag as arguments, the first operator returning an LTM-item, the second a Task-item. The result of the translation is a production in which the identifier of a retrieved object that meets the predicate tests is stored in the given tag.
A predicate is a comparison between operands. Comparisons in G2A are limited to tests of equality or non-equality of symbols. The first operand to the comparison is a property of the object to be retrieved. The second is either a literal, a literal stored in a tag, or a property associated with an object stored in a tag. In the first two cases, the production that is generated includes a pattern that makes a direct comparison between the object property and the literal or tag. In the third case, an additional buffer is added to the production that allows the retrieval of the chunk corresponding to the identifier stored in the tag and access to its properties. Predicates are processed recursively, one at a time, until the list of predicates is exhausted. The last action carried out by the production is to store the identifier of the retrieved object in the tag slot of the current Goal chunk. Figure 2 shows a typical result for a retrieval with a single predicate.

Look-for-object-whose/search takes a list of predicates and a store tag as arguments. The translation generates productions that search through the set of visual objects present until the desired object is found. These productions use the standard ACT-R idiom of find/attend/harvest for visual acquisition. The harvest production is augmented via the same predicate processing as with Recall-LTM-item-whose.
If the visual module could be guaranteed to find an object that meets all of the predicate tests on its first try, then the translation would be similar to Recall-LTM-item-whose. Lacking this guarantee, however, productions must be generated for each of the predicates that may fail. In case of failure, these secondary productions cause the find production to be fired again, to visit another not-yet-attended visual location. When a visual object is found that passes all the predicate tests, its identifier is recorded in the tag slot of the current Goal chunk.

```
;;; Get-task-item-whose ((is name [task-name])) [current-task]
(p production86
   =goal>
      isa              goal
      [task-name]      =[task-name]
      ?state           edit-document-3
   =match88>
      isa              task-item
      ?id              =temp87
      name             =[task-name]
==>
   =goal>
      [current-task]   =temp87
      ?state           edit-document-4)
```

Figure 2: Sample generated production.

Look-for-object-whose/direct is another *A*-type translation. In many situations, an exhaustive visual search is not carried out, because the user either knows the location of a visual object already, or because the predicates rely on pop-out properties of the object. For such situations, the translation generates a production that simply moves the attention to a specific visual location (randomly generated by default, but recorded for later use, as with Point-to.)

Store generates a production that records a given value (either a literal value, the contents of a tag, or the property the contents of a tag) in a target tag. There is one special case for translation of Store (and Delete, below), as dictated by GOMSL. When these operators occur in the same step as other operators, their action is merged with the other operators, which means that they have no independent duration.

Delete generates a production that sets the contents of a given tag to a null (Empty) value.

Verify/last generates a production that moves attention to the visual location that was most recently visited by a Look-for-object-whose or a Point-to operator. Verify/none generates no production, taking no time.

The translations we have defined have minor limitations, which we are working to remove. We have also begun to examine different sets of translations; for example, we have reproduced several of the models of Salvucci and Lee (2003), using the ACT-Simple modeling language rather than GOMSL as a source language.

Evaluation

Comparing GOMSL and ACT-R models. We can directly demonstrate the accuracy of the GOMSL to ACT-R translation by choosing a GOMSL model, translating it using G translations for each of its methods, and running a comparison of execution times. To simplify our comparison here (and in the rest of this section), ACT-R models are executed using default values for all parameters, with no base-level learning.

For our comparison we used the Edit Document model described earlier. This GOMSL model includes four top-level methods describing activities in a mouse-based word processing environment, executed sequentially with no delays between them. Copy word involves selecting and copying a single word and pasting it elsewhere in the document. Copy arbitrary is a comparable task for a sequence of words. Delete word involves selecting a word and pressing the delete key. Move arbitrary is similar to Copy arbitrary. All editing actions are executed through selections from a pull-down menu. These lower-level activities are accomplished by the methods given in the Method column of Table 1.

The GOMSL column shows the durations per method as given by the GOMSL model (Kieras, 1999, p. 58), and the G2A/M_g column shows the corresponding values for the ACT-R model that G2A generated, using a G-type translation for every GOMSL operator, averaged over 20 runs. The numbers are very close, $r = 0.999$, with small discrepancies due to parallel execution of visual and motor actions in ACT-R, plus the automatic addition to the ACT-R model of hand movements between keyboard and mouse that are implicit in the GOMSL model (specifically, in Select-word and Select-insertion-point). On average, the predictions of the ACT-R translations are within 5% of the GOMSL method durations, and the overall task duration prediction is within 1%. This is a straightforward but non-trivial result: it shows that the hierarchical relationships between GOMSL methods, flow of control within the methods, and transfer of information between methods are captured in executing ACT-R productions, even if production durations are fixed.

Predicting user performance. An obvious next step, staying within the same task domain, is to see how well these models predict actual user performance. If we time users carrying out the Edit Document task in a standard word processing application, we find that it takes 20 seconds or less. This value is far from the predictions of the GOMSL or M_g model. If we believe that these models, despite their performance, correctly represent the basic structure of the task, we can develop explanations for the observed behavior in a systematic way with G2A.

The translation alternatives in the previous section can be thought of as ACT-R idioms for representing particular activities. We can then view G2A as facing the same issues as a human cognitive modeler in constructing an accurate model. Though G2A lacks actual domain knowledge, it can rely on the GOMSL model of the task, and in cases involving analysis of an existing system, G2A may have access to user data. In other words, G2A can carry out a model-fitting process, exploring different elaborations of the GOMSL task structure.

Table 1: Model predictions for the Edit Document task.

Method	GOMSL	G2A/M_g	Error
Select-insertion-point	3.60	4.14	15%
Select-word	4.40	4.71	7%
Erase-text	6.40	7.25	13%
Select-arbitrary-text	6.70	6.52	3%
Issue-command	9.05	8.55	6%
Paste-selection	12.80	12.84	0%
Copy-selection	14.85	14.36	3%
Cut-selection	16.20	15.79	3%
Copy-text	28.85	28.55	1%
Move-text	30.80	29.98	3%
Edit-document	101.20	100.63	1%

To do this, G2A treats the alternative translations of GOMSL operators as a search space. By varying the translations that are activated in generating ACT-R models, G2A explores this search space, using hill-climbing to identify the best translation. In hill-climbing, an evaluation function f is applied to a current state s. The function f is applied to each of the states neighboring s, and if any of these successor states produces a better value, the best of them becomes the new current state. The process repeats until no successor state produces an improvement in f.

A state for the G2A search is a set of translations, each chosen from the set of possible translations for a single operator: Look-for-object-whose/direct + Verify/none + Click/G + ... Each translation set produces a unique model. Successors to a translation set are those that differ in the translation of one GOMSL operator. For f, G2A executes the model corresponding to the current translation set 20 times, collects predictions of the total duration of the Edit Document task, and computes the difference from a target duration. (Notice that we cannot directly compare method execution times, because method boundaries are only implicit in user behavior—all we have access to is external events.)

To test this idea, we conducted a small user study. We implemented a simple, instrumented text editing application roughly equivalent to Microsoft Notepad that supports the Edit Document tasks. A pilot subject ran ten trials of four tasks (Copy word, Copy arbitrary, Delete word, and Move arbitrary) that the Edit Document model also performed. We then ran six users through the same procedure.

We used the mean total duration of the pilot user's trials as a target for the G2A search. G2A evaluated about 50 models in its search. The best model found, M_{td}, relies on A translations for all mental and visual activities and most motor actions, performs no visual search, directing attention to known locations, and does no verification. Comparing M_{td}'s predictions of total duration with the performance of the six users of the study shows good results. The grand mean duration over the six users was 18.85 seconds, with a standard deviation of 1.82 over the six user means. M_{td}'s prediction of 19.18 seconds gives an error of about 2%.

This result is also worth noting: the ACT-R model automatically produced by G2A reflects straightforward decisons that a human modeler might make to capture human performance in this domain, and its prediction is much more accurate (and obtained with no further modeling effort) than that of the original GOMSL model. M_{td} has about the same predictive power as comparable generated ACT-R models described in the literature (John et al., 2004; Salvucci & Lee, 2003).

Refining a model for more detail. M_{td} has obvious built-in limitations. In particular, the duration of its mouse movements and visual attention shifts are based on random locations, rather than an actual environment, and thus its detailed predictions of user actions are unlikely to be reliable. Using the data from the pilot subject, however, it is possible to carry out a more detailed analysis by refining M_{td}.

We defined a new search evaluation function that measures the intervals between keyboard/mouse events (i.e., stripping out the cumulative duration to each event), and computes the mean squared error with the corresponding interval durations for our pilot user. We also altered the model generation process to use actual locations of objects, as measured in our instrumented application, rather than random values. The best model produced by the search, M_{id}, is almost identical to M_{td}: it does no visual search and no verification, and it relies on A translations for all mental operators. It varies only in that Double-click is implemented by a single action translation, rather than a sequence of mouse clicks.

M_{id}'s prediction of overall task duration is about the same as that of M_{td}, 19.03 seconds. (In general, we would not expect models generated by different search evaluation functions to be so similar, but in this domain it appears that several plausible models are clustered in the same region of performance.) The more detailed predictions of M_{id} are shown in Table 2, along with the mean intervals between mouse click and keystroke events for the six users in the study. Unfortunately, these predictions of M_{id} are not as good as we might have hoped for. The correlation between user intervals and M_{id} predictions is 0.754, and the average error in the predictions is 35%. There is no obvious pattern in the error values shown in Table 2, but if the two largest could be reduced, this would bring the model's error down to 24%—still high, but more respectable.

There are two ways we might improve M_{id}. First, as discussed in the previous section, some low-level mouse actions, such as Double-click, Hold-down, and Release, have no direct representations in ACT-R; our substitutions are inevitably inaccurate. We expect that once such actions are developed, validated, and added to the ACT-R architecture, we will see better results. Second, it is possible that the structure imposed on ACT-R productions by GOMSL forms (e.g., GOMSL methods entail bookkeeping overhead due to AG and RGA operators) degrades the accuracy of predictions of a generated model, compared with the predictions of a native ACT-R model. Translations that combine or parallelize GOMSL operators more effectively may help.

Table 2: User performance on the Edit Document task.

Action	User	M_{id}	Error
Double-click	0.25	0.20	20% (0.05 s)
Select Copy	1.48	1.40	5% (0.08 s)
Set insertion	0.53	1.01	91% (0.48 s)
Select Paste	1.28	1.18	8% (0.10 s)
Select sentence	2.16	2.02	6% (0.14 s)
Select Copy	1.81	1.16	36% (0.65 s)
Set insertion	0.76	1.01	33% (0.25 s)
Select Paste	1.62	1.18	27% (0.44 s)
Double-click	1.35	1.74	29% (0.39 s)
Press Delete	0.64	1.39	117% (0.75 s)
Select sentence	2.92	2.00	32% (0.92 s)
Select Cut	1.82	1.13	38% (0.69 s)
Set insertion	0.74	0.96	30% (0.22 s)
Select Paste	1.49	1.18	21% (0.31 s)

Despite the limited predictive power of M_{id} at a detailed level, it constitutes another interesting result. M_{id} was produced as a refinement of M_{td}, in a largely automated process that called for only the effort of supplying environment information. Otherwise the process depended mainly on the GOMSL model for domain knowledge and pilot user data to guide model construction. No existing work on model translation has tested model predictions at the same level of detail as M_{id}, and given this lack of experience we find its performance adequate.

Limitations. One important limitation in our evaluation of M_{td} is that it is among the fastest models in the search space explored by G2A. In other words, if we had chosen an evaluation function that simply minimized execution duration, our results would have been about the same. Our combination of task and subject pool is thus less informative than it might be, because our target performance lies at an extreme, allowing little flexibility in model structure.

We have also neglected one of the most interesting opportunities for G2A: exploring possible translations of the Think-of operator. John et al. (2004) have shown that the placement of mental operators in a keystroke-level model, a difficult task for human modelers, can be automated. An elaboration of our search process might insert ACT-R productions into a model to better match user durations, but determining exactly what these productions should do is a much harder problem that requires domain-dependent reasoning. This is an open issue for future research.

Discussion

G2A is a work in progress, but it is sufficient to demonstrate the promise of the approach. We expect that with further development and refinement by ourselves and others, G2A can benefit the research community.

There are several avenues worth pursuing. One is bracketing (Kieras & Meyer, 2000), a tactic that can guide the iterative modeling process. A fastest possible model and a slowest reasonable model (based on optional or inefficiently executed task components) are derived from a base strategy; results can then help a designer decide, for example, whether the performance demands of a system are likely to outstrip human capabilities. G2A could contribute to bracketing by generating different models given a base strategy represented as a GOMSL model. If the fastest and slowest models are within the search space of G2A's translations—an important consideration, but a reasonable expectation in some domains—then these might be identified automatically. Work along these lines could lead to a useful extension to G2A, in the form of translations of higher-level GOMSL methods as well as primitive operators.

An idea closely related to bracketing is that modelers might associate classes of translations with different classes of users. In the same way that a programmer sets compiler flags to generate faster or safer executable code, sets of translations can be identified with expert or novice behavior. These translations might go further than the ones described above, varying internal ACT-R parameters rather than only model structure. In such cases, the expert modeler would simply bypass G2A search to settle directly on a specific model.

Another possible benefit is the use of G2A's abstractions in teaching modeling concepts to computer scientists. Operator translations generate sequences of productions comparable to those in ACT-R tutorial modules, though G2A's productions tend to be much more specialized. The abstraction of productions into methods can be thought of as providing a simple, high-level programming language for model components. Productions from different methods do not interact with each other except at the beginning and end of their translation sequences; information is passed through named, global slots. These factors make processing in a generated model much easier to understand. G2A might be extended to allow user queries (e.g., "What does an exhaustive visual search look like in ACT-R?") by showing how specific translations are performed.

A final potential benefit relates to model size and reuse. The largest models that G2A generates for the Edit Document task contain about 100 productions, and nothing prevents us from generating much larger models for more complex tasks. It is time-consuming and error-prone to generate models of this size by hand even if the modeled behavior is completely understood. Adding methods to a GOMSL model and examining their translation is straightforward in G2A and easier, in our experience, than writing ACT-R productions directly. If a task appears to be too large to be handled with conventional ACT-R development tools, or if model requirements outgrow GOMS (e.g., a model of visual processing or learning behavior may be needed), or if a high-level, procedural tasking language is required to control an extended experiment, then G2A offers a starting point.

Any potential benefits of G2A will involve tradeoffs that can only be determined by further research. For example, it may happen that abstract tuning of models via high-level compiler-like directives does not allow sufficient precision, but on the other hand it may provide more control. It may be that increasing model modularity, which makes larger models more feasible, involves a reduction in accuracy. This is suggested by our model refinement efforts, but it is not clear that this is a necessary implication of G2A processing. Still, we believe that G2A represents a promising approach. G2A advances the state of the art in cognitive model generation and points to several important areas for further research.

Acknowledgments

Thanks to Thomas Horton, who built the editing application and ran the user study. Thanks to Mike Byrne and two anonymous reviewers for suggesting significant improvements to this paper. This research was supported by NSF award IIS-0083281 and by ONR award N00014-02-1-0021. The information in this paper does not necessarily reflect the position or policies of the U.S. government, and no official endorsement should be inferred.

References

Anderson, J. R., Bothell, D., Byrne, M. D. & Lebiere, C. (in press). An integrated theory of the mind. *Psychological Review.*

Anderson, J., & Lebiere, C. (1998). *The Atomic Components of Thought.* LEA, Mahwah, NJ.

John, B. E., Prevas, K., Salvucci, D. D., & Koedinger, K. (2004). Predictive human performance modeling made easy. *Proceedings of CHI '04*, 455–462. ACM.

Kieras, D. E. (1999). *A guide to GOMS model usability evaluation using GOMSL and GLEAN3.* Technical Report, University of Michigan, Ann Arbor, MI.

Kieras, D. E., & Meyer, D. E., (1997). An overview of the EPIC architecture for cognition and performance with application to human-computer interaction, *Human-Computer Interaction, 12*(4): 391-438.

Kieras, D. E., & Meyer, D. E., (2000). The role of cognitive task analysis in the application of predictive models of human performance. In *Cognitive Task Analysis*, Schraagen, J. M., Chipman, S. F., & Shalin, V. L. (eds.) Lawrence Erlbaum, Mahwah, NJ.

Ritter, F. E., Shadbolt, N. R., Elliman, D., Young, R., Gobet, F., & Baxter, G. D. (2003). *Techniques for modeling human performance in synthetic environments: A supplementary review.* Wright Patterson AFB: Human Systems Information Analysis Center. WWW: //iac.dtic.mil/hsiac/S-docs/SOAR-Jun03-Front.pdf

St. Amant, R., Horton, T. E., & Ritter, F. E. (2004). Cognitive modeling for cell phone menu evaluation. *Proceedings of CHI '04*, 343–350. ACM

Salvucci, D. D., & Lee, F. J. (2003). Simple cognitive modeling in a complex cognitive architecture. *Proceedings of CHI '03*, 265-272. ACM.

Yost, G. R. (1993). Acquiring knowledge in Soar. *IEEE Expert, 8*(3), 26-34.

Symposium

Competitive Modeling Symposium: PokerBot World Series

Christian Lebiere (clebiere@maad.com)
Micro Analysis and Design, Inc.
6800 Thomas Blvd., Pittsburgh, PA 15208 USA

Dan Bothell (db30+@andrew.cmu.edu)
Department of Psychology, Carnegie Mellon University
Baker Hall 446J, 5000 Forbes Ave, PA 15213 USA

The Cognitive Challenges in Poker

Game playing seems to satisfy a basic craving of human cognition by exercising its fundamental abilities in a competitive setting. Therefore, it provides an excellent benchmark to study and evaluate cognitive models in tractable yet naturalistic settings that are simple and formal yet reproduce much of the complexity of real life. Poker is probably the most widely played card game, with endless variations played by millions of adherents from casual players gambling pennies to professionals competing in million-dollar tournaments. Unlike other games that emphasize one particular aspect of cognition, poker involves a broad range of cognitive activities, including:

- Reasoning under uncertainty (opponents' cards)
- Dealing with probabilistic outcomes (future cards)
- Decision-making with multiple options (chips used for bets)
- Individual differences (different styles of play)
- Inference of intent (from opponents' bets)
- Intentional deception (bluffing, sandbagging)
- Pattern recognition (detecting trends from flow of game)
- Social and emotional aspects (dealing with winning and losing)
- Economic behavior (factoring impact of amount of bets)

Because of the range of cognitive activities involved, poker provides a broader and more challenging test for cognitive modeling than other games such as chess that focus on a more restricted range of mechanisms (e.g. search). Despite the complexity of aspects involved, it remains a highly tractable domain, partly because it abstracts away from computationally demanding perception and interaction problems. Poker is increasingly being played in online gaming communities where the need for challenging, cognitively plausible agents is increasing. Poker therefore provides a challenging domain at the intersection of fundamental research questions and potential mass application.

PokerBot at ICCM2004

Open competitions such as Robocup (http://www.robocup.org/) and the DARPA grand challenge (http://www.darpa.mil/grandchallenge/) have recently provided a promising mechanism to make progress on long-term research issues while focusing on a practical, measurable goal. To that end, a competitive modeling symposium took place at ICCM-2004 with the ultimate goal of producing a fully functional, human-like poker agent. The competition took place in two parts, emphasizing the dual aspects of cognitive modeling: the production of accurate models of human cognition that are not mere post-hoc descriptions of human performance but can perform actual tasks in a fully functional manner. The first part involved multiple runs of head-to-head, winner-take-all tournament play between models to test their functionality. The second part involved a live evaluation of the cognitive plausibility of the models during the conference. The best models were presented at a symposium during the conference where progress on research issues was discussed.

Tournament Rules

Because of the open-ended nature of the game that favors the exploration of the research issues, the variant of the game selected for the competition was No Limit Texas Hold'em a popular, fast-paced game that is the centerpiece of the World Series of Poker (http://www.harrahs.com/wsop/index.html).

The initial bankroll given each player at the start of the game was $10,000, with initial blinds of $10 and $20 that doubled every 100 hands to bring the game to a timely conclusion. To fairly distribute resources, the response time and memory space for each player was sharply limited. The winner of each tournament was the last player left, with as many tournament runs as possible played during the week between the submissions and the announcement of the results, with the players restarted to their initial state between each game. The complete logs of all games played were made available after the tournament to allow anyone to review them for fairness and accuracy and to serve as the basis for a database of games for further cognitive modeling.

Models of Eye-Movement Control During Reading

Erik D. Reichle (reichle@pitt.edu)
Department of Psychology, University of Pittsburgh
LRDC, 3939 O'Hara St., Pittsburgh, PA 15260 USA

Ronan Reilly (Ronan.Reilly@may.ie)
Department of Computer Science, National University of Ireland
Maynooth, Co. Kildare, Ireland

Eike M. Richter (richtere@rz.uni-potsdam.de)
Department of Psychology, University of Potsdam
Postfach 60 15 53, 14415 Potsdam, Germany

Shun-Nan Yang (syang3@ski.org)
The Smith-Kettlewell Eye Research Institute
2318 Fillmore St., San Francisco, CA 94115 USA

Eye-Movement Control During Reading

Models of eye-movement control during reading can be classified along two theoretical dimensions: (1) the degree to which cognitive processes (e.g., word identification) influence the moment-to-moment decisions about when to move the eyes; and (2) whether attention is allocated serially, from one word to the next, or allocated in parallel, across more than one word. This symposium will provide a forum to review four of the models that represent four distinct positions along these theoretical dimensions. The models that will be discussed include: (1) the *Competition-Inhibition model* (Yang & McConkie, 2001); (2) *E-Z Reader* (Reichle, Rayner, & Pollatsek, 2003); (3) *Glenmore* (4); and *SWIFT* (Engbert, Longtin, & Kliegl, 2002). A comparative analysis of the models will then examine how these models account for a variety of independent variables that are known to affect eye movements during reading, including: word length, word frequency, word predictability, orthographic regularity, morphological complexity, and strategic modulation of eye movements (e.g., careful reading vs. skimming). This comparative analysis will examine how the models account for each of the preceding variables and their effects on a number of dependent measures, including: various commonly used word-based measures (e.g., first-fixation duration, gaze duration, probability of skipping, number of fixations), measures of saccade amplitude, and fixation locations. Finally, the comparative analysis of the models will also include a discussion of various phenomena (some of which are currently disputed to exist) that have been predicted by and used to evaluate the models of eye-movement control. These phenomena include: (1) the general shape of the landing-site distributions and how they are affected by saccade length (i.e., systematic range error) and launch-site fixation duration; (2) the effect of the optimal-viewing position on the probability of making a refixation; (3) the inverted optimal-viewing position effect on first-fixation durations; (4) the effects of parafoveal preview on fixation durations; (5) the attenuation of the parafoveal preview effect due to foveal processing difficulty; (6) parafovea-on-fovea preview effects; (7) the lagged (spill-over) effects of word difficulty; and (8) the fixation duration costs that have been associated with word skipping. This discussion will focus on how—rather than how well—the models explain these phenomena, and will emphasize unique predictions so as to provide points of maximum contrast among the models.

References

Engbert, R., Longtin, A., & Kliegl, R. (2002). A dynamic model of saccade generation in reading based on spatially distributed lexical processing. *Vision Research*, *42*, 621-636.

Reichle, E. D., Rayner, K., & Pollatsek, A. (2003). The E-Z Reader model of eye-movement control in reading: Comparisons to other models. *Behavioral and Brain Sciences*, *26*, 445-526.

Reilly, R. & Radach, R. (2003). Glenmore: An interactive activation model of eye movement control in reading. In J. Hyönä, R. Radach, & H. deubel (Eds.), *The mind's eye: Cognitive and applied aspects of eye movement research.* Amsterdam: Elsevier.

Yang, S.-N. & McConkie, G. W. (2001). Eye movements during reading: A theory of saccade initiation times. *Vision Research*, *41*, 3567-3585.

Presented Papers

Streak Biases in Decision Making: Data and a Memory Model

Erik M. Altmann (ema@msu.edu)
Bruce D. Burns (burnsbr@msu.edu)
Department of Psychology
Michigan State University
East Lansing, MI 48824

Abstract

Streaks of past outcomes, for example of gains or losses in the stock market, are one source of information for a decision maker trying to predict the next outcome in the series. We examine how prediction biases based on streaks change as a function of length of the current streak. Participants experienced a sequence of 150 flips of a simulated coin. On the first of a streak of heads, participants showed positive recency, meaning that they predicted heads for the next outcome with a greater-than-baseline probability. As streak length increased, positive recency first decreased but then increased again, producing a quadratic trend. We explain these results in terms of outcome-prediction processes that are sensitive to the historical frequency of streak lengths and that make heuristic assumptions about changes in bias of the outcome-generating process (here, the coin). An ACT-R simulation captures the quadratic trend in positive recency, as well as the baseline heads bias, in two experimental conditions with different coin biases. We discuss our memory-based model in relation to a model from the domain of economics that posits explicit representation of an "urn" from which events are sampled without replacement.

Introduction

A streak of repeated outcomes can be an important source of information for decision makers trying to predict the outcome of the next event. For example, asked to predict whether the global average temperature will increase or decrease next year, a decision maker who has access to historical records may predict an increase simply on the basis of past trends. A bias of this form, in which a streak of past outcomes is taken as evidence that the next outcome (or measurement) will be in kind, is often referred to as *positive recency*. In terms of the decision-maker's causal reasoning, one could imagine that a long streak of temperature increases induces a belief that an underlying causal mechanism is at work (warming due to greenhouse gases, for example). This causal mechanism will then govern changes in next year's measurement as well (modulo extraneous variance).

Of course, whether the decision maker makes this or the opposite prediction will depend on precisely how he or she represents the mechanism producing the streak. If the decision maker happens to work for the current Bush administration, he or she may well understand historical increases in the global temperature in terms more similar to the well-known gambler's fallacy, in which a gambler takes a streak of undesirable outcomes as evidence that his or her luck will soon change. More generally, a bias of this form, in which a streak of outcomes is taken as evidence that the next outcome will be opposite, is often referred to as *negative recency*. In the case of gambling, each gamble is an independent event, so there is no causal mechanism linking the outcomes (hence the fallacy). However, there are situations in which negative recency is a rational bias, namely when outcomes are sampled without replacement. For example, a rat in a maze, reflecting instincts of foraging animals, usually has a bias against returning to the location where it found food on the previous trial. The rat's causal model is, presumably, that food at a given location is depleted before it is replaced; if this model is correct, then negative recency is adaptive.

In this paper we examine how recency biases change as a function of streak length. That is, we are interested in how the length of a streak, up to and including the most recent outcome, affects the decision maker's prediction concerning the next outcome. An experimental paradigm appropriate for addressing such issues involves a two-choice prediction task (Estes, 2002; for simulations, see, e.g., Lebiere, Gray, Salvucci , & West., 2003; Lovett, 1998). In the standard experiment with this kind of paradigm, the participant is asked on each trial to predict the outcome of an event such as a coin flip. The "coin" typically has a bias toward one outcome or the other, of which the participant is not informed. The question of interest often has to do with probability learning – how the participant's bias to predict one outcome or the other changes over time. The usual finding is that participants "match" rather than "maximize", meaning that over many trials their bias tends to asymptote at the level of the bias in the event generator; for example, if the "coin" is biased to produce 75% heads, then participants will, by the end of a session, predict heads on roughly 75% of trials. Under a maximizing strategy, participants would come to predict heads 100% of the time, once they detected a bias, so a matching strategy is to difficult to explain using the simplest rules of rational choice.

Probability learning studies have thus shown that people are to some extent sensitive to base rates and changes in base rates, and adjust the frequency of their predictions accordingly, if sub-optimally. Base rates, though related to streaks, are a distinct source of information, with different dynamics that may make them more or less appropriate to a given decision-making scenario. Thus, a probability learning experiment involving a biased coin might track changes in the bias to predict heads as experience with the biased coin grows. We are interested in tracking changes in the bias to predict heads as a streak of heads increases in length, from one head (following a tail), to two consecutive heads, and so on. Thus, in terms of the gambler's fallacy, we are interested in how the strength of the gambler's bias might change as a function of number of losses. Similarly, in terms of the hot-hand heuristic (Burns, 2004), in which streaks of successes serve as an adaptive allocation cue, we are interested in how the strength of the team's bias to give the ball to one shooter is affected by that shooter's recent success at scoring.

Apart from the empirical question of whether decision makers respond to streaks, there is also an important theoretical question relating streaks to base rates. Both reflect any biases toward one outcome or the other in the outcome-generating process; the base rate of that outcome will be higher, and the occurrence and length of streaks of that outcome will be more frequent. However, if the bias in the outcome-generating process happens to change, as a function of a shift in environmental characteristics, for example, the base rate will change only gradually in response. In contrast, the probability of a streak of a given length will change much more quickly. For example, a sudden increase in the heads bias of a virtual coin can immediately produce a streak of heads that is much longer than any that the decision maker experienced under the old bias; the base rate, which represents the overall historical frequency of heads, will change much more slowly. Thus, streak statistics are the more sensitive measure of change in the environment, and may play an important role in the decision making of adaptive organisms, particularly when there is variability in the bias of the outcome-generating process (Burns, 2004).

In the first section below, we describe an experiment in which participants experienced a series of coin flips and were asked after each flip to predict whether or not the next outcome would be heads. Dependent measures were (1) the overall frequency of heads predictions, and (2) the frequency of heads predictions conditional on length of heads streak (the frequency of heads predictions after one head, after two heads, and so on). In the second section, we present a model that accounts for the resulting patterns of conditional heads predictions in terms of simple memory-based decision processes; the model has available a strategy in which it asks itself whether it has seen a streak of this length before, and bases its prediction based on what is retrieved from memory. The discussion relates our memory-based model to a formal model from the economics literature.

Experiment

In this experiment, each participant experienced a series of 150 coin flips, and was asked after each flip to predict the outcome of the next flip. Aiming for replication with some generality, we manipulated the coin's bias towards heads, between subjects. Participants were told that there might be a bug in the computer program generating the coin flips, such that the outcome of one flip might influence the outcome of the next; the nature of the bug was not specified, but the goal was to invite participants to view flips in causal terms, such that some underlying mechanism might be responsible for streaks of heads. Burns (2002) found this instruction to be effective in manipulating participants' beliefs about the randomness of the outcome-generating process. Finally, participants were asked periodically to report a count of heads since their last report, to draw their attention to the bias in the coin.

Method

Participants were 71 students recruited from the Michigan State University subject pool. They were randomly assigned to one of two conditions that differed in the heads-bias of the coin being flipped (75% or 60%).

Participants first read instructions indicating that they would observe a virtual coin being flipped, and would be asked after each flip to predict the next outcome:

> In this experiment you will observe the result of a series of penny tosses. Before each toss, you will be asked to try to predict what you think the result of each toss will be. You can take as long as you like for this, but there is nothing to be gained by waiting. The result of each toss should be random, but the coin may be biased towards heads or tails, and there may be a bug in the program such that one flip may influence the result of the next flip. After selecting "heads" or "tails", you will see the result of the coin toss. Every so often, we will ask you some questions about the task, including how many "heads" have been flipped since we last asked. (You can't keep a written record, so you have to remember this.)

After asking any questions they had, participants then predicted the result of a training flip, and observed the flip itself – a computer animation that unfolded over 3 sec.

Participants experienced then 150 flips, without knowing how many flips there would be. Every 30 flips they were interrupted and asked four questions:

(1) How many times has the coin come up heads?

(2) How well do you think you are doing at the task? Responses were given using a scale from 1 (very poorly) to 7 (very well).

(3) How random do you think have been the flips produced by the program? Responses were given using a scale from 1 (not at all random) to 7 (completely random).

(4) If you have an idea about any error in how the program generates flips, please write it below. Responses were entered in a free-form text box.

Questions (3) and (4) were meant to focus participants on the possibility that some underlying mechanism could be responsible for producing streaks of heads.

In the analyses below, the first 50 trials were excluded, as is standard in probability learning experiments (Estes, 2002).

Results and Discussion

Figure 1 shows the baseline bias to predict heads, measured over all trials (filled bars). The difference in baseline prediction bias across conditions mirrors the change in bias of the coin.

Figures 2 and 3 show conditional prediction bias for 75% and 60% conditions, respectively. The x-axes represent

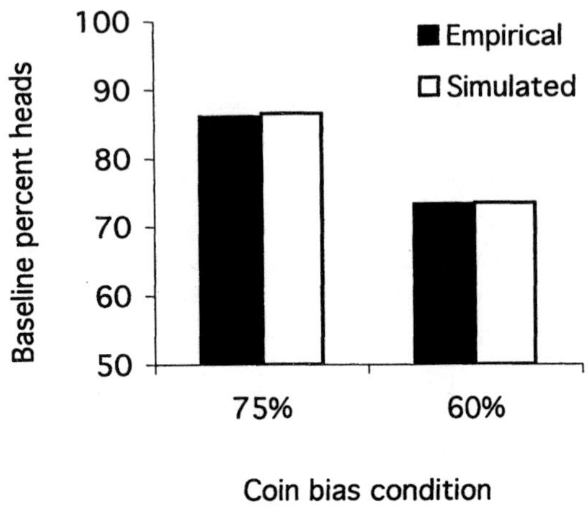

Figure 1: Baseline heads predictions.

streak length; 1 means one head after a tail, 2 means two consecutive heads after a tail, etc. The scale of the x-axes differs because we restricted our analysis to streak lengths represented at least once in every participant's session; longer streaks are more frequent with a more biased coin, so the scale is wider for the 75% condition. The y-axes represent recency biases, expressed in terms of deviation from baseline prediction bias. For example, a value of 2 for streak length 1 would indicate that the bias to predict heads after a streak of one head was 2% greater than the baseline bias to predict heads. Positive values thus indicate positive recency – a greater-than-baseline tendency to predict that a streak will continue – and negative values indicate negative recency.

Analyses of variance confirmed the quadratic trends for length of heads streak apparent in Figures 2 and 3, $t(140)=2.6$, $p<.02$ for the 60% condition and $t(238)=2.4$, $p<.03$, for the 75% condition.

A Model

We explain the U-shaped pattern in Figures 2 and 3 in terms of use of the historical frequency of streak lengths as a heuristic. In particular, we assume that if the decision maker is experiencing a streak, and can recall encountering a longer streak in the past, he or she exhibits positive recency, on the assumption that the past predicts the future. If, on the other hand, the decision maker is unable to recall any streak at all, he or she assumes that the current streak reflects a change in the environment that has made a streak of the current length more probable – and again exhibits positive recency. This logic is common in decision-making models predicated on representativeness and related heuristics. In particular, Rabin (2002) argues that decision makers "overinference" based on streaks, meaning that a streak that is unusually long in their experience leads them to overestimate the bias toward that outcome in the outcome-generating process, relative to the estimate that would be produced by pure Bayesian updating (with its infinite historical window).. Viewing this overinferencing in adaptive terms, it may make sense in a changeable environment to factor a new experience into one's beliefs about the frequencies with which different outcomes are generated, and perhaps experiment with revised beliefs in future decisions.

Our model is implemented in the ACT-R cognitive theory (version 4.0; Anderson & Lebiere, 1998), which incorporates in a variety of ways the notion that the past predicts the future; we exploited the declarative memory mechanism, in which the activation and availability of memory elements is linked to their statistical patterns of use in a given task environment. The model fits are shown in Figures 2 and 3. The empirical data are noisy, and replications currently under way will help to smooth out the curves, but the figures show that the model qualitatively captures the quadratic trends.

Our model-fitting strategy was to fit the baseline and conditional prediction biases from the 75% condition (Figure 1, left, and Figure 2), then to vary the smallest possible number of parameters (which turned out to be one)

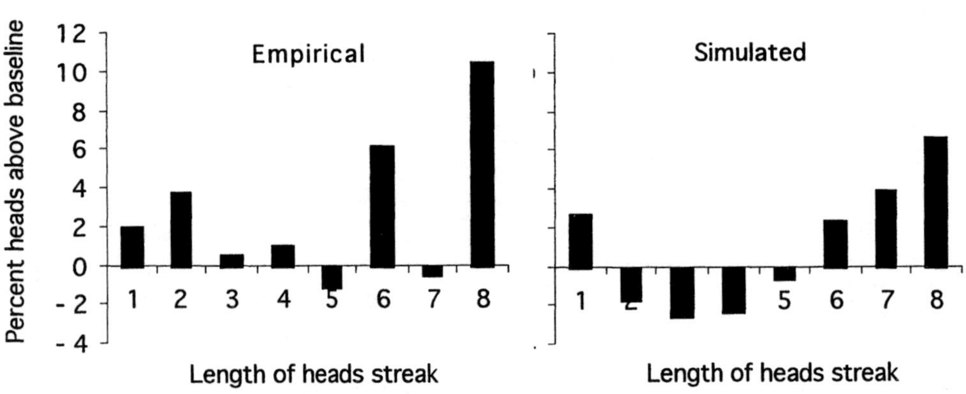

Figure 2: Percent heads predictions above baseline for the 75% condition (see text), conditional on length of heads streak.

to fit the baseline prediction bias from the 60% condition (Figure 1, right). Encouragingly, this one parameter change concurrently produced the fit to the conditional prediction biases in the 60% condition (Figure 3), suggesting that the model is capturing basic sources of variation in the human response to the different conditions.

A Process-Level View

Here we describe the processes that execute on individual trials, as the model observes an outcome (heads or tails) and makes a prediction for the next outcome.

On most trials, the model stochastically selects one of three processes to predict the next outcome. Two of these processes are simple; one guesses heads and one guesses tails (regardless of the current outcome). Guess-heads is biased to be selected more frequently than guess-tails, reflecting the bias of the coin being flipped; the probability learning that produces this bias is of secondary interest here, and we currently do not attempt to model it.

The third process, *heuristic-predict*, is more complex, implementing the frequency-based heuristics sketched above. If this process is selected on a given trial, if spans the next several trials, exiting when the current streak terminates; while this process is active, guess-heads and guess-tails are locked out. The intuition is that the model can wave in and out of using frequency heuristics to guide its decision making, but when it waves in, it remains focused on that heuristic until the current streak ends. While active, the frequency-heuristic process is responsible for updating a count of the current streak in the model's mental focus of attention, retrieving past streak lengths from memory as guidance for predicting the next outcome, and, when the streak is over, encoding the length of the just-ended streak in episodic memory. All such streak lengths are represented as distinct items in ACT-R's declarative memory.

On a given trial, having first updated its count of the length of the current streak, the heuristic-predict process then tries to retrieve a streak length from memory; that is, it tries to retrieve any streak length encoded by heuristic-predict in the past. The retrieval is primed by the length of the current streak; thus, if the model has tracked four heads in a row, memories for streaks of length four will be more active, other things being equal, than memories for streaks of other lengths. This associative activation diminishes with integer distance from the current count, with length-four memories receiving more than length-three and length-five memories, which in turn receive more than length-two and length-six memories.

Three decision rules govern how the heuristic-predict process maps the outcome of an attempted memory retrieval to a prediction for the next coin flip. Successful retrieval is not guaranteed, but if a streak length is retrieved, the model compares the length it represents to the length of the current streak. If the retrieved length is longer than the current length, the model takes this as reason to expect the current streak to continue, so it predicts a repeat of the current outcome. If the retrieved length is the same as or shorter than the current length, the model takes this as reason to expect the current streak to end, so it predicts the opposite of the current outcome. Finally, if retrieval fails altogether, the model takes this as evidence that the current streak is unusually long, and again predicts a repeat of the current outcome. Recapping the logic, the notion is that it may make sense, in a changeable environment, for the decision maker to be open to the possibility that the rate at which an outcome is generated has changed. Here, although there is no explicit revision of beliefs about the environment, there is a decision rule that interprets a retrieval failure as evidence of a novel circumstance.

Each of the three prediction processes – guess-heads, guess-tails, and heuristic-predict – is associated with a probability of being selected on a given trial (when heuristic-predict is not already active). These three parameters were among those adjusted to fit the 75% data. Having fit those results, we held all parameters constant in fitting the 60% data, except the probability of selecting guess-tails. This was increased, reflecting the higher bias toward tails, given the drop to a 60% heads bias in the coin.

An Implementation-Level View

Here we briefly describe the implementation of the model, in terms of ACT-R constructs. The model code is available for downloading at http://www.msu.edu/~ema/streaks.

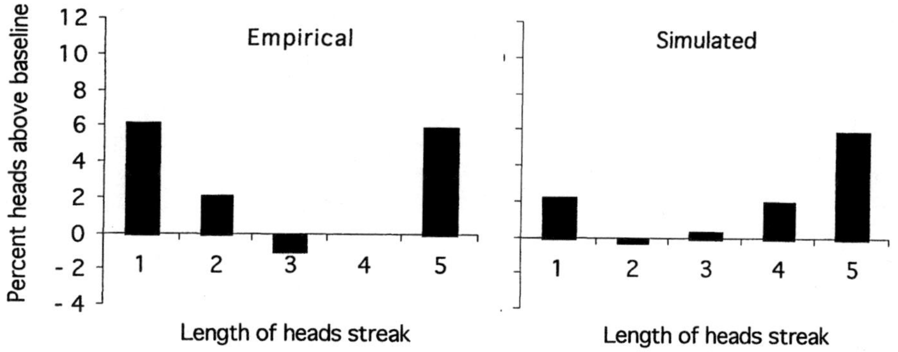

Figure 3: Percent heads predictions above baseline for the 60% condition (see text), conditional on length of heads streak.

The simpler prediction processes – guess-heads and guess-tails – are each represented by one ACT-R production. The selection probability of each production is represented using ACT-R's production utility parameters, which are the inputs to the conflict resolution scheme in this particular production system (selection probabilities are therefore represented only indirectly, in terms of utilities).

The heuristic-predict process is represented by a set of productions operating on a variety of memory representations. The process is initiated by a *start-count* production, which competes in conflict resolution with guess-heads and guess-tails. If start-count is selected on a given trial, it enables an *increment-count* production on the following trials, and eventually a *stop-count* production, which is selected in response to the outcome that ends the streak. Stop-count is responsible for encoding a declarative memory element representing the length of the streak when it ended. On trials on which heuristic is active – that is, on trials after start-count has fired to activate the process but before stop-count has fired to terminate it – guess-heads and guess-tails are excluded from conflict resolution.

The predictive component of the heuristic-predict process begins with a *retrieve-streak* production, triggered once the increment-count production has fired on the current trial. Retrieve-streak attempts to retrieve some streak length encoded by stop-count in response to a past streak. This retrieval attempt can fail, if no potential target's activation level is above a threshold that is a system parameter. The retrieval attempt can succeed with any streak length in memory that is active enough, with the most active streak length at that instant being the one that satisfies the retrieval request. Therefore, activation dynamics, reflecting decay and priming, govern which streak length is retrieved. Decay, characterized by ACT-R's base-level learning mechanism, means that older streaks have less activation and thus are less likely to be retrieved; this suggests, as a test of the model, engineering particular distributions of streak lengths within a session to see if model and humans respond in similar ways.

Priming (associative activation) flows from the current streak length, stored in the system's mental focus of attention, to all streak lengths in memory, weighted by their integer distance from the current streak length. Thus, as described above, if the current streak is of length four, past streak lengths of four receive the most associative activation, past streak lengths of three and five the next most, etc. This priming gradient plays a critical role in producing the quadratic trend in conditional heads predictions. Longer streaks are less frequent than shorter ones, so as the current streak grows, the probability of retrieving any streak length from memory decreases, because fewer streak lengths receive the maximum amount of priming (which may be necessary to bring a target above threshold). Failure to retrieve a streak length triggers one of the decision rules described above, namely to predict continuation of the streak, on the assumption that the bias of the outcome-generator may have suddenly changed. If the next outcome is as predicted, the heuristic-predict process will continue to be active, and will ultimately encode a long streak in memory. Thus, the second and later instances of a long streak will be less likely to trigger this particular decision rule, given the model's prior experience.

Note that the heuristic-predict process is generic with respect to outcome, meaning that in principle another test of the model would be against the empirical biases in tails predictions. We have not yet conducted this analysis.

The model performs the same 150 trials as do human subjects, so in this sense its task environment is veridical; however, we did not try to match the timing parameters of the experiment, so, for instance, there is currently no account of three-second duration of the animated coin flip. We would expect, though, that the changes required accommodate the actual time course of the experiment would be absorbed by existing model parameters.

The parameters adjusted to fit the 75% data (baseline and conditional predictions) include the utilities of guess-heads, guess-tails, and start-count; the activation threshold for retrievals (ACT-R's :rt); and two parameters (peak and slope) that determine the gradient with which priming flows from the current streak length to the same and other streak lengths in memory. To then fit the 60% data, we adjusted only the guess-tails utility; this utility was increased, consistent with the decrease in the heads bias of the coin. The ACT-R decay parameter (:bll) was set to its default value of 0.5, and the activation noise (:ans) and utility noise (:egs) parameters were both set to 0.3.

General Discussion

Previous attempts to reconcile positive and negative recency heuristics in a single model (Rabin, 2002) have appealed to the law of small numbers. This "law" is a belief, named by analogy to the law of large numbers, that the frequency of events in a sample should match the frequency of those events in the population (Tversky & Kahneman, 1971). The law of small numbers affords one explanation of the gambler's fallacy (negative recency), in which, for instance, a series of red outcomes on a roulette wheel leads the gambler to bet on black next. The black outcome is thought to be necessary to "even out" the preceding streak of red outcomes in the current (relatively small) sample.

Although the law of small numbers is a fallacy in the context of gambling, where events are independent and sampling is with replacement, it has some validity when events are sampled without replacement, even if they are independent (Rabin, 2002). If a red ball is drawn from an urn containing red and black balls and not replaced, then the probability of the next ball being black is higher than it was before the red ball was drawn. For large urns this effect is minimal, but a decision maker applying the law of small numbers effectively sets the urn size to be quite small, such that each new outcome in a streak warrants a relatively large update in expectations concerning the next outcome. Thus, the longer a streak of reds becomes, the further the decrease in the subjective probability that the next ball will red. Thus, the law of small numbers, applied to an outcome generator in which events are sampled from a (small) urn, offers one explanation of the decrease in positive recency across shorter streak lengths in Figures 2 and 3.

The law of small numbers may also explain the upward trend in positive recency that follows the initial downward trend. Rabin (2002) proposes that as a streak grows, there is not only Bayesian updating of the probability of the next ball being red or black, but also updating of beliefs about the relative proportions of reds and blacks in the urn. As more and more red balls are drawn, the decision maker comes to believe that there are more red balls in the urn than he or she previously thought. This effect should be greater for decision makers applying the law of small numbers, because the smaller the (mental) urn, the less likely the subjective probability of a streak of a given length. The smaller the urn, therefore, the more a streak indicates that the decision maker's beliefs about relative proportions of red balls and black balls in the urn need revision. Thus, as more and more red balls are drawn, the expectation that the next ball will be red should begin to increase again, at the point where the decision maker revises his or her subjective probability of reds and blacks in the urn. Rabin's is not a process model, so how these different forms of Bayesian updating may interact is unclear. Our model thus builds on his work by offering a precise formulation of how positive and negative recency interact.

Rabin (2002) draws support for his model from evidence that investors tend to under-react to a firm's financial prospects in the short term and to over-react in the long term. Empirically, stock prices tend to auto-correlate positively in the short term (a period of months), which can be interpreted to mean that investors insufficiently react to good or bad news. This is a form of positive recency, to the extent that it reflects a willingness to bet that recent trends in stock price will continue – for example, that a firm with a low stock price will continue to have a low stock price, despite a recent money-making breakthrough. On the other hand, stock prices tend to auto-correlate negatively in the medium term (a three- to five-year horizon), suggesting, for example, that such breakthroughs are factored only belatedly into investor decisions. This is a form of negative recency, to the extent that it reflects a willingness to accept that less recent trends in stock price may now reverse themselves. Qualitatively, then, there appears to be a mapping from Rabin's analysis of investor data, couched in terms of his urn model, and our U-shaped curve (positive recency, then negative recency), which we explain in terms of memory processes.

As a model, the law of small numbers by itself is problematic in that it can equally well explain positive and negative recency, and thus has no predictive power (Burns & Corpus, in press). Similarly, it seems unlikely that people really think in terms of "urns" (Rabin, 2002), although this is a representational hypothesis that remains to be tested. In our model we have tried to explain the U-shaped curve in terms of basic cognitive processes, linked ultimately to memory. A mapping could be made between urn size and memory capacity, and from different types of Bayesian updating to different reactions to different recall events. Thus, our model can be viewed as putting cognitive processing flesh on previous verbal theories. Whether its memory processes and simple decision rules generalize to account for use of recency biases in other decision-making tasks, and across longer time spans, are important questions for future research.

Acknowledgements

We thank the conference reviewers for their helpful suggestions for improvement.

References

Anderson, J. R. & Lebiere, C. (1998). *The atomic components of thought.* Hillsdale, NJ: Erlbaum

Burns, B. D. (2002). *Does the law of small numbers explain the gamblerís fallacy?* Paper presented at the Forty-Third Annual Meeting of the Psychonomic Society, Kansas City, MO.

Burns, B. D. (2004). Heuristics as beliefs and as behaviors: The adaptiveness of the "hot hand". *Cognitive Psychology, 48,* 295-331.

Burns, B. D., & Corpus, B. (in press). Randomness and inductions from streaks: "Gambler's fallacy" versus "Hot hand." *Psychonomic Bulletin & Review.*

Estes. W. K. (2002). Traps in the route to models of memory and decision. *Psychonomic Bulletin & Review,9. 3-25.*

Hintzman, D. L. (1988). Judgments of frequency and recognition memory in a multiple-trace memory model. *Psychological Review, 95,* 528-551.

Hintzman, D. L. (1984). MINERVA 2: A simulation model of human memory. *Behavior Research Methods, Instruments and Computers, 16,* 96-101.

Lebiere, C., Gray, R., Salvucci , D., & West., R. (2003). Choice and learning under uncertainty: A case study in baseball batting. In R. Alterman & D. Kirsh (Eds.), *Proceedings of the Twenty-fifth Annual Meeting of the Cognitive Science Society.* Hillsdale, NJ: Erlbaum.

Lovett, M. (1998). Choice. In J. R. Anderson & C. Lebiere (Eds.), *Atomic components of thought* (pp. 255-296). Hillsdale, NJ: Erlbaum.

Rabin, M. (2002). Inference by believers in the law of small numbers. *Quarterly Journal of Economics, 117,* 775-816.

Tversky, A., & Kahneman, D. (1971). Belief in the law of small numbers. *Psychological Bulletin, 2,* 105-110.

OPTIMIST: A New Conflict Resoution Algorithm for ACT–R

Roman V. Belavkin (R.Belavkin@mdx.ac.uk)
School of Computing Science, Middlesex University
London NW4 4BT, United Kingdom

Frank E. Ritter (ritter@ist.psu.edu)
School of Information Sciences and Technology,
The Pennsylvania State University, University Park, PA16802, USA

Abstract

Several studies have suggested recently that a more dynamic conflict resolution mechanism in the ACT–R cognitive architecture (Anderson & Lebiere, 1998) could improve the decision–making behaviour of cognitive models. This part of ACT–R theory is revisited and a new solution is proposed. The new algorithm (OPTIMIST) has been implemented as an overlay to the ACT–R architecture, and can be used as an alternative mechanism. The operation of the new algorithm is tested in a model of the classical Yerkes and Dodson experiment on animals' learning. When OPTIMIST is used, the resulting model fits the data better than the previous model (e.g. R^2 increases from .85 to .93 in one example).

Introduction

Conflict resolution is an important part of many intelligent systems, and from a cognitive science perspective it represents a model of a decision–making mechanism in the brain. In this paper, we introduce a new conflict resolution algorithm that can be used as an alternative to the standard mechanism in the ACT–R cogntivie architecture (Anderson & Lebiere, 1998). The new algorithm is called OPTIMIST (it stands for 'Optimism' plus 'Optimisation'), and recently it has been introduced as a search method (Belavkin, 2003).

Although OPTIMIST can, indeed, be used as a general purpose search strategy, its roots come from ACT–R models of cognitive development (Jones, Ritter, & Wood, 2000) and the effect of emotion on learning and decision–making (Belavkin & Ritter, 2003). These works exposed where to improve the well–established cognitive architecture.

The standard conflict resolution mechanism of ACT–R, its achievements and problems will be discussed in the first section of the paper. Then, the underlying theory of the new method will be explained, and the new algorithm will be presented. This section will repeat some results of the previous paper (Belavkin, 2003). The third section will demonstrate how the new algorithm works in a model. Early results suggest that a model with the OPTIMIST conflict resolution matches the data better than with the standard implementation of ACT–R.

The ACT–R Conflict Resolution

The symbolic level of ACT–R is organised as a goal–directed production system with declarative and procedural knowledge encoded in the form of chunks and production rules respectively. The chunks representing the current goal, some facts currently retrieved from the long term memory, and the states of perceptual and action buffers are compared with the patterns in the left–hand sides of the production rules. Then, after a set of all the rules that match the current working memory pattern has been created (the conflict set), a single rule has to be selected from this set and fired. This last step is called *conflict resolution*, and it is important how this rule selection occurs because it controls which 'decisions' the model makes and affects the search of the problem space.

In ACT–R, the conflict resolution uses subsymbolic information associated with the rules. During the model run the number of successes and failures of each rule (decision) is recorded by the architecture. In addition, ACT–R records the efforts (e.g. time) spent after executing the rule and actually achieving the goal (or failing). This information is used to estimate empirically the probability of success P_i and the average cost C_i of each rule

$$P_i = \frac{\text{Successes}_i}{\text{Successes}_i + \text{Failures}_i} \quad (1)$$

$$C_i = \frac{\text{Efforts}_i}{\text{Successes}_i + \text{Failures}_i} . \quad (2)$$

Here, Efforts_i is the sum of all costs, associated with previous tests of the ith rule: $\text{Efforts}_i = \sum_{j=0}^{k} C_{ij}$, where $k = \text{Successes}_i + \text{Failures}_i$ is the number of previous tests of rule i. For example, if cost is measured in time units, then equation (2) calculates the average time for exploring particular decision path. This way, probabilities and costs of rules are learned by the architecture.

When several rules compete in the conflict set, ACT–R calculates their ***utilities*** by the following equation

$$U_i = P_i G - C_i + \xi(\sigma^2) . \quad (3)$$

Here, G is called the ***goal value***, and it is measured in the same units as the cost (e.g. time); ξ is a random number taken from a normal distribution with zero mean and variance σ^2. Thus, the rational parts of the rules' utilities ($P_i G - C_i$) are corrupted by noise ξ of variance σ^2. Finally, the rule is selected according to utility maximisation: $i = \arg\max U_i$. Below is the summary of conflict resolution in ACT–R:

1. Set the goal value G and noise variance σ^2
2. Calculate P_i, C_i and $P_i G - C_i$ of the matched rules
3. Add noise $\xi(\sigma^2)$ to the utilities U_i
4. Fire rule $i = \arg\max U_i$

One can see that ACT–R learning equations (1) and (2) provide a kind of Bayesian estimation of rules' utilities. However, this mechanism only estimates the mean values $(P_iG - C_i)$ of distributions from which the random utilities are drawn. The variances σ^2 remain the same for all rules and do not change. This issue will be addressed in the new mechanism.

The utility equation (3) has allowed ACT–R to model successfully some important properties of human and animal decision–making:

Probability matching. The choice in humans and animals decision–making is proportional to the probability of success. The use of P_i in the utility has allowed ACT–R to model the data of many probability matching experiments (e.g. see Anderson & Lebiere, 1998 for models on Friedman et al., 1964).

Stochasticity. The nondeterministic (irrational) property of choice behaviour is achieved by adding noise to the utility, and different variance σ^2 values were needed to simulate various experimental data (see Anderson & Lebiere, 1998). For example, the somewhat irrational behaviour of children could be simulated by a model of an adult with increased noise in conflict resolution (Jones et al., 2000). Moreover, it has been suggested that risk–taking behaviour characteristic to choice involving losses and negative emotions (Tversky & Kahneman, 1981; Johnson & Tversky, 1983) can be simulated by higher noise variance values, while low noise variance is better for simulating the risk–aversive behaviour associated with choice involving gains and positive emotions (Belavkin & Ritter, 2003).

Levels of stimulation. The reward (or penalty) values are known to influence choice. For example, higher pay–off leads to preferences towards decisions with higher success probabilities (Myers, Fort, Katz, & Suydam, 1963). This effect was modelled by using higher goal values G (Lovett & Anderson, 1995; Anderson & Lebiere, 1998). Also, it was shown that G can be used to represent different levels of aversive stimulation and even different levels of arousal (Belavkin & Ritter, 2003).

Recently, however, several problems in models' performance have been associated with the limitations of the ACT–R conflict resolution mechanism. In particular, it was noticed that ACT–R models usually produced more errors in the final stages of experiments than subjects. This effect was especially noticeable in models of tasks with incremental learning, such as the Tower of Nottingham (Jones et al., 2000) or the Yerkes and Dodson experiment (Belavkin & Ritter, 2003). Figure 1 shows such an example: The model matches the data well during the first five simulated days,[1] but produces more errors after day 5. Using smaller values of noise variance σ^2 could eliminate the problem, but would lead to a higher discrepancy with the data in the earlier stage of the curve. A similar lack of convergence was noticed by other researchers (Taatgen, 2001; Lebiere, 2003).

It has been suggested that noise variance σ^2 should not remain constant, but should gradually decrease. Taatgen used an exponential decay of σ^2 as a function of time and achieved better results. However, it was argued that noise variance should be an inverse function of success rate and should not

[1]Days A and B denote the preference series before training.

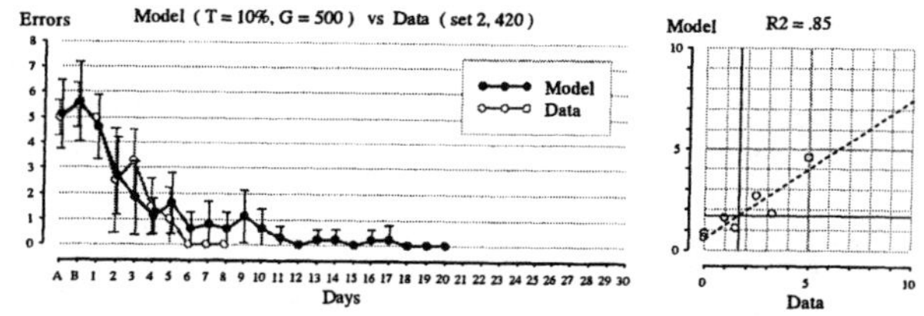

Figure 1: A model of the Yerkes and Dodson experiment compared with the data. Left: Error curves. Right: Regression plot of errors per day.

necessarily always decrease, but may increase if more failures occur (Belavkin, 2001). This would not only improve the models' match with the data, but also optimise the decision–making in a way similar to a simulated annealing heuristic. An alternative method was proposed to control noise variance by the entropy of success parameter (Belavkin & Ritter, 2003). Indeed, uncertainty of achieving success, estimated by the entropy, decreases as a result of learning, but may increase locally if more failures occur. The experiments demonstrated consistently that models with such a control matched the data better. In this interpretation, noise can be seen as compensation for missing information about the utilities of rules. The idea that noise should be proportional to the uncertainty (lower expertise) may explain also why children were simulated better by a model with higher noise variance (Jones et al., 2000).

The increase of expertise is not only reflected in the form of statistical information about the production rules. An ACT–R model may learn new rules using the production compilation mechanism. Taatgen proposed that noise should affect these new rules more than the 'older' rules in the system. This would provide a smooth transition in a model from the use of old to the more recently learned rules.[2] Again, this is not possible in the current ACT–R implementation, because σ^2 is a global parameter which does not depend on rules' creation times.

Another concern expressed is regarding the goal value parameter G, which is used as a constant in the current implementation. In the real–world situations, however, the value of the goal may change due to various reasons: Environmental change, re–evaluation of the efforts required, change of motivation due to boredom or anxiety and so on. Moreover, it was shown that G controls the problem space search strategy, and an increase of G from small to high values implements the best–first search heuristic (transition from breadth–first to depth–first) that can greatly optimise the search (Belavkin, 2001).

Unfortunately, the current ACT–R theory does not account for such dynamics. Let us summarise the new properties desirable for the conflict resolution algorithm:

1. Noise variance should be rule specific.

2. Noise variance should be inversely proportional to the rate of success, and should decrease on average with time.

[2]This effect has been achieved by using the production strength parameter and strength learning mechanism.

3. The goal value should be dynamic and increase on average.

In the next section, a new algorithm that implements the above properties is introduced.

The OPTIMIST Conflict Resolution

The OPTIMIST algorithm (Belavkin, 2003) has been derived in attempt to address the issues discussed in previous section. In the first part of this section, we present some theoretical background that helped derive the new algorithm, and in the second part, we describe the algorithm and its properties.

Theoretical background

It is well–known that many problems can have several solutions. Moreover, in the real world, applying even the same solution to one problem several times may produce slightly different outcomes. For example, using the same strategy to reach the goal in some task in several experiments may take slightly different amounts of time due to slightly different initial conditions or other uncontrolled factors in the environment. In view of this, it is natural to consider the cost C (e.g. time) needed to achieve the goal as a random variable, and the expected cost is thus

$$E\{C\} = \sum_C C\,P(C) \quad \left(\text{or} \quad E\{t\} = \int_0^\infty t\,\varphi(t)\,\mathrm{d}t\right) ,$$

where $P(C)$ is the probability that the goal will be achieved exactly at cost C, and the summation is made across all possible values of C (or an integral on $t \in [0, \infty)$ if C is continuous, such as time). Note, that in this notation, probability distribution $P(C)$ (or probability density $\varphi(t)$) defines the probability of achieving success on any time interval on $[0, \infty)$, which is different from the ACT–R notation.

Knowledge of distribution functions $P_i(C)$ for different alternative decisions $i \in [1, \ldots, N]$ would allow us to calculate their expected costs $E_i\{C\}$, and to choose the best rule. Indeed, better decisions should have smaller expected costs. For example, we can use several strategies to assemble a Rubik's cube puzzle. One such strategy can be a random rotation of edges of the cube, and it may eventually assemble the puzzle, but it will probably take much longer than by using some more sophisticated rules. Thus, one can choose the rule by minimising the expected cost: $i = \arg\min E_i\{C\}$ (optimisation).

The problem is, of course, that usually there is little information about $P_i(C)$, especially when making a decision for the first time, and in order to estimate the expected cost even for one decision one would have to apply this decision several times to get a sample estimate: $\bar{C} = \frac{1}{k}\sum_{j=1}^{k} C_j \approx E\{C\}$, where k is the number of tests.

We suggested to use Poisson distribution to approximate $P_i(C)$, if the cost is continuous, such as time (Belavkin, 2003). Indeed, if the expected cost of some strategy was known, then one could repetitively solve a problem by this strategy and would expect to observe the goal state at a rate $\lambda \equiv 1/\theta$, where $\theta \approx E\{C\}$ is the average waiting time. The probability of observing $n = 0, 1, 2, \ldots$ number of successes by the time t in such a process is given by the Poisson distribution

$$P(n \mid \lambda) = \frac{(\lambda t)^n}{n!} e^{-\lambda t} , \qquad n = 0, 1, 2, \ldots . \tag{4}$$

Here, $\lambda = 1/\theta$ is called the *mean count rate*. Note, that for $\lambda \to 0$ (or $\theta \to \infty$) the probability (4) becomes zero. The corresponding waiting times until the success events (costs) are distributed according to Gamma distribution with mean $\mu = n\theta$ and variance $\sigma^2 = n\theta^2$. Thus, a strategy with a higher success rate λ should not only have a smaller expected cost, but also the variance of costs is smaller.[3]

Equation (4) describes the conditional probability of n successes on time interval $[0, t]$ for a known λ. However, in our case λ is unknown, and the expected cost $E\{C\} \approx \theta$ is what we are trying to estimate after observing $n = 0, 1, \ldots$ successes on time interval $[0, t]$. This can be done using posterior probability density $\varphi(\lambda \mid n)$, which can be obtained from Bayes' formula

$$\varphi(\lambda \mid n) = \frac{P(n \mid \lambda)\varphi(\lambda)}{P(n)} .$$

In the worst case scenario when no information about prior $\varphi(\lambda)$ is available, we should assume that all λ are equally probable (the maximum entropy principle). We can use the following exponential function $\varphi_\varepsilon(\lambda) = \varepsilon e^{-\varepsilon\lambda}$ with $\varepsilon \to 0$ for such an unbiased estimate. Using this prior and Poisson likelihood $P(n \mid \lambda)$, one can show that

$$\varphi(\lambda \mid n) = t\,P(n \mid \lambda) .$$

Note, that after some finite period of time, if no successes have been observed, the distribution of λ becomes a decreasing function with high rates having smaller probabilities. Now, the posterior mean estimate of λ is:

$$E\{\lambda\} = \int_0^\infty \lambda\,\varphi(\lambda \mid n)\,\mathrm{d}\lambda = \frac{n+1}{t} \quad \left(E\{C\} \approx \frac{t}{n+1}\right)$$

Here, t and $n+1$ correspond to the Efforts$_i$ and Successes$_i$ parameters in ACT–R equations (1) and (2).[4] Note that we can use the above estimate even when $n = 0$ (i.e. when no successes have occurred). This property is very important, because it means that we do not have to explore all the solution paths in full trying to succeed. Indeed, in our probabilistic interpretation of cost, any decision or strategy may eventually lead to the desired goal (optimistic approach), although the chance may be very small. An illustration of this idea can be the classical example from the probability theory of a monkey randomly typing on a keyboard. The probability that it will come up with a literature text, such as *War and Peace*, is in fact non–zero. Therefore, it is desirable for such 'impractical' decision paths to be explored only partially, and after accepting a failure ($n = 0$) the system should give up and try another decision (or strategy).

---

[3] We use the term *success rate* to describe the number of successes per time, and term *success ratio* as the number of successes per attempts.

[4] In ACT–R, Successes$_i$ $= 1, 2, \ldots = n + 1$ with 1 being the default value.

We showed using the maximum entropy principle that the optimal moment to make new estimation of the expected cost and its posterior probability density $\varphi(\theta \mid n)$ is at $t = \theta \approx E\{C\}$ (see Belavkin, 2003). If after the new estimation another rule has smaller expected cost, then this also would be the best moment to give up and try another alternative. The following recursive procedure can be used to estimate $E\{C\}$ of one decision

$$\Delta t_0 = C_{\min} \,, \quad \Delta t_{k+1} = \theta_k = \frac{\sum_{i=0}^{k} C_i}{n+1} \,.$$

Here, k is the cycle number, and Δt_k is the time (or cost) interval, on which, after the decision has been made, we expect to achieve a success. If success does not occur before the end of Δt_k, then a failure is accepted. The number of successes (n) in this case does not change, but the efforts (t) increase by $C_k = \Delta t_k$. Thus, on failures the estimate increases. If the success occurs before Δt_k, then n increases by one, and efforts increase by $C_k < \Delta t_k$. Thus, on successes the estimate decreases. Figure 2 shows an example of $E\{C\}$ estimation over 20 cycles: After increasing above the $E\{C\}$ level, its estimate quickly converges to $E\{C\}$.

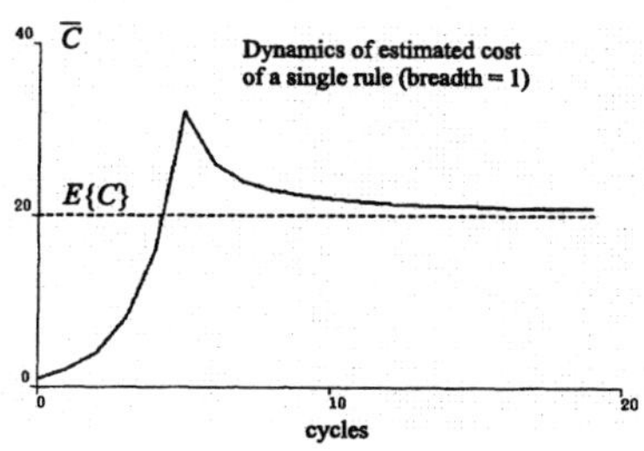

Figure 2: Estimated cost of one rule (vertical axis) as a function of test cycles (horizontal axis). Estimated cost converges to the expected cost $E\{C\}$ with cycles $k \to \infty$.

Now, if several alternative decisions (rules) are being considered, the choice can be made by selecting the decision with the smallest estimate.

Algorithm Description

The OPTIMIST algorithm uses the same subsymbolic information as that of the standard ACT–R implementation — the number of successes and the overall efforts associated with each rule. However, instead of calculating probabilities P_i and average costs C_i (equations (1) and (2)), OPTIMIST estimates the expected costs of achieving the success by each rule

$$\theta_i = \frac{\text{Efforts}_i}{\text{Successes}_i} \approx E_i\{C\} \,. \qquad (5)$$

Next, the estimates θ_i of all the rules in the conflict set are replaced by random numbers ξ_i, which we call *random estimated costs.* Ideally, ξ_i should be drawn from Gamma distributions with parameters θ_i that have just been estimated. As has been mentioned earlier, these distributions have means $\mu_i = \theta_i$ and variances $\sigma_i^2 = \theta_i^2$ (we are expecting only the fist success, or $n = 1$). Thus, ξ_i represent a sample approximating the distribution of expected costs for several rules in the conflict set. Finally, the rule choice is made by minimisation of random estimated costs:

$$i = \arg\min \xi_i \,.$$

Below is the summary of the OPTIMIST algorithm:

1. Calculate the estimates θ_i of rules' expected costs
2. Replace θ_i by corresponding random ξ_i
3. Fire rule $i = \arg\min \xi_i$

For computationally efficiency, instead of Gamma distributions, we used the following uniform distributions to generate the random estimated costs

$$\varphi(\xi_i) = \begin{cases} \frac{1}{2\theta_i} & \text{if } |\xi_i - \theta_i| < \theta_i \\ 0 & \text{otherwise} \end{cases} \,. \qquad (6)$$

These distributions have means $\mu_i = \theta_i$ and variances $\sigma^2 = \theta^2/3$. In addition, the algorithm can be made sensitive to the successes per attempts ratio by using the following functions to generate ξ_i:

$$\xi_i = \frac{k_i\,\theta_i + \text{rand}(2\theta_i)}{k_i + 1} \,, \qquad (7)$$

where $k_i = \text{Successes}_i + \text{Failures}_i$ is the number of attempts.

Note that the algorithm does not use the goal value parameter G. However, to some extent it is identical to the expected cost estimation θ (or more precisely its minimum $\min\theta_i$). Thus, the goal value in OPTIMIST is dynamically learned through equation (5). Figure 3 shows the dynamics of $\min\theta_i$ for twenty rules in an example conflict set.

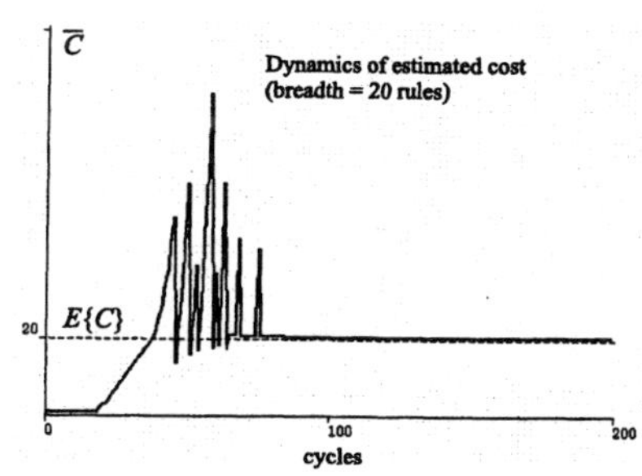

Figure 3: Dynamics of the smallest estimated cost for a conflict set of 20 rules as a function of test cycles.

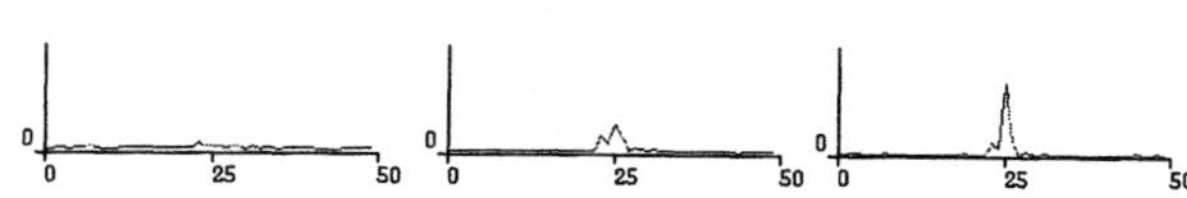

Figure 4: Dynamics of choice proportion (vertical axis) for different rules (horizontal axis) as a function of time. From left to right: The choice concentrates on more successful rules.

Also, one has been mentioned earlier, θ_i controls the variance of the costs' distributions. Thus, the noise variance in the conflict set also increases on failures and decreases on successes. In addition, the variance may also decrease with k_i — the number of times a rule has been used (see Eq. 7). Figure 4 shows from left to right the dynamics of choice proportion between fifty rules (horizontal axis), with the best rule placed in the middle. One can see that the choice quickly concentrates on the best rule.

Finally, because both θ_i and k_i are rule specific parameters, the randomness is different for all the rules in the system. In general, the less successful rules in the system (smaller rates $1/\theta_i$) as well as newer rules (smaller k_i) are more 'noisy' than rules with higher success rates and rules used more frequently. One can see from above that OPTIMIST possesses all three desired properties stated in the previous section: Noise variance is rule specific, dynamic and proportional to the success rate; Goal value is also dynamic and increases on average. Table 1 provides a comparison between the current ACT–R mechanism and OPTIMIST.

Table 1: Comparison of ACT–R and OPTIMIST conflict resolution mechanisms.

	ACT–R	OPTIMIST
n. of successes	$n_i + 1$	$n_i + 1$
n. of attempts	k_i	k_i
efforts spent	t_i	t_i
success probability	$P_i = \frac{n_i+1}{k_i}$	$P_i(n_i \mid \lambda_i)$
expected cost	$C_i = \frac{t_i}{k_i}$	$\theta = \frac{t_i}{n_i+1}$
goal value	$G = \text{const}$	$\min \theta_i$
noise variance	$\sigma^2 = \text{const}$	$\sigma_i^2 \sim \theta_i^2$
utility	$U_i = P_i G - C_i$	θ_i
conflict resolution	$\max U_i$	$\min \theta_i$

A Model Example and Additional Parameters

The OPTIMIST conflict resolution algorithm was put into a test in a model of the Yerkes and Dodson experiment. In this classical animal learning task, mice were trained over several days to escape a discrimination chamber (a box with two doors) through one particular door. Ten tests per day were performed with each mouse, and the number of errors was recorded. Figure 5 shows one example of distribution of errors, produced by the model with the OPTIMIST algorithm and compared with the experimental data. Horizontal axis represents the day numbers, and vertical axis shows the number of errors per day.

These first tests demonstrated that the new algorithm works, and the model produces behaviour comparable to both the data and a model with the standard conflict resolution mechanism. Moreover, because noise variance in OPTIMIST version has decreasing dynamics, the models with the new algorithm do not suffer from lack of convergence discussed earlier in the paper (see Figure 1). In fact, when using the weighted average version of the algorithm (7), the convergence is sometimes too fast. Therefore, a parameter has been introduced to enable OPTIMIST to retain some level of noise.

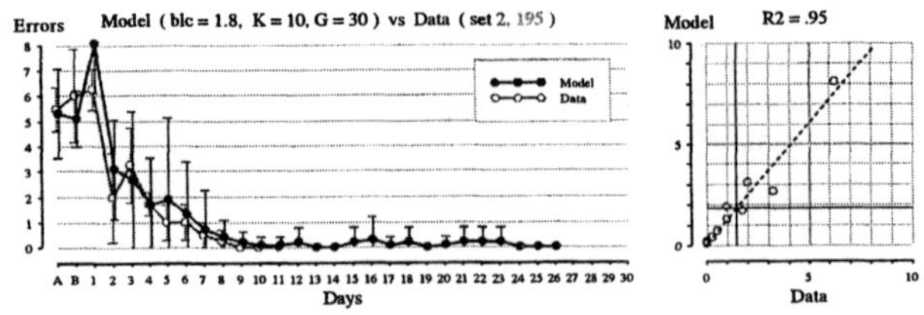

Figure 5: A model with OPTIMIST conflict resolution compared with experimental data (Yerkes & Dodson, 1908). Left: Error curves. Right: Regression plot of errors per day.

In the current implementation, this is achieved by limiting the number k used in equation (7).

Another adjustment to the algorithm concerns different levels of stimulation. In ACT–R, different values of pay–off can be represented by the goal value parameter G. In OPTIMIST, if the cost is only measured by time, then there is no way of distinguishing between different levels of a pay–off (i.e. values of reward or penalty). Indeed, the time spent on choosing an option with a prize worth \$10 is the same as for \$100. In order to account for these effects, the current OPTIMIST implementation uses reinforcement mechanism, which can modulate the costs of particular outcomes:

- If a rule fired has explicit `:failure` flag, then *penalty* value increases the cost of of the outcome and hence increases the expected cost of the rule associated with the failure.
- If a rule fired has `:success` flag, then the cost of the outcome is reduced by the *reward* amount.

The values of penalty and reward are defined by the corresponding variables in the system, and in fact they describe characteristics of the environment and interaction of a cognitive model with the environment, rather than internal state of the model. Moreover, this implementation allows a modeller to define several different rewards and penalties in various places of the simulated environment (unlike global G).

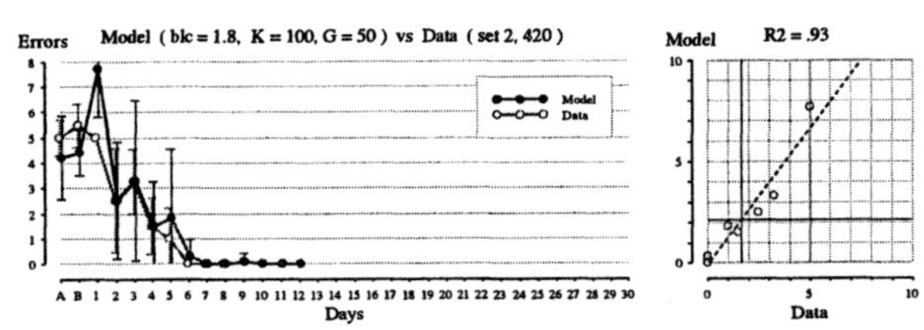

Figure 6: The effect of reinforcement: A model and data for an experiment with higher level of stimulation (Yerkes & Dodson, 1908).

Figure 6 shows the results of a model of the Yerkes and Dodson experiment with higher value of stimulation (in the original experiment it was an electrical stimulus). The model uses penalty value that modulates the costs of rules that are choosing the wrong door leading to an error. As a result, the model learns faster. Note that the data set the model is compared with is the same as shown on Figure 1. One can see that the OPTIMIST model matches the data better than the model with standard mechanism of ACT–R version 5 (R^2 has increased from .85 to .93), which indicates in favour of the new algorithm.

Discussion and Conclusions

In this paper, we challenged one of the most important mechanisms of a well–established cognitive architecture — the conflict resolution of ACT–R. Many cognitive models in ACT–R rely on and use the utility equation (3) and its parameters. One of such models, mentioned in this paper, is the model of the Yerkes and Dodson experiment. The limitations of the current ACT–R implementation, as encountered by this model, has become the main motivation for the new algorithm. For example, we have shown that a model with dynamic control of noise variance by means of entropy reduction improves significantly the match between the model and data (Belavkin & Ritter, 2003). As has been discussed earlier in this paper, similar concerns have been expressed by other researchers.

The new algorithm uses some elements of statistical decision–making theory and estimates expected costs of production rules using a Poisson distribution. Interestingly, several studies on kinetics of choice in animals learning have suggested earlier that estimation of the Poisson rate λ (or equivalently $\theta \equiv 1/\lambda$) may explain animals' choice behaviour. In particular, Myerson and Miezin (1980) used a Poisson distribution to explain the change of response frequency in rats (see also Mark & Gallistel, 1994). Moreover, an attempt to incorporate this into the ACT–R theory has been already made in a form of the *events discounting* mechanism (Lovett & Anderson, 1996; Lovett, 1998). Unfortunately, this mechanism suffers from computational overhead and turns out to be impractical for complex models. The new algorithm, introduced in this paper, directly estimates the rate of a Poisson process. In addition, the algorithm is computationally efficient and uses the standard subsymbolic information of ACT–R, so it is relatively easy to implement.

The first implementation of the algorithm as an overlay to ACT–R has been created, and it is available for download on

`http://gold.mdx.ac.uk/~rvb/software/optimist/`

The operation of the algorithm has been demonstrated on a model. Early results are in favour of the new algorithm and suggest that it indeed may improve the match of some cognitive models to data. However, more tests in different models and on other data sets still have to be done. It is, therefore, suggested to use the new algorithm in addition to the standard to provide valuable feedback for further development.

References

Anderson, J. R., & Lebiere, C. (1998). *The atomic components of thought.* Mahwah, NJ: Lawrence Erlbaum.

Belavkin, R. V. (2001). The role of emotion in problem solving. In C. Johnson (Ed.), *Proceedings of the AISB'01 Symposium on Emotion, Cognition and Affective Computing* (pp. 49–57). Heslington, York, England: AISB. (ISBN 1-902956-19-7)

Belavkin, R. V. (2003). Conflict resolution by random estimated costs. In D. Al-Dabass (Ed.), *Proceedings of the 17th European Simulation Multiconference (ESM2003)* (pp. 105–110). Nottingham, UK. (ISBN 3-936150-25-7)

Belavkin, R. V., & Ritter, F. E. (2003). The use of entropy for analysis and control of cognitive models. In F. Detje, D. Dörner, & H. Schaub (Eds.), *Proceedings of the Fifth International Conference on Cognitive Modelling* (pp. 21–26). Bamberg, Germany: Universitäts–Verlag Bamberg.

Friedman, M. P., Burke, C. J., Cole, M., Keller, L., Millward, R. B., & Estes, W. K. (1964). Two–choice behaviour under extended training with shifting probabilities of reinforcement. In R. C. Atkinson (Ed.), *Studies in mathematical psychology* (pp. 250–316). Stanford, CA: Stanford University Press.

Johnson, E., & Tversky, A. (1983). Affect, generalization, and the perception of risk. *Journal of Personality and Social Psychology, 45*, 20–31.

Jones, G., Ritter, F. E., & Wood, D. J. (2000). Using a cognitive architecture to examine what develops. *Psychological Science, 11*(2), 93–100.

Lebiere, C. (2003, April). *Personal communication.*

Lovett, M. C. (1998). Choice. In *The atomic components of thought* (pp. 255–296). Mahwah, NJ: Lawrence Erlbaum.

Lovett, M. C., & Anderson, J. A. (1995). Making heads or tails out of selecting problem solving strategies. In *Proceedings of the 17th Annual Conference of the Cognitive Science Society* (pp. 265–270). Hillsdale, NJ: Lawrence Erlbaum.

Lovett, M. C., & Anderson, J. A. (1996). History of success and current context in problem solving: Combined influences on operator selection. *Cognitive Psychology, 31*, 168–217.

Mark, T. A., & Gallistel, C. R. (1994). Kinetics of matching. *Journal of Experimental Psychology, 20*(1), 79–95.

Myers, J. L., Fort, J. G., Katz, L., & Suydam, M. M. (1963). Differential monetary gains and losses and event probability in a two–choice situation. *Journal of Experimental Psychology, 77*, 453–359.

Myerson, J., & Miezin, F. M. (1980). The kinetics of choice: An operant systems analysis. *Psychological Review, 87*(2), 160–174.

Taatgen, N. A. (2001, July). Production compilation. In *Eigths annual post–graduate summer school.* act-r.psy.cmu.edu.

Tversky, A., & Kahneman, D. (1981). The framing of decisions and the psychology of choice. *Science, 211*, 453–458.

Yerkes, R. M., & Dodson, J. D. (1908). The relation of strength of stimulus to rapidity of habit formation. *Journal of Comparative Neurology and Psychology, 18*, 459–482.

Good Enough But I'll Just Check: Web-page Search as Attentional Refocusing

Duncan P. Brumby (BrumbyDP@Cardiff.ac.uk)
Andrew Howes (HowesA@Cardiff.ac.uk)
School of Psychology, Cardiff University,
Cardiff CF10 3YG, Wales, UK

Abstract

When people search a set of Web labels for links that are relevant to their information goal, they attend to the labels and estimate the likelihood that the link will lead to the goal. Recent findings indicate that people sometimes, but not always, assess only a subset of the links available. We report an ACT-R model of Web-page search that was inspired by Young's (1998) rational account of exploratory choice, but which was also sensitive to the psychological constraints encoded in the ACT-R theory of the human cognitive architecture. The behavior of the model differs substantially from previous ACT-R models of Web navigation, at least for single-page search. We describe an experimental test of the models behavior and qualitative and quantitative fits of the model to the data.

Introduction

Consider searching a newly encountered Web page for links that are relevant to the achievement of some search goal. During this activity, people focus on the labeled links in order to derive a subjective assessment of the likelihood that the selection of a given link will lead to the achievement of the current information goal. It is known that people tend to select links to Web sites with labels that are more relevant to their current goal (Pirolli & Fu, 2003). In a simplified single page menu search task, however, Brumby and Howes (2003) observed that people do not always assess all of the items available in the choice set prior to the selection of an item, and that they re-fixate a smaller and smaller subset of these items prior to selection. In the current paper, we present a model of menu search behavior developed with the ACT-R framework (Anderson & Lebiere, 1998). The model makes use of ACT-R's architectural assumptions in order to produce behavior that (1) provides both qualitative and quantitative fits across a range of performance measures including eye-movement data, and (2) is broadly consistent with a rational analysis of exploratory choice (Young, 1998). We also report an experimental test of the model.

In a simplified single-page Web search task, Brumby and Howes (2003) asked participants to search single-level Web pages in pursuit of an information goal. The pages contained a single goal, or *target*, link and the rest of the links were *distracters*. An eye tracker recorded participant's eye movements while they searched the labeled links. The relevance of the distracter links to the goal was varied while the quality of the target link was held constant.

As we have said, Brumby and Howes found that participants did not always fixate all of the links available in the choice set prior to selection, and that they re-fixated a smaller and smaller subset of these links prior to selection. These signature behaviors are consistent with qualitative observations of people freely exploring a novel device interface (Rieman, 1994) and with the search of early keyboard-driven menus in which only a single choice can be seen at a time (MacGregor, Lee, & Lam, 1986).

Brumby and Howes found that participants selected the target link immediately after its first fixation, i.e. they satisficed, on 31% of trials and repeatedly rescanned a smaller subset of the links prior to selection on 69% of trials. They also found that participants sometimes fixated spatially adjacent links consecutively in time and sometimes skipped links, as observed by Byrne (2001) and Hornof (in press). Interestingly, participants more frequently skipped links after they had fixated the item that they would eventually select.

Further, Brumby and Howes found that the presence of lower relevance distracters resulted in fewer items being fixated. This was because participants who were given lower relevance distracters, and target links of the same relevance, were more likely to satisfice, i.e. to select a target link immediately after fixating it for the first time.

While it is known that the relationship between eye-movements and cognitive processing, in this case assessment, can be complex, we assume that the fact that Brumby and Howes' participants were more likely to satisfice if the distracters were of lower quality is evidence that participants made assessments of fewer links. This finding therefore supports the idea that people may adjust an independent assessment of the relevance of a link, in order to derive an estimate that is interdependent with the quality of the other links in the choice set. Consequently, people may make implicit assumptions about the value of items that they *have not assessed* on the basis of generalization from those that they have assessed.

A number of models of Web navigation and menu search have previously been reported, including models constructed in ACT-R (Byrne, 2001; Salvucci, 2001) and also in EPIC (Hornof, in press; Kieras & Meyer, 1997). These models have focused on the basic perceptual/motor components involved in simple, routine menu search and are not able to model the semantically laden assessment process that characterizes complex menu search, typical of Web navigation.

SNIF-ACT (Pirolli & Fu, 2003) makes an important contribution to the development of models that can simulate users searching the World Wide Web for information relevant to an unfamiliar information goal. The model provides a high-level characterization of Web search behavior, focusing on link selection and site leaving behaviors. Importantly, the model uses ACT-R's declarative memory module in order to derive assessments of a menu labels relevance to the information goal (what Pirolli & Card, 1999, term *information scent*). The model accurately predicts that users will select menu labels that have high information scent. However, SNIF-ACT lacks a plausible model of how people search an individual page. This over simplification is non-trivial, if people do not always assess all items on a menu page prior to selection.

While some previous models do not capture the menu search behavior observed by Brumby and Howes (2003), Young (1998), and more recently Cox and Young (submitted), have presented a rational analysis of exploratory choice, which provides an important theory from which to understand people's menu search behavior. These models extend Anderson's (1990) rational analysis of problem solving to the uncertain environment inherent in menu search. In these models, the underlying choice at each cycle is between selecting the item judged most likely to lead to goal immediately, or performing another (re)assessment. The decision between action and assessment is based upon a simple utility function. In other words, items are assessed so long as the expected gain of making another assessment exceeds the cost of the assessment.

Young's (1998) model is particularly interesting because it is sensitive to the implications of the structure of the task environment in many menu search tasks. When searching a menu that contains a given choice set of menu items ($item_1$ … $item_n$), typically only a single item will lead to the achievement of the information goal. Given some probability estimate of this likelihood it can be assumed that the sum of estimates across all items in the choice set must be equal to one. This *normalization assumption* "reflects real cross-relationships between the judgments about choices made by a person, and cannot be avoided … the reality is that people are often forced to make rapid and radical revisions of their estimates of the correctness of particular options as they work their way through [the options available]" (Young, 1998, p. 474). A novel prediction to emerge from the normalization assumption is that the relevance of both the target *and* the distracters will affect the decision of whether to select or continue assessment.

Web-page search as attentional refocusing

We present an ACT-R model of our previous menu search data (Brumby & Howes, 2003). The model is partially constrained by Young's (1998) rational analysis of exploratory choice, and in addition is consistent with the memory constraints imposed by ACT-R (Anderson & Lebiere, 1998). Our aim was that the model should at least demonstrate the observed search behaviors: Participants typically do not fixate all of the items in the choice set, and often items are fixated on multiple passes prior to selection.

In order to capture these behaviors it is necessary to model how people choose between selection of an item and further assessment of items. A candidate mechanism is ACT-R's conflict resolution mechanism. Competition between productions for control of behavior is governed by a mechanism derived from Anderson's rational analysis of choice. However, as Young and Cox (2000) point out, although the production rule system in ACT-R takes account of past experience in operator selection (Lovett & Anderson, 1996), the parameters associated with a given production rule change relatively slowly in order to reflect the long-term learning of production utility. Consequently, our approach to this problem was to further explore ACT-R's declarative memory retrieval mechanism. This approach is consistent with that employed in previous ACT-R models of Web navigation (Pirolli & Fu, 2003) but makes a novel use of the mechanism by which source-activation models the focus of attention.

The key features of the model were:

1. The ACT-R goal chunk included *n* slots, one for each label link and each of which, initially, had a value of *unassessed*. We call these *assessment* slots. The goal also has two additional slots, one for the current information goal and the other for the current attended visual location.
2. Assessment of label links was achieved by repeated attempts to retrieve chunks from declarative memory (Figure 1). Chunks represent world knowledge. The successful retrieval of a chunk was assumed to indicate that there was positive information linking the label and the goal, and resulted in the replacement of *unassessed* values on the goal with the retrieved value. An unsuccessful retrieval resulted in the replacement of a slot value with "not relevant", i.e. equivalent to setting the slot value to nil.

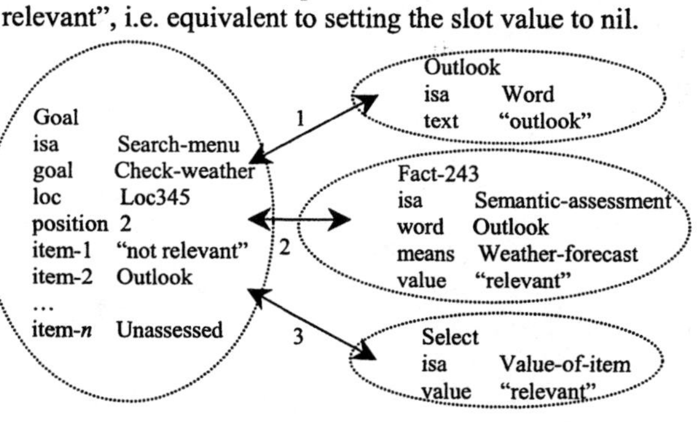

Figure 1: An example of assessment chunks for a menu item. The goal chunk contains slots for $item_1$ … $item_n$ that can take the values *Unassessed, Word-n*, or *"not relevant"*. An iconic representation of the currently attended menu item is passed from the visual buffer (*"outlook"*) to signal a retrieval request for a syntactic assessment chunk (arrow 1), which if retrieved leads to requests for further assessment chunks to be retrieved (e.g. arrow 2 → arrow 3). Decision to select the item is signaled if all assessment chunks are retrieved. Attention is shifted to another menu item if there is a failure to retrieve any of the assessment chunks.

3. Whether or not a chunk was retrieved depended in part on its activation. In ACT-R the activation is determined by a combination of base-level activation B, spreading activation $W_j \cdot S_{ji}$, and a transient noise.

$$A_i = B_i + \sum W_j \cdot S_{ji} + \varepsilon \quad (1)$$

The former was set to zero in our model and played no role in capturing the desired behaviors. Spreading activation played a crucial role (see bullets 4 and 5 below).

4. The amount of spreading activation received by a chunk depended in part on the attentional focus. The more label links that were in the choice set (as represented by the assessment slots) the lower the amount of source activation W received by the chunk. Conversely, the fewer the number of items in the choice set the greater the amount of source activation W received by the chunk.

$$W_j = 1 / (n + 2) \quad (2)$$

5. The amount of activation received by a chunk also depended on the strength of association S between the goal and the chunk. Following Budiu & Anderson (2004) S_{ji} reflects the similarity between chunk *i* and *j* and was defined as

$$S_{ji} = C + M \cdot \sigma(j, i) \quad (3)$$

Where, the σ(j, i) component reflects the input measure of semantic similarity between chunks *j* and *i*, and C is a negative quantity that serves as a base of associative strength and M is a positive multiplier. Moreover, this definition of the similarity between two chunks reflects a simple linear mapping of similarity that varies between 0 and 1, where values closer to 1 reflect greater similarity (see Table 1).

6. Different chunk-types were used to represent different types of assessment (e.g. syntactic or semantic). Importantly, the model could sometimes retrieve some of the assessment chunks for an item, while other assessment chunks for the item would fail to be retrieved (see Figure 1). In this case the item was judged partially relevant to the information goal. Given some change in the value of source activation W, future assessment of the item may warrant selection. Thus, the model maintained pointers on the goal chunk to such partially relevant menu items in order to allow the item to be reassessed.

7. The model (re)assessed another item when there was a failure to retrieve an assessment chunk for the currently attended item. In deciding which item to assess next, production rules for assessing an unassessed item (nearest the current visual location) and reassessing a previously relevant item (see bullet 6 above), compete in a stochastic selection process. The utility of the production rule for assessing an unassessed item was greater than the utility of the production for reassessing an item.

8. The model chose to select an item if it had been judged to be highly relevant to the current information goal, indicated by retrieval of all chunk-types related to the item.

In summary, if the assessed distracter links were not relevant to the current information goal, then the model was more likely to make a positive assessment of a labeled link, and furthermore was more likely to select a highly relevant link immediately. This behavior is consistent with Young's (1998) normalization assumption and reflects a constraint imposed by the structure of the task environment.

Menu Search Experiment

The models performance was compared with data collected from a menu search experiment that extended our previous work (Brumby & Howes, 2003). In the study participants searched single-page menus. All menus contained a single goal item and the participant's task was to select the goal item as quickly and accurately as possible. Participants did not progress to the next trial until they had selected the goal item. Independent ratings were taken of label quality. The menus differed in terms of the quality of the goal item (very good or moderately good) and the quality of the distracter items (moderate, bad, or very bad). The quality of a label was determined from participant's ratings of the degree to which an item was relevant to the achievement of the search goal. Participants (n = 20) were native English speakers. Eye tracking was performed using an ASL Pan/Tilt optic eye-tracking system.

Results and Discussion

It was found that the choice between continued assessment and selection in menu search was affected by the quality of both the goal item that was selected and quality of the assessed distracter items. Figure 2 demonstrates that (1) participants rarely fixate all of the items available in the choice set prior to the selection of an item, and (2) they re-fixated a smaller and smaller subset of these items prior to the selection of an item. In particular, when the quality of the goal item was held constant, participants fixated fewer of the items fewer times prior to selection of an item when the quality of the distracter items that were assessed were less attractive ($F(2, 38) = 3.66$, $p<.05$). Indeed, if we

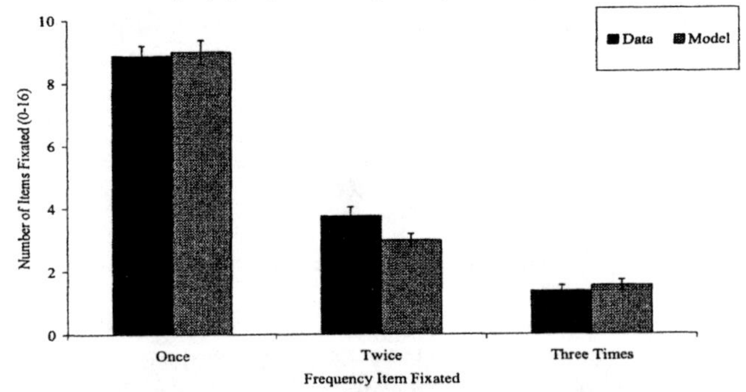

Figure 2: Data and model fits for the number of items fixated at least once, twice, and three times (collapsed across condition)

consider only the number of items fixated at least once, then this effect amounted to participants on average fixating two items (approx. 12.5%) fewer when the distracters were very bad in quality compared to moderate in quality. This key finding provides evidence of interdependence in menu choice assessment: whether people even bother to fixate some items depends on the quality of the distracters that they have assessed. There was also significant effects of quality of distracter item on time to selection ($F(2, 38) = 8.71$, $p<.001$), accuracy of selection ($F(2, 38) = 107.54$, $p<.001$), the number of items fixated after the initial fixation of the goal item ($F(2, 38) = 7.75$, $p<.005$), and a significant trend for the proportion of trials in which participants selected the goal item immediately after fixating it for the first time ($F(2, 38) = 3.18$, $p<.1$).

Furthermore, we found that when the selected goal item was more relevant to the current information goal, participants were quicker to select that item ($F(1, 19) = 258.24$, $p<.001$) and were more accurate with that selection ($F(1, 19) = 174.63$, $p<.001$), and fixated fewer items after the initial fixation of the selected item ($F(1, 19) = 84.37$, $p<.001$). Indeed participants were more likely to select the goal item immediately after fixating it for the first time ($F(1, 19) = 46.69$, $p<.001$). Although intuitively obvious, it is not clear that previous models of menu search would predict some of these findings, e.g. SNIF-ACT (Pirolli & Fu, 2003).

Model Fitting

The primary aim in evaluating the validity of the model was to match the models performance across a range of dependent variables used in the menu search experiment. Given that the primary focus of our study was on eye movement data, we made the assumption that ACT-R's movements of visual attention to items in the menu can be taken to broadly match a participants eye movement fixation centered around a menu item in our menu search task. To this end, the model used Salvucci's (2001) EMMA system to provide a more detailed theory of visual encoding. The ACT-R models output was therefore subject to the same analysis as a participant's eye movement fixation sequence.

The ACT-R model interacted with a menu that was the same as that searched by participants in the experiment. Although, we gathered participant ratings of the degree to which menu labels from the experiment were relevant to the achievement of the search goal. It was not possible to use these ratings as input to the model. Participant's ratings were provided on a discrete scale, whereas the model required a probability-like quantity (see bullet 5). Furthermore, as described in bullet 6, it was assumed that labeled links were evaluated in stages, involving several assessments of the label (Young, 1998). The input similarity values were therefore estimated for labels of different qualities and were essentially free parameters in the model (see Table 1 for the estimated values). In Table 1 the label quality 'moderate distracter', for example applied to all distracter labels in the experimental condition where the distracters were of moderate relevance to the search goal.

Table 1: Estimated Similarity Values

Label Quality	Chunk-Type: Word	Semantic-Assessment	Value-of-Item
Very Good Goal	.80	.45	.25
Moderate Goal	.50	.20	.18
Moderate Distracter	.25	.15	.10
Very Bad Distracter	.20	.10	.05

In fitting the model to the data, we systematically attempted to maximize the fit of the models performance across three of the four[1] experimental conditions on the main dependent variables (number of items fixated at least once, twice, and three times; accuracy of selection; percentage of self-terminating searches; time to selection; percentage of items fixated after the initial fixation of the goal item).

In order to estimate the input similarity values for each label of the same quality (see Table 1), we went through each experimental condition in a piece meal fashion, and obtained input values that maximized the models fit, over 100 model runs, across the range of dependent variables. Once an input similarity value was estimated, we moved to the next condition and obtained a best fitting model by varying input similarity values for the chunk-types associated with only a single label quality. For example, taking the *very good goal, very bad distracter items* condition, input similarity values for the chunks associated with very good goal chunks and very bad distracter items chunks were estimated which provided a good fit with the data. Next, fitting the model to the *moderately good goal, very bad distracter items* condition, we held constant the estimated input similarity values for the chunks associated with very bad distracter items, and varied the value of the chunks associated with the goal item only.

Importantly, all of the models free parameters were estimated from fitting the model to three of the four experimental conditions. Thus, the models performance on the final condition (*very good goal, moderate distracter items*) was predictive (in the sense that the model was not iteratively fitted to the data).

The model provided a good fit with the data across most of the dependent variables. The model demonstrated the observed signature behaviors: (1) the model rarely attended all of the items available in the choice set prior to the selection of an item, and (2) re-attended a smaller and

[1] The actual design included six experimental conditions, however, post-hoc analysis revealed non-significant pairwise differences between very bad quality and bad quality distracter items across quality of goal. For simplicity we focused on the contrast between moderate and very bad quality distracters across very good quality goal and moderately good quality goal (i.e. we did not model the two bad quality distracter conditions).

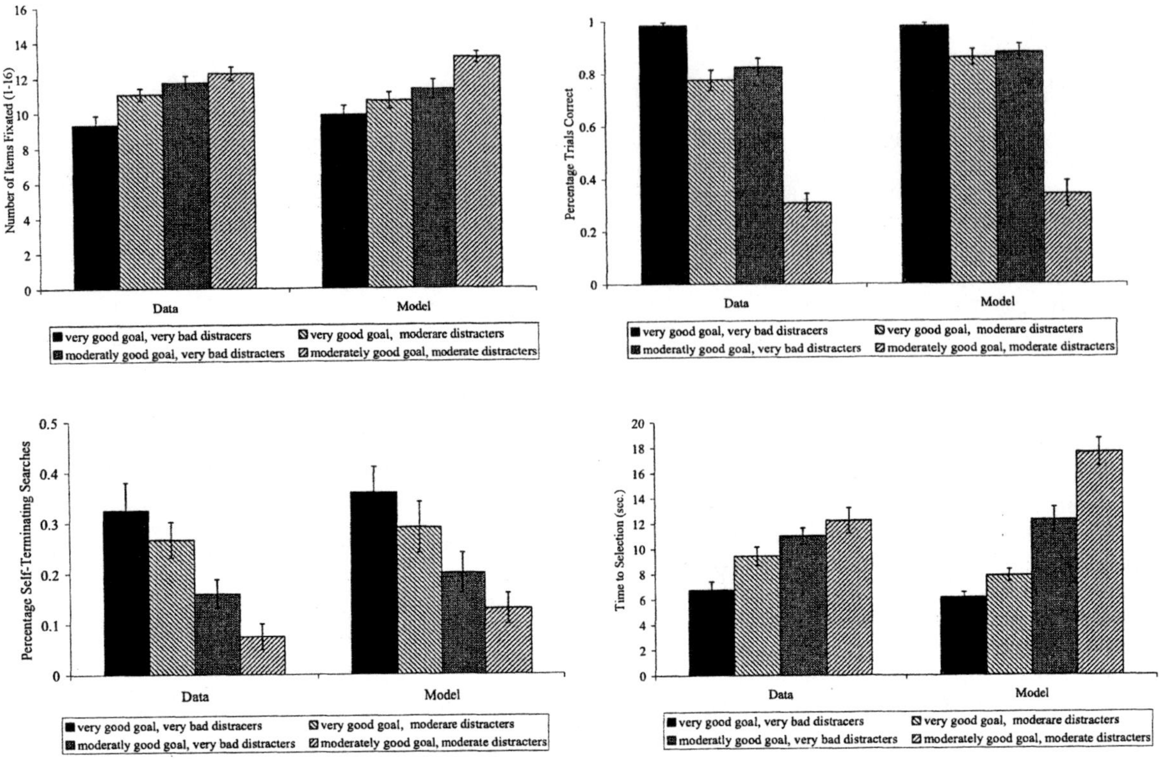

Figure 3: Data and model fits for effect of quality of goal and quality of distracter items on the (1) number of items fixated; (2) percentage of trials correct; (3) percentage of self-terminating searches; and (4) time to selection

smaller subset of these items prior to the selection of an item (see Figure 2 for aggregate model fit). Across conditions, the model was able to account for shifts in this behavior (r^2 = .96), i.e. attending fewer items when the distracter items were very bad in quality. This correlation was highly significant ($F(1, 10) = 214.08$, $p<.001$). In particular, Figure 3 shows that the model provided very good fits to the data for (1) the number of items fixated; (2) percentage of trials correct; (3) percentage of self-terminating searches; and (4) time to selection. The model did not provide a good fit for the number of items fixated after an initial fixation of the goal item, however.

General Discussion

We have presented a model of how participants searched single simplified Web pages in order to find links that were relevant to an information goal. In contrast to previous models of Web navigation (e.g. SNIF-ACT, Pirolli & Fu, 2003), but consistent with the data, the reported model sometimes selected a link before evaluating all links. The model was built in ACT-R and took advantage of the fact that in the architecture a fixed amount of source activation is distributed among the declarative chunks that are associated with the goal. The search process was therefore modeled as attentional refocusing. The probability of retrieving knowledge that associated a label with the goal was therefore dependent on the number of other labels in the attended choice set. In other words, the estimated likelihood that any single label would lead to the desired information was partly dependent on the values of other items. This interdependency reflects Young's (1998) normalization assumption.

The model was evaluated by comparing its predictions to a range of measures. In an experiment, the relevance of the goal item and the distracter items to the information goal was manipulated. The data were consistent with the prediction that people would fail to assess all of the items in the choice set, and that they would repeatedly reassess a smaller and smaller subset of potential candidate items prior to selection.

The model was inconsistent with the data in at least one respect. It did not provide a good fit to the number of items fixated after the initial fixation of the eventually selected item. This discrepancy may be partially explained by the systematic search strategy used in the model. Consistent with previous models of routine menu search (e.g. Byrne, 2001; Salvucci, 2001), the model attended each item in turn in the menu in a top-to-bottom fashion, with the exception of choosing to reassess an item. Our analysis of participant's eye movement fixation sequences revealed that the distance between contiguous fixations was typically in the region of

1.5 items, suggesting participants regularly skip over items. Our model does not capture this item-skipping strategy.

Hornof (in press) proposed a maximally efficient foveal sweep strategy, in which multiple items within the fovea (defined as one degree visual angle) are assessed with a single fixation. Interestingly, in our data the foveal sweep strategy (for fixating previously unfixated items) was moderated by whether or not the participants had already fixated an item that was attractive. However, it is not clear that participants in our experiment were assessing multiple items with a single fixation, because items were approximately one-degree visual angle in height and the distance between items was also approximately one-degree visual angle. Further theoretical work is required.

The model does not capture all of our previous data (Brumby & Howes, 2003). In particular, we have found that the history of information search moderates the local search strategy. After participants had completed trials in which they were more likely to select an incorrect item (because the distracters had been made more attractive), they were more cautious about selection. That is, they assessed more of the items in the choice set and were less likely to select an item immediately following an initial fixation of that item. Extending the current model to account for these findings should be relatively straight forward, because ACT-R's production rule learning mechanism is well suited to modeling the influence of history of successes on operator selection (Lovett & Anderson, 1996).

The idea that Web-page search is attentional focusing may seem counter-intuitive. In our model the goal of assessing unassessed items reduces the probability of retrieving information about the currently fixated item. Although, this mechanism predicts the observed behavior it seems counter-intuitive because, given that the goal is presumably under strategic control, an implication is that participants deliberately reduced the probability of retrieval of information associating an item with the goal (at least initially) in order to achieve the desired overall search strategy. Further data is required.

In summary, the aim of the work reported in this paper was to build a model of single-page Web-search that was constrained by ACT-R architectural assumptions and by Young's (1998) rational analysis of exploratory choice. The model was supported by the results of an experiment that tested the consequences of distracter-quality and goal-quality for eye fixations during the search of simplified Web pages.

Acknowledgments

This work was partially funded by an EPSRC studentship to Duncan Brumby. An earlier version of this work was presented at the 1st UK ACT-R Learning Group held at UCL, London UK, January 2004. We thank Richard Young and Anna Cox for comments on this work, and three anonymous reviewers.

References

Anderson, J.R. (1990). *The adaptive character of thought.* Hillside, NJ: Erlbaum.

Anderson, J.R., & Lebiere, C. (1998). *The atomic components of thought.* Mahwah, NJ: Erlbaum

Brumby, D.P., & Howes, A. (2003). Interdependence and past experience in menu choice assessment. In the *Proceedings of the 25th Annual Conference of the Cognitive Science Society,* Boston, MA, 2003

Budiu, R., & Anderson, J.R. (2004). Interpretation-based processing: a unified theory of semantic sentence comprehension. *Cognitive Science, 28 (1),* 1 – 44

Byrne, M.D. (2001). ACT-R/PM and menu selection: applying a cognitive architecture to HCI. *International Journal of Human-Computer Studies, 55,* 41 – 84

Cox, A.L., & Young, R.M. (submitted). A rational model of the effect of information scent on the exploration of menus. *6th Internal Conference on Cognitive Modeling,* Pittsburgh, PA, 2004

Hornof, A.J. (in press). Cognitive strategies for the visual search of hierarchical computer displays. To appear in *Human-Computer Interaction*

Kieras, D.E., & Meyer, D.E. (1997). An overview of the EPIC architecture for cognition and performance with application to human-computer interaction. *Human-Computer Interaction, 12,* 391–438

Lovett, M.C., & Anderson, J.R. (1996). History of success and current context in problem solving: combined influences of operator selection. *Cognitive Psychology, 31,* 168 – 217

MacGregor, J., Lee, E., & Lam, N. (1986). Optimizing the structure of database menu indexes: a decision model of menu search. *Human Factors, 28(4),* 387–399

Pirolli, P., & Card, S.K. (1999). Information Foraging. *Psychological Review, 106,* 643–675

Pirolli, P., & Fu, W-T.F. (2003). SNIF-ACT: a model of information foraging on the world wide web. In *Proceedings of the Ninth International Conference on User Modeling,* 2003

Rieman, J. (1994). *Learning strategies and exploratory behaviour of interactive computer users.* PhD dissertation, University of Colorado, Boulder, CL

Salvucci, D.D. (2001). An integrated model of eye movements and visual encoding. *Cognitive Systems Research, 1(4),* 201– 220

Young, R.M. (1998). Rational Analysis of exploratory choice. In M.Oaksford & N.Chater (Eds.). *Rational Models of Cognition.* Oxford: Oxford University Press

Young, R.M., & Cox, A.L. (2000). A new rational framework for modelling exploratory device learning … but does it fit with ACT-R? In the proceeding of the *Seventh Annual ACT-R Workshop and Summer School,* Pittsburgh, PA, 2000

Routine Procedural Isomorphs and Cognitive Control Structures

Michael D. Byrne, David Maurier, Chris S. Fick, Philip H. Chung
{byrne, dmaurier, cfick, pchung}@rice.edu
Department of Psychology
Rice University, MS-25
Houston, TX 77005

Abstract

A major domain of inquiry in human-computer interaction is the execution of routine procedures. We have collected extensive data on human execution of two procedures which are structurally isomorphic, but not visually isomorphic. Extant control approaches (e.g., GOMS) predicts they should have the same execution time and error rate profiles, which they do not. We present a series of ACT-R models which demonstrate that control of visual search is likely a key component in modeling similar domains.

Introduction

Every day, people execute countless procedures which are more or less routine. Many of these are uninteresting, but many of these occur in contexts such as emergency rooms and command-and-control centers where failures of speed or correctness can have serious consequences. Thus, understanding how humans execute routine procedures is critical in at least some domains. Card, Moran, and Newell (1983) and John and Kieras (1996) define a routine cognitive skill as one where the person executing the skill has the correct knowledge of how to perform the task and simply needs to execute that knowledge. Roughly speaking, that can be thought of as the point where people are no longer problem solving, but rather applying proceduralized knowledge to a relatively familiar task.

This level of skill has been the focus of attention for an entire family (the GOMS family; John & Kieras, 1996) of techniques for analysis and execution time prediction. This is largely due to the fact that such a wide array of situations fall under this classification, from occasional but not infrequent programming of VCRs to situations involving highly-motivated people in safety-critical situations, such as commercial pilots and medical professionals. As noted, GOMS, which stands for goals, operators, methods, and selection rules, is one of the primary techniques for predicting human performance under these conditions, and the empirical success of GOMS is well-documented (again, see John & Kieras, 1996). A typical GOMS analysis is based on a hierarchical goal decomposition and then a listing of the primitive operators needed to carry out the lowest-level goals. Thus, GOMS analyses are highly sensitive to the goal-based task structure and the number of primitive operations required.

What such an analysis predicts is that two tasks with the same goal/method/operator structure should produce identical performance. While this may be true in a great many cases, it is not universally true. We will present data from two tasks which would yield equivalent GOMS models (which we term "GOMS-isomorphic") but produce significantly different profiles in terms of the time of execution for each step, as well as the error rates at each step. This is not intended as a criticism of the GOMS modeling approach, but rather as the identification of an opportunity for improvement.

This presentation will focus on performance in a series of laboratory experiments in which participants were trained on a number of relatively simple computer-based tasks and then in a subsequent session, returned to perform those tasks along with a concurrent memory-loading task. This paradigm is essentially the same as that used in Byrne and Boviar (1997), which focused on a particular type of procedural error, the postcompletion error. This line of research is primarily concerned with errors made in the task, but to fully understand the errors made, we felt it would first be necessary to understand the cognitive control structures which would produce execution times similar to those we found in the lab. In order to understand these experiments, a relatively thorough understanding of the tasks is required.

The Tasks

Common Procedures

The two tasks under examination were both set in a fictional Star Trek setting to encourage engagement of the undergraduate participants. Participants came in for two sessions spaced roughly one week apart. The first session was training, in which participants were given a description for each task and a manual, walked through the task once with the manual in hand, and then had to repeat each task until they performed it without error three times. In the second session, participants performed the tasks on which they were trained in the first session, along with a concurrent memory-loading task. In this task, they had to monitor a stream of spoken letters which was occasionally interrupted with a beep, after which they responded with the last three letters heard. Participants earned points for correctly executed steps, lost points for errors, received bonus points for rapid performance, and lost points for incorrect answers to the memory probes. High scorers received additional compensation.

While participants were trained on several tasks, not all of which were the same from experiment to experiment, the current research is focused on two tasks, called the Phaser and the Transporter. These two tasks are isomorphic in that they have the same number of steps which were grouped in the training manuals in the same subgoals. The names of those goals, and the names of the buttons and some of the displays and actual controls, however, were different between the two tasks.

The displays for the two tasks appear in Figures 1 and 2 and the list of subgoals and steps appears in Table 1. The main goal in the Phaser task is to destroy the hostile Romulan vessel; the main goal in the Transporter task is to energize it to return some crewmembers to safety. One of the immediately obvious visible differences between the two layouts is that the controls for the Transporter are visually grouped according to subgoal while in the Phaser they are not.

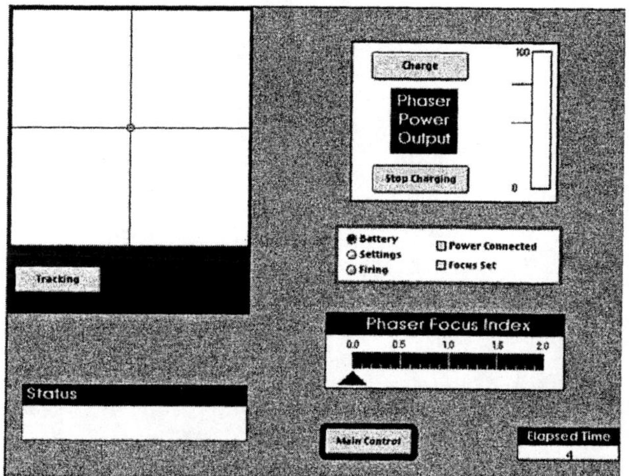

Figure 1. Phaser task display

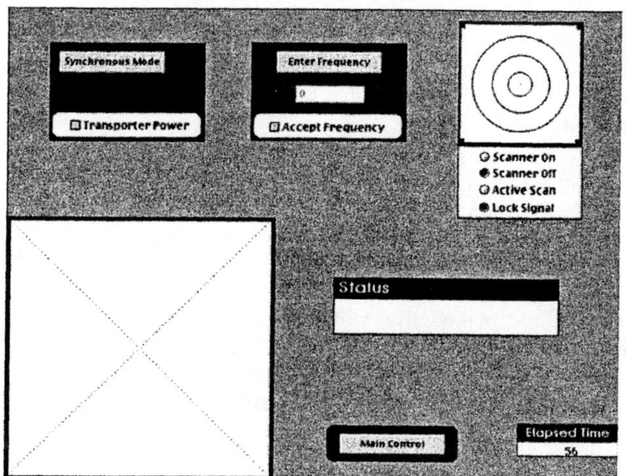

Figure 2. Transporter task display

There are some other important features to note as well. After Step 3 in both tasks, the participants had to wait until the display reached an acceptable state before clicking the next button. Step 6 in both tasks involved, or could involve, multiple actions: multiple drag adjustments to the slider in the case of the Phaser and multiple keystrokes in the case of the Transporter. Step 10 in both procedures involved a somewhat extended tracking task, done with the arrow keys for the Phaser and with the mouse for the Transporter. All of the other steps required the simple clicking of a button.

The exact responses of the display and some of the task structure did differ between the two tasks for Steps 11 and 12 as part of manipulations concerned with postcompletion errors, so those steps were excluded from all present analyses.

The major dependent variables of interest here were step completion time and error frequency Step completion time is measured as the time between clicks. That is, the time for Step 2 in the Phaser is the time elapsed between the click on "Power Connected" and the click on "Charge." For the first step, the start time was the start of the trial. Steps on which errors were made were excluded from the time analysis. Times for steps that include other actions (waiting, tracking) were be excluded from the analysis because this other time is difficult to factor out.

Error frequency was also measured. This hinges on the definition of what counts as an error. Each step can be considered a sequential choice (Ohlsson, 1996), so the definition was based on the step, not the action. If any incorrect action was taken at a step, that step counted as an error, regardless of the number of incorrect actions taken. For example, if a participant is at Step 4 in the Phaser task, and they click on the "Settings" button and then the "Firing" button, only one error was recorded because an error was made at that step. Frequency was calculated as the number of error-containing steps divided by the total number of steps executed.

Step #	Phaser	Transporter
First subgoal		
1	Power Connected	Scanner On
2	Charge	Active Scan
3	Stop Charging	Lock Signal
4	Power Connected	Scanner Off
Second subgoal		
5	Settings	Enter Frequency
6	<slider>	<type>
7	Focus Set	Accept Frequency
Third subgoal		
8	Firing	Transporter Power
9	Tracking	Synchronous Mode
10	<tracking task>	<tracking task>
Fourth subgoal		
11	Tracking	Synchronous Mode
12	Main Control	Main Control

Table 1. Steps in the two task isomorphs

Because these tasks are essentially isomorphic, there is no *a priori* reason to necessarily expect different performance on the two tasks (through Step 10), except perhaps slightly longer step completion times for those steps where the mouse has further to go. Nor was assessing such differences the original purpose of the three experiments we will report; those experiments were primarily focused on postcompletion errors in the Phaser task.

Results

While three separate experiments were run, these experiments differed from each other in detail only. Experiment 1 actually included subsequent sessions with a variety of between-subjects manipulations; Experiment 2 added visual cueing at the postcompletion step of the Phaser; Experiment 3 used a cue and a mode indicator to attempt to mitigate postcompletion effects in the Phaser at step 11; the exact point system used in the three experiments differed slightly; etc. However, none of these surface dissimilarities made much difference; the results are nearly identical for all three experiments (inclusion of "experiment" as a between-subjects variable reveals no main effects or interactions involving that variable). Across the three experiments, data from a total of 164 participants were used. Figure 3 shows the results for step completion time for the Phaser and the Transporter, while Figure 4 presents the error frequency. Steps 3, 6, and 10 are excluded from Figure 3 because those steps involve other processes (e.g., tracking) or possibly multiple actions, as described above. Note that both graphs also include the 95% confidence intervals (non-pooled error).

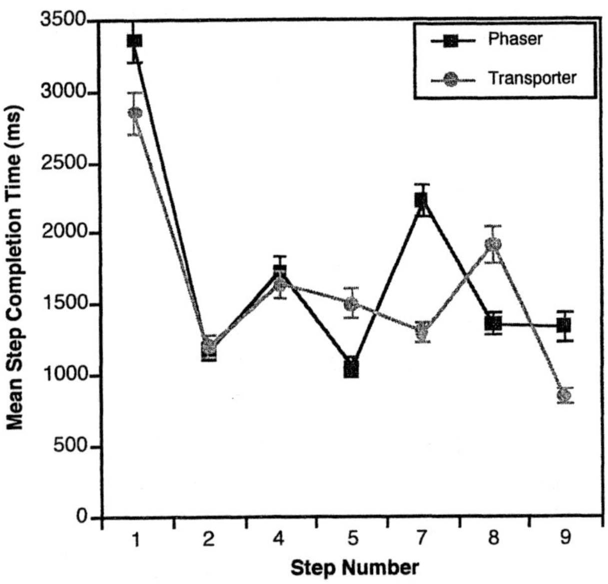

Figure 3. Mean step completion times by task and step

So, while the tasks are isomorphic in terms of subgoals and steps, they produce clearly different step completion time profiles. This runs clearly counter to any account which relies entirely on the GOMS-level structures.

Similarly, if one assumes that the same failure mechanisms are in operation in each task, the two tasks should produce identical error rate profiles as well. This is also obviously not the case. While both tasks share a spike in error rate at step 4 in the procedure (this is, in fact, a postcompletion error), the Phaser shows other spikes at steps 1 and 6 while the Transporter only shows another spike at step 8. Note that these spikes in error rate are not particularly linked to exceptionally large or small step times, either; for example, step 7 in the Phaser is particularly slow, but is not especially error-prone. Step 1 is slow in both tasks, but only markedly error-prone in the Phaser.

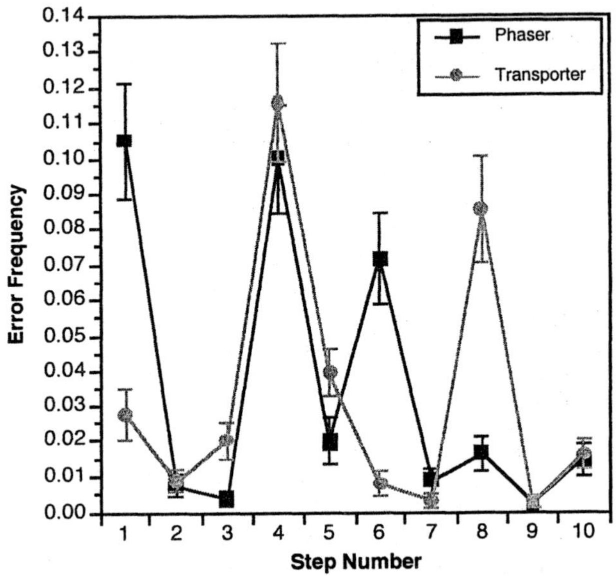

Figure 4. Mean error frequency by task and step

Discussion

These data are obviously problematic for any account which relies solely on the goal-subgoal-method structure for predicting execution time. It is hard to know how a GOMS-style account might accommodate these data. Subjects were probably not at the level of skill where extreme interleaving of cognitive, perceptual, and motor operations is required to model their peformance, thus it is not clear that the CPM variant of GOMS (e.g., Gray, John, & Atwood, 1993) would be appropriate. This is not to say that motor operators are unimportant; there are some differences in going from button to button in terms of pointing time as predicted by Fitts's law, but these differences are relatively small (as will be shown later).

One possibility that seems straightforward is that each of these buttons has to be visually located in order for the mouse to be moved to the button and a click registered. However, there is no single "visual search" operator in GOMS (or ACT-R or Soar for that matter) which would obviously capture the differences here. Each button on the display is at least approximately equal in terms of visual salience; while one might argue that the larger gray pushbuttons are more salient and should thus be found faster, there is little difference between steps 8 and 9 of the Transporter, one of which is a large gray button and the other is simply a labeled checkbox. Furthermore, consider step 7 in the Phaser is markedly slower than step 7 in the Transporter and yet both are simple check boxes with two-word labels. So, if the difference is simply in a "visual search" operator, this operator must itself be driven by something substantially more sophisticated than what is present in a typical GOMS analysis. Furthermore, if the only difference between the two tasks is in their visual search latencies, the source of the differential error spikes remains a mystery.

This obviously raises the question of what kind of control structure could account for the differences between these two tasks? Accounting for the error profiles seems extremely difficult with any model at this point; generative theories of error are in their infancy at best (though that is ultimately our goal, see also Byrne, 2003). Thus, we entered into a modeling exploration with the modest goal of trying to understand what drove the step completion times.

Modeling

We constructed a number of models of this task using ACT-R 5.0 (Anderson, et al., in press). This was done not so much because of a strong commitment to any particular mechanism in ACT-R, but rather because ACT-R contains the full suite of perceptual, motor, and cognitive functionality required for these tasks. It is likely that some version of Soar or EPIC would have served equally well for present purposes but we are much more familiar with ACT-R (and further suspect we will need the subsymbolic mechanisms for future error modeling).

We constructed four models of each task. It was our hope that this way we might "bracket" performance (Kieras & Meyer, 2000; Gray & Boehm-Davis, 2000) and see if the models could provides reasonable predictive bounds. The four models represented a crossing of two dichotomies:

Goal organization. The first dichotomy was whether the model used a hierarchical representation of the goal structure, with intermediate subgoals (e.g., "charge the phaser") or a "flat" goal structure where 12 low-level goals were simply executed in sequence. There is reason to believe that even well-practiced experts do not entirely flatten their goal hierarchies (Kieras, Wood, & Meyer, 1997) and that, in fact, often times fairly slow retrieval-based strategies are appropriate (Altmann & Trafton, 2002). The hierarchical goal strategy is noted with "Hier" in the model label and the flat with "Flat."

Visual search. We implemented two very simplistic visual search strategies here: one in which the location of each button had to be determined through serial visual examination with a tendency to search near the current focus of visual attention (Fleetwood & Byrne, 2003) and one in which the model is assumed to have declarative knowledge of the locations of the buttons which must be retrieved for each button. Various ACT-R models (Ehret, 2002; Anderson, et al. in press) have shown that this kind of learning is a key component of skill development in similar interfaces. The unguided serial search strategy is noted with "DS" (for dumb search) and the alternate with "RL" for "retrieve location."

It should be noted that in ACT-R, these dichotomies may interact. ACT-R's visual system has a memory for which locations (though not explicitly which objects) have been viewed recently, but this memory decays over time (we used 1.5 seconds for this decay time; the models are indeed sensitive to this parameter but in unusual ways which are beyond the scope of this presentation). Thus, additional time spent in traversing the goal hierarchy can result in the loss of this information, which may affect the time course of the serial visual search.

ACT-R also embeds Fitts's law for prediction of mouse movement times. We used ACT-R to calculate the expected movement time between the various buttons to make clear the movement time contribution to the results. We did not compute it for step 1 because the initial location of the cursor was not recorded; informal observation of the participants indicated that many of them moved the mouse around before clicking anyway.

Finally, these models are all stochastic. Time for memory retrievals and perceptual-motor operations in ACT-R can be made noisy and ACT-R chooses randomly between options in various subsystems in cases of ties, so each run of the model is not identical to the last. We present the mean model-generated times for 100 runs of each model.

Model Results

Figure 5 presents the data and the model predictions, as well as the Fitts's Law time, for the Phaser task. Figure 6 presents the same for the Transporter.

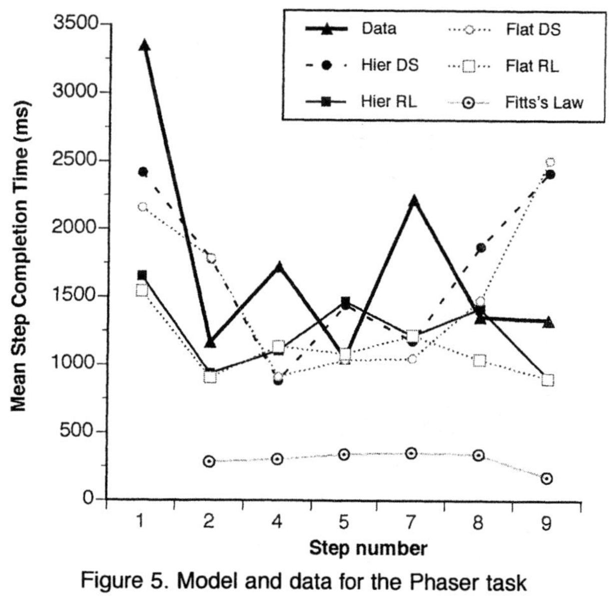

Figure 5. Model and data for the Phaser task

None of the four models provides a particularly good fit; which model is the "best" model by fit metric depends on which metric of fit one uses: by r-squared, the best model is Flat RL at 0.73; by RMSD, the best model is the Hier DS at 640 ms; by mean absolute deviation (MAD) the best model is the Hier RL model at 26%. These are fairly fine distinctions since RMSD ranged from 640–739 ms and MAD ranged from 26–33%. Note that the model variant here which is most similar to a GOMS-style model is the Hier RL model. This model uses hierarchical goal decomposition as per GOMS, and essentially has a fixed time "find-on-screen" operator (the retrieval of the location). This model is generally good, if a bit too fast, for the Transporter, but is a poor model for the Phaser.

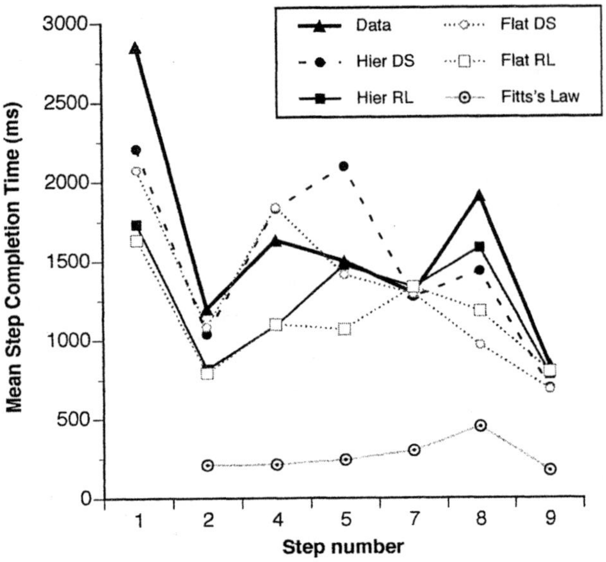

Figure 6. Model and data for the Transporter task

While the models do not provide outstanding levels of fit, they do provide some important insights. First, Fitts's Law alone provides an r-squared of 0.28 for those steps where it is applicable. Obviously time is grossly under-predicted by Fitts's Law—it is hardly surprising that more is going on here than simple motor movement, though it is clearly a contributor.

In general, the ordinal effects one would expect from the basic construction of the models held: the Flat models were faster than the Hier models and the DS models were generally faster than the RL models. Note that in general, the RL models' performance on the two isomorphs was quite similar. This is a reflection of their isomorphic task structure. The DS models, on the other hand, reflect differences between the tasks. This is consistent with the notion that it is the visual aspects of the display—the DS models are sensitive to button location with the RL models are only in the Fitts's law sense—that drives the differences between the two tasks.

However, there were a few cases where the DS and RL models were rougly equivalent, and even one case where the DS models were faster (Phaser step 4, Power Connected). There were some degenerately bad performances by the DS model, notably Phaser step 9, Tracking, and the Hier DS on Transporter step 5, Enter Frequency. Both of these cases involve a visual shift to a location which has a lot of competition from items much closer to the starting attentional focus.

On the other hand, the slower DS resulted in better fit in some instances, namely step 1 for both tasks, and steps 2 and 3 for the Transporter. Step 1 is particularly problematic for all four models; this is a very slow step for both the models and the participants, but more so for the participants. We suspect this is due to some kind of initial orienting or goal construction on the part of the participants which was not well-represented in the model, but may be partially represented by the DS behavior of taking an initial visual survey of the display. This plays into the next insight we gained from these models.

In general, the RL models were slightly better than the DS models. What this suggest to us is that participants in this case are at an intermediate point in their learning of the locations of the objects on the interface. Our next model will likely not start with the locations explicitly encoded in declarative memory but will instead use the strategy of attempting to retrieve them from memory, but this time from the memories created as a by-product of visual searches conducted along the way.

Comparisons between the Flat and Hier models are also revealing. These models differed primarily at steps where either they interacted with the visual search process (Transporter step 5, Enter Frequency is a good example of this) or there was a delay for additional goal traversal (steps 1, 5, and 8. This additional time appears correct for both tasks for steps 1 and 8, but step 5 indicates something else going on. Both Hier models are too slow for step 5 in the Phaser, but the Hier RL model is right on target for step 5 for the Transporter.

Finally, some points were fairly strategy-insensitive. Transporter step 7 (Accept Frequency) was fit equally well by all four models. This is an interesting case for two reasons: [1] the DS visual search strategy will almost always search the correct location first here because of visual proximity to where the model is looking prior to this step, [2] it is the last subgoal within the second goal, and thus not differentially affected by the goal organization, and [3] the completion time for the similar Phaser step is radically different. None of the models captured this deviant time in the Phaser at all.

Discussion

While it may appear that our goal was to somehow falsify or criticize GOMS models, but that was not our intent. Instead, we wanted to explore where and why models based purely on the GOMS-style structure would misfit, not for the purposes of finding fault, but to find opportunities for improvement. One of the primary things GOMS-style models lack is a consideration of the visual task faced by interface users. This was certainly reasonable when most users faced command-line tasks which were indeed primarily cognitive, but the shift to increasingly visual interfaces has raised the importance of systematically addressing the problem of how the visual and cognitive systems are integrated. While this has certainly been a big topic for some cognitive scientists for many years (see Pylyshyn, 1999 and the associated commentary for an excellent discussion), it has not been a prominent theme in computational modeling of human-machine interfaces until fairly recently and in cases where the task is clearly defined as primarily a visual search task (e.g., Fleetwood & Byrne, 2003; Everett & Byrne, 2004; Hornof & Kieras, 1997, 1999) . Our research suggests this may be an important part of routine procedure execution even when visual search may *not* appear to be a dominant factor. Furthermore, it appears that it is neither the case that the most optimistic assumption (users memorize the location of all controls) or the most pessimistic assumption (users search randomly every time) is an appropriate representation of user behavior, at least at this level of skill. This suggests that more research is needed on the integration of cognitive mechanisms such as representing and traversing goal structures with visual-cognitive mechanisms such as search strategies. While we doubt anyone would have denied that this is an important domain in a general sense, we suspect that most researchers in this area would underestimate the impact such considerations might have on execution of routine procedures.

To end on a speculative note, consider the hint provided by Phaser step 5 (settings). In that case, the Hier models are too slow and the Flat models are spot-on, suggesting that the goal traversal performed by the model is not being done by the participants. This might be an indicator that the participants have re-configured their internal representation of the task structure to match the visual structure of the interface! This suggests a possibly important role for the match between the task structure and the visual layout of an interface, something clearly not predicted by extant GOMS-class models.

Acknowledgements

We would like to acknowledge the support of the Office of Naval Research under grant number N00014-03-1-0094 to the first author. The views and conclusions contained herein are those of the authors and should not be interpreted as necessarily representing the official policies or endorsements, either expressed or implied, of ONR, the U.S. Government, or any other organization.

References

Altmann, E. M., & Trafton, J. G. (2002). Memory for goals: An activation-based model. *Cognitive Science, 26*, 39–83.

Anderson, J. R., Bothell, D., Byrne, M. D., Douglass, S., Lebiere, C., & Quin, Y. (in press). An integrated theory of the mind. To appear in *Psychological Review*.

Byrne, M. D. (2001). ACT-R/PM and menu selection: Applying a cognitive architecture to HCI. *International Journal of Human-Computer Studies, 55(1)*, 41–84.

Byrne, M. D. (2003). A mechanism-based framework for predicting routine procedural errors. In R. Alterman & D. Kirsh (Eds.) *Proceedings of the Twenty-Fifth Annual Conference of the Cognitive Science Society*. Austin, TX: Cognitive Science Society.

Byrne, M. D., & Bovair, S. (1997). A working memory model of a common procedural error. *Cognitive Science, 21*, 31–61.

Card, S. K., Moran, T. P., & Newell, A. (1983). *The psychology of human-computer interaction*. Hillsdale, NJ: Lawrence Erlbaum Associates.

Ehret, B. D. (2002). Learning where to look: location learning in graphical user interfaces. In *Human Factors in Computing Systems: Proceedings of CHI 2002* (pp. 211-218). New York: ACM.

Everett, S. P., & Byrne, M. D. (2004). Unintended effects: Varying icon spacing changes users' visual search strategy. *Human Factors in Computing Systems: Proceedings of CHI 2004* (pp. 695–702). New York: ACM.

Fleetwood, M. D. & Byrne, M. D. (2003). Modeling the visual search of displays: A revised ACT-R/PM model of icon search based on eye-tracking and experimental data. In F. Detje, D. Dörner, & H. Schaub (Eds.) *Proceedings of the Fifth International Conference on Cognitive Modeling* (pp. 87–92). Bamberg, Germany: Universitas-Verlag Bamberg.

Gray, W. D., & Boehm-Davis, D. A. (2000). Milliseconds matter: An introduction to microstrategies and to their use in describing and predicting interactive behavior. *Journal of Experimental Psychology: Applied, 6*, 322-335.

Gray, W. D., John, B. E., & Atwood, M. E. (1993). Project Ernestine: A validation of GOMS for prediction and explanation of real-world task performance. *Human-Computer Interaction, 8*, 237–309.

Hornof, A. J., & Kieras, D. E. (1997). Cognitive modeling reveals menu search is both random and systematic. In *Human Factors in Computing Systems: Proceedings of CHI 97* (pp. 107–114). New York: ACM Press.

Hornof, A. J., & Kieras, D. E. (1999). Cognitive modeling demonstrates how people use anticipated location knowledge of menu items. In *Proceedings of ACM CHI 99 Conference on Human Factors in Computing Systems* (pp. 410-417). New York: ACM.

John, B. E., & Kieras, D. E. (1996). The GOMS family of user interface analysis techniques: Comparison and contrast. *ACM Transactions on Computer-Human Interaction, 3*, 320-351.

Kieras, D. E., & Meyer, D. E. (2000). The role of cognitive task analysis in the application of predictive models of human performance. In J. M. Schraagen & S. F. Chipman (Eds.), *Cognitive task analysis* (pp. 237-260). Mahwah, NJ: Erlbaum.

Kieras, D. E., Wood, S. D., & Meyer, D. E. (1997). Predictive engineering models based on the EPIC architecture for multimodal high-performance human-computer interaction task. *Transactions on Computer-Human Interaction, 4*(3), 230-275.

Ohlsson, S. (1996). Learning from performance errors. *Psychological Review, 103*, 241–262.

Pylyshyn, Z. (1999). Is vision continuous with cognition? The case of impenetrability of visual perception. *Behavioral & Brain Sciences, 22*(3), 341-423.

Why Soccer Players Yell: Using RoboCup to Model the Advantage of Signaling

Sanjay Chandrasekharan (schandra@sce.carleton.ca)
Babak Esfandiari(babak@sce.carleton.ca)
Institute of Cognitive Science,
Carleton University, Ottawa, Canada, K1S 5B6

Neal Arthorne(narthorn@connect.carleton.ca)
Department of Systems and Computer Engineering,
Carleton University, Ottawa, Canada, K1S 5B6

Abstract

We report an experiment where the RoboCup simulation environment was used to study the cognitive advantage provided by signals, which we view as *task-specific* structures generated in the environment to improve decision-making. We used the passing problem in RoboCup as our test problem and soccer-players' 'yells' of their 'passability' as the task-specific signals. We found that yells improved accuracy– agents using the yells to decide the best player performed much better than agents computing the best pass themselves. The accuracy advantage derives from the task-specific nature of the yell, and such task-specific structures (signals) are used by organisms across species. From this, we reason that player yells are an instantiation of an implicit, evolved, and adaptive strategy, rather than an explicitly reasoned-out process.

Many organisms generate stable structures in the world to reduce cognitive complexity, for themselves, for others, or both. Wood mice (*Apodemus sylvaticus*) distribute small objects, such as leaves or twigs, as points of reference while foraging. They do this even under laboratory conditions, using plastic discs. Such "way-marking" diminish the likelihood of losing interesting locations during foraging (Stopka & MacDonald, 2003). Red foxes (*Vulpes vulpes*) use urine to mark food caches they have emptied. This marking acts as a memory aid and helps them avoid unnecessary search (Henry, 1977, reported in Stopka & MacDonald, 2003). The male bower bird builds colorful bowers (nest-like structures), which are used by females to make mating decisions (Zahavi & Zahavi, 1997). Many birds advertise their desirability as mates using some form of external structure, like colorful tails, bibs etc (Bradbury & Vehrencamp, 1998). Other animals have signals that convey important information about themselves to possible mates and even predators (Zahavi & Zahavi, 1997).

Such *epistemic structures* (Chandrasekharan & Stewart, 2004), usually termed signals, form a very important aspect of animal life across biological niches. These structures allow the organisms to hive off part of their cognitive load to the world. How much cognitive advantage do such structures provide in noisy, dynamic and adversarial environments? Where do the advantages come from? What are its components? These are the problems we address in this paper. We used the RoboCup simulation environment to study the cognitive advantage provided by epistemic structure strategy.

Signaling is generally studied as communication (considered a good thing), and most approaches do not focus on the computational advantages provided by signaling. To understand the computational advantage provided by signaling, consider the peacock's tail, the paradigmatic instance of an animal signal. The tail's function is to allow female peacocks (peahens) to make a mating judgment, by selecting the most-healthy male (Zahavi & Zahavi, 1997). The tail reliably describes the inner state of the peacock, that it is healthy (and therefore has good genes).

To see the cognitive efficiency of this mechanism, imagine the peahen having to make a mating decision without the existence of such a direct and reliable signal. The peahen will need to have a knowledge base of how the internal state, of health, can be inferred from behavioral and other cues. Let's say "good dancing", "lengthy chase of prey", "long flights" (peacocks fly short distances), "tough beak" and "good claws" are cues for the health of a peacock. To arrive at a decision using these cues, first the peahen will need to "know" these cues, and that some combinations of them imply that the male is healthy.

Armed with this knowledge, the female has to sample males for an extended period of time, and go through a lengthy sorting process based on the cues (rank each male on each of these cues: good, bad, okay). Then it has to compare the different results, keeping all of them in memory, to arrive at an optimal mating decision. This is a computationally intensive process.

The tail allows the female peacock to shortcut all this computation, and go directly to the most-healthy male in a lot. The tail provides the peahen a single, chunked, cue, which it can compare with other similar ones *perceptually* (i.e. without computation) to arrive at a decision. The tail is a *task-specific* structure. It exists just for the peahen to make the mating decision. The other cues (like tough_beak etc.) do not exist for this purpose, they are *task-neutral* structures, *which have to be synthesized by the pea-hen into a task-specific structure*, to help with the mating decision. The tail, being a task-specific structure, allows the pea-hen to short-cut this synthesizing process. The tail 'fits' the peahen's task, and provides a standardized way of arriving at a decision, with the least amount of computation. The peacock describes its system state using its tail. Such self-description is one of nature's ways of avoiding long-winded sorting and inference.

The peacock example (and others above) shows that the reduction of others' cognitive complexity using task-specific external informational structures (what we term epistemic structures) is very common, and it can be considered one of the building blocks of nature. Note that while we develop the notion of epistemic structure using stable (or quasi-permanent) structures added to the world like markers and tails, task-specificity is a property common to all signals, including transient environment structures (like vocal signals for mating, warning etc.). In our view, transient signals are an adaptation of the basic epistemic structure theme (of adding task-specific structures to the world), to suit the highly dynamic or adversarial nature of such decision-making environments. In other words, stability is not the crucial property for being an epistemic structure/signal, task-specificity is. To understand the advantage provided by signaling, we have to understand the efficiency provided by task-specific structures.

A Taxonomy of Agent-Environment Relations

Even though signaling is a basic structure of cognition, it has received very little attention *as a cognitive strategy*. In the following section we develop a framework to understand how signaling, or the epistemic structure strategy (where the environment is changed in a way that it contributes task-specific structures for decision-making), fits in with other agent-environment relationships. We categorize agent-world relations into four strategies. To illustrate these strategies, we use the design problem of providing disabled people access to buildings. There are four general strategies to solve this problem.

Strategy 1: This involves building an all-powerful, James Bond-style vehicle that can function in all environments. It can run, jump, fly, climb spiral stairs, raise itself to high shelves, detect curbs etc. This design does not incorporate detailed environment structure into the vehicle, it is built to overcome the limitations of all environments.

Strategy 2: This involves studying the vehicle's environment carefully and using that information to build the vehicle. For instance, the vehicle will take into account the existence of curbs (and them being short), stairs being non-spiral and having rails, level of elevator buttons etc. So it will have the capacity to raise itself to short curbs, climb short flight of straight stairs by making use of the rails etc. Note that the environment is not changed here.

Strategy 3: This involves adding structure to the environment. For instance, building ramps and special doors so that a simple vehicle can have maximum access. This is the most elegant solution, and the most widely used one. Here structure is added to the environment, the world is "doped", so that it contributes to the agent's task. Our analysis will focus on this approach.

Strategy 4: This strategy is similar to the first, but here the environment is all-powerful instead of the vehicle. The environment becomes "smart", and the building detects all physically handicapped people, and glides a ramp down to them, or lifts them up etc. This solution is an extreme case of strategy III, we will ignore it in the following analysis.

The first strategy is similar to the centralized AI one, which ignores the structure provided by specific environments. The environment is something to be overcome, it is not considered a resource. This strategy tries to load every possible environment on to the agent, as centrally stored representations. The agent tries to map the encountered world on to this internal template structure.

The second strategy is similar to the situated AI model promoted by Rodney Brooks (1991). This strategy recognizes the role of the environment as a resource, and analyses and exploits the detailed structure that exists in the environment to help the agent. Notice the environment remains unchanged, it is considered a given.

The third strategy is similar to one aspect of distributed cognition, where task-specific structures are generated in the environment, allowing the agent to hive off part of the computation to the world. Kirsh (1996) terms this kind of "using the world to compute" active redesign. This strategy underlies many design techniques to minimize complexity. At the physical level, the strategy can be found in the building of roads for wheeled vehicles. Without roads, the vehicles will have a hard time, or all vehicles will need to have tank wheels. With roads, the movement is a lot easier for average vehicles. This principle is also at work in the "intelligent use of space" where people organize objects around them in a way that helps them execute their functions (Kirsh, 1995). Kitchens and personal libraries (which use locations as tags for identifying content) are instances of such use of space in cognition.

Another application of task-specific structures is bar coding. Without bar coding, the checkout machine in the supermarket would have to resort to a phenomenal amount of querying and object-recognition routines to identify a product. With bar coding, it becomes a simple affair. The Semantic Web enterprise is another instance. The effort is to generate task-specific structure in an information environment (the Web) so that software and human agents can function effectively in it. This principle is also at work in the Physical Markup Language effort, which tries to develop a common standard to store information in low-cost Radio-frequency Identification (RFID) tags. These tags can be embedded in products, like meta-tags in web pages. Such tagged objects can be easily recognized by agents fitted with RFID readers (for instance, robots in a recycling plant).

The epistemic structure strategy is applied at the social level as well, especially in instances involving trust. Humans add structures to the environment to help others make trust decisions. Formal structure created for trust includes credit ratings, identities, uniforms, badges, degrees, etc. These structures serve as reliable signals for people to make trust decisions. Less reliable, and informal, structure we create include standardized ways of dressing, talking etc.

Using RoboCup to Study Epistemic Structure

Given this broad range of current and potential applications of this strategy, it is important to understand how the epistemic structure strategy works, how it evolved, how efficient it is compared to other strategies, in what conditions it works, and where it breaks down.

We have modeled organisms learning to add task-specific structures to the world, across generations and within their lifetime, using genetic algorithms and Q-learning (Chandrasekharan & Stewart, 2004). In both cases, we used an environment with no adversaries and almost no noise. The results show that the advantage of using the epistemic structure (ES) strategy is quite significant, agents spend 58% of their time generating such structures. But since most organisms live in noisy, dynamic, adversarial environments, static environments do not provide a sense of the comparative advantage of the ES strategy over others.

The RoboCup simulation environment, which simulates a soccer game, provides an interesting dynamic and adversarial environment to study the efficiency of the epistemic structure strategy. Briefly, the simulation environment consists of a standard central server and two teams of decision-making agents (usually 11 to a team) that connect to the server. Researchers around the world develop agent teams to study multi-agent systems, while the server is maintained by the RoboCup administration as a standard test-bed for such systems.

In a game, the server sends the agents field information (like current coordinates, coordinates of opponents or teammates seen, coordinates of the ball etc.). The agents use this information to update their world model, and analyze this information to send action commands back to the server (like kick, dribble, turn, turn_neck etc.), which are then 'executed' by the server, thereby changing the state of the agents and what they can perceive. This process also changes the configuration of the field in a very dynamic fashion, akin to a soccer game. The agents can communicate with each other using a narrow channel broadcast by the server (each agent can hear utmost 2 messages in a cycle), but every message by every agent is heard by every other agent, an approximation to player yells in a real soccer field.

Being a game environment, Robocup does not provide much scope to add *stable* task-specific structure to the environment. However, it does allow transient task-specific structure to be added, these are 'yells', or signals from teammates. We used this structure to study the advantage provided by task-specific structures, using the passing problem (i.e. how an agent in control of the ball can decide whom to pass the ball) as our test decision-making problem. For our study, we developed three RoboCup teams (11 agents each) that used three different approaches to passing. The teams were based on the publicly available UvA TriLearn 2002 team (Kok, 2002), which implements some basic low-level skills (like dribbling, kicking etc.).

Team 1: Centralized Passing

This team (A1) uses strategy 1 in our agent-world taxonomy. A1 does all computations centrally, and does not depend on task-specific information from other agents. In A1, when an agent has possession of the ball (i.e., the ball is within a kickable margin), it calculates the pass suitability (passability) for each teammate, and passes the ball to the teammate with the highest passability. If no teammate has passability above a fixed threshold value, the agent will dribble the ball toward the opponent goal.

The goalie in this team is based on the original UvA algorithm, except for one modification: in a goal kick or free kick, the goalie will use A1 to calculate the best receiver for a pass and kick the ball to that teammate. This differs from the UvA standard behavior of the goalie kicking the ball straight down the field. The A1 passing algorithm is described below:

```
A1: Centralized Passing
   Input(s):  None.
   Output(s): Best pass receiver

// set the minimum passability
   Pb <- passabilityThreshold

// initialize best pass receiver to none
   receiver <- none

   for each Teammate except goalie
        Pt <- calculatePassability( agent, Teammate
   ) // see P1
        if ( Pt > Pb ) then
                Pb <- Pt
                receiver <- Teammate
        end if
   end for

return receiver
```

The following section describes P1, the algorithm that computes the suitability of an agent to receive a pass, or what we term passability.

```
P1 Calculate Passability

   Input(s):   source   -   the   agent   who   has
   possession of the ball
   target - the target player whose passability is
   to be calculated
   Output(s):  passability   -   a   real   number
   indicating pass suitability of target player

   posSource <- global position of source
   posTarget <- global position of target

   // draw a line between source and target
   Line     L     <-     Line::makeLineFromTwoPoints(
   posSource, posTarget )
   sumOfDistances <- 0.0;


   // for each opponent, add their distance to the
   line to the
   // sum of distances
   for each Opponent
        oppDistToLine   <-   L.getDistanceWithPoint(
   position of Opponent )
        // only add opponents that are close to the
   line
        if ( oppDistToLine < 15.0 ) then
                sumOfDistances += oppDistToLine
        end if
   end for
```

```
passability <- sumOfDistances

// modify passability to favour forward passing
if ( angle to opponent goal -
     angle to posTarget < 50 ) then
     passability *= 1.3
else
     passability *= 0.4
end if

// modify for congestion
if ( target is congested ) then
     passability *= 0.5
end if
if ( source is congested ) then
     passability *= 0.5
end if
// modify to prevent long passes
if ( distance to target > 20.0 ) then
     passability *= 0.5
end if
```

The passability values used above are based on test games, where different pass situations and passabilities were tested. The algorithm modifies the original UvA player's decision-making algorithm and is used by all our agents.

Team 2: Passing with Yells

This team (A2) is an implementation of the epistemic structure approach. Here every agent calculates its own passability using P1. This calculation is done for every cycle a teammate has control of the ball. The fastest player in a set who can reach the ball is determined to have control of the ball. Once the passability value is calculated, each player uses the 'say' command to signal this value to teammates.

When updating the world model, every agent uses these 'yells' from teammates to track the best passability at a given time. If a message arrives announcing a higher passability, then the sender of the message becomes the new best pass receiver. Every five cycles, the best passability is reset to the minimum threshold, to ensure that old information is not used to make the passing decision.

As in centralized passing, the goalie uses A1 to calculate the pass receiver, but unlike its teammates, the goalie uses the centralized approach with no input from teammates. This ensures that the goalie always passes to someone.

Team 3: Passing with Filtered Yells

This team (A3) is also an implementation of the signaling strategy, but it has some properties of the Brooksian approach, because it takes into consideration the limitation of the communication channel, which is a significant property of the environment. A3 also uses P1, and in the same way as the A2 algorithm. However, instead of agents yelling their passability every cycle, here agents listen to others' yells and compare their passability with the ones they hear. That is, they compare their passability with the current best value, and announce their passability only if it is better. This lowers the load on the communication channel, by allowing only the best messages through. Once again, the goalie uses the centralized approach to passing.

Experiment

To test the efficiency provided by the task-specific structure strategy, our three teams were pitted against the original UvA team. Each team played 10 games. Logs of individual agents' decision-making were collected and analyzed to extract the successful and unsuccessful passes, and the passability values. Note that even though A1 uses centralized decision-making to pass, the other agents in A1 calculate their own passabilities and store these values. In effect, all agents in all the three conditions calculate their passabilities using P1 when a team mate has the ball. In A2 and A3, this information was 'yelled', and the passing agent's decision to pass was based *entirely* on this information. In A1, there was no yelling by individual agents, they just stored their passability values. The passing agent here used the P1 algorithm in a centralized manner, to calculate the passability for everyone else.

Results

Since we are interested in understanding the performance of the strategies in making the passing decision, we analyze only the completion of passes, and not the goals scored, which is affected by many factors other than passing.

Pass Completion: Pass completion is a measure of the ability of a player to pick the correct pass recipient. Although pass completion primarily depends on the effectiveness of the passability function P1, it can also show the relative effectiveness of the three algorithms with respect to each other. We analyzed the log files of games played with the three passing strategies, and checked who next kicked or caught the ball after a player made a pass. If it was the intended recipient, the pass was completed, otherwise the pass failed. Table 1 shows the results of running our three teams against the original UvA team, and testing over ten games for each team. The centralized approach achieved the best results with A2 and A3 achieving a similar percentage of pass completion.

Table 1: Number of passes completed

Team	Completed passes	Total Passes	Percentage
A1	1110	2416	45.94%
A2	804	2384	33.72%
A3	1324	3455	38.32%

Correct Passing: To understand the effectiveness of the three algorithms in deciding the right player, we determined the number of 'correct' passes, which is defined as passes where the agent in possession of the ball passes to the best player (the agent with the highest passability value). This view assumes that an individual player's judgment of its passability is the 'correct' value. The validity of this view is tested in the next section.

For A1, this analysis provides a sense of how often the centralized algorithm agreed with the individuals' assessment of their own passabilities. For A2 and A3, which depend entirely on signaled structures to decide on passing, a correct pass indicates that the message from the most suitable player got through to the passing agent. As the communication channel bandwidth is low, the agent may not always know the best passability in the team, and therefore will make an incorrect pass. Thus, for A2 and A3, the ratio of correct passes to incorrect passes reveals the effectiveness of the communication channel, irrespective of the performance of the passing algorithms.

The results (in Table 2) show that the centralized algorithm makes the 'correct' judgment around 39 percent of the time. A3, the filtered yell model, is more effective than A2 in allowing agents to know the teammate with the highest passability in any given cycle. Overall, the A3 strategy is better in choosing the player with the highest passability, even though the signaling channel is feeble.

Table 2: Number of passes where the player with the highest passability was chosen

Team	Best Player chosen	Total Passes	Percentage
A1	942	2416	38.99%
A2	803	2384	33.68%
A3	1454	3455	42.08%

Correct and Completed Passes: The above tables give us our preliminary data. Given the narrow communication channel, there is no direct way of gauging the effectiveness of the epistemic structure strategy, because only a third of the signals get through to the decision-making agent. To understand the effectiveness of the ES strategy, we need to create a 'what if' scenario, i.e. what if all the messages get through? To find out this, we have to first find out the effectiveness of knowing the best player, i.e. what percentage of time did knowing the best player result in a completed pass? To get this value, we extracted the intersection of the two tables above -- the completed passes when the best player was chosen. This is given in Table 3.

Table 3: Number of passes completed when the best player was chosen

Team	Passes completed	Best player chosen	Percentage
A1	510	942	54%
A2	344	803	42.8%
A3	668	1454	45.9%

Averaging for the three teams, this means nearly 48% of the time, when the best player (the player with the highest passability, according to his own estimate) is chosen, the pass is completed. We consider this quite impressive for a simple passing algorithm. This is the maximum advantage provided when an agent uses task-specific structures for decision-making in the Robocup environment.

What about the mirror of this, how many times was the pass completed when the player chosen was *not* the best player? Table 4 below provides these values.

Table 4: Passes completed when the best player was not chosen

Team	Passes completed	Best player not chosen	Percentage
A1	512	1474	34.7%
A2	400	1581	25.3%
A3	563	2001	28.1%

Comparing with the earlier table, this means knowing the passability value (task-specific structure) can provide around 20% improvement in passing for A1 and 18% for A2 and A3. This means individuals' calculation of their own passability is more accurate than others' calculation of the same value. This justifies the assumption we made about 'correctness' above. Once computed, signaling of this task-specific structure to a decision-making agent can result in around 19% improvement in success rate overall, in a dynamic and adversarial environment like robocup.

At the strategy level, this illustrates why signaling is a preferred strategy in making mating decisions – *because an agent revealing his/her own system state is always more accurate than another agent's judgment of the same state.* We hypothesize that this is also the reason why soccer players yell. Anecdotal evidence (from soccer-players) indicates that yells are used heavily by players while making passing decisions, suggesting an implicit understanding of this accuracy advantage. Given that it is the task-specificity of the yell that is contributing to its accuracy and use, and this task-specificity is shared by signals across species, it is also reasonable to assume that yelling is not an explicitly thought-out process, but an instantiation of an evolved, general, adaptive strategy. Even novice and beginner players yell. In tasks or functions involving cooperation (like mating, or hunting in a group), adding task-specific structures about your system (status, intention, perspective) to the world has adaptive value.

On the receiver side, a decision-maker would be basing her decision on more accurate information when she is using such structures. Of course, this is complicated by the problem of judgement, reliability (mimicry in organisms) and 'eavesdropping', where other agents 'listen in' and use the signal to further their own interests. It is interesting to note that this complexity has not lead to organisms dropping the signaling strategy altogether. Instead, they have evolved ways to counter mimicry and eavesdropping. This means the adaptive advantage provided by signaling is quite high.

The above analysis shows the improvement in performance for individual players. What about team performance? To understand this, we extracted the number of times the home team got the pass when the best player was chosen.

Table 5: Home team got the ball when the best player was chosen

Team	Home team got ball	Best player chosen	Percentage
A1	653	942	69.3%
A2	560	803	69.7%
A3	976	1454	67.1%

When the best player was chosen, the ball stayed with the home team nearly 70% of the time. This is probably a result of the passability algorithm taking into account the number of opponent players in the pass trajectory, or the movement of the agents towards the pass trajectory. Either way, the passability calculations by individuals provide an overall advantage for the team.

The individual and team performance in retaining control of the ball (when individual passabilities are known) shows that the epistemic structure strategy is quite effective. The environment contributing to cognition provides significant advantages in improving accuracy.

Limitations and Future Work

The current experiment investigated one aspect of decision-making based on the ES strategy, the advantage in accuracy, i.e. identifying the best player to pass. In the next set of experiments, we are planning to investigate a second aspect of the task-specific structure strategy, the advantage in processing load, by measuring the time taken for passability calculations by agents with different perspectives. The efficiency of a cognitive strategy is a combination of both accuracy and processing efficiency. We are interested in both, and also in the way they interact.

One of the major limitations of the study is the indirect way of assessing the effectiveness of the epistemic structure strategy. This is a direct result of the narrow communication channel. If the server parameters had allowed us to manipulate the number of messages the agents can hear to beyond 2 messages per cycle, it would have been possible to judge the effectiveness of the strategy better. The ability to vary the size of the communication channel would have also provided a way to better understand the relationship between channel-width and signal effectiveness in a dynamic environment. Such freedom to change parameters, and a more user-friendly way of doing this, could lead to the RoboCup environment being used more widely by disciplines like cognitive science and ethology.

In this study, the opponent team was the same in all the games. Even though this could be considered as providing a standardization for the results reported here, it is desirable to test a cognitive strategy in a variety of situations.

A further limitation is that the opponent team was not designed to intercept the passability messages, or to manipulate them. So the adversarial nature of the environment was limited to pass interception. In future work, we plan to use different teams against our teams.

We also plan to investigate how unreliable messages affect decision-making based on the ES strategy. This is the equivalent of mimicry in biological systems. Another interesting study would be to examine how centralized decision-making can be combined with ES-based strategies, and in what conditions such combinations are effective. Varying the noise parameters for different combinations of strategies may provide insight into how the structure of the environment can lead to different decision-making strategies.

References

Alcock, J. (1998). *Animal Behavior: An evolutionary approach*, Massachusetts, Sinauer Associates.

Bradbury, J.W. & Vehrencamp, S.L (1998). *Principles of Animal Communication*. Sunderland, Mass: Sinauer Associates

Brooks, R. A. (1991) Intelligence without Representation, *Artificial Intelligence*, 47:139-160

Chandrasekharan, S. and Stewart, T. (2004). Reactive Agents Learn to Add Epistemic Structures to the World. Carleton University Cognitive Science Technical Report 2004-01. http://www.carleton.ca/iis/TechReports/

Henry, J.D. (1977). The use of urine marking in the scavenging behaviour of the red fox (Vulpes vulpes). *Behaviour*, 62:82-105.

Kirsh, D., & Maglio, P. (1994). On distinguishing epistemic from pragmatic action. *Cognitive Science*, 18, 513-549.

Kirsh, D.(1995) The Intelligent Use of Space. *Artificial Intelligence*. 73: 31-68

Kirsh, D. (1996). Adapting the environment instead of oneself. *Adaptive Behavior*, Vol 4, No. 3/4, 415-452.

Kok, J.R. (2002) UvA TriLearn 2002 Source Code, University of Amsterdam (UvA), Faculty of Science Intelligent Autonomous Systems Group, http://carol.wins.uva.nl/~jellekok/RoboCup/2002/index_en.html

Noda, I. (1995). Soccer Server: A Simulator for RoboCup, JSAI AI-Symposium 95: Special Session on RoboCup

Stopka, P. & Macdonald, D. W. (2003) Way-marking behaviour: an aid to spatial navigation in the wood mouse (Apodemus sylvaticus). *BMC Ecology*, published online, http://www.biomedcentral.com/1472-6785/3/3

Zahavi, A., & Zahavi, A. (1997). The Handicap Principle: A missing piece of Darwin's puzzle. Oxford: Oxford University Press

A Connectionist Model of the Attentional Blink Effect During a Rapid Serial Visual Presentation Task

Sylvain Chartier (chartier.sylvain@courrier.uqam.ca)
Département de psychologie, Université du Québec à Montréal
C.P. 8888, succ. Centre-ville, Montréal, H3C 3P8 CANADA

Denis Cousineau (denis.cousineau@umontreal.ca)
Département de psychologie, Université de Montréal
C.P. 6128, succ. Centre-ville, Montréal, H3C 3J7 CANADA

Dominic Charbonneau (dominic.charbonneau@umontreal.ca)
Département de psychologie, Université de Montréal
C.P. 6128, succ. Centre-ville, Montréal, H3C 3J7 CANADA

Abstract

We introduce a connectionist model that reproduces the attentional blink effect during a rapid serial visual presentation task. The model is composed of two layers, a competitive one that acts as an identification layer and a recurrent one that acts like short term memory where the main mechanisms is the presence of an inhibition gate and a neural fatigue. Simulations show that the model generates data that are as variables as the one obtained from human participants and the mean performance is identical to the performance obtained by human participants.

Introduction

During visual scanning, the eyes gaze at a given area then make a saccade to another area. Between the saccades, they receive brief images. How the brain can process such a flow of information is studied in the laboratory using the Rapid Serial Visual Presentation (RSVP) task. In this experimental setting, stimuli are presented in rapid succession, usually, 6 to 20 stimuli per second, at a same spatial location. Within the stream, there is usually a target which is marked by a different attribute (e.g. color) and the task is to identify it. Thus, the RSVP can be seen as a visual search without saccade.

The attentional blink (AB) effect is characterized by a decrease in performance for a second target when a first target has been identified (T2|T1). The performance for recalling a second target will decrease if it is presented within 200 to 500 milliseconds of the first. However, if the second target is presented next to the first target, there is no decrease in performance. This phenomenon is called lag of 1 sparing effect. Figure 1 shows a typical AB curve with human participants obtained from our lab.

Many independent models have been postulated to explain the AB phenomenon.The first explanation is given by Raymond, Shapiro & Arnell (1992) and is based on the postulate of an inhibition process. When a first target is perceived, the perceptual system is inhibited to avoid confusion with subsequent items. However, the inhibition is slow to start, so that if a second target is next to the first, it will be processed along the way.

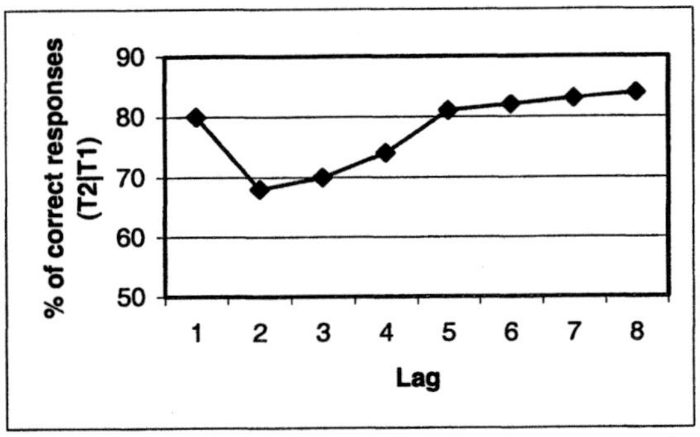

Figure 1: Results with human participants in the AB task from our lab.

Shapiro, Ward & Duncan (1997) challenged this explanation by showing that the second target could facilitate the processing of a third target. If perception was inhibited, there should be no facilitation, according to Raymond & al. (1992).

Another explanation was proposed by Chun & Potter (1995) who postulated a two-stage model. The first stage operates rapidly and decides which of the stimuli is sent to the second stage for encoding. The second stage is slow and can process only one item at a time. Meanwhile, the second target is left in a waiting stage. Thus, as time passes the probability of encoding the second target correctly decreases. However, if the second target is presented next to the first target, it will be encoded

Finally, the most recent explanation comes from Jolicoeur (1998). He proposed a modified version of the two-stage model by adding a third stage. According to Jolicoeur, the first and last stages are the perception and recall stage respectively. The second stage is used to encode the targets in short term memory and is characterize by its limited capacity (1 item at a time). This bottleneck delays the processing of a second target. The probability of correctly encoding the second target decreases as a function of the delay.

These three explanations seem to capture the data in a qualitative manner. However, few formal simulations were ever proposed (Bowman, Wyble and Barnard, 1994). Without form simulation it is difficult to evaluate them quantitatively. This paper introduces a connectionist model of perception and short-term memory that accounts for the AB effect. The paper is divided as follow: the first section presents the general methodology used with human participant to obtain the AB effect. The second section presents the model, and gives the methodology used for the simulations. The third section compares the results obtained by the model with the one obtain by human participants. Finally, the last section gives a brief conclusion.

Description of the RSVP Task

The typical task used to obtain the AB effect consists in presenting a RSVP stream of 16 items. Each item is a number between 0 and 9. A target item is red whereas a distractor is green. An item is presented during a period of 100 milliseconds. What is controlled is the position of the second target relative to the first target, a variable named lag. For example, a lag of 1 indicates that the second target is presented immediately after the first target and a lag of 2 indicates that there is one distractor between the first and the second target. The participant performances are obtained by calculating the recall percentage of the second target when the first target is correctly detected. Figure 2 shows a typical RSVP used to obtain AB effect.

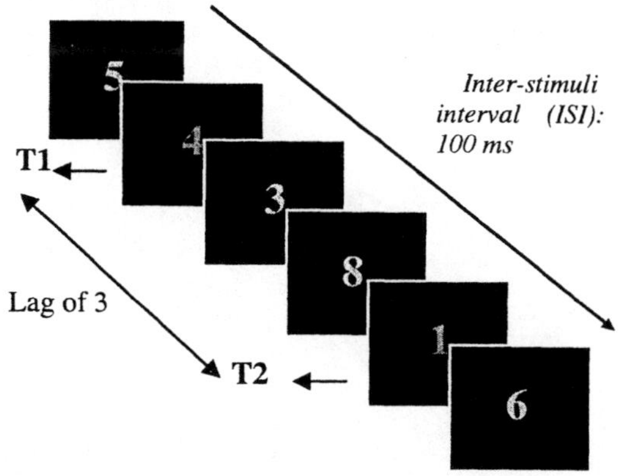

Figure 2: Six frames of an RSVP stream showing the two targets (4 and 1), the second occurring at a lag of 3.

Description of the AB Model

The proposed model is based partially on Chun & Potter (1995) and Jolicoeur (1998) models, where the first stage identifies the stimulus and the second stage stores the input into a short term memory (STM). The overall model is illustrated in Figure 3. However, as seen latter, the model also has the characteristic of Raymond, Shapiro and Arnell (1992) model, an inhibition process.

As seen in Figure 3, the input is distributed and is decomposed into its attributes. The identification stage is composed of 2 competitive networks (Kohonen, 1982, 1984). One competitive network identifies which letter is presented and the second network identifies which color is presented. In this modelization, the color determines if the stimulus is a target or a distractor. This information is thus the decision criterion used to decide if a given letter must proceed to STM. The STM memory is represented by an autoassociative layer (Hopfield, 1982).

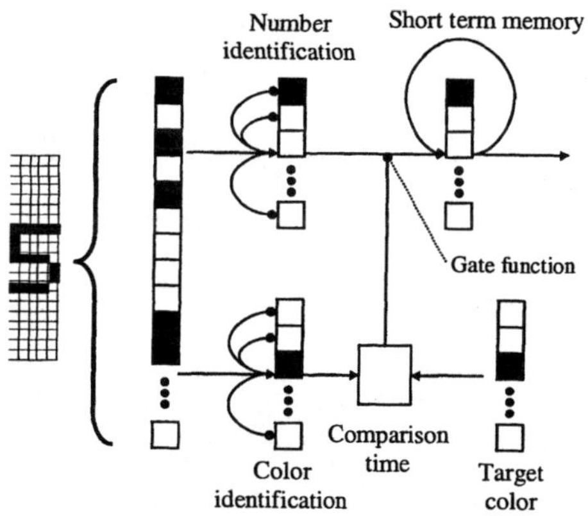

Figure 3: Model architecture

The main characteristic of the model is the presence of a dual mechanism that accounts for the decrease in the overall performance and the lag of 1 sparing effect. The first mechanism is the comparison process: once a target has been identified, it is compared to the desired attributes. If the stimulus attribute corresponds to the desired one, two things occur: a) a gate to STM is open so that the stimulus starts being encoded; b) an inhibitive process takes place making further comparison harder for a given period of time. This can be viewed as a psychological refractory period (PRP) as described by Pashler (1994). In addition, the comparison time is also involved to close the gate to STM.

The second mechanism takes place at the STM layer. At this stage, while learning of the first target occurs, a neural fatigue takes place diminishing the learning strength over time (Tsodyks & Markam, 1997). These mechanisms are illustrated in Figure 4 for a target presented at lag of 1, 2 and 3. The figure shows that the memory trace is function of both learning strength and comparison time. Thus, the amount of learning is given by the trapezoidal area.

The responses of the model are the two items that are

the most learned in STM. Yet, the memory is not perfect: because the comparison time is slow, not only the targets are memorized but also the distractor following a target. As a consequence the model predicts intrusion effects (the participant reports the distractor following a target instead of the target) as well as inversion effects (the participant reports the two targets in reversed order).

Model Description

In this section, we give the details of the identification network, implemented by a competitive model, and the short-term memory, implemented by an autoassociative model. As for any artificial neural network, these networks are entirely described by their architectures, their transmission rules and their learning rules.

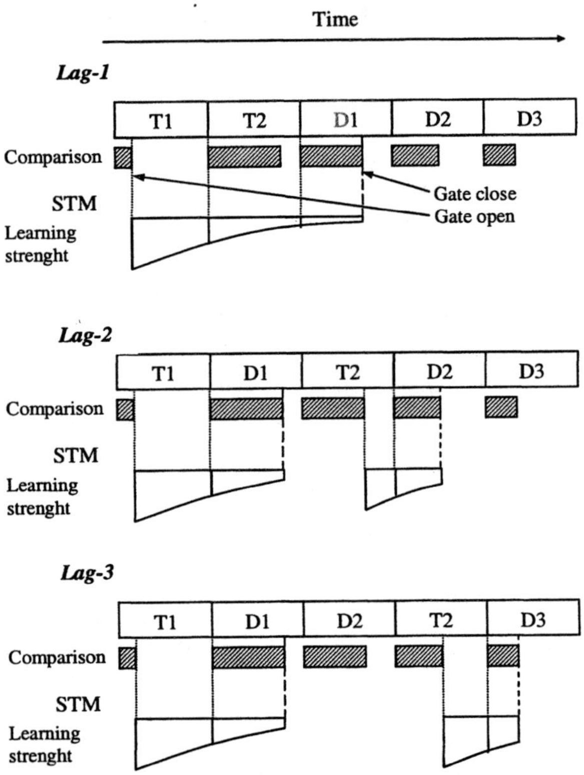

Figure 4: Illustration of the model main mechanisms

Competitive Model of Identification

Figure 5 illustrates the architecture of a typical competitive model (Kohonen, 1982). As we can see, each unit inhibits its neighbors. Thus after some times only one unit will remain active. This unit will be the one whose weights are closer to the input. As a consequence the network's output is localist (Page, 2000): each unit is responsible for the perception of a given input. In our simulations, we used the digits from 0 to 9 so that there are 10 units.

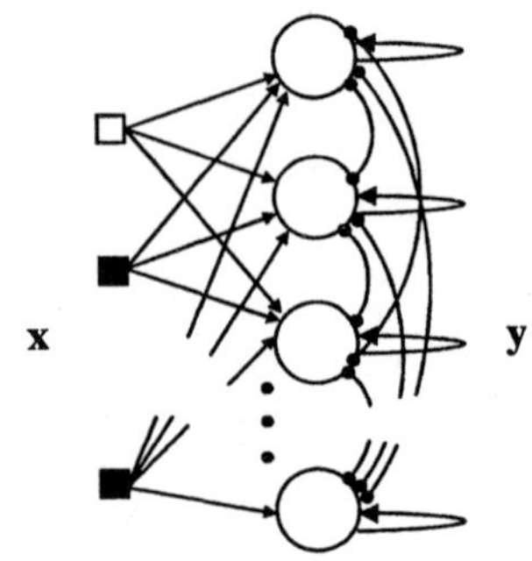

Figure 5: architecture of the competitive layer

The network's transmission rule is given by the following equation:

$$\mathbf{a} = \mathbf{W}\mathbf{x} \tag{1}$$

where **a** represents the activation vector, **W** the weight matrix, and **x** an input vector. The inhibitive function is usually modeled by the Mexican hat function. However, we can use the following Winner-Take-All (WTA) simplification: the most active unit is set to 1 and all the others are shut down to zero, so that:

$$\mathbf{y} = WTA(\mathbf{a}) \tag{2}$$

where **y** represents the output of the network. Learning in the model is usually carried out by an unsupervised algorithm (Kohonen, 1982). However, in our case, learning at this layer is not a preoccupation because the participant should already be able to identify a letter or a color. Consequently, for simplification, the network was trained to identify colors (7 possible colors) and identity (10 possible identities) before it was submitted to the RSVP task using the supervised delta rule (Widrow & Hoff, 1960) described by the following expression

$$\mathbf{W}_{[k+1]} = \mathbf{W}_{[k]} + \eta(\mathbf{z} - \mathbf{y})\mathbf{x}^{\mathrm{T}} \tag{3}$$

where η represents the learning parameter, **z** the desired output, T the matrix transposition operator, and k the learning trial.

Autoassociative Model of Short-Term Memory

This network is usually used to model unsupervised memorization process (Hopfield, 1982, Anderson, Silverstein, Ritz and Jones, 1977). Figure 6 shows the network's architecture. The output vector **y** computed from the identification network is the original input used for the autoassociative memory, so that $\mathbf{x}_{[0]} = \mathbf{y}$. Because the memory is fed by the output of the competitive network, the STM also has 10 units. As we can see, the model is recurrent and the input is associated with its-self.

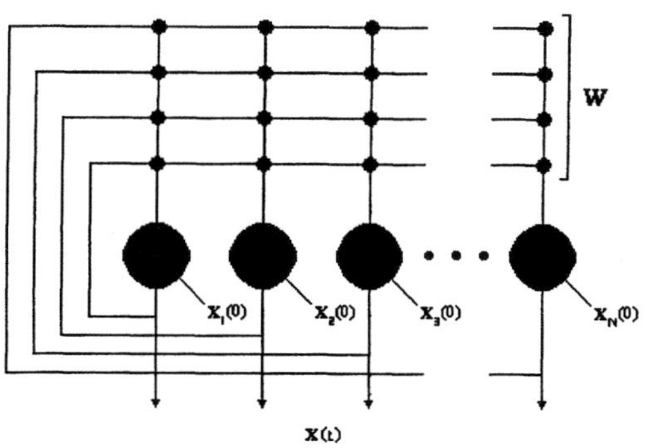

Figure 6: Architecture of the autoassociative layer

Memorizing an item was carried out by the following equation:

$$\mathbf{W}_{[k+1]} = \mathbf{W}_{[k]} + \eta(\mathbf{x}_{[0]}\mathbf{x}_{[0]}^{\mathrm{T}} - (\mathbf{W}\mathbf{x}_{[t]})(\mathbf{W}\mathbf{x}_{[t]})^{\mathrm{T}}) \quad (6)$$

We used this learning rule instead of a strictly hebbian learning rule found in (Silverstein, Ritz and Jones, 1977) for the simple reason that the hebbian does not converge whereas this one converge in a finite number of learning trial (Chartier & Proulx, 2001). Because the inputs to be learned are not linearly dependant, this autoassociative model can memorize up to 10 items with different memory strength for each, depending on the study time.

Methodology

Stimulus Composition

The inputs were simplified digits made from a 5 × 7 array of pixels. However, to model the presence of color, we used one such array for each color channel (red, green, blue). Therefore, the full input is composed of 3 arrays of 5 × 7. This input is separated in two sections. The first section consisted of the pixel information. This stimulus is thus an array of 5 × 7 given a 35 dimensions input vector. The second section consisted of the color information. The color information is thus a 3-dimensions input vector, corresponding to the RGB channels. An example of a RSVP stream is illustrated at the Figure 7.

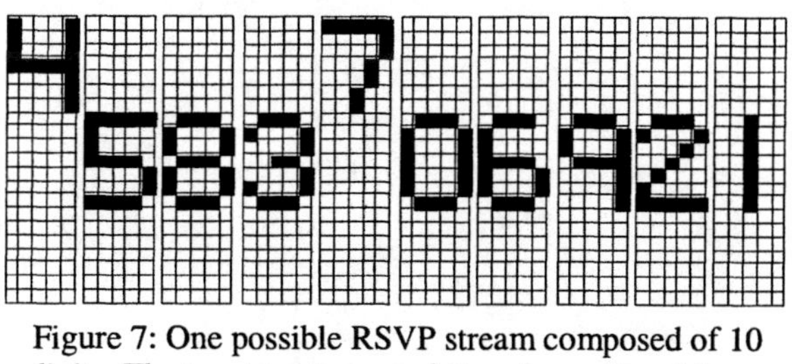

Figure 7: One possible RSVP stream composed of 10 digits. The two targets are red (top channel) and the remaining distractors are green (middle channel).

Variables

The independent variable, were the learning strength and the comparison time, which is implemented by an inhibition time function. The initial learning strength was varied from 0.001 to 0.02, whereas the inhibition time function varied from 0 to 100 time step. It is noted the inhibition function decrease following a linear function

$$inhib_{[t+1]} = inhib_{[t]} - 1 \quad (7)$$

whereas the learning strength decrease following a simple exponential function described by

$$\eta_{[t+1]} = \eta_{[t]} * \gamma \quad (8)$$

where t represents the time step and γ a general constant (γ was set to 0.997 for all the simulation). The dependant variables were the memory strength in STM for each stimulus, the average model performance and the probabilistic output. Moreover, to generate data that are closely related to human performance simple uniformed noise were introduced to both learning strength and comparison time. The random component was 30% of a given parameter initial value. To have a good idea of the variability obtained in the model's response, we performed 100 simulations. From those outputs, we compute the mean to see if, the network was able to reproduce mean performance obtained by the participants.

Results

Figure 8 shows the different AB curve obtained from the variation of the learning strength and the inhibition time. We can see that the learning strength is responsible for the vertical movement of the curve, higher the learning strength is, higher the overall performance. The inhibition time is responsible for the depth of the blink effect. Higher is the amount of inhibition, lower will be the recall performance, creating a bigger blink.

From these results, we selected a learning strength of 0.017 and inhibition value of 85 to be used for the probabilistic network. Figure 9 shows the results obtained from the simulations and the one obtained from human participants. We can see that the simulations yield similar results as the one obtained by human. Moreover if we look at Figure 10 we can see that the mean performance gives the same as the one obtained by human participants (Figure 1).

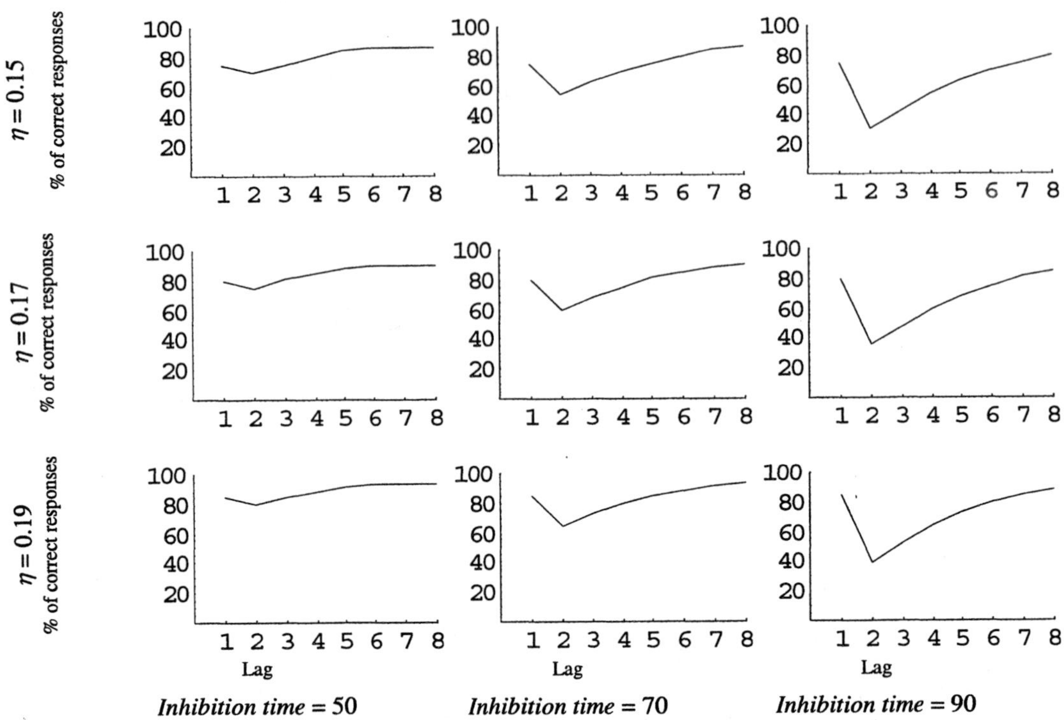

Figure 8: Various performances obtained by varying the parameters.

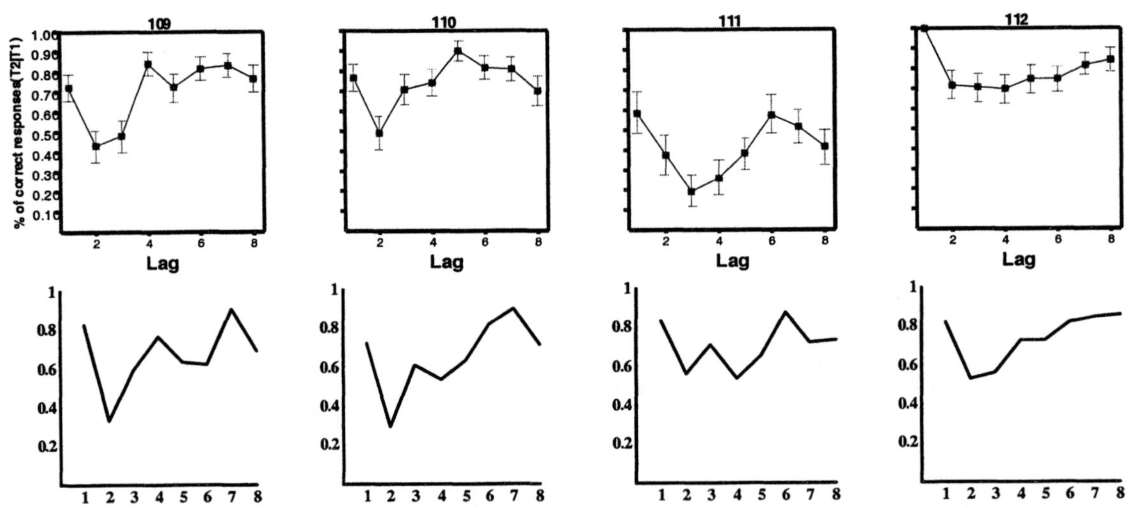

Figure 9: Examples of individual data from 4 participants (top row) and from the simulations (bottom row).

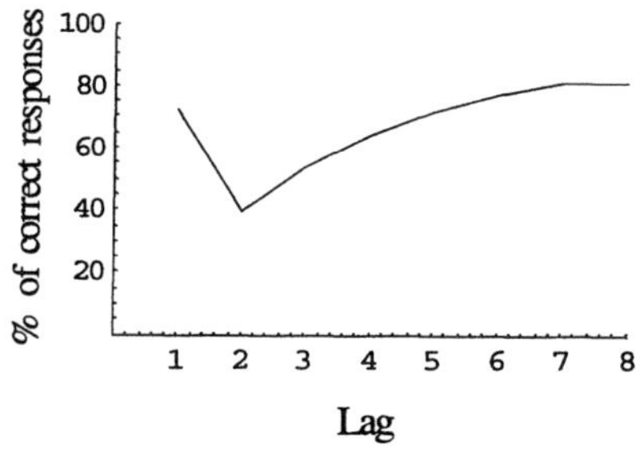

Figure 10: Mean results over 100 simulations

Conclusion

Experiments conducted with human participants showed the presence of an attentional blink if stimuli are presented within a rapid serial visual task. Many authors have proposed different mechanism that could account for the data but none have tested their affirmations using simulations. We proposed a two stage model composed of an identification stage and a memorization stage where the mains characteristic are an inhibition gate and neural fatigue. The important point is that the two-stage model alone or the inhibition model alone cannot account for the standard results.

We performed several simulations that shown that the model can reproduce the average performance of the human subjects. Moreover, under its probabilistic version the model generates data that are as variables as the one obtained from human participants. If we average those data we have shown that mean performance is the same as the one obtained by human participant. It is thus concluded that model can describe accurately the AB effect during a RSVP task.

References

Anderson, J. A., Silverstein, J. W., Ritz, S. A. & Jones, R. S. (1977). Distinctive features, categorical perception, and probability learning: Applications of a neural model. *Psychological Review, 84*, 413-451.

Bowman, H., Wyble, B. and Barnard, P.J. (2004). Towards a Neural Network Model of the Attentional Blink. In H. Bowman and C. Labiouse, editors, *Proceedings of the Eighth Neural Computation and Psychology Workshop, Connectionist Models of Cognition and Perception II, volume 15 of Progress in Neural Processing*, 178 - 187, Singapore.

Chartier, S. & Proulx, R. (2001). A new online unsupervised learning rule for the BSB model, *Proceeding of the International Joint Conference on Neural Networks (IJCNN'01)*, 448 - 453, Washington D.C.

Chun, M. M. & Potter, M. C. (1995). A two-stage model for multiple target detection in rapid serial visual presentation. *Journal of Experimental Psychology: Human Perception and Performance, 21*, 109-127.

Hopfield, J. J., (1982). Neural networks and physical systems with emergent collective computational abilities. *Proceedings of the National Academy of Sciences, 79*, 2554-2558.

Jolicoeur, P. (1998). Modulation of the attentional blink by on-line response selection : Evidence from speeded and unspeeded Task1 decisions. *Memory & Cognition, 26*, 1014-1032.

Kohonen, T. (1982). Self-organized formation of topologically correct feature maps. *Biological Cybernetics, 43*, 59-69.

Kohonen, T. (1984). Self-organization and associative memory. Springer Verlag.

Page, M. (2000). Connectionist modelling in psychology: A localist manifesto. *Behavioral and Brain Sciences, 23*, 443-512.

Pashler, H. (1987). Detecting conjunctions of color and form: Reassessing the serial search hypothesis. *Perception and Psychophysics, 41*, 191-201.

Raymond, J. E., Shapiro, K. L. & Arnell, K. M. (1992). Temporary suppression of visual processing in an RSVP task: An attentional blink?. *Journal of Experimental Psychology: Human Perception and Performance, 18*, 849-860.

Shapiro, K., Driver, J., Ward, R. & Sorensen, R. E. (1997). Priming from the attentional blink: A failure to extract visual tokens but not visual types. *Psychological Science, 8*, 95-100.

Tsodyks, M. V., & Markam, H. (1997). The neural code between neocortical pyramidal neurons depends on neurotransmitter release probability. *Proceedings of the National Academy of Science, 94*, 719-723.

Widrow, B. & Marcian E. H. (1960) Adaptive switching circuits, *IRE WESCON Convention Record*, New York: IRE, pp. 96-104.

A Model of Human Behavior in Coalition Formation Games[1]

Alex K. Chavez (achavez@sas.upenn.edu)
University of Pennsylvania; 4110 Baltimore Avenue
Philadelphia, PA 19104 USA

Steven O. Kimbrough (kimbrough@wharton.upenn.edu)
University of Pennsylvania; 3730 Walnut Street; Suite 500, Room 565
Philadelphia, PA 19104 USA

Abstract

Coalition formation is a type of mixed-motive game in which n players strategically negotiate to secure positions in advantageous contracts. In this study, we find that systems of agents which learn by a simple linear updating rule successfully can model the outcomes of human players across five coalition formation games studied experimentally by Kahan and Rapoport (1974). "Greedy" agents, which are deterministic and maximizing in their selection of whom to offer to, achieve outcomes on par with humans within a few hundred trials. In comparison, "Matching" agents which use *probability matching*[2] for selecting whom to offer to achieve overall outcomes qualitatively similar to those of humans, but not as closely as the Greedy agents do.

Introduction

There recently has been interest in modeling human learning in strategic games, both in economics (Roth and Erev, 1995; Mookherjee and Sopher, 1997; Erev and Roth, 1998; Camerer and Ho, 1999) and in cognitive science (e.g., Ritter and Wallach, 1998; Lebiere et al., 2003). Previous efforts have focused on modeling learning in repeated, extensive and normal form games. Here, we extend the literature by modeling the outcomes reported in an experimental study of a characteristic function form game (Kahan and Rapoport, 1974).

Our goals in this paper are to see if agents endowed with a minimal learning rule can model the play by humans in the coalition formation game of Kahan and Rapoport (1974), and to compare the two selection rules–Greedy, and Matching–to be described in detail later. Because the human data available is limited, our approach is not to produce entire trajectories of play, but to see if the overall behavior of our agents corresponds to that of humans.

Previous work of Dworman et al. (1995a,b,c, 1996) used genetic programming to explore coalition games similar to those in this paper, where agents were explicit strategies in the space of offers and players. However, such agents cannot cannot be said to have a cognitive component, as each agent represents a single rule. In contrast, our agents represent players of low or minimal rationality, who must make decisions on how much to offer and to whom. Finally, while characteristic function games have been investigated widely in terms of outcomes or solutions through both experimental and mathematical approaches (see Kahan and Rapoport, 1984; Uhlich, 1990, for overviews), less attention has been paid to the learning processes associated with such games.

In the next section of our paper, we describe the coalition formation game, and also present the experimental study of Kahan and Rapoport (1974). We then describe our model, and finally present simulation results. While we only explore three-person games with a certain characteristic function, we can hope our results generalize to games with larger numbers of players.

Coalition Formation

Description

In the coalition formation game with a set of n players P, any of the players can join together to form a *coalition* $S \subseteq P$. Such a coalition can attain a guaranteed payoff of $v(S)$, called the *coalition value*, where v is defined over all coalitions and is known as the *characteristic function* of the game; along with P, this defines the coalition game. While it generally is advantageous for a player to be included in some coalition,[3] it is up each member of the coalition to secure a portion of $v(S)$ for herself or himself. Once a coalition S forms an agreement of how to split $v(S)$, then this agreement is enforced. While some coalitions may have greater guaranteed payoffs than others, individual players should be drawn to those coalitions where they can attain the greatest individual reward.

To provide an example, imagine a situation where there are three researchers (players), A, B, and C, each of whom has some resource needed to run a study. No researcher can run the study alone (so that $v(A) =$

[1] Correspondence may be addressed to Alex Chavez.

[2] *Probability matching* holds that humans and other animals choose between different alternatives in ratios that matches those of their respective rewards. The rule has been used successfully to model dynamic learning in strategic contexts (Erev and Roth, 1998, e.g.,). However Sarin and Vahid (2001) have contested that *non-probabilistic selection*, where the agent always selects the action that it currently perceives will yield the highest payoff, can produce results as good as those obtained under the Erev and Roth (1998) model. The latter corresponds to our Greedy agents, and the former to Matching agents.

[3] If the condition $v(AB) > v(A) + v(B)$ holds for all players (where AB is shorthand for the coalition $\{A, B\}$), then v said to have the property of superadditivity. Such a set of relations specifies that players achieve a higher joint payoff in a coalition compared to the sum of their payoffs when acting alone, and represents comparative advantages. Superadditivity holds for all games studied here.

$v(B) = v(C) = 0$), but any two of them can collaborate to run the study. Assume that the combined resources of A and B permit them to run the "best" study (say $v(AB) = 95$); that A and C can run the next "best" one (let $v(AC) = 90$); and B and C the worst one (let $v(BC) = 65$). Assume further that the coalition ABC does not yield any value, so that $v(ABC) = 0$ (there is a limit on the number of researchers who can be involved). Thus, the characteristic function v has been defined completely, so the situation constitutes a coalition game.

Now, researcher A must ask the question, "With whom should I propose a coalition, and how should I propose to allocate the resources assigned by v?" For example, A might be greedy and propose the split $(65, 30, 0)$, where A receives 65 (say, in units of recognition), B receives 30, and C is not included in the *winning coalition AB* (note $65 + 30 = 95 = v(AB)$). But realizing that C would be better off receiving even a small payoff, B might then propose the allocation $(0, 40, 25)$ for the coalition BC, where both B and C do better than they would under A's proposal. The bargaining might continue, with C trying to increase her payoff by proposing the allocation $(60, 0, 30)$ (excluding B from the coalition). This last proposed allocation is special because neither player in the coalition (e.g. A) can make an offer to the excluded player (B) without the excluded player (B) being able to make a counteroffer to the remaining player (C) in which they both do better than under the original allocation; in this sense, the proposed split is *stable* (Aumann and Maschler, 1964). For example, say A was not satisfied with the proposed allocation $(60, 0, 30)$, and decides to propose a coalition with B with the split $(61, 34, 0)$. In this case, B could offer $(0, 35, 30)$ to C, which is another stable split. Along with $(0, 0, 0)$ (where no players form a coalition) and $(60, 35, 0)$, these stable allocations form the solution concept of the "bargaining set" of Aumann and Maschler (1964) (See also chapters 3 and 4 of Kahan and Rapoport, 1984). As Kahan and Rapoport (1974) point out, the bargaining set solution does not predict *which* of the four above allocations will emerge.

An Experimental Study

In their study, Kahan and Rapoport (1974) used human subjects in a computerized experiment designed to test behavior in situations similar to the one described above. 48 undergraduate male subjects were divided into groups of 16 and participated in three separate experiments. In the first experiment, messages were public, so that all players were aware of the others' offers. Subjects had to send messages publicly in a fixed order (as opposed to being able to speak at will). In the second experiment, messages could be private, but again were sent in order. In the last experiment, messages could be private, but were sent at will. The 16 players in each experiment were broken up into 4 quartets, each of which played five 3-person characteristic function games for 4 iterations. For each game, one member of the quartet would sit out as an observer – this procedure was employed to allow subjects to reflect upon the task, and to increase

	Game: I	II	III	IV	V
Char. Function					
$v(AB)$	95	115	95	106	118
$v(AC)$	90	90	88	86	84
$v(BC)$	65	85	81	66	50
Quota Values					
ω_A	60	60	51	63	76
ω_B	35	55	44	43	42
ω_C	30	30	37	23	8

Table 1: Characteristic Function and Quota Solutions by Game

the validity of the assumption of independence between games. Order of play between and within games was randomized subject to the condition that no player would be observer in two consecutive rounds. Subjects were given an extended practice session. The 5 games are shown in Table 1.

Quota Values

The type of characteristic function game considered here is a special case known as the *quota games* (Kahan and Rapoport, 1974, 1984), where the conditions $v(ABC) = v(A) = v(B) = v(C) = 0$ and $v(AB), v(AC), v(BC) > 0$ hold. Such games have *quota solutions*, which are generally accepted by cooperative game theory, and are given by the following equation for player i:

$$\omega_i = .5 \sum I(j,k) * v(jk) \quad \forall j,k \in P, j \neq k \qquad (1)$$

where $I(j,k) = 1$ if $i = j$ or $i = k$, and equals -1 otherwise (note that $\omega_i + \omega_j = v(ij)$). So for example, player A's quota simply is calculated by $.5(v(AB) + v(AC) - v(BC))$. These quotas represent normative predictions; when players follow a set of weak rationality conditions (e.g. of the bargaining set (Aumann and Maschler, 1964)), they will arrive at the quota values. The quota solution is a solution concept for characteristic function form games, just as the Nash equilibrium is one for normal form games. Table 1 shows the quota values for players A, B, and C in all 5 games.

Kahan and Rapoport (1974) consider for each player *the mean reward as a member of the winning coalition* (MRAC). Because solution concepts such as the quota make predictions about how much players will get *given* that they are in a coalition, MRAC is the appropriate measure if one wishes to test the theory. Kahan and Rapoport (1974) report that human subjects' overall deviations from the quotas are not significantly different from zero, for each of the experimental conditions; this reinforces that idea that the quota is an important theoretical notion. Table 2 displays the human data in an aggregate form.

	I	II	III	IV	V
Experimental MRAC, Averaged across Experiment					
A	57.43	63.00	53.73	62.07	71.6
B	38.90	54.40	43.40	45.20	45.70
C	29.53	26.93	34.67	19.27	18.17
Experimental Frequency of Coalition Structures					
A,B,C	.0208	.0000	.0000	.0000	.0000
AB,C	.5625	.5208	.3542	.7292	.8750
AC,B	.3125	.1667	.3333	.1458	.1250
BC,A	.1050	.3125	.3125	.1250	.0000

Table 2: Human Data (Kahan and Rapoport, 1974)

While MRAC is the appropriate measure for comparing human performance with theoretical predictions, if one wishes simply to measure how well the subjects do in terms of wealth extracted from the game, then the mean reward (MR) is more appropriate. MRAC and MR can differ greatly; if a stubborn player s always refuses to accept any amount below his quota + 20, for example, then he may be included in the winning coalition once or twice out of a hundred trials, so that his MRAC would be $\omega_s + 20$, but his MR close to zero. The difference arises because in rounds when a player is not in the winning coalition, he or she receives a payoff of zero, which the MRAC ignores. We report both MRAC and MR below.

The Model

We wish to consider the behavior of agents using a simple learning rule in the context of the five coalition formation games above, and to see if these agents achieve outcomes close to those of humans in the experimental study of Kahan and Rapoport (1974). The model assumes that each player updates a belief about how much payoff it can expect from every other agent. We shall refer this value as player i's *aspiration level*, following Macy and Flache (2002), to player j at a given time t, or $A_i^j(t)$ for short. Aspiration levels are updated over time by adding a fraction of the difference between actual reward received from the environment, and the payoff level expected, as given by the equation (e.g., Sutton and Barto, 1998; Macy and Flache, 2002):

$$A_i^j(t) = A_i^j(t-1) + \alpha[r_i^j(t) - A_i^j(t-1)] \qquad (2)$$

where $\alpha \in (0,1)$ represents a recency constant, and $r_i^j(t)$ is the reward received by player i from being in a coalition with player j at time t, as specified by their agreement. For all results reported here, we set $\alpha = 0.2$, a typical value in the reinforcement learning literature. In simulations not reported here, α values between 0.1 and 0.4 yielded results similar to those in this paper. In addition to the updating rule given in Equation (2), we considered two methods for agents to decide whom to offer to. They are given in Table 3[4].

	Probability of i Offering to j
Greedy	1 if $j = \arg\max_{p \in P} A_i^p(t)$ 0 else
Matching	$A_i^j(t) / \sum_{p \in P} A_i^p(t)$

Table 3: Selection Rules

In addition, each agent makes offers *at* its aspiration levels, and not below. So for example, if agent A has aspiration levels of 50 to agent B and 65 to agent C, then it would offer $v(AC) - 65$ to agent C with probability 1 under the Greedy rule. Under the Matching rule, it would offer $v(AB) - 50$ to agent B with probability 65/115 and $v(AC) - 65$ to agent C with probability 50/115. While Sarin and Vahid (2001) note that Greedy (i.e. non-probabilistic) action selection is in line with the traditional economic precept of choice as maximization over beliefs, others have found that humans and other animals use probability matching to select between actions associated with a reward (e.g. Gallistel, 1990). Thus, we investigate the performance of both types of selection rules.

Also, our agents are myopic subjective maximizers in their offer behavior, in that they make offers based on the maximum they currently "think" they can get, without considering the possible ramifications of their offer behavior on the future state of the system. Such agents represent players who vastly simplify their objective environment, collapsing the available history of offer behavior for each other player into a single real value, A_i^j (Sarin and Vahid, 2001). Again, we wish to explore the behavior of agents representing players of minimal rationality, and to see if they can model the overall results of humans. The simple agents described in this section satisfy such a condition.

Finally, simulations were run as follows. At the start of a simulation, each agent i's initial aspiration level to j was initialized from the uniform distribution on $[0, v(ij)]$[5]. Next, for each episode, an agent was selected at random as the initial offerer. This agent made an offer to a receiving agent, according to the rules described above. The receiving agent accepted if the offer amount was greater than or equal to its current aspiration level to the offering agent. If it declined, then the receiving

[4] In simulations not reported here, other relatively exploitative selection methods drawn from reinforcement learning literature, such as ϵ-greedy and Softmax selection (Sutton and Barto, 1998), produced results similar to those reported below for the Greedy agent.

[5] We found that setting $A_i^j(0) = v(ij)/2$ – that is, to half of the coalition value between i and j – did not affect MRAC or MR in expectation (it resulted in less variance).

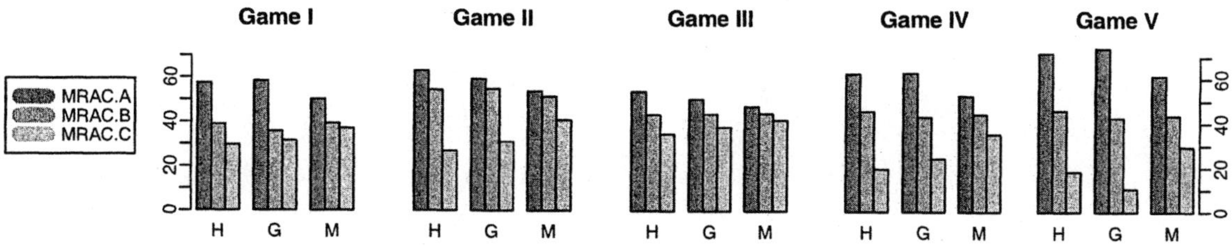

Figure 1: Mean Reward as Members of Winning Coalition for <u>H</u>uman Subjects, <u>G</u>reedy Agents, and <u>M</u>atching Agents

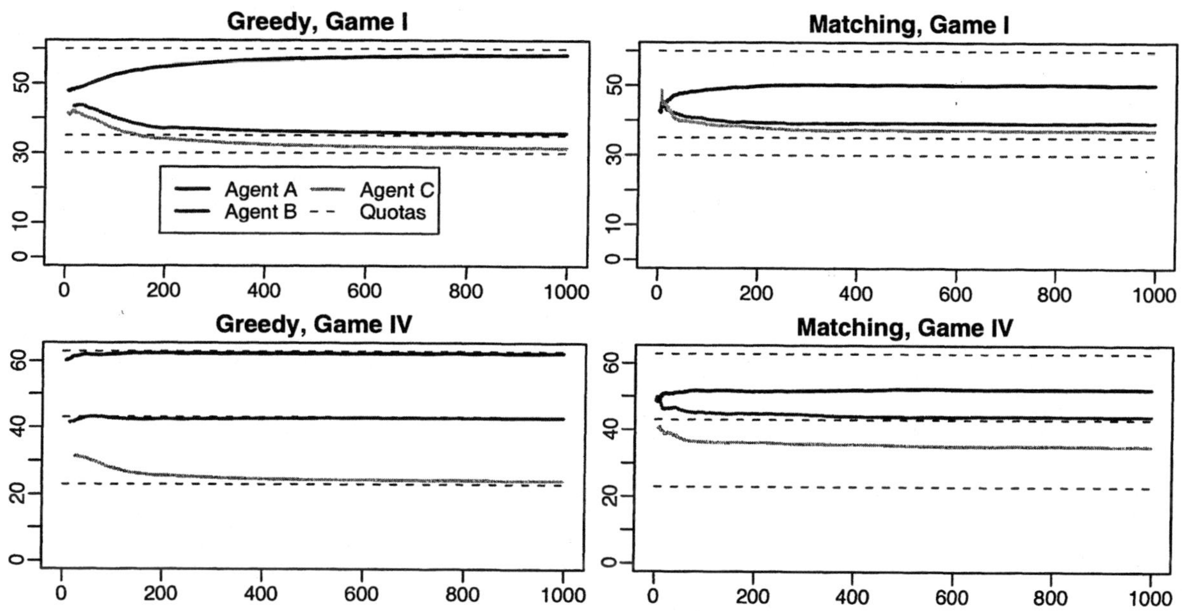

Figure 2: MRAC over Time in Games I and IV, for Greedy and Matching Agents

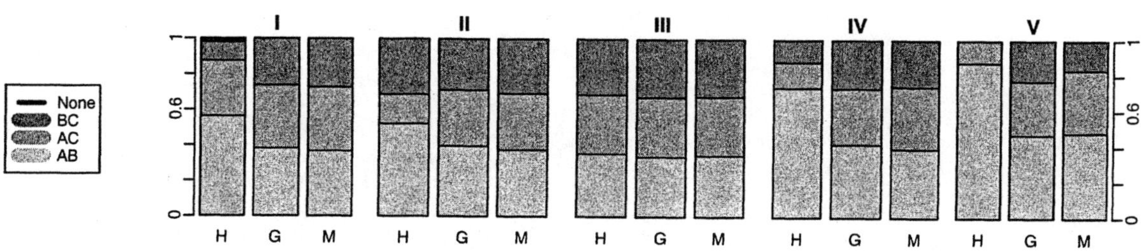

Figure 3: Frequency of Coalition Formation for <u>H</u>uman Subjects, <u>G</u>reedy Agents, and <u>M</u>atching Agents

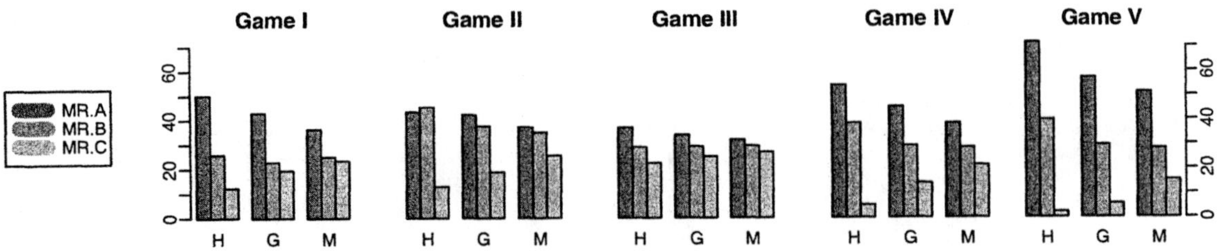

Figure 4: Mean Reward for Human, Greedy, and Matching Agents

agent would become the new offering agent, and so on. The process continued until a receiving agent accepted an offer, which always happened within 6 rounds. When an episode ended, agents' aspiration levels were updated, and the next episode was started. This process continued until a maximum number of episodes was reached, whereupon the simulation would terminate.

Results

The MRAC values for the data from Kahan and Rapoport (1974) for human subjects are presented along with results for Greedy and Matching agents (20 simulations of 1,000 episodes each) in Figure 1. As Figure 1 shows, the MRAC values of the agents, especially for the Greedy type, closely fit those of human subjects ($r = .98$ for the Greedy agents, and $r = .96$ for the Matching agents). Figure 2 shows the rate of convergence of MRAC values for Greedy and Matching agents in Games I and IV (Games II, III, and V are not displayed for space reasons). As can partially be seen from the figure, MRAC values for Greedy agents rapidly approach the quota solutions in Games II, III, and IV, more slowly in Games I and V, and never for Matching agents (this is explained later). Figure 3 shows frequencies of coalition structures for human subjects and for simulations. The model does not predict some important differences in these frequencies. Namely, player A is included in the winning coalition more often than in the human data than in the simulations, and player C less often (this can be seen for player A by looking at the AC and AB blocks together, and for player C by looking at the BC and AC blocks together).

As discussed previously, to gauge the overall performance of players, it is useful to consider the measure of mean reward (MR) of players. MR values for human subjects, Greedy agents, and Matching agents are shown in Figure 4. The MR values for both agent types fit the data of human subjects well, with Greedy agents again doing better than the Matching agents ($r = .97$ vs. $r = .95$). However, there are some noticeable departures. For example, player A has lower MR values and players C higher, for agents vs. humans. This is attributable directly to the differences in the frequencies of coalition formation (Figure 3), where player A is in the winning coalition less often (and player C less more) for agents vs. humans.

We would like to make two statements here about our findings. The first is that systems of relatively exploitative agents (e.g. Greedy agents) converge to neighborhoods of quota solutions, and that this outcome is robust to variations of initial aspiration levels. The second is that the difference between an agent's initial aspiration level and its quota value is related to its speed of convergence in MRAC value (the smaller the difference, the faster the convergence). This may be important because although humans subjects arrived at the quota solution in only 4 iterations of play, their initial offers appear to have been very close to their quota values[6]. Whether this occurred as a result of practice sessions, transfer between games or from acting as the observer, or from deliberation is not clear. We merely point out that if the initial aspiration levels of the agents are close to their quota values, then their MRAC values converge almost immediately to the quotas (see e.g. agents A and B in Game IV of Figure 2).

Conclusion

There is now an extensive literature in behavioral game theory that aims at describing, modeling, and explaining human behavior in games (see Camerer, 2003; Kagel and Roth, 1995, for literature reviews)). This literature has discovered that a variety of simple learning models provide reasonably good predictions of human behavior in laboratory games. A number of authors have found that reinforcement learning models supplemented with additional heuristic rules (e.g., pertaining to the specifics of the game) achieve good predictive power. Moreover, other simple models of learning, e.g. using activation-based recall (Lebiere et al., 2003), have also provided good accounts of human behavior in games.

In the present work, we found that systems of agents using a simple learning rule can model human data reasonably well, and that their performance converges to the theoretical predictions of quota values in a class of 3-player, coalition formation games. The MRAC values for both Greedy and Matching agents were very close to those of humans, while their MR values showed some systematic differences from the human data. Although we do not claim the learning rule presented in Equation (2)

[6]Kahan and Rapoport (1974) report that first offers made to the winning coalition had an average deviation over games and coalitions of -2.95, e.g..

fully accounts for the way people play these games, we submit that it is a fruitful and practicable way of investigating strategic environments. While much remains to be done, it is intriguing that the agents' behavior so closely matches the human data, much more so than has been reported for simple reinforcement learning models in non-cooperative games.

The results of both agent and human bargaining results are interesting, but perhaps the most important questions regarding bargaining games pertain to their dynamics. When will (and when should) a player accept an offer? If a player chooses to reject, how much and to whom should he/she offer? The answers to these questions may depend not only on the coalition values and other initial parameters of the game, but also on the history of offers and round number. For example, if only 2 rounds are left before the game is ended by default, can player A (with the highest total coalition value) extract more reward by making an aggressive offer? or, perhaps he/she must accept to avoid forfeiting any positive reward for the game. To answer these questions, different approaches such as classifier systems are needed, for example to find bargaining rules such as (for Agent A): "If round is 4 and offeringAgent is B and amount is at least aspirationLevel$-0.1 * v(AB)$, then accept."

In sum, the prospects for mutual benefit between the cognitive modeling community and the behavioral game theory community are bright indeed.

Acknowledgements

The work reported here was supported in part by NSF grant number SES-9709548 and funding from the University Scholars Program at the University of Pennsylvania. We thank two anonymous reviewers, Frank Lee, and Dario Salvucci for helpful comments.

References

Aumann, R. J. and Maschler, M. (1964). The bargaining set for cooperative games. In Dresher, M., Shapley, L. S., and Tucker, A. W., editors, *Advanced in Game Theory*. Princeton University Press.

Camerer, C. and Ho, T. (1999). Experience-weighted attraction learning in normal form games. *Econometrica*, 67:827–874.

Camerer, C. F. (2003). *Behavioral Game Theory: Experiments in Strategic Interaction*. Russell Sage Foundation and Princeton University Press, New York, NY and Princeton, NJ.

Dworman, G. O., Kimbrough, S. O., and Laing, J. D. (1995a). Bargaining in a three-agent coalitions game: An application of genetic programming. In *Working Notes: AAAI-95 Fall Symposium Series, Genetic Programming*, pages 9–16, Boston, MA, November 10-12, 1995. AAAI. File: gpconf08.doc.

Dworman, G. O., Kimbrough, S. O., and Laing, J. D. (1995b). On automated discovery of models using genetic programming: Bargaining in a three-agent coalitions game. *Journal of Management Information Systems*, 12(3):97–125.

Dworman, G. O., Kimbrough, S. O., and Laing, J. D. (1995c). On automated discovery of models using genetic programming in game–theoretic contexts. In Nunamaker, Jr., J. F. and Sprague, Jr., R. H., editors, *Proceedings of the Twenty-Eighth Annual Hawaii International Conference on System Sciences, Volume III: Information Systems: Decision Support and Knowledge-Based Systems*, pages 428–438, Los Alamitos, CA. IEEE Computer Society Press.

Dworman, G. O., Kimbrough, S. O., and Laing, J. D. (1996). Bargaining by artificial agents in two coalition games: A study in genetic programming for electronic commerce. In Koza, J. R., Goldberg, D. E., Fogel, D. B., and Riolo, R. L., editors, *Genetic Programming 1996: Proceedings of the First Annual Genetic Programming Conference, July 28-31, 1996, Stanford University*, pages 54–62. The MIT Press.

Erev, I. and Roth, A. E. (1998). Predicting how people play games: Reinforcement learning in experimental games with unique, mixed strategy equilibria. *The American Economic Review*, 88(4):848–881.

Gallistel, C. R. (1990). *The Organization of Learning*. The MIT Press, Cambridge, MA.

Kagel, J. H. and Roth, A. E., editors (1995). *The Handbook of Experimental Economics*. Princeton University Press, Princeton, NJ.

Kahan, J. P. and Rapoport, A. (1974). Test of the bargaining set and kernel models in three-person games. In Rapoport, A., editor, *Game Theory as a Theory of Conflict Resolution*, pages 119–160. D. Reidel, Dordrecht, The Netherlands.

Kahan, J. P. and Rapoport, A. (1984). *Theories of Coalition Formation*. Lawrence Earlbaum Associates, Hillsdale, NJ.

Lebiere, C., Gray, R., Salvucci, D., and West, R. (2003). Choice and learning under uncertainty: a case study in baseball batting. In *Proceedings of the 25th Annual Conference of the Cognitive Science Society*.

Macy, M. W. and Flache, A. (2002). Learning dynamics in social dilemmas. *Proceedings of the National Academy of Science (PNAS)*, 99(suppl. 3):7229–7236.

Mookherjee, D. and Sopher, B. (1997). Learning and decision costs in experimental constant sum games. *Games and Economic Behavior*, 19(1):97–132.

Ritter, F. E. and Wallach, D. P. (1998). Models of two-person games in act-r and soar. In *Proceedings of the 2nd European Conference on Cognitive Modelling*.

Roth, A. E. and Erev, I. (1995). Learning in extensive-form games: Experimental data and simple dynamic models in the intermediate term. *Games and Economic Behavior*, 8:164–212.

Sarin, R. and Vahid, F. (2001). Predicting how people play games. *Games and Economic Behavior*, 34:104–122.

Sutton, R. S. and Barto, A. G. (1998). *Reinforcement Learning: An Introduction*. The MIT Press, Cambridge, MA.

Uhlich, G. R. (1990). *Descriptive Theories of Bargaining*. Springer-Verlag, Berlin, Germany.

Fast Fitting of Convolutions Using Rational Approximations

Denis Cousineau (denis.cousineau@umontreal.ca)
Université de Montréal, C.P. 6128, succ. Centre-ville
Montréal (Québec), H3C 3J7 Canada

Abstract

Models of response time distributions are powerful models because they capture mean response times as well as variability and asymmetry. Some of these models are expressed with a closed-form equation (e.g. the Weibull distribution). However, a large number of others are not available in closed form, including convolutions. Fitting such distributions with the likelihood method is still possible with the help of numerical integration techniques. However, the time required to fit a single subject is a matter of days. We present an alternative to numerical integration based on rational approximations. This approach is 100 times faster so that fitting a single subject can be done in less than an hour. As shown with simulations, the approach can be fully automatized, and is both reliable and accurate.

Introduction

Response times (RT) give very precious information on elementary mental processes. Psychological models can be used to fit mean RT alone, or means and standard deviations of RT simultaneously (Cousineau & Larochelle, in press). However, a more stringent test is imposed when a model must fit the whole distribution of RT (Cousineau & Shiffrin, in press). Some models predict simple distributions and therefore are easy to fit to the RT data using the maximum likelihood method (described next). This is the case for the diffusion models which predict a Wald distribution (Ratcliff, Van Zandt, & McKoon, 1999). However, such models are the exceptions: a large number of models predict that the observed RT distributions will conform to a convolution of two or more simple distributions which cannot be simplified into a closed-form equation.

A simple psychological model that predicts a convolution is one which assumes two stages, for example one stage for encoding the stimulus and one stage for making a motor response. The observed RT is the sum of the two processing times. In this situation, the mean RT is the sum of the two mean processing times. The variance in RT is also the sum of the variance of the two processes. However, the RT distribution **is not** the sum of the two processes distributions. It is obtained by an operation called a convolution of the two processes distributions.

Because two-stage models are very frequent, convolutions occur naturally in psychology.

Convolutions

A convolution is the distribution of the sum of two or more random values. Formally, let $\boldsymbol{T}$ be a random deviate and $\boldsymbol{T}_1$ and $\boldsymbol{T}_2$ be the processes that contribute to $\boldsymbol{T}$, such that $\boldsymbol{T} = \boldsymbol{T}_1 + \boldsymbol{T}_2$. Often, only $\boldsymbol{T}$ is observable. We can estimate the probability that on a given trial, $\boldsymbol{T}$ will take the value t with the equation:

$$\Pr(\boldsymbol{T} = t) = \Pr(\boldsymbol{T}_1 = s \,\&\, \boldsymbol{T}_2 = t - s)$$

for all s (because s and $t - s$ cannot be negative response times in psychology, we have the constraint that $0 < s < t$). Assuming that the second processing time is independent of the first, we can write:

$$\begin{aligned}\Pr(\boldsymbol{T} = t) &= \Pr(\boldsymbol{T}_1 = s) \times \Pr(\boldsymbol{T}_2 = t - s)\ \forall s \\ &= \int_0^t \Pr(\boldsymbol{T}_1 = s) \times \Pr(\boldsymbol{T}_2 = t - s)\, ds\end{aligned} \quad (1)$$

Equation 1 is the basic equation for convolutions. When assumptions are given on the distributions of $\boldsymbol{T}_1$ and $\boldsymbol{T}_2$, Eq. 1 can sometimes be solved in closed form. Let $f_1(t \mid \theta_1)$ and $f_2(t \mid \theta_2)$ be two probability density functions (pdf) for $\boldsymbol{T}_1$ and $\boldsymbol{T}_2$ with parameters θ_1 and θ_2 respectively. The convolution is often denoted using the * so that $(f_1 * f_2)(t)$ yields the density that the sum of the two processes equals t.

As an example, if f_1 is the normal distribution with parameters $\{\mu_1, \sigma_1^2\}$ and f_2 is also normal with parameters $\{\mu_2, \sigma_2^2\}$, Eq. 1 is easy to solve, and the result of $(f_1 * f_2)(t)$ is normal with parameters $\{\mu_1 + \mu_2, \sigma_1^2 + \sigma_2^2\}$ (Cramér, 1947).

Another example often seen in psychology relates to the Poisson race model (Townsend & Ashby, 1983). This model assumes that when n spikes reach a certain neuron, this neuron is triggered. Suppose that under given conditions, a spike occurs every τ milliseconds on average. If the time between two spikes is random following a Poisson process (an exponential distribution), then the total time before the neuron fires is given by $\boldsymbol{T} = \boldsymbol{T}_1 + \boldsymbol{T}_2 + \ldots + \boldsymbol{T}_n$. The distribution of $\boldsymbol{T}$ is a convolution of n exponential distributions. Again, this convolution can be solved and yields a gamma distribution with two parameters $\{n, \tau\}$ (Luce, 1986).

On other occasions, the convolution cannot be solved in closed form but good approximations exist for its computation. This is the case for the exGaussian distribution (Ratcliff, 1979). It is based on the assumptions that the decision process is modeled by an exponential distribution and that all the other processes, owing to the theory of errors (Gauss, 1809/1864), are normally distributed. The observed time is therefore a sum of a normal component and an exponential component. Using Eq. 1, we find:

$$\Pr(T=t)=\int_0^t \frac{e^{-\frac{(s-\mu)^2}{2\sigma^2}}}{\sqrt{2\pi}\sigma}\times\frac{e^{-\frac{(t-s)}{\tau}}}{\tau}ds$$

$$=\frac{1}{2\tau}e^{\frac{\mu-RT}{\tau}+\frac{\sigma^2}{2\tau^2}}\left(1+erf\left(\frac{RT-\mu}{\sigma}-\frac{\sigma}{\tau}\right)\right)$$

where $erf(t)$, often called the error function, is given by

$$erf(t)=\int_0^t e^{-\frac{s^2}{2}}ds$$

This integral is not available in closed form. However, good approximations exist and are quite efficient in terms of computation time. The most commonly used approximation was obtained by using a ratio of two cubic polynomials (Kennedy and Gentle, 1980). Sometimes, by subdividing the curve in sub intervals, better approximations can be found. The *erf* function implemented on most programming platform is subdivided in three intervals, and for each, a rational approximation was found. As an example, the following is the rational approximation for the $erf(x)$ function within 0 and 0.5:

$$\frac{3.2027\times10^{-7}+1.1283x+34.011x^2-1.0401x^3}{1+30.140x-0.56796x^2+9.9358x^3}$$

When no approximation is available, an equation containing an integral can still be evaluated using numerical integration techniques (mostly based on Monte Carlo and Markov chain techniques). To estimate $(f_1 * f_2)(t)$ at a single point t, these techniques use an approach analogous to throwing a large number of darts and finding the proportion of them that are below the curve. Hence, to evaluate the function at one point, hundreds or thousands of computations have to be done, resulting in very slow computations.

This paper aims at presenting a solution to speed up fitting of convolutions. When fitting a convolution to RT data, the function has to be estimated for each datum and each parameter set explored. Because the convolution is so slow to estimate, finding the best-fitting parameters typically requires from two to five days per subject on a 2.0 GHz PC. The approach described next will reduce this time by a factor of 100, so that fitting one subject is possible in about an hour.

Distribution Fitting

Before presenting the approach, we briefly describe the maximum likelihood method used to find best-fitting parameters (Cousineau & Larochelle, 1997). It consists in finding the parameters that make the data the most likely. Let the likelihood L of the data $\{t_i, i = 1 .. n\}$ given parameters θ be:

$$L(t_i,\theta)=\Pr(T=t_1 \,\&\, T=t_2 \,\&\ldots\&\, T=t_n)$$

where n is the number of data to be fitted. When the data are independent, we can simplify the equation to:

$$L(t_i,\theta)=\prod_{i=1}^{n}\Pr(T=t_i)$$

$$=\prod_{i=1}^{n}f(t_i\mid\theta)$$

using f to denote the pdf underlying the data which is a function of the parameters θ. The best-fitting parameters are those that make the overall probability L closest to 1. To avoid underflow, it is often more convenient to use the log of the probabilities, so that the log likelihood LL is given by:

$$LL(t_i,\theta)=\log(\prod_{i=1}^{n}f(t_i\mid\theta)$$

$$=\sum_{i=1}^{n}\log(f(t_i\mid\theta)) \qquad (2)$$

The best-fitting parameters θ^* are those that maximize $LL(t_i, \theta)$ or equivalently, that minimize minus $LL(t_i, \theta)$:

$$\underset{\theta\in\Theta}{Max}\, LL(t_i,\theta)$$

Maximizing a function is an iterative process that requires considering many possible parameters. Some methods are based on gradient descents (Chandler, 1965), others are based on geometric methods (simplex, Nelder and Mead, 1965). Typical code in *Mathematica* is given in Listing 1.

Note that the programs in Listings 1 and 2 are not optimized for speed. By using Apply and Map instead of the summation, the code will operate much faster:

```
-Apply[Plus, Map[Log[f[#,{p1,p2,p3}]]&,data]]
```

In the simulations reported next, the code was optimized for speed.

Objectives

Distribution fitting of a convolution can be accomplished using numerical integration. However, this approach is too slow to be useful in any practical application. We propose in this paper to systematize the use of approximations. There exists nowadays algorithms that can find a good approximation to any function within a given interval $[a, b]$ by evaluating this

```
<< Statistics`ContinuousDistributions`
data = ReadList[file, Real];
n = Length[data];
f[x_, θ_] := PDF[WeibullDistribution[θ[[3]], θ[[2]]], x - θ[[1]]] /; x ≥ θ[[1]]
f[x_, θ_] := 10^-50 /; x < θ[[1]]
FindMinimum[
  -Σ_{i=1}^{n} Log[f[data[[i]], {p1, p2, p3}]],
  {p1, 250, 300},
  {p2, 50, 150},
  {p3, 1.8, 2.2}
]
```

Listing 1. A typical Mathematica code for finding the best-fitting parameters. The distribution f in the example is a Weibull distribution (built-in) with three parameters θ[[1]] (the position), θ[[2]] (the scale) and θ[[3]] (the shape).

function at only a few selected positions. As soon as the data set is sufficiently large ($n > n_0$), it is faster to find an approximation and use it *n* times that to estimate the original function *n* times using numerical integration. In the following, we explore one type of approximation, the rational approximation. We will examine how reliable the process is, how accurate the resulting approximations are, and how much faster fitting using approximations is.

Finding Rational Approximations

A rational approximation to an equation $f(x)$ in the interval $[a, b]$ is another equation composed by the ratio of two polynomials of degree *m* and *n* (*m* can be equal to *n*). Let $r_{m,n}(x)$ denote such an approximation:

$$r_{m,n}(x) = \frac{a_0 + a_1 x^1 + \ldots + a_m x^m}{b_0 + b_1 x^1 + \ldots + b_n x^n}$$

in which the coefficients $a_0, .., a_m$ and $b_0, .., b_n \in \mathbb{R}$ have to be chosen so that the error

$$\varepsilon(x) = \left| f(x) - r_{m,n}(x) \right|$$

is small (Jackson, 1930). If *f* depends on parameters θ, every time θ are changed, a new approximation has to be found.

It is possible to demonstrate (Petrushev & Popov, 1987) that for any continuous function $f(x)$, *x* being restricted to a domain $[a, b]$, there exists a single set of coefficients $\{ a_0, .., a_m, b_0, .., b_n\}$ so that the difference is the smallest. Further, there exists an algorithm, the Remez algorithm (described in Petrushev and Popov, 1987) that can find these coefficients rapidly (the algorithm is said to be quadratically convergent).

The Remez algorithm is not particularly complex to implement and is already provided with some software (such as *Mathematica*, by loading the *NumericalMath`Approximations`* library). Fitting a distribution is therefore a two-step process: For a tested θ, find an approximation, then compute *LL* using the approximation instead of the original equation. Listing 2 shows an example of code in *Mathematica* which automatically finds a rational approximation $r(x)$ of degree 10, 10.

Reliability of the Approach

We tested the approach under three aspects: its reliability, its accuracy and its speed. The following simulations were run using Weibull distributions. It was used because of its relations to psychological models (Cousineau, Goodman & Shiffrin, 2002). Its pdf is given by:

$$f(x \mid \alpha, \beta, \gamma) = \gamma \beta^{-\gamma} (x - \alpha)^{\gamma - 1} e^{-\left(\frac{x-\alpha}{\beta}\right)^{\gamma}}$$

where α is the shift parameter, β is the scale parameter, and γ is the shape parameter. When the shape is 1, the Weibull reduces to a shifted exponential distribution. Unless the shape parameter is 1, a convolution of two Weibull distributions never yields a closed-form equation.

Reliability

Reliability measures whether the approach always returns an approximation. To test this, we generated random parameters for two Weibull components that were convolved. We then looked whether an

```
<< Statistics`ContinuousDistributions`
<< NumericalMath`Approximations`
data = ReadList[file, Real];
n = Length[data];
f[x_, θ_] := PDF[WeibullDistribution  [θ[[3]], θ[[2]]], x - θ[[1]]] /; x ≥ θ[[1]]
f[x_, θ_] := 10^-50 /; x < θ[[1]]
FindMinimum[
 r[x_] = RationalInterpolation[f[x, {p1, p2, p3}], {x, 10, 10},
   {x, Min[data], Max[data]}];
   n
 - ∑ Log[If[r[data[[i]]] ≤ 0, 10^-16, r[data[[i]]] ]],
  i=1

 {p1, 250, 300},
 {p2, 50, 150},
 {p3, 1.8, 2.2}
]
```

Listing 2. A typical Mathematica code for finding the best-fitting parameters using an approximation. The approximation is found within the interval [Min(t_i), Max(t_i)]

approximation was found.

The parameters were as follow: α_1 and α_2 were normal with mean 250 and standard deviation 40; β_1 and β_2 were normal with mean 60 and standard deviation 10; γ_1 and γ_2 were normal with mean 2 and standard deviation 0.2. The means were chosen so that the resulting distributions look like an RT distribution. $\alpha_1 + \alpha_2$ is a lower bound so that $\Pr(T < \alpha_1 + \alpha_2) = 0$.

The approximations requested in all the following simulations were ratios of polynomials of degree 10 on both the numerator and the denominator. The approximation had to be valid within the interval $[\alpha_1 + \alpha_2, \alpha_1 + \alpha_2 + 6(\beta_1 + \beta_2)]$, going well beyond four standard deviation above the mean.

Repeated over a thousand replications, the process never failed, always returning an approximation. Figure 1 shows one such approximation. As seen the approximation is visually quite good. However, it is sometimes hovering around zero in the tails. The curve should not go beyond zero since it is suppose to approximate probabilities. Further, later on, when we proceed to distribution fitting, computing the log of a negative value stops the minimization algorithm. To avoid these problems, we added a condition that if the approximation returns a value below zero, 10^{-16} should be returned instead (see Listing 2). Note that the true probability of sampling a RT in these areas is quite small (once every million samples approximately).

Accuracy

Accuracy being crucial, we tested it from three different points of view.

Area Under the Curve The maximum likelihood method assumes that a probability density function is used in Eq. 2. Because the area under a pdf is 1, the area under the approximation should also be 1. We generated the parameters of a convolution randomly as above and then generated an approximation $r(x)$ over the interval $[\alpha_1 + \alpha_2, \alpha_1 + \alpha_2 + 6(\beta_1 + \beta_2)]$. Afterward, the area under the curve was estimated.

Repeated over a thousand replications, the largest absolute deviation to 1 was in the order of 10^{-10} which is of the same magnitude as the numerical integration error.

Likelihood Result The fitting procedure relies on the $LL(t_i, \theta)$ quantity to guide the search for the optimal parameters. Therefore, this quantity must be very accurate. To test this, we first generated random parameters as above, followed by the generation of a sample containing 1200 deviates. One deviate was obtained by generating two Weibull deviates with parameters $\{\alpha_1, \beta_1, \gamma_1\}$ and $\{\alpha_2, \beta_2, \gamma_2\}$ respectively and adding them. Finally, the quantity LL was computed using the true parameters, once with the convolution equation (and numerical integration), and once with the rational approximation. The rational approximation was

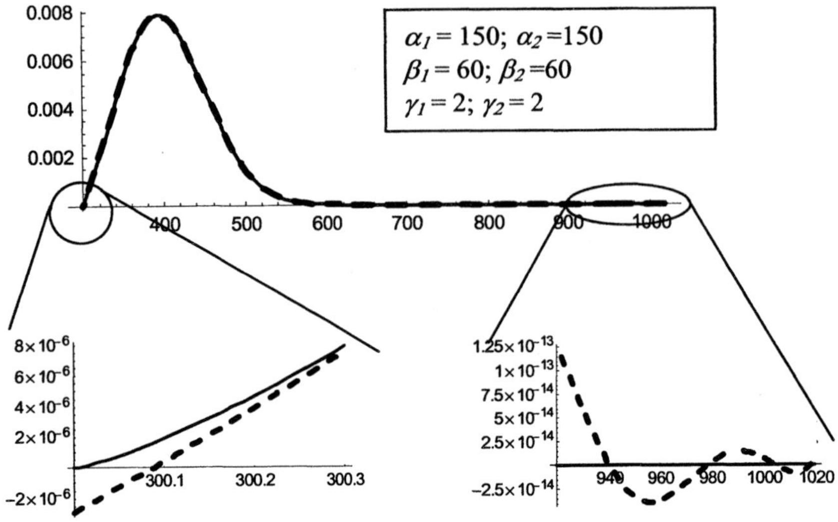

Figure 1. An example of approximation with magnification of the tails. The full line represents the true function $(f_1 * f_2)(t)$ and the dash line represents its approximation $r(t)$

computed over the interval [Min(sample), Max(sample)].

Repeated over a thousand replications, the log likelihood using the convolution was on average -6299.566. The absolute deviation between the *LL* of the convolution and the *LL* of the approximation was on average 0.011, that is, an error in the order of 10^{-6}.

Parameter Estimation We also checked the error in the best-fitting parameter estimated. Because fitting parameters using a convolution is so slow (see next), we used a single Weibull distribution instead of a convolution. We generated random parameters α, β, γ in the same manner as previously, followed by the generation of a sample containing 1200 deviates. The best-fitting parameters of the sample were first found using the true Weibull pdf equation, returning θ_f* and second, using the rational approximation to it, returning θ_r*. To assess the difference between the two sets of parameters, we looked at the Euclidian distance between them:

$$\left\| \theta_f * - \theta_r * \right\|$$

Repeated over a thousand replications, the average distance between the two best-fitting solutions was 0.00014, an error in the order of 10^{-8}.

The efficiency of the likelihood method was explored in Cousineau, Brown & Heathcote (in press). They found that the error between parameters from two samples generated with the same parameters was on average 0.2. This is larger than the error between the two methods as reported here by a factor of about 10^2.

The second and third tests demonstrate that the error arising from using a rational approximation is totally negligible, being smaller than the error that would occur if numerical integration was used, and being much smaller than the sampling error.

At this point, the two approaches are equivalent. They will diverge on the time they require.

Computation Times

To assess computation times, we generated random parameters for a convolution as above and from these, a sample containing 1200 deviates. Then, we measured the time taken to compute *LL* once with the true parameters and the convolution equation. We did not try to find the best-fitting parameters because the minimization process needs to compute *LL* from a hundred to a thousand times to converge onto the optimal parameters, slowing down the simulations accordingly. Using the same parameters and the same sample, we also generated an approximation $r(x)$ to the convolution and computed *LL* using the approximation. We recorded the time taken to find the approximation and the time taken to compute *LL* with the approximation.

Repeated over a thousand replications, the average time to compute *LL* using the convolution required 164 seconds on a 2.0 GHz PC with *Mathematica* 4.1. The times varied between 60 and 600 seconds.

As for the second method, the average time to find an approximation $r(x)$ was 0.96 second (varying from 0.50 to 1.60 seconds). The variations in the time to find an approximation came from the Remez algorithm requiring one, two or three iterations to find the best rational approximation. The average time to compute *LL* with the approximation was nearly constant at 0.59 seconds (the variations came from the operating system, Windows 2000).

On average, the two-step method (finding an approximation, using it 1200 times) is 106 times faster than using a convolution 1200 times.

Of course, *Mathematica* is notably known to be slow (but see version 5). Using, for example, *Matlab* would speed-up these computations by about a factor of 20. However, all the computations would equally benefit from this general speed-up so that the 100 to 1 ratio in favor of using an approximation would still be present.

Conclusion

The method presented here relies on brut computational power to find an approximation every time a new parameter set is tested. It does speed up computation times considerably. In fact, we can estimate from the above that as soon as the sample size is 8 or more, it is preferable to use rational approximations. In addition, the whole process can be automatized using the Remez algorithm which is already implemented in *Mathematica*.

Yet, the process might be made even faster. For example, the exGaussian distribution is based on an approximation. However, it was possible to isolate the approximation from the parameters so that the same approximation (*erf*) is always used whatever the parameters θ. As such, the initial time to find the approximation occurred only once, and is no longer part of the computation. In the above simulations, it would eliminate the 0.96 second interval to find the approximation, resulting in a near 300 times speed-up in computation. It remains to be demonstrated whether convolutions (of two Weibull distributions for example) could be reduced to a single approximation independent of the parameters.

Acknowledgments

This research was supported in part by the Fonds pour la formation de Chercheurs et l'Aide à la Recherche and the Conseil de Recherches en Sciences Naturelles et en Génie du Canada.

References

Chandler, P. J. (1965). Subroutine STEPIT: An algorithm that finds the values of the parameters which minimize a given continuous function [computer program]. Bloomington: Indiana University, Quantum chemistry.

Cousineau, D. Larochelle, S. (1997). *PASTIS: A Program for Curve and Distribution Analyses.* Behavior Research Methods, Instruments, & Computers, 29, 542-548.

Cousineau, D. Larochelle, S. (in press). *Visual-Memory search: An integrative perspective.* Psychological Research.

Cousineau, D., Brown, S., Heathcote, A. (in press). *Fitting distributions using maximum likelihood: Methods and packages.* Behavior Research Methods, Instruments, & Computers.

Cousineau, D., Goodman, V. Shiffrin, R. M. (2002). *Extending statistics of extremes to distributions varying on position and scale, and implication for race models.* Journal of Mathematical Psychology, 46, 431-454.

Cousineau, D., Shiffrin, R. M. (in press). *Termination of a visual search with large display size effect.* Spatial Vision.

Cramér, H. (1946). Mathematical Methods of Statistics. Princeton: Princeton University Press.

Gauss, C. F. (1809/1864). Theoria Motus Corporum Coelestium in Sectionibus Conicis Solem Ambientum. Paris: A. Bertrand.

Jackson, D. (1930). The theory of approximation. New York: American Mathematical Society.

Kennedy, W. J. Gentle, J. E. (1980). Statistical computing. New York: Marcel Dekker inc.

Luce, R. D. (1986). Response times, their role in inferring elementary mental organization. New York: Oxford University Press.

Nelder, J. A. Mead, R. (1965). *A simplex method for function minimization.* The computer journal, 7, 308-313.

Petrushev, P. P., Popov, V. A. (1987). Rational approximation of real functions. Cambridge: Cambridge University Press.

Ratcliff, R. (1979). *Group Reaction Time Distributions and an Analysis of Distribution Statistics.* Psychological Bulletin, 86, 446-461.

Ratcliff, R., Van Zandt, T. McKoon, G. (1999). *Connectionist and diffusion models of reaction time.* Psychological Review, 106, 261-300.

Townsend, J. T. Ashby, F. G. (1983). Stochastic Modeling of Elementary Psychological Processes. Cambridge, England: Cambridge University Press.

A Rational Model of the Effect of Information Scent on the Exploration of Menus

Anna L Cox[1] (a.cox@ucl.ac.uk) & Richard M Young[2] (r.m.young@herts.ac.uk)
[1]UCLIC, Dept of Psychology, University College London, London, UK
[2]Dept of Psychology, University of Hertfordshire, Hatfield, UK

Abstract

This paper describes a computational model based on rational analysis which is compared to experimental data and enables us to explain the number and order of assessments that participants make on the items on a web-page. The model provides both a good fit to the scan paths of the participants and an explanation for why a particular item is chosen. The strength of this research is that it provides a qualitative account of why particular patterns of scent result in particular scan patterns. The research provides further support for the rational framework as an explanation of exploratory behaviour.

Introduction

In the real world people often do not read the instructions for a novel system either through choice, or because they are not available. Instead, they learn how to use the novel system by actually trying to use it, drawing on a combination of prior knowledge, information from the interface itself such as the semantic similarity between the labels and their task description, and problem solving skills. This phenomenon is described as exploratory learning.

When exploring an interface, people might select a particular option for one of two reasons: because they believe that it will lead them to the goal (focused exploration); or because they believe that it will provide them with some information about the system they are interacting with (free exploration) (Cox 2002).

Our earlier research (Cox 2002) led to the development of a rational framework for modelling this kind of exploratory learning. We have already used it to explain empirical findings such as why people do and learn different things from free as against focused exploration (Young & Cox, 2000). We are now using it as a framework to build models of single-level menu exploration which we compare to data gained from empirical studies.

The framework and how it works

To begin, we will consider the behaviour of a model built within the framework when engaged in free exploration. At any moment, the model has a number of possible things it can do. These might be clicking on a hyperlink, considering some feedback that has just been received, reading a label, formulating a hypothesis to test, etc. These are referred to as Exploratory Acts or EAs. EAs differ in the number of stages they consist of and therefore in the amount of time and effort required to carry them out. For example, pressing a button or clicking on a hyperlink would require a single quantum of effort, whereas interpreting feedback from the interface might require a number of quanta: one to read the display; another to realise that the information has changed; and a third to understand the type of change.

Each EA has a cost (C) and a value (ΔI) determined by the costs and values of its constituent quanta. Each quantum has a cost reflecting the time it would take to perform it, and a value related to the estimated increase in information that would be gained. The cost of the EA that proposes a single button press therefore is likely to be low as it only requires one quantum of processing, but the value may be high or low depending on what is known about the button already. The cost of the EA that proposes interpreting some feedback would be higher than that of the button press as this EA requires more quanta of processing (three in this example) and again the value would vary depending on the amount of information expected to be gained.

We assume that which of the proposed EAs is chosen in any given situation is determined by rational analysis. This theory suggests that when trying to learn about the device, the only reason to choose an EA will be the one that is believed to elicit the highest amount of information (ΔI) for the least cost (C): $\Delta I/C$.

During focused exploration, the model has a specific goal to achieve and therefore an EA could be chosen for one of two reasons. Firstly, in the same way as for free exploration, an EA might be chosen because it will elicit information that is not known, or secondly, it might be chosen if it is expected to lead to the goal. In this situation, therefore, the efficiency of some EAs is calculated as the expected amount of information elicited about the device per unit of cost, and for some others the efficiencies are calculated as the probability (P) that they will lead to the completion of the goal (G) minus the cost (C) of getting to the goal (PG-C). This results in the behaviour suggested by Pirolli & Card's scent following theory (1999).

The framework describes a cycle of three stages. In stage one, the efficiency of all the EAs possible at that moment are calculated. The efficiency of an EA is equal to the expected amount of information gained by the EA, divided by the cost of executing it. In stage two, the EA with the highest efficiency is chosen. The model will therefore choose whichever EA proposes the highest information gain (ΔI) per unit of cost (C) or the greatest probability of reaching to goal (PG) minus the cost (C). Finally, in stage three, the chosen EA is executed.

Menu Search Data

In response to an earlier model of menu search (Young, 1998), Brumby and Howes (2003, 2004) set out to test the hypothesis that people would assess fewer poor distracter items than mediocre distracter items. Participants were shown a single page on which a series of hyperlinks were arranged in a vertical list and asked to identify which of the menu items they would choose in order to find the goal information. Eye-tracking was performed using an ASL Pan/Tilt optic eye-tracking system.

Results

Average performance across menus. Rieman (1994) noted that users often appear not to assess all possible items before making their selection. Initial analysis of the empirical data supported this observation and suggested that on average, participants looked at about half the items in the menu before making their selection (Brumby & Howes 2003).

Performance on Individual Menus. More detailed analysis of the traces provided by the eye-tracker shows that in fact there are two kinds of behaviour that can be observed. The first is where a participant appears to continue scanning the menu in search of the goal item after fixating on the item eventually selected. This was the most common behaviour and occurred on 69% of trials. The second is the opposite of this, when the participant selects the goal item immediately after fixating on it (self-terminating behaviour). This occurred on 31% of the trials.

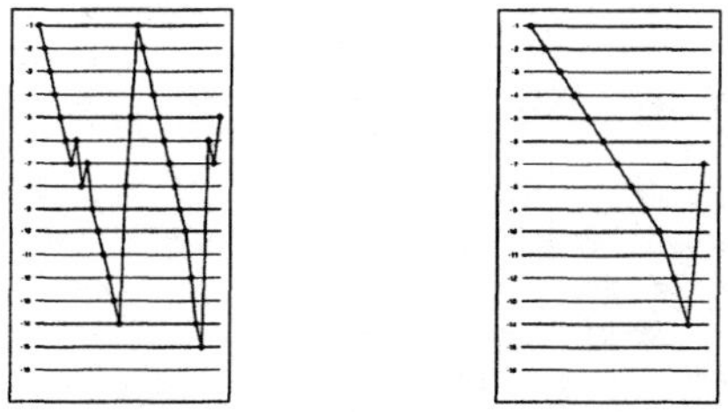

Figure 1: Illustrations of continued scanning of a menu.

The illustrations in figures 1 and 2 show a simplified version of the scan paths of participants in the experiment. The menu items are arranged in order downwards on the vertical axis, while time runs to the right. The scan paths show which items were fixated and in what order.

The first illustration in figure 1 shows an example of continued scanning of the menus. This pattern was seen in approximately 32% of the continued scan cases (22% of the total). The second illustrates an example where the participant continues to fixate on the menu items below the goal item before returning to that item and selecting it. This pattern was seen in approximately 68% of the continued scan cases (47% of the total).

Although the eye-tracking data suggests that sometimes the participants do not fixate on all the items in the list, but, as in figure 1, stop short of the end of the menu, we suggest that this is not sufficient evidence to suggest that the last two items in the list are not being assessed. Byrne et al (1999) provide evidence from eye-tracking studies of visual search in which participants tend to start at, or near, the top of a list and search downwards sometimes skipping items. This behaviour is explained in Salvucci's (2001) model of menu search in which he noted that although the model did not fixate on some items, it did attend to and encode them.

Figure 2 illustrates the other scan pattern from the eye-tracking data: the participant assesses each of the items in the menu until he locates the goal item, and then selects it immediately, without fixating on any other items in the menu.

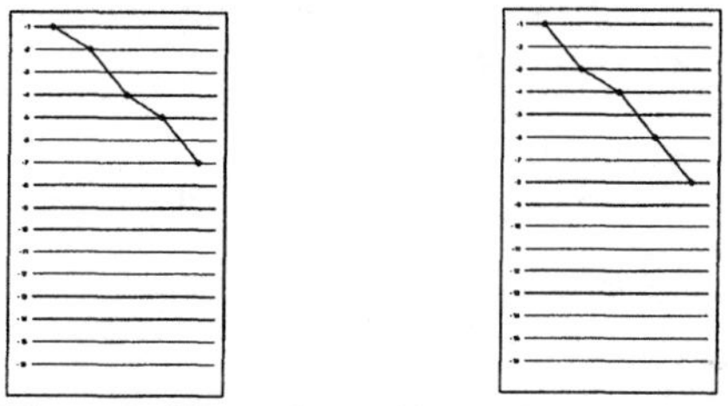

Figure 2: Illustrations of immediate selection following fixation on the goal item (self terminating behaviour).

The Model

In this paper we report the behaviour of a model built within the framework described above. Previous research suggests it is not straightforward to implement the framework in ACT-R (Young & Cox 2000) and therefore the framework has been built in Lisp. It is a model of the cognitive processes involved in an agent's *assessment* of a novel menu presented on a web-page. The agent has the goal of selecting the item that will lead to goal completion and therefore this is focused exploration. However, as the menu presented is novel, the first thing that the model has to do is to gain some information about the menu that will then enable it to conduct focused exploration (i.e. use a scent *following* strategy such as PG-C). The model presented here stops short of the scent *following* part of the task (i.e. selecting the item that will lead to the goal) and as a consequence it is not necessary for this model to make use of the two measures of efficiency. Instead, this model incorporates only the ΔI/C efficiency measure. In order to extend the model to include scent *following*, an additional type of EA (SELECT) would need to be incorporated. The efficiency value for this EA would need to be calculated using PG-C rather than ΔI/C.

The model interacts with a representation of a single level menu that consists of sixteen items as used by Brumby & Howes (2003). Each menu is associated with a particular goal: e.g. Find out who stars in the film "The Lord of the Rings".

The model includes two types of EA: 'assess information SCENT' and 'ANTICIPATE the result of selecting this item'. The SCENT EA should be thought of as being an amalgamation of perceiving the label, reading the label (at a lexical level) and considering the semantic similarity between the label and the current task. The ANTICIPATE EA should be thought of as some additional cognitive effort that considers whether the label is likely to lead to the goal. For example, given the goal of finding an armchair on a furniture website, the ANTICIPATE EA models the agent's consideration of whether the label 'home' is likely to lead to the homepage of the site, or to a list of home furnishings. Each of these EA types has a cost associated with it with the ANTICIPATE EA type being more expensive in mental effort than the first type. There is also a fixed cost of moving attention from one item in the menu to the next.

Brumby and Howes (2003) asked participants to rate the relevance of each item to the goal, on a five point scale. We consider the median ratings from the participants to be a good indication of the information scent of each menu item to the relevant goal. These ratings have been used as the judgments made by the model regarding the amount of information scent from a particular item. These are used to represent the value of the outcome of the SCENT EA in the model.

The ratings provided by participants were also used as the basis upon which to calculate the conditional probabilities of an item leading to the goal or not, given each scent rating. Before assessing any items, the model 'knows' the number of items in the menu. Each of these items is considered to be equally (ir)relevant to completing the task. The scent ratings provided by the participants are also used as the basis for determining the new relevance (R) value of an item following an assessment. On each page, the set of relevances R_i are mapped into a set of probabilities P_i by the transformation $P_i = \text{odds}(Ri)/\sum \text{odds}(R_j)$, where odds(R) is defined in the standard way as $\text{odds}(R) = R/(1-R)$. Note that $\sum P_i = 1$, reflecting the fact that exactly one option on the page leads to the goal.

Results

The model was run on the same menus that were used in the empirical data collection (Brumby & Howes 2003). The goal item in each menu had a high scent value (1) and appeared in various positions within the menus. The menus varied in terms of the quality of the distracter items. They were either low quality (all distracters had a scent value of 5) or mediocre quality (mean scent value of the distracter items was 3). The output trace from the model identifies which EA fired on each of the cycles. Average performance of the model across menus is briefly considered together with performance on individual menus. The performance of the model is compared to that of human participants.

Fitting the model. No substantial effort has been made to manipulate the parameters of the model in order to ensure a good fit between the quantitative data from the empirical study and that from the model. However, even without this, the model appears to perform similarly to Brumby & Howes' participants and their ACT-R model (Brumby & Howes 2004) in so far as exhibiting two tendencies:

I. The model tends not to assess all of the items in the choice set.
II. Items are often assessed on multiple passes before selection.

Looking more closely at the model runs on individual menus shows that the model behaves similarly to the participants. The following sections describe the model's behaviour on the menus in detail.

Good goal, poor distracters. Just over half (8/14) of the menus used in Brumby and Howes' (2003) experiment were made up of a good goal (scent value 1) and 15 poor distracters (scent value 5). These eight menus can be thought of as being identical to each other except for the position of the high scent item. Of these menus, two were completely identical as far as the model was concerned as the high scent item was in the same position in each: position 1. On these menus the model chooses to do a SCENT assessment on the first item, and then does the ANTICIPATE assessment on that item immediately and then stops. (If scent *following* were also to be implemented in the model, the model would choose to SELECT this item at this point.) In concordance with the eye-scanning data, the model predicts that this pattern of behaviour is most likely to occur on menus where there is (at least) one menu item rated very high (1), and all the distracters encountered prior to the highly rated item are rated low. The ANTICIPATE EA can have one of three outcomes (yes, no, maybe) which can increase, decrease or leave unchanged, the relevance of the item assessed. Figure 3 shows the sequence of assessments made by the model on these menus. Bold text has been used to highlight the high scent item.

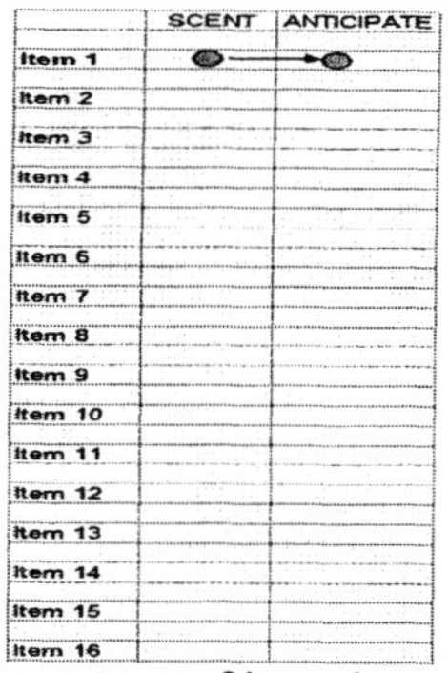

Figure 3: Assessments of items in a menu when the high scent item is in position 1 and the distracters are poor.

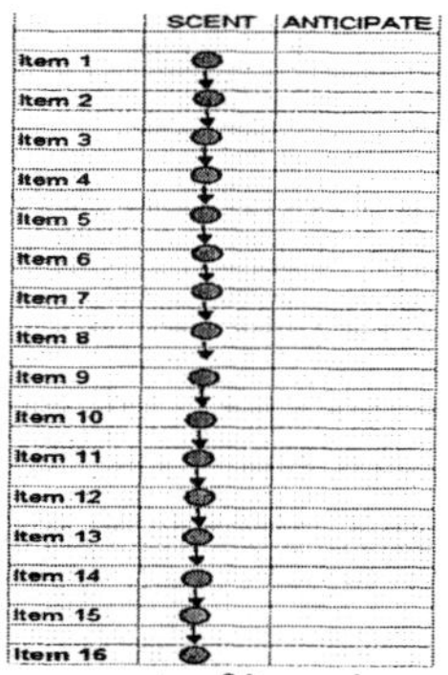

Figure 4: Assessments of items in a menu when the high scent item is in position 16 and all the distracters are poor.

A further two of the eight menus have the high scent item in the last position (16th) on the menu. On these menus the model chooses to do a SCENT assessment on each item and then, on reaching the bottom of the menu, stops immediately without conducting any ANTICIPATE assessments. This behaviour is illustrated in figure 4.

The second pattern in figure 1 shows an example of a participant assessing (nearly) all the items in the menu before refocusing on a previously assessed item and then selecting it. One might imagine that the agent considers the investment in the extra effort required to assess the rest of the list as worthwhile when considered against the probability that the high scent item is not the goal item. However, this is not an accurate way in which to conceptualise the behaviour of the model. Instead, this behaviour is most likely to occur when the previously encountered items have been discounted as irrelevant (scent = 5, R=0).

For the remaining four of these eight menus, the high scent goal item is in either position 7, 12, 14 or 15 in the menu while all the distracters have a low scent score. On these menus the model choose to do a SCENT assessment on every item in the menu before doing the ANTICIPATE assessment and stopping. This therefore provides us with an example of continued scanning before selection (figure 5).

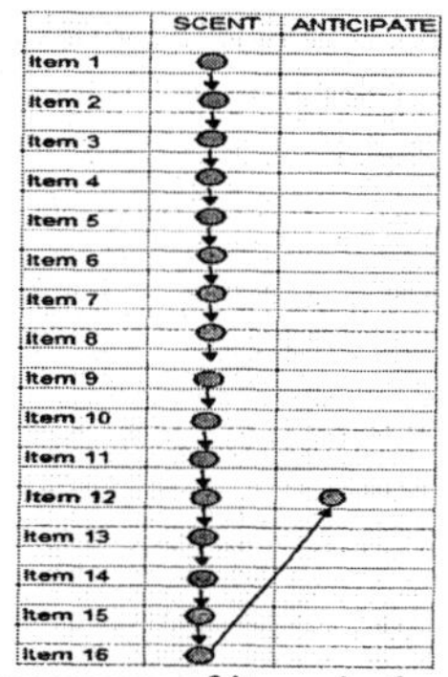

Figure 5: Assessments of items in the menu when the high scent item is in position 12.

Why the position of the goal item results in different behaviour. In order to understand how the position of the high scent item can result in these two different types of behaviour we will compare the behaviour of the model when the high scent item is in position two (as an example of occurring early in the menu) and in position twelve (as an example of occurring late in the menu) in more detail. In both examples, initially, all sixteen menu items are rated equally and all have a relevance (R) value of 0.06. In the first cycle, the EA that proposes assessing the scent of the first item in the menu is rated as having the highest efficiency due to it having the lowest cost. Consequently, the model assesses the first item which gets rated as a 5 (i.e. very low scent). As a result, the new R value of this item is set at 0. On the next cycle, the EA that proposes SCENT assessment on the second item in the list is the most efficient (due to the lower cost) so this item gets assessed. This behaviour continues until the model assesses the high scent item.

In menus where the high scent item occurs early on in the menu, the second item in the menu gets a R value of 0.5097 which raises the probability that this item will lead to the goal to 0.6220. On the following cycle (see figure 6), the R value of the high scent item (the black diamond) leads to an efficiency of 0.008 (the value of the low solid curve in figure 6 at 0.5097 on the x-axis) whilst the second best item (an item yet to be assessed – shown as a black square) has an R of 0.06 which results in an efficiency of 0.006 (the value of the dashed curve at 0.06 on the x-axis). Although the efficiencies of the two EAs are very similar, on is larger than the other rand this is what determines which EA is chosen. The grey diamonds in figures 6 & 7 show the relevance values of the other items in the menu. In figure 6, one item (that in position 1) has had its relevance reduced to zero and therefore lies at zero on the x-axis. All the other grey diamonds lie at 0.06 on the x-axis. Their height against the y-axis is immaterial.

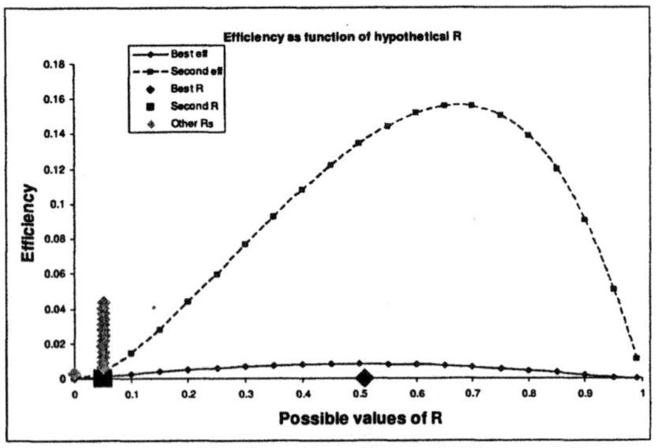

Figure 6: The R values for each item and the efficiency curves for best and second best EAs immediately after assessing the high scent item that occurs early in the menu.

In our example of a menu where the high scent item occurs late on in the menu, the relevance of each of the low scent items that have already been assessed falls to zero. When the model assesses the twelfth item, its relevance gets a value of 0.5097 which raises the probability that this item will lead to the goal to 0.6220. On the following cycle (see figure 7), the R value of the high scent item only has an efficiency of 0.005 (the value of the dashed curve in figure 7 at 0.5097 on the x-axis) whilst the item with the best efficiency (an item yet to be assessed) has an R of 0.05 which results in an efficiency of 0.006 (the value of the dashed curve at 0.06 on the x-axis). The result is that the model continues to assess each item in the menu until it reaches the bottom because the efficiency of conducting a SCENT assessment of a new item is greater than the efficiency of conducting the ANTICIPATE assessment on the high scent item in position twelve. This has the effect of slowly increasing the probability of the item in position twelve leading to the goal.

The reason this is different to the outcome shown in figure 6 is because the shapes of the two curves in figure 7 are slightly different due to the effect of the distracter items (or lack of them). The fact, in figure 6, that there are more distracter items still competing (i.e. with an R>0) has resulted in the peak of both curves shifting to the right which results in a flatter curve at the left-hand end and consequently lower efficiencies for items with low R values.

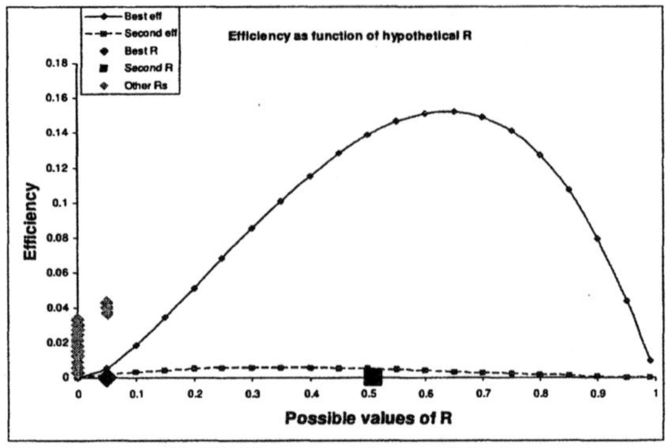

Figure 7: The R values for each item and the efficiency curves for best and second best EAs immediately after assessing the high scent item that occurs late in the menu.

Good goal, mediocre distracters. The remaining menus used in the empirical study (6/14) consisted of menus that were far more varied in terms of the pattern of scent. The goal item had a high scent rating (1) while the ratings of the distracters varied, both within and between menus, from 2 to 5 with a median of 3. On two of these six menus the high scent item is in position one on the menu. As in the previous examples where this is the case the model conducts a SCENT assessment of this item followed by an ANTICIPATE assessment and then stops. On a further menu, where the high scent item is in position three, the model conducts SCENT assessments of the first three items before conducting the ANTICIPATE assessment on the third item and then stops. Here again then we see examples of self-terminating behaviour. For these menus, the model does not 'know' that the rest of the menu has a more vaired scent pattern than those menus with the poor distracters (where all distracters were rated 5).

On the three remaining menus the high scent item is in position twelve (twice) or fourteen and the model produces patterns of repeated scanning similar to those of the participants. The following description of the model's behaviour closely matches that of the participants' scan pattern shown in figure 1. The model conducts a scent assessment on each of the items in the menu in turn (see figure 8). This is then followed by ANTICIPATE assessments on the highest scent items (those with a scent level of 2) on the upward scan. The model conducts another downward scan this time conducting ANTICIPATE assessments on those items assessed to have a scent level of 3. The final upward scan (shown by a dotted line) is inferred from the fact that, in order to complete the task, a model that included scent following would refocus its attention on the item in position 5 which had a scent level of 1 before selecting it.

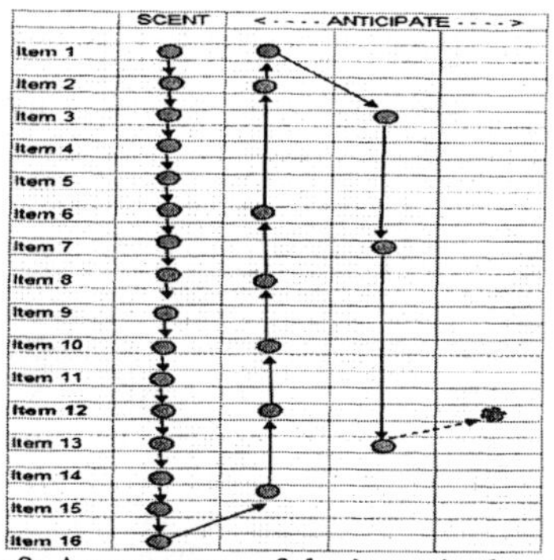

Figure 8: Assessments of the items in the mediocre menu when the high scent item is in position 12.

General Discussion

Comparisons of the traces of the model with the empirical data suggest that the model provides a good explanation of the cognitive processes involved in this task. This suggests that participants make an assessment of the relevance of a label to the current goal and then, together with the estimated relevance of previous items, choose to either i) select that item as the one that will lead to the goal, ii) conduct some further assessment of the current item, or iii) move on to another item and assess that. Which of these EAs is chosen is driven by the pattern of information scent that has been experienced so far.

The model provides us with an explanation of how and why the position of the goal and the quality of the distracter items affect the behaviour of the participants on the task. Regardless of the pattern of scent of the menu, our model predicts that the agent will tend to stop exploring the menu as soon as it comes across a menu item that has high information scent (self-terminates) if this is encountered early in the menu. On menus where there is one high scent item amongst a set of low scent items and the high scent item occurs later in the menu, the agent continues to assess the other items in the menu before conducting further assessment of the high scent item and finally selecting it. The model enables us to explain why we see these different patterns of behaviour on menus which have such similar patterns of information scent. This is due to the effect of the interdependence of the probability that each of the items will lead to the goal. The actual point on the menu at which the model swaps from one behaviour to the other is sensitive to a number of factors such as the length of the menu and the costs of the EAs. It would appear therefore that it is in the nature of exploratory behaviour that there are close calls which suggests that people can rationally do either behaviour and that a number of factors have an effect on the behaviour of participants exploring real menus.

When there are a number of distracter items that receive a similar level of scent assessment to that of the goal item (i.e. there are mediocre distracters) it is more difficult to identify the best item from amongst the distracter items and therefore additional cognitive effort is required in order to determine which item is most likely to lead to the goal. As a consequence, the agent is more likely to exhibit multi-scanning behaviour on these menus.

Although one might expect the data to show continued scanning of menus and multi-scanning one would not necessarily expect participants to self-terminate. The value of this research is that the model exhibits all three patterns of behaviour from a single set of parameters This suggests that no alternative strategies are necessary to explain variations in human behaviour on this task. The success of this model in providing an explanation of the eye-tracking data adds further support to the rational framework as a general model of exploration.

Acknowledgements

Thanks are due to Duncan Brumby and Andrew Howes for access to their menus and data and for comments on an earlier draft.

References

Brumby, D.P., & Howes, A. (2003). Adaptive decision making in menu search: The role of interdependence and past experience on link selection. *Proceedings Volume 2 of the* Conference *HCI 2003: Designing for society,* Bath, UK, 133-134.

Brumby, D.P. & Howes, A. (2004). Good Enough But I'll Just Check: Web-page Search As Attentional Refocusing. *Proceedings of the 6th International Conference on Cognitive Modelling*, Pittsburgh, USA.

Byrne, M. D., et al. (1999). Eye tracking the visual search of click-down menus. *Human Factors in Computing Systems: CHI 99*, New York: ACM Press.

Cox, A.L. (2002) *Exploratory Learning of Interactive Devices: What People Do and What People Learn.* Unpublished PhD, University of Hertfordshire, UK.

Pirolli, P. & Card, S., (1999) Information Foraging, *Psychological Review*, 104 (4), 645-675

Rieman, J. (1994) *Learning strategies and exploratory behaviour of interactive computer users.* PhD dissertation, University of Colorado, Boulder, CL.

Salvucci, D. D. (2001). An integrated model of eye movements and visual encoding. *Cognitive Systems Research* 1(4): 201-220.

Young, R.M. & Cox, A.L., (2000) A New Rational Framework for Modelling Exploratory Device Learning ... but does it fit with ACT-R? *Proceedings of the 7th ACT-R Workshop,* Carnegie Mellon University

Properties of the Decision Environment and Display Organization in Choice

Fabio Del Missier (delmisfa@units.it)
Department of Cognitive and Educational Sciences, University of Trento
Via Matteo del Ben, 5/B, I-38068 Rovereto, Italy

Abstract

This paper describes two experiments and a simulation, aimed at investigating the impact of specific features of the decision environment and of the organization of the information display on choice processes and outcomes. The first experiment compared a matrix organization with a Cartesian graph, while also manipulating the distribution of the decision weights and the dominance structure of the problems. It was conducted to test the predictions of models derived from different choice frameworks, but none of the models were able to provide a full account of the results. Therefore, an alternative explanation was put forward, relying on the simulation of a variant of the lexicographic heuristic (LEX-2). The second experiment supported the existence of LEX-2 and explained a counter-intuitive display-related effect brought out by the first study.

Theories and Models of Choice

The first experiment presented in this paper takes into account both the influence of the display and the role of two significant properties of the decision environment: the distribution of importance weights and the dominance structure of the decision problems. Two prominent choice frameworks were selected and their predictions are clearly set out. The theories will be described in general terms in the following sections, while their specific predictions will be presented further on in the paper.

The Adaptive Decision Maker Framework

The adaptive decision maker framework (Payne, Bettman, & Johnson, 1988, 1993) regards choice as a constructive process. According to this theory, the decision maker (henceforth DM) is sensitive to the specific properties of the decision environment and is able to construct a choice strategy or to select it from memory in an adaptive way. The basic assumption of the framework is that the DM will use a heuristic that assures a satisfying trade-off between the anticipated accuracy of the decision and the anticipated effort required in the choice process, *given the specific features of the decision environment*. This framework has recently included in the main decision goals the need to avoid negative emotions and the desire to justify the decision (Bettman, Luce, & Payne, 1998).

The properties and the performance profiles of different choice strategies have been analyzed through simulations, and the predictions of the framework have received a good degree of experimental support (Payne et al., 1993). This research effort has provided us with knowledge about the effects of different task and context variables on strategy selection. The dominance structure of the decision problem is one of the variables that has been taken into account. It was shown that the presence of dominated options in the choice set allows noncompensatory heuristics to perform reasonably well (Payne et al., 1993). Another significant variable is the distribution of the DM's importance weights. These weights represent the importance that the DM gives to the various features of the available options. The weight distribution, together with the value functions describing the DM's subjective evaluation of the attribute values, defines the DM's subjective utility function.

Kleinmuntz and Schkade (1990, 1993) analyzed the impact of the information display variables on choice tasks. Their cost-benefit proposal complements the adaptive decision maker framework. According to them, strategy selection is also sensitive to the features of the display: "*This cost-benefit perspective implies that information displays define a cognitive incentive system for decision makers, whereby displays influence the effort and accuracy of each available strategy and, therefore, induce decision makers to use different strategies*" (Kleinmuntz & Schkade, 1993, p. 224). Kleinmuntz and Schkade (1994) highlighted the role of three features of the display: the organization of the information items, their form, and their sequence.

The information display may affect choice strategies through the reduction of the cognitive effort required. For instance, if comparisons are simplified by the display, the effort involved in comparison-based heuristics will be reduced, increasing the likelihood that a (sufficiently accurate) procedure of this kind will eventually be adopted. Alternatively, some strategies could be more prone to error with some specific displays. In these cases, accuracy concerns may lead the DM to avoid potentially unsound procedures. The effects of the display variables on strategy selection are mediated by the anticipation of the effort required and the accuracy of the different strategies. It has been shown that these estimates (in particular the accuracy anticipation) may be incorrect (Fennema & Kleinmuntz, 1995). The cost-benefit approach predicts that the DM will learn to select the best strategy from experience, using the feedback on the experienced effort and accuracy collected during previous choices. However, despite the evidence for adaptive decision behavior, there are also indications that the DMs are not very sensitive to accuracy feedback (Brehmer, 1980).

The Search for a Dominance Structure Theory

Montgomery (1983, 1989) proposed the Search for a Dominance Structure theory (SDS), postulating a tight linkage between decision and action. A decision is seen in the context of an intention to act, which is defended and supported by the DM. Therefore, it is important that the DM is able to find good reasons for his/her choices. Being able to identify a dominance structure provides a particularly strong justification for choice: if an alternative dominates the other options in the set, this choice will be very easy to defend. Therefore, the DM will be motivated to find a dominance structure and, if this does not exist, he/she will restructure the problem, in order to bring about such a structure out. According to the SDS theory, the dominance structure found by the DM may be close or far from a strict dominance relationship: this distance will depend on the type of problem presented and on the specific type and sequence of restructuring operations applied by the DM.

In the process that leads to the definition of a dominance structure the DM will go through the following stages (Montgomery, 1983): (a) pre-editing, (b) search for a promising alternative, (c) dominance testing, and (d) dominance structuring. Different known decision rules may have a local function in this process, depending on the stage in which they are used. In the pre-editing stage, the DM selects alternatives and attributes to set up the problem representation. The less important attributes and the less valuable options may be discarded. During the search for a promising alternative, the DM identifies an option that could have some chance of dominating the others (or to be better). This promising option is then evaluated through an absolute or relative test. The absolute test may be carried out using a procedure similar to a satisficing heuristic (Simon, 1955). The relative test is based on pairwise comparisons of the attribute values of different alternatives, and it can be considered as a real test of dominance. If the promising alternative receives a good evaluation (i.e. the dominance test succeeds), the DM will choose the current option. If the dominance test fails, the DM will enter the dominance structuring stage. In this stage a decision problem can be restructured in many ways. The de-emphasizing operation is used to reduce the importance of a disadvantage or of a difference between alternatives concerning the promising option. The bolstering operation increases the advantages of the promising option or decreases the advantages of the other options. The cancellation operation rules out a disadvantage through the association with a (logically connected) advantage. Finally, the collapsing operation ties together different attributes in a single dimension (for instance, using a time or money scale). If the dominance structuring process is not able to find a satisfying solution, the DM starts again from the pre-editing stage or searches for a new promising alternative.

Experiment 1

According to Kleinmuntz & Schkade (1993), the comparisons of graphs and tables typically confounded the manipulation of form and organization, contrasting graphic organization with pictorial form and tabular organization with numeric form. Therefore, in this experiment the form of the values is kept constant (numeric) and only their organization is varied. A 2D Cartesian graph and a table were compared, starting from the assumption that a Cartesian representation may facilitate the detection of dominance. Two contextual factors were manipulated: the presence of dominated options in the choice set and the relative importance of the attributes.

Method

Participants and Procedure Sixty undergraduates (30 males and 30 females; mean age = 24) were required to perform a series of multiattribute choices. They were asked to imagine being a travel agent who must select a flight for a series of customers with different preferences. The task was to examine the information provided on the computer screen and to select the best option for each specific customer. The customers' preferences were represented by attribute weights. The decision problems were presented through the PsyScope program. All the choices and the decision times were recorded.

Experimental Design The experimental design was a nested design. Two within-subjects variables were fully crossed: the distribution of weights (uniform, two-level), and the presence of dominated options in the choice set (dominated present, dominated absent). The third variable, the type of display, was varied between-subjects and nested within the distribution of weights (Table 1). Sixty participants were randomly assigned into four groups.

Table 1: Nesting of the display type within the distribution of weights.

	Uniform weights	Two-levels weights
Group 1	neutral graph	compatible graph
Group 2	neutral graph	incompatible graph
Group 3	neutral table	compatible table
Group 4	neutral table	incompatible table

Stimuli The choice problems were of moderate complexity, being composed of four alternatives (available flights) and four attributes (price of the ticket, travel duration, connections, and comfort). The attribute values were presented on a five-point numeric scale, common to all the attributes. The lowest value of the scale represented the best subjective value for the potential customer. All of the attribute values were randomly generated from a normal distribution with a mean of 3 and a standard deviation of 1.

Depending on the experimental condition, some problems contained (or did not contain) dominated options, according to the definition of strict dominance. Also the attribute weights varied across conditions, but they were always presented in percentage form. In the uniform condition, the weights were equal for all the attributes (.25). In the two-level condition, two weights were high (.40) and the other two were low (.10). The association between the weights and the specific attributes in the two-level conditions was randomized. Seven problems for each of the four 'within' cells were generated. They were presented to the participants in 4 blocks, randomizing both the block order and the sequence of problems within each block.

Three types of graph were constructed: 'compatible', 'incompatible' and 'neutral'. The alternative flights are always represented by capital letters (A, B, C, D) placed within the graph and the attribute weights are always presented in a table in the upper part of the graph. The values of two of the attributes are represented by the placing of the letter within the graph, and the names of these attributes are printed on the corresponding axes. The values of the other two attributes are represented within the graph, by attribute-value strings (like TIME=3) which are printed near the letter. In the neutral graphs (with uniform weights) two of the attributes are randomly associated with the axes while the other two are displayed through attribute-value strings. In the compatible graphs the two most important attributes (weight=.40) are represented on the axes and the remaining two by attribute-value strings. In the incompatible graph condition the situation is completely reversed.

Three types of tables were constructed: neutral, compatible and incompatible. In all of the tables the alternatives (capital letters) are represented on the rows and the attributes on the columns. The first row of the table contains the attribute names and the second one the attribute weights, in percentage form. The other rows display the attribute values associated with the different alternatives. In the 'neutral' table with uniform weights there is a random attribute-column association. When the weights are not uniform and the table is compatible, those attributes with the greatest weights are associated with the first two columns of data. On the contrary, if the table is incompatible, the values of the most important attributes are contained in the last two columns. Figure 1 presents the same decision problem through a compatible table and a compatible graph.

Hypotheses

Several hypotheses have been formulated, deriving from models taken from the adaptive decision maker framework (henceforth ADM) and from the SDS theory.

ADM 1. According to the ADM framework, the participants should adopt a variant of the lexicographic strategy (LEX-2) in the conditions with two-level weights and an equal weight strategy (EQW) in the conditions with uniform weights. LEX-2 chooses the option with the best values on the two most important attributes. This would represent an adaptive response to the environment demands and assure a good effort-accuracy trade-off. Simulations carried out in similar environments (Payne et al., 1993) lead to the prediction that accuracy will always be high and the effort (response time) will be greater in the conditions with uniform weights (EQW requires more effort than LEX). The most difficult condition may be the one without dominated options and with uniform weights.

ADM 2. This model is a variant of the previous one, and it is based on the assumption that the most difficult condition is the one without dominated alternatives and with two-level weights. This may be due both to the lack of clearly inferior alternatives and to the need to consider a more complex weighting scheme. Therefore, if these problems are perceived as being particularly difficult, it is possible that the participants will try to adopt a more complex and accurate strategy, like the weighted additive rule (WADD). Due to the use of WADD, both the accuracy and the response time (RT) will be higher in this specific condition.

ATTRIBUTI	COSTO	CLASS	CAMBI	TEMPO
PESO	40%	40%	10%	10%
A	3	1	3	2
B	4	4	3	1
C	3	3	5	3
D	4	3	3	3

ATTRIBUTI	COSTO	CLASS	CAMBI	TEMPO
PESO	40%	40%	10%	10%

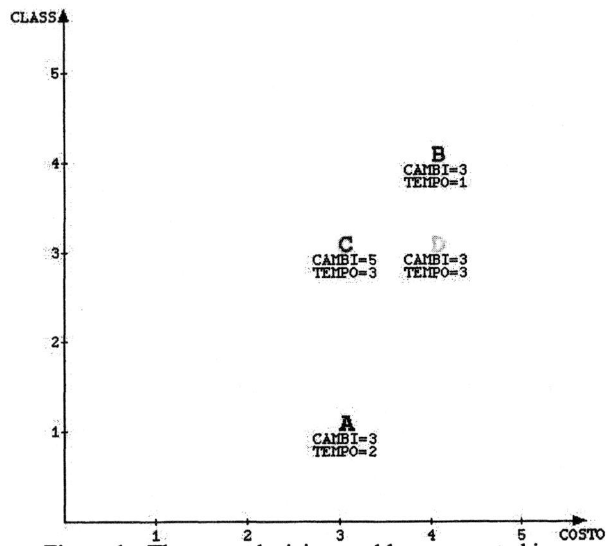

Figure 1: The same decision problem presented in a compatible table and in a compatible graph.

SDS. Starting from the Search for a Dominance Structure theory it is possible to specify different models, which can be used to generate quite different predictions. In the flight choice task it is assumed that the pre-editing stage will be skipped, given that the problems are already presented in a structured form. In the search for a promising alternative, when the weights are not uniform the participants will use a selective, noncompensatory heuristic. When the weights are uniform they will adopt a less selective heuristic. This is in compliance with the weak constraints put forward by SDS on this stage. According to the SDS theory, the dominance test can be of two different types. Therefore, two simple models have been specified (SDM-sat and SDM-dom). Finally, it was hypothesized that only if the dominance test fails will the participants enter the dominance structuring stage. The SDM-sat model predicts only a main effect of the dispersion of weights on the RT (due to the different heuristics used in the search for a promising alternative). The SDS-dom model also predicts a main effect of dominance: if there are dominated alternatives in the choice set, the dominance testing phase will take less time and the restructuring process will be completely avoided. According to the theory, accuracy may be lower if a clear dominance structure is absent, as the use of bolstering or de-emphasizing operations may lead to a sub-optimal choice.

Display-related predictions. The main prediction of a cost-benefit framework regarding the display manipulation is that in the two-level dominance possible environment, the participants using the compatible graph will make faster decisions. This will happen because they will apply a 'perceptual' variant of the lexicographic strat egy on the two most important attributes (represented on the axes). In other words, the participants will be able to take advantage of the display in this specific condition, applying an efficient strategy that requires them to use only simple perceptual operations (to locate the alternative closest to the origin).

Results

No significant difference was found between the conditions with a table display, therefore these data were pooled.

Accuracy A relative loss score was computed from the participants' choices. This score is the difference between the utility of the best alternative in the choice set and the utility of the selected alternative, divided by the difference between the best and the worst alternatives in the set. Therefore, the lower the loss score, the greater the accuracy of the participants' choices. The analyses were conducted on the means of loss scores for the different blocks. No effect of the display type was found. A significant interaction between the dispersion of weights and the dominance structure was highlighted by the ANOVA (Figure 2; $F(1,57)=36.278$, $MSE=.011$, $p<.0001$).

The condition with two-level weights and no dominated options in the choice set was significantly more difficult. Participants were very accurate in all of the other conditions. This pattern of results was not predicted by any model.

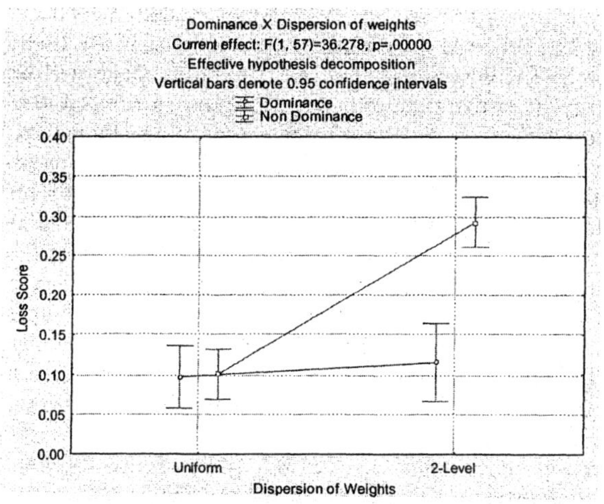

Figure 2: Interaction between dominance structure and dispersion of weights on the loss score.

Response Time A significant interaction between the dispersion of weights and the dominance structure was shown by the ANOVA (Figure 3; $F(1,57)=11.46$, $MSE=10968$, $p<.01$). The condition with uniform weights and the absence of dominated options was significantly slower than the others. Again, n one of the models were able to predict this specific interaction.

Finally, an ANOVA on the conditions with two-level weights showed a main effect of the display type ($F(2,57)=10.545$, $MSE=45025$, $p<.001$). However, the slowest display was the one that was expected to foster the most efficient perceptual strategies. The compatible graph ($M=20125$ ms) was slower than the table ($M=13281$ ms, $p<.001$) and the incompatible graph ($M=14814$ ms, $p<.01$).

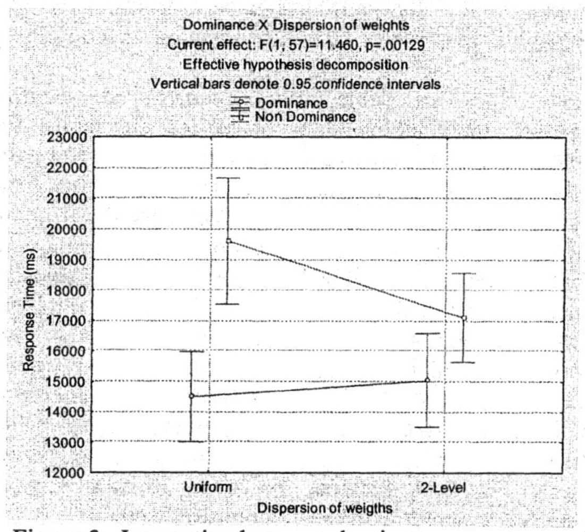

Figure 3: Interaction between dominance structure and dispersion of weights on the response time.

Discussion

There was a strong facilitating effect of dominance, which interacted with the dispersion of importance weights in an unpredicted way.

Indications of both adaptive behavior and potential failures in adaptivity were obtained. In the condition without dominated options and uniform weights the participants were able to make good choices with a significant increase in effort. On the contrary, in the conditions without dominated options and with two-level weights, participants presumably used faster but less accurate strategies.

Finally, the 'adaptive' hypothesis on the di splay type in the conditions with two-level weights is not supported.

Two issues have been opened up by this experiment: (a) what kind of choice processes could have produced the weights by dominance interactions? (b) why a potentially useful display is not able to elicit an adaptive behavior?

Choice Simulation

The interactions obtained in the first e xperiment can be explained by a deeply revised version of the ADM 1 model. The participants may have used the LEX-2 heuristic with two-level weights, but *they may have failed to apply it in an accurate way in the dominance absent environment.* Different explanations for this behavior could be put forward. For instance, while applying LEX-2 in a negatively correlated environment, participants might not be able to properly handle the trade-offs needed to make a choice (Hogarth, 1987), perhaps resorting to LEX. In the environments with uniform weights, ADM 1 assumed that participants will have used the EQW strategy. This prediction seems compatible with the data collected in the dominance absent condition, but it is clearly not supported by the data in the dominance present condition. Therefore, it may be assumed that, in the latter condition, *the participants were able to exploit a very fast 'perceptual' detection of dominance to make their choice*. The revised model is presented in Figure 4.

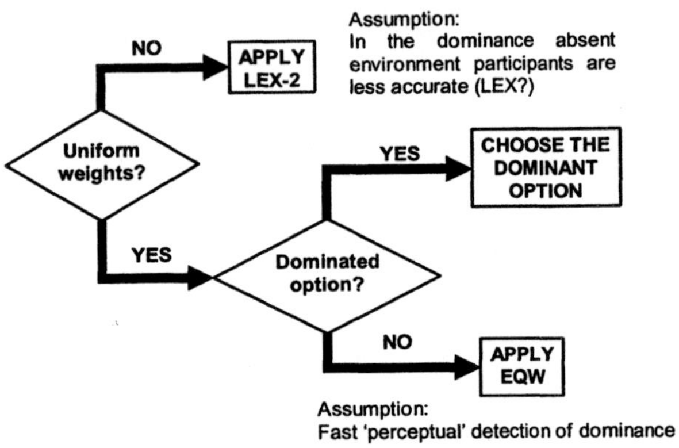

Figure 4: The revised ADM model.

The new assumptions made in the revised model can be tested experimentally or by simulation. The remaining part of this paper focuses on the first assumption; future work may investigate the decision maker's capacity to detect and exploit the dominance relationships. Testing the first assumption requires proving the following: (a) LEX-2 is accurate in the two-level environments; (b) DMs use it in these conditions; (c) DMs are less accurate in applying LEX-2 when dominance is absent. In this paper evidence supporting the theoretical accuracy of LEX-2 and its real use in the two-level dominance present condition will be provided. Further experiments are necessary to show that this heuristic is also used in the dominance absent condition and to shed light on the reasons for its low real accuracy.

A set of LISP procedures simulating the choices of different heuristics (WADD, EQW, LEX, LEX-2, DOM) was written. The WADD procedure always picks out the best option and it is used as a reference for the evaluation of the performance of the other heuristics. The procedures were applied to two groups of 200 problems. Each group of problems was randomly generated from the same normal distribution and scale used for the construction of the stimuli in the first experiment. The first group of problems contained dominated options, which were instead not present in the second group. In the simulation, two attributes were randomly assigned the .40 weights while the other two were assigned the .10 weights. The dominance rule (DOM) was applied only to the problems with dominated options, using both uniform and two-level weights. The loss score and the percentage of problems in which only the best option in the set was selected are presented in Table 2.

Table 2: Choice simulation results. *SE* in parentheses.

	Dominance Present		**Dominance Absent**	
Strategy	Optimal Choice	Loss Score	Optimal Choice	Loss Score
LEX-2	73%	.035 (.006)	71%	.047 (.007)
LEX	35%	.193 (.015)	38%	.249 (.02)
EQW	68%	.078 (.009)	43%	.214 (.018)
Dominance Strategy (DOM)				
Environments	Optimal Choice	Loss Score		
Two-level weights	55%	.151 (.016)		
Uniform weights	57%	.117 (.016)		

As can be seen, the LEX-2 strategy is quite accurate in both the environments. Given that it also requires less effort than EQW and WADD, this heuristic represents a good solution. It should be noted that the LEX-2 loss score in the dominance present condition is slightly lower than the real (experimental) loss score, and this may be due to some minor errors in the application of the strategy. On the contrary, the LEX-2 loss score in the dominance absent condition is much lower than the real one (that is close to the LEX score). To explore the viability of an explanation based on the difficulty in handling trade-offs while using LEX-2 in the two-level environments, the frequency of problems in which the strategy had to face a trade-off between attributes was counted. As expected, the proportion of trade-offs was significantly higher in the dominance absent condition (.80 vs. .71, p=.018 one-tailed). Finally, the performance of the DOM strategy properly agrees with the data collected in the uniform condition.

Experiment 2

A brief summary of a second experiment will now be presented. This experiment was designed to try to explain the counter-intuitive display effect obtained in the first experiment and to provide, at the same time, more direct evidence for the existence of the LEX-2 heuristic.

The basic hypothesis was that in the Cartesian graph conditions participants do not rely uniquely on perceptual operations: they explicitly encode the numeric values from the axes in order to perform further comparisons between alternatives (Benbasat, Dexter, & Todd, 1986). If the most important attributes are on the axes and the participants rely on these, this unnecessary encoding and the following comparisons will increase the RT.

Sixty-four undergraduates (50 males and 14 females; mean age = 23) were randomly assigned to four groups (compatible graph, semi-compatible graph, incompatible graph, and table). To provide the strongest test of the hypothesis, a single decision environment was employed in which the adoption of a perceptual strategy could be maximally advantageous: all of the decision problems had two-level weights and dominated options.

Participants underwent through 3 training trials. Then a block of 5 trials with verbal protocols and another five-trial block without protocols were executed. The block order was counterbalanced within each group. For each participant, each block consisted of problems which were randomly selected from the same set of basic problems, and generated using the same criteria as in the first experiment. Before the block of trials with verbal protocols, participants were given instructions, warm-up exercises (Ericsson & Simon, 1993), and one training trial with verbalization.

Despite slight differences, the accuracy was always very high. There was also a global speed up of response times, probably due to the longer practice peridod in a single environment. The display effect obtained in the previous experiment was substantially replicated: the pattern of the RT means in the trials without protocols was as predicted (compatible graph: 13806 ms; semi-compatible graph: 11866 ms; incompatible graph: 10362 ms; table: 11681 ms) and there was a significant difference between the compatible and the incompatible graph ($t(27)=1.73$, $p<.05$, one-tailed). The analysis of verbal protocols showed that the prevailing acquisition strategy in both the conditions is mainly limited to the two most important attributes (in more than 75% of participants). This pattern is compatible with the LEX-2 strategy. Furthermore, the proportion of statements indicating a numeric encoding of the values on the axes (like "*A has 3 for the price and 2 for the quality*") was significantly higher in the compatible graph group than in the incompatible graph group (.73 vs. .20, $p=.009$). More detailed analyses showed that, in the compatible group, perceptual comparisons (like "*B is the flight closest to the origin*") were prevailing in only five participants. Furthermore, only in the protocols of two participants was it possible to find exclusively perceptual operations, without any other kind of comparison. These indications suggest that participants in the compatible graph condition frequently used numerical encoding of values and non-perceptual comparisons.

Conclusions

The first main result of this study is that participants are not always accurate, even if their overall choice performance is good. In one condition (two-level dominance absent environment) their choices appeared to rely on fast but inaccurate heuristics. This may be due to an accuracy-effort trade-off, to the application of an inaccurate strategy, or to the incorrect or partial application of a valid decision procedure (LEX-2). According to the experiments and to the simulation, the last explanation seems more likely, but only further experiments may provide a conclusive answer.

The results on the display organization suggest that decision makers rely on number-based comparisons even if the information is presented in graphs and only perceptual operations are required. A lack of confidence in the use of Cartesian graphs for choice may explain this effect. If this were to be the case, display design theories should try to combine effort-accuracy principles and theories of attitudes and expertise in situated decision making.

Acknowledgments

This research was funded by Provincia Autonoma di Trento (INDIRECS project). Many thanks to Vera Falcon, (who carried out the data collection) and to Donatella Ferrante.

References

Benbasat, I., Dexter, A. S., & Todd, P. (1986). The Influence of Color and Graphical Information Presentation in a Managerial Decision Simulation. *Human-Computer Interaction*, *2*, 65-92.

Bettman, J., Luce, M. F., & Payne, J. (1998). Constructive consumer choice processes. *Journal of Consumer Research*, *25*, 187-217.

Brehmer, B. (1980). In one word: Not from experience. *Acta psychologica*, *45*, 223-241.

Ericsson K. A., & Simon, H. A. (1993). *Protocol analysis: Verbal reports as data*. Cambridge, Mass: MIT Press.

Fennema, M. G., & Kleinmuntz, D. N. (1995). Anticipations of effort and accuracy in multiattribute choice. *Organizational Behavior and Human Decision Processes*, *63*, 21-32.

Hogarth, R. M. (1987). *Judgement and choice: The psychology of decision*. (2nd edition). Chichester, England: John Wiley & Sons.

Kleinmuntz, D. N., & Schkade, D. A. (1990). Cognitive processes and information displays in computer supported decision making: Implication for research. Unpublished working paper, Sloan School of Management, MIT. Cambridge, MA.

Kleinmuntz, D. N., & Schkade, D. A. (1993). Information displays and decision processes. *Psychological Science*, *4*, 221-227.

Montgomery, H. (1983). Decision rules and the search for a dominance structure: Towards a process model of decision making. In P. C. Humphreys, O. Svenson, & A. Vari (Eds.), *Analyzing and aiding decision processes* (pp. 343-369). North Holland: Amsterdam.

Montgomery, H. (1989). From cognition to action: The search for dominance in decision making. In H. Montgomery & O. Svenson (Eds.), *Process and structure in human decision making* (pp. 23-49). New York: Wiley.

Payne, J. W., Bettman, J. R., & Johnson, E. J. (1988). Adaptive strategy selection in decision making. *Journal of Experimental Psychology: Learning, Memory, and Cognition*, *14*, 534-552.

Payne, J. W., Bettman, J. R., & Johnson, E. J. (1993). *The adaptive decision maker*. New York, NY: Cambridge University Press.

Schkade, D. A., & Kleinmuntz, D. N. (1994). Information displays and choice processes: Differential effects of organization, form, and sequence. *Organizational Behavior and Human Decision Processes*, *57*, 319-337.

Simon, H. A. (1955). A behavioral model of rational choice. *Quarterly Journal of Economics*, *69*, 99-118.

Resolving Ambiguities in the Extraction of Syntactic Categories through Chunking

Daniel Freudenthal (DF@Psychology.Nottingham.Ac.Uk
Julian Pine (JP@Psychology.Nottingham.Ac.Uk
School of Psychology, University of Nottingham
University Park, Nottingham, NG7 2RD, UK

Fernand Gobet (Fernand.Gobet@Brunel.Ac.Uk)
Department of Human Sciences, Brunel University
Uxbridge, Middlesex, UB8 3PH, UK

Abstract

In recent years, several authors have investigated how co-occurrence statistics in natural language can act as a cue that children may use to extract syntactic categories for the language they are learning. While some authors have reported encouraging results, it is difficult to evaluate the quality of the syntactic categories derived. It is argued in this paper that traditional measures of accuracy are inherently flawed. A valid evaluation metric needs to consider the well-formedness of utterances generated through a production end. This paper attempts to evaluate the quality of the categories derived from co-occurrence statistics through the use of MOSAIC, a computational model of syntax acquisition that has already been used to simulate several phenomena in child language. It will be shown that derived syntactic categories which may appear to be of high quality will quickly give rise to errors which are not typical of child speech. A solution to this problem is suggested in the form of a chunking mechanism which serves to differentiate between alternative grammatical functions of identical word forms. Results are evaluated in terms of the error rates in utterances produced by the system as well as the quantitative fit to the phenomenon of subject omission.

Introduction

In recent years, several authors have argued that co-occurrence statistics can provide powerful cues that may aid children in extracting syntactic categories for the language they are learning. Redington, Chater and Finch (1998) analysed large corpora of child directed speech and performed a cluster analysis on vectors describing the lexical context in which words occurred. They found that words that occurred in linguistically similar contexts (tended to be preceded and followed by the same words) had a high likelihood of belonging to the same syntactic class.

Mintz (2003) expanded on the work of Redington et al. Rather than analysing vectors describing lexical context, Mintz's unit of analysis was a frame: two jointly occurring words with one word in between. Mintz restricted his analysis to the 45 most frequent frames that occurred in a large corpus.

While both Redington et al. and Mintz showed that their procedure resulted in apparently good syntactic categories, there is an inherent difficulty with the use of co-occurrence statistics to derive syntactic categories. As Pinker (1987) points out, words that occur in similar contexts may not be of the same category. Pinker argues that a distributional learning mechanism faced with utterances 1a, b and c, would produce an ungrammatical utterance like 1d.

1a. John ate fish
1b. John ate rabbits
1c. John can fish
1d. *John can rabbits

Mintz (2003) claims that 'in children's actual input, these problems do not significantly undermine the informativeness of distributional patterns' (p. 92). He also suggests that 'although problematic environments may exist, there is nonetheless enough "signal" in the distributional patterns compared to the noise created by the problematic environments that categorization from distributional patterns is not intractable' (p. 93).

There is, however, an inherent difficulty with the approach taken by Mintz and Redington et al., which may obscure the extent of the problem identified by Pinker. Mintz and Redington et al. evaluated the quality of the extracted categories using criteria of accuracy and completeness. Accuracy was computed by classifying every word-pair within a category as a hit (same syntactic class), or miss (different syntactic class). Where the grammatical class of a word was unclear, the corpus was consulted to disambiguate and label the word. Mintz used two types of labeling. In standard labeling, all nouns and pronouns were classed as nouns, and all verbs (lexical verbs, auxiliaries and the copula) were classed as verbs. In expanded labelling, nouns and pronouns were labeled as distinct categories, as were lexical verbs, auxiliaries and the copula. While Mintz achieved high levels of accuracy with both types of labelling, closer inspection of his categories reveals that they may not be as accurate as his analyses suggest. One of Mintz's verb categories contains verbs in present tense and past tense as well as progressive particles, verbs that can and cannot be used in an imperative frame, and verbs such as *do* and *have* that can be used both as a main verb and as an auxiliary.

This heterogeneity of the derived word classes may not appear problematic since neither Mintz nor Redington et al. concern themselves with production (Mintz views the process of extracting distributional categories as a precondition for a (relatively unspecified) process of bootstrapping into a parametrized universal grammar). When one considers how the extracted categories might be used in production, however, it quickly becomes apparent that heterogeneous word classes will result in utterances that deviate considerably from child speech. The simplest way in which a child producing speech could use the categories arrived at through a distributional analysis of the input is by considering the members of a category as equivalent. That is, if words *a* and *b* occur in the same category, the child may simply substitute *a* for *b* in a context where it knows *b* has occurred. Taking the words *do*, *have* and *put* (which were classed together in Mintz's analysis) as an example, such a substitution mechanism will result in (clearly incorrect) utterances such as *Do you got an ice-cream* and *Put you want a drink.*

It is argued here that when syntactic categories derived from co-occurrence statistics are used to generate speech, more subtle problems emerge that are not apparent with the use of an evaluation metric based on a researcher's intuition about a word's syntactic class. These problems become especially apparent in detailed quantitative simulations of child data, where seemingly correct substitutions may drastically affect the fit to actual child data. This became clear when Freudenthal, Pine & Gobet (2002a) used MOSAIC, a computational model of syntax acquisition which utilizes co-occurrence statistics to substitute phrases that occurred in similar contexts, to simulate the phenomenon of subject omission and the associated verb phrase length effect (Bloom, 1990). This phenomenon revolves around the fact that there is a stage in development where children produce subjectless utterances such as *Want a cookie*. While the model simulated the general pattern of results, it tended to overestimate the levels of subject omission. One of the reasons for this was that, in order to identify ungrammatical subjectless utterances, the analyses were restricted to utterances containing 'non-imperative verbs'. Since MOSAIC tended to substitute non-imperative verbs for imperative verbs it generated a relatively high number of subjectless utterances. The reason why these verbs were substituted was that both verb types were linked because they both occur in non-imperative frames. While their substitution in imperative frames did result in child-like utterances, the substitution rate was too high to allow a good quantitative fit to the data. This type of problem is not apparent in an approach that simply extracts syntactic categories and does not use a production end to generate utterances.

Thus, the main cause of problematic substitutions is that a substitution that is correct in one context is incorrect in another context. This paper aims to show that one possible solution to this problem is to compute co-occurrence statistics over longer units. Redington et al. considered longer contexts (two or three words preceding and following the target word), and found this did not improve the quality of their syntactic categories. This paper investigates a different approach, inspired by the well-established chunking theory (Chase & Simon, 1973; Gobet et al., 2001).

A new version of MOSAIC has been developed which incorporates a novel chunking mechanism[1] which results in frequent phrases being treated as one unit. One consequence of this is that single words that have been chunked up will no longer be substituted unless when substituted as part of a chunk. It will be shown that this mechanism decreases the amount of unwanted as well as incorrect substitutions, resulting in a decreased overall error rate as well as a better fit to the phenomenon of subject omission.

The remainder of this paper is organized as follows. Firstly, MOSAIC and its chunking mechanism will be described. MOSAIC will be trained on corpora of child directed speech while the parameter governing chunking frequency is manipulated. In order to provide an evaluation of the quality of the output, a sample of generated utterances is judged against criteria of 'well-formedness'. The output is also compared to actual child speech, which is analysed with respect to the phenomenon of subject omission.

Simulating Language Acquisition in MOSAIC

MOSAIC has already been used to simulate several phenomena in child speech. Earlier versions have been used to simulate the Verb-Island phenomenon (Jones, Gobet & Pine, 2000; negation errors (Croker, Pine & Gobet, 2003) the Optional Infinitive phenomenon in Dutch, Spanish and English (Freudenthal, Pine & Gobet, 2002a, 2003, in preparation), as well as phenomena related to subject omission in English Freudenthal, Pine & Gobet 2002b). Whilst the version

[1] Earlier versions of MOSAIC employed a chunking mechanism as well. The novel chunking mechanism differs from this in that chunks can now occur at the primitive level.

used for the simulations discussed here has changed from the earlier simulations, the main theoretical underpinning of the model remains the same. The basic tenet of the model is that the learning of language is a performance-limited process which is heavily weighted towards the most recent elements in the speech stream (i.e., which has an utterance final bias). Several authors have argued that children are better at learning material that occurs towards the end of the utterance (Shady & Gerken, 1999; Wijnen et al., 2001).

MOSAIC learns from orthographically coded input, with whole words being the unit of analysis. The model is a simple discrimination net (an n-ary tree) which is headed by a root node. At the start of learning the discrimination net consists of just the root node. More nodes (encoding words or phrases) are added as the model is shown more utterances. An important requirement for nodes to be added is that whatever follows the word to be encoded in the input must already have been encoded in the model. That is, the model will only learn a new word when it has already encoded the rest of the utterance. Thus, while the model processes utterances from left to right, it builds up its representation of the utterances it receives by starting at the end of the utterance, and slowly working its way to the beginning[2]. The probability of creating a node in MOSAIC is given by the following formula:

$$NCP = \left| \frac{1}{1 + e^{m \,|\, u/c}} \right|^{\sqrt{d}}$$

where: NCP = Node Creation Probability

m = a constant, set to 20 for these simulations.
c = corpus size.
u = total number of utterances seen.
d = distance to the end of the utterance.

The formula results in a basic sigmoid curve (when plotted as a function of the number of utterances the model has seen). The formula contains the size of the corpus and the total number of utterances seen. The size of the corpus is included because the size of the available input corpora differs considerably. The use of the term (m – u/c) ensures that after n presentations of the complete input corpus the Node Creation Probability is identical for corpora of different sizes. The 'distance to the end of the utterance' in the exponent causes material that occurs near the beginning of the utterance to have a lower likelihood of being encoded than material that occurs near the end. This effect decreases as the model sees more input. Since learning in MOSAIC is slow, the input corpus is fed through the model several times, so that output of increasing average length can be generated after consecutive exposures to the input corpus.

[2] Earlier versions of MOSAIC simulated such an utterance final bias by restricting production to utterances that had appeared in sentence final position.

Production of Novel Utterances

Utterance production in MOSAIC involves outputting all the utterances the model has encoded. However, the output that MOSAIC produces consists of more than the input it has seen. MOSAIC has a mechanism for linking words or phrases that have occurred in similar contexts. All nodes being traversed when processsing input are deposited into a buffer of limited size reflecting the most active/recently encountered input. The nodes in the buffer are then compared with respect to their preceding and following context. When the overlap between two nodes is sufficiently high (more than 20% of both the context that preceded and followed the target node are the same), a *generative link* is created between them. The contents of nodes that are linked can be substituted for each other when the model produces output. This mechanism allows MOSAIC to produce utterances that were not present in the input.

Chunking in MOSAIC

MOSAIC employs a chunking mechanism, which results in frequent multi-word phrases being treated as one unit. Nodes in the network contain a frequency slot, the value of which is increased every time that node is traversed when the net sorts an input utterance. The frequency of a node at one of the lower levels (non-primitive nodes[3]) in the tree encodes the frequency of the entire phrase leading up to that node. Thus, if a node for *you* occurs underneath *do*, the frequency of that node encodes the number of times the phrase *do you* has been encountered. When the frequency of a non-primitive node exceeds a pre-determined value, the node is chunked up with the node above it: the two nodes are merged into one node at the primitive level. Thus, in the above example the two nodes encoding the phrase *do you* will be merged into one node at the primitive level. The chunk is then propagated through the network; all occurrences of the phrase *do you* are chunked up. Nodes encoding chunks can be linked to other nodes encoding chunks (or words) in the same way that nodes encoding individual words are linked. When two nodes are chunked, it is no longer possible to substitute words for the individual words making up the chunk. Thus, the chunk *Do I* may be substituted for the

[3] The distinction between nodes directly underneath the root node (primitive nodes) and those at lower levels (non-primitive nodes) is an important one. Due to the structure of the discrimination net, primitive nodes encode all the context the word or phrase has been seen in (the 'global context'). Non-primitive nodes encode 'local context'.

chunk *Do you*, when they share sufficient context. However, should the words *I* and *you* be linked, they can be substituted in unchunked contexts, but not in chunks. In this way, chunking serves to differentiate different grammatical functions of the same word form: if the dummy modal *Do* is chunked with the subject *you*, it will no longer be substituted by verbs that are linked to *Do* by virtue of its occurrence as a main verb.

Chunking affects the substitution of words in two ways. Firstly, chunks themselves are deposited into the buffer, making phrases the target for the creation of generative links. Secondly, the context preceding and following a target node may be chunked up. Thus if the phrase *He goes into the house* contains the chunks *he goes* and *the house*, the context for the word *into* will be *he goes* and *the house* rather than *goes* and *the*. Chunking thus serves to increase the length of items considered for a generative link, as well as increase the context considered in the creation of a generative link.

The chunking mechanism also affects learning. If the model receives a novel input utterance containing a phrase that has been chunked up earlier, it will treat the phrase as a unit, rather than attempting to encode its constituent words separately.

The Simulations

Simulations were run using two corpora of English (maternal) child-directed speech (those of Anne and Becky) taken from the Manchester corpus (Theakston, Lieven, Pine & Rowland, 2001) available through the CHILDES data-base (MacWhinney, 2000). The size of the input sets is approximately 33,000 and 24,000 utterances. Simulations were run using different levels of chunking. The models' output was analysed with respect to error rates and levels of subject omission.

Error rates

For the first simulation, models were trained with and without chunking for both children. For the chunked model, the chunking threshold (frequency required for a node to be chunked up) was set at 1/4 times the square root of the number of nodes in the net. The chunking threshold was expressed relative to the square root of the nodes in the net to ensure that the chunking rate was relatively constant over the development of the model. For all simulations, an output file was selected at an MLU (Mean Length of Utterance) of approximately 3.5

Next, a sample of 500 utterances from each of the output files was coded by two independent raters for the presence or absence of syntactic errors. Syntactic errors were defined as cases in which one or more of the substitutions made by the model resulted in an utterance that was grammatically incorrect (e.g. *Pegs find fallen down* from *Pegs have fallen down*). Note that this definition of syntactic errors is designed to exclude cases in which the model substituted a grammatically correct word into a sentence fragment (e.g. *My toys out* v *Your toys out*) and cases in which the substitutions made by the model were semantically but not syntactically anomalous (e.g. *Shall I cut them with the puzzle?* v *Shall we cut them with the knife?*). The vast majority of the errors identified in this way fell into one of the following categories: word-class errors (e.g. *To vest on his tummy* v *To lie on his tummy*); subject-verb agreement errors (e.g. *They am sitting* v *I am sitting);* missing argument errors (e.g. *Putting the story* v *Reading the book*); errors involving the use of a verb with the wrong particle (e.g. *Shall we use her t-shirt off* v *Shall we take his dungarees off*); errors involving the use of a verb form with the wrong auxiliary (e.g. *I've just finish that off* v *You've just taken that off*) and errors involving the use of a particular type of noun with the wrong determiner (e.g. *Put it on a sand* v *Put it on the sand*). Interestingly, virtually all of the errors falling into these categories seemed to involve either the substitution of a word from the wrong syntactic category for a word that is a member of two or more syntactic categories (e.g. the use of *vest* as a verb instead of *lie* which can be both a noun and a verb) or the substitution of a word from the correct syntactic category into a context in which that particular instance of the category is not permitted to occur (e.g. the use of the indefinite article *a* with a mass noun instead of the definite article *the* which can be used with both mass and count nouns in English). Note that these are precisely the kinds of errors that are likely to be hidden by the kind of evaluation metrics used in previous research using distributional learning mechanisms.

Agreement between the raters was high, at .93 (Kappa = .74). The results are shown in Table 1. Error rates are lower for the chunking models (X^2 = 40.70, p < .001 for Anne, and X^2 = 5.42, p < .05 for Becky).

Table 1: Syntactic error rates for Anne and Becky's simulations at two levels of chunking.

	No chunking	Chunking
Anne	.21	.07
Becky	.24	.18

However, a potentially confounding factor is that the unchunked models simply generate more novel utterances. It could therefore be argued that any mechanism that restricts the generativity of the model will reduce the error rate. In order to test this possibility, the generativity of the unchunked models was reduced by increasing the overlap parameter governing the creation of generative links to .25. This resulted in the proportion of novel utterances being similar to that in the chunked models. The error rate for Anne's model was reduced to .16, less than for the high generativity model X^2 = 4.15, p < .05, but more than for the chunked version X^2 = 19.90, p < .001. For Becky's model, the error rate was .23, not significantly different from the high generativity version X^2 < 1, p < 1, and still higher than the chunked version X^2 = 4.44, p < .05.

Thus, the reduced error rates in the chunking version are not just a result of chunking reducing the proportion of generated utterances, but rather of chunking reducing unwanted substitutions. The reason why error rates remain higher than in the chunked version is that even at a high overlap percentage, some links may remain which give rise to errors. As the overall rate of generativity decreases, these undesirable links may even gain weight, and could conceivably even increase the error rate.

Subject omission

A second analysis assessed whether chunking can decrease the levels of subject omission. For these simulations, the chunking threshold was set to three levels: 4, 1, and .25 times the square root of the number of nodes in the net. These simulations are referred to as low, medium and high chunking, respectively.

In order to match the models' output to child speech, models were trained iteratively to match the MLU of the children at two points in time. The models' output was then compared against children's output with respect to the phenomenon of subject omission.

The analysis of the levels of subject omission was performed in the same way as in Bloom (1990), and Freudenthal, Pine and Gobet (2002b). Utterances were limited to those that Bloom identified as non-imperative (though the verb *see* was excluded from this list as it was not considered non-imperative). The analysis was restricted to declaratives. Double verb constructions and utterances containing the words *don't*, *no*, or *not* were excluded from the analysis.

The remaining utterances were scored with respect to the inclusion of a subject. Figure 1 shows the results for the children and the six simulations at two different MLUs. Model MLUs were matched as closely as possible to the children's MLUs.

Figure 1 shows that the fits for the simulations increase as the chunking rate increases. While the overall fits are not particularly good, the chunking mechanism appears to have been successful in avoiding unwanted substitutions, with the high chunking model providing the best fit at high MLU (particularly for Becky). It may be worth stressing that only one parameter in the chunking mechanism has been manipulated. Future work may suggest manipulations that result in a better fit.

Generative Chunks

While it could be argued that the main effect of the chunking mechanism is to reduce error by avoiding incorrect substitutions, it is worth pointing out that the chunking mechanism leads to different types of substitutions (and errors) as well. The reason for this is that chunks themselves can be linked (both to single words and other chunks). While a full analysis of the role of linked chunks is beyond the scope of this paper, some interesting examples can be given. In some of the high chunking models phrases like *I can* were linked to *Can I*, thus allowing the model to generate declaratives off questions and the other way round. Similarly, in one of the models the phrase *I wouldn't* was linked to *I don't want to*. The chunking mechanism thus aids in linking phrases as well as words that fullfil a similar grammatical role. The chunking mechanism resulted in some interesting errors as well. One of the simulations substituted *don't want to* for *want to*. While this resulted in some grammatical utterances, it also resulted in phrases like *Do you don't want to*. This is clearly a syntactic error. However, it is a type of error that children do occasionally make.

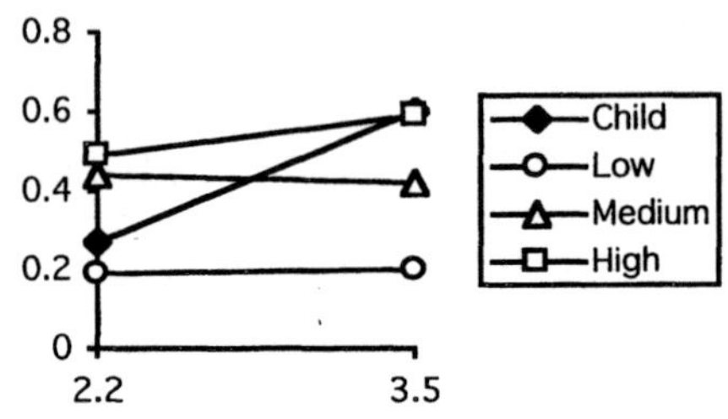

Figure 1a: Levels of subject provision for Anne and simulations

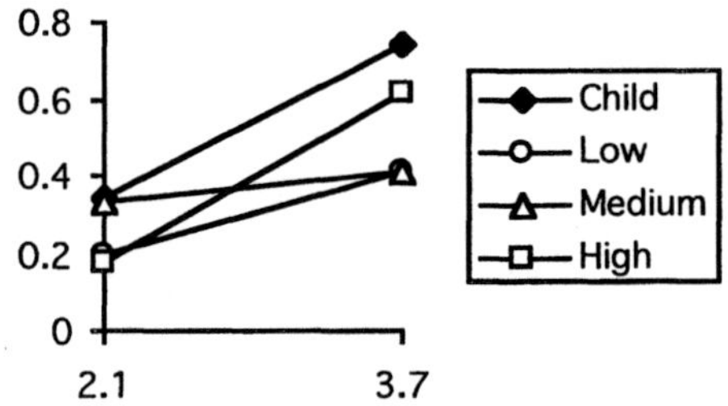

Figure 1b: Levels of subject provision for Becky and simulations.

Conclusions

Several conclusions can be drawn from the simulations reported here. Firstly, in a global analysis of generated utterances, clear word class errors do occur, but not at very high rates. The problem identified by Pinker (1987) therefore does not appear to be particularly significant. However, when the analysis is restricted to a subset of the data (such as utterances containing non-imperative verbs), it becomes apparent that the fact that a simple distributional analysis does not pick up subtle differences between different verb classes can greatly affect the fit to child data. It was shown that the chunking mechanism was able to reduce the overall error rates as well as prevent the substitution of similar words in incorrect contexts.

It should be stressed that chunking does not simply cut generativity in all contexts (as increasing the overlap parameter does). Rather, chunking restricts the contexts in which two words may be substituted. Thus, two single words that share a generative link may be substituted in unchunked contexts, but not in contexts where the word is chunked up (unless of course the chunk itself has a generative link). The chunking mechanism is thus able to cut generativity selectively. Besides diminishing unwanted generativity, chunking also adds to generativity by substituting phrases rather than words.

It is important to bear in mind that the only parameter manipulated in these simulations is the chunking threshold. There is clearly a range of parameters that can be manipulated in conjunction with chunking threshold. At present, the chunking threshold is a function of the square root of the number of nodes in the net. Variations of this formula may affect the chunking rate differentially for different stages of development, thus affecting more detailed fits to child data. We are not committed to the fits and specific implementation used in these simulations, but rather stress the fact that chunking can be a powerful tool in resolving ambiguities in the extraction of syntactic categories.

On a more general level, these analyses illustrate two strengths of MOSAIC as an approach to modelling language acquisition: the use of realistic child-directed speech, and the production of utterances that can be compared with child speech. The use of child-directed speech is important because it ensures a realistic frequency distribution. As all distributional analyses are frequency sensitive, a realistic frequency distribution in the input is crucial for obtaining good fits to detailed phenomena in child language.

The use of a production end has shown that traditional measures of accuracy are insufficient to evaluate the quality of syntactic categories derived from co-occurrence statistics, as accuracy not only depends on the researcher's intuitions regarding a word's syntactic class, but also on the context in which the word is used. Researchers should therefore be careful about relying on measures of accuracy to evaluate the quality of syntactic categories as the addition of a production end may show the accuracy to be considerably lower than it appeared.

Acknowledgments

This research was funded by the ESRC under grant number RES000230211

References

Chase, W.G. & Simon, H.A. (1973). Perception in chess. *Cognitive Psychology*, *4*, 55-81

Croker, S., Pine, J. M., & Gobet, F. (2003). Modelling children's negation errors using probabilistic learning in MOSAIC. In F. Detje, D. Dörner & H. Schaub (Eds.) *Proceedings of the Fifth International Conference on Cognitive Modeling* (pp. 69-74). Bamberg: Universitäts-Verlag.

Freudenthal, D., Pine, J. & Gobet, F. (2002a). Modelling the development of Dutch optional infinitives in MOSAIC. In W. Gray & C. Schunn (Eds.), *Proceedings of the 24th Annual Meeting of the Cognitive Science Society* (pp. 322-327). Mahwah, NJ: LEA.

Freudenthal, D., Pine, J. & Gobet, F. (2002b). Subject omission in children's language: The case for performance limitations in learning. In W. Gray & C. Schunn (Eds.), *Proceedings of the 24th Annual Meeting of the Cognitive Science Society* (pp. 328-333). Mahwah, NJ: LEA.

Freudenthal, D. Pine, J. & Gobet, F. (2003). Modelling syntax acquisition in MOSAIC. Paper presented at the 8th International Cognitive Linguistics Conference, Logroño, Spain.

Freudenthal, D., Pine, J. & Gobet, F. (in preparation). Modelling the development of children's use of optional infinitives in English and Dutch using MOSAIC.

Gobet, F., Lane, P. C. R., Croker, S., Cheng, P. C.-H., Jones, G., Oliver, I., & Pine, J. M. (2001). Chunking mechanisms in human learning. *Trends in Cognitive Sciences, 5*, 236-243.

Jones, G., Gobet, F. & Pine, J. M. (2000). A process model of children's early verb use. *Proceedings of the Twenty Second Annual Meeting of the Cognitive Science Society* (pp. 723-729). Mahwah, NJ: LEA.

MacWhinney, B. (2000). *The CHILDES project: Tools for analyzing talk. Third Edition.* Mahwah, NJ: LEA.

Mintz, T. (2003). Frequent frames as a cue for grammatical categories in child directed speech. *Cognition*, *90*, 91-117.

Pinker, S. (1987). The bootstrapping problem in language acquisition. In B. MacWhinney (Ed.). *Mechanisms of language acquisition*. Hillsdale, NJ: LEA.

Redington, M., Chater, N., & Finch S. (1998). Distributional Information: A powerful cue for acquiring syntactic categories. *Cognitive Science*, *22*, 435-469.

Shady, M. & Gerken, L. (1999). Grammatical and caregiver cues in early sentence comprehension. *Journal of Child Language*, 26, *163-176*.

Theakston, A.L., Lieven, E.V.M., Pine, J.M. & Rowland, C.F. (2001). The role of performance limitations in the acquisition of verb-argument structure: An alternative account. *Journal of Child Language*, *28*, 127-152.

Wijnen, F., Kempen, M. & Gillis, S. (2001). Root infinitives in Dutch early child language. *Journal of Child Language*, *28*, 629-660.

Learning From Real-Time Over-The-Shoulder Instructions in a Dynamic Task

Wai-Tat Fu[1], Daniel Bothell[1], Scott Douglass[1], Craig Haimson[2], Myeong-Ho Sohn[1], John Anderson[1]
[1]Carnegie Mellon University, Pittsburgh, PA 15213, USA
[2]Aptima, Inc, Washington, DC 20005, USA

Abstract

We extended the work of Anderson et al. (in press) by modeling how people learn from "over-the-shoulder" instructions – instructions given immediately after actions were executed – while participants were working on a Anti-Air Warfare Coordinator (AAWC) task. Specifically, we modeled the incremental top-down influence of instructions on the visual search process. We constructed a model that first responded to and converted the over-the-shoulder instructions into declarative memory chunks. These declarative chunks of instructions were strengthened with repeated exposures to these instructions. The model then incrementally learned to improve the selection of next track as the strengths of these declarative chunks of instructions increased, and performance declined as the instructions decayed over time. The model was able to fit the data well, suggesting that the model was able to capture the effects of the instructions on learning and performance in this dynamic task.

Introduction

Anderson et al. (in press) describe a system in the ACT-R cognitive architecture (Anderson & Lebiere, 1998) that takes the initial set of instructions for a dynamic task and converts the set of instructions into a declarative representation, which is interpreted by a generic set of production rules. Through the production compilation mechanism, a set of specific productions that directly perform the task are generated as the task is practiced. We extended this effort by constructing a model that learns from both initial instructions and instructions given by the system while participants are doing the task. These "over-the-shoulder" instructions are given to provide real-time feedback on participants' actions, so that eventually the initial set of instructions can be strengthened, complemented, or overridden. These instructions are given based on a cognitive tutor, a technology that has been used successfully in a variety of educational applications to facilitate learning by doing (e.g. Anderson, Corbett, Koedinger, & Pelletier, 1995).

Learning from Cognitive Tutors

Cognitive tutors were computer-based instructional technology to help students learn from instructions. The idea is that instructions should be given based on a cognitive model of the competence that the student is being asked to learn. In other words, the cognitive model should incorporate the underlying skills that allow the model to perform the task the student was expected to perform. In ACT-R, the skills will be represented as a set of production rules (procedural knowledge) for executing the correct actions under the right conditions, and as a set of declarative memory elements (declarative knowledge) for the appropriate domain knowledge. Based on the model, the tutor is able to monitor actions of the student and infer the intentions of the student. Immediate feedback can then be given to the student to facilitate learning. A major challenge in building the model is to understand how people respond and learn from immediate feedback on their actions from the tutor, and how it influences the existing declarative and procedural knowledge. In this paper, we describe a model of how people learn from immediate feedback from the tutor. The model allows better understanding of the impact of over-the-shoulder instructions on learning and performance.

The Task: The Anti-Air Warfare Coordinator

The task is the same as the one described in Anderson et. al (in press), which was constructed based on the Georgia Tech Aegis Simulation Program (GT-ASP, Hodge, Rothrock, Kirlik, Walker, Fisk, Phipps, & Gay, 1995). GT-ASP is a tactical decision-making computer game that simulates tasks facing an anti-air warfare coordinator (AAWC) on board a US Navy cruisers and destroyers. A participant assumes the role of an AAWC, which includes monitoring a radar screen for unknown aircraft, requesting and collecting information regarding the unknown aircraft, and updating the identity of the aircraft. The task we used is a simplified version of the GT-ASP, and a model-tracing[1] tutor was constructed to monitor participants' actions. We refer to it as the CMU-ASP task henceforth.

The radar screen of the CMU-ASP task (Figure 1) consists of three major areas. First, the radarscope shows various air tracks. Vectors emanating from the aircraft indicate speed and course. The AAWC is on a ship at the center of the radarscope (called ANZIO). The AAWC moves the mouse within the scope and "hooks" a target airplane by clicking the mouse button. This hooking is necessary whenever the AAWC tries to update identity of unknown aircraft. Second, there is a group of information boxes on the left of the screen where the participant can get information on tracks. Third, the menu panel shows the

[1] Owing to the complexities and the dynamic nature of the task, the tutor used a grammar-based representation of procedural knowledge instead of the traditional production-based representation (Douglass, 2004). The grammar representation was able to capture more variations of different courses of actions more succinctly than the traditional production representation. However, the basic tracing methodology was the same.

currently bindings of the function keys (F1 to F12 on the computer keyboard) that are used to issue commands. The AAWC spent the majority of the time in identifying the intent (friendly or hostile) of tracks on the screen and their air type (e.g., strike, or commercial). It is this identification task that the experiment focuses on. We will focus on the first unit task in the identification task – how participants select the next track -- with and without the over-the-shoulder instructions. This unit task involves a selection and a visual search component, both of which are subject to significant improvement with the over-the-shoulder instructions. We will elaborate on this aspect below.

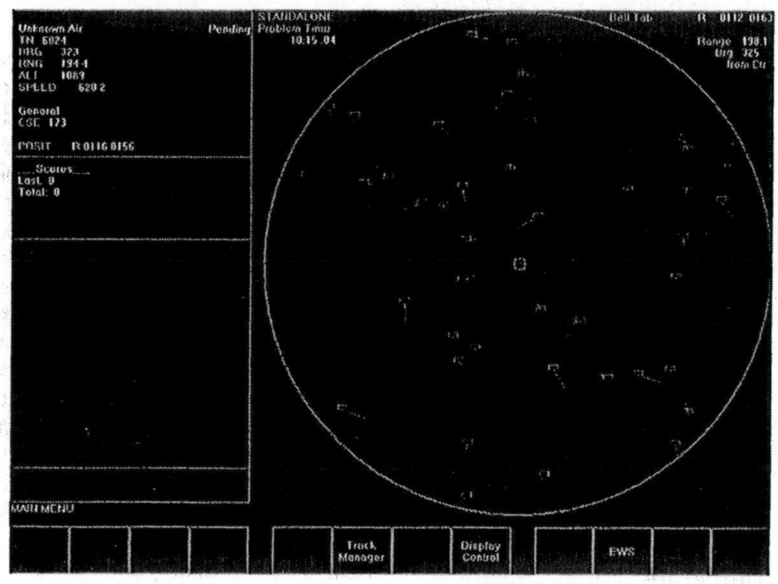

Figure 1. The display in the CMU-ASP task.

Instructions

Before the experiment participants were asked to memorize a set of initial instructions that were sufficient to finish the task (details in Anderson, et. al., in press). Half of the participants were assigned to a group where over-the-shoulder instructions were given (instruction group) and the other half did not receive any over-the-shoulder instructions (no-instruction group). The over-the-shoulder instructions were given right after the participants finished a unit task (e.g. after hooking a track or after identifying a track as a commercial profile). In this paper, we focus on the over-the-shoulder instructions that were given right after participants had hooked a track. Participants were told that their scores depended on the importance of the tracks classified, but the initial set of instructions did not explain how the importance of a track could be evaluated. After a track had been identified, a score would be displayed in the middle information box on the left of the screen as shown in Figure 1. The score was calculated by the following equation (not known to the participants): *Identification score* = (*speed* + 3 * (512 – *range*)) * *at-me*. Hence, the score was high if the *speed* was high, if it was close to the center of the screen (*range* was small), or if it was flying towards the ship. The *at-me* value ranges from 1.0 to 3.0, and it depends on the difference between the course of the track and the bearing of the track measured from the ANZIO. A small difference implies that the track is flying towards the ship, and the *at-me* value will be high. Since the score decreases linearly over time (from 100% to 25% during the 6-min trial) and there is often insufficient time to classify all tracks in 6 minutes, the final scores (calculated as the sum of all time-weighted identification scores) depend critically on whether participants can classify tracks in the order of importance.

At any point in time, the identification scores of all unidentified tracks were calculated and tracks were ranked according to their scores. In the instruction group, after the participant hooked a track, if the hooked track was less important than 20% of all unidentified tracks the tutor would be triggered to give an audio instruction. The most important track on the screen would be highlighted, and instruction would be given to inform the participant why the highlighted track was more important (e.g. it was fast and directly approaching the ship). Participants could then learn from the instructions and improve future performance by selecting the most important unclassified tracks.

ACT-R 5.0

The basic architecture of ACT-R 5.0 consists of a set of modules, each devoted to processing a different kind of information. Coordination in the behavior of these modules is achieved through a central production system. This central production system is not sensitive to most of the activity of these modules but rather can only respond to a limited amount of information that is deposited in the buffers of these modules. The core production system can recognize patterns in these buffers and make changes to these buffers – as for instance, when it makes a request to perform an action in the manual buffer, or a request to move attention in the visual field (e.g. to a track) or to an audio signal (e.g. an over-the-should instruction).

The Experiment

Thirty-two participants were recruited for a two-day experiment. Half the participants were assigned to the instruction group and the other half to the no-instruction group. On the first day they were given the set of initial instructions that taught them how to hook and identify a track, and how to use the F-keys to input the classifications of the tracks. On the first day participants memorized the instructions and were tested on 10 6-minute scenarios. On the second day they were tested on 10 more. Each scenario involved 40 tracks randomly placed on the screen. In the instruction group, over-the-should instructions were given except the first two and the last two scenarios given on each of the two days. In other words, in the instruction group, participants were given 2 no-instruction trials, followed by 6 instruction trials, then 2 more no-instruction trials on each of the two days of experiment.

The Model

The model uses the same set of parameters as described in Anderson et al (in press), and uses the same declarative-to-procedural system to process the initial set of instructions. The current extension processes over-the-shoulder instructions in a similar fashion, i.e. instructions are initially represented as declarative knowledge and are retrieved when needed. The over-the-shoulder instructions are also subject to the same declarative-to-procedure conversion as the initial set of instructions. However, the strength of the over-the-shoulder instructions is assumed to be much weaker than that of the initial set of instructions. To preview our results, we did find that the learning of the over-the-shoulder instructions were slower. Besides, when the tutor was turned off, performance declined quickly, suggesting that the instructions were weakly encoded and not yet proceduralized as the initial set of instructions.

To model the track selection strategies, we asked participants to freely write down the criteria they used to select the next track. The top three criteria were shown in Table 1. It shows that participants were able to learn to attend to the right features in the instruction group. It is interesting that even with no instruction, participants tended to choose tracks that were closer to the ANZIO (range), and apparently some of them learned that at-me and speed were important features.

	Instruction	No-instruction
At-me	15	4
Range	12	10
Speed	15	4

Table 1. The number of participants reported that they used the criteria to select the next track in each group.

We represented each of the above criteria as a "search-factor" and they were created as declarative chunks when the first over-the-shoulder instruction was given. The search-factors provided constraints to the visual search process. For example, a "fast" search-factor commands a search for a track that has a long vector emanating from the track, and a "range" search-factor commands a search for a track that was close to the ANZIO, and so on.

When the task began, since no search-factor was available, the model started to select the track closest to the ANZIO. This assumption was based on the apparent tendency for participants to select tracks closer to the ANZIO in both groups (see Table 1). Another assumption was that participants would tend to minimize the search time required to select a track, so that they could classify as many tracks as possible in each trial. Under this assumption, when no additional information is available, the best strategy is to select the track closest to the track that has just been classified, i.e., a nearest-neighbor strategy. This will minimize both the visual search time and the mouse movement time.

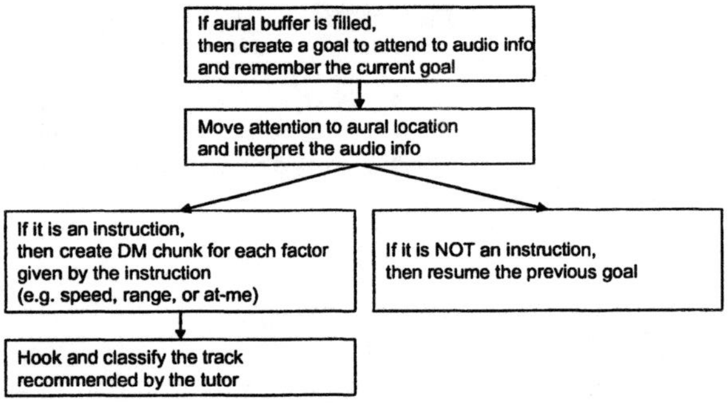

Figure 2. How the model responds to the over-the-shoulder instruction through the aural buffer. (DM = Declarative Memory)

When an over-the-shoulder instruction is given, the instruction will be stored automatically in the buffer of the aural module. A production will fire whenever there is an audio signal in the aural buffer (see Figure 2). This production will also create a new goal of moving attention to this audio signal, and stores the current goal as part of the new goal so that in case the model chooses not to follow the instruction, it can resume the current goal. After attention is moved to the audio signal, the content of the audio signal can be interpreted. If the audio signal is an instruction given by the tutor, the model will interpret the instruction (we choose not to model the interpretation, instead the interpretation is done by a lisp function). The interpretation process generates search-factors that explain why the track highlighted by the tutor is important. There are three search-factors given by the over-the-shoulder instruction: speed, range, and at-me. For example, if the instruction is: "This fast track is coming at the ANZIO and needs attention", then the interpretation process will generate the "fast" and "at-me" search-factors. For each of the search-factors generated, the model will create separate chunks representing them in declarative memory. If there is already an identical chunk in declarative memory when the new chunk is created, the chunks will be merged and the activation of the chunk will be increased and its likelihood of being retrieved in the future increases.

After the model repeatedly processes the over-the-shoulder instructions, the activation of the search-factor chunks will be strengthened and the chunks will eventually be retrieved. These search-factors affect the selection of the next track through repeated cycles of retrieval and visual search (see Figure 3). When the model starts to select a track, it will first try to retrieve a search-factor. If a search-factor, for example, speed, is retrieved, the model will search for a fast track on the radar screen. At the same time, retrieval for the second search-factor is initiated. The visual search and the retrieval compete against each other. If the visual search process finds a track before retrieval finishes, the model will satisfice on the track and stop searching. However, if retrieval finishes first, a new visual search will be initiated based on the two search-factors retrieved. For example, if the first search-factor is speed and the second factor is range, then the model will search for a track that is both fast and close to the ANZIO. At the same time, retrieval for the third search-factor will be initiated. If, again, a track is found based on the two search-factors before the retrieval for the third search-factor finishes, the model will proceed to hook and identify the track. If retrieval finishes first, a new search will be initiated based on the three search-factors. At this point, since all factors have been retrieved, the model will proceed to hook the track when a track is found. The flow diagram in Figure 3 therefore shows how top-down influence of instructions can be incrementally combined with the visual search process.

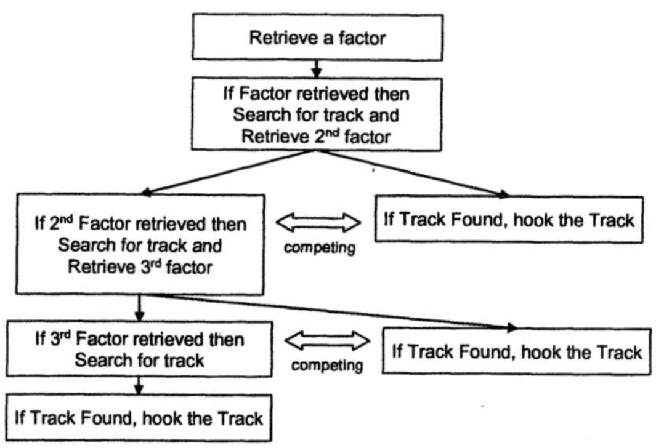

Figure 3. The model selects the next track retrieving factors that guides the visual search.

The competition of the retrieval and visual search process allows the model to incrementally improve the selection of tracks. Perhaps more importantly, it allows performance to decline gradually as over-the-shoulder instructions were dropped (e.g., see Figure 7 below). This performance decline will be difficult to be modeled by, for example, a strategy-competition approach.

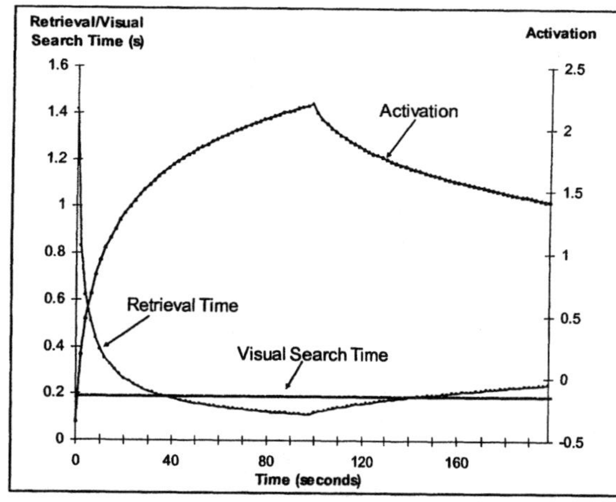

Figure 4. The activation of a chunk encountered every two seconds and its noiseless retrieval time. The visual search time is fixed at 0.185s. The encounter stops at 100s.

In ACT-R, the retrieval time (T) is calculated as T=F*exp(-A) where F is a scaling factor that defaults to 1, and A is the activation of the chunk. The calculation of the activation is more complicated, but for the current purpose, if a chunk is created t_1 before (i.e. a lag of t_1), and a new chunk was merged to the chunk at lags t_2, t_3, ... t_n, then the activation of the chunk will be

$$A = \ln(\sum_{j=1}^{n} t_j^{-d}) + \varepsilon$$

where d is the decay parameter which defaults to 0.5, and ε represents normally distributed noise. Figure 4 shows how the retrieval time (without noise) decreases with time, during which the same chunk is encountered (i.e. in this case, merging of chunks) every 2 seconds. Figure 4 also shows the relationship between activation and retrieval time. After 50 encounters (i.e. after 100 seconds), strengthening of the chunk stops (i.e. the chunk is not encountered anymore). We can see that after this point, the activation of the chunk decays with time (the rate of decay is controlled by the value of d), and the retrieval time increases. The visual search time is fixed as 0.185 seconds[2]. We can see that in this simple case, as the number of previous encounters is more than 20 (i.e. after 40 seconds of regular encounters), the retrieval time is faster than the visual search time. However, when the encounter stops, the retrieval time eventually becomes larger than the visual search time.

Results

Figure 5 shows the final scores obtained after each scenario by the participants and the model in both the instruction and no-instruction groups. The final scores are the total scores

[2] ACT-R assumes that it takes 85ms to encode the features of a visual object. However, it often requires the firing of two productions, one for finding the location of the visual object, the second for initiating the visual encoding. Each production takes 50ms. Therefore the total time for a visual search is 185ms.

obtained for all tracks identified during each 6-min scenario. The model was able to fit the data well (R^2 = 0.92). However, the model did start off with slightly better scores in the first two scenarios than the participants. Apparently some participants might have picked less important tracks than the model in the beginning trials. However, given the good fit between the model and data at later trials, we consider this a good tradeoff by constructing only a single simple strategy (i.e. the nearest-neighbor strategy) to capture performance without making assumptions on the various strategies that might have been used by the participants at different stages of learning.

Participants improved steadily across trials. The final scores dropped slightly at scenario 11, when the participants came back on the second day. In general, the improvements of final scores of the model were similar to those of the participants.

Figure 5 shows that there was no significant difference in total scores between the instruction and no-instruction groups. However, Figure 6 shows that the number of identification made in each scenario was actually higher for the no-instruction group than the instruction group. Apparently the over-the-shoulder instructions had slowed down the participants. This is most obvious in scenario 3, where the over-the-instructions were first given. Combining the results from Figure 5 and Figure 6, participants in the no-instruction group had fewer scores per each track they identified. In other words, participants in the instruction group had identified more important tracks than the no-instruction group. It suggests that although participants (and the model) were slower performing the task, they were able to choose more important tracks and obtained final scores as high as participants in the no-instruction group.

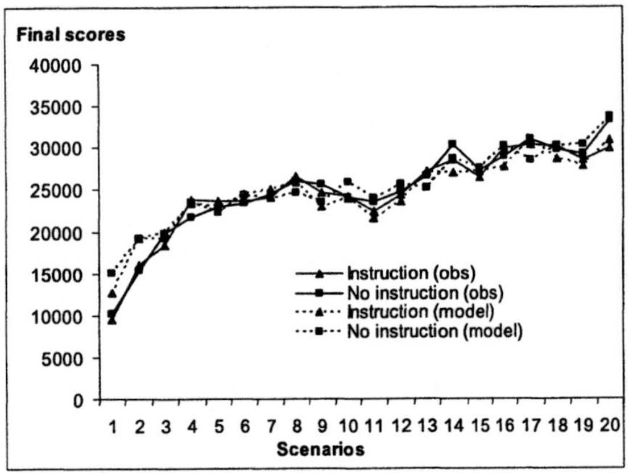

Figure 5. Final scores obtained by the participants and the model after each scenario. Note that the second day starts on scenario 11. (obs = observed data from participants)

Participants sped up steadily across scenarios, except again in scenario 11 when participants came back on the second day. The difference between the instruction and no-instruction group also became smaller with practice. It seemed that participants in the no-instruction group reached asymptotic performance earlier than the instruction group. From the analyses in Anderson et al (in press), during the early stages of learning, time to identify a track depended mostly on cognitive components. At later stage of learning, the time to identify a track depended mostly on perceptual-motor components. The difference between the two groups indicates that the over-the-shoulder instructions may require more cognitive operations to process the instructions. The smaller difference at later stage of learning indicates that the processing of instructions had also sped up. The model was able to capture the patterns of data well in both the instruction and no-instruction groups (R^2=0.96).

To further investigate the effect of the over-the-shoulder instructions, we derived a measure of how well participants were able to identify the tracks in the order of their importance (i.e., identify the most important first and least important track last). In each scenario, we calculated the correlation between the scores of the tracks and the order of their identification. For example, if tracks A, B, and C are identified one after the other and the scores obtained are 900, 800, and 600 respectively, then the correlation between the two arrays (900, 800, 600) and (1, 2, 3) is calculated to be -0.98. However, if the tracks were identified in the order of A, C, and B, then the correlation between the two arrays (900, 600, 800) and (1, 2, 3) will be -0.33[3]. In other words, the more negative the correlation, the better the participant is able to identify the tracks in the order of their importance.

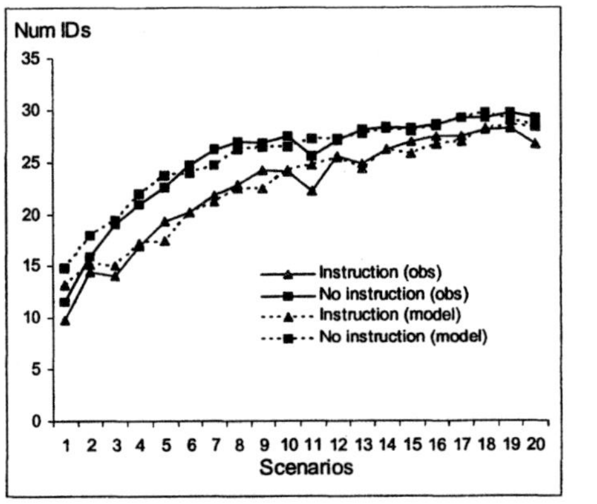

Figure 6. The number of identifications made in each scenario. (IDs = identifications)

Figure 7 shows the correlations across the scenarios for both groups. The correlations for the instruction group were lower than the no-instruction group throughout the 20 scenarios. One may wonder whether the fewer number of identifications may have contributed to the lower correlations (more negative) in the instruction group. However, after checking the data we found that the differences were too big to be caused by the fewer number of identifications (except perhaps in the first two scenarios).

[3] In the actual task, the scores will not be identical when tracks were identified in different orders because scores decrease linearly with time.

In fact, the difference in the number of tracks identified between the two groups was very small (see Figure 6) in the second day. We conclude that participants in the instruction group were better at identifying the most important tracks first, and the least important tracks last.

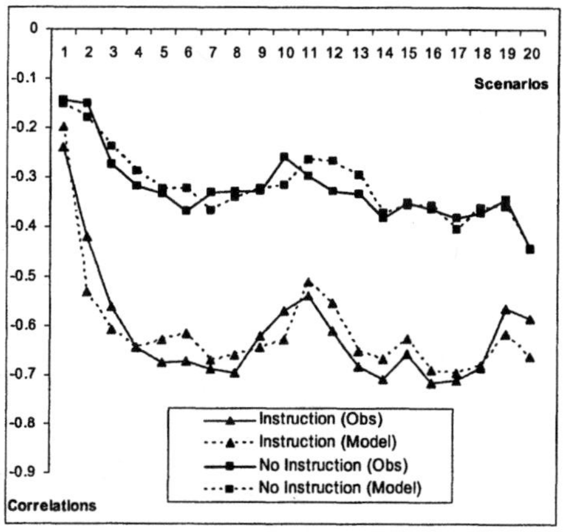

Figure 7. The correlations between scores of the track identified and their order of identification. The more negative the correlation, the better the identification was in a decreasing order of scores.

The effect of instructions can also be assessed from Figure 7 at scenarios 9 to 12, and scenarios 19 and 20 for the instruction group, during which the tutor was turned off. The instruction group shows some reduction in the magnitude of their correlations but still show stronger (more negative) correlations than the uninstructed group.

The model captures the data well ($R^2 = 0.95$). The model exhibited the same increases of correlations when the tutor was turned off as the data did. This shows that the model was able to respond and learn from the over-the-shoulder instructions as the participants did.

The nearest-neighbor strategy also did a good job capturing the small learning curve in the no-instruction group. Since the strategy always started with the track closest to the ANZIO, it captured the general tendency to classify tracks closer to the ANZIO first as reported by the participants. Since tracks close to the ANZIO tended to have a high importance, the general tendency therefore led to the sharp decrease of correlation during early trials as the number of identification increased. However, after tracks close to the ANZIO were identified, the order of identification became more random than systematic, therefore the correlation asymptote at a much less negative value than that of the instruction group.

Conclusions and Discussions

We found that participants were able to learn from the over-the-shoulder instructions and paid attention to more important tracks. To understand how people learned from these instructions, we extended previous work by Anderson et al. (in press) by constructing a mechanism that allows the model to learn from over-the-shoulder instructions. Instructions were represented as declarative memory traces. Repeated instructions strengthened these memory traces to make them more available when needed. A mechanism is constructed that allows visual search to be incrementally improved and declined based on the strengths of these memory traces of these top-down instructions. The ability to search for important tracks improved as the strengths of these memory traces increased. The strengths of these memory traces of instructions decayed when the tutor was turned off, leading to the decline in performance.

The treatment of the over-the-shoulder instructions is basically the same as the initial instructions, except that (a) they take time away from the processing, and (b) we realistically model the slow learning of the over-the-shoulder instructions whereas we assume that the initial instructions are well encoded. Intuitively, one might speculate that over-the-shoulder instructions might be easier to learn as they are presented right in the exact same situation. As least from our data, although the over-the-shoulder instructions apparently had a large impact on performance, they did take time away from the task and their effect seemed to decay quite rapidly. This may suggest that over-the-shoulder instructions may not have the privileged status (as compared to practice) as believed.

Cognitive tutors have been successfully applied in relatively static task domains such as algebra or geometry. The current line of work attempts to transfer the technology of cognitive tutors out from the protected classroom or laboratory to a real-world, complex, and dynamic task. The task requires all forms of learning and all buffers in ACT-R to interact and thus is a good test bed for the architecture. Our goal is to have the model eventually handles more complicated over-the-shoulder instructions, thus allowing automatic evaluation of a real-time instructional system for dynamic tasks.

Acknowledgments

This research is supported by the ONR grant N00014-99-10097.

References

Anderson, J. R., Bothell, D., Byrne M. D., Douglas S., Lebiere, C. & Qin, Y. (in press). An Integrated Theory of the Mind. *Psychological Review.*

Anderson, J. R., Corbett, A. T., Koedinger, K. R., & Pelletier, R. (1995). Cognitive Tutors: Lessons learned. *The Journal of the Learning Sciences,* 4, 167-207.

Anderson, J. R., Lebiere, C. (1998) *Atomic components of thought*. Mahwah, NJ: Erlbaum

Douglass, S. A. (2004). Using the PLASTIC framework to augment scenario-based training systems with instructional agenst. In *Proceedings of the 48th Annual Meeting of the Human Factors and Ergonomics Society.*

Hodge, K., Rothrock, L., Kirlik, A., Walker, N., Fisk, A., Phipps, D., & Gay, P. (1995). *Trainings for tactical decision making under stress: Towards automatization of component skills.* Atlanta, GA: Georgia Institute of Technology.

Memory, Emotion, and Rationality: An ACT-R interpretation for Gambling Task results

Danilo Fum (fum@units.it)
Andrea Stocco (stocco@units.it)
Dipartimento di Psicologia, Università di Trieste
via S. Anastasio 12, I-34134, Trieste, Italy

Abstract

A new interpretation for the results of several experiments carried out using the Gambling Task paradigm is presented that differs from that put forward by Damasio (1994). While Damasio grounds his analysis on the somatic markers hypothesis, we propose a perspective based on the functional integration of emotion with memory that is sympathetic with the view of the prefrontal cortex endorsed by Rolls (1999, 2000). Our interpretation is supported by the development of a computational model implemented in the ACT-R cognitive architecture. The model highlights some limitations of the architecture that arise when dealing with situations charged with emotional significance. In particular, the model—when developed according to the standard ACT-R theory—produces a perseverating behavior that replicates the performance exhibited by participants with lesions in the orbitofrontal cortex. An adjustment to the ACT-R equations describing the activation of elements in declarative memory is suggested that allows modeling the performance of both normal controls and orbitofrontal patients.

Introduction

The idea that emotions could play a critical role in some cognitive processes—like reasoning, planning, and decision making—that were traditionally considered as "cold" has been put forward in the last years by several researchers (LeDoux, 1996; Panksepp, 1998; Rolls, 1999, 2000). The arousal of interest for these topics should be possibly credited to Antonio Damasio who, in his *Descartes' error* (Damasio, 1994), strongly emphasized the link between emotion and rational action, and highlighted the functional role played by the former in determining the latter.

According to Damasio, completely rational decisions (i.e., emotionless decisions based on cost-benefit analysis, and on the maximization of subjective utility) are insufficient when dealing with personal and social affairs. Making a good decision in such domains means in fact to select an option that will prove advantageous for the survival of the organism, and for the quality of such a survival. Making a good decision often means to decide fast, because the environment asks for quick choices. In the best cases, deciding on a purely rational basis will require too much time. In the worst cases, it will lead to bad choices, or to the impossibility of making any choice, because the cognitive resources brought to bear in the decision process do not allow to take into account all the information needed for an optimal conclusion.

While authors who have acknowledged these limitations have spoken of *bounded rationality* (Simon, 1978), *heuristics and biases* (Kahneman, Slovic, & Tversky, 1982) or *adaptive decision making* (Payne, Bettman, & Johnson, 1993), and tried to solve the problem of sub-optimal reasoning within a purely cognitive approach, Damasio takes a different path.

His basic idea is that, in considering the alternatives for a course of action, we base our decision on the representation of similar events experienced in the past. These memories, however, are not neutral entities but are charged with emotional dispositions resulting from the positive or negative outcomes they have been associated with. In contrast with what happens in more abstract domains, when the topic at issue has personal significance, our reaction is not purely intellectual but involves a physical component. According to Damasio, affective memories generate *somatic markers*, bodily sensations activated by situations analogous to the current one.

Somatic marker are thus triggers of (positive and negative) emotions that have been associated, through learning, with the outcomes obtained in previously experienced situations. The markers make the decision process more precise and efficient. For instance, when a negative somatic marker is associated with a particular outcome, it plays the role of an alarm signal. By making the reasoner dump immediately the option associated with the outcome, it allows the discarding of unacceptable choices to promote a rational, in-depth analysis of more promising courses of action.

The acquisition of the somatic markers is mediated, according to Damasio, by the orbitofrontal cortex (henceforth OFC). This area plays, in fact, a critical role "in regulating our abilities to inhibit, evaluate, and act on social and emotional information" (Gazzaniga, Ivry, & Mangun, p. 547). It receives signals from all the sensitive regions of the brain, and it is directly involved in every chemical or motor response. A damage in this area causes an impairment in personal and social decisions in spite of the fact that intellectual capacities are generally preserved. In particular, patients with lesions in the OFC seem oblivious to the consequences of their actions, and guided only by immediate prospects. They are generally unable to learn from their mistakes, and persist in making decisions that lead to negative conseguences. According to Damasio, OFC lesions can damage the emotional processing of affective memories, disrupt the delivery of somatic markers and, consequently, worsen the quality of decision making.

The best empirical corroboration for the somatic markers hypothesis comes from a group of experiments (Bechara, Damasio, Damasio, & Anderson, 1994; Bechara, Damasio, Damasio, & Lee, 1999; Bechara, Damasio, Tranel, & Damasio, 1997; Bechara, Tranel, & Damasio, 2000) carried out with the so-called Gambling Task (henceforth, GT) paradigm. This task simulates some of the features of daily-life decisions, that occur in real time, have personal consequences, are based on uncertain premises, and could end up with rewards or punishments. While normal participants in this task are able to show an adaptive behavior, patients with OFC lesions repeatedly engage in decisions leading to negative consequences. The difference in performance between these two groups, according to Damasio, makes a case for the role of somatic markers in shaping rational behavior.

In the paper we offer an interpretation for the GT findings that avoids the somatic markers hypothesis and that is based on the functional integration of emotion with memory. Our interpretation is supported by the development of a computational model implemented in the ACT-R (Anderson & Lebiere, 1998) cognitive architecture. The following section illustrates in detail the GT paradigm, presents the data to be modeled, and contrasts Damasio's interpretation with other positions. The essential assumptions of our approach are presented in section 3, where the main features of the model are also described. The model highlights some limitations of the ACT-R architecture that arise when dealing with situations charged with emotional significance. Section 4 shows how the model—when developed according to the standard theory—produces a perseverating behavior that replicates the performance exhibited by participants with lesions in the OFC. An adjustment to the ACT-R equations describing the activation of elements in declarative memory is suggested that allows modeling the performance of both normal controls and OFC patients. The final section illustrates the significance of our work for the ACT-R architecture and for neuropsychological data modeling.

The Gambling Task

The paradigm

In a typical experiment with the GT paradigm, participants are put in front of four decks of cards, labeled A, B, C, and D, respectively. Each deck is composed of 40 cards and, because the backs of the cards all look the same, the decks are distinguishable only by their position. Participants are initially given an amount of play money, and they are said their goal is to maximize the money they will hold at the end of the game. Participants are then asked to choose one card at a time from any of the four decks until told to stop. They are free to switch from any deck to another, but they are not informed about how many selections they have to make. Unbeknown to participants, the experiment is stopped after 100 trials.

Once a card is chosen, it is turned over and it allows participants to gain some money whose amount varies with the deck. After turning some cards, however, the participant is both given money *and* asked to pay a penalty whose amount is revealed only after the card is turned, and that also varies with the deck—and with the card position within the deck—according to the fixed schedule reported in Figure 1.

Turning any card from deck A or deck B yields $100 to the participant; turning any card from deck C or deck D yields $50. However, the net yielding of each deck varies because the penalties also vary with the decks. In particular, decks A and B assure high wins but are also associated with higher losses (up to $1250); decks C and D are associated with both small wins and small penalties (for a maximum of $100). The payoff sequence, however, is unknown to the participants, and is arranged in a manner that makes it difficult to learn the long term effects of each deck.

For instance, after turning ten cards from deck A, the participants earn $1000 but they also receive five punishments—ranging from $150 to $350—that will bring the total cost to $1250, resulting in a net loss of $250. Ten cards from deck B will gain $1000, but they will produce a penalty of $1250. Decks A and B are thus equivalent in terms of the overall net loss over trials. The difference is that in deck A the punishment is more frequent but of smaller magnitude, whereas in deck B the punishment is less frequent but of higher magnitude. On the other hand, after turning ten cards from decks C or D, participants earn $500 but the total of their punishments is only $250 (ranging from $25 to $75 in deck C, and resulting from a single $250 loss in deck D, respectively), bringing a net gain of $250. Again, one deck (C) leads to more frequent and lower magnitude punishments while the other (D) produces the opposite pattern.

Taking the overall effects into account, however, it is clear that decks A and B are disadvantageous, and that choosing from them will lead in the long run to financial disaster. On the other hand, decks C and D are advantageous because they eventually result in an overall gain.

The results

Experiments carried out with the GT paradigm produce findings that exhibit essentially the same pattern. As the task progresses, normal participants gradually concentrate their choices on the good decks avoiding the bad ones. Patients with OFC lesions, on the other hand, fail to demonstrate this shift, and persevere in choosing from riskier decks even though these decks lead to negative consequences. Card selection profiles reveal that controls initially sample all decks and make quite a few selections from the bad ones. Eventually, however, they make more and more selections from good decks, with only occasional returns to the alternative options. OFC patients behave like normal controls only in the very first trials while in the long run, even though they make occasional selections from good decks, they stick more frequently and more systematically to bad choices.

Figure 2 presents the number of cards selected from each deck by normal controls and OFC patients in the experiment carried out by Bechara et al. (1994). Statistical analysis revealed that the number of cards selected by normal controls from bad decks was significantly less than the number of cards selected by OFC patients from the same decks. On the contrary, the cards selected from good decks by normals were significantly more numerous than those selected by patients.

Normal controls and OFC patients differed also in the pattern of results shown at the SCR (skin conductance response), an effect mediated by the autonomic nervous system that is used as an index of somatic state activation. On turning over a card, both groups displayed a transient increase in SCR, and hence an autonomic response to reward and penalty. After encountering a few losses (usually by card 10), however, normal participants began to generate anticipatory SCR to bad decks: i.e., their SCR showed a peak immediately before choosing a card from these decks. None of the OFC patients, however, revealed this anticipatory reaction, their only SCR responses being exclusively reactive.

	1	2	3	4	5	6	7	8	9	10	11	12	13	14	15	16	17	18	19	20
A (+100)			-150		-300		-200		-250	-350		-350		-250	-200		-300	-150		
B (+100)									-1250					-1250						
C (+50)			-50		-50		-50		-50	-50		-25	-75				-25	-75		-50
D (+50)										-250										-250

	21	22	23	24	25	26	27	28	29	30	31	32	33	34	35	36	37	38	39	40
A (+100)		-300		-350		-200	-250	-150			-350	-200	-250				-150	-200		
B (+100)	-1250											-1250								
C (+50)				-50	-25	-50			-75	-50				-25	-25		-75		-50	-75
D (+50)									-250						-250					

Figure 1: The payoff schedule used by Bechara et al. (1994).

The interpretations

According to Damasio, the anticipatory SCRs show evidence for a process of nonconscious signaling (the somatic markers) produced by traces of previous experiences, and their associated emotional states. Damages to OFC act by precluding the access to these individual experiences and, consequently, by disabling the possibility for emotions to bias the decision making process. The performance offered by OFC patients, with their repeated engagement in decisions leading to negative consequences, emphasizes thus the role of somatic markers as mediators for the influence of emotions on cognition.

This claim has been questioned by Tomb, Hauser, Deldin, & Caramazza (2002) who observed how GT findings could be explained by simply sticking to the rewards and punishments schedule. Because, on the average, the amount of money that is both win and lost is greater for the disadvantageous than for the advantageous decks, the anticipatory SCRs produced by normal participants could be due to the higher activation aroused by the former. The authors built an experimental situation in which the *advantageous* decks were the ones associated with higher rewards and higher punishments. (Normal) participants in their experiment not only picked more cards but also showed higher anticipatory SCRs for these decks. Although this result is not necessary incompatible with Damasio's hypothesis (that posits the existence of both negative and positive somatic markers) it could be reconciled with it only by assuming that the modified task produced an inversion of the marker signal.

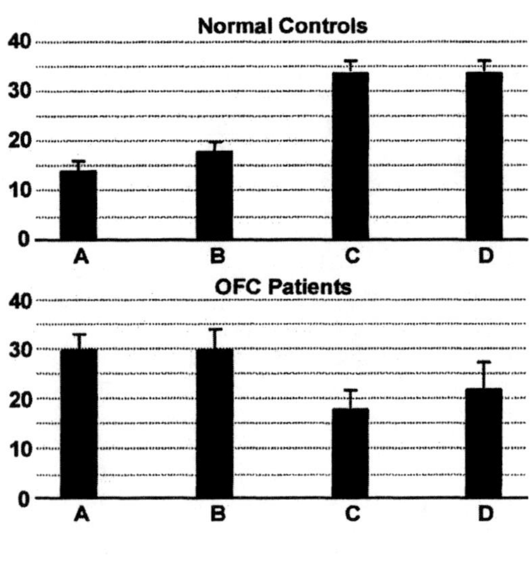

Figure 2: Number of cards chosen from the different decks (after Bechara et al., 1994).

A different perspective has been put forward by Rolls (1999, 2000). According to his point of view, the OFC is critical for a rapid, on-line evaluation of the reinforcement properties of a stimulus or of the outcome associated with a response. The point is that these properties may change as we interact with the environment, and therefore a correction of the inappropriate associations may be required. Rolls (2000, p. 284) noticed that humans with frontal lobe damage can show impairment in a number of tasks in which an alteration of behavior is required as a response to a change in environmental reinforcement strategies. For example, in the Wisconsin Card Sorting Task frontal patients have difficulty in shifting to a second sorting principle when required to do so. In stylus maze tasks it is hard for them to change direction when a sound indicates that the correct path has been left. In the same vein, it is tough for patients with OFC damage to move from disadvantageous to advantageous decks.

In summary, the idea that memories charged with emotional associations could help in guiding our actions is uncontroversial, and equally well accepted is the idea that patients with lesions in the OFC, who generally experience emotional problems, show an impairment in their decision making. The fundamental problem, however, consists in determining whether their defective behavior depends on the incapacity to link a stimulus or an action to their reinforcing properties (Rolls' point of view), or to the incapacity to generate adequate somatic markers (Damasio's position).

An Alternative Approach

The interpretation we suggest for the results of the GT deals with the problem of emotion within the general framework of cognition. It avoids taking into account the concept of somatic markers, and proposes instead a functional integration of emotion with memory.

We start by the rather uncontentious assumption that participants, following the instructions they have received, try to pursue the options that will maximize their gains and minimize their losses. In doing so, they rely on their past experiences to inform their future choices. From this point of view, the GT could be considered as a memory-based task: when participants figure out which card to pick up, they try to remember the typical outcomes associated with each deck, and restrict their choices to the most promising alternatives.

The details of this process are specified in a computational model framed in the ACT-R cognitive architecture. ACT-R (Anderson & Lebiere, 1998) is grounded on the idea that cognition depends on the interaction between two knowledge sources, i.e, declarative knowledge—expressed through *chunks*, frame-like structures composed of labeled slots with their associated filler values—and procedural knowledge—represented by *productions*, rule-like elements that subordinate the execution of their actions to the existence of particular conditions represented through declarative chunks. Productions retrieve and transform declarative knowledge.

Another assumption is that the kind of memory that plays a critical role in the GT is the declarative one. We believe that the performance of normal participants in the GT experiments, who gradually shift their choices from disadvantageous to more advantageous decks, reflects a type of learning that could not be assimilated to the acquisition of more effective procedural strategies, but that is grounded on the memory of previous events, i.e. on remembering the outcomes associated with the various decks.

The reasons why we think it would be unacceptable to model the GT in ACT-R by relying mainly on the procedural knowledge are essentially two. First, as reported by Bechara et al. (1997), all normal participants—and no orbitofrontal patient—are able to reach, approximately at the middle of the experiment, a phase in which they begin to express the "hunch" that decks A and B are riskier than C and D, and most of them are able to explicitly verbalize, before the end, why it is so. The only explicit knowledge in ACT-R is that represented through declarative chunks that can be retrieved and inspected. Procedural knowledge (and the subsymbolic processes that control the retrieval of declarative chunks) are not accessible to consciousness. The fact that participants are able to verbalize the reasons underlying their choices constitutes therefore a case for the declarative nature of the elements their behavior is based upon.

Second, ACT-R assumes that productions are implemented, at the neural level, by basal ganglia and associated connections. The quality of the performance obtained in pure implicit learning experiments, like those employing the serial reaction time paradigm, is correlated with the integrity of these structures. Patients with damages at the basal ganglia, e.g., suffering from Parkinson's disease, fail in these as well as in other tasks identified as "habit learning" (Knowlton, Mangels & Squire, 1996). Apparently, however, these patients behave like normal controls in the GT (Stout, Rodawalt & Siemers, 2001), a fact that supports the idea that the main responsible for their performance is the declarative component.

In ACT-R the key factor determining the accessibility of the declarative knowledge is represented by chunk activation: the more active a chunk, the higher its probability of being retrieved, and the higher the speed of its retrieval. The activation A_i of a generic chunk i is determined by two components:

$$A_i = B_i + \sum_j W_j S_{ji} \quad (1)$$

where B_i constitutes the base component of the activation while the other term constitues the contextual, or associative, component. The base component reflects the general usefulness of a chunk: its value is higher the more times the chunk has been retrieved in the past and it decays as a function of the time passed by the last retrieval. Frequency and recency of use are thus the elements that control the base activation.

The associative component reflects the probability that a given chunk will be used in the current context. It depends on the multiplicative combination of two factors. The first, is represented by a parameter, W_j, expressing attentional weighing, i.e., the variable degree of attention different individuals are able to dedicate to the elements of the current context. The second, S_{ji}, reflects the degree in which the given chunk is related or associated to the elements of the current context (more precisely: to the different attributes, or *slots*, of the chunk representing the goal being currently pursued). The associative strength S_{ji} defines thus an estimate of the degree in which the presence of an item j in a slot of the goal chunk increases the utilization probability of the chunk i.

Banning some fluctuations due to stochastic noise, when an attempt is made to access a chunk by specifying some of its feature, the most active chunk sharing those features is retrieved from declarative memory. The activation of a chunk is determined, as we have seen, essentially by frequency-based constraints that are not affected by the content of the item being retrieved, nor by any other non cognitive factor, such as the emotional dispositions associated with the chunk.

A model of the GT based on the ACT-R theory, in trying to establish the typical outcome of each deck, will retrieve the most active chunk associated with it. Because the most frequently retrieved chunks are those with the highest activation, it is possible to predict that the disadvantageous decks A and B will remind of the numerous and frequently repeated positive-only outcomes, being the punishments relatively rare and not representative of the typical result obtained by choosing a card from those decks. Moreover, because the value of the positive-only outcomes of the disadvantageous decks outweighs that of the advantageous ones, the model, trying to be rational, will concentrate its choices mainly on the former decks, not taking into account the fact that, in the long run, they will lead to an awful performance. In other words, a typical ACT-R model will behave like an OFC patient.

To be able to replicate the performance of normal participants, taking advantage from the emotional effect aroused by receiving a punishment, it is necessary to allow the chunks encoding these relatively rare, but emotionally salient, events to have an activation sufficient to be retrieved, and to be considered as the typical outcomes associated with a given deck.

According to our point of view, the emotional effect of an event should be restricted to the contextual part of the activation. While the base component reflects the statistical structure of the environment, as revealed by the outcomes of past events, the contextual one could be considered as mirroring the biological, or evolutionary, usefulness of a given memory, helping to establish the cases in which a given context could profitably facilitate its retrieval. The emotion contributes to this process by charging declarative memories with biologically significant pleasant or unpleasant feelings, thus influencing their contextual activation, and therefore enhancing their memorability. An ACT-R model having its activation equation modified to take into account these emotional aspects should therefore be able to model the behavior of normal participants in the GT.

An alternative approach to take into account the fact that memory retrieval is influenced not only by the frequency of an event but also by the quality of the outcome associated with it would be to devise a mechanism capable of explicitly evaluating, and updating, the deck values by computing the average outcome associated with each deck. Even if it could be possible to build such a mechanism adhering strictly to the principles of the ACT-R theory, and even if the mechanism would replicate the shift from disadvantageous to advantageous decks shown by normal participants, it would certainly fail in explaining why OFC patients keep on choosing from the disadvantageous decks when the total yielding has become negative.

Implementation and results

Relying on the above reported considerations, we developed two versions of a GT model. To keep a better control of the memory retrieval processes, and of the parameters underlying them, we did not use the customary release of the ACT-R architecture but we reimplemented its required parts.

The first version of the model was based on the standard ACT-R retrieval mechanisms, as described in Anderson & Lebiere (1998). In our implementation we adopted the equation of Altmann & Trafton (2002) which sets the basic activation B_i of a chunk i to:

$$B_i = \log\left(\frac{n_i}{\sqrt{T_i}}\right) \tag{2}$$

where n_i is the number of past retrievals of the chunk and T_i is the time passed since the creation of the chunk in memory.

The association between the chunk representing the outcome of a card (C_i) and the deck it comes from (D_j) was implemented in the link S_{ji}. The value of S_{ji} is computed as the log ratio between the conditioned probability of the outcome, given that the card has been selected from the deck, and the unconditioned probability of the outcome:

$$S_{ji} = \log\left(\frac{P(C_i|D_j)}{P(C_i)}\right) \tag{3}$$

S_{ji} represents thus the increment in the likelihood that a particular outcome associated with a deck will be retrieved as a consequence of trying to retrieve the outcomes of that deck.

The model utilizes the most active outcome associated with a deck as an estimate of the typical outcome that could be obtained from that deck. At each trial the model then selects, following a Boltzmann soft-max rule, the card on the top of the deck with the best estimate. This version of the model was used to simulate the performance of OFC patients.

To replicate the performance of normal participants, we modified the contextual component of the activation equation by introducing a new parameter η taking into account the emotional impact of an event. The basic idea is that different situations could be more or less emotionally charged. For instance, if participants in the GT would gain and loss real money instead of playing with simulated dollar bills, the emotional effect of an outcome on memory would be enhanced, and this fact would be reflected in a higher value of the η parameter. In most cases, however, the emotional impact of an event is related also to its magnitude. For instance, losing \$1250 (of both real and simulated money) would provoke a more negative feeling that losing only \$150. This fact is captured by a a normalized factor V representing a situation-independent evaluation of the magnitude of an emotional effect. Being the representation of numerical magnitudes subjected to distortion effects, we represent the magnitude of an outcome i through its logarithm. To obtain values of V_i comprised between 0 and 1 we normalized the absolute value of each outcome i on the absolute value of the maximum possible outcome:

$$V_i = \frac{\log(|i|)}{\log(|\max(\mathbf{i})|)} \tag{4}$$

The value thus obtained is then multiplied by η, and added to the original value of S_{ji} [1]

In summary, the activation equation (1) in the case of normal participants becomes now:

$$A_i = B_i + \sum_j W_j \bar{S}_{ji} \tag{5}$$

where:

$$\bar{S}_{ji} = S_{ji} + \eta V_i \tag{6}$$

Figure 3 reports the comparison between the results of the simulations with the two versions of the model (the normal version using a value of η = 2), and the data of Bechara et al. (1994).

[1] More precisely, the original likelihood ratio of S_{ji} is multiplied by $e^{\eta V_i}$. Taking the logarithm of this value, the two terms are simply added.

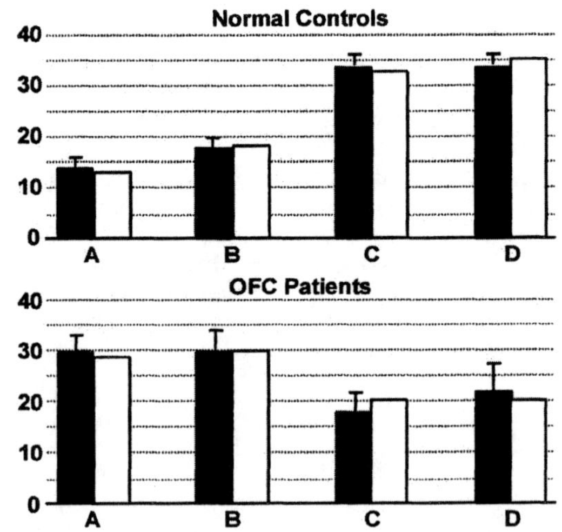

Figure 3: A comparison between participants (black) and model (white) choices.

Conclusions

In the paper we presented an ACT-R model capable of replicating the behavior of normal and OFC patients in the GT. In addition to clarifying the findings obtained with that paradigm, our work presents some implications for the interpretation of the role of the frontal cortex and for the development of the ACT-R cognitive architecture.

At the best of our knowledge, the present model is the first computational account of the GT. Previous investigations, in fact, were focused more on the discussion of the neurological issues than on providing a precise explanation for the experimental results. The only approach comparable to ours is that of Busemeyer & Stout (2002) who, however, pursue purely statistical modeling and are able to fit their models to data but cannot use them to make predictions about the behavior of participants in untested situations. After fixing the relevant parameters, on the other hand, our model is able to anticipate the participants performance for practically every variation of the GT.

The model takes a stance on the role of the OFC, and on the effects of injures in this area. It follows essentially Rolls' (1999, 2000) position according to which OFC patients are unable to exploit the emotional effect associated to outcomes because they are unable to associate an event with the emotion it produces.

A non trivial consequence of this position is that OFC patients should not differ from normal participants in experiencing emotive reactions that follow immediately their actions, while they should be disadvantaged when the task requires further processing that depends on the retrieval of emotional outcomes. An example of this distinction is given by the difference between *disappointment* (i.e, the simple reaction to an unpleasant outcome) and *regret* (that requires a form of counterfactual thinking comparing what was obtained to what would have been obtained by making a different decision). We predict that OFC patients should be impaired with their regret reactions but not with disappointment. Giorgio Coricelli from the University of Siena (personal communication, March 11, 2004) has tested the difference between disappointment and regret in frontal patients, and his data are congruent with our approach.

The present work has some implications for the ACT-R architecture, too. As far as we know, our model constitutes the first attempt to apply ACT-R to the explanation of neuropsychological results, being previous ACT-R neuroscience related work (e.g., Fincham, Carter, van Veen, Stenger, & Anderson 2002) focused exclusively on the interpretation of fMRI data. It should be noted that the behavior of the OFC patients could be considered as rational because, regarding the most frequent results as the typical outcomes of a deck, they stick to them ignoring the less frequent events that are kept in shadow by an inadequate contextual activation. To replicate the performance of normal participants in the GT, we propose a change to the ACT-R retrieval mechanism that takes into account the role of emotions in memory. The proposed modification makes ACT-R more sensible to this important situational factor, and is essentially backword compatible with previous modeling work that has dealt mainly with emotionally neutral tasks.

References

Altman, E. M. & Trafton, J. G. (2002). Memory for goals: An activation-based model. *Cognitive Science, 26*, 39–83.

Anderson, J. R. & Lebiere, C. (1998). *The atomic components of thought.* Mahwah, NJ: Erlbaum.

Bechara, A., Damasio, A. R., Damasio, H., & Anderson. S.W. (1994). Insensitivity to future consequences following damage to human prefrontal cortex. *Cognition, 50*, 7–15.

Bechara, A., Damasio, H., Damasio, A. R., & Lee. G.P. (1999). Different contributions of the human amygdala and ventromedial prefrontal cortex to decision-making *Journal of Neuroscience, 19*, 5473–5481.

Bechara, A., Damasio, H., Tranel, D., & Damasio, A.R. (1997). Deciding advantageously before knowing the advantageous strategy. *Science, 275*, 1293–1295.

Bechara, A., Tranel, D., & Damasio, H. (2000). Characterization of the decision-making deficit of patients with ventromedial prefrontal cortex lesions. *Brain, 123*, 2189–2202.

Busemeyer, J. R. & Stout, J. (2002). A contribution to cognitive decision models to clinical assessment: Decomposing performance on the Bechara Gambling Task. *Psychological Assessment, 14*, 253–262.

Damasio, A. R. (1994). *Descartes' error: Emotion, reason, and the human brain.* New York: Grosset/Putnam.

Fincham, J. M., Carter, C. S., van Veen, V., Stenger, V. A., & Anderson, J. R. (2002). Neural mechanisms of planning: A computational analysis using event-related fMRI. *Proceedings of the National Academy of Sciences, 99*, 3346–3351.

Gazzaniga, M. S., Ivry, R. B., & Mangun, G. R. (2002). *Cognitive neuroscience: The biology of the mind.* (2nd. ed.) New York: Norton & Company.

Kahneman, D., Slovic, P., & Tversky, A. (1982). *Judgment under uncertainty: Heuristics and biases.* New York: Cambridge University Press.

Knowlton, B. J., Mangels, J. A., & Squire L. R. (1996). A neostriatal habit learning system in humans. *Science, 273*, 1399–1402.

LeDoux, J. E. (1996). *The emotional brain.* New York: Simon & Shuster.

Panksepp, J. (1998). *Affective neuroscience.* New York: Oxford University Press.

Payne, J. W., Bettman, J. R., & Johnson, E. J. (1993). *The adaptive decision maker.* New York: Cambridge University Press.

Rolls, E. T. (1999). *The brain and emotion.* Oxford, UK: Oxford, University Press.

Rolls, E. T. (2000). The orbitofrontal cortex and reward. *Cerebral Cortex, 10*, 284–294.

Simon, H. A. (1978). Rationality as a process and as a product of thought. *American Economic Review, 68*, 1–16.

Stout, J. R., Rodawalt, W. C., & Siemers, E. R. (2001). Risky decision making in Huntington's disease. *Journal of the International Neuropsychological Society, 7*, 92–101.

Tomb, I., Hauser, M., Deldin, P., & Caramazza, A. (2002). Do somatic markers mediate decisions on the gambling task? *Nature Neuroscience, 5*, 1103-1104.

Learning to Choose the Most Effective Strategy: Explorations in Expected Value

Wayne D. Gray, Michael J. Schoelles, & Chris R. Sims
Cognitive Science Department
Rensselaer Polytechnic Institute
[grayw, schoem, simsc] @rpi.edu

Abstract

Small variations in how a task is designed can lead humans to tradeoff interaction-intensive for memory-intensive strategies. In this paper we introduce one such task, Blocks World, and present empirical data that shows such tradeoffs. Our attempts to model the acquisition of these tradeoffs using the default ACT-R conflict resolution mechanisms have met with failure. We have now run our model using four different methods for tallying a strategy's success and failure. For each run, we discuss the formulation of expected value used for conflict resolution and the reasons why the model does or does not match the human data.

Introduction

Few, if any, tasks are so new as to require the invention of strategies that have never been used by the task performer. Hence, in most situations, settling on a strategy or set of strategies for performing a task is not so much a matter of learning new strategies as it is learning which strategy, out of a set of already acquired strategies, works best in the current task environment.

This paper is motivated by our attempts to model strategy selection in Blocks World using ACT-R. First we introduce Blocks World and the empirical phenomena we seek to model. Second, in ACT-R, the *expected value* equation (Anderson, Bothell, Byrne, & Lebiere, in press; Anderson & Lebiere, 1998) determines which of two or more alternative strategies will be selected. We present data from a model that uses the default expected value equation and discuss why we found this mechanism inadequate for modeling Blocks World. Third, we layout and discuss three alternative bases for calculating expected value and present data from the original model run with each of these alternatives. We discuss how each alternative influenced model behavior as well as its fit or misfit to the empirical data. Fourth and finally, we summarize our work and draw conclusions regarding both the Blocks World task specifically, as well as our three alternative bases for calculating expected value.

Blocks World

Blocks World is a simple task that has been used to study the trade-off between interaction-intensive and memory-intensive strategies (Ballard, Hayhoe, & Pelz, 1995; Fu & Gray, 2000; Gray & Fu, 2000). The task is to copy a pattern of colored blocks shown in the *Target window* to the W*orkspace* window, using the colored blocks in the R*esource* window (for our version see Figure 1).

The Blocks World Studies

Each trial begins with a random placement of 8 colored blocks into empty spaces (defined by an invisible 4 x 4 grid) in the Target window. Unlike Figure 1, during the study all three windows are covered by gray windows. In our studies the gray windows that cover the Resource window and the Workspace window vanish as soon as the cursor enters those windows. The between-Ss manipulation varies how effortful it is to uncover the Target window. Across three studies (a brief description of the first is available as Fu & Gray, 2000, the other two are not published) we have varied difficulty "intuitively", by varying the Fitts' Law Index of Difficulty, and by lockout time.

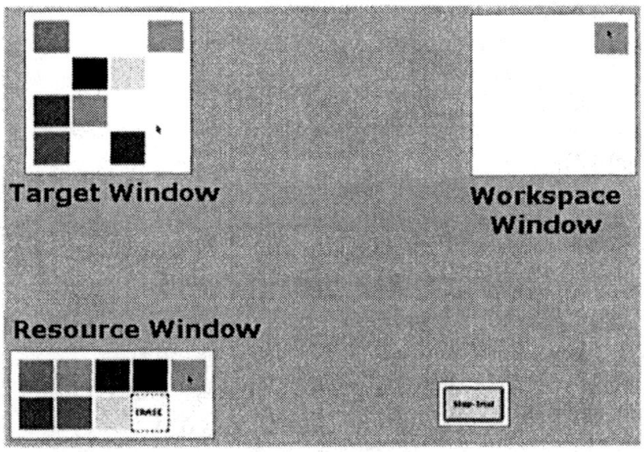

Figure 1. The blocks world task at the start of a new trial. In the actual task all windows are covered by gray boxes and at any time only one window can be uncovered. (The labels do not appear in the actual task. The Start/Stop button is shown at the lower right.)

Subjects were asked to do 40 (E1) or 48 (E2-3) trials. Each trial continued until they had correctly duplicated in the Workspace window the pattern (color and location) of blocks shown in the Target window.

The human and model data reported here are based on the version of Blocks World used in our third study. In that study the costs of opening the Target window were varied by increasing the lockout time (i.e., the delay in uncovering the Target window after the cursor had been moved into it). The three conditions reported have a 0 (0-Lock), 400 (400-Lock), and 3,200 (3200-Lock) millisecond lockout time.

Strategies

To access the information in the Target window subjects could adopt either an interaction-intensive or a memory-intensive strategy. An extreme interaction-intensive strategy would entail uncovering the Target window to obtain color information for a single block, obtaining that block from the

Resource window, another uncovering of the Target window to obtain the block's position information, followed by placing the block in the Workspace window. In contrast, an extreme memory-intensive strategy would entail one look at the Target window to encode both color and position for all 8 blocks.

We did not expect to find either extreme strategy to be popular with our subjects. However, as the cost of accessing information in the Target window increased, we expected to find that subjects shifted from more interaction-intensive strategies to more memory-intensive ones.

The Blocks World Results

For the current report, our measure of performance is the number of blocks correctly placed after the first, but before the second, uncovering of the Target window. At the time of the first uncovering, each of the lockout conditions has 8 blocks that have to be placed. Our empirical data shows that the number of blocks placed on the first uncovering varies significantly between conditions. Hence, on subsequent uncoverings, the number of remaining to-be-placed blocks differs between conditions.

Likewise, as it takes some time for the models and humans to settle on stable strategies, we only report data for trials 25-48. The process of "settling in" is interesting but beyond the scope of this short report.

There were 18 subjects in each of the three conditions. For these subjects, Figure 2 shows that as lockout time increases, the number of blocks placed in the Workspace window increases. Human subjects are clearly trading off interaction-intensive for more memory-intensive strategies.

Failure to Pick a Good Strategy: Issues in Credit Assignment and Expected Value

In data reported in Fu and Gray (2000), we showed that as the costs of opening the Target window increased subjects spent more time with the window open before going off to place the blocks. As the number of blocks placed also increased, the obvious inference is that the increased time spent with the Target window open, reflects increased time spent encoding a larger number of blocks.

To capture human adaptation to the cost of opening the Target window, we implemented a set of 8 *DO-strategies*. These strategies, DO-1 through DO-8, varied in the number of blocks they encoded per opening of the Target window. After each round of encoding, the model would go to the Resource window and attempt to retrieve the memory (declarative memory element or DME) of an encoded, but not-yet-placed block. If a DME was retrieved, a block of that color was picked up from the Resource window and placed in the Workspace window. After placing a block in the Workspace window the model tried to retrieve another DME of another encoded, but not-yet-placed block. When no more DMEs of not-yet-placed blocks could be retrieved, the model picked a new DO-strategy according to its expected value and another round began. A trial ended with all 8 blocks correctly placed in the Workspace window.

Model Details

The above description generally characterizes our modeling approach. This section provides further details on the construction and operation of our model.

Limits on DO-strategies. On reflection, it will be clear that all DO-strategies could encode their full range of blocks on the first uncovering of every trial, but not thereafter. For example, if on the first round of encoding, DO-4 fired, encoded 4 blocks, and placed 3, on the next round only 5 to-be-placed blocks would remain. Hence, on round 2, DO-5, DO-6, DO-7, and DO-8 would all encode 5 blocks. At best this would blur the distinction between DO-strategies. At worst, it seems cognitively implausible that, for example, people would fire a strategy to encode 8 blocks when only 1 block remained to be placed. To avoid this problem we wrote our model so that a DO-strategy would compete only if the number of to-be-placed blocks was greater than or equal to the strategy's *DO-number* (i.e., if 4 blocks remained, only DO-4, DO-3, DO-2, and DO-1 would be in the conflict set).

Calculating Expect Value. The DO-strategies compete with each other based on their expected value. ACT-R's expected value equation is:

$$EV = PG - C +/- noise \qquad \text{Equation 1}$$

P reflects the probability that a production has been successful in the past. P is simply calculated as the ratio:

$$P = \frac{successes}{(successes + failures)} \qquad \text{Equation 2}$$

G is a constant expressed in units of time. G is loosely thought of as the number of seconds that a person would be willing to pursue a given goal. The default value of G is 20.

C is a ratio of the sum of all past efforts attributed to the production divided by all past uses:

$$C = \frac{efforts}{(successes + failures)} \qquad \text{Equation 3}$$

Finally, noise adds variability to the expected value, but rather than constituting unexplained variability it seems to be an essential element. Too little noise leads the system to prematurely settle on strategies that gain an early advantage in P and C. Too much noise prevents the model from settling on any strategy, regardless of the values of P and C.

Credit Assignment. The credit assignment issue is "when" – when are the parameters in the expected value equation updated? These quantities could be updated for all productions once per trial; that is, after all 8 blocks are placed. However, as our model interacts with the same software as our humans interact with, many hundreds of productions fire on each trial. Indeed, we counted 762 productions firings on a randomly sampled trial that took 128 seconds of ACT-R time to complete. (This count includes many refirings by some productions.) As all trials ended successfully, each production fired on a trial would have the value of its successes updated by one. (If it fired multiple times, it would receive multiple updates.) Each production fired on a trial would have its efforts incremented by the difference in ACT-R time between when it was selected and the end of the trial. (If it fired multiple times, its efforts would be updated for each firing by the difference between firing time and trial end time.)

Perhaps more to the point, placing 8 blocks entails a number of different DO-strategies firing a number of different times. This is the problem of structural credit assignment. Given a number of competing strategies, to what extent should each be credited with contributing to the final success of the trial? Updating all productions at the end of each trial would make it extremely difficult for credit assignment to properly credit the success, failure, and cost of any given DO-strategy. (Note that due to forgetting, the higher DO-strategies seldom, if ever, placed their complete allotment of encoded blocks.)

Rather than updating credit assignment once per trial, we updated it once per firing of a DO-strategy. Credit assignment time began ticking when a DO-strategy was selected. Time ended when the model could no longer retrieve the DME of a not-yet-placed block.

Model Runs. One model was run with four different schemes for updating successes and failures. For convenience, we refer to the model when it is running a particular updating scheme as, e.g., "the Success-Weighted model." However, each of these "models" used the same production rules, the same DMEs, and the same settings for all ACT-R parameters. The only change between models is in the updating of successes and failures that are discussed in the next section.

With two exceptions, all ACT-R parameters were left at their defaults. Specific parameters[1] important to our model include *enable subsymbolic computations* (:esc t); enable randomness (:er t); optimize learning (:ol t); parameter learning (:pl t); and base level learning (:bll 0.5). Although we make special mention of these parameters, this set is required by any model in which expected value and declarative memory activation is learned. They are all set to their default values. Our two exceptions do not have definite default values. We set activation noise (ans) to 0.23 and expected gain noise (egs) to 0.3. Activation noise is the noise added and subtracted to the activation of a DME on each retrieval attempt. The value we picked is within the normal range of this parameter and is one that we have used in other studies. Expected gain noise is the noise added and subtracted to the expected value of a production each time it appears in a conflict set (see Equation 1). The value we picked is within the range that we typically use in models (0.25 to 0.50). The setting of both egs and ans were done *a priori* – neither were tuned to the particular results of our models.

Problems in Updating Successes and Costs

The default scheme for calculating expected values based on successes, failures, and costs does not reproduce the data. The number of blocks placed by the model that used ACT-R's default scheme for updating successes and failures (the *Vanilla model*) is plotted at the bottom of Figure 2. It can be seen that this model differs quantitatively (RMSE = 2.26) as well as qualitatively from the human data. By the 25th trial, only one block is being placed in each lockout condition. As this is preliminary work, we are more concerned with the qualitative mismatch in trends across lockout conditions than with the quantitative mismatch in absolute number of blocks placed. Why does the number of initial blocks placed not increase for the higher lockout conditions?

Unfortunately, the answer to our question is as obvious as it is basic to the ACT-R calculation of expected value. As each round ends with a success, the value of P stays at 1.0. The expected value is driven entirely by the costs. As we go from DO-1 to DO-8 the model spends more time encoding blocks, more time getting blocks from the Resource window, and more time placing blocks in the Workspace window. For the higher DO-strategies the costs soar and the expected value plummets (see Table 1).

Table 1: Expected values after 48 trials for DO-strategies 1-8 by lockout condition in the Vanilla ACT-R model

Lockout	1	2	3	4	5	6	7	8
0	8.4	4.8	3.8	1.9	0.4	-1.0	-6.0	-4.7
400	8.1	5.1	4.0	3.7	-1.0	-1.5	-6.3	-6.7
3200	5.3	2.5	0.6	1.4	-1.0	-2.4	-9.8	-9.5

We believe that the failure to capture the qualitative trends reflects a fundamental flaw with the default ACT-R mechanism for credit assignment in tasks involving a sequence of steps over a time period of secs to 10s of secs. We elaborate this argument in the next section.

Weighting Successes and Failures

As success in Blocks World is defined as correctly placing 8 blocks in the Workspace window, the Blocks World paradigm allows us to define partial success in terms of the number of blocks placed. Hence, if a DO-strategy places one block into the Workspace window it is less successful than a DO-strategy that places four blocks.

Table 2: Three weighting schemes for changing ACT-R's system for calculating expected value.

	P		C	
Weighting	Success	Failure	Success	Failure
Success-Weighted	Yes	No	Yes	No
All-Weighted	Yes	Yes	Yes	Yes
Mixed-Weighted	Yes	Yes	Yes	No

Prior models of ACT-R have apparently not had to deal with such nuances. In thinking about how to overcome this limit to ACT-R we generated a number of schemes that weight the updating of the successes and failures parameters by the number of blocks placed. Three of these schemes are discussed below and shown in Table 2.

Implementing these schemes required adding a hook to ACT-R's *parameters learning* function to bypass the normal updating of the successes and failures parameters with the updates required by each scheme. The hook

[1] The only detailed discussion of ACT-R parameters that we know of is in Anderson and Lebiere (1998). Updated documentation, ACT-R5parameters.doc, can be retrieved from http://act-r.psy.cmu.edu/tutorials/.

function is called by the parameters learning function for each production in the sequence.

Success-Weighted

The most basic change is to vary the count of successes to reflect the number of blocks correctly placed. This update is shown in the first row of Table 2.

As the current model almost always successfully places at least one block, the Success-Weighted update is equivalent to dropping "failures" from the calculation of P (compare Equation 2 and 4) and C (compare Equation 3 and 5). This change has the effect of setting P to one.

$$P = \left(\frac{successes}{successes}\right) \quad \text{Equation 4}$$

$$C = \frac{efforts}{successes} \quad \text{Equation 5}$$

The effect of our change is to increase the denominator of C. Rather than adding one to successes each time a DO-strategy has fired, our change adds in the number of blocks that have been successfully placed. (For example, if DO-3 places 3 blocks successes will be incremented by 3. Hence, the cost in terms of the additional time required to encode and place multiple blocks is amortized over the number of blocks actually placed.

All-Weighted

An alternative update would be to weight both the number of successes and the number of failures that an update returns. This alternative is shown in the 2nd row of Table 2.

For *All-Weighted*, the equations for P and C are the same as the default equations shown in Equations 2 and 3. However, All-Weighted differs from the default in two ways. First, both successes and failures can be updated on a given round. Second, the number of successes and failures is weighted by the amount of the goal accomplished or attempted. As per the Success-Weighted scheme, All-Weighted increases the denominator of C and P by the number of blocks correctly placed (columns 2 and 4 of Table 2). However, unlike Success-Weighted, failures are also credited (columns 3 and 5). Failures are defined as the difference between the number of blocks encoded versus the number of blocks placed (DO-number minus number-placed). If, for example, DO-8 fires and encodes 8 blocks but places only 3, then DO-8 will be credited with 3 successes and 5 failures.

Unlike Success-Weighted, All-Weighted affects the value of P by differentially changing both the numerator and denominator (as per Equation 2). A DO-strategy that encodes more blocks than it can retrieve from memory will be severely punished by a decrease in P (the denominator increases faster than the numerator).

On the other hand, regardless of the number of successes and failures, for a given DO-strategy, All-Weighted equally increments the denominator of C (see Equation 3). For example, if DO-6 fires, encodes 6 and places 6, 6 successes will be added to the denominator for C. If the next time DO-6 fires it encodes 6 but places 3, the denominator will again be incremented by 6 (3 successes + 3 failures).

It would be one thing if All-Weighted were neutral with respect to the effect of success and failure on C; however, it seems to reward failure. If 6 blocks are encoded and only one is placed, then the time (and therefore effort) between initiating the strategy and finishing the strategy is less than if 6 blocks were placed, but the effect on the denominator is the same. Counterintuitively, for the same DO-strategy, costs are reduced more by an early failure than by an eventual complete success.

Mixed-Weighted

Mixed-Weighted is an alternative to All-Weighted that simply drops the count of failures from the denominator of costs. The expected value equation for Mixed-Weighted borrows its calculation of P from Equation 2 and its calculation of C from Equation 5. As per All-Weighted, if a DO-strategy promises more than it can deliver, then it is punished by a reduction in P. As per Success-Weighted, costs are reduced in proportion to the amount of the goal accomplished. Credit is not given for promises, only for results.

Model Data: Comparing Weighting Schemes[2]

The four models differ only in the scheme they use for counting successes and failures. In all other respects, in terms of productions, DMEs, and all other parameters, the models are identical.

Success-Weighted Model. As Figure 2 shows, unlike Vanilla ACT-R, the Success-Weighted model is not absolutely flat and consistently overshoots human performance. However, it is fair to say that Success-Weighted is a poor fit both qualitatively and quantitatively to the human data (RMSE = 1.42).

Table 3: Expected values after 48 trials for DO-strategies 1-8 by lockout condition in the Success-Weighted model

Lockout	1	2	3	4	5	6	7	8
0	7.1	8.8	10.0	10.2	10.3	10.4	10.2	9.9
400	6.3	8.1	9.4	10.2	10.4	10.4	10.1	9.8
3200	3.6	6.1	7.7	9.0	9.4	9.5	9.4	9.1

Across all three lockout conditions (see Table 3) the expected values of the smallest DO-strategies, DO-1 and DO-2, is much below that of the other DO-strategies. Post-hoc comparisons show that the comparison of DO-1 and DO-2 versus DOs3-8 was significant [$F(1, 105) = 1816$, $p < .0001$] and accounted for 83% of the variance due to DO-Strategy. DO-3 is close to the higher DOs for 0-Lock, it begins diverging slightly for 400-Lock, and by 3200-Lock it is still close, but 1.3 units of expected value away from the next highest expected value. Hence, the three lockout conditions are relying on essentially the same pool of DO-strategies with the slight increase in number placed for 400-Lock and 3200-Lock due to the less frequent use of DO-3 in favor of a slightly increased use of the higher DO-strategies.

All-Weighted Model. Compared to the Vanilla and Success-Weighted models, the All-Weighted model is a much better fit. As shown by Figure 2, this is the first model

[2] For each of the three models reported here, the model was run six times for each of the three lockout conditions.

that comes close to capturing the qualitative and quantitative (RMSE = 0.60) trends in the human data.

Table 4: Expected values after 48 trials for DO-strategies 1-8 by lockout condition in the All-Weighted model

Lockout	1	2	3	4	5	6	7	8
0	6.4	8.1	7.8	9.0	7.9	7.5	7.0	6.8
400	5.5	7.8	8.9	8.5	7.4	7.1	7.3	6.3
3200	2.6	4.9	6.8	8.1	8.0	7.6	6.6	6.7

Across the DO-strategies the difference between the maximum and minimum expected value varied from 2.6 for 0-Lock, to 3.4 for 400-Lock, and 5.5 for 3200-Lock (see Table 4). Post-hoc comparisons showed that DO-1 and DO-2 had a much lower expected value for 3200-Lock than did the other DO-strategies [$F(1, 35) = 67.7$, $p < .0001$] with this comparison accounting for 79% of the variance due to DO-strategy. This same comparison accounted for 15% of the variance for 400-Lock and 4% of the variance for 0-Lock. Hence, in contrast to the Success-Weighted model, it is clear that for the All-Weighted model a different mix of DO-strategies was favored across the three lockout conditions.

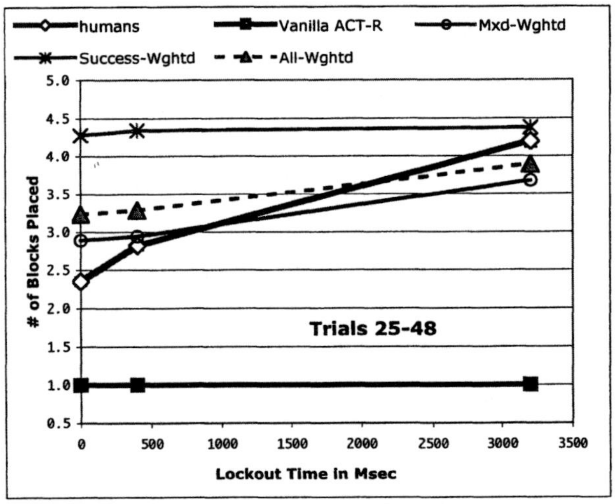

Figure 2: Blocks placed following the first uncovering of the Target window for Humans versus four ACT-R models. Except for the use of different systems for weighting successes and failures, all models use the same parameters and same productions.

Mixed-Weighted Model. The Mixed-Weighted model is the best fitting of the three both qualitatively and quantitatively. Quantitatively it has the smallest RMSE (0.44). Qualitatively, this model shows the greatest increase in blocks placed across lockout conditions. The difference between number of blocks placed at 0-Lock versus 3200-Lock is 1.84 for humans (see Figure 2), 0.79 for Mixed-Weighted, 0.66 for All-Weighted, and 0.09 for Success-Weighted.

In terms of expected value, performance in the 3200-Lock condition is dominated by DOs3-6 (see Table 5). The expected value of these DO-strategies were quite similar. The next closest DO-strategy was 0.50 expected value units below this range. In contrast, for 400-Lock the both DO-2 and DO-8 were within the same range of expected value as DOs3-6. For 0-Lock, DO-2 fell within the range of values shown across DOs3-6. Hence, compared to Success-Weighted and All-Weighted, for the Mixed-Weighted model as lockout time increases the extreme DO-strategies (high as well as low) are less likely to be selected.

Table 5: Expected values after 48 trials for DO-strategies 1-8 by lockout condition in the Mixed-Weighted model

Lockout	1	2	3	4	5	6	7	8
0	6.7	8.5	9.1	8.4	8.3	8.0	7.0	6.5
400	5.5	7.7	8.9	7.9	8.3	7.2	7.0	7.2
3200	2.7	6.0	6.9	7.1	6.9	6.9	6.4	6.3

Comparing Models. Across the four models the number placed gets closer and closer qualitatively and quantitatively to the human data, and the pattern of expected values for the DO-strategies begins to seem like a reasonable reflection of what humans must be doing.

The default, or vanilla, ACT-R model simply cannot handle these data. A strategy is either a success or a failure. If it places at least one block it is successful. The vanilla model was entirely driven by costs to an exclusive use of DO-1 over all runs of the model for trials 25-48. (Note that this model was so consistent that we only did three runs for each condition. In contrast, each of the other models was run six times per condition.)

The Success-Weighted model reduced the costs of the higher DO-strategies by the number of blocks they successfully placed. This cost reduction sufficed to boost the expected value of all higher DO-strategies. The expected values for DOs4-8 fell within 0.46 expected value units of each other for 0-Lock, within 0.56 for 400-lock, and within 0.52 for 3200-lock. Hence, the number placed was much higher than for the Vanilla model, but the number placed did not vary between lockout conditions.

The All-Weighted model punished strategies that encoded more than they placed by lowering their P value, but worked against itself by reducing costs based solely on the number encoded. This bias in reducing costs actually worked to favor strategies that encoded a lot but placed little. (More detail on this aspect has to await a fuller report in which we examine and report changes in P and C across models.) This all worked to favor DO-strategies in the range of DO-3 on up.

Like the All-Weighted model, the Mixed-Weighted model punished strategies that encoded more than they placed by lowering their P value. Unlike that model, it only reduced costs for the number of blocks actually placed. This combination worked to favor DO-strategies in the range of DO-4 to DO-6 over both the lower and higher strategies. The Mixed-Weighted model provided the best qualitative and quantitative fit to the empirical data.

Discussion and Conclusions

We divide this section into a brief discussion of alternative changes to expected value computations, conclusions about our work on the expected value equation, and conclusions about our model of Blocks World.

Other Changes to the Expected Value Equation

Rather than changing how successes and failures are calculated, our initial instinct was to change G – the value of the goal. It made much sense to us that if the value of completing a trial was worth, say 24, then the value of placing each block would be worth 3. (A strategy that placed 3 blocks would have its G incremented by 9 units.)

Whatever the merits of this scheme, unlike P and C, G does not accumulate separately for each production. Rather, G is a global value that is applied equally to calculate the expected value for each item in the conflict set. To experiment with G would require learning G. This would require more extensive changes to the current version of ACT-R (5.0) than the changes reported in this paper.

Other than our work, the only work we know that explores changes in ACT-R's calculation of expected value is that of Belavkin and colleagues. Whereas our work focuses on discriminating among strategies, Belavkin's work is focused more on changes in the range of expected values considered as the model gains expertise within a domain (Belavkin & Ritter, 2003) or on when to give up on a strategy (Belavkin, 2003).

Conclusions for calculations of expected value

This research into the parameters of the ACT-R expected value equation arose out of failure in trying to fit a model to the Blocks World data. As soon as we ran the model with ACT-R's default expected value settings (the Vanilla model) we realized that unless we could amortize costs and punish strategies that encoded more than they could place that we could not hope to fit the human data. The problem stemmed from the sparse reinforcement in the Blocks World environment, as well as the binary nature of success or failure in ACT-R. This dilemma prompted us to explore the space of expected values equations generated by changes in how successes and failures were counted and accumulated.

Our Mixed-Weighted model provides the best fit to our data and we believe it makes the most intuitive sense. For problems such as Blocks World where eventual success can be easily quantized into smaller units, it makes sense to us to reward and punish strategies based on how much of the problem they succeed in solving.

Conclusions for Blocks World

Although we believe the Mixed-Weighted model is a general solution to similar problems, we do not believe that we have adequately modeled the Blocks World data.

Under the Mixed-Weighted scheme, a key to a strategy's success or failure is its ability to retrieve from memory the items it has encoded. Currently we have run our models with optimized learning on (:ol t). This is the default for ACT-R models. However, in related research (Sims & Gray, 2004) we have come to believe that optimized learning overestimates the amount that can be retrieved in situations like the Blocks World where there is a long interval between an item's early encoding and rehearsal versus its later retrieval.

Indeed, for the larger DO-strategies there is a sizable period of time between the encoding and rehearsal of the first encoded block and the encoding and rehearsal of the last block on that round; this period, of course, precedes the long placement period. We believe that more realistic forgetting might work to drive the 0-Lock condition to rely more on lower DO-strategies so that the number placed for 0-Lock declines to somewhere closer to the human data (see Figure 2). Contrarily, more forgetting may drive the higher lockout conditions (such as 3200-Lock) to use fewer of the larger DO-strategies and more of the middle DO-strategies. We expect that this shift would have the counterintuitive effect of boosting number placed for 3200-Lock from its current 3.7 blocks to someplace closer to the 4.2 blocks of the humans in this condition.

Acknowledgments

The work reported was supported by a grant from the Office of Naval Research ONR #N000140310046. Additional support was provided by a subcontract to Rensselaer Polytechnic Institute from contract #MDA-904-03-C-0408 to Booz Allen Hamilton from the Advanced Research and Development Activity (ARDA). Thanks to Wai-Tat Fu for running human subjects as well as many other contributions to this project.

References

Anderson, J. R., Bothell, D., Byrne, M. D., & Lebiere, C. (in press). An integrated theory of the mind. *Psychological Review.*

Anderson, J. R., & Lebiere, C. (Eds.). (1998). *Atomic components of thought.* Hillsdale, NJ: Lawrence Erlbaum Associates.

Ballard, D. H., Hayhoe, M. M., & Pelz, J. B. (1995). Memory representations in natural tasks. *Journal of Cognitive Neuroscience, 7*(1), 66–80.

Belavkin, R. V. (2003). Conflict resolution by random estimated costs. In D. Al-Dabass (Ed.), *Proceedings of the 17th European Simulation Multiconference* (pp. 105-110). Nottingham, UK.

Belavkin, R. V., & Ritter, F. E. (2003). The use of entropy for analysis and control of cognitive models. In F. Detje, D. Dorner & H. Schaub (Eds.), *Fifth International Conference on Cognitive Modeling.*

Fu, W.-T., & Gray, W. D. (2000). Memory versus perceptual-motor tradeoffs in a Blocks World task. In L. R. Gleitman & A. K. Joshi (Eds.), *Twenty-second Annual Conference of the Cognitive Science Society* (pp. 154–159). Hillsdale, NJ: Lawrence Erlbaum Associates.

Gray, W. D., & Fu, W.-T. (2000, November). *Memory versus perceptual-motor tradeoffs in a Blocks World task.* Conference of the Psychonomics Society, New Orleans.

Sims, C. R., & Gray, W. D. (2004). Episodic versus semantic memory: An exploration of models of memory decay in the serial attention paradigm. In M. C. Lovett, C. D. Schunn, C. Lebiere & P. Munro (Eds.), *6th International Conference on Cognitive Modeling–ICCM2004.* Pittsburgh, PA.

Beyond Cognition: Modeling Emotion in Cognitive Architectures

Eva Hudlicka (evahud@earthlink.net)
Psychometrix Associates, Inc.
Blacksburg, VA, USA

Abstract

Recent research in psychology and neuroscience has identified both the critical role of emotion in decision-making and social interaction, and some of the mechanisms mediating the functioning of emotion. Yet the majority of cognitive architectures do not include models of emotion. In this paper I first motivate the need for including emotion in cognitive architectures, and then describe a generic methodology for modeling the effects of emotion within a symbolic cognitive architecture, as well as some of the associated representational requirements. I then present preliminary results, and conclude with an outline of key issues and future work in emotion modeling.

Introduction

When is it necessary to include emotion in cognitive architectures? The answer depends entirely on the purpose of the architecture, which also defines the constraints on the modeling effort (e.g., level of resolution, which emotions to include, which emotion effects to represent, which processes to model, etc.) (Hudlicka, 2003b). There are a number of reasons why emotion may be necessary, including: 1) research-motivated emulation of human decision-making and appraisal processes, to better understand their mechanisms; 2) enhancing agent and robot realism (e.g., for training and educational purposes, or assistive roles); 3) developing user and operator models for improved adaptive HCI. Depending on the specific purpose, an explicit model of emotion within the architecture may or may not be necessary. For example, it may be sufficient to 'fake' distinct affective states to enhance the effectiveness of tutoring systems, without including any of the underlying mechanisms that actually generate these states within the agent.

While much has been accomplished in the affective computing area since Picard coined the term in 1997, (Picard, 1997), most agent and cognitive architectures do not address the 'emotion issue'. There seems to be a degree of polarization within the field, with some researchers avidly embracing the need for emotion, at times perhaps even uncritically so, while others dismiss the need for modeling emotion. To make an informed decision regarding emotion modeling within a particular agent, a deeper understanding is necessary of: (1) the contexts that require emotion models, (2) the methods available to model emotions, and their effects, within cognitive architectures, and (3) the appropriateness of alternative modeling strategies, as a function of the specific context.

The purpose of this paper is twofold. *First,* to provide background information about recent emotion research, and its relevance to cognitive modeling. *Second,* to describe a generic methodology for modeling emotion effects within symbolic architectures, in terms of parametric manipulations of the architectural processes and structures, as well as the associated cognitive-affective architecture that implements this methodology – the MAMID architecture, and present preliminary results.

Background: Emotion Research

What are emotions?

There are many terms in the emotion research literature describing what we generally refer to emotions (Ekman and Davidson, 1994). *Affective states* generally refer to high-level stimulus evaluations yielding a positive or negative valence, and leading to correspondingly high-level behavioral directives such as approach or avoid. *Emotions* refer to transient states described by the familiar terms such as joy, fear, anger, etc. Some researchers divide these further into *basic emotions* (e.g., joy, sadness, frustration, anger, fear and anxiety), and more *cognitively-complex and self-oriented emotions* (e.g., guilt, pride, shame, jealousy). There are also *dimensional characterizations* of emotions, which identify sets of two or three dimensions defining spaces within which the individual emotions can be placed (e.g., valence and arousal, hedonic and tense arousal (Matthews & Deary, 1998)). Any and all of these may be candidates for modeling within a cognitive architecture, depending on the objective of the model.

Roles of Emotions

Three primary roles of emotion have been proposed, both interpersonal and intrapsychic: (1) *Emotions as Interpersonal Communication Mechanisms,* serving to communicate intentions and behavioral tendencies (e.g., imminent attack or withdrawal, pleasure vs. displeasure), thereby improving social behavior coordination; (2) *Emotions as Internal Goal Management Mechanisms,* required to coordinate physical and mental activities aimed at satisfying agent's multiple goals in an uncertain and unpredictable environment, and monitoring and regulation of goal-directed behavior; and (3) *Emotions as Behavior Preparation Mechanisms,* with distinct emotions linked to distinct desired behavior, improving the organism's chances for survival (Oatley & Johnson-Laird, 1987; Frijda, 1986).

Appraisal

Key components of emotional processing are the mechanisms that evaluate a current situation (internal and external), in terms of an affective label (affective state or emotion). This process is referred to as *appraisal*, (also *cognitive appraisal*), and has been studied extensively over the past 15 years (e.g., Scherer, 2003; Ellsworth & Scherer, 2003).

Earlier studies focused on descriptive characterizations of the phenomenon, a type of black-box, I/O view, attempting, for example, to identify types of elicitors required for particular emotions. More recently, attempts have been made to identify the mechanisms mediating these processes, as exemplified by the work of Smith and colleagues (Smith & Kirby, 2001), who elaborated earlier models of Lazarus into mechanistically-oriented process models, which lend themselves to computational implementations.

Appraisal processes are typically divided into two stages, an *automatic appraisal*, generating a high-level initial assessment, followed by a slower, more *deliberate appraisal*, with more idiosyncratic components and frequently including assessment of the individual's coping potential.

Emotion effects

A critical aspect of understanding and modeling emotions concerns their effects on attention, perception and cognitive processing. The specific effects on attention and cognition of a number of affective states have been studied extensively (e.g., anxiety and fear, anger and frustration, positive and negative affect, etc.). These effects include *altering the nature of attentional processing* and *working memory* (e.g., changes in capacity and bias) and helping to activate (or inhibit) particular perceptual and cognitive schemas that enhance (or inhibit) the perception and processing of specific stimuli. These include the following: *perceptual categorization biases* towards threats; *memory encoding and recall* effectiveness and biases; and a variety of additional *influences on reasoning, judgment, and decision-making* (Williams et al., 1997; Isen, 1993; LeDoux, 1989). Effects thus exist at both the low levels of processing (attention and working memory speed and capacity), and the higher levels (situation assessment, decision making, planning, learning and judgment).

Emotion mechanisms

The theories of emotion mechanisms have evolved over the past 100 years, with recent neuroscience findings stimulating much progress by providing concrete data, and demystifying earlier theories. The recent consensus regarding emotions is that they play a pervasive neuromodulatory role across the brain circuitry, replacing the earlier views of isolated 'emotion' modules (e.g., the amygdala) or emotion circuits (e.g., the limbic system) (Fellous, 2004).

Requirements for Modeling Emotion

How then do we model emotion within cognitive architectures? What specific structures and processes are necessary? What representational formalisms and inferencing strategies are most appropriate? Again, the answers depend on the ultimate purpose of the modeling effort.

In some cases, researchers explore the fundamental architectural features and configurations capable of producing emotion (e.g., Sloman, 2003), providing what we might call *deep emotion models*. In other cases, it is sufficient to reason 'about' emotion, focusing on *shallow emotion models*. In either case, a number of available symbolic (as well as non-symbolic) representational and inferencing mechanisms are appropriate. For example, in a shallow model of emotion, rules or belief nets can represent patterns such as "If user exhibits behavior xxx, then s/he is likely experiencing emotion yyy".

A middle ground approach, such as the MAMID architecture described below, assumes that certain structures and processes are necessary to adequately model both the *effects of emotions*, and some of the *core processes* involved (e.g., appraisal). Specific *representational requirements* include the explicit representation of the following: particular mental constructs (e.g., situations, expectations, goals); the self and self-relevant stimuli and states; particular attributes of these constructs (e.g., threat level, valence, desirability); and long-term and working memories. Specific *processing requirements* include explicit representation of the processes mediating the mapping of external and internal stimuli onto particular affects, via distinct stages, and the processes whereby specific affects exert particular influences on attention, perception and cognitive processing, including situation assessment, goal management, decision-making, and action selection.

Below we briefly describe one approach to modeling emotions within a symbolic cognitive architectures: the MAMID methodology and architecture.

MAMID Methodology and Architecture

The key components of the MAMID emotion modeling approach are: (1) an architecture capable of deriving an affective state via an appraisal process; and (2) a means of representing the effects of the resulting affective state on processing. The latter is implemented via a generic methodology capable of representing the effects of emotions (and other individual differences, including personality traits) within a symbolic cognitive architecture via parametric manipulations of the architecture processes and structures. We describe both components below.

MAMID Cognitive Architecture

The cognitive architecture implements a sequential see-think-do processing sequence (figure 1), consisting of the following modules: *sensory pre-processing*, translating incoming data into task-relevant cues; *attention*, filtering incoming cues and selecting a subset for processing; *situation assessment*, integrating individual cues into an overall situation assessment; *expectation generation*, projecting current situation onto possible future states; *affect appraiser*, deriving the affective state (both valence and four of the basic emotions) from a variety of external and internal elicitors, both static and dynamic; *goal selection*, selecting critical goals for achievement; and *action selection*, selecting the best actions for goal achievement.

These *modules* map the incoming stimuli (cues) onto the outgoing behavior (actions), via a series of intermediate internal representational structures (situations, expectations, and goals), collectively termed *mental constructs*. This mapping is enabled by long-term memories (LTM) associated with each module, represented in terms of belief nets or rules.

Mental constructs are characterized by their attributes (e.g., familiarity, novelty, salience, threat level, valence, etc.), which influence their processing, by determining their rank and the consequent likelihood of being processed within a given execution cycle; (e.g., cue will be attended, situation derived, goal or action selected). (Note that the availability of the mental constructs from previous frames of the execution cycle allows for dynamic feedback among constructs, and thus departs from a strictly sequential processing sequence.)

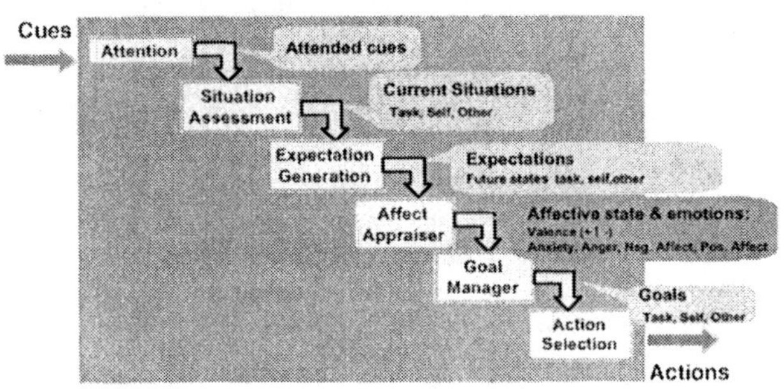

Figure 1: MAMID Cognitive Architecture: Modules & Mental Constructs

Affect Appraisal Process

The Affect Appraiser module (figure 2) incorporates elements of several recent appraisal theories: *multiple-levels* (Leventhal & Scherer, 1987; Sloman, 2003; Smith & Kirby, 2001), and *multiple stages* (Scherer, 2003), and incorporates empirical data.

The Affect Appraiser module generates both a *low-resolution assessment* of the current set of stimuli, in terms of a valence, and a *higher-resolution categorical assessment*, in terms of four of the basic emotions: anxiety/fear, anger, sadness, happiness. Its multiple stages use both *universal elicitors* (e.g., novelty, threat level, (un)pleasantness, unexpectedness), to generate a valence, using an *automatic appraisal* (roughly corresponding to the largely 'hardwired', 'primitive' appraisal components), and more *cognitively-complex and idiosyncratic elicitors* (e.g., individual history, expectation- and goal-congruence), to generate a categorical assessment using an *expanded appraisal*.

The MAMID Affect Appraisal module consists of three stages: *automatic appraisal*, *expanded appraisal*, and *current state modulator*. *Automatic appraisal* emphasizes the stimulus properties to calculate the valence state (positive or negative); specifically, unexpectedness ("is situation part of current expectations"), novelty ("is situation part of individual history"), and the situation's intrinsic threat and "pleasantness" levels. Trait effects are included via a bias factor, reflecting the agent's temperamental predisposition toward negative or positive states. For example, high extraversion and low neuroticism individuals tend to be predisposed towards positive affective states, whereas low extraversion and high neuroticism individuals tend towards negative affective states.

Expanded appraisal emphasizes the influence of the agent's internal motivational context, by taking into consideration the congruence of the current situations and expectations with the agent's goals, the general level of success in achieving the current goals (e.g., number of goals met vs. failing), and the individual idiosyncratic effects of particular elicitors (e.g., previous experiences with a specific stimulus). This latter factor captures the domain-, task-, and individual-specific emotion-eliciting stimuli, situations or expectations. These are encoded in a set of belief nets, associated with a particular emotion, for each agent type.

The belief nets vary in type and structure, reflecting the differences in individual histories (e.g., previous negative / positive experience in a particular situation), sensitivity to particular factors (e.g., considerations of own sense of competence or control in affect appraisal), and responsiveness (e.g., magnitude of particular affect generated in response to a set of situations or expectations). For example, a trait-anxious agent considers own competence and sense of control during appraisal to determine the level of anxiety, whereas a non-trait anxious individual does not consider these factors.

Trait effects are included as above, with specific trait combination biasing towards a particular emotion (e.g., high neuroticism - low extraversion predisposes towards anxiety and negative affect (Matthews & Deary, 1998)). The current valence also influences the expanded appraisal, by contributing to the intensity of the valence-congruent emotions; thus negative valence increases the intensity of anxiety, anger and negative affect, while positive valence increases the intensity of positive affect.

The expanded appraisal produces a vector of intensities for each of the four represented emotions. This representation, along with the multi-level appraisal structure, supports the representation of mixed, ambiguous, and possibly conflicting emotions, which are quite common in real life, but have not been adequately explored in models (Scherer, 2003). We are just beginning to explore the behavioral consequences of these complex representational possibilities.

Both valence and emotion intensities are calculated via linear functions of the weighted eliciting factors. The weights controlling the contributions of the individual eliciting factors can be modified interactively by the model developer. This allows model tuning to reflect emerging empirical data and alternative theories regarding the mechanisms of appraisal and state / trait effects, as well as modeling of a wide range of individual differences (e.g., differences in sensitivity to the valence produced by the automatic appraisal can be explored by modifying the weight of the valence component in the expanded appraisal functions).

The *Current State Modulator,* the final stage of the appraisal process, consists of modulating the newly-derived valence and emotion values by the valence and emotions generated in the previous execution cycle, thereby assuring smooth transitions among states. Traits exert an effect on this stage by influencing the ramp-up and decay rates of individual emotions, as well as their maximum intensities.

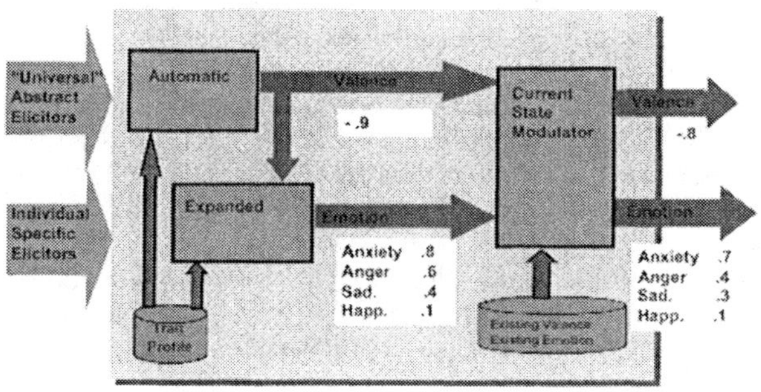

Figure 2: Affect Appraisal Model

The resulting affective states then influence processing in several ways. *First*, they are used directly in the rules selecting the agent's goals and actions. *Second*, they influence the speed and capacity of the architecture modules – a parameter-controlled effect analogous to the system-wide neuromodulatory role of emotions. *Third*, they influence mental construct ranking, thus determining whether a specific cue or situation is processed, or specific goal selected. The last two effects have been a particular focus of this effort, and aim to emulate some of the empirically-identified mechanisms of emotion effects.

Generic Modeling Methodology

We used a previously described methodology for modeling state and trait effects within a cognitive architecture (Hudlicka, 2002; 1998), which consists of mapping particular state / trait profiles onto specific architecture parameter values (figure 3). These parameters then control processing within the individual architecture modules.

Functions implementing these mappings were constructed on the basis of the available empirical data. For example, reduced attentional and working memory (WM) capacity, associated with anxiety / fear, are modeled by dynamically reducing attention and WM capacity of the architecture modules, which then reduce the number of constructs processed (fewer cues attended, situations derived, expectations generated, etc.). Attentional threat bias is modeled by higher ranking of threatening cues, thus increasing their likelihood of being attended, and by higher ranking of threatening situations and expectations, thus increasing the chances of a threatening situation / expectation being derived. Trait-linked structural differences in LTM are supported by allowing the flexible selection of alternative LTM clusters, reflecting distinct personality traits. Traits also influence the dynamic characteristics of the emotional responses (ramp up, decay, and maximum intensities).

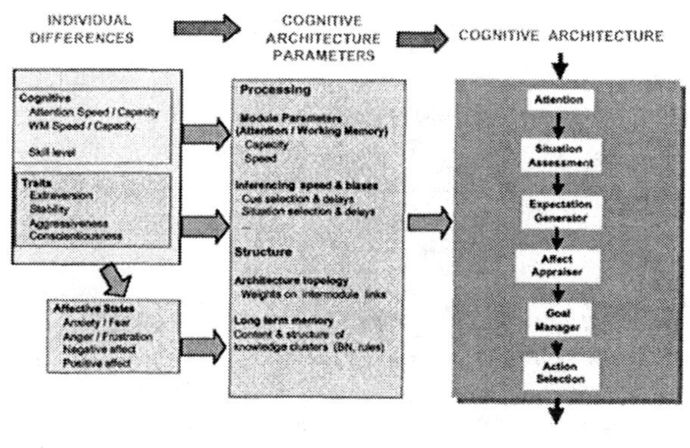

Figure 3: Parametric State / Trait Modeling Methodology

A key objective of the methodology was to provide flexibility regarding the *types of factors* selected for inclusion in a model, the *nature of their influence* on cognitive, perceptual, and decision-making processing, and the *degree of this influence.* This approach is necessitated by the fact few definitive theories exist regarding the exact mechanisms of the emotion influences, particularly with respect to the more complex, internal processes of situation assessment, expectation generation, and goal management and planning. In addition, new data regarding these effects continue to emerge and need to be rapidly accommodated.

The MAMID methodology, and the architecture and testbed that implement it, were designed with the explicit purpose of facilitating a rapid accommodation of these emerging findings. The methodology achieves this flexibility through a high-degree of parameterization of the processes and structures that comprise the distinct architecture modules, and the analyst's interactive access to these parameters and their weights. The distinct individual differences factors are mapped onto distinct configurations of the cognitive architecture parameters, which in turn produce different processing within the model, and ultimately produce different behavioral outcomes for the associated simulated agent.

There are several advantages of this methodology for modeling both emotions, and a broad range of individual differences (behavior moderators). *First,* it facilitates rapid modeling of a broad range of distinct individual profiles. *Second*, the rich architecture parameterization allows the definition of additional individual characteristics (e.g., additional traits can be introduced into the model by identifying a mapping between that trait and the corresponding set of architecture parameters). *Third*, it provides a means of integrating (possibly conflicting) effects of multiple, interacting emotions and traits, much as these influences interact in humans. *Fourth*, it is psychologically grounded, with both the emotions and their corresponding architecture parameters selected on the basis of empirical data and psychological theory. This is in contrast to some recent efforts that have adopted this approach, but use parameters such as global system noise to degrade system performance, which has no direct counterpart in specific psychological processes (e.g., Ritter et al., 2002). While such high-level manipulations may be sufficient to model the surface manifestations of some emotion effects at the shallow level, they tell us little about the mechanisms that mediate these effects.

Preliminary Results

The MAMID architecture and methodology were evaluated by modeling agent decision-making in the context of a peacekeeping scenario. Several simulated commander types (anxious, aggressive, normal) encountered a series of 'surprise' scenario situations (e.g., destroyed bridge, hostile crowd), designed to elicit different reactions as a function of their state and trait differences. The agents' affective states were dynamically generated by the Affect Appraisal module, in response to each surprise situation.

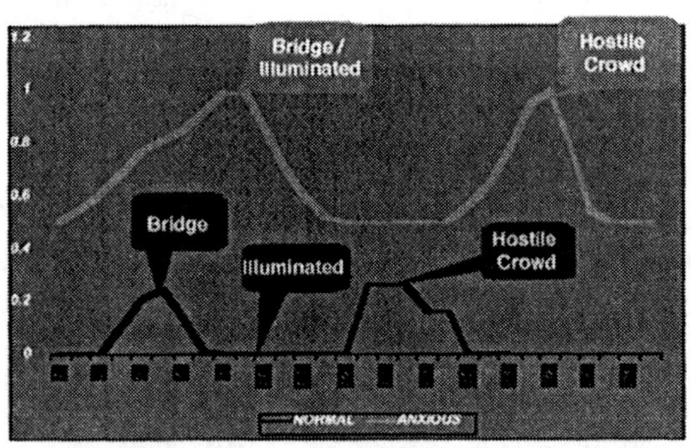

Figure 4: Anxiety Fluctuations Over Time for Normal and Anxious Commanders

The resulting emotions then influenced processing within each of the architecture modules, as outlined above (e.g., contributed to lower or higher processing capacities, threat bias, etc.). Figure 3 shows the fluctuating anxiety levels of a normal and a trait-anxious commander, during the course of the simulation scenario. The different emotions, together with the trait-related differences in both processing and memory, then resulted in differences in behavior, in response to the identical set of external circumstances produced by the scenario simulation (figure 4). Thus, for example, an anxious commander used inappropriate force against a hostile crowd, moved more slowly, and spent more time in situation assessment than his 'normal' counterpart.

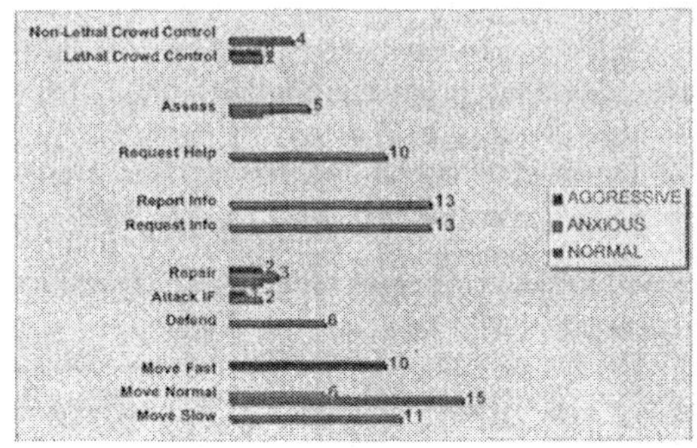

Figure 5: Distinct Behavior of Different Agent Types

Related Work

A number of models addressing emotion have been developed in cognitive science and AI. These models range from individual processes to integrated architectures, and explore several of the emotion roles outlined above. The most frequently modeled process has been *cognitive appraisal.* Several alternatives have been proposed for these processes in the psychological literature (Ortony et al., 1998; Frijda, 1986; Lazarus, 1991; Smith and Kirby, 2001; Scherer, 2003). A number of these models have been implemented, both as stand-alone versions, and integrated within larger agent architectures (e.g., Scherer, 2003; Bates et al., 1992; Elliot et al., 1999; Andre et al., 2000). Other emotion model implementations include models of emotions as goal management mechanisms (Frijda and Swagerman, 1987), models of interaction of emotion and cognition (Araujo, 1993; Hudlicka, 2002), and effects of emotions on agent's belief generation (Marsella & Gratch, 2002). Examples of *integrated architectures* focusing on emotion include most notably the work of Sloman and colleagues (Sloman, 2003). There are also recent efforts to integrate emotion effects in Soar (Jones et al., 2002), and ACT-R (Ritter et al., 2002), the latter using in-part a parametric-manipulation approach analogous to MAMID.

Conclusions and Future Work

The model described above has merely begun to 'scratch the surface' of emotion modeling. Many extensions are possible, and necessary, to develop a more accurate model of the complex appraisal processes and emotion effects on decision-making. One of the most challenging issues is validation of these computational models, whereby specific empirical data are compared with model performance over time. To date, this has typically been done at the I/O level, comparing model input/output on some particular task with those of humans.

Validating deep emotion models at the process level, by matching the mechanisms involved rather than just the input/output, is particularly challenging, since the types of internal data required about the detailed structures and processes are not available.

Some of the potential future directions for this work include: additional processing levels; parallelism and interaction among processing levels; more complex elicitors; inclusion of coping potential in elicitors (Lazarus' deliberate appraisal); additional emotions (basic and complex); more complex functions calculating the architecture parameters and emotion dynamics; and explicit models of meta-cognition.

Acknowledgments

The research described in this paper was supported in part by US ARI Contracts DASW01-98-M-1985 & DASW01-00-C-3000. We would like to acknowledge continued support of the COTRs, Drs. B. Witmer & J. Psotka, and the contributions of the following individuals: Prof. G. Matthews, Prof. W. Revelle, Mr. Ted Fichtl, Dr. G. Zacharias, & the Psychometrix software team: J. Pfautz, M. Turnbull, L. Buomano, C. Ganoe, J. Helms., and B. Staderman.

References

Andre, E., Klesen, M., Gebhard, P., Allen, S.,Rist,T. (1999). Integrating Models of Personality and Emotions in Lifelike Characters. In *Proceedings of IWAI*. Siena, Italy.

Araujo, A.F.R. (1993). Emotions Influencing Cognition In *WAUME '93: Workshop on Architectures Underlying Motivation and Emotion*. Birmingham, UK: The University of Birmingham.

Bates, J., Loyall, A.B., and Reilly, W.S. (1992). Integrating Reactivity, Goals, and Emotion in a Broad Agent. In *Proceedings of the 14th Meeting of the Cognitive Science Society*.

Davidson,R.J.,Scherer,K.R.,Goldsmith,H.H.(2003). *Handbook of Affective Sciences*. NY: Oxford.

Ekman,P. & Davidson,R.J. (1994). *The Nature of Emotion*. Oxford: Oxford.

Elliot, C., Lester, J., & Rickel, J. (1999). Lifelike pedagogical agents and affective computing: An exploratory synthesis. In M. Wooldridge and M. Veloso, (Eds.). *AI Today*. Lecture Notes in AI. NY: Springer-Verlag.

Ellsworth,P.C. & Scherer, K.R. (2003). Appraisal Processes in Emotion. In *Handbook of Affective Sciences*. R.J. Davidson, K.R. Scherer, H.H. Goldsmith (Eds.). NY: Oxford.

Fellous, J-M. (2004).From Human Emotions to Robot Emotions. In *Proceedings of the AAAI Spring Symposium-Architectures for Modeling Emotion*, TR SS-04-02. Menlo Park, CA: AAAI Press.

Frijda, N.(1986). *The Emotions*. New York: Cambridge.

Frijda, N.H. and Swagerman, J. (1987). Can Computers Feel? Theory and Design of an Emotional System. *Cognition and Emotion*, 1 (3), 235-257.

Gratch, J. & Marsella, S. (2002). Modeling the influence of emotion on belief for virtual training simulations. *Proceedings 11th CGF Conference*, Orlando, FL.

Hudlicka, E. (1998). Modeling Emotion in Symbolic Cognitive Architectures. In *Proceedings of the AAAI Fall Symposium-Tangled Knot of Cognition*. TR FS-98-03. Menlo Park, CA: AAAI Press

Hudlicka, E. (2002). This time with feeling: Integrated Model of Trait and State Effects on Cognition and Behavior. *Applied AI*, 16:1-31.

Hudlicka, E. (2003a). Modeling Effects of Behavior Moderators on Performance. In *Proceedings of BRIMS-12*, Phoenix, AZ.

Hudlicka, E. (2003b). To Feel or Not to Feel: The Role of Affect in Human-Computer Interaction. *Int. Journal of Human-Computer Studies*, 59 (1-2).

Hudlicka, E.(2004). Two Sides of Appraisal. In *Proceedings of the AAAI Spring Symposium-Architectures for Modeling Emotion*, TR SS-04-02. Menlo Park, CA: AAAI

Isen, A.M. (1993). Positive Affect and Decision Making. In *Handbook of Emotions*, J.M. Haviland, M. Lewis, (eds.). NY: Guilford.

Lazarus, R.S. (1991). *Emotion and Adaptation*. NY: Oxford.

Jones, R., Henninger, A. & Chow, E. (2002). Interfacing Emotional Behavior Moderators with Intelligent Synthetic Forces *Proceedings 11th CGF Conference*, Orlando, FL.

LeDoux, J.E. (1989). Cognitive-Emotional Interactions in the Brain. *Cognition and Emotion*, 3(4), 267-289.

Leventhal, H. & Scherer, K.R. (1987). The relationship of emotion to cognition. *Cognition and Emotion*, 1, 3-28.

Matthews, G., and Deary, I.J. (1998). *Personality Traits*. Cambridge, UK: Cambridge.

Oatley, K. & Johnson-Laird, P. (1987). Towards a cognitive theory of emotion. *Cognition and Emotion*, 1, 51-58.

Ortony, A., Clore, G.L., & Collins, A. (1988). *The Cognitive Structure of Emotions*. NY: Cambridge.

Picard, R. (1997). *Affective Computing*. Cambridge, MA: The MIT Press.

Ritter, F , Avramides, M., & Councill, I. (2002). Validating Changes to a Cognitive Architecture to More Accurately Model the Effects of Two Example Behavior Moderators *Proceedings of 11th CGF Conference*, Orlando, FL.

Scherer, K.R. (2003). Cognitive Components of Emotion. In *Handbook of Affective Sciences*. R.J. Davidson, K.R.,Scherer, H.H.Goldsmith(Eds.). NY: Oxford.

Sloman, A. (2003). How many separately evolved emotional beasties live within us? In *Emotions in Humans and Artifacts*, R.Trappl, P. Petta, & S. Payr (Eds.). Cambridge, MA: The MIT Press.

Smith, C.A. & Kirby, L.D. (2001). Toward Delivering on the Promise of Appraisal Theory. In *Appraisal Processes in Emotion*. K.R. Scherer, A.Schorr, T. Johnstone (Eds.). NY: Oxford.

Williams, J.M.G., Watts, F.N., MacLeod, C., & Mathews, A. (1997). *Cognitive Psychology and Emotional Disorders*. NY: John Wiley.

Explaining Eye Movements in the Visual Search of Varying Density Layouts

Tim Halverson (thalvers@cs.uoregon.edu)
Anthony J. Hornof (hornof@cs.uoregon.edu)
Department of Computer and Information Science, 1202 University of Oregon
Eugene, OR 97401-1202 USA

Abstract

This research advances computational cognitive modeling of visual search, and the synergistic relationship between cognitive modeling and eye tracking. The paper presents cognitive models of the perceptual, cognitive, and motor processing involved in the visual search of words in structured layouts that vary in density. The layouts are all-sparse, all-dense, or mixed. A principled approach is taken to account for eye movement data, specifically the mean number of fixations per trial and mean fixation durations. A random search strategy without replacement is used as a base model. The best-fitting model assumes that people examine two to three items per fixation regardless of the density. A new implementation of the EPIC cognitive architecture is used to build the models in this study. Modeling adjustments necessary to account for the data are discussed.

Introduction

Cognitive modeling is useful to the field of human-computer interaction because it reveals patterns of human performance at a level of detail not otherwise available to analysts and designers (as in Gray, John, & Atwood, 1993). The ultimate promise for cognitive modeling in human-computer interaction is that it provides the science base needed for predictive analysis tools and methodologies (Card, Moran, & Newell, 1983). This article reveals patterns of human performance in visual search, and contributes to predictive analysis of visual search.

The density of items in a display is one factor that has been shown to affect visual search. Bertera and Rayner (2000) varied the density of a fixed number of characters by varying the spacing between characters in a search task and found that search time decreased and the estimated number of letters processed per fixation increased as the density increased. Bertera and Rayner concluded that the effective field of view, the visual space from which information is perceived in a fixation, did not vary with the stimuli density. Ojanpää, Näsänen, and Kojo (2002) studied the effect of spacing on the visual search of word lists, and found that as the vertical spacing between words increased (i.e. as density decreased), search time also increased. In general, research examining the effect of density on visual search has found that more dense stimuli are searched faster per object, with a decrease in the number of fixations required to find the target being the largest factor influencing search time.

The modeling presented here focuses on the issues raised by previous research on density, e.g., the number of items perceived per fixation, and other fundamental perceptual and ocular motor issues of visual search. Previous modeling has used data from eye tracking to inform the development of models with respect to the order of search (e.g. Byrne, 2001; Hornof & Halverson, 2003). Here we use fixation duration and number of fixations to inform the development of other aspects of the models.

This paper presents models of a task that investigates the effect of local density on visual search. The purpose of these models is to determine the perceptual and ocular-motor constraints that are required to explain eye movement data collected with this local density task. Other aspects of the data, such as fixation order, are left for future research.

The Visual Search Experiment

The task modeled in this paper is the visual search of a known target among words in structured layouts. Figure 1 shows a sample layout from a trial. All layouts contained six groups of left-justified, vertically-listed black words on a white background. The groups were arranged in three columns and two rows. There were two types of groups of different densities: *Sparse* groups contained five words of 18 point Helvetica font. *Dense* groups contained 10 words of 9 point Helvetica font. Both group types subtended the same vertical visual angle.

There were three types of layouts: *sparse*, *dense*, and *mixed-density*. Sparse layouts contained six sparse groups. Dense layouts contained six dense groups. Mixed-density layouts contained three sparse groups and three dense groups. Figure 1 shows one such layout. The arrangement of the group densities in the mixed-density layouts was randomly determined for each trial.

Target and distractors items were words selected randomly from a list of 765 nouns generated from the MRC Psycholinguistic Database (Wilson, 1988). No word appeared more than once per trial. The words in the list were selected as follows: three to eight letters, two to four phonemes, above-average printed familiarity, and above-average imagability. Participants were precued with the target word before each layout appeared.

Each trial proceeded as follows: The participant studied the precue; clicked on the precue to make the precue disappear and the layout appear; found the target word; moved the cursor to the target word; and clicked on it.

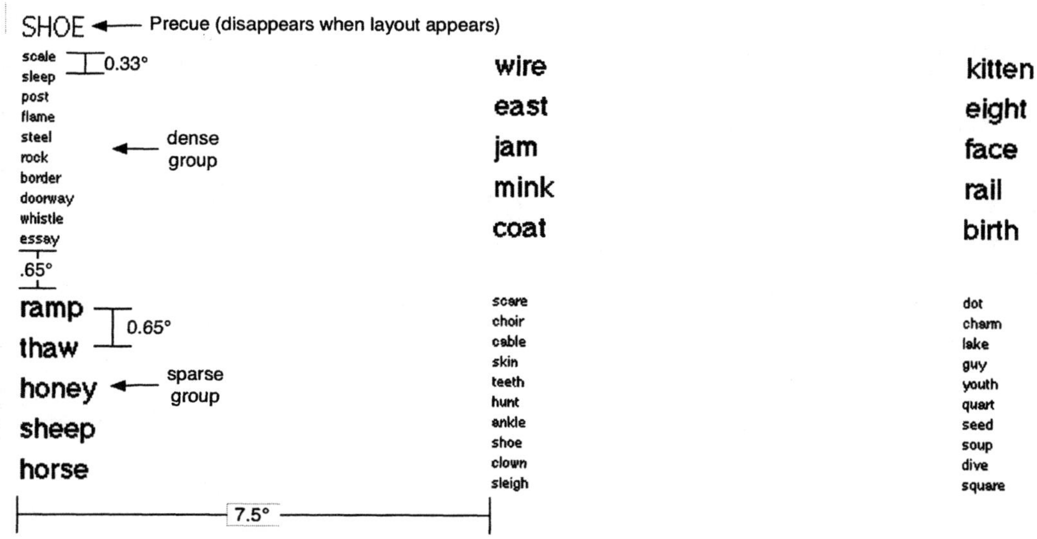

Figure 1. A mixed-density layout. All angle measurements are in degrees of visual angle.

The trials were blocked by layout type. Each block contained 30 trials, preceded by five practice trials. The blocks were counterbalanced using the Latin square technique. Eye movements were recorded using an LC Technologies Eyegaze System, a 60 Hz pupil-center/corneal-reflection eye tracker.

The Observed Data

The solid line plotted in Figure 2 shows the observed search time per trial, in Figure 3 the observed mean number of fixations per trial, and in Figure 4 the observed mean fixation duration. These data were collected from 24 participants (see Halverson & Hornof, 2004 for details). The mean number of fixations per trial and mean fixation duration are used for comparison with the model data for two reasons. First, they were both shown to vary significantly by layout type, $F(2,46) = 60.17$, $p < .01$ and $F(2,46) = 61.82$, $p < .01$, respectively. Second, the majority of the shift in search time per trial across the different layout types may be accounted for by the number of fixations and fixation duration. The search time per trial shows to what extent adjusting the number of fixations per trial and fixation duration can account for the search performance. The observed search time was also found to vary significantly with layout type, $F(2,46) = 127.80$, $p < .01$.

The data of primary interest are the eye movement data from the sparse and dense layouts. Previous work (Halverson & Hornof, 2004) suggested that the observed behavior for the mixed density layouts was affected by global strategies, which are not the primary concern in the present discussion. The important features in the observed data are that all three measures in this study, search time, number of fixations per trial, and fixation duration, increased when the local density increased.

It is well established that search time varies with distractor numerosity. However, the number of distractors cannot account for all of the shift in search time observed across the three layout types. If we normalize for the number of words per layout, the between-layout differences in mean search time per word and mean number of fixations per word are also significant, $F(2,46) = 14$, $p < .01$ and $F(2,46) = 3$, $p = .05$, respectively. However, search time per trial and the number of fixations per trial will be compared with the predicted data since they are more directly interpretable.

EPIC Cognitive Architecture

The models were constructed using the new C++ implementation of the EPIC (Executive Process Interactive Control) cognitive architecture (Kieras & Meyer, 1997). EPIC captures human perceptual, cognitive, and motor processing constraints in a computational framework that is used to build cognitive models. Into EPIC, we encoded (a) a reproduction of the task environment, (b) the visual-perceptual features associated with each of the screen objects, such as the text feature, and (c) the cognitive strategies that guide the visual search, encoded as production rules. These components were added based on task analysis, human performance capabilities, previous visual search model, and parsimony.

After these components are encoded into the architecture, EPIC executes the task, simulates the perceptual-motor processing and interactions, and generates search time and eye movement predictions. EPIC simulates ocular-motor processing, including the fast ballistic eye movements known as *saccades*, as well as the fixations during which the eyes are stationary and information is perceived.

The Models

Three models are presented in this section. Each improves on the last, correcting for a particular shortcoming in the prediction generated from the previous model. A good fit of the mean search time is the primary goal. However, while the mean search time is analyzed for each model, the models are improved incrementally based on eye movement data.

All models are based on a purely random, without-replacement, search strategy. While we do not necessarily assert that people move their eyes from item to item randomly, it may be that a random search strategy is a good first approximation for predicting *mean layout search time* without complicated strategies or visual features beyond the locations of objects. Such a strategy has the added benefit for *a priori* engineering models, as each object need be encoded with only one directly-extractable feature, its location. Subtler features, such as visual prominence or group-inclusion are not needed. Hornof (in press) found that a purely random search model with two to three items examined per fixation was good predictor for mean layout search time, even though it was a marginal predictor of search time per position.

Each model was run for 2,520 trials per layout type. The predictions generated by each model are discussed next.

Purely Random Base Model

The first model examined – the base model – was a purely random search model with all of EPIC's perceptual properties left at the default values. This base model is the standard to which we can compare subsequent models. In general, each word in the layout had an equal probability of becoming the destination of the next saccade. For this and following models, *text* is only available within one degree of visual angle (dov) from the center of fixation. We assume a without-replacement search. Any object for which the *text* had been perceived was excluded from being the destination point of future saccades.

The base model was overall a poor predictor of human performance. As seen in Figures 2 and 4, the predicted search times and fixation durations are incorrect in value and trend. It might seem that search time should decrease with dense layouts because more items fit into the fovea, but dense layouts also had more items to search.

Figure 3 shows the mean number of fixations. While the trend is incorrect for the predicted number of fixations as well, the prediction for the sparse layouts is quite good. This is promising and suggests that the purely random search model is a good starting point for modeling the characteristics of participant eye movements. As the predicted fixation durations have the greatest error of the two eye movement measurements, the next model will focus on improving the fixation durations produced by the model.

Wait-for-*text* Model

People appeared to adopt a search process that increased the duration of fixations on smaller, denser text. This could be

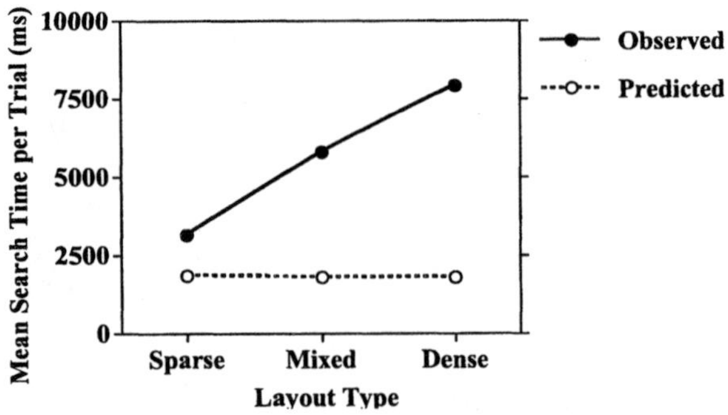

Figure 2. Mean search time per trial observed (solid line) and predicted (dashed line) by the purely random base model. Average absolute error (AAE) = 62.1%

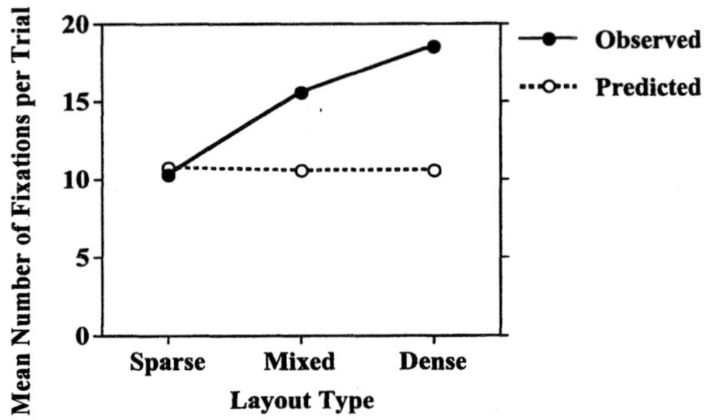

Figure 3. Mean number of fixations per trial observed (solid line) and predicted (dashed line) by the purely random base model. AAE = 26.7%

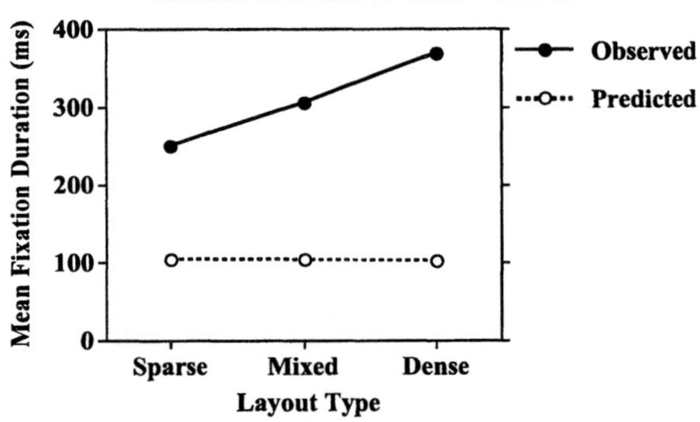

Figure 4. Mean fixation durations observed (solid line) and predicted (dashed line) by the purely random base model. AAE = 65.5%

achieved a number of ways in the model. One brute force approach would be for the production rules to directly set the fixation duration, though EPIC provides no such facility. Another would be to hold back each saccade until a certain amount of information is gathered from the currently fixated stimuli. This is straightforward to model in EPIC by (a) modulating the recoding time for a visual property (as in, dense text takes longer to recode) and (b) using a strategy that holds back each saccade based on this modulated

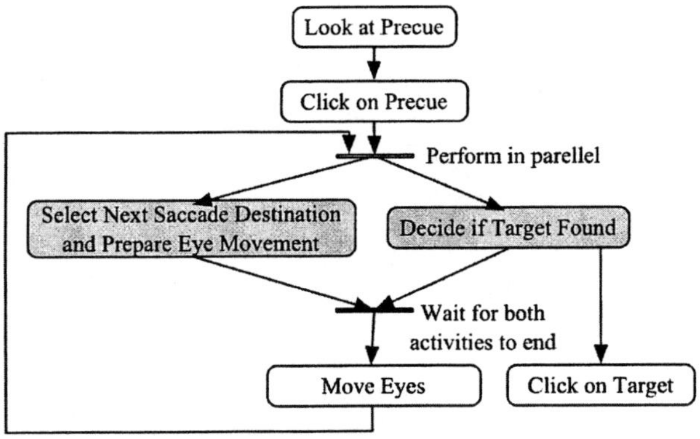

Figure 5. Flow chart of the Wait-for-TEXT strategy. Universal Modeling Language, a diagrammatic standard in software engineering, is used here.

feature. The Wait-for-*text* model uses such a strategy. Figure 5 shows a flow-chart based on the production rules.

The EPIC's default recoding time for the *text* feature is a constant 100 ms. This was modified when trying to explain the human data. As shown in Figure 4, the mean observed fixation duration was over 100 ms longer in the dense layouts than in the sparse layouts. To model this a stepped recoding function was introduced to calculate the perceptual time for a feature based on the proximity of adjacent items. If an object's closest neighbor was closer than 0.15 dov (a dense object), the *text* recoding time was 150 ms. Otherwise the *text* recoding time was 50 ms.

The base model initiated a saccade to the next randomly chosen object as soon as the previous saccade was complete. We hypothesized that humans may not initiate a saccade to another object before the text of the currently fixated object had been perceived, but that they may select another object and prepare the eyes to move to that object. Based on this belief, the procedure used for initiating saccades became: (1) Select an object at random as a saccade destination and prepare the eyes to move to that object. (2) Wait for the *text* of the currently fixated objects to become available before performing the prepared saccade. A similar prepare-then-perform strategy was used successfully in the modeling of another visual search task with EPIC (Kieras, 2003).

The predicted mean search time improved slightly with these modifications. As seen in Figure 6, there is now a very slight upward trend in the search time. However, the slope of the predicted search time line is not steep enough. The lack of a sharp increase between sparse and dense layouts correlates with the continued poor fit of the predicted number of fixations per trial. The model still does not make more fixations in layouts with dense objects. Further, the overall mean number of fixations has dropped in comparison to the base model. Though the mean fixation duration is now explained very well.

Detailed traces of the models revealed that the drop in the mean number of fixations was due to the prepare-then-perform strategy. The base model initiated approximately three additional fixations after the target had been fixated but before the *text* property for the target had become available. This resulted in roughly three more fixations per trial than the prepare-then-execute strategy, which inhibited additional fixations until the *text* property was perceived.

A large improvement was found in the predicted fixation durations. The inhibition of saccades and an increased recoding time for dense objects resulted in a differentiation in fixation durations similar to that in the observed data. The predicted data could have been further improved by reducing the *text* recoding time for sparse objects further, as

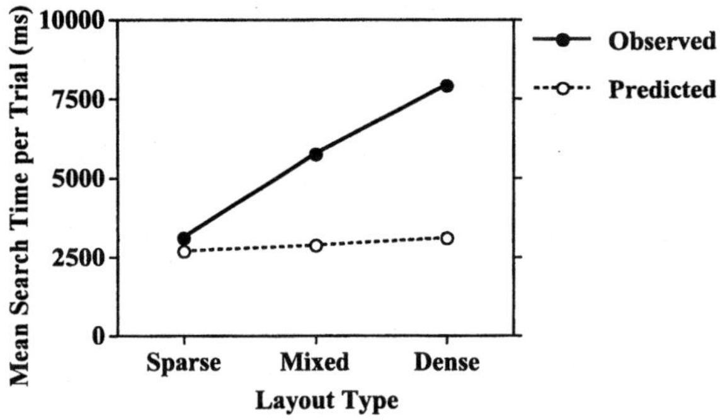

Figure 6. Mean search time per trial observed (solid line) and predicted (dashed line) by the Wait-for-the-TEXT model. AAE = 41.7%

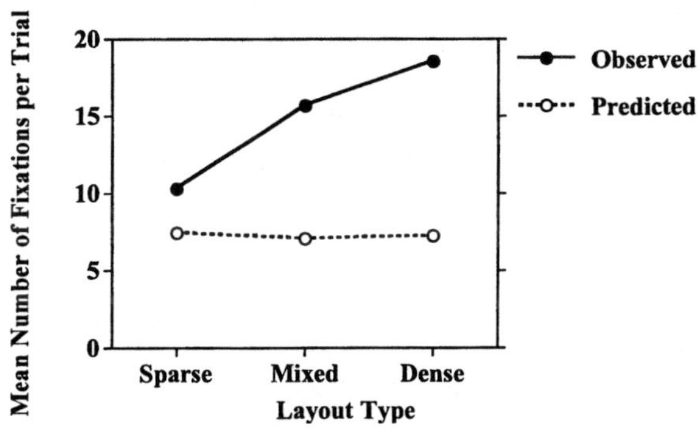

Figure 7. Mean number of fixations per trial observed (solid line) and predicted (dashed line) by the Wait-for-the-TEXT model. AAE = 48.1%

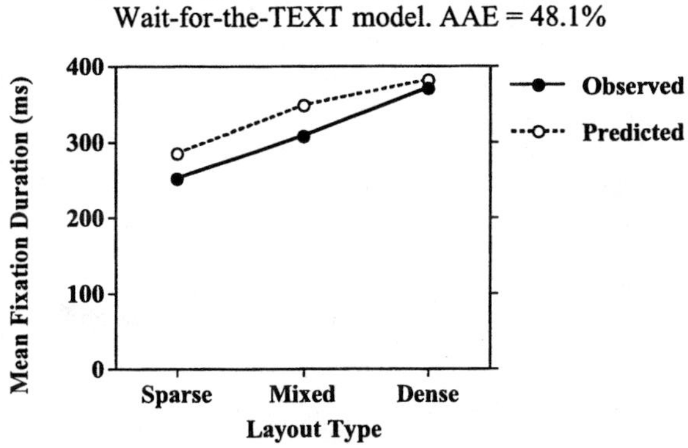

Figure 8. Mean fixation durations observed (solid line) and predicted (dashed line) by the Wait-for-the-TEXT model. AAE = 10.0%

the majority of the error in the predicted data lies in the sparse and mixed layouts. However, the purpose of this modeling was to approximate the ocular-motor behavior in the observed data, so further fine-tuning of fixation durations was not performed. Rather, since the greatest error now lies in the predicted number of fixations per trial the next model will focus on improving the number of fixations produced by the model.

Reduced *Text* Availability Model

A simplifying assumption in previous models directly influenced the predicted number of fixations. The assumption was that all text within the fovea (1 dov) is perceived for every fixation. This results in the model perceiving two to three sparse objects or five to seven dense objects in each fixation. Consequently, the model was able to perceive all *text* in a layout with an equal number of fixations, regardless of the layout density. The observed data suggests that humans do not do this. People require more fixations for dense text. An increase in the number of fixations required for dense objects can be achieved in a number of ways. One way is to reduce the region within which dense *text* can be perceived. Another is to reduce the probability of correctly perceiving *text* based on the size or spacing of the text. Both methods were tested in the models.

It was found previously that processing two to three items per fixation accounts well for the observed number of fixations in a search task (Hornof & Halverson, 2003). Here, we first limited the *text* availability to two to three items per fixations by decreasing the region of *text* availability for dense words to 0.5 dov. The default settings already limited sparse words to two or three per fixation. Perceiving two to three words per fixations resulted in a much better fit for the predicted number of fixations per trial. However, the model was still under-predicting the number of fixations per trial in all layouts.

A second availability function was created to vary the probability of perceiving the *text* property dynamically based on the distance to the closest neighboring object. This is one of several ways to measure density. One advantage of this measure is that it only requires the position of each item on the screen. If an object's closest neighbor was less than 0.15 dov away (a dense object), the probability of perceiving the *text* was 50%. Otherwise, the probability of perceiving the *text* was 90%. These probabilities were chosen because they would result in two to three items, on average, perceived per fixation across densities.

As seen in Figure 9, the predicted mean search time per trial improved considerably. Examining the ocular-motor behavior, we see that the predicted fixation durations improved slightly compared to the Wait-for-*text* Model, and the number of fixations per trial now closely approximates the observed data. The average absolute errors for the three measures range between 6.5% and 8.8%. This has been achieved by only modifying perceptual parameters and fundamental aspects of the strategy (i.e. prepare-then-perform).

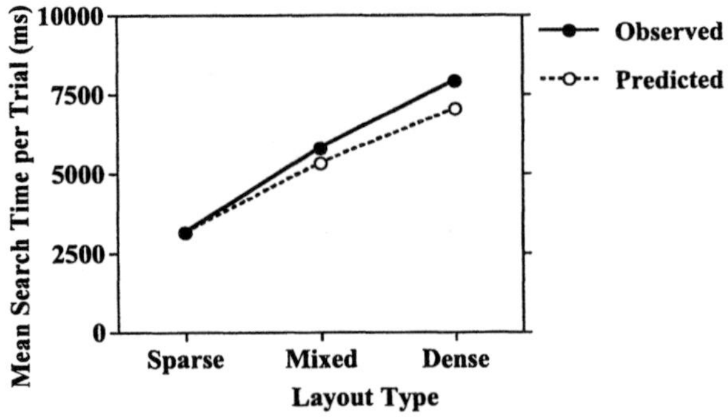

Figure 9. Mean search time per trial observed (solid line) and predicted (dashed line) by the Reduced Text Availability model. AAE = 6.5%

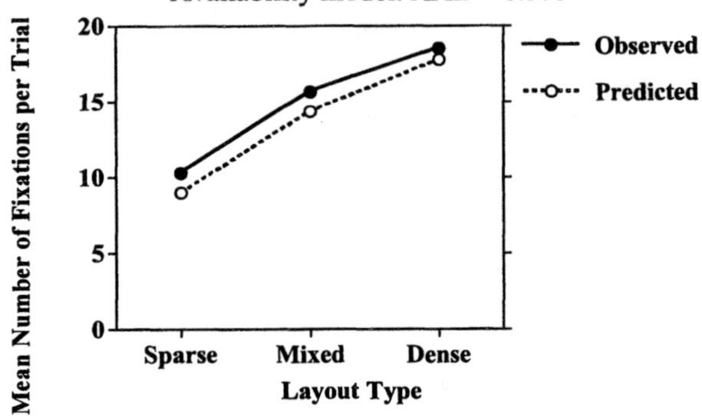

Figure 10. Mean number of fixations per trial observed (solid line) and predicted (dashed line) by the Reduced Text Availability model. AAE = 8.8%.

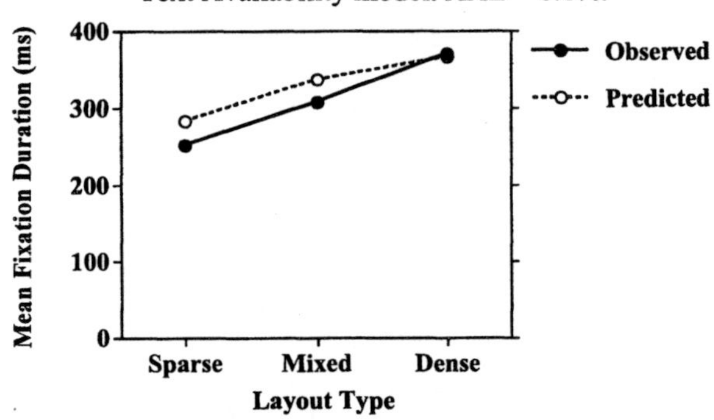

Figure 11. Mean fixation durations observed (solid line) and predicted (dashed line) by the Reduced Text Availability model. AAE = 7.8%

Discussion

Implications for cognitive modeling of visual search

A random search strategy is a reasonable first approximation that allows the analyst to focus on other fundamental ocular-motor activity that affects visual search. If an analyst can initially account for fundamental perceptual and ocular-motor activity with such a parsimonious model, the model may help to constrain the search space of more elaborate strategies.

One means of accounting for the number of fixations in a visual search of words is to limit the number of words perceived per fixation to two to three on average. Hornof (in press) found in that limiting the number of objects perceived per fixation to two to three items helped predict observed search times. The same assumption here helped to predict search time and number of fixations. However, it was not found that reducing the region of perception for denser objects was the best predictor, but rather maintaining the same region of perception and reducing the probability of perceiving denser objects so that an average of two to three items was perceived per fixation.

A straightforward and parsimonious means to account for fixation durations in a visual search of words is to use a prepare-then-perform strategy. The improvement in the predictions resulting from the use of this strategy suggests that future models that account for other aspects of the observed data should constrain the manner in which saccade destinations are selected. The saccade destination should be selected, and the eye prepared to move there, before the currently fixated text is perceived and the current fixation is complete. Otherwise, the predictions for saccade durations will probably be worse than was found here.

Implications for the theory of visual search

Bertera and Rayner (2000) concluded that the effective field of view did not decrease as density increased. The findings here work within that conclusion and expand upon it. It was found that if we tried decreasing the region in which *text* could be perceived (i.e. the effective field of view), that our model under-predicted the number of fixations required to find the target. However, if we left the region in which *text* could be perceived the same size, regardless of density, and changed the probability of perceiving *text* within that region, the models could better account for the observed data. The task modeled in this research differed from that used by Bertera and Rayner. In the current task, density was manipulated by varying the size of text and spacing (which is arguably more ecologically valid, see Halverson & Hornof, 2004). Still, similar conclusions were reached. Future work is required to study the effects of density where text size and spacing vary independently.

Conclusion

This paper presents models for a task that investigates the effect of local density on the visual search of words. A principled approach was used to account for the observed eye movement data. We started with a model that used the default perceptual parameters of EPIC and a random, without-replacement search strategy.

We found that a random search model that examines an average of two to three words per fixation predicts the observed number of fixations per trial data well. Further, it was found that a fixed region for *text* perception and a probability of perceiving *text* that varied with density was a better prediction than a variable region for *text* perception.

We also found that a model that waits for *text* to become available before initiating a saccade, but that prepares the next saccade in parallel with examining the *text*, is a good predictor of the observed fixation durations.

Further work is in progress to account for other observed data, such as the order of visitation and strategy shifts in mixed density layouts. The research presented here has established fundamental strategies and perceptual parameters that will constrain further modeling of the task.

Acknowledgments

This research is supported by the Office of Naval Research grant N00014-02-10440 and the National Science Foundation grant IIS-0308244. Both grants are to the University of Oregon with Anthony Hornof as the principal investigator.

References

Bertera, J. H., & Rayner, K. (2000). Eye movements and the span of effective stimulus in visual search. *Perception & Psychophysics*, 62(3), 576-585.

Byrne, M. D. (2001). ACT-R/PM and menu selection: Applying a cognitive architecture to HCI. *International Journal of Human-Computer Studies*, 55, 41-84.

Card, S. K., Moran, T. P., & Newell, A. (1983). *The Psychology of Human-Computer Interaction.* Hillsdale, NJ: Lawrence Erlbaum Associates.

Gray, W. D., John, B. E., & Atwood, M. E. (1993). Project Ernestine: Validating a GOMS analysis for predicting and explaining real-world task performance. *Human-Computer Interaction,* 8, 237-309.

Halverson, T., & Hornof, A. J. (2004). Local Density Guides Visual Search: Sparse Groups are First and Faster. *Proceedings of the Human Factors and Ergonomics Society 48th Annual Meeting*, New Orleans, LA.

Hornof, A. J. (in press). Cognitive Strategies for the Visual Search of Hierarchical Computer Displays. *Human-Computer Interaction.*

Hornof, A. J., & Halverson, T. (2003). Cognitive strategies and eye movements for searching hierarchical computer displays. *Proceedings of the Conference on Human Factors in Computing Systems*, Ft. Lauderdale, FL.

Kieras, D. E. (2003). Modeling Visual Search in the EPIC Architecture. *Presented at the meeting of Office of Naval Research Grantees in the Area of Cognitive Architectures,* University of Pittsburgh.

Kieras, D. E., & Meyer, D. E. (1997). An overview of the EPIC architecture for cognition and performance with application to human-computer interaction. *Human-Computer Interaction*, 12(4), 391-438.

Ojanpää, H., Näsänen, R., & Kojo, I. (2002). Eye movements in the visual search of word lists. *Vision Research*, 42(12), 1499-1512.

Wilson, M. D. (1988). The MRC Psycholinguistic Database: Machine Usable Dictionary, Version 2. *Behavior Research Methods, Instruments, and Computers*, 20, 6-11.

Integrating models and tools in the context of driving and in-vehicle devices

Bonnie E. John (bej@cs.cmu.edu)
Human-Computer Interaction Institute, Carnegie Mellon University, 5000 Forbes Ave.
Pittsburgh, PA 15213 USA

Dario D. Salvucci (salvucci@cs.drexel.edu)
Department of Computer Science, Drexel University, 3141 Chestnut St.
Philadelphia, PA 19104 USA

Peter Centgraf (centgraf@cmu.edu), Konstantine Prevas (gusp@cmu.edu)
Human-Computer Interaction Institute, Carnegie Mellon University, 5000 Forbes Ave
Pittsburgh, PA 15213 USA

Abstract

The integration of cognitive models offers interesting theoretical and practical benefits to the development of complex models, but at the same time brings up new challenges that the modeling community is only beginning to address. In this paper we present our experiences integrating models and tools for modeling,. Specifically, on the modeling end, we integrated an ACT-R model of driving with an independently-built model of cell phone dialing. To do this, we integrated the driving tool (a minimal-graphics driving simulator that can be controlled by ACT-R) with a modeling-by-demonstration tool, CogTool, that also used ACT-R as its cognitive engine. We found that both the models and tools required re-working because of the integration. The models required new motor movements and methods for switching between tasks. The tools required new specification techniques for placing prototyped devices in the vehicle and demonstration techniques for indicating points of task-switching. Our experiences indicate that while the integration of cognitive models is indeed an approach worthy of pursuit, the field may not understand all the ramifications of such integration until many more integrated models and tools are developed.

Introduction

People often integrate separate skills in the course of their everyday lives. For example, a person learns how to drive separately from learning to use a cell phone, and integrates these two skills when they use a cell phone while driving. Likewise, to make building computational cognitive models of complex domains tractable, a reasonable approach is to integrate independently-built models of different behavior. (Here, we are defining "integration"' of models to mean the joining together of independently-built models of tasks at approximately the same task size, as opposed to the "composition" of models from smaller tasks that are components of a larger task, e.g., the composition of templates at the level of mouse movements or keystrokes into large tasks (John & Gray, 1992; John et al., 2002).)

The integration approach has advantages over constructing complex task models from scratch or even from smaller components. For example, an expert in a particular domain can invest time and expertise in producing a veridical model of an important behavior, and it can be integrated with models of different behavior by other people who are not expert in that domain. The first author took this approach in the early 1990s when modeling the tasks of the NASA Test Director (the person responsible for assuring that all tests are conducted on the Space Shuttle before launch). Because the NASA Test Director's job involves substantial verbal communication, we integrated an independently-developed, in-depth and well-verified model of natural language comprehension (NL-Soar, Lewis, 1999) with a hierarchical goal decomposition model of the NTD's task knowledge (John, Remington & Steier, 1991; Nelson, Lehman, & John, 1994a,b). Even when working within a single cognitive architecture like Soar or ACT-R, the integration approach is not without difficulties (discussed in Nelson, Lehman, John, 1994a), because independently-built models often do not exchange information or pass control in a way that interleaves the tasks appropriately.

Because this approach is relatively new in computational cognitive modeling, each new attempt teaches lessons in how we might proceed with models, methods, and tools to make integration more successful and easier. This paper describes our most recent work in integrating stand-alone Keystroke-Level Models (KLM, Card, Moran & Newell, 1980) of skilled users of information devices with a model of driving an automobile (Salvucci, Boer & Liu, 2001). In this work we integrated both models and modeling tools, to the betterment of both the models and the tools. We will briefly describe the separate components that contributed to this work, then describe our experiences integrating the tools, the resulting models, and an iteration on both the models and the tools.

Separate Contributing Components

The work reported integrates two models, an ACT-R model of driving an automobile (Salvucci, et.al., 2001) and a KLM of dialing a cell phone, and two tools, a "minimally-graphic" simulated driving environment[1] (Salvucci et al., 2001) and CogTool (John et al., 2004). CogTool is a tool for

[1] Salvucci maintains two versions of his driving model for operating a simulator with textured road graphics (see animation at http://hcil.cs.drexel.edu/projects/drivermodel.html) or the "minimally-graphic" simulator used here and shown in Figure 2.

cognitive modeling by demonstration focused on HCI tasks, which itself is an integration of several tools: Dreamweaver™, the Behavior Recorder (Koedinger, Aleven & Heffernan, 2003), ACT-Simple (Salvucci & Lee, 2003), and the ACT-R environment (Bothell, 2004).

ACT-RPM and Salvucci's Model of Driving

The model of driver behavior is an ACT-R model that integrates control, monitoring, and decision making to navigate highway environments with traffic. It uses the full ACT-RPM cognitive architecture, with simulated eyes, hands, and feet as well as ACT-R's cognitive processor. For control, the model employs a two-level model of steering that uses a "far point" on the road to guide predictive steering and a "near point" on the road to center the vehicle. For monitoring, the model encodes its surrounding environment using ACT-R's simulated eyes to maintain situation awareness. For decision-making, the model checks the current situation and decides when to perform maneuvers such as lane changes. Thus, the driver model incorporates both lower-level perception and action for vehicle guidance and higher-level cognition for awareness and decision-making. This driver model has been shown to account for a number of aspects of human highway driving, including nearing the inner curb during curve negotiation and switching gaze to the destination lane at the start of a lane change (see Salvucci et al., 2001).

When the driver model is run in ACT-R, a window opens with a simulated driving environment. This window contains a picture of the simulated environment (i.e., the roadway and other vehicles) from the point of view of the driver (model). The far point and near point are indicated and the model's point of attention and eye fixation are displayed. As the model drives, the road appears to move past the car and the eye fixation point moves as it assesses the driving situation.

CogTool

The CogTool is a new tool for creating KLMs by demonstration (John et al., 2004). The KLM technique uses a very simple framework for modeling skilled performance on a computer-based task. It was originally described in Card, Moran, and Newell (1980) and later in Card, Moran, and Newell (1983, Ch. 8). The KLM makes several simplifying assumptions that provide more constraint than other modeling frameworks in cognitive science. For example, the analyst must specify the method used to accomplish the particular task of interest, which typically entails modeling specific task instances. Furthermore, the specified method is limited to containing only a small set of pre-established keystroke-level primitive operators. Given the task and the method, the KLM uses duration estimates of these keystroke-level operators to predict the time a skilled user will need to execute the task. The original KLM included six types of operators: K to press a key or button, P to point with a mouse to a target on a display, H to home hands on the keyboard or other device, D to draw a line segment on a grid, and R to represent the system response time during which the user has to wait for the system. Each of these operators has an estimate of execution time, either a single value, a parameterized estimate (e.g., K is dependent on typing speed and whether a key or mouse button click, press, or release is involved), or a simple approximating function (e.g., Fitts's Law estimates for P). A final operator, M, represents the pauses sometimes observed between users' actions, presumably to mentally prepare to do that action or a closely-related series of primitive actions. Using a single mental operator instead of decomposing it into different types for different perceptual and cognitive activities was a deliberate simplification. Card, Moran and Newell (1983) explored many different levels of granularity and simplification and determined that this single M was sufficient for many HCI design problems. The KLM also includes a set of five heuristic rules for placing mental operators to account for mental preparation time. An analyst constructing a KLM by hand starts with Rule 0 which places Ms before many Ks or Ps, then applies Rule 1 through Rule 4 to remove many of the Ms just placed. If applied rigorously and consistently, these rules produce a model that matches skilled execution time to well within 20%.

Despite its simplicity, some user interface (UI) designers in Human-Computer Interaction see KLM as a relatively difficult technique to learn and use. Novices creating KLMs by hand often forget to include all physical operators and have difficulty placing Ms rigorously and consistently (John, 1994). CogTool was created in response to these problems, to make it possible for UI designers to create KLMs through demonstration on HTML storyboards. CogTool includes Macromedia Dreamweaver™ extensions that provide a palette of widgets commonly used in interactive devices that can be dragged-and-dropped onto a canvas for WYSIWYG construction of HTML storyboards. In addition, an image can be used as the basis for a storyboard and hotspots inserted in that image to simulate buttons and other interactive devices (see Figure 1). Existing HTML storyboards that were not constructed with our widget palette can be converted to work with modeling-by-demonstration through an "instrument all widgets" command.

After creating the HTML storyboard, an analyst opens the storyboard in Netscape, and demonstrates a task while CogTool's Behavior Recorder (Koedinger et al., 2003) watches the demonstration. An export command then creates a KLM of that task expressed in ACT-Simple, a simplified language that compiles into ACT-R productions (Salvucci & Lee, 2003). The export command inserts Ms (mental operators) automatically, using new rules based on specific widgets to implement Card, Moran and Newell's heuristics (John et al. 2004). The analyst can then load this KLM into ACT-R and run the model. ACT-R operates the HTML storyboard through the Behavior Recorder and produces a time-stamped trace of its activities that predict skilled execution of the task.

Although this particular combination of systems into a single tool for modeling by demonstration is new, it borrows heavily from previous work (CRITIQUE by Hudson et al., 1999 and WorldBuilder reported in Remington et al., 2002). It has been shown to produce valid KLMs for the skilled use

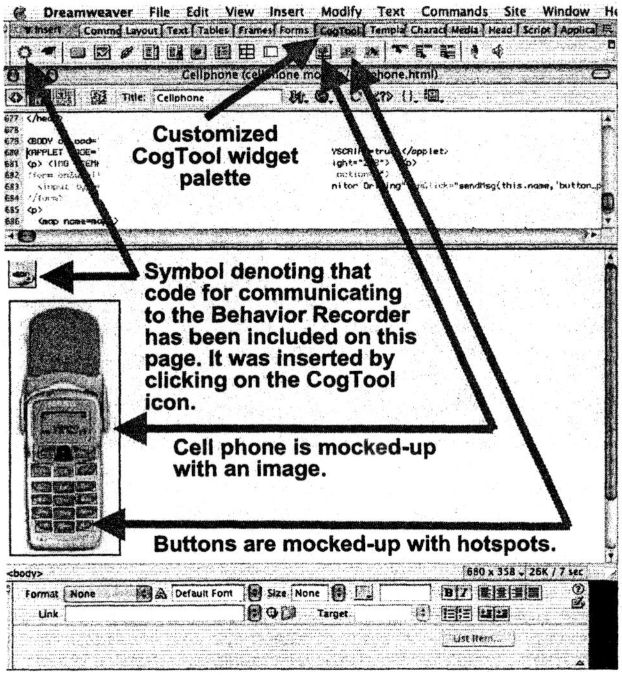

Figure 1: Using CogTool's Dreamweaver™ extensions to mock up a cell phone.

of interactive devices when the CogTool widget palette is used (John et al., 2004).

Storyboards of physical devices typically use an image of the device and hotspots to indicate arbitrary interactive mechanisms, e.g., buttons, knobs and toggle switches. Tasks on such storyboards are more difficult for the CogTool to model because images and hotspots representing different types of mechanisms are all simply tagged as HTML images and hotspots, not as the devices they represent. The CogTool cannot distinguish between them, nor does it possess knowledge about how KLM's rules for inserting mental operators apply to these arbitrary devices. In the absence of such knowledge, CogTool places a mental operator in front of each click on a hotspot, which is likely to over-predict skilled performance times compared to human data for arbitrary devices. To address this problem, we have expanded the CogTool widget set to include some common UI devices that are not standard HTML widgets, e.g., pull-down, cascading menus. Work is continuing to expand valid modeling by demonstration to other common interactive mechanisms found both on computer-based devices and physical devices.

We used CogTool to create an HTML storyboard of the cellular phone used to explore driver distraction in Salvucci, 2001. Demonstrating dialing a telephone number on this image-and-hotspot storyboard produced a stand-alone model of dialing with too many mental operators to match human data, as expected. However, the CogTool produces an intermediate product, namely ACT-Simple code in a text file, which can easily be edited by hand to rectify this problem. The resulting independently-developed model of cell phone dialing is similar to Salvucci's (2001) hand-crafted ACT-R production system.

Integrating the Tools and Models

To make mock-ups of in-vehicle devices as easy to build as mock-ups of desktop interfaces, we enhanced the CogTool so that the models it produces can be integrated seamlessly with the model of driving. Since the CogTool already uses the ACT-R environment, only a few modifications were needed to integrate it with the simulated driving environment.

First, since in-vehicle devices typically use the driver's finger, but the HTML mock-up is demonstrated with a computer mouse, we created an option to export the model as a non-computer-based device. When exporting the model as neither a mouse-based device nor a keyboard-based device, every point-and-click in the demonstration is translated into a simple point based on Fitts's Law, without the click. The original Fitts's Law was derived with measurements where a person tapped a target with a stylus (Fitts, 1954), very similar to a finger tapping the short-travel buttons or touch-screen commonly used on in-vehicle devices. This option is also applicable to modeling interaction with a PDA like a PalmPilot™.

Second, moving to in-vehicle devices required new motor commands in both ACT-R and ACT-Simple. Previously, both languages had operators that homed the hands between keyboard and mouse (KLM's H), pointed with a mouse (KLM's P), and typed with the fingers (KLM's K). To drive while operating an in-vehicle device, we had to add homing between the steering wheel and the device and pressing buttons with the fingers as described above. Though arguably not theoretically interesting, these additions illustrate that new integration and new domains push the development of modeling architectures to cover new classes of human behavior.

Third, an HTML mock-up of a device can be drawn to any scale, but the actual device needs to be placed in the context of the car's dashboard so that eye- and hand-movements can be modeled. Therefore, we needed to record the actual dimensions and location of the in-vehicle device in the storyboard to transmit to the simulator. We added a standard "first-page" to the HTML storyboard with labeled text fields into which the analyst enters the size and location of the device. CogTool reads this first page when the demonstration is exported, creating code in the beginning of the ACT-Simple file that tells the ACT-R environment to place a wireframe of the device in the driving simulation window when the integrated model is run. Figure 2 shows the wireframe of the cellphone's numeric keypad and function buttons on the dashboard console in the lower right.

Resulting Models and Iteration of the Models

Using the integrated tool resulted in an integrated model that ran in ACT-R, displayed the screen shown in Figure 2 so the integrated behavior could be viewed, and predicted relevant measures such as average lateral deviation from the center of the lane and the total time to perform the secondary task on the in-vehicle device. Unfortunately, when examining the traces and the data produced, the first set of models was insufficient to match human data, or even to be plausible. However, several reasons for this failure immediately presented themselves, which were easy to fix and feed back into the design of the tools.

First, the straightforward integration of the models inherited a decision-making portion of the driving model that switched control from driving to the secondary task (cell phone dialing), when the vehicle's position on the road was reasonably close to the center of the lane and stable. This ability to shift tasks had been included in the driving model when Salvucci hand-crafted the task knowledge to dial a cell phone (Salvucci, 2001). This mechanism worked "out of the box" with the CogTool model. However, the independently-developed model of cell phone dialing did not have any mechanism for switching back to driving. The result was a model that switched to dialing when the driving

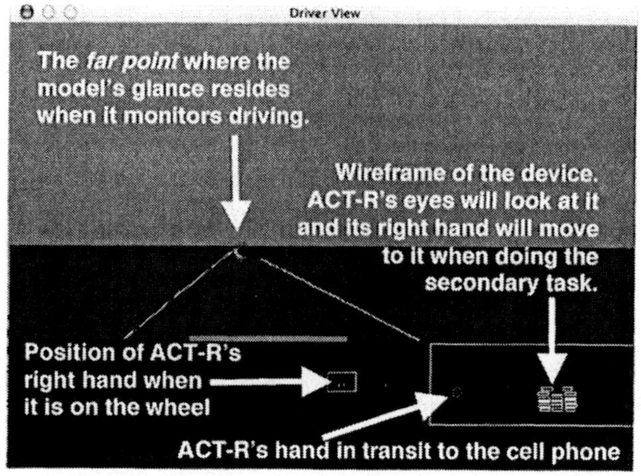

Figure 2. View of ACT-R driving while operating a cell phone mocked up in CogTool.

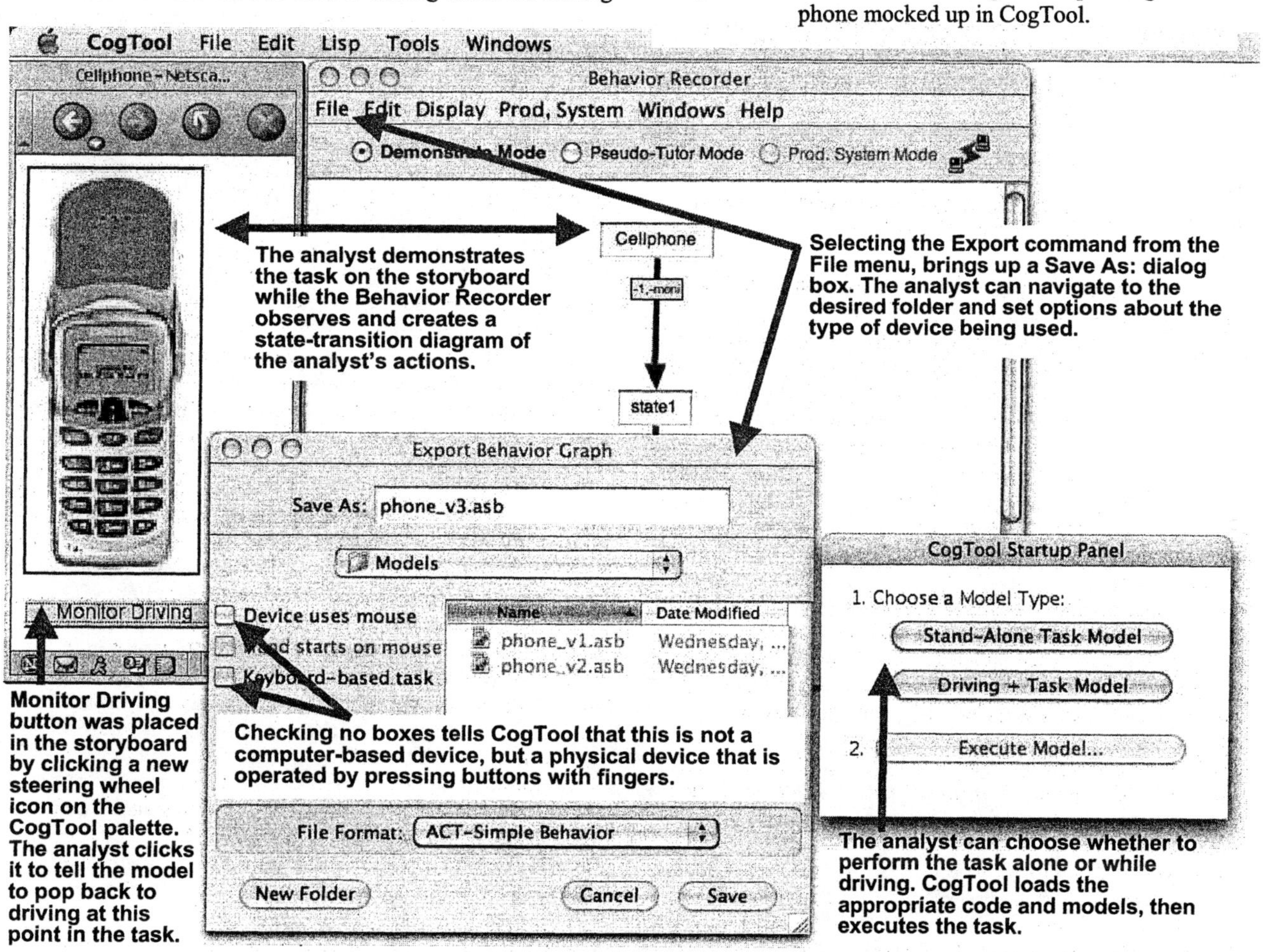

Figure 3. Current version of CogTool being used to demonstrate dialing a phone number on a cell phone. The leftmost window shows the HTML mock-up running in Netscape. The partially-hidden center window shows the Behavior Recorder running and the state transitions it observed in the demo. The window labeled "Export Behavior Graph" in the center is a save dialog box that appears when the behavior observed in a demo is exported to ACT-Simple code (.asb=ACT-Simple Behavior). The rightmost window appears when the Model Launcher portion of CogTool is invoked to let the analyst easily run the model as either a stand-alone task or an in-vehicle information task while driving.

was stable, worked exclusively on that task, and drove off the road! The knowledge of how to interleave secondary tasks with driving in the general case (e.g., Kurosu, 1994) is still under-examined in the field of driver distraction and the subject of further research. However, in response to this problem, we added a "popping" mechanism in ACT-Simple that dictates when the secondary model "pops" back to driving; specifically, the "pop" switches ACT-R's current goal to the driving goal, which runs at least one iteration of vehicle control (i.e., one update of steering angle and pedal depression, taking roughly 200 ms) and then returns switches the goal back to the secondary task. The "pop" can be considered part of a customized executive control process similar to that used by EPIC to switch between tasks in a dual-task environment (see Kieras et al., 2000), or it could be control knowledge that results from learning in a dual-task domain. Both approaches result in behavior similar to our "pop" and match human data in the Wickens' task, a simple dual task (Lallement & John, 1998).

The initial model's lack of knowledge of when to return to driving was compounded by the relatively large mental operators placed by the CogTool. Card, Moran and Newell (1980) used a single type of mental operator, M, with a duration of 1350 ms to denote all operations a skilled user performed that resulted in pauses between keystrokes or mouse movements. Thus, M included visual search, recall, decision-making, verification, and other unspecified perceptual and mental operations. This simplification provided a good fit to keystroke data and was sufficient for an engineering tool to differentiate between different human-computer interaction designs. The models produced by the CogTool implement this M by occupying ACT-R's cognitive processor for 1200 ms and usually launching an eye movement that takes 150 ms (totaling Card, Moran and Newell's 1350 ms). This is again an approximation that is sufficient for evaluating different interfaces when they are simulated as the only task a user is doing. However, when CogTool models are integrated with the driving model, this M takes ACT-R's cognitive processor away from driving far too long for it to maintain a good match to human data. We explored breaking this large mental operator up into smaller pieces that each pop back to driving, working under the notion that people can interleave their mental operations with driving, rather than treating the mental operator as one atomic unit. In this exploration we found that splitting M into three segments of 400 ms each produced a better fit to the human driver data (Salvucci et al., 2004).

Iteration of the Tools

After manipulating the models directly in both ACT-Simple and ACT-R to get a good fit to human data, we can now feed this information back into the design of the tools. For example, the CogTool Dreamweaver™ tool palette now includes an icon that looks like a steering wheel. When this icon is clicked, it places an HTML button labeled "Monitor Driving" at the bottom of the page, outside the boundaries of the mock-up (Figure 3). When demonstrating a task on the mock-up, the analyst can click on the Monitor Driving button to indicate that the device model should relinquish control to driving at this point in the task. This gives the analyst an easy way to control the interleaving of driving and the secondary task.

We have also already included the shorter duration mental operators in CogTool. Now when an M is automatically inserted in a model, it is also automatically broken into three 400 ms segments with a pop back to driving between the segments, as was found to fit best to empirical data (Salvucci et al., 2004).

Because the data on how humans choose to break up secondary tasks and return to driving is not yet conclusive, we wish to make it easy for an analyst to explore many possibilities of when a model pops back to driving from the secondary task. The Monitor Driving button and the sequence of three 400 ms mental operators are two mechanisms already implemented in CogTool for making this easy. However, we intend to make it possible to automatically explore many possible combinations of popping and mental operation durations. Automatic placement of these operations combined with facilities for running and analyzing multiple instances of a task should help researchers quickly explore the space of modeling styles and parameters.

Finally, when more models and data have yielded sufficient information with sufficient confidence, we intend to build heuristics into the CogTool that automatically create models of in-vehicle devices that interleave with driving in a plausible manner.

Conclusions

We conclude from this exercise that integrating independently-built computational models is indeed an approach worthy of pursuit. However, researchers taking this approach in the near future will most likely encounter unforeseen difficulties. Until many more independently-built models are integrated, the field will not understand all the ramifications of integration, nor will it be able to build integrated tools that support this style of modeling in general.

In both the NASA Test Director model mentioned in the introduction and the in-vehicle models described here, the issue of how control is passed from one model to the other arose in unforeseen ways. This issue is likely to continue to be a major challenge until either a theory of general task integration is developed, validated, and incorporated into integration tools, or we convince ourselves that such a general theory is not viable and many examples of domain-specific solutions populate our field. As a third possibility, models may be developed of how people learn to integrate two well-practiced tasks, as was investigated by Chong (1997). In any case, the integration approach to cognitive modeling clearly has both scientific and engineering implications, and will continue to spur improvement and development of theoretical cognitive architectures and also application to real-world complex tasks.

Acknowledgments

This research was supported by grants from the Office of Naval Research (N00014-03-1-0086 & N00014-03-1-0036) to the first and second authors (respectively), a research contract between General Motors and the first author, and a gift from Ford Motor Company to the second author. The views and conclusions contained in this document are those of the authors and do not represent the official policies, either expressed or implied, of General Motors, Ford Motor Company, the Office of Naval Research or the U. S. Government.

References

Anderson, J. R., & Lebiere, C. (1998). *The atomic components of thought*. Hillsdale, NJ: Erlbaum

Bothell, D. (2004) *ACT-R Environment Manual*. Available for download at http://act-r.psy.cmu.edu/software/

Byrne, M. D. & Anderson, J. R. (1998). Perception and action. In J. R. Anderson & C. Lebiere (Eds.) *The Atomic Components of Thought,* (pp. 167-200). Mahwah, NJ: Lawrence Erlbaum.

Card, S., Moran, T., & Newell, A. (1980). The Keystroke-Level Model for user performance time with interactive systems. *Communications of the ACM, 23*, 393-410.

Card, S. K., Moran, T.P. and Newell, A. *The Psychology of Human-Computer Interaction.* Lawrence Erlbaum Associates, Hillsdale, NJ, USA (1983).

Chong, R. & Laird, J. (1997). Identifying dual-task executive process knowledge using EPIC-Soar. *Proceedings of the nineteenth annual conference of the Cognitive Science Society*, 107–112.

Fitts, P. M. (1954). The information capacity of the human motot system in controlling amplitude of movement. *Journal of Experimental Psychology, 47*, 381-391.

Hudson, S. E., John, B. E., Knudsen, K., and Byrne, M.D. (1999). A tool for creating predictive performance models from user interface demonstrations. *UIST'99: Proceedings of the ACM Symposium on User Interface Software and Technology, CHI Letters 1*(1), 93-102.

John, B. E. (1994) Toward a deeper comparison of methods: A reaction to Nielsen & Phillips and new data. In *Proceedings Companion of CHI, 1994* (Boston, MA, April 24-28, 1994) ACM, New York, 1994. pp. 285-286.

John, B. E. & Gray, W. D. *GOMS Analyses for Parallel Activities*. Tutorial materials, presented at CHI, 1992 (Monterey, California, May 3- May 7, 1992), CHI, 1994 (Boston MA, April 24-28, 1994) and CHI, 1995 (Denver CO, May 7-11, 1995) ACM, New York.

John, B. E., Prevas, K., Salvucci, D. D., & Koedinger, K. (2004). Predictive human performance modeling made easy. To appear at *CHI 2004*.

John, B. E., Remington, R. W., & Steier, D. M. (1991) *An analysis of space shuttle countdown activities: Preliminaries to a computational model of the NASA test director*. Carnegie Mellon University School of Computer Science Technical Report No. CMU-CS-91-138.

John, B., Vera, A., Matessa, M., Freed, M., & Remington, R. (2002) Automating CPM-GOMS. *Proceedings of CHI, 2002* (Minneapolis, April 20-25, 2002) ACM, New York.

Kieras, D. E., Meyer, D. E., Ballas, J. A., & Lauber, E. J. (2000). Modern computational perspectives on executive mental processes and cognitive control: Where to from here?. In S. Monsell & J. Driver (Eds.), Control of Cognitive Processes: Attention and Performance XVIII (pp. 681-712). Cambridge, MA: MIT Press.

Koedinger, K. R., Aleven, V., and Heffernan, N. (2003). Toward a rapid development environment for cognitive tutors. *Artificial Intelligence in Education: Shaping the Future of Learning through Intelligent Technologies, Proceedings of AI-ED 2003* (pp. 455-457). IOS Press.

Kurosu, M. (1994). Dual task model: An evaluation model for the complex operation. In *Proceedings of ACM CHI'94 Conference on Human Factors in Computing Systems* (pp. 125-126). New York: ACM Press.

Lallement, Y., & John, B. E. (1998). Cognitive architecture and modeling idiom: An examination of three models of the Wickens' task. Proceedings of the Twentieth Annual Conference of the Cognitive Science Society, August 1998.

Lewis, R. L. (1999). Specifying architectures for language processing: Process, control, and memory in parsing and interpretation. In Crocker, M., Pickering, M., and Clifton, C. Jr., (editors) *Architectures and Mechanisms for Language Processing*. Cambridge University Press.

Nelson, G., Lehman, J. F., John, B. E. (1994a) Experiences in interruptible language processing, In Proceedings of the 1994 AAAI Spring Symposium on Active Natural Language Processing, 1994.

Nelson, G. H., Lehman, J. F., & John, B. E. (1994b) Integrating cognitive capabilities in a real-time task. *Proceedings of the Sixteenth Annual Conference of the Cognitive Science Society*, August 1994. pp. 353-358.

Remington, R. John, B. E., Vera, A., Matessa, M., Freed, M., Dalal, M., Harris, R., & Dahlman, E. (2002) *Apex/CPM-GOMS: Modeling human performance in applied HCI domains.* Tutorial materials presented at the Twenty-Fourth Annual Conference of the Cognitive Science Society, (George Mason University, Fairfax VA. August 7, 2002).

Salvucci, D. D. (2001). Predicting the effects of in-car interface use on driver performance: An integrated model approach. *International Journal of Human-Computer Studies, 55*, 85-107.

Salvucci, D. D., Boer, E. R., & Liu, A. (2001). Toward an integrated model of driver behavior in a cognitive architecture. *Transportation Research Record,* 1779, 9-16.

Salvucci, D. D., John, B. E., Prevas, K., & Centgraf, P. (2004) *Interfaces on the road: Rapid evaluation of in-vehicle devices.* Manuscript in preparation.

Salvucci, D. D., & Lee, F. J. (2003). Simple cognitive modeling in a complex cognitive architecture. In Human Factors in Computing Systems: *CHI 2003 Conference Proceedings* (pp. 265-272). New York: ACM Press.

Salvucci, D. D., & Macuga, K. L. (2002). Predicting the effects of cellular-phone dialing on driver performance. *Cognitive Systems Research,* 3, 95-102.

A Preliminary Model of Pronoun/Verb Co-occurrences in Child-Directed Speech

Aarre Laakso (alaakso@indiana.edu)
Linda B. Smith (smith4@indiana.edu)
Department of Psychology, 1101 E. 10th Street
Bloomington, IN 47405 USA

Abstract

Given the selectional restrictions on the kinds of subjects and objects that a verb may take, it seems likely that children learn verbs partly by exploiting statistical regularities in co-occurrences between verbs and nouns. This paper explores the role of pronouns in this process. Although pronouns are semantically "light," they dominate nouns in the speech children hear and systematically partition important classes of verbs. We show that a statistical learner can exploit these regularities to constrain the possible verbs that might fit in a simple syntactic frame.

Introduction

Pronouns stand for central elements of adult conceptual schemes—as Quine pointed out, pronouns "are the basic media of reference" (Quine, 1980, p. 13). Most syntactic subjects in spontaneous spoken adult discourse are pronouns (Chafe, 1994), and preferred argument structure analyses have suggested that, for information processing reasons, pronouns *should* be the most common subjects of transitive and copular clauses, because they maintain the topic (DuBois, Kumpf, & Ashby, 2003). When addressing their children, English-speaking mothers often begin with a high-frequency pronoun — *you* and *I* occur most frequently (Valian, 1991). Parents use the inanimate pronoun *it* far more frequently as the subject of an intransitive sentence than of a transitive one (Cameron-Faulkner, Lieven, & Tomasello, 2003). As Cameron-Faulkner et al. note, this suggests that intransitive sentences are used more often than transitives for talking about inanimate objects. It also suggests, we would note, that the use of the inanimate pronoun might be a cue for the child as to some aspects of the verb (i.e., whether it is transitive or intransitive). Similarly, Childers & Tomasello (2001) have suggested that pronouns may form the fixed element in lexically-specific frames acquired by early language learners—so-to-speak "pronoun islands," something like Tomasello's (1992) "verb islands."

What has not been studied is how these "basic media of reference" may help children learn other words by virtue of systematic co-occurrences. We address this issue by measuring the statistical regularities among the uses of pronouns and verbs in a large corpus of parent and child speech and by modeling them with a connectionist network.

Experiment 1

The first experiment consisted of a corpus analysis to identify patterns of co-occurrence between pronouns and verbs in the child's input.

Method

Parental utterances from the CHILDES database (MacWhinney, 2000) were coded for syntactic categories, then subjected to cluster analysis. The target children in the transcripts were aged approximately 1;4 – 6;1.

Materials The following corpora were used: Bates, Bliss, Bloom 1970, Brown, Clark, Cornell, Demetras Working, Gleason, Hall, Higginson, Kuczaj, MacWhinney, Morisset, New England, Post, Sachs, Suppes, Tardiff, Valian, Van Houten, Van Kleeck and Warren-Leubecker.[1] Coding was performed using an Internet application that randomly selected transcripts, assigned them to coders as they became available, collected coding input, and stored it in a MySQL database. The application occasionally assigned the same transcript to all coders, in order to measure reliability. Five undergraduate coders were trained on the coding task and the use of the system.

Procedure Each main tier line was coded for speaker, addressee, and syntactic frame (no verb, question, passive, copula, intransitive, transitive or ditransitive). Each word was then coded for its syntactic category in that utterance (subject, auxiliary, verb, direct object, indirect object and oblique — others were ignored). In total, 59,977 utterances were coded from 123 transcripts. *All* of the coders coded 7 of those transcripts for the purpose of measuring reliability. Average inter-coder reliability (measured for each coder as the percentage of items coded exactly the same way they were coded by each other coder) was 86.1%.

We only considered parental child-directed speech (PCDS), defined as utterances where the speaker was a parent and the addressee was a target child. Clauses with no verb, questions, passives and copulas were excluded from further analysis; thus, the analysis was conducted using only

[1] Citations to original sources for the corpora are omitted here due to space constraints. The full references for each corpus may be found in (MacWhinney, 2000).

clauses that were intransitives, transitives, or ditransitives, a total of 12,377 clauses.

From these clauses, we formed 2 matrices: a verbs-by-subjects matrix and a verbs-by-objects matrix. The verbs-by-subjects matrix contained only verbs used with an overt subject; its size was 621 verbs by 317 nouns (subjects). The verbs-by-objects matrix contained only verbs used with a direct object; its size was 524 verbs by 907 nouns (objects). Each cell of each matrix contained the proportion of times that verb was used with that noun (as subject or object) in a coded clause.

We then performed 4 cluster analyses. First, we took the 50 nouns most commonly used as objects and clustered them according to their proximity in verb space, i.e., the space formed by considering each verb as a dimension. Each noun was placed along each dimension according to the proportion of times it was used with the corresponding verb. Hence, a noun never used as the object of a particular verb would be at 0, and a noun always used as the object of a particular verb would be at 1. Second, we clustered the 50 most common subject-nouns in verb space. Third, we took the 50 verbs most commonly used with objects and clustered them according to their proximity in noun space (defined analogously to verb space). Finally, we clustered the 50 most common verbs-with-subjects in noun space. As another means of understanding the structure of the data, for each of the four cluster diagrams, we also plotted the corresponding words in the principal components of the relevant space.

Results

We cannot show all of the cluster diagrams or principal components plots here due to space limits. However, the plot of verbs in the principal components of the space of syntactic objects is shown in Figure 3 on the next page, and the corresponding cluster diagram is shown in Figure 4, also on the next page

Overview The most frequent nouns in the corpus — both subjects and objects — are pronouns, as shown in Figures 1 and 2. The objects divided the most common verbs into three main classes: verbs that take the pronoun *it* and concrete nouns as objects, verbs that take complement clauses, and verbs that take specific concrete nouns as objects. This may be observed in Figures 3 and 4 — approximately the lower 1/3 of the verbs in Figure 4 take complement clauses instead of nominal objects, whereas the middle 1/3 take *it* as an object, and the upper 1/3 take primarily concrete nouns. This may also be observed in Figure 3, in which the verbs that take clauses as objects are clumped in the lower right, the verbs that take *it* as an object are clumped in the lower left, and the verbs that take a variety of concrete nouns are scattered across the rest of the figure. The subjects divided the most common verbs into four main classes: verbs whose subject is almost always *I*, verbs whose subject is almost always *you*, verbs whose subject is almost always either *I* or *you*, and other verbs. The verbs divided the most common object nouns into a number of classes, including objects of telling and looking verbs, objects of having and wanting verbs, and objects of putting and getting verbs. The verbs also divided the most common subject nouns into a number of classes, including subjects of having and wanting verbs, and subjects of thinking and knowing verbs.

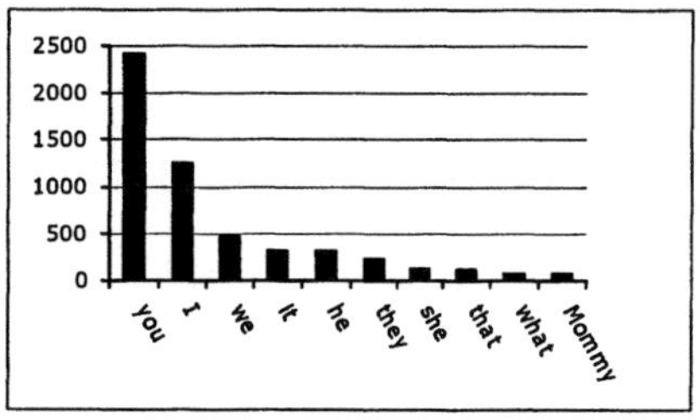

Figure 1: The 10 most frequent subjects in PCDS by their number of occurrences

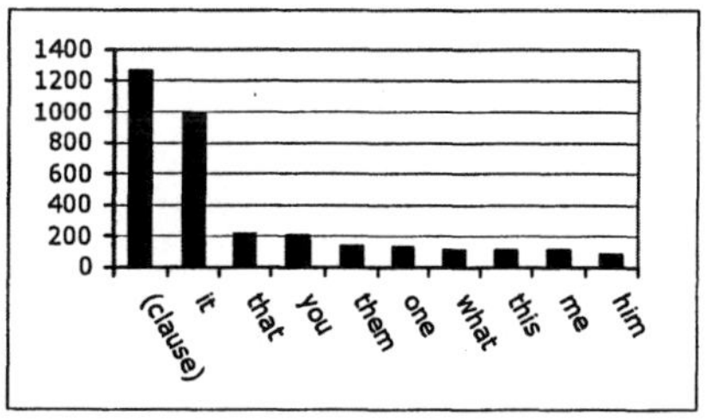

Figure 2: The 10 most frequent objects in PCDS by their number of occurrences.

Verbs that take *it* as an object The verbs that take *it* as their most common object include verbs of motion and transfer, as shown in Table 1.

Table 1: Verbs most commonly used with object *it*.

Verb	Total	it (#)	it (%)
turn	56	33	58.9
throw	36	20	55.5
push	25	13	52.0
hold	42	19	45.2
break	36	16	44.4
leave	27	12	44.4
open	36	15	41.7
do	256	105	41.0
wear	25	10	40.0
take off	24	9	37.5
put	276	93	33.7
get	348	74	21.3
take	106	22	20.8
put on	42	8	19.0
buy	50	9	18.0
give	85	14	16.5
have	340	26	7.6

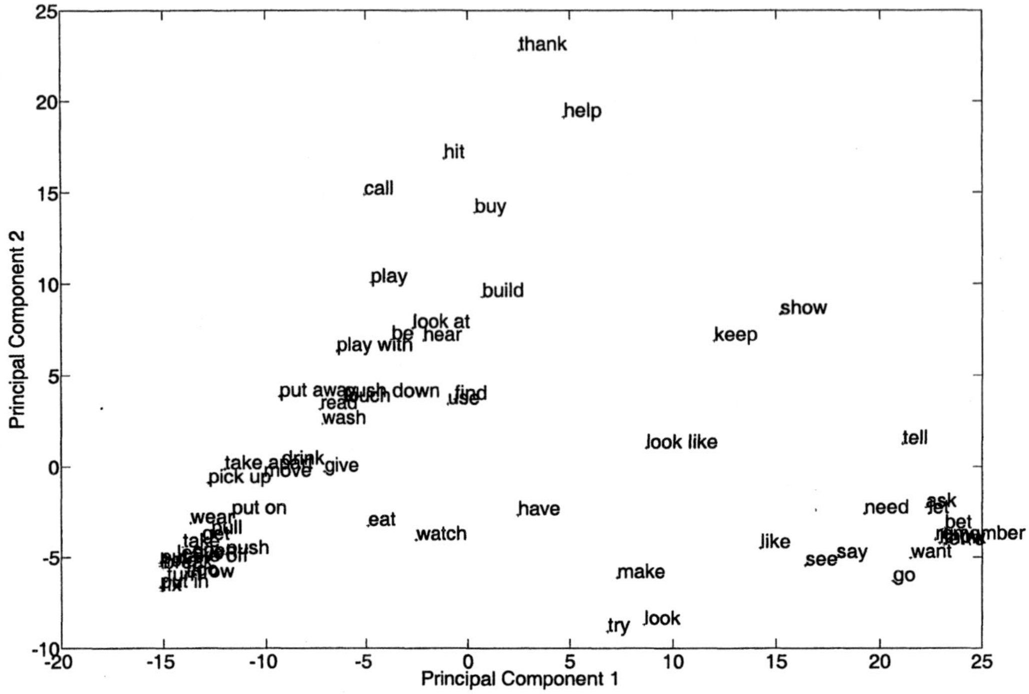

Figure 3: Verbs plotted in the first two principal components of syntactic object space.

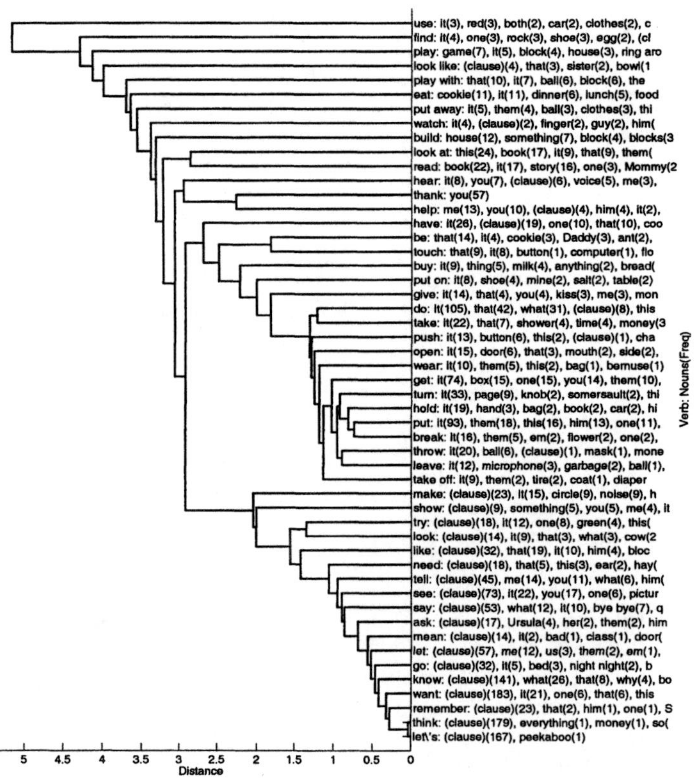

Figure 4: Cluster diagram of verbs in syntactic object space.

Verbs that take complement clauses Most verbs that did not take *it* as their most common object instead took complement clauses. These are primarily psychological verbs, as shown in Table 2.

Verbs that take *I* as a subject Verbs whose most common subject is *I* include *bet* (23 out of 23 uses with a subject, or 100%), *guess* (21/22, 95.4%), *think* (212/263, 80.6%), and *see* (95/207, 45.9%). Parents were not discussing their gambling habits with their children — *bet* was being used to indicate the epistemic status of a subsequent clause, as were the other verbs.

Table 2: Verbs most used most with complement clauses.

Verb	Total	<clause> (#)	<clause> (%)
think	187	179	95.7
remember	31	23	74.2
let	78	57	73.1
know	207	141	68.1
ask	29	17	58.6
go	55	32	58.2
want	317	183	57.7
mean	25	14	56.0
tell	115	45	39.1
try	51	18	35.3
say	175	53	30.3
look	48	14	29.2
need	64	18	28.1
see	266	73	27.4
like	123	32	26.0
show	36	9	25.0
make	155	23	14.8

Table 3: Some verbs commonly used with subject *I* or *you*.

Verb	Total	I (#)	I (%)	you (#)	you (%)
bet	23	23	100	0	0
guess	22	21	95.4	0	0
think	263	212	80.6	38	14.4
see	207	95	45.9	50	24.1
mean	32	15	46.9	12	37.5
know	360	159	44.2	189	52.5
remember	23	9	39.1	12	52.2
like	134	20	14.9	86	64.2
want	270	34	12.6	192	71.1
need	65	5	7.7	33	50.8

Verbs that take *you* as a subject Verbs whose most common subject is *you* include *like* (86 out of its 134 total uses with a subject, or 64.2%), *want* (192/270, 71.1%), and *need* (33/65, 50.8%). These verbs are being used to indicate the deontic status of a subsequent clause, i.e., the subject's disposition or inclination, volition, or compulsion with respect to the proposition expressed by the complement.

Verbs that take *you* or *I* as a subject Verbs that take *I* and *you* more or less equally as subject include *mean* (15 out of 32 uses, or 46.9%, with *I* and 12 of 32 uses, or 37.5%, with *you*), *know* (*I*: 159/360, 44.2%; *you*: 189/360, 52.5%), and *remember* (*I*: 9/23, 39.1%; *you*: 12/23, 52.2%).

Subjects of *think* and *know* The subject *I* appeared most frequently with the verbs *think* and *know*.

Discussion

Although pronouns are semantically "light," their particular referents determinable only from context, they may nonetheless be potent forces on early lexical learning by statistically pointing to some classes of verbs as being more likely than others. The results of Experiment 1 clearly show that there are statistical regularities in the co-occurrences of pronouns and verbs that discriminate between broad classes of verbs. The verb clusters identified in Experiment 1 share more than their associations with pronouns — each cluster corresponds roughly to a broad class of verbs with similar semantic aspects. Specifically, when followed by *it*, the verb is likely to describe physical motion, transfer, or possession. When followed by a relatively complex complement clause, by contrast, the verb is likely to attribute a psychological state. If the subject is *I*, the verb is likely to have to do with thinking or knowing, whereas if the subject is *you*, *she*, *we*, *he*, or *they*, the verb is likely to have to do with having or wanting. This regularity most likely reflects the ecology of parents and children — parents "know" and children "want" — but it could nonetheless be useful in distinguishing these two classes of verbs.

The results thus far show that there are potentially usable regularities in the statistical relations between pronouns and verbs. However, they do not show that these regularities can be used to cue the associated words.

Experiment 2

To demonstrate that the regularities in pronoun-verb co-occurrences in parental speech to children can actually be exploited by a statistical learner, we trained a connectionist network to auto-associate subject-verb-object "sentences" from the input, then tested it on individual verbs and pronouns.

Method

Data The network training data consisted of the subject, verb, and object of all coded utterances that contained the 50 most common subjects, verbs and objects. There were 5,835 such utterances. The inputs used a localist coding wherein there was one and only one input unit out of 50 activated for each subject, and likewise for each verb and each object. Absent and omitted arguments were counted among the 50, so, for example, the utterance "John runs" would have 3 units activated even though it only has 2 words — the third unit being the "no object" unit. With 50 units each for subject, verb and object, there were a total of 150 input units to the network. Active input units had a value of 1, and inactive input units had a value of 0.

Network Architecture The network consisted of a two-layer 150-8-150 unit autoassociator with a logistic activation function at the hidden layer and three separate softmax activation functions (one each for the subject, verb and object) at the output layer—see Figure 5. Using the softmax activation function, which ensures that all the outputs in the bank sum to 1, together with the cross-entropy error measure, allows us to interpret the network outputs as probabilities (Bishop, 1995). The network was trained by the resilient backpropagation algorithm (Riedmiller & Braun, 1993) to map its inputs back onto its outputs. It is well known that this sort of network performs nonlinear dimensionality reduction at its hidden layers, extracting statistical regularities from the input data.

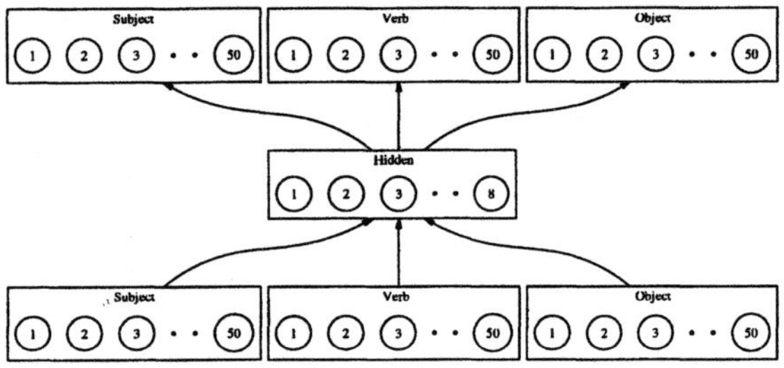

Figure 5: Network architecture

Training The data was randomly assigned to two groups: 90% of the data was used for training the network, while 10% was reserved for validating the network's performance. Starting from different random initial weights, five networks were trained until the cross-entropy on the validation set reached a minimum for each of them. (Multiple networks were used in order to ensure that the results were not idiosyncratic.) Training stopped after approximately 150 epochs of training, on average. At that point, the networks were achieving about 81% accuracy on correctly identifying subjects, verbs and objects from the training set. Near perfect accuracy on the training set could have been achieved by further training, with some loss of generalization. We decided that it was better to avoid over fitting.

Testing After training, the networks were tested with incomplete inputs corresponding to isolated verbs and pronouns. For example, to see what a network had learned about *it* as a subject, it was tested with a single input unit activated — the one corresponding to *it* as subject. The other input units were set to 0. Activations at the output units were recorded. The results presented below report averages over all five networks.

Results

The networks learn many of the simple co-occurrence regularities observed in the data, as well as higher-order co-occurrences. When tested on the object *it*, the most activated verbs are *get*, *hold*, *take* and *have*, which are among the most common verbs with *it* in PCDS (see Table 1). Similarly, *tell*, *make* and *say* are the most activated verbs when networks are tested with the *clause* unit activated in the object position, and they are also among the verbs most commonly associated with a *clause* in the input (Table 2).

However, the networks do not merely learn the relative frequencies of pronouns with verbs. For example, the verbs most activated by the subject *you* are *have* and *get*, neither of which appears in Table 3. The reason for this, we believe, is that the subject *you* is strongly associated with the object *it*, and the object *it*, as mentioned in the previous paragraph, is strongly associated with the verbs *have* and *get*. The difference may be observed most clearly when the network is prompted simultaneously with *you* as the subject and *clause* as the object. In that case, the verb *want* is strongly preferred and, though *get* still takes second place, *tell* and *know* rank third and fourth, respectively — consistent with the results in Table 1. This demonstrates that the network model is sensitive to high-order correlations among words in the input, not merely the first-order correlations between pronoun and verb occurrences.

Conclusion

We have shown that there are statistical regularities in co-occurrences between pronouns and verbs in the speech that children hear from their parents. We have also shown that a simple statistical learner can exploit these regularities, including subtle higher-order regularities that are not obvious in a casual glance at the input data. Furthermore, it can use these regularities to predict the verb in an incomplete sentence.

There are several ways in which this might help children learn verbs. First, pronouns may 'highlight' verbs by consistently bracketing the verb with simple, frequent markers, making it easier to segment the verb from the speech stream. Second, *which* pronouns are used may indicate the animacy, gender, and number of the participants in the action or event that an utterance describes, and the *order* of the pronouns may further indicate temporal sequence or causal direction. Finally, one set of verb-pronoun co-occurrences may lead to another, in the following sense: once the child has learned at least one verb and its pattern of correlations with pronouns, when she hears another verb being used with the same or a similar pattern of correlations, she may hypothesize that the unknown verb is similar to the known verb in some respects. In a sense, this is what our network does — although it does not learn new verbs by this method, it does identify the broad class of a missing verb purely based on its co-occurrences with pronouns and other high-frequency nouns.

We do not claim that the network that we used in Experiment 2 is, by itself and in its current state, an adequate cognitive model. We do claim, however, that it is a suggestive first step. In part, we used the model to study some statistical properties of the data itself, namely higher-order associations. That was not the only purpose of introducing the model, though — the other purpose was to demonstrate that it is possible for a simple statistical learner to use the regularities in the data to predict the class of a missing verb. We would agree that it is very unlikely that children autoassociate simple word patterns; on the other hand, we would insist that it is highly likely that children somehow keep track of lexical co-occurrence patterns. The autoencoder, while assuredly nowhere close to an exact model of the child's processing of parental speech and subsequent lexical learning, is a first step in that direction, to be refined by subsequent research.

The next step in our research is to show that children actually pick up on these regularities. We could be wrong that the lexical co-occurrences are significant — perhaps what children really attend to are referential co-occurrences. While the words *I* and *you* will generally correspond with the referents *me* and *Mommy* / *Daddy* in PCDS, the referents of other pronouns will be more variable. To the extent that children do attend to the lexical regularities, we predict, they should, at a minimum, use pronouns and verbs together with roughly the same frequencies that they hear in their parents' speech. Measuring frequencies is the subject of research in progress using the coded corpus data from Experiment 1 to compute the correlations in the patterns of co-occurrence between verbs and pronouns in parental speech and children's speech.

A further step would be to show that children not only pick up on the regularities, and exhibit them in production, but also use them to comprehend verbs. That is, children should better comprehend ordinary but relatively infrequent verbs when they are presented in frames that are consistent with regularities in the inputs, as opposed to when they are presented in frames that are inconsistent with those regularities. This is a strong test of the network model — to the extent that the model is at all suitable, children should be able to perform at least as well as it does at guessing known, but missing or garbled verbs. In particular, they should be sensitive to higher-order correlations, as our model predicts.

The third and most stringent test is whether children not only exhibit these regularities in production and use them to comprehend known verbs, but also use them to learn novel verbs. Thus, future experiments are planned to measure children's sensitivity to higher-order correlations by testing their preferences for different scenes when primed with sentences containing nonce verbs and pronouns, such as "You blick it" and "I moop him." If children are better able to learn novel verbs when they are presented in pronoun frames consistent with the regularities we have observed, this provides strong support for our hypothesis that pronoun-verb co-occurrences help children learn verbs.

Work is also underway to collect comparable crosslinguistic data from both Japanese and Tamil, relatively verb-heavy languages with frequent argument dropping and case-marked pronouns referring to various levels of social status. Finally, we intend to refine our model along the way, to incrementally approach a fuller and more accurate account of the child's learning processes.

Acknowledgements

This research was supported by NIMH grant number ROI MH 60200. Additional thanks go to our coders, to members of the Cognitive Development Laboratory at IU for useful discussions of these results, and to several anonymous reviewers for helpful comments on an earlier draft.

References

Bishop, C. M. (1995). *Neural Networks for Pattern Recognition.* Oxford: Oxford University Press.

Cameron-Faulkner, T., Lieven, E. V. M., & Tomasello, M. (2003). A construction-based analysis of child directed speech. *Cognitive Science, 27,* 843-873.

Chafe, W. L. (1994). *Discourse, Consciousness and Time: The Flow and Displacement of Conscious Experience in Speaking and Writing.* Chicago: University of Chicago Press.

Childers, J. B., & Tomasello, M. (2001). The role of pronouns in young children's acquisition of the English transitive construction. *Developmental Psychology, 37*(6), 739-748.

DuBois, J. W., Kumpf, L. E., & Ashby, W. J. (Eds.). (2003). *Preferred Argument Structure: Grammar as an Architecture for Function.* Amsterdam: John Benjamins Publishers.

MacWhinney, B. (2000). *The CHILDES Project: Tools for Analyzing Talk* (3rd ed. Vol. 2: The Database). Mahwah, NJ: Lawrence Erlbaum Associates.

Quine, W. V. O. (1980). On what there is. In W. V. O. Quine (Ed.), *From a Logical Point of View* (2nd ed.). Cambridge, MA: Harvard University Press.

Riedmiller, M., & Braun, H. (1993). *A direct adaptive method for faster backpropagation learning: The Rprop algorithm.* Paper presented at the IEEE International Conference on Neural Networks 1993 (ICNN 93), San Francisco, CA.

Tomasello, M. (1992). *First Verbs: A Case Study of Early Grammatical Development.* Cambridge: Cambridge University Press.

Valian, V. (1991). Syntactic subjects in the early speech of American and Italian children. *Cognition, 40*(1-2), 21-81.

Comparing Agent-Based Learning Models of Land-Use Decision Making

Tei Laine (telaine@cs.indiana.edu)
Indiana University
Computer Science Department and the Cognitive Science Program
150 S. Woodlawn Ave.
Bloomington, IN 47405-7104 USA

Jerome Busemeyer (jbusemey@indiana.edu)
Indiana University
Psychology Department
1101 E. 10th Street
Bloomington, IN 47405-7007 USA

Abstract

An agent-based model, incorporating a small set of primarily agent-based variables, was designed to explain private land-use decision making. Agents are land-owners, who allocate their labor and land for different uses in regular time intervals. The goal is to understand what kind of spatial patterns emerge from different agent characteristics, and decision and learning mechanisms. Landscapes produced by two different learning models are compared to actual land-cover data. By calculating a set of spatial metrics from the simulated land-cover and comparing them to the metrics calculated from the actual land-cover data, the role of agent preferences for different land-uses is explored. The preliminary results suggest that the models capture relatively well the quantitative patterns of land-cover changes but they are poor in predicting the location of changes.

Land-use Decision Making

Land-use decision making is complex, multi-asset, real world decision task. The land-owner has to consider which activities (land-uses) he wants to implement on his land and decide where on that land to implement them. The decision maker's task is to find an effective way of using his assets — size and quality of land, technology, education and experience — in allocation of available resources — labor and land — to different uses. The number of factors to be considered range from the suitability of the land, dictated by various bio-physical variables, to the expected monetary or non-pecuniary returns from the uses. The optimal or good decision does not depend solely on the careful consideration of the above factors, but also on the decisions of neighboring owners and the use of their land.

General goal of modeling land-use decision making is to understand global environmental change and human role in bio-ecological systems. Specific goals are several; for instance, to explain the factors that drive people's land-use decisions, e.g., the role the preferences play in decision making; to understand the nature of spatial patterns emerging from learning and interaction of land-owners; to predict the location and rate of land-cover change (Schneider & Pontius, 2001); finally, the knowledge obtained in modeling studies may be used in environmental policy making (Daalen, Dresen, & Janssen, 2002) or to educate land-owners about sustainable land management practices.

Schneider & Pontius (Schneider & Pontius, 2001) model deforestation caused by residential development in the watershed in North-eastern Massachusetts. They predict the quantity and the location of deforestation between 1985 and 1991 using suitability maps calibrated with the real changes between 1971 and 1985. When comparing the model predictions of the rate of changes to the real changes 1985-1991, the model performs well, but fails to predict the sites of changes, which are relatively evenly spread over the study area. The model predicts suitability maps for residential development in the watershed, but does not account for individual decision makers' behavior.

Ligtenberg *et al.* combine multi-actor approach with cellular automata in modeling spatial planning (Ligtenberg, Bregt, & Lammeren, 2001). They apply the model with several types of decision making agents in a single-use framework to predict the spread of urban areas in Netherlands. Axtell *et al.* propose a multi-agent model of growth and collapse of Anasazi settlement in North-eastern Arizona from 1800 B.C. to 1300 A.D. (Axtell et al., 2002). Agents represent households whose primary source of nourishment is growing crops. By incorporating a number of agent-based variables (e.g., household size, age, nutritional needs, behavioral rules) to the archaeological and environmental data, agents' periodical choices for the locations of their residence and agricultural activities are simulated. The model closely reproduces the actual size of the population, its gradual migration north, and eventual abandonment of the area, while it occassionally fails to predict exact locations of household residences.

Hoffman, Kelley and Evans (Hoffman, Kelley, & Evans, 2002) and Kelley and Evans (Kelley & Evans, Under review) propose an agent-based agricultural land-use model and compare it to the actual data in order to explain land-cover changes in Southern Indiana. Individual agents represent households, who make decisions about how to use their land or whether to seek off-farm employment. They approach the decision process as a expected utility-maximizing behavior. They explore how spatial patterns of land-uses are influenced by land-owners' preferences and land suitabilities. Their finding is that land-owners preference heterogeneity plays at least as important role in emerging land-use patterns as the land suitability.

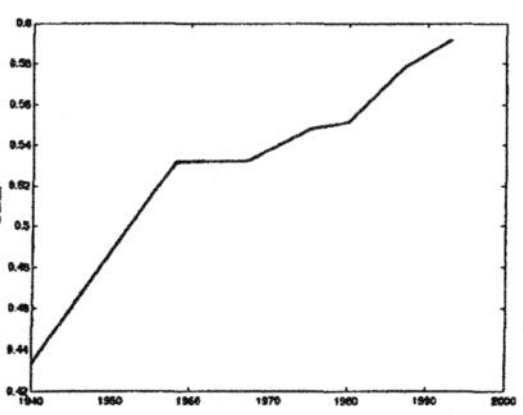

Figure 1: Changes in forest cover percentage in Indian Creek township from 1940 to 1993.

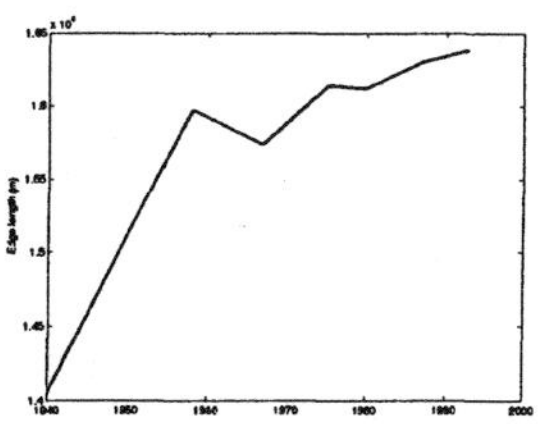

Figure 2: Changes in forest edge length in meters in Indian Creek township from 1940 to 1993.

Study Objective

Although this study concentrates on specific land-uses and a particular geographical area, the general principles used here are applicable to variety of cases of modeling private agricultural land-use decision making. The interest of this study is in the forest cover changes in Indian Creek township in South-central Indiana between the years 1940 and 1993. The available land-cover data indicates that there has been a significant increase in forest cover within the first 15 years of the study period and after that a modest, but gradual increase. The overall increase of forest cover is about 16%. The Figure 1 shows the monotonic increase of the percentage of forest in the area. Another metric that characterizes land-cover configuration is the length of the forest edge; the total length of the border between forest cells and non-forest cells. The non-monotonic increase in forest edge in meters in Indian Creek is shown in the Figure 2.

Theories in land and agricultural economics assume that land-use decision preferences are primarily formed by comparing expected financial benefits from different activities to the potential monetary costs of carrying out these activities. Koontz conducted an interview study among South-central Indiana land-owners in which he tried to explicate their motivations in land-use decision making (Koontz, 2001). The survey results suggest that non-monetary benefits also play a significant role, especially if the land is not the land-owner's primary source of income.

The goal of the current modeling study is to explain spatial patterns in South-central Indiana by postulating a small set of individual characteristics and learning mechanisms for decision makers and the payoff scheme that combines both monetary and non-monetary benefits. Simulating land-owners' yearly land-use decisions model makes predictions about the set of alternative land-uses, from which the changes of forest cover patterns are predicted.

Landscape Data

Three types of actual data are used in the modeling enterprise: forest-cover data, slope data and land ownership data. The time series of land-cover data were acquired from the historical areal photographs of the years 1939, 1958, 1967, 1975, 1980, 1987 and 1993. The slope data was extracted from the topographic maps and the ownership information from the historic parcel maps from the modeled period (Kelley & Evans, Under review). However, since the current model does not account for parcelization, only the parcel boundaries of the year 1957 are used.

These data were encoded into layered raster GIS (Geographic Information System) representation. In this representation the landscape is divided into a grid of cells of equal size and each layer records one type of information for each cell. In the current study the cell size is 50m × 50m and the landscape consists of 195 × 189 cells. A group of cells belonging to an individual landowner is called a *parcel*. There are 190 landowners in the modeled area, and their parcel sizes vary between 12 and 1936 cells, the average being 185 cells. For modeling purposes some additional landscape variables were introduced and encoded as layers in the GIS representation: soil quality and the number of years the cell has preserved its current use.

Agent-based Model

An agent-based model (see (Janssen, 2004) and (Parker, Manson, Janssen, Hoffman, & Deadman, 2003) for a review of agent-based models) is proposed that simulates land-owners' land-use decision making. The basic components of the model are the *landscape*, a rectangular area of land divided into cells corresponding to the actual landscape, and *agents*, the landowners whose primary source of income is the land they own. Like in the actual landscape data, the cell determines the resolution in which bio-physical information of the land is encoded. The cell is also the basic decision-making unit. Figure 3 represents the general model architecture.

Since the only available land-cover information is whether the cell is covered in forest or not, the initial landscape is constructed so that it allows more elaborate land-use decisions. This is accomplished by introducing two new land-uses — farming and abandoned land — and initializing the original landscape so that the shallow slope non-forest cells are marked as farmed cells, while the cells with steeper than average slope are marked as abandoned. After the initialization, the agents make two kinds of decisions each year; first, how to allocate their available labor between different activities, and secondly, on which parts of their land to apply these activities. The available activities are farming, off-farm employment, cutting trees and growing trees. The possible land-covers resulting from these activities are farmland, abandoned and forested land.

The specific agent actions are to decide whether to farm their non-forest land or seek off-farm employment, in which case they leave part of the land unused, and to decide whether to harvest trees or let them grow. The only information they use in these decisions is the payoff received from these activities in the past. In the farming decision the agents compare last year's hourly income from farming to the income from off-farm employment; if farming was more profitable, they decide to farm more

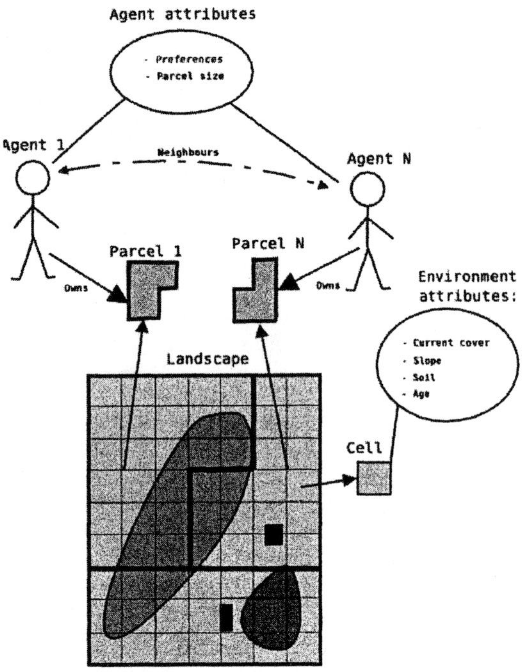

Figure 3: The basic components of the agent-based land-use decision model. Different shades of gray represent different land-uses.

in the future. In case farming was less profitable, they decide to work more off-farm and leave more cells unused. The cells left unused revert back to forest after certain number of years. Similarly, the decision between harvesting more trees or letting them grow depends on the payoffs from past decisions. The harvested areas start growing trees back. The tree growth follows a logistic function that is weighted by the average distance to other cells growing trees.

Agents

Agents represent private land-owner households, who make decisions about the use of their own land. They are assumed to have different individual characteristics; besides that their land varies in size, location and quality, the agents may vary in age, wealth, education, family size, experience, risk attitude, goals and subjective preferences.

Since we are interested in agent heterogeneity and its role in emerging land-use patterns, but most of the above data is not available, we postulate for each agent the subjective preference for farming (α_{farm}) and for growing trees, i.e., aesthetic enjoyment of having trees around (α_{trees}). Instead of estimating these parameters individually for each agent, the property size is used as the primary variable to infer the distribution of preference parameters. In other words, the agent j's preference for the use $i \in$ {*farm, trees*} is assumed to be a linear function of the property size s^j: $\alpha_i^j = b_i s^j + c_i$, where parameters b_i and c_i are estimated from the data. For the other land-uses α's are assumed to be 1.

Two additional parameters, fitted to the data, are the strength of externality effects for farming (ϵ_{farm}) and for growing trees (ϵ_{trees}). The externality effect in this context means either a positive or negative effect the cover of the neighboring cells have on the cover of the cell, or the effect of land-uses across parcel borders. For instance, the cell may be more likely to grow trees if, there are trees on the neighboring cells, or the agent may be more likely to farm the part of its land that borders neighbor's farmed land rather than forest. The strength of the externality effect can be conceived as agent's preference for large continuous areas of homogeneous land-use rather than scattered pattern of different uses.

Learning and Decision Strategies

Two different types of decision making agents are implemented and compared in the same land-use decision task. Both types of agents use the reward received from past decisions to infer if the decision was successful or not. The reward combines both monetary and non-pecuniary profits, and depends on the number and location of cells allocated to the activities, the suitability of these cells for the activities, unit returns from the activities and agent's subjective preferences for the activities. The payoff from the land-use i ($\in$ {*farming, off-farm employment, cutting trees, growing trees*}) is calculated by:

$$P_i = \sum_j \alpha_i \rho_i \sigma_i^j \delta_i^j \epsilon_i^j, \qquad (1)$$

where j goes over all the agent's cells that are allocated to use i, α_i is the general preference for the land-use i, ρ_i is the unit return from the use i [1], σ_i^j is the suitability parameter for the use i on the cell j, δ_i^j is the average distance from the cell j to other cells with use i, and finally ϵ_i^j is strength for externality effect of use i on cell j. Finally, the P_i's are summed up to get the agent's total payoff.

The suitability represents a discount factor that penalizes land-uses in unfavorable conditions. For farming and tree harvesting it depends linearly on soil quality and inversely on slope: $\rho = \frac{soil}{1+slope}$, where the soil quality degrades linearly over time with continuing use of the same kind. For other land-uses the suitability is assumed to be 1.

Learning and Resource Allocation Rieskamp *et. al* (Rieskamp, Busemeyer, & Laine, 2003) compared two learning algorithms to data obtained in the resource allocation laboratory experiment, in which subjects repeatedly allocated a fixed amount of money between three assets with either certain or uncertain returns. The payoff function was designed so that it had two maxima;

[1] The crop prices are aggregated from corn and soybean prices per produced unit, while timber prices are aggregated from certain hardwood species commonly harvested in the modeled time period. The minimum wage is used as an indicator of the changes in off-farm income rates in the modeled period.

two local and one global. The goal of the study was to find out, if the subjects were able to learn the location of these maxima from the numerical feedback they received after each allocation.

The two algorithms are based on different learning principles. The local adaptation model (LOCAD) makes small adjustments to the allocation depending on the feedback received from the trial one time step before, while the global search model (GLOS) keeps track of all possible allocations and probabilistically samples the areas in allocation space, which have proved profitable earlier.

The models' performance was compared to individual subjects' performance along various dimensions, e.g., the average allocation across trials, and the success in increasing payoff through learning. The main result is that the LOCAD model explains subjects' learning behavior better than the GLOS.

These promising experimental results give us a guideline to how to approach modeling of decision making and learning behavior in real-world resource allocation task that is much more complex. The proposed models for land-use decisions do not assume expected utility maximizing behavior, but the agents are expected to adapt to changes in the environment, for instance, to degradation of soil quality or changes in market prices, and adjust their decisions accordingly. The two learning models, discussed next, have different adaptation mechanisms.

Local adjustment learning agent considers the payoff received from one time step before and makes small adjustments to labor allocation in the next time step. The amount of labor it allocates to certain use reflects the outcome of the previous year's allocation so that, if the allocation was success, the new allocation is adjusted upwards, and if it was a failure, it is adjusted downwards. The success and failure are defined relative to the payoffs received from other uses. The decision is not deterministic; there is a slight chance that an opposite decision is made. For instance, even if the farming was more profitable than off-farm employment in the last period, the amount of farmed land may be decreased rather than increased in the next period.

The local adjustment learning agent uses *mixed* learning and decision strategy; each agent starts as a *individual learner*, i.e., it only considers its own payoff from the previous decision, but changes its type to *social learner*, if the decision turns out to be poor. This learner type considers the payoffs of all the agents when making decisions. Depending on the success or failure of the subsequent decisions the agent changes its type between these two types. Success is defined as an increase in the total payoff in two consecutive years, while failure is defined as decrease in the payoff.

The local adjustment agent is assumed to track small changes in market prices or environmental conditions relatively well. However, it is prone to get stuck in local minimum or in case of larger yearly changes, to oscillate between two allocations.

Experience-based agent has a perfect memory of all labor allocations it has tried before and the payoffs they have resulted in. Instead of making a small adjustment to the previous labor allocation, it samples its memory of allocations and chooses the one that produced the highest payoff. The choice is stochastic, proportional to the payoff received from the allocation, so that the allocations with higher payoffs are more likely chosen. After finding a good allocation from the memory, in order to assure a some level of exploration, the agent adjusts it slightly, i.e., it either increases or decreases the percentage of labor allocated to one or more activities.

This agent is assumed to repeat the early decision several times in the beginning, but to shift to more radical changes later on. It is not sensitive to changes in the environment, but chooses good allocations from the past, which may not be profitable in a new situation.

Results

The models' prediction accuracy is assessed by fitting the free parameters to the actual data in order to minimize the sum of square error (see equation 2) between the spatial metrics produced for the actual landscapes and the simulated landscapes. The spatial metrics are landscape-level forest cover percentage and forest edge length.

The model fit is assessed with respect to the null model, the landscape before the simulated period, to test if the proposed model predicts the land-cover changes better than the model that assumes no changes occurred (Kelley & Evans, Under review). Since the changes in the actual data are relatively small, it is expected that the initial landscape produces pretty accurate predictions.

$$\begin{aligned} SSE &= \sum_i (y_i - \hat{y}_i)^2 \\ SSE_{null} &= \sum_i (y_i - \hat{y}_{null})^2 \\ R^2 &= 1 - \frac{SSE}{SSE_{null}} \end{aligned} \tag{2}$$

Figures 4 and 5 present forest cover percentage and edge length changes for real and modeled landscapes, when preference parameters and strengths of externality effects are estimated as described above. The experience-based model fails to predict the data both qualitatively and quantitatively. The local adjustment model is relatively accurate in forest cover percentage, but is not able to account for the non-monotonic trend in the increase of the forest edge length.

Since both models use the same tree growth function, the experience-based model either harvests too many trees or farms extensively. In order to test the former hypothesis, one more parameter is added and fitted to the data: the frequency in which the logging decision is made. This parameter does not have any effect in the

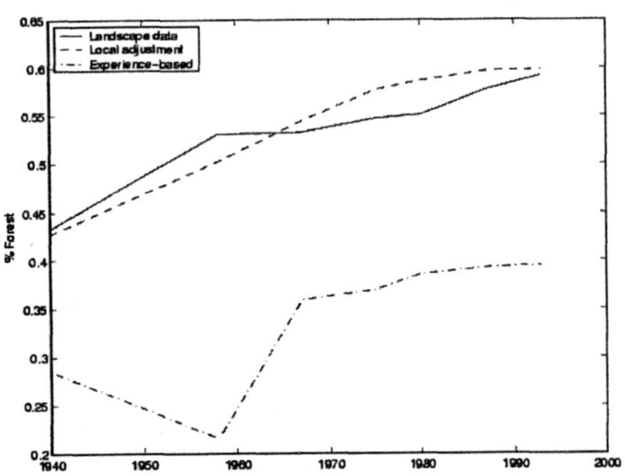

Figure 4: Changes in forest cover percentage over time. The parameters fitted are the linear preference parameters and the strength of externality effects.

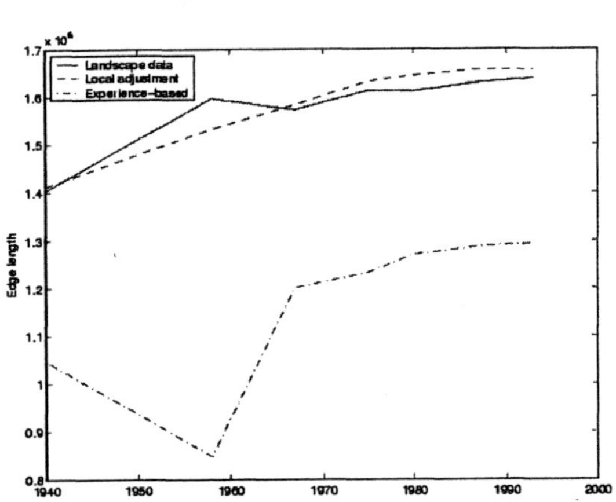

Figure 5: Changes in edge length over time. The parameters fitted are the linear preference parameters and the strength of externality effects.

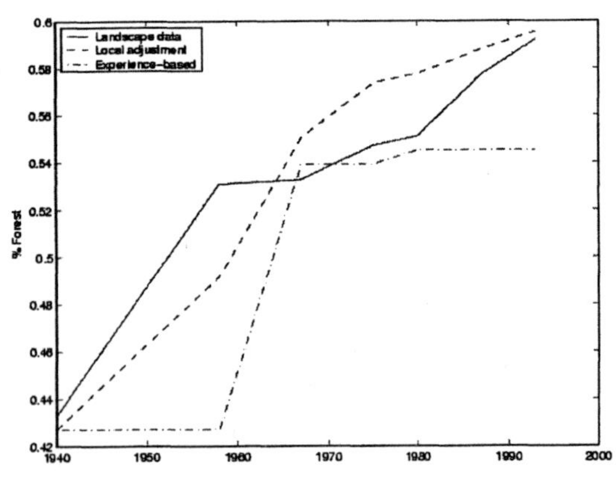

Figure 6: Changes in forest cover percentage over time. In addition to the preference parameters, the frequency of logging decision is fitted to the data.

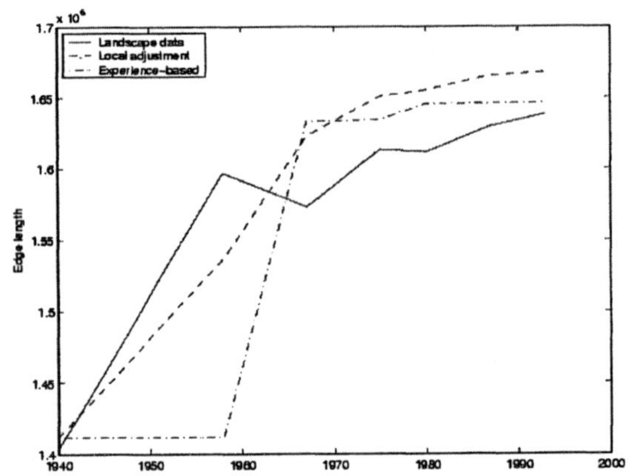

Figure 7: Changes in edge length over time. In addition to the preference parameters, the frequency of logging decision is fitted to the data.

performance of the local adjustment model, but clearly changes the other model's behavior in the beginning. Although it is still unable to predict the early significant increase in the cover, it captures some of the trend later. The results of these simulations are presented in Figures 6 and 7.

All the above results are from simulation runs in which the local learning model uses a mixed learning and communication strategy. In these runs a little more than half of the agents eventually converge to social learners. In order to test the effect of agent communication the possibility of changing the type was disabled and the agents were all set to either individual or social learners.

The trends of forest cover changes for the real landscape and the three types of local adjustment models are presented in the Figure 8[2]. If the agents share the information of everyone's payoffs, the model predicts an excess of forest cover starting early in the period, while the agents learning only from their own experience predict too conservative changes. As expected, the more complex model, with one extra parameter fitted to the data, predicts the data more accurately.

The model performance was also tested by comparing the metrics on the individual decision-maker level, i.e., forest cover percentage and edge length were calculated for each parcel separately. The preliminary results of these runs suggests that neither of the models does better than the null model in predicting individual parcel metrics. Actually, both models do slightly worse. They fail to explain why certain land-owner parcels have regrowth of forest while others do not. This may be because the same parameters are used to predict all the stages of forest-cover increase. However, the parameters get mostly calibrated to the early period in which most of the change occurs and to the parcels that change

[2]The pattern of the edge length change is pretty similar, and therefore not presented here.

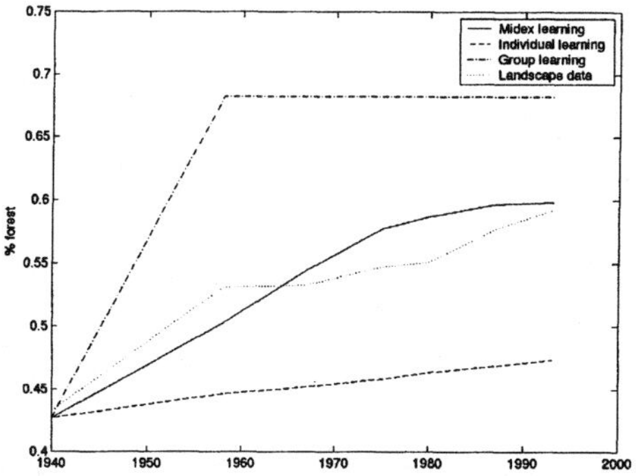

Figure 8: Changes in forest cover percentage over time. Results for different types of local adjustment models.

the most. Consequently, they are unable to capture the smaller changes occurring later.

Discussion

Two agent-based learning models, the local adjustment model and the experience-based model, were proposed to explain private land-owners' agricultural land-use decision making. The agents differ in the amount of information they use in decisions; the local adjustment learning agent utilizes the payoff information of only one time step behind, while the experience-based learning agent uses the information of all past decisions.

The local adjustment model predicts the increase in forest cover percentage relatively accurately, while the experience-based model fails to account for the significant increase in the beginning of the modeled period. This may be a direct effect of the strategy used; it has very little experience at early stages, so it basically repeats the same decision, or almost the same decision, over and over again. Neither of the models is able to predict the non-monotonic increase in the edge length.

The models do not do very good job in predicting the locations of changes on the parcel level. This may be because of the parameters do not capture the essential agent characteristics, i.e., the size of the parcel is not a good indicator of land-use preferences. However, Koontz's survey among Indiana land-owners indicated that there is a relationship between the parcel size and motivations for land-use activities (Koontz, 2001): on larger parcels the activities were associated with monetary benefits, while on smaller parcels the motivations were driven by non-monetary benefits. In the current study the aesthetic enjoyment of having trees around is assumed to be driven by non-monetary benefits. The model does not distinguish between financial and non-pecuniary motivations, and therefore the size of the parcel may not have a real effect in the preferences of different activities.

Finally, unlike the two learning models designed to explain the laboratory resource allocation behavior, the models designed for the real-world resource allocation task make both qualitatively and quantitatively different predictions when the amount of changes is considered. The models do not differ significantly in their accuracy of predicting the location of changes; they both fail equally.

Acknowledgments

This research was supported by NSF grant SES-008351. The authors are grateful for the helpful comments received from the members of the CIPEC, Biocomplexity project throughout the modeling process.

References

Axtell, R., Epstein, J., Dean, J., Gumerman, G., Swedlund, A., Harburger, J., Chakravarty, S., Hammond, R., Parker, J., & Parker, M. (2002). Population growth and collapse in a multiagent model of the Kayenta Anasazi in Long House Valley. In *Proceedings of the National Academy of Sciences of the U.S.A.* (Vol. 99: Suppl. 3, p. 7275-7279).

Daalen, C. E. van, Dresen, L., & Janssen, M. A. (2002). The roles of computer models in the environmental policy life cycle. *Environment Science & Policy*(238), 1-11.

Hoffman, M., Kelley, H., & Evans, T. (2002). Simulating land cover change in South-central Indiana: Complexity and ecosystem management. In M. E. Janssen (Ed.), *The theory and practice of multi-agent systems.* Edward Elgar, Cheltenham, U.K.

Janssen, M. A. (2004). Agent-based modelling. In J. Proops & P. Safonov (Eds.), *Modelling in ecological economics.* Edward Elgar.

Kelley, H., & Evans, T. (Under review). The relative influences of land-owner and landscape heterogeneity in an agent-based model of land-use. *Journal of Economic Dynamics and Control.*

Koontz, T. M. (2001). Money talks — But to whom? Financial versus nonmonetary motivations in land use decisions. *Society and Natural Resources, 14,* 51-65.

Ligtenberg, A., Bregt, A. K., & Lammeren, R. van. (2001). Multi-actor-based land use modelling: spatial planning using agents. *Landscape and Urban Planning, 56,* 21-33.

Parker, D., Manson, S., Janssen, M., Hoffman, M., & Deadman, P. (2003). Multi-agent system models for the simulation of land-use and land-cover change: A review. *Annals of the Association of American Geographers, 93*(2), 316-340.

Rieskamp, J., Busemeyer, J., & Laine, T. (2003). How do people learn to allocate resources? Comparing two learning models. *Experimental Psychology: Learning, Memory & Cognition, 29*(6).

Schneider, L. C., & Pontius, R. G. (2001). Modeling land-use change in the Ipswich watershed, Massachusetts, USA. *Agriculture, Ecosystems and Environment, 85,* 85-94.

A Constraint-based Approach to Understanding the Composition of Skill

Richard L. Lewis (rickl@umich.edu)
Department of Psychology, University of Michigan, Ann Arbor, MI 48109-1109.

Alonso Vera (avera@arc.nasa.gov)
NASA Ames Research Center, MS 262-4 Moffet Field, CA 94035.

Andrew Howes (HowesA@cardiff.ac.uk)
School of Psychology, Cardiff University, Cardiff, Wales, UK CF10 3YG.

Abstract

A hallmark of human cognition is the ability to compose novel behaviors from an existing repertoire of skills Newell (1990). These compositional processes range from search-based problem solving to the rapid, smoothly meshed perceptual-motor coordinations of well-practiced device interaction. In this paper we describe an approach to partially automating the composition of both semi-routine and highly skilled interactive behaviors. This approach, called Cognitive Constraint Modeling (CCM), is characterized by three principles: (a) descriptions of behavior are derived via constraint satisfaction over explicitly declared architectural, task, and strategy constraints; (b) the details of behavioral control (and therefore behavior composition) emerge in part from optimizing behavior with respect to objective functions intended to capture general strategic goals (e.g., go as fast as possible); and (c) the architectural building blocks are based on a simple ontology of resource-constrained cascaded processes. We show that these three principles jointly support modeling two important aspects of an interactive task: the overlapping and anticipatory behavior of highly skilled performance, and the hierarchical control of behavior evident earlier in practice.

Introduction

A hallmark of human cognition is the ability to compose novel behaviors from an existing repertoire of skills (Newell, 1990). These compositional processes range from search-based problem solving to the rapid, smoothly meshed perceptual-motor coordination of well-practiced device interaction. In this paper we describe an approach to predicting the composition of both semi-routine and highly skilled interactive behaviors. The design of interfaces for complex systems affects both task efficiency and error rates, but the consequences of design decisions are at best difficult to work out by hand and, at worst, counter-intuitive. For example, there have been cases where the introduction of automation, a putatively straightforward way to decrease task completion time, has instead had the opposite effect of increasing task time (Ballas, 1992). Similarly, increasing the efficiency of individual interface operations does not necessarily increase overall task efficiency, if those operations are not on the critical path that determines overall time. The a priori prediction of performance time and accuracy is difficult because human performance depends on many complex, interacting constraints that derive both from the task environment and from the constraints on the human cognitive, perceptual, and motor processing architecture. Skilled performance of a routine task usually involves the execution of a number of parallel but interdependent streams of activity. For example, one hand may move to a mouse, while the other finishes typing a word; the eyes track a target while a command label is retrieved from long-term memory. The details of how these cognitive, perceptual and motor processes are scheduled and the nature of their dependencies have significant consequences for the overall time requirement and error rate. Unfortunately, there are currently no modeling approaches that yield a priori prediction of human performance on semi-routine tasks that do not require the modeler to either go through complex simulations of the learning process or to hand-code the intricate sequencing of every type of action that comprises the behavior. In the next section, we briefly sketch the requirements for an effective modeling approach.

Requirements for compositional modeling

Effective tools for modeling semi-routine behavior must support at least the following features: (a) reuse of existing composed tasks; (b) automatic composition; (c) parametric flexibility; and (d) easy manipulation and inspection of theoretical/task assumptions.

We discuss each of these in turn, summarizing the technical challenge associated with each.

Compositional reuse and specification

People can compose new behaviors out of existing component skills. One approach to modeling composition is with *template* modeling. Templates corresponding to common interactive behaviors such as using a mouse-driven pull-down menu can be reused without specifying again all their internal details.

Programmable cognitive architectures such as ACT-R (or Soar) provide considerable reuse of architectural-level components, but relatively little support for the flexible recombination of existing strategies at the task level. There are two key reasons for this. First, there are often theoretically irrelevant incompatibilities in existing task models that prevent their straightforward reuse and composition into new tasks. Second, architectural approaches lack an explicit theory of task-level behavior composition to guide the modeler in composing new tasks

P1: Cognitive modeling proceeds by deriving descriptions of behaviors from (possibly underspecified) declarative statements of architectural, strategic, and task constraints. These constraints are an explicit, formal statement of the psychological theory.

P2: The architectural constraints are based on an ontology of cascading, communicating processes in which the resource consumption of both processes and their communication channels can be reasoned about.

P3: The modeler does not specify all of the control details; rather, details of behaviors are predicted by optimizing algorithms that maximize or minimize objective functions (such as total time of behavior) specified by theorist. This embodies the hypothesis that skilled behavior is the optimal solution to a constraint satisfaction problem.

P4: Explicit goal hierarchies provide (a) the structures out of which new tasks are composed; (b) the representation over which learning parameters are specified, and (c) the cognitive control processes that organizes the model's behavior.

Table 1: Technical Principles

from existing parts---the composition theory is implicit in the architecture and learning theory.

Encapsulation

The principal advantage of composing new behaviors out of existing subcomponents is the ability to work at higher levels of abstraction without worrying about the details of how to achieve the component tasks. Unfortunately, this is difficult to capture in models of skilled human behavior. Consider the following simple example: Suppose we want to compose two behaviors A and B in sequence. Behavior A might be a keyboard button press, and behavior B might be a mouse-and-click. It is insufficient to specify a serial order on the two behaviors, expand them into their detailed subcomponents, and then impose that ordering constraint on the subcomponents (as is typical in software engineering). Human skilled behavior is not so neatly encapsulated---for example, the eye-movements associated with the mouse-movement of behavior B might begin as early as the start of behavior A. What is specified as serial at a high level might be executed in nearly complete parallelism at the lower levels, and this can have significant implications for overall performance time and accuracy. In short, the problem is the seamlessness of human behavior. To first approximation, skilled human behavior does not respect task boundaries---certainly not the potentially arbitrary boundaries introduced by a modeling decomposition. In particular, *anticipatory acts*, in which aspects of a later behavior intrude on or modulate an earlier one, are a hallmark of high degrees of skilled, and can be seen in behaviors ranging from coarticulation in speech to aggressive eye-movements in computer interaction. The result is that it is not possible to be concerned only with the "inputs" and "outputs" and temporal relationships of component behaviors---the internal details, and how they mesh, matter.

Parametric flexibility

As explained above, most realistic task scenarios require a mix of skilled and novel components. The modeling approach must therefore support flexibility in specifying the degree of learning of subcomponents of the overall task. Current approaches permit either programming only at a fixed skill level (e.g., Epic), or use implemented learning mechanisms, such as production compilation, to simulate the progression through various stages of learning (ACT-R). Although this approach has significant theoretical advantages by providing explanatory accounts, in practice it makes modeling difficult, and it becomes especially cumbersome when the goal is to model mixes of skilled and novel components within a single task. It also puts the specific learning mechanisms, which themselves are evolving hypotheses, on the critical path.

Easy manipulation of theory and task specification

Architectural assumptions should be as explicit, inspectable, and modifiable as task specifications. This is important for both the applied HCI practitioner and the cognitive scientist.

Unfortunately, cognitive architectures embody architectural assumptions in underlying program code, and are not easy to change. This would not be a problem if the details of an architectural theory were stable and comprehensive enough to be applied to a wide range of tasks. But in the forseeable future many modelers will find it valuable to easily manipulate and add architectural assumptions. Examples of such assumptions may range from relatively low level commitments such as the distributions of times for specific kinds of perceptual processes, to constraints on short term memory retention of specific kinds of task information, to major architectural choice points such as the existence of a central response selection bottleneck.

Addressing the requirements

In the remainder of this paper we describe a modeling approach characterized by the three principles of Cognitive Constraint Modeling (CCM) and a fourth principle of composition (see Table 1) (Vera et al. 2004).

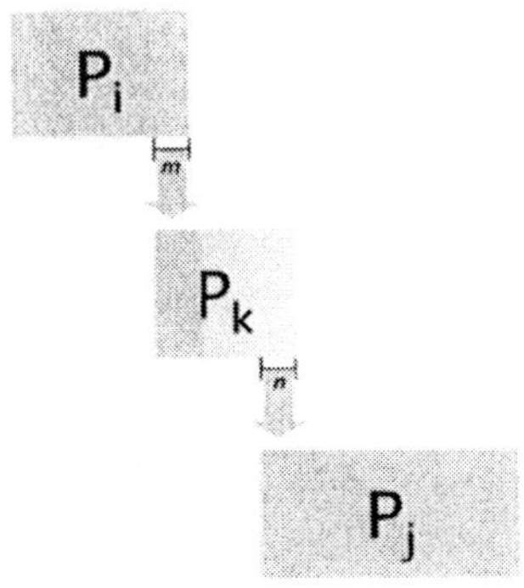

Figure 1: Two processes joined in cascade. See text for a formal specification.

The first principle defines the nature of the cognitive modeling infrastructure; the remaining principles constitute a set of hypotheses about the nature of the human cognitive architecture. In the next sections we explain these principles, and how they address the significant modeling issues above, by demonstrating them in a prototype modeling tool. We also briefly summarize new empirical evidence for the psychological reality of hierarchies in controlling behavior.

A cascade-based process ontology

The arithmetic constraints defined above give a formal definition of a cascade-theory of process communication (McClelland, 1979). Rather than one process following the next, as in discrete-stage theories, processes overlap in time. The temporal overlap is sufficient to allow the information required by process Pi to be communicated to process Pj through buffer process Pk. Pi overlaps with Pk which overlaps with Pj.

Deriving behavior descriptions from constraints

Constraints on behavior can be specified in terms of predicate calculus statements relating entities in the environment, tasks, and psychological processes. An entity can be represented as a set of elements where each element is an ordered pair. For example, the following reads, there exists a cognitive process called initclick with a start and a duration.

$$\leq P_i\text{: \{ (isa,process) (name,initclick) (resource,cognition) (start, Si), (duration, Di) \}} \leq P_i \qquad (1)$$

Each pair in the above statement consists of an attribute and a value. A set must only contain a single element with a particular attribute (e.g. there must only be, at most, one pair that matches the pattern (name,_)). Further features may complete the specification of this process (see Howes et al. submitted for a full description). Relationships between the start times and durations of processes, for example between Pi and Pk in Figure 1, can be represented with simple integer-arithmetic inequalities. The following represents the assumption that a motor process is a necessary consequence of an initialization process, that a motor process must start before the end of its initialization process, and that the maximum temporal gap between the two processes is 300ms. This constraint must hold irrespective of the task.

$$\leq P_j\text{: \{ (isa,process) (name,initclick) (start,}S_j\text{) (duration,}D_j\text{) \}} \leq P_j$$
$$\leq$$
$$\leq P_i\text{: \{ (isa,process) (name,click) (start,}S_i\text{) \}} \leq P_i$$
$$\leq S_j + D_j \leq S_i$$
$$\leq S_i - (S_j + D_j) \leq 300 \qquad (2)$$

Statements of this form can be used to represent theoretical assumptions about the task environment, about instruction taking, about the strategies that people deploy, and about the human cognitive architecture. Universally quantified constraints specified in a predicate calculus are not production rules. The constraints may appear to possess a similar surface form to production rules but, in fact, the semantics are very different. Unlike production rules, these declarative statements of theory are statements of what must be true irrespective of context. They are not elements of a procedure that generates the description. The constraint must hold for every circumstance where its antecedent is met. The generation of a model with these constraints is entirely monotonic and the order of expansion can be (and often is) different to the predicted order of behavior.

Optimal constraint satisfaction

In contrast to Anderson (1990), Simon (1992) emphasizes the need for cognitive scientists to uncover optimal solutions given constraints on task environment, strategies, knowledge and human cognitive architecture.

Given these constraints, it is possible to generate a prediction of optimal performance (Vera et al., 2004). First the implications of the theory are derived. Subsequently, using a constraint satisfaction engine such as CLP FD, a prediction of the optimal behavior can be calculated by finding a set of variable bindings that are consistent with the defined constraints. A branch-and-bound algorithm, for example, can be used to generate a schedule with the greatest utility. The particular algorithm is irrelevant to the

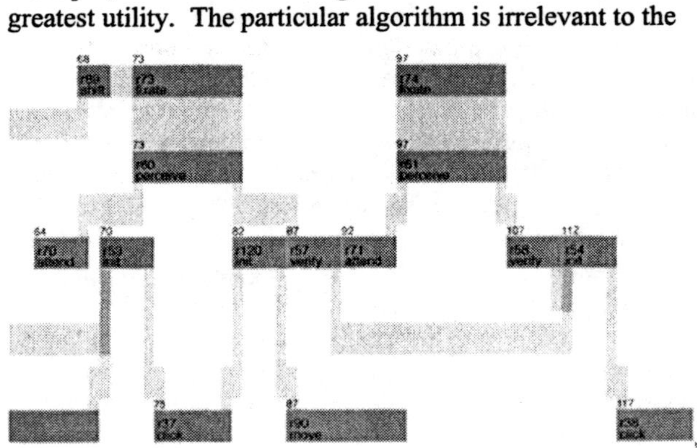

Figure 2: Interleaved mouse move-click tasks.

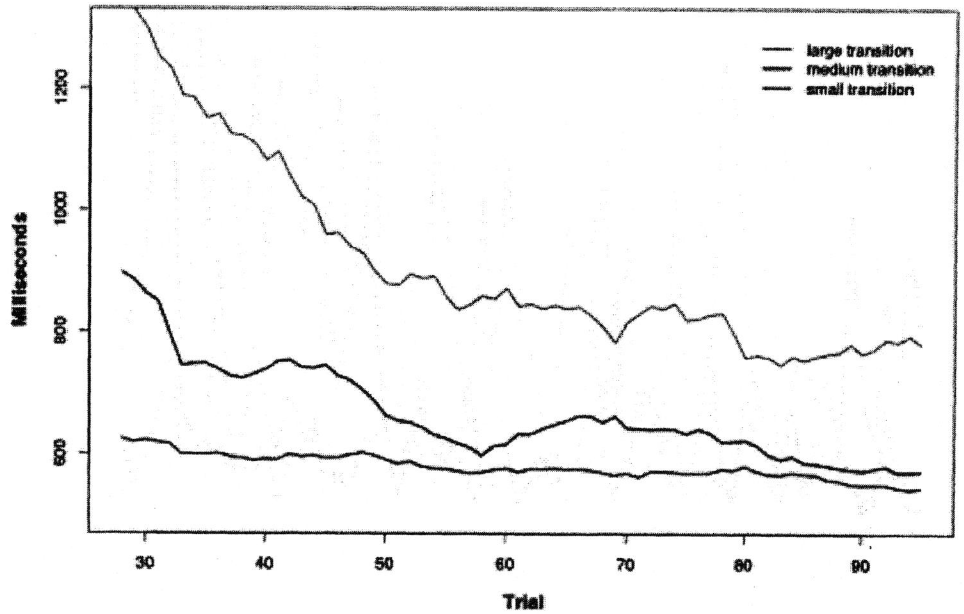

Figure 3: Reaction times for button presses in a simulated ATM-task. Buttons are grouped into three groups According to the size of the transition required in the task hierarchy (Lewis, Vera & Remington, in preparation).

theory, what is important is the objective function and the fact that its value can be minimized or maximized. In the examples reported in this paper the objective is to minimize time.

Example 1: Composing highly skilled behavior

We have built a prototype modeling tool, called CORE (Vera, Howes, Lewis McCurdy, 2004), that takes as input a set of mathematically stated constraints on behavior and outputs a prediction. One of the formats for the output, a behavior-graph, similar to a CPM-GOMS-like Pert chart, is illustrated in Figure 2. The prediction in the Figure is for a pair of move-click mouse tasks (one represented in light gray and the other dark). Each box represents a process. In the Figure, time is represented on the horizontal access and each row represents a different resource or processor, perception at the top, through cognition, to motor actions.

The example illustrates how the automatic composition of highly skilled behavior can be achieved with a combination of optimization and cascaded information-processes. Cascades prevent cognitively implausible process orderings that are legal when the relationships between processes are described in terms of dependencies. For example, the process ordering init(x), init(y), click(y), click(x) is legal if the relationship between an init and a click is specified as a dependency but cognitively implausible because it assumes no cost to buffering information between the cognitive intention and the motor action.

Optimization can lead to the generation of counter-intuitive schedules. It can lead to both anticipatory behavior, as in Figure 2, and to strategic deferment. For example, Howes et al. (submitted) describe how, under certain architectural assumptions, the minimum-time cost schedule for a dual-task behavior required the strategic deferment of response retrieval.

Example 2: Hierarchical composition at multiple skill levels

But how is behavior organized earlier in practice? We hypothesized that the hierarchical organization of behavior early in practice would result in longer response times at subtask boundaries. We distinguish fixed task hierarchies from subgoaling during problem solving (e.g. as in the Tower of Hanoi). Fixed task hierarchies must be retrieved, one goal at a time, from memory and are not generated dynamically in response to novel problem solving contexts.

Empirical evidence for hierarchical control structures

Using a simulation of an Automated Teller Machine (ATM), we designed a set of studies to analyze the effects of task hierarchy on the acquisition of skill. We hypothesized that, for example, participants who were told that they must enter a 4-digit PIN in order to access all of their accounts and that they must then enter the specific account number for the desired transaction (e.g., withdrawal from checking versus savings) would form a different task representation than those told to enter their PIN and account number together in order to access all of the functions. The sequence of numbers in each case could be exactly the same and yet it would be possible that, because of a potentially different hierarchical representation there would be a short pause between the two numbers in one case but not the other.

15 University of Michigan undergraduate students participated in the study. The ATM simulation interface,

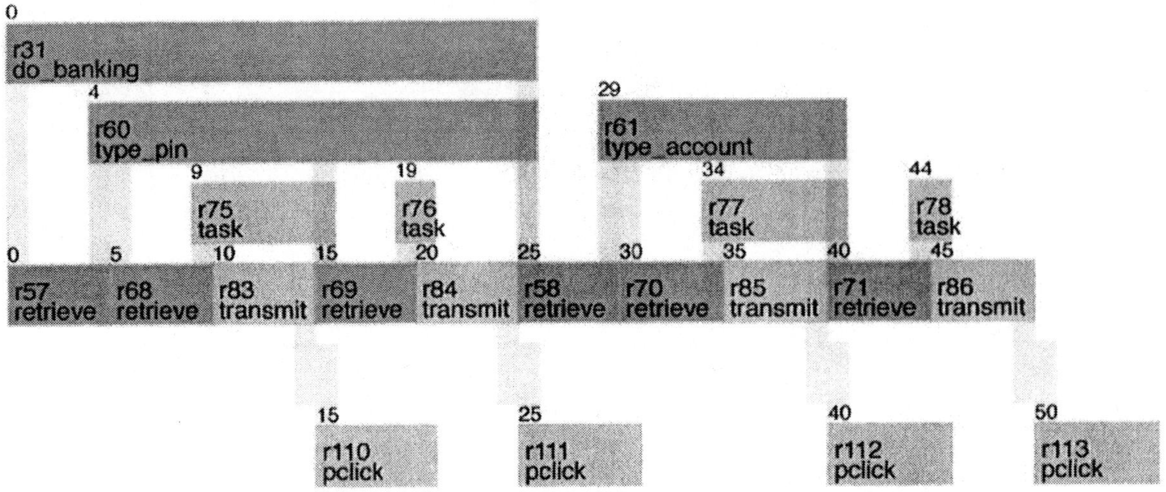

Figure 4: Prediction of novice performance on the ATM task. There is a greater duration between click r111 and r112 Than between r110 and r111 because of the time required to retrieve the 'type_account' subtask.

had five elements: a numeric keypad, function buttons to each sided of the numeric keypad (e.g. "OK", "Cancel"), a display screen, buttons at the sides of the screen and two slots (one for inserting the card and the other for retrieving money/receipt). The entire task was mouse-driven, including clicking on the card and money/receipt slots.

Each subject performed the same task 200 times. Subjects were instructed on the interface and given a set of of actions to perform: enter card, enter PIN, enter account, enter amount, etc. Subjects were instructed to complete each trial as quickly as possible. Errors in the sequence were indicated by the interface and subjects had to go back and complete the correct sequence before proceeding.

Results: As shown in Figure 3, we found evidence for a multi-tiered hierarchical representation of the task through the first 100 trials of performance. We analyzed all the task transitions where we hypothesized a three-level hierarchy traversal (e.g., from enter PIN to enter account), all two-level traversals, and all one-level traversals separately. ,Figure 4 shows a clear differentiation for each type of transition. Actions requiring three-level transitions took 200-400ms longer over the first 100 trial than mouse clicks requiring only two-level transitions. Similarly, the difference between two-level an one-level traversals was in the order of 100-200ms over the first 100 trials. Two additional interesting effects are that the curves begin to converge between 100-200 trial but do not fully overlap by 200 trials (the three-level traversal curve remains 50-100ms above the other two) and that the one-level traversal curve remains relatively flat throughout (i.e., participants start out fast on those transitions and remain fast).

A model of hierarchical control and learning

We specified a cognitive constraint model of Lewis et al.'s data focusing on the effect of hierarchy on inter-mouse-click latencies. Two constraints were of particular importance:

1. There is a time cost to retrieving subtasks from memory. The average time cost of retrieving a single subtask is fixed but the traversal of multiple layers of the hierarchy requires multiple retrievals. More retrievals therefore requires more time.
2. Learning consists of incrementally removing intermediate layers in the task hierarchy from bottom to top. This assumption reflects the idea that subtasks are chunked together into new more specialized but more efficient (in terms of memory and time costs) procedures.

Figure 4 shows the model's predicted behavior for novice performance. This behavior graph is an optimal solution that was derived, using CORE, from a set of mathematically specified constraints. Time is represented from left to right. First note that we have assumed, for now, that no perceptual processing is required to achieve the task. Therefore the cognitive processing consists of the retrieval of task information and the initiation (transmit) of motor responses.

Above the cognitive processor in the Figure are the buffers that represent the goal information. At the top is the top level task (do banking). It is predicted to occupy the buffer for at least as long as is required to cue the retrieval of its two subtasks type_pin and type_account. Type_pin and type_account each cue the retrieval of further subtasks. Eventually the correct sequence of motor responses is retrieved and initiated. Most importantly for our current purposes, the contiguous retrieval of task r61 and its first subtask r77 results in a longer latency between r111 and r112 than between r110 and r111. I.e. consistent with the human data, the model predicts longer latencies at subtask boundaries.

Figure 5 illustrates the model's predicted behavior for expert performance. The behavior graph has been generated from task knowledge that was derived from the specified novice task knowledge using a set of five universally

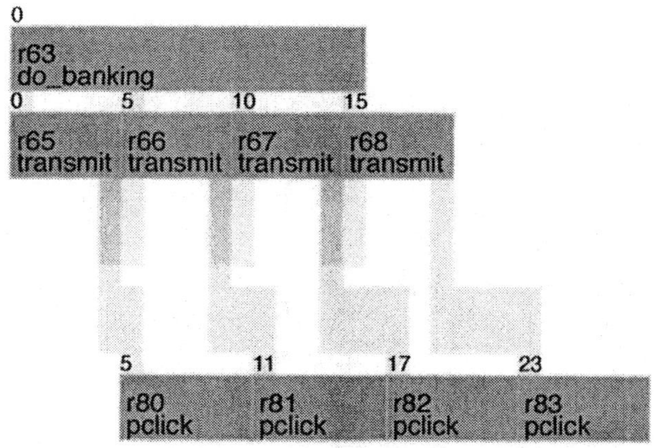

Figure 5: Prediction of skilled performance on the ATM task. Due to the composition of the hierarchy, the duration between the clicks is both less than and more homogeneous than in the novice schedule (see Figure 3).

quantified predicate calculus statements. The statements encode a theory of how procedures are composed with practice. They specify how to infer the structure of an expert task hierarchy from a given novice task hierarchy and how to ensure that the task ordering is maintained once the control information provided by the hierarchy has been removed. In Figure 5 the hierarchy has been replaced by a buffer (represented below cognition) that chains the transmit processes together.

Discussion

We have described the application of a tool for making inferences about the implications of formally specified theories of behavior to the problem of understanding the composition of skill. The tool uses a constraint logic-programming environment to support inference given a specification of the constraints on the task environment, on perception, on cognition, and on action.

A particular model constructed with the tool was used to illustrate the way in which composition can be understood as incremental replacement of layers in the task hierarchy with process chaining. The behavior predicted by the model was consistent with human performance on a simulated ATM task. Most importantly, predicted inter-move latency corresponded to the task boundaries.

Though the work described in this article is still in progress the framework and model directly address the requirements for compositional modeling that we articulated in the introduction: (a) existing composed tasks were reused while avoiding the restricted-interleaving associated with encapsulation (Example 1); (b) automatic composition was achieved by separating constraints from scheduling and then using an optimal scheduling algorithm to predict the optimal behavior given constraints on task environment and strategy and architecture (see Simon, 1992); (c) parametric flexibility was achieved by specifying the declarative relationship between task knowledge structures at different levels of skill, rather than by simulating the incremental transitions of the human cognitive learning algorithm; (d) Specifying constraints on cascading information processes in terms of mathematical relationships between their start times and durations supported rapid specification and manipulation of theory (only five rules were required to articulate a simple theory of composition).

Acknowledgments

This work was supported by ONR Grant number N0001404IP2002 and by NASA through the HMP/HADMT Program.

References

Anderson, J.R. (1990). *Rational Analysis.* LEA.

Anderson, J.R. and Lebiere, C. (1998). *The Atomic Components of Thought.* Mahwah, NJ: Erlbaum.

Ballas, J. A., Heitmeyer, C. L., & Perez, M. A. (1992). Evaluating two aspects of direct manipulation in advanced cockpits. *Proceedings of ACM CHI 92 Conference on Human Factors in Computing Systems* (pp. 127-134). NewYork: ACM.

Card, S.K., Moran, T.P., Newell, A. (1983). *The Psychology of Human Computer Interaction.* NJ: Erlbaum.

Howes, A., Vera, A.H., Lewis, R.L & McCurdy, M. (submitted). Cognitive constraint modeling: A formal approach to supporting reasoning about behavior. Paper submitted to Cognitive Science Conference.

McClelland, J.L. (1979). On the time relations of mental processes: An examination of systems of processes in cascade. *Psychological Review, 86,* 287-330.

Simon, H.A. (1992). What is an "explanation" of behavior? Psychological Science, 3, 150-161.

Vera, A.H., Howes, A., McCurdy, M. & Lewis, R.L. (2004). A constraint satisfaction approach to predicting skilled interactive cognition. In *Proceedings of the Conference on Human Factors in Computing Systems CHI'04,* Vienna, April 24-29, 2004.

A Queuing Network Model for Eye Movement

Ji Hyoun Lim (limjh@umich.edu)
Department of Industrial and Operations Engineering, University of Michigan
1205 Beal Ave, Ann Arbor, MI 48109 USA

Yili Liu (yililiu@umich.edu)
Department of Industrial and Operations Engineering, University of Michigan
1205 Beal Ave, Ann Arbor, MI 48109 USA

Abstract

Eye movement is a basic human behavior that offers a valuable means to explore human cognitive processes. This article introduces two modeling studies of eye movement. First, random menu search was modeled using a queueing network approach and second, a reinforcement learning algorithm was used to generate various eye movement patterns. Random menu search is a task component involved in many human-machine interfaces and has been modeled with several cognitive models including ACT-R and EPIC. Based upon review of empirical data in menu search, and strengths and limitations of existing models, this article proposes a queueing network model, which has been successfully applied in some other task domains (e.g., response time, driver performance). The queueing network model for random menu search was implemented and evaluated through model simulation. In contrast to existing models that rely on four task-specific strategies to account for data, the queueing network model accounted for the same data using only one strategy already employed in cognitive modeling. To extend this parsimonious, "minimal task strategy" modeling approach, Q-Learning, one of the reinforcement learning methods, was adopted to generate different patterns in eye movement. The same strategy from random menu search was used to generate eye movement, and the simulated eye movements were qualitatively compared to the human eye movement.

I. Introduction

Computational modeling of menu search and eye movement has both practical and theoretical significance. Practically, menu search is almost universally required in computer interfaces. Most computer interfaces contain menus for users to select functions from. Selection of a menu is then followed by a visual search. Theoretically, eye movement is an integral component of cognitive performance. Many experimental studies (Nilsen, 1991; Byrne, Anderson, Douglass, & Matessa, 1999) and computational modeling efforts (Byren, 2001; Hornof, 1999; Anderson, Matessa, and Lebiere, 1997) have been conducted to investigate random menu search and eye movement.

Computational cognitive models are useful because they can integrate results of various experiments, provide understanding of perceptual and cognitive processes underlying observable behaviors, make quantitative predictions for scenarios not yet tested, and provide a precise common language for description of phenomena of interest. We examined existing computational cognitive models of menu search and eye movement and proposed a new computational cognitive model rooted in queueing network theory (Liu, 1996, 1997) that overcomes the limitations of the existing models.

Two studies are introduced here. The first focuses on the development of a computational cognitive model for menu search with a minimal number of eye movement strategies. Performance of the queueing network model is compared to existing models involving complex strategies. The second study investigated the potential performance of our model to generate eye movement patterns. Results show that our model generated various types of eye movement patterns based on the same simple strategy used in the first study. To explain the underlying rules for different eye movement patterns, reinforcement learning algorithm was adopted.

II. Random Menu Search Models

Random menu search is the simplest type of search task because it requires only one-dimensional eye movement (up or down). A menu search task consists of control of aimed movements and visual search of menus. Nilsen conducted a series of menu search experiments using sets of randomly ordered and vertically listed items as the stimuli for menu search (1991). The subject's task was to find a given target from a set of items. Nilsen collected only response time in this experiment. Nilsen's data have been used to validate existing models (Anderson, 1997; Hornof, 1999).

In addition to response time data, Byrne, Anderson, Douglass, and Matessa (1999) considered the use of eye movement data as validation of computational cognitive models for random menu search. According to their eye movement records of a random menu search task, there are two typical phenomena in eye movement for menu search: *varying lengths of saccade* and *overshooting*. A saccade is one type of eye movement characterized as a ballistic movement with a fixed destination. The records of eye movement showed that the length of saccade is not constant. Overshooting is a phenomenon in which the eye moves beyond the location of the target. Byrne's data revealed the existence of negative lengths on the last saccade in menu search, a finding that can be interpreted as overshooting.

II. 1. ACT-R and EPIC Models

Anderson, Matessa, and Lebiere (1997) showed that the ACT-R model could reproduce menu search response time by incorporating a critical production called the *hunt-feature*. The hunt-feature production moves attention by hunting for objects that have a target feature. The hunt-feature production determines the next fixation position, which determines the length of the following saccade.

Hornof (1999) developed the *dual strategy varying distance hybrid model* (DSVDH model) based on EPIC. He addressed the difference between local and global control of eye movement in visual search. In the DSVDH model, he assumed four different search strategies and their fractions of use in their global control of eye movement.

Byrne (2001) compared the performance of ACT-R (Anderson et al., 1997) and DSVDH (Hornof, 1999) in random menu search using eye movement data (Byrne et al., 1999). ACT-R successfully generated the varying length of saccade with the hunt-feature production (Anderson et al., 1997). However, the fundamental serial cognitive processing assumption of the ACT-R architecture did not allow the eyes to overshoot. Since subsequent fixation location is determined before the eye moves in ACT-R, there is no reason for the eyes to move further than the target location.

Unlike ACT-R's serial cognitive processing assumption, the EPIC architecture assumes parallel cognitive processing. Thus, Hornof's (1999) DSVDH model allows multiple target items to be compared simultaneously. Hornof's assumption of multiple strategies allows DSVDH to reproduce eye overshooting and varying length of saccade, although the DSVDH model itself provides insufficient explanation of how and why they occur.

II. 2. Queueing Network Model for Menu Search

The *queueing network* modeling approach was proposed by Liu (1996, 1997) and has been applied successfully to integrate a large number of influential mathematical models in psychology (Liu, 1996), in modeling multitask performance (Liu, 1997), and in modeling driver performance (Tsihmoni and Liu, 2003).

The queueing network model consists of four components: customers, servers, arrivals and routes. The *stimulus*, either outside or inside of human body, is encoded as customers, which need to be processed by servers. Each *server* in a queueing network represents a fundamental functional unit in human information processing. The *arrival* of a customer at a particular server is an event that initiates information processing at that server, and the order of the servers though which a customer should go is the *route* traversed by that customer.

To model a random menu search task with the queueing network, three types of stimuli required in random menu search tasks were defined as customers: precue stimulus, reference stimulus (the target), and a list of menu items containing one target. Nine servers form the architecture of the queueing network model: four servers for the visual perceptual sub-network, three servers for the cognitive sub-network, and two servers for the motor sub-network. Figure 1 shows the layout of servers and routes.

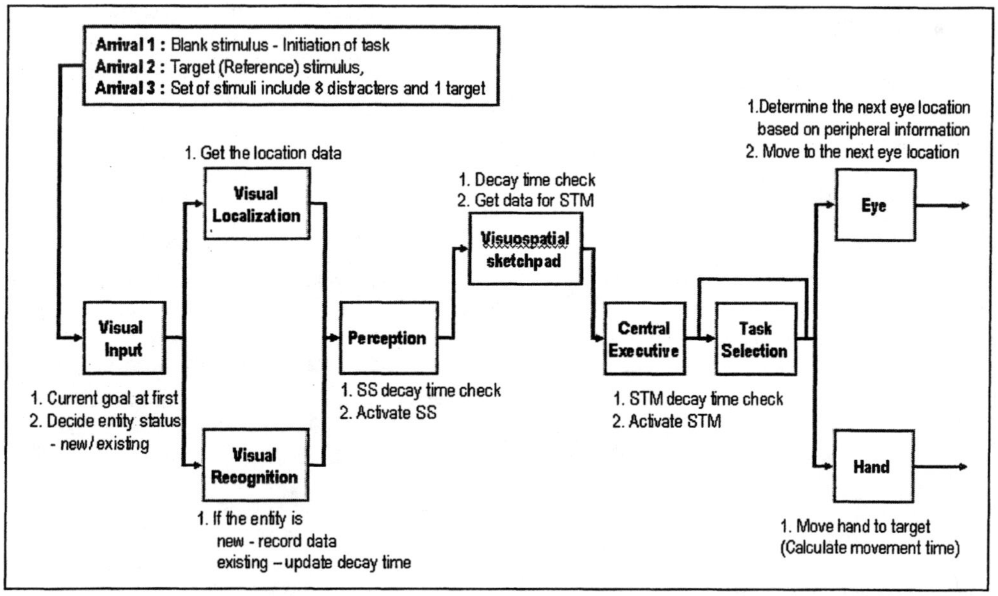

Figure 1. The layout of the servers and routes of the queueing network model for random menu search

The queueing network model for menu search has both serial and parallel routes. The servers located at the same horizontal location in Figure 1, for example *Visual Localization* server and *Visual Recognition* server, can be processed in parallel. The queueing network model relies on two distinct assumptions for menu search task. First, the model uses ACT-R's *hunt-feature* production (Anderson et al., 1997) as the only strategy for eye movement. Second, *separate routes* are suggested for attended and unattended visual information based on the preattentive location encoding theory (Wright and Word, 1998).

The queueing network model was implemented and simulated with Pro-Model—a commercially available simulation software widely used for manufacturing and operational applications. The stimuli consisted of nine randomly ordered numbers, one of which was the target. The location of the target was varied randomly from top to bottom.

The queueing network model was evaluated in two conditions—separate-routes versus same-route. The former had separate-routes for attended and unattended information, allowing partial parallel processing. The latter had the same-route for all information, which was expected to behave like ACT-R due to the shared assumption of serial information processing. The attended information was regarded as being achieved from the center zone of fovea (two degrees of visual angle) on the retina and the unattended information from outer area of fovea.

II. 3. Results and Discussion

Model simulation reproduced both the varying length of saccade and overshooting eye movement phenomena in random menu search task.

The relationship between the number of fixations and the location of target is presented in Figure 2. In the separate-routes condition, the queueing network model generated slightly more fixations because of the eye overshooting.

The major difference in the simulation results between the separate- routes condition and the same-route condition was the presence of eye overshooting. In the separate-routes condition, 30% of the menu selection trials generated overshooting eye movement. In contrast, there was no overshooting eye movement in the same-route condition. Byrne's eye tracking data showed that overshooting existed in visual search (about 32%).

Varying length of saccade was reproduced via use of the hunt-feature production. If the target stimulus was *out of* current unattended vision, the next destination of the saccade would be determined at the very end of the current unattended vision area. This condition generated a longer distance of eye movement than when the target stimulus was *within* an unattended vision area. Since the hunt-feature production used attended and unattended visual information to determine the next location of eye fixation and this was considered a unique rule of eye movement, multiple competing strategies were not required for each saccade.

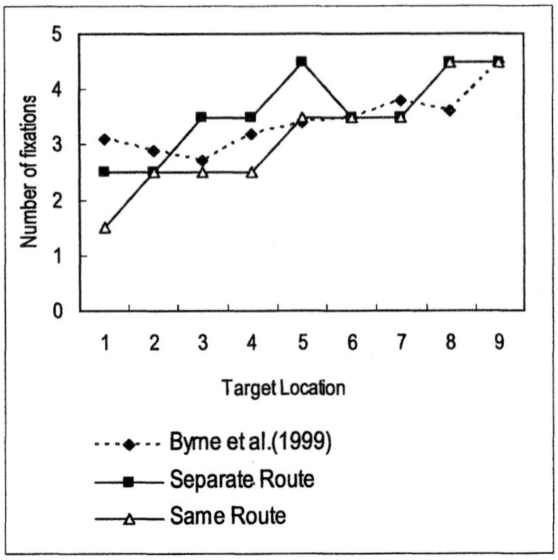

Figure 2. Comparison of the queueing network simulation result to Byrne et al.'s eye tracking data

This study showed that both varying length of saccade and eye overshooting could be generated by a single strategy rather than multiple competing strategies. Furthermore, the unique strategy used in the queueing network model was not proposed here, but was originally incorporated by the ACT-R model (Anderson et al., 1997). This shows a potential strength of the queueing network structure compared to other computational cognitive models.

Based on these initial findings, an interesting question was raised. Would this simple, single strategy be sufficient to model more complex eye movement patterns? The following study shows our effort to answer this question.

III. Single Strategy Eye Movement

In 1967, Yarbus conducted an interesting experiment in which a subject was shown a picture and then asked seven different questions about the picture. The subject's eye movement was recorded for each question and the results showed that the subject moved his/her eyes over the picture very differently depending on the questions.

Modeling eye movement requires considering various patterns in eye movement. If we need multiple strategies to move our eyes along only one-dimension (up and down), as EPIC (Hornof, 1999) and ACT-R/PM (Byrne, 2001) do, it is difficult to estimate the number of strategies required to move the eyes in a two-dimensional space. Furthermore, explaining the different patterns in eye movements would very likely require even more strategies.

This speculation was confirmed in Hornof's recent study on visual search of hierarchical menus, in which additional strategies were needed to produce the varying length of saccade (Hornof, 2003). One of the questions addressed in our modeling work is whether there exists a modeling approach that relies on intrinsic behaviors of a cognitive architecture and is more parsimonious than depending on a large and growing number of task-specific strategies.

Our study shows that a single strategy can explain various patterns in human eye movements by adopting a Q-learning algorithm, which is task independent and is one type of method in reinforcement learning.

III.1. Reinforcement Learning and Eye Movement

This study is not the first attempt to adopt learning methods to visual search or eye movement study. Reward maximization has been used to model human eye movement (Sprague and Ballard, 2003). The ACT-R/PM model (Byrne, 2001) adopted learning method of conflict resolution for multiple competing strategies. Their study focused on 'where' an eye looked at (fixating location) rather than 'what' guided the movements of eye because each strategy is associated with the next eye location. The present study focuses not simply on the location of eye fixations, but on what guides the eye movement

The difference in the research approach between the queueing network model and other models is not only the number of strategies, but also the method used to guide the eyes. The hunt-feature production, which is used in queueing network model for menu search as a unique strategy for eye movement, guides eyes to the next location depending on the information in the unattended visual zone. In other words, the strategy is not directly related to the next location of eye fixation, while each strategy used in the other models explicitly tells the eyes where they should go next.

In the hunt-feature production rule originally used in ACT-R (Anderson, 1997), feature selection is assumed to be random. Here, we suspect that the randomness may be associated with the type of eye movement patterns. Therefore, instead of assuming numerous different strategies to explain different patterns of eye movements, a *reinforcement learning algorithm* is proposed to explain different kinds of patterns by adjusting the probability of selecting a feature used in hunt-feature production rule.

If we sketch eye movement as a problem of location, it becomes a complex problem in a continuous two-dimensional space. When we formulate the various patterns of eye movement into the feature selection problem, we can reduce the problem space into a finite number of alternative features rather than a continuous two-dimensional space problem. The picture used in Yarbus (1967)'s experiment was also used for this study.

III. 2. Q-Learning Model for Eye Movement

Our approach to modeling eye movement was to consider each eye movement as a Markov decision process and then to find an underlying policy using reinforcement learning method. A Markov decision process is described by four attributes: $\boldsymbol{S}$ is the state space, $\boldsymbol{A}$ is the action space, $\boldsymbol{T(s, a, s')}$ is the transition function that indicates the probability of arriving in state $\boldsymbol{s'}$ when action $\boldsymbol{a}$ is taken in state $\boldsymbol{s}$, and the reward function $\boldsymbol{R(s, a, x)}$.

The objective of the reinforcement learning algorithm is to discover an optimal policy $\boldsymbol{\pi^*(s)}$ (Sprague and Ballard, 2003) that includes which action should be taken in a certain state to maximize discounted long term reward.

One way to find the optimal policy $\boldsymbol{\pi^*(s)}$ is to obtain the optimal value function $\boldsymbol{Q(s, a)}$. This function represents the expected discounted return if action $\boldsymbol{a}$ is taken in state $\boldsymbol{s}$ and the optimal policy is followed thereafter. There are many algorithms (Sutton and Barto, 1998) for learning $\boldsymbol{Q(s, a)}$. For the feature selection in eye movement, the on-line $\boldsymbol{Q}$-learning update rule is used.

$$Q(s, a) \leftarrow Q(s, a) + \alpha \left[r + \gamma \max_{a'} Q(s', a') - Q(s, a) \right]$$

Here, $\boldsymbol{\alpha}$ is a learning rate parameter, and $\boldsymbol{\gamma}$ is a discounting factor of future reward (Sutton and Barto 1998). In this study, the initial value of learning rate parameter $\boldsymbol{\alpha_0}$ was 0.3 and it decreased by following rule: $\boldsymbol{\alpha_n = 0.999\ \alpha_{n+1}}$. The discounting factor of future reward $\boldsymbol{\gamma}$ is set as 0.5 and does not change.

To apply a reinforcement learning method to our eye movement study, the four attributes such as agent, environment, states, action and reward, were defined in the context of Yarbus' experiment as follows.

The *Agent* was defined as an eye that has two kinds of visual zones: attentional and unattentional visual zones. In Figure 3, the small square indicates attended visual zone and the large square indicates unattended visual zone. The Environment (*E*) was the picture shown to the Agent. The *State* ($\boldsymbol{s \in S}$) was a feature selected by the agent and in this case we assumed there were three alternative features which determined the next eye location. Depending on the state (selected feature), the *Location* ($\boldsymbol{x \in E}$ and $\boldsymbol{x}$ is two-dimensional vector) of eye fixation was determined as the location of attended visual zone.

The *Action* ($\boldsymbol{a \in A(s)}$) was a selection of the next hunting feature. There were three alternative criteria of hunting feature: the most similar contrast, the most different contrast, or the most edge information. Though the next eye location was not directly selected by the action, a selected feature determined the next location $\boldsymbol{x}$ of eye as described in Figure 3. Once the next location of eye has been determined, the eye moves to the location in the following saccade.

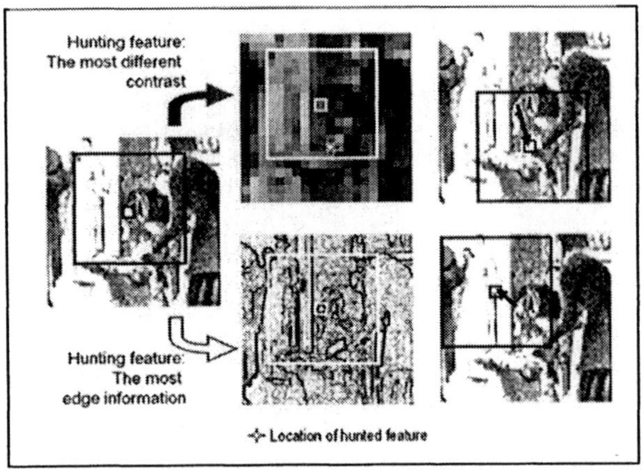

Figure 3. Different location is selected for next eye fixation depend on hunting feature used

Figure 4 presents the nine transition functions. The three states are the three alternative features currently used, and each action can keep the current state or move to other state.

The *Reward* (***R***(*s*, ***a***, ***x***)) was information given in a certain location ***x*** which was determined by an action ***a*** taken in the state ***s***. In this case, the amount of information was encoded as High (10), Low (5), and None (0). The location of reward was assumed to be different according to the goal. For example, under 'give the ages of the people', the rewards were assumed to be distributed mainly at people's face in the picture whereas under 'what family had been doing', the rewards were assumed to be distributed at people's posture. For different reward conditions, a different optimal policy could be obtained by Q-Learning method.

Following the different optimal policy, the simulated eye movement data were collected and they were qualitatively compared to the record of actual human eye movement patterns from Yarbus' experiment.

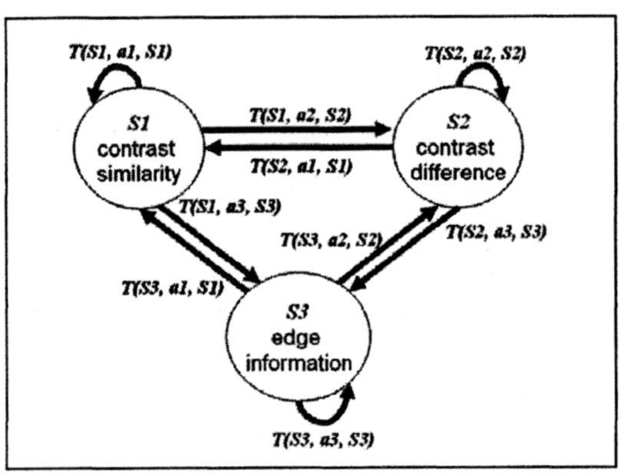

Figure 4. The nine transitions with the three states and the three actions

III. 3. Results of Q-Learning and Discussion

Using 800 randomly selected actions, the value of each ***Q***(***s***, ***a***) was recorded. Two different reward arrays were used for the 'Give the ages of the people' condition and 'What the family had been doing' condition. The results are summarized in Table 1. The proposed policy would be: *In the state s, the action **a** is selected with probability **F**(s, **a**).* Two different conditions showed the most different policy in state 1.

Table 1. The summary of different optimal policies for two different conditions

a. Give the ages of the people	
T(s, a, s')	*F(s, a)*
T(S1, a1, S1)	0.43
T(S1, a2, S2)	0.25
T(S1, a3, S3)	0.32
T(S2, a1, S1)	0.31
T(S2, a2, S2)	0.43
T(S2, a3, S3)	0.26
T(S3, a1, S1)	0.34
T(S3, a2, S2)	0.32
T(S3, a3, S3)	0.34

b. What family had been doing?	
T(s, a, s')	*F(s, a)*
T(S1, a1, S1)	0.24
T(S1, a2, S2)	0.29
T(S1, a3, S3)	0.47
T(S2, a1, S1)	0.33
T(S2, a2, S2)	0.35
T(S2, a3, S3)	0.32
T(S3, a1, S1)	0.33
T(S3, a2, S2)	0.34
T(S3, a3, S3)	0.33

S1: the most similar contrast
S2: the most different contrast, and *S3*: the most edge information

Using the two policies above, the viewing tasks were simulated. Figure 5 shows the simulated eye movement of 100 saccades and the human eye movement patterns from Yarbus' experiment.

Due to the lack of actual data from Yarbus' experiment, it was impossible to compare the simulated and the human eye movements quantitatively. However, the simulated eye movement patterns showed striking difference between the different conditions, and also showed similarity with the human eye movement patterns of the same conditions (the questions asked).

The only difference in modeling the two conditions—the 'Give the ages' and 'What the family had been doing'—was the *location of rewards*. All other parameters, input data, and strategy for eye movement were exactly same for both cases.

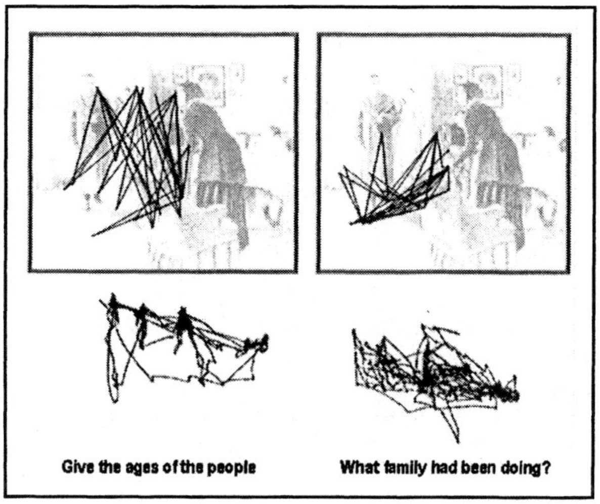

Figure 5. The simulated eye movements (top) and human eye movement (bottom)

In the two studies reported in this article, patterns of eye movement emerged mainly as the behavior of the underlying queueing network with far less reliance on task specific performance strategies than other models. The two studies demonstrated some of the unique aspects of the queueing network approach for modeling eye movement in specific and human performance in general.

ACKNOWLEDGMENTS

This article is based upon work supported by the National Science Foundation under Grant No. NSF 0308000. However, any opinions, findings and conclusions or recommendations expressed in this article are those of the authors and do not necessarily reflect the views of the National Science Foundation (NSF).

REFERENCES

Anderson, J. A., Matessa, M., & Lebiere, C. (1997). ACT-R: A theory of higher level cognition and its relation to visual attention. *Human-Computer Interaction*, 12, 439-462.

Byrne, M. D. (2001). ACT-R/PM and menu selection: applying a cognitive architecture to HCI. *Int. J. Human-Computer Studies*, 55.

Byrne, M. D., Anderson J. R., Douglass, S., & Matessa, M. (1999). Eye tracking the visual search of click-down menu. *Human Factors in Computer Systems: Proceedings of CHI*, 402-209.

Hornof, A. J. (1999). *Computational Models of the Perceptual, Cognitive, and Motor Processes Involved in the Visual Search of Pull-down Menus and Computer Screens*. Ph. D. dissertation in computer science engineering, The University of Michigan, Ann Arbor, Michigan.

Hornof, A. J., & Halverson, T. (2003). Cognitive strategies and eye movements for searching hierarchical computer displays. *ACM CHI 2003: Conference on Human Factors in Computing Systems,* New York: ACM, 249-256.

Liu, Y. (1996) Queuing network modeling of elementary mental processes. *Psychological Review*, 103, 116-136.

Liu, Y. (1997). Queuing Network Modeling of Human Performance of Concurrent Spatial and Verbal Tasks. *IEEE Transactions on system, Man, and Cybernetics – Part A: Systems and Humans.* 27(2), 195-207.

Nilsen, E. L. (1991). *Perceptual-Motor Control in Human-Computer Integration*. Ph.D. dissertation in Psychology, The University of Michigan, Ann Arbor, Michigan..

Sprague, N. and Ballard, D (2003), Eye Movements for Reward Maximization, *Advances in Neural Information Processing Systems*, 15.

Sutton, R. S. and Barto, A. G. (1998), *Reinforcement* ambridge, Massachusetts, London, England

Tsihmoni, O., and Liu, Y. (2003). Modeling steering using the Queueing Network-Model Human Processor (QN-MHP). *In Proceedings of the 2003 Annual Conf. of the Human Factors and Ergonomics Society.*

Wright, R. D., & Ward, L. M. (1998) The control of visual attention. Wright, R. D. (ed.), *Visual Attention*, Oxford; Oxford University Press.

Yarbus, A. L. (1967). Eye movements during perception of complex objects, in L. A. Riggs, ed., *Eye Movements and Vision*, Plenum Press, New York, chapter VII, pp. 171-196.

Combining learning approaches for incremental on-line parsing

Deryle Lonsdale (lonz@byu.edu)
Brigham Young University Department of Linguistics; 3186 JKHB
Provo, UT 84602 USA

Michael Manookin (mbm5@email.byu.edu)
Brigham Young University Department of Linguistics; 2129 JKHB
Provo, UT 84602 USA

Abstract

This paper discusses the integration of two different machine learning approaches to modeling language, NL-Soar and analogical modeling (AM). The resulting hybrid system is capable of functionality that is not possible when using only one of the systems in isolation. After a brief introduction of each system, an explanation is given of how AM is used to provide information useful to NL-Soar for two tasks. Examples are given, and related issues are outlined.

Introduction

Ongoing investigation in computational language modeling involves assessing the relative strengths and weaknesses of various approaches, for example symbolic versus subsymbolic, or rule-based versus exemplar-based. In this paper we show how an exemplar-based method is able to provide two types of crucial information that otherwise might not be available to a symbolic cognitive modeling system.

In the following section we sketch two substantially different machine learning approaches to language modeling. Next, we mention two well-studied natural language learning tasks: named entity recognition and prepositional phrase attachment resolution. The subsequent section discusses how two systems, NL-Soar and Analogical Modeling, have been combined in a way that brings together their relative strengths in novel and interesting ways involving these two tasks. In the last section we present conclusions, observations, and ideas for future work.

Language modeling: symbolic and exemplar-based

We begin by discussing two heretofore unrelated systems that have traditionally been used to model different language use phenomena: NL-Soar and analogical modeling (AM). Their complementarity motivates this integration: the former provides cognitive-level control, and the latter gives robust low-level instance-based matching.

Natural-language Soar

Natural-language Soar (NL-Soar) is an agent-based, hierarchical, goal-directed machine learning system that is based on the Soar cognitive modeling approach [Newell, 1990]. It has been used to model language use in a variety of modalities—comprehension [Lonsdale and Rytting, 2001], generation [Lonsdale, 2000], discourse [Green and Lehman, 2002]—in a variety of communicative task settings. A rule-based system, its basic knowledge repository is a set of if-then productions. Probabilistic reasoning is not a core feature of the basic architecture, and this introduces various challenges when addressing language-related tasks (among others).

The system receives lexical input word-by-word, and lexical access is performed for each word in turn. During lexical access WordNet [Fellbaum, 1998] provides relevant morphological, syntactic, and semantic information for all of the senses and homographs of the word in question [Rytting and Lonsdale, 2001]. The system then attempts to integrate the incoming words incrementally into linguistic models: a syntactic X-bar parse tree, and a semantic lexical-conceptual structure. All potential and possible syntactic and semantic material is considered in piecing together licit constructions. Constraints operate to rule out attachments that do not follow standard principles. In certain cases, some types of limited structure can be undone and reformulated when ongoing hypotheses prove untenable in the presence of new incoming words.

The semantic conceptual primitives are based on WordNet's lexical filenames and senses, which constitute (respectively) course-grained and fine-grained categories such as `v-body` for body verbs (e.g. sneezed, tripped) and `n-plant` for rhododendron[1]. Given the high degree of lexically-based ambiguity in English, much processing in NL-Soar involves determining compatibility between words and phrases at the syntactic and semantic levels.

The system, which is symbolic in its functionality, relies on a set of hand-coded rules and on-line learning as it performs the task. Its limited backtracking abilities allow modeling such processes as local ambiguity resolution, syntactic garden pathing, and complexity-induced breakdown in human parsing performance [Lewis, 1993].

Analogical modeling

Analogical modeling (AM) is a data-driven, exemplar-based approach to modeling language [Skousen, 1989] and other types of data. It has no rule-based component, either explicit or implicit, requires no explicit knowledge representations beyond the set of exemplars, and is a flexible and robust language modeling paradigm. Several linguistic applications have been reported using analogical modeling as the basic approach involving phonology, morphology, word sense disambiguation, speech processing, and lexical selection. So far

[1] In reality the WordNet codes are `verb.body` and `noun.plant` but this paper uses abbreviated names as shown.

it has only modeled low-level individual tasks, precluding its use as a comprehensive modeling framework.

The system operates as follows. A set of exemplars that address and illustrate a particular linguistic phenomenon is prepared; each instance has a fixed-length feature-vector encoding that represents salient (and perhaps nonsalient or questionable) properties for that instance. Each instance is labelled with an outcome that is used by the system to output how that instance behaves with respect to the phenomenon in question. At run time, the user inputs to the system a set of queries in the form of similarly encoded feature vectors. The system matches each input query with the exemplar base, and generates one or more probabilistically weighted outcomes. The system is able to tolerate noisy or incomplete data and behaves differently than other approaches in that it takes into consideration so-called "gang effects" that are problematic for the more traditional machine-learning languge-modeling methods. More details are available elsewhere concerning the system's application to language [Skousen, 1989], its statistical foundations and processing metrics [Skousen, 1992], NLP applications [Jones, 1996], and recent comparative work [Skousen et al., 2002].

Relevant tasks

In developing and scaling up the NL-Soar system, several issues arose that were not solvable using traditional symbolic methods. In this section we survey two such problems and how solutions were achieved from recent research in the natural language learning community.

Proper-noun semantics

Proper nouns are a crucial component of natural language, but in previous versions of NL-Soar they were not addressed. Using WordNet as the primary lexical resource allows for the retrieval of some of the more common proper nouns that are contained therein, such as "Virginia" and "France". For such words WordNet also provides semantic information; for example, different senses of "Washington" are encoded as `n-group`, `n-location`, and `n-person`.

Of course, only a small fraction of proper nouns are included in WordNet, which is problematic for NL-Soar's processing. A simple approach for syntax is to assume that any non-sentence-initial capitalized word encountered in text and absent from WordNet acts syntactically a proper noun[2]. Following this assumption has been straightforward and successful for handling the syntax of proper nouns. On the other hand, determining the semantics of proper nouns not included in WordNet has been more problematic.

Fortunately, this so-called named entity recognition (NER) problem has recently undergone extensive study by the natural language learning community [Tjong Kim Sang and De Meulder, 2003]. Though it has not been discussed in previous conferences on the topic, AM, like other modeling approaches, has been successfully used for named entity recognition. Using standardized data sets released from previous CoNLL shared tasks[3], AM researchers have been able to achieve state-of-the-art results for English, Dutch, and Spanish. Exemplar vector encodings used such features as the lexical item itself, its part-of-speech information, shallow-parsed constituent information, and the standard IOB codes for semantic classification.

Providing NER data for the semantics of noun phrases has been one motivation for integrating NL-Soar and AM processing. Before discussing and exemplifying this integration, we first mention another ideal application where a hybrid approach is advantageous.

PP attachment

The prepositional phrase attachment (PP-attachment) problem is an important and widely studied issue in natural language processing. Determining syntactic PP-attachment is even problematic for humans, as different attachment sites lead to multiple semantic interpretations (e.g. "I saw the man with the telescope."). Psycholinguistic research shows that human strategies for resolving PP-attachment ambiguities include the use of argument relations in the sentence [Schelstraete, 1996, Schuetze and Gibson, 1999], prosodic cues [Schafer, 1998, Straub, 1998], lexical constraints [Boland and Boehm-Jernigan, 1998], and context [Ferstl, 1994]. Others [Spivey-Knowlton and Sedivy, 1995] have demonstrated that lexical bias and contextual information have a strong effect. NL-Soar, as a rule-based symbolic system, has traditionally inferred PP-attachments primarily using lexical subcategorization information (i.e. WordNet verb frames). Thus leveraging the complement/adjunct distinction is based largely on data provided by WordNet.

Figure 1 reflects parses of two sentences: (a) "The minister warned the president of the danger.", and (b)-(c) "The minister warned the president of the republic of the danger." During parsing of the latter sentence, "...of the republic" is first temporarily linked as the PP-complement of "warned" (cf. "...of the danger" in (a)). When the second preposition is encountered, though, NL-Soar removes "...of the republic" from the verb's complement position and remakes the structure by adjoining the PP to the noun "president". The second "of" is then linked in as the PP-complement of "warned" (b). The parse then completes (c).

Though the system has been capable of handling relatively complex constructions like the one just discussed, a large class of PP-attachment scenarios could not be processed by the system. In particular, problems arose when the attachment decision was determined, not by subcategorization information on the matrix verb, but rather via lexical semantic information contained in the oblique object of the preposition. This leads to familiar structural ambiguities, which are sometimes ambiguous even for humans:

I saw the man with a beard/telescope.

Here attachment is determined by the PP object ("beard" or "telescope"), not by the subcategorization of "saw". Hence subcategorization information alone is insufficient to make PP-attachment decisions in many contexts; another approach was needed by NL-Soar to deal with ambiguities of this type. The next section presents the solution to this problem.

Past approaches to PP-attachment disambiguation have focused on statistical [Hindle and Rooth, 1993, Ratnaparkhi et al., 1994, Collins and Brooks, 1995] or rule-based [Brill and Resnik, 1994] methods. The statistical

[2]For the purposes of this paper we do not discuss words like "eBay", or "and" in multi-word proper-noun expressions.

[3]See http://lcg-www.uia.ac.be/conll200{2,3}/ner.

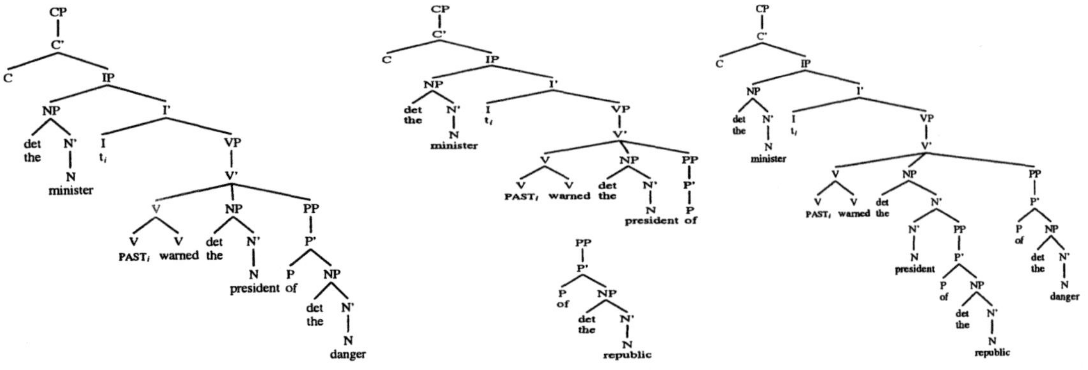

(a) (b) (c)

Figure 1: Sentence showing subcategorized PP complement 1(a); similar sentence with second preposition after reanalysis of the first PP now detached 1(b); final two-PP sentence including reattachment of the first PP 1(c).

approaches generally involve mining large annotated corpora and treebanks for determining the probability of an unknown attachment, usually based on the environment's lexical content. For "I saw the man with the telescope," a stochastic parser might assert an 84% probability for attachment to the verb "saw" and 16% probability for attachment to "man".

Since, for both of these tasks, exemplar-based methods are more appropriate, we describe their integration within the previously described symbolic framework.

Integration: Hybrid on-line processing

Having surveyed the NL-Soar and AM systems, we are in a position to follow how their integration allows for treatment of the two questions raised above. From a technical standpoint, composing the systems was straightforward; AM is written in C and Perl, and NL-Soar is an agent-based system written in C with a Tcl interface. It is therefore possible to call the AM system from NL-Soar via Tcl.

In the rest of this section more detail is provided about each of the integrations, and how hybrid processing unfolds.

Proper-noun semantics

Consider how NL-Soar processes the sentence: "Pendleton homered in the third inning."[4] The word "Pendleton" is not found in WordNet, and since it is also capitalized the system assumes that it is a proper noun. This initiates projection to an NP in the syntax. In order to ascertain the semantics of the word the system, during lexical access, calls the AM system with a named-entity query. The data consists simply of the word, putative part-of-speech category, and associated constituency information (i.e. initial part of an NP).

The AM system is invoked, calling the NER task with one test item—the word "Pendleton" and its associated features. The instance base consists of over 200,000 exemplars from the CoNLL-2003 shared task English NER training data. Two results are returned from the system: I-PER and I-LOC representing a person and a location, respectively. These codes are are translated into corresponding WordNet semantic classes: `n-person` and `n-location`. Syntactic and semantic nodes are created; since this is the first word of the sentence, nothing more can be done at this point.

Then the next word, "homered", arrives into the system and is attended to. This is clearly a verb, so the system's syntactic processing component projects a VP and an IP and then links in the NP "Pendleton" into the subject position. Semantically, this word is a competition verb (i.e. the semclass `v-competition` is provided by WordNet). NL-Soar, via a set of corpus-mined selectional restrictions [Rytting, 2000], is able to determine that agents of such verbs are much more likely to be of class `n-person` than of class `n-location`: people rather than places typically perform competitive actions[5]. Figure 2 shows the incremental syntactic (a) and semantic (b) representations for the sentence after the second word has been processed. Note the the LCS in (b) shows a `v-competition` node with an external argument consisting of the `n-person` construal of the proper noun; the other meaning `n-location` remains unlinked and will eventually disappear.

Thus we see that calling an exemplar-driven, corpus-based AM task for crucial NER information is useful in providing semantic information for NL-Soar. To be sure, other resources could also be accessed to find such information (e.g. online gazeteers or onomasticons), but the exemplar-based approach provides a level of generalization and robustness for processing novel items that a direct-lookup method cannot assure.

PP attachment

In similar fashion, we address the PP-attachment problem by combining NL-Soar with analogical modeling. NL-Soar calls

[4] This means that a baseball player with the surname Pendleton hit a home run in the third inning of a baseball game.

[5] Teams referred to as locations in a figurative sense may also serve as the agent in such verbs, as in "France won the World Cup."; consideration of this type of non-literal construction cannot be addressed within the scope of this paper.

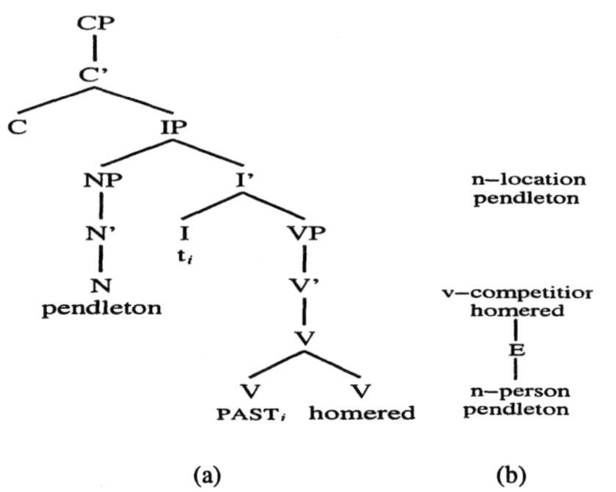

Figure 2: Syntactic and semantic incremental parses after processing the second word of the sentence.

AM when a preposition is encountered after a transitive verb and passes the verb, direct object noun, and preposition to AM. AM, in turn, runs the test item against a set of exemplars and returns the result to NL-Soar: whether the preposition should attach to the verb or to the noun.

Exemplars for this task are taken from the Brill/Resnik PP-attachment corpus[6] against an analogical set (of 12,266 examples) mined from the Penn Treebank.

Consider the sentence: "The military protects the regime against resistance." NL-Soar incrementally parses the sentence until the preposition "against" arrives; at this point there is no straightforward way for the system to decide the attachment of "against." By default, when a transitive verb in WordNet (e.g. "protect") does not subcategorize for a prepositional phrase complement, NL-Soar will adjoin the incoming PP to the direct object NP. Consequently in our sentence the PP "against resistance" will adjoin to "regime," not to "protects" (contra expectations).

In the NL-Soar/AM integrated framework, though, when "against" is encountered, a system callout queries AM for attachment preference information: the collocation "protect regime against" is sent to AM as a test item. AM computes the degree of (dis)agreement between the test item and exemplars from the previously mentioned PP-attachment corpus. The binary decision regarding the proposed attachment is returned via Tcl to NL-Soar; if licensed, the suggested attachment is implemented. If a verbal attachment is preferred, the proposed adjunction of the PP to the noun is rejected, and the preposition is instead adjoined to the verb.

Figures 3 and 4 demonstrate that integrating AM with NL-Soar generates different PP-attachments for the syntactically similar sentences "The military protects the regime against resistance." and "The military protects communication between foreign diplomats." Classified as unproblematic for humans, this type of structural ambiguity occasionally involves minor local readjustment of the parse and is relatively straightforward [Pritchett, 1992].

Figure 3(a) illustrates some of the processing involved during the parsing that creates 3(b). During cycles 259-263 NL-Soar attempts to link "against" as a complement of "protects", but fails since subcategorization requirements do not permit this attachment. Next (cycles 265-268) NL-Soar attempts to adjoin "against" to "regime"; this action is blocked by AM in cycle 268. Finally, the system attempts to adjoin "against" to "protects"; this action passes all constraints and the link is built (cycles 270-275).

A similar sequence of events occurs in Figure 4(a). The preposition "between" fails as a complement of "protects", but unlike in the previous example AM licenses[7] adjunction of "between" to "communication". Hence the link is built (cycles 253-260). The final result is shown in 4(b).

A final observation concerning learning should be made at this point. NL-Soar, through its hierarchical subgoaling architecture, learns as it performs tasks. This is true for the hybrid system described here: the system, once it has learned an attachment decision or constraint, chunks up this knowledge and retains it for future use. Thus, for example, when running the sentence in Figure 3(b) recognitionally (i.e. with all learned chunks available) vs. deliberately (i.e. with none available), the system uses one-third the CPU time (7 sec. vs. 2.3 sec.), one-half the decision cycles (500 vs. 250), one-eighth the rule firings (20,500 vs. 2500), one-tenth the number working memory changes (47,000 vs. 4700), and almost one-half the working memory size (3500 vs. 1900).

Discussion and conclusions

The work described in this paper is subject to the same criticisms that are often leveled at hybrid systems: functionality is gained, but only at the cost of abandoning a pure theoretical basis. On the other hand, much research has focused on the hypothesis that cognition (including the language faculty) could in fact be hybrid in nature. Another consideration when evaluating this tradeoff is the purpose for the system: whereas in the past NL-Soar has been used for cognitive modeling only, these problems have been encountered while trying to scale up coverage for more general NLP applications.

While only two exemplar-based callouts have been presented here, we anticipate that others will be useful in the future. The complement/adjunct distinction extends beyond PP's, and the determination of where to attach more complicated structures such as untensed clauses might profit from corpus-based exemplar approaches. Another difficult problem for on-line parsing is disambiguating gerund/participle environments: "My favorite cousin was singing." versus "My favorite activity was singing." Whereas this sentence pair is relatively straightforward, many cases are not as clear.

The NER data used in this paper comes off-the-shelf from a previous shared task training corpus, so the available information was limited to its contents. Only three categories (LOC, PER, ORG) have been implemented, so more thorough coverage of the semantics of proper nouns (e.g. determining gender

[6]The corpus is freely available at the following website: www.mit.edu/afs/athena/course/6/6.863/Lovecraft/PP.

[7]Since the actual exemplar is in the instance base, AM's score is 100%. Even if the instance is not used, AM approves this linkage by a margin of 72.6% to 27.4%.

```
250: O: O221 (u-constructor46)
251: ==>S: S196 (operator no-change)
252:     ==>S: S197 (state no-change)
259:         O: O227 try(link(protects.v--comp-->against.p)
260:         ==>S: S199 (operator no-change)
261:             O: C514 (check form)
262:             O: C512 (check subcat-feature)
263:             O: C510 (check subcat)
264:         O: O231 try(link(regime.n--adjoin-->against.p)
265:         ==>S: S200 (operator no-change)
267:             O: C529 (check subcat-bead)
268:             O: C527 (check amppattach)
269:         O: O229 try(link(protects.v--adjoin-->against.p)
270:         ==>S: S201 (operator no-change)
272:             O: C538 (check subcat-bead)
273:             O: C536 (check receiver-follows)
274:             O: O237 (constraint-success)
275:     O: C487 link(protects.v--adjoin-->against.p)
276:     ==>S: S202 (state no-change)
277:         O: O239 (exhausted)
278: O: O241 (return-operator)
```

(a)

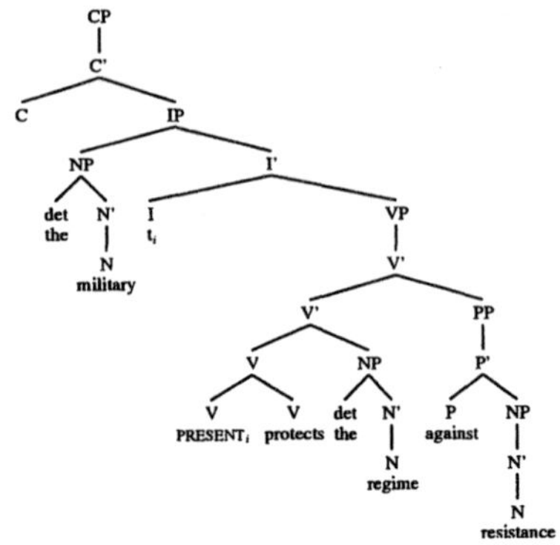

(b)

Figure 3: Trace of processing during call to AM (a) and the resulting syntactic structure (b). Processing cycles in (a) are indicated on the left; note that in cycle 268 a call to AM (i.e. amppattach) blocks adjunction of the preposition to the noun.

```
238:  O: O211 (u-constructor86)
239:  ==>S: S199 (operator no-change)
240:     ==>S: S200 (state no-change)
247:        O: O217 try(link(protects.v--comp-->between.p)
248:        ==>S: S202 (operator no-change)
249:           O: C513 (check form)
250:           O: C511 (check subcat-feature)
251:           O: C509 (check subcat)
252:        O: O222 try(link(communication.n--adjoin-->between.p)
253:        ==>S: S203 (operator no-change)
255:           O: C528 (check subcat-bead)
256:           O: C526 (check amppattach)
257:           O: C524 (check verb-saturated)
258:           O: C522 (check receiver-follows)
259:           O: O227 (constraint-success)
260:     O: C488 link(communication.n--adjoin-->between.p)
261:     ==>S: S204 (state no-change)
262:        O: O229 (exhausted)
263:  O: O231 (return-operator)
```

(a)

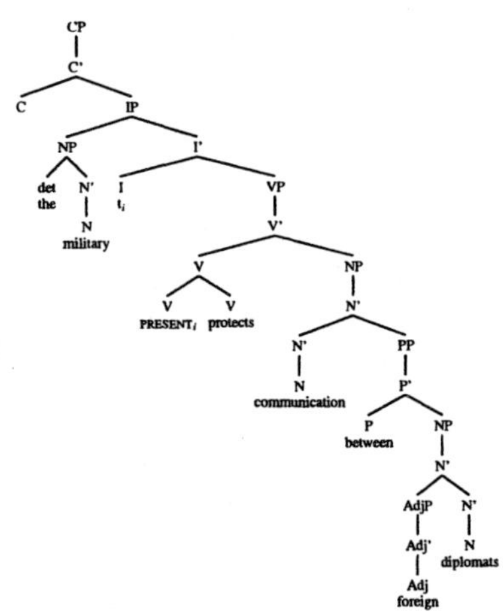

(b)

Figure 4: Trace of processing during call to AM (a) and the resulting syntactic structure (b). Processing cycles in (a) are indicated on the left; note that in cycle 256 a call to AM permits adjunction of the preposition to the noun.

of people from their names) might require more systematic corpus processing and exemplar base development.

Though empirical and theoretical issues remain, the integration of NL-Soar and AM has demonstrated how exemplar-based input can be used by a symbolic system in determining constraints, whether syntactic or semantic, for linguistic model construction. This in turn obviates the need for developing complex rule sets that likely cannot describe some phenomena in as perspicuous and robust a way as a well-constructed exemplar base.

References

[Boland and Boehm-Jernigan, 1998] Boland, J. and Boehm-Jernigan, H. (1998). Lexical constraints and prepositional phrase attachment. *Journal of Memory and Language*, 39(4):684–719.

[Brill and Resnik, 1994] Brill, E. and Resnik, P. (1994). A transformation-based approach to prepositional phrase attachment disambiguation. In *Proceedings of the Fifteenth International Conference on Computational Linguistics (COLING-1994)*.

[Collins and Brooks, 1995] Collins, M. and Brooks, J. (1995). Prepositional phrase attachment through a backed-off model. In *Proceedings of the Third Workshop on Very Large Corpora*, pages 27–38, Cambridge, MA.

[Fellbaum, 1998] Fellbaum, C. (1998). *WordNet: An electronic lexical database*. MIT Press, Cambridge, MA.

[Ferstl, 1994] Ferstl, E. (1994). Context effects in syntactic ambiguity resolution: The location of prepositional phrase attachment. In *Proceedings of the Sixteenth Annual Conference of the Cognitive Science Society*, pages 295–300. Lawrence Erlbaum Associates.

[Green and Lehman, 2002] Green, N. and Lehman, J. F. (2002). An integrated discourse recipe-based model for task-oriented dialogue. *Discourse Processes*, 33(2).

[Hindle and Rooth, 1993] Hindle, D. and Rooth, M. (1993). Structural ambiguity and lexical relations. *Computational Linguistics*, 19(1):103–120.

[Jones, 1996] Jones, D. (1996). *Analogical Natural Language Processing*. Studies in Computational Linguistics. University College of London Press Limited, London, England.

[Lewis, 1993] Lewis, R. (1993). An architecturally-based theory of human sentence comprehension. Technical Report CMU-CS-93-226, CMU.

[Lonsdale, 2000] Lonsdale, D. (2000). Leveraging analysis operators in incremental generation. In *Analysis for Generation: Proceedings of a Workshop at the First International Natural Language Generation Conference*, pages 9–13. Association for Computational Linguistics.

[Lonsdale and Rytting, 2001] Lonsdale, D. and Rytting, C. A. (2001). An operator-based account of semantic processing. In *The Acquisition and Representation of Word Meaning*, pages 84–92. European Summer School for Logic, Language, and Information.

[Newell, 1990] Newell, A. (1990). *Unified Theories of Cognition*. Harvard University Press.

[Pritchett, 1992] Pritchett, B. (1992). *Grammatical Competence and Parsing Performance*. University of Chicago Press.

[Ratnaparkhi et al., 1994] Ratnaparkhi, A., Reynar, J., and Roukos, S. (1994). A maximum entropy model for prepositional phrase attachment. In *Proceedings of the Human Language Technology Workshop*, pages 250–255, Plainsboro, N.J. ARPA.

[Rytting, 2000] Rytting, C. A. (2000). Semantic class disambiguation in Natural Language Soar. Brigham Young University Department of Linguistics. Unpublished honors thesis.

[Rytting and Lonsdale, 2001] Rytting, C. A. and Lonsdale, D. (2001). Integrating WordNet with NL-Soar. In *WordNet and other lexical resources: Applications, extensions, and customizations*, pages 162–164. North American Association for Computational Linguistics.

[Schafer, 1998] Schafer, A. (1998). *Prosodic parsing: The role of prosody in sentence comprehension*. PhD thesis, University of Massachusetts, Amherst.

[Schelstraete, 1996] Schelstraete, M. (1996). Definiteness and prepositional phrase attachment in sentence processing. *Current Psychology of Cognition*, 15(5):463–486.

[Schuetze and Gibson, 1999] Schuetze, C. and Gibson, E. (1999). Argumenthood and English prepositional phrase attachment. *Journal of Memory and Language*, 40(3):409–431.

[Skousen, 1989] Skousen, R. (1989). *Analogical modeling of language*. Kluwer, Dordrecht.

[Skousen, 1992] Skousen, R. (1992). *Analogy and structure*. Kluwer, Dordrecht.

[Skousen et al., 2002] Skousen, R., Lonsdale, D., and Parkinson, D., editors (2002). *Analogical modeling: An exemplar-based approach to language*. John Benjamins, Amsterdam.

[Spivey-Knowlton and Sedivy, 1995] Spivey-Knowlton, M. and Sedivy, J. (1995). Resolving attachment ambiguities with multiple constraints. *Cognition*, 55(3):227–267.

[Straub, 1998] Straub, K. (1998). *The production of prosodic cues and their role in the comprehension of syntactically ambiguous sentences*. PhD thesis, University of Rochester.

[Tjong Kim Sang and De Meulder, 2003] Tjong Kim Sang, E. F. and De Meulder, F. (2003). Introduction to the conll-2003 shared task: Language-independent named entity recognition. In Daelemans, W. and Osborne, M., editors, *Proceedings of CoNLL-2003*, pages 142–147. Edmonton, Canada.

Abductive Proofs as Models of Students' Reasoning about Qualitative Physics

Maxim Makatchev (maxim@pitt.edu), Pamela W. Jordan (pjordan@pitt.edu), Umarani Pappuswamy (umarani@pitt.edu) and Kurt VanLehn (vanlehn@pitt.edu)
Learning Research and Development Center
University of Pittsburgh
3939 O'Hara Street, Pittsburgh, PA 15260 USA

Abstract

In this paper we describe a part of the Why2-Atlas tutoring system that models students' reasoning in the domain of qualitative physics. The main goals of the model are (1) to evaluate correctness of the student's essay, and, in case the essay contains errors, (2) to direct remedial tutoring actions according to plausible errors in the student's reasoning. To meet these goals, a backchaining theorem prover generates a set of assumptions and a chain of reasoning (a proof) that plausibly led the student to write the observed essay. A proof can include correct as well as buggy reasoning steps and assumptions. After a proof is generated, it is analyzed for correctness and the analysis is used to generate appropriate feedback to the student. We describe the weighted abductive theorem proving framework, outline previous and upcoming evaluations and discuss possible future directions.

Question: Suppose a man is in a free-falling elevator and is holding his keys motionless right in front of his face. He then lets go. What will be the position of the keys relative to the man's face as time passes? Explain.

Explanation: The keys are affected by gravity which pulls them to the elevator floor, because the keys then have a combined velocity of the freefall and the effect of gravity. If the elevator has enough speed the keys along with my head would be pressed against the ceiling of the elevator, because the acceleration of the elevator car along with me and the keys would overwhelm the gravitational pull.

Figure 1: The statement of the problem and an example explanation.

Introduction

The Why2-Atlas tutoring system is designed to encourage students to write their answers to qualitative physics problems as essays that include explanations of their arguments (VanLehn, Jordan, Rosé, Bhembe, Böttner, Gaydos, Makatchev, Pappuswamy, Ringenberg, Roque, Siler, & Srivastava, 2002). If the essay is incomplete or incorrect, the system generates elicitation or remediation feedback, respectively. For the purpose of evaluating completeness and correctness of the essay, a deep understanding of its contents is necessary. In the domain of qualitative physics, a deep understanding of the essay involves reasoning about logical relationships between the statements in the essay. The theorem proving approach that we use in Why2-Atlas provides the means to recover a structure of logical dependencies that connects the propositions representing (a) the essay text, (b) the problem statement, and (c) plausible student reasoning steps not explicitly stated in the essay.

Previously, formal methods for analyzing natural language text have encountered a number of challenges, such as the difficulty of obtaining propositional representations for sentences and the need for large amounts of common sense knowledge in order to interpret the many concepts expressed. Consider, for example the qualitative physics problem presented in Figure 1 along with an actual student explanation. To address all the errors in the essay, the propositional representation of the essay must include a significant amount of common sense knowledge, for example that an elevator has a ceiling, and that a human has a head. In addition, even if the system is able to represent the respective statements, it also needs to be able to reason about correctness of logical relations between these statements (a statement b can be false in the context of the problem, while the statement $a \to b$ can be true). Such reasoning would also be desirable for the purpose of providing a more substantive feedback to the student, e. g. excessive argumentation. An informal example of a possible chain of reasoning that student used to arrive at the statement "The keys would be pressed against the ceiling of the elevator" is shown in Figure 2.

An abductive theorem proving approach allows one to cope with propositions that cannot be proven due to lack of applicable rules by assuming such propositions are true without a proof (the operation is also referred to in the literature as *abducing*) when such assumptions allow a proof to be completed. The fewer the assumptions made, the better the proof. Weighted abduction takes this approach farther and assigns a cost to the set of assumptions depending on their individual weights and the chain of reasoning that led to the generation of these assumptions. A proof can include correct as well as buggy reasoning steps and assumptions.

In this paper we describe a part of the Why2-Atlas tutoring system that models students' reasoning of qualitative physics. The theorem prover, called Tacitus-lite+, is a derivative of SRI's Tacitus-lite (Hobbs, Stickel, Martin, & Edwards, 1988, p. 102) that, among other extensions, incorporates sorts. First we provide an overview of the

Step #	Proposition	Justification
1	before the release, the keys have been in contact with the man, and the man has been in contact with the elevator	given
2	at the moment of release, velocity of the keys is equal to velocity of the elevator	bodies in contact over a time interval have same velocities
3	after the release, nothing is touching the keys	given
4	after the release, the keys are in freefall	if there is no any contact then the body is in freefall
5	after the release, the keys' acceleration is not equal to the elevator's acceleration	*elevator is not in freefal
6	after the release, the keys' velocity is not equal to the elevator's velocity	if initial velocity is the same and accelerations are different the final velocities are different
7	the keys touch the ceiling of the elevator	if the keys' velocity is smaller than the elevator's velocity, the keys touch the ceiling

Figure 2: An informal proof of the excerpt "The keys would be pressed against the ceiling of the elevator" (From the essay in Figure 1). The buggy assumption is preceded by an asterisk.

knowledge representation. Next we describe the abductive theorem proving framework and the heuristics we developed that aim at maximization of the plausibility of the proof as a model of the student's reasoning and the utility of the proof for the tutoring system. The measure of plausibility is evaluated with respect to (a) the misconceptions that were identified as present in the essay by the prover and by a human expert, and (b) the proof as a whole. The utility for the tutoring task can be interpreted in terms of relevance of the tutoring actions (triggered by the proof) to the student's essay, whether the proof was plausible or not. We also discuss the assumptions of cognitive economy and concept-level consistency that we make about the student in relation to the plausibility of the model. Next we summarize previous evaluations of the Why2-Atlas system (VanLehn et al., 2002) and of an early version of the abductive reasoning engine (Jordan, Makatchev, & VanLehn, 2003). Finally we conclude with a section on our future work.

Knowledge Representation for Students' Reasoning about Qualitative Physics

Envisionment and idealization

Generating an internal (mental) representation plays a key role for both novice and expert problem solving (Ploetzner, Fehse, Kneser, & Spada, 1999; Reimann & Chi, 1989). (Reimann & Chi, 1989) describes the internal representation in terms of "objects, operators, and constraints, as well as initial and final states." This notion of internal representation overlaps with *envisionment*, which is defined in qualitative physics problem solving (de Kleer, 1990) as a sequence of events described in the problem or implied by the description. A further step, translating the envisionment into the domain terminology (bodies, forces, motion properties) is referred to as *idealization* in (Makatchev, Jordan, & VanLehn, 2004a).

For the problem in Figure 1, for example, a possible envisionment is: (1) the man is holding the keys (elevator is falling); (2) the man releases the keys; (3) the keys move down with respect to the man and hit the elevator floor. The idealization would be:

Bodies: Keys, Man, Elevator, Earth.

Forces: Gravity, Man holding keys

Motion: Keys' downward velocity is larger than the velocity of the man and the elevator.

Many misconceptions that students have are rooted in the envisionment and idealization (Ploetzner, 1999). To make the task of representing possible correct and erroneous envisionments feasible we restrict ourselves to problems with few plausible envisionments. The rules of mechanics, which rely mostly on the formal domain terminology, are augmented by rules for reasoning about most common envisionments, which use a looser language. Further we briefly describe these representations.

Qualitative mechanics ontology

The ontology for the subset of qualitative mechanics that the system addresses consists of bodies (e. g., keys, man), agents (air), phenomena (e. g., gravity, friction), and conventional physical quantities (e. g., force, velocity, position). To adequately represent justifications, we also have representations for physics laws (Newton's First Law) and basic algebraic expressions ($F = ma$). While internally the reasoning is done within a coordinate system that is fixed for each problem (for example, horizontal axis x directed to the right and vertical axis y directed up), a student's reasoning can be independent of coordinate system choice, operating instead in relative terms (up, down, in front of). The representation and corresponding translation rules are described in the following section.

Logical constants and variables, corresponding to bodies, agents, and quantities are associated with a sort symbol. Sorts are partially ordered by a natural subset order. Domains of the predicate symbols are restricted to certain sorts (so that each argument position has a corresponding sort symbol). These associations and constraints constitute an *order-sorted signature* (Walther, 1987).

Description	Sort
quantity	Quantity1b
identifier	Id
body (or two bodies in case of force)	Body
axial component or not	Comp
qualitative derivative of the magnitude	D-mag
quantitative derivative of the magnitude	D-mag-num
zero or non-zero magnitude	Mag-zero
quantitative magnitude	Mag-num
sign for axial component	Dir
quantitative direction	Dir-num
qualitative derivative of the direction	D-dir
beginning of time interval	Time
end of time interval	Time

Table 1: Slots of a vector quantity of sort Quantity1b.

Time is represented using time instants as basic primitives. Time intervals are denoted as a pair (t_i, t_j) of instants. This and the order relation before on time points enables us to reason about a reasonably rich subset of the mechanics domain.

Argument slots and an order-sorted signature for a predicate representing a vector quantity that involves a single body (for example velocity, total-force) are shown in Table 1.

A number of relation predicates are used to specify various algebraic and logical relations between physical quantities (see Table 2).

Two bodies can also be related via a state of contact with possible fillers detached, attached, and moving-contact (for the case of relative motion between bodies in contact).

Rules

The rules cover correct reasoning at the formal domain level of an idealized problem ("zero acceleration implies constant velocity"), buggy reasoning at this same level ("zero force implies decreasing velocity"), and some common relevant aspects of the idealization and envisionment stages ("if axis y is directed upward and velocity is vertical and positive then velocity is upward").

The rules are represented as *extended Horn clauses*, namely the head of the rule is an atom or a conjunction of multiple atoms. Further details of the knowledge representation are covered in (Makatchev et al., 2004a).

Weighted Abductive Theorem Proving

Order-sorted abductive logic programming framework

Similar to (Kakas, Kowalski, & Toni, 1998) we define the *abductive logic programming framework* as a triple $\langle T, A, I\rangle$, where T is the set of *givens* and *rules*, A is the set of abducible atoms (potential hypotheses) and I is a set of integrity constraints. Then an *abductive explanation* of a given set of sentences G (goals) is (a) a subset Δ of abducibles A such that $T \cup \Delta \vdash G$ and $T \cup \Delta$ satisfies I, and (b) the corresponding *proof* of G. The set Δ is *assumptions* that explain the goals G. Since an abductive explanation is generally not unique, various criteria can be considered for choosing the most suitable explanation (see Section "Proof search heuristics").

An *order-sorted abductive logic programming framework* $\langle T', A', I'\rangle$ is an abductive logic programming framework with all atoms augmented with the sorts of their argument terms (so that they are sorted atoms) (Makatchev et al., 2004a). Assume the following notation: a *sorted atom* is of the form $p(x_1, \ldots, x_n) : (\tau_1, \ldots, \tau_n)$, where the term x_i is of the sort τ_i. Then, in terms of unsorted predicate logic, formula $\exists x\ p(x) : (\tau)$ can be written as $\exists x\ p(x) \wedge \tau(x)$. For our domain we restrict the sort hierarchy to a tree structure that is naturally imposed by set semantics and that has the property $\exists x\ \tau_i(x) \wedge \tau_j(x) \rightarrow (\tau_i \preccurlyeq \tau_j) \vee (\tau_j \preccurlyeq \tau_i)$ where $\tau_i \preccurlyeq \tau_j$ is equivalent to $\forall x\ \tau_i(x) \rightarrow \tau_j(x)$.

Tacitus-lite+ uses backward chaining with the order-sorted version of modus ponens:

$$\frac{\begin{array}{l} q(x', z') : (\tau_5, \tau_6) \\ p(x, y) : (\tau_1, \tau_2) \leftarrow q(x, z) : (\tau_3, \tau_4) \\ \tau_5 \preccurlyeq \tau_3,\ \tau_6 \preccurlyeq \tau_4 \end{array}}{p(x', y') : (\min(\tau_5, \tau_1), \tau_2)}$$

Proof search heuristics

The aim of the proof search heuristics is to quickly find a proof that optimizes a measure of utility of the proof for tutoring applications and a measure of plausibility of the proof as a model of a student's reasoning. A highly plausible proof has a high value for its utility measure since it potentially allows the tutoring system to generate feedback that is more relevant to the student's actual mental state. However a less plausible proof would have the same utility measure if it results in the same tutoring action as a more plausible proof. In fact, we would prefer a less plausible proof over the more plausible proof, their utility measures being same, if the former takes less time to compute.

The plausibility measure is based on two cognitive assumptions. The first assumption, *cognitive economy*, can be interpreted in the context of the abductive proofs as a preference for a simpler proof structure (for example a smaller proof) and a smaller cost for the propositions that have to be assumed. The second assumption, *concept-level consistency*, is based on the fact that even young children are unlikely to make mistakes in tasks involving taxonomic categories (Chi & Ceci, 1987). Thus we assume that, while proofs can have errors, errors in categorical and taxonomic reasoning are less plausible. For example, the consistency constraints that we enforce for proofs prevent propositions such as "velocity of the keys is increasing" and "velocity of the keys is constant" from appearing within the same proof.

A proof is considered sufficiently cheap if the total cost of its assumed atoms is below a certain threshold. The cost is computed for each proposition of the proof via the following procedure. First, costs are uniformly assigned to the goal atoms (observations), namely the propositional representation of the student's essay. Conjunct atoms p_i in the body of a rule have pre-assigned weights

Relation	1st and 2nd arguments	3rd argument	4th argument
non-equal	any terms		
before	Time		
rel-position	Body	Rel-location	
compare	Mag-num or D-mag-num of any scalar or vector quantity	Ratio	Difference
compare-dir	Dir-num of any vector quantity	Rel-dir	
dependency	any terms	Rel-type	time interval

Table 2: Relations.

w_i (Stickel, 1988):

$$p_1^{w_1} \wedge \cdots \wedge p_m^{w_m} \rightarrow r_1 \wedge \cdots \wedge r_n.$$

If this rule is used to prove a goal g by unifying it with atom r_j, then the cost of assuming p_i, $1 \leq i \leq m$, is computed according to the following cost propagation formula: $cost(p_i) = cost(g) \cdot w_i$. The cost of the proof is the total cost of all assumed atoms.

A weighted abductive proof for student's statement "The keys would be pressed against the ceiling of the elevator" is shown in Figure 3. Total cost of the proof is 0.15, the cost of its only assumption. Incidentally, the proof indicates a possible presence of the wrong assumption "The elevator is not in freefall," which is made by the student likely due to a wrong interpretation of a problem given.

Since the cost of a proposition is a penalty for assuming it without a proof, it can also be interpreted as a degree of disbelief in the proposition. This interpretation suggests that more general existentially quantified propositions should be cheaper to assume than more specific propositions. The mechanism for such cost adjustment is implemented in the most recent version of the theorem prover.

Various rule choice heuristics have the aim of finding a sufficiently cheap proof of a small size. Generally, if atoms in the head of the rule are unifiable with a subset of goals then application of such a rule will result in the subset of atoms being removed from the goal list. If a rule has atoms in its body that are unifiable with the goals, then the new subgoals will be *factored* with the unifiable goals, namely only the most specific of the unifiable atoms will be left on the goal list. These nuances imply that proving via rules that have heads and bodies that are unifiable with larger subsets of goals lead to a faster reduction of the goal list and consequently a smaller resultant proof.

In addition, a set of atoms can be cross-referenced via shared variables. The cross-reference graph encodes a large amount of semantics for the proposition corresponding to the respective set of atoms. One of the major rule choice heuristics implemented in the theorem prover is based on the similarity between the graph of cross-references between the propositions in a candidate rule and the graph of cross-references between the set of goals. The metric for the match between two labeled graphs is computed as the size of the largest common subgraph using the decision-tree-based algorithm proposed in (Shearer, Bunke, & Venkatesh, 2001). For further details on the proof search heuristics we refer the reader to (Makatchev, Jordan, & VanLehn, 2004b).

Evaluation

Although students in a baseline evaluation of the Why2-Atlas system showed significant learning gains (VanLehn et al., 2002), the sentence-level representations of the students' essays produced by the system, that are the input to Tacitus-lite+, were too sparse for any misconceptions to be correctly identified. To evaluate Tacitus-lite+ we developed a test suite of 45 student generated essays in which we manually corrected and completed the input generated by the system for input to Tacitus-lite+ and annotated the misconceptions expressed in each essay that Tacitus-lite+ should identify. The student essays were randomly selected from those collected during the baseline evaluation and from subsequent experiments with students and human tutors. In the 45 essays of the test suite, three essays have two misconceptions each, eight essays have one misconception each, and the rest of the essays don't have any misconceptions from the list of 54 misconceptions that could arise for the training problems according to our physics experts.

There are two types of evaluations of interest to us for the abductive theorem prover: (1) the accuracy of the misconceptions revealed by the proofs and (2) the accuracy of the proofs as models of the students. We summarize here the results of both for an earlier version of Tacitus-lite+, as described in (Jordan et al., 2003), and plan to repeat both in the near future for the newer version described in this paper.

To assess the accuracy of the misconceptions identified by the theorem prover, we compare the misconceptions revealed by the proofs of each essay to those annotated for each test suite essay. We accumulated the number of true positives TP, false positives FP, true negatives TN, and false negatives FN for each essay; and from this computed recall TP/(TP+FN), precision TP/(TP+FP), and positive false alarm rate FP/(FP+TN). In addition, we calculated these measures for the theorem prover's results at various proof cost thresholds to see how the performance changes as we move closer toward building a complete proof. The results are shown in Figure 4.

The recall increases from 0 at a proof cost of 1 (where everything is assumed without proof) to 62% at a proof cost threshold of 0.2. As the recall increases, the precision degrades but then levels off. These results mean that the earlier theorem prover can help to reveal up to 62% of the misconceptions that a human would recognize, but at the cost of identifying some misconceptions that are not justified by the essays. We consider recall to be the more important measure for misconceptions since it is important to find and address the misconceptions

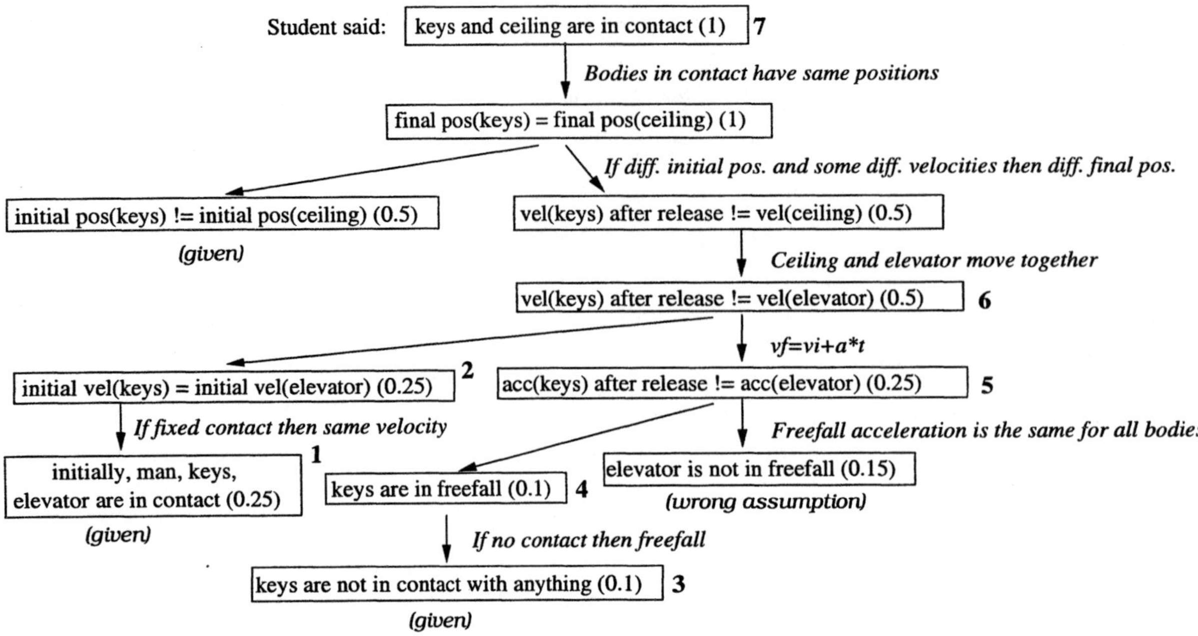

Figure 3: A weighted abductive proof of the proposition representing the excerpt "The keys would be pressed against the ceiling of the elevator." Rule names are in italics; arrows are in the direction of abductive inference; costs of the propositions are in parenthesis; the references to the steps from Figure 2 are in bold. Total cost of the proof is 0.15.

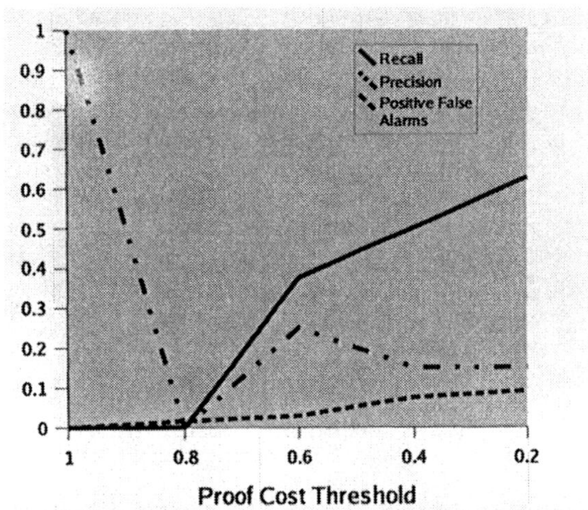

Figure 4: Recall, precision and false alarm measures as proof cost threshold decreases.

that are expected to be obvious to a human tutor. The positive false alarm is quite low and although our goal is to reduce this value as close to 0 as possible, we consider a high recall to be a higher priority as we expect that it is more important not to miss misconceptions. On the other hand, some possible drawbacks of not also trying to lower the positive false alarms are inadvertently strengthening the reasoning that leads to a misconception and a loss of student motivation and cooperation if the student perceives the system is too frequently giving inappropriate feedback.

While these results are encouraging, we expect that the recent improvements we've made to Tacitus-lite+, along with additional testing and fine-tuning of rules, will further improve the results. In addition, an evaluation of misconceptions revealed is only a coarse measure of the quality of the proofs generated. To determine the fitness of the theorem prover's modeling to support assessments of completeness, we must also consider the accuracy of the proof structure generated. Assessing the accuracy of the proof structure is more difficult because the proofs must be hand verified. It is difficult to create a reliable gold standard against which to evaluate the accuracy of proofs for essays and the reasons for any inaccuracy. This is because, in general, language in context gives rise to many inferences (Austin, 1962; Searle, 1975). For this assessment we judged whether the lowest cost proofs generated for 15 of the test suite essays was a plausibly good, satisfactory or bad model of the student essay. As shown in Table 3, as the proof cost threshold decreased and consequently the number of assumptions made fell, the number of good proofs increased and the number of bad ones fell to 0.

Conclusions and Future work

In this paper we described an approach to modeling of a student's reasoning about qualitative physics problems by treating the student's essay as an observation, the problem statement as a set of given facts, and us-

Threshold	0.8	0.6	0.4	0.2
good	7	7	10	11
satisfactory	4	4	4	4
bad	4	4	1	0

Table 3: Evaluation of plausibility of proofs generated for different proof cost thresholds.

ing an abductive proof of this observation as a plausible approximation of the student's reasoning. Abductive proofs provide an intuitively natural representation for logical relations between the arguments of the essay. The problem of insufficient coverage of the domain and common-sense knowledge—one of the difficulties that formal methods face when applied to natural language text analysis—is alleviated by allowing proofs to include assumptions, namely propositions that cannot be proven. Weighted abduction provides a facility to rate such proofs by assigning costs to their respective sets of assumptions. The weighted abductive theorem prover has been implemented and evaluated with respect to plausibility of proofs as models of students' reasoning.

There are a number of challenges still to address. One is to handle various degrees of formalism in the input representations of the student's language. For example, if a student says "throw," the current representation input to Tacitus-lite+ is "apply an upward vertical force." But the student's actual lexical choices need additional reasoning relative to the model of the student in order to determine whether the correct formal representation is plausible for the student. Otherwise, the student is credited with understanding more about physics than may be plausible. So the goal is to take over more of the natural language semantic interpretation process within Tacitus-lite+. The favorable evaluation results we have obtained so far make it more promising that such a move will be successful. Other challenges include increasing the coverage of the rule-base, further improving the efficiency of the theorem prover, and further improved consistency checking.

Acknowledgments

This work was funded by NSF grant 9720359 and ONR grant N00014-00-1-0600. We thank the entire Natural Language Tutoring group, in particular Michael Ringenberg and Roy Wilson for their work on Tacitus-lite+, and Brian 'Moses' Hall and Michael Böttner for their work on knowledge representation and rules.

References

Austin, J. L. (1962). *How to Do Things With Words.* Oxford University Press, Oxford.

Chi, M. T. H., & Ceci, S. J. (1987). Content knowledge: Its role, representation and restructuring in memory development. *Advances in Child Development and Behavior*, *20*, 91–142.

de Kleer, J. (1990). Multiple representations of knowledge in a mechanics problem-solver. In Weld, D. S., & de Kleer, J. (Eds.), *Readings in Qualitative Reasoning about Physical Systems*, pp. 40–45. Morgan Kaufmann, San Mateo, California.

Hobbs, J., Stickel, M., Martin, P., & Edwards, D. (1988). Interpretation as abduction. In *Proc. 26th Annual Meeting of the ACL, Association of Computational Linguistics*, pp. 95–103.

Jordan, P., Makatchev, M., & VanLehn, K. (2003). Abductive theorem proving for analyzing student explanations. In *Proceedings of International Conference on Artificial Intelligence in Education*, pp. 73–80, Sydney, Australia. IOS Press.

Kakas, A., Kowalski, R. A., & Toni, F. (1998). The role of abduction in logic programming. In Gabbay, D. M., Hogger, C. J., & Robinson, J. A. (Eds.), *Handbook of logic in Artificial Intelligence and Logic Programming*, Vol. 5, pp. 235–324. Oxford University Press.

Makatchev, M., Jordan, P. W., & VanLehn, K. (2004a). Abductive theorem proving for analyzing student explanations to guide feedback in intelligent tutoring systems. *To appear in Journal of Automated Reasoning, Special issue on Automated Reasoning and Theorem Proving in Education.*

Makatchev, M., Jordan, P. W., & VanLehn, K. (2004b). Modeling students' reasoning about qualitative physics: Heuristics for abductive proof search. In *Proceedings of Intelligent Tutoring Systems Conference*, LNCS. Springer. To appear.

Ploetzner, R., Fehse, E., Kneser, C., & Spada, H. (1999). Learning to relate qualitative and quantitative problem representations in a model-based setting for collaborative problem solving. *The Journal of the Learning Sciences*, *8*, 177–214.

Reimann, P., & Chi, M. T. H. (1989). Expertise in complex problem solving. In Gilhooly, K. J. (Ed.), *Human and machine problem solving*, pp. 161–192. Plenum Press, New York.

Searle, J. R. (1975). Indirect Speech Acts. In Cole, P., & Morgan, J. (Eds.), *Syntax and Semantics 3. Speech Acts.* Academic Press. Reprinted in *Pragmatics. A Reader*, Steven Davis editor, Oxford University Press, 1991.

Shearer, K., Bunke, H., & Venkatesh, S. (2001). Video indexing and similarity retrieval by largest common subgraph detection using decision trees. *Pattern Recognition*, *34*(5), 1075–1091.

VanLehn, K., Jordan, P., Rosé, C., Bhembe, D., Böttner, M., Gaydos, A., Makatchev, M., Pappuswamy, U., Ringenberg, M., Roque, A., Siler, S., & Srivastava, R. (2002). The architecture of Why2-Atlas: A coach for qualitative physics essay writing. In *Proceedings of Intelligent Tutoring Systems Conference*, Vol. 2363 of *LNCS*, pp. 158–167. Springer.

Walther, C. (1987). *A many-sorted calculus based on resolution and paramodulation.* Morgan Kaufmann, Los Altos, California.

Toward a Comprehensive Computational Model of Emotions and Feelings

Robert P. Marinier III (rmarinie@umich.edu)
University of Michigan, 1101 Beal Avenue
Ann Arbor, MI 48109-2110 USA

John E. Laird (laird@umich.edu)
University of Michigan, 1101 Beal Avenue
Ann Arbor, MI 48109-2110 USA

Abstract

We describe a computational framework for emotions and feelings that combines biological, cognitive, and social influences. The framework, Soar-Emote, combines ideas from Damasio and Gratch & Marsella. From Damasio we include physiological influences, ideas about the differences between emotions and feelings, and the direction of causality. From Gratch we include cognitive and social influences, appraisal theory and coping. We give results that show that these systems influence agent behavior for a simple interactive game.

Introduction

If one thing is clear in the field of computational models of affective reasoning, it is that the exact phenomena and behavior that falls under the rubric of "emotion" or "affective reasoning" is very murky. Some theories of emotion attempt to model low-level physiological effects of specific emotions in isolation, whereas others attempt to model the interaction between of emotion and cognition, while others attempt to model the impact of emotional responses on social interactions, typically without attempting to model any of the underlying physiological processes. An analysis by Biddle et al. (2003) of a sampling of current computational models showed that several systems attempt to model affective behavior in one or two of these areas, but no systems attempts to model behavior across all. Our hypothesis is that affective reasoning is not an isolated phenomenon and that to understand it we need to build computational models that have impact across all of these areas. Further, such models must be intertwined with a theory of cognition – emotion is not a separate "chip" (as often portrayed in science fiction AI systems) but is integral to cognitive processing.

This paper describes a computational framework for affective reasoning. We chose two theories that together cover biological, cognitive and social influences on emotions and feelings. Damasio (1994, 2003) describes the role of the body in emotions and feelings as well as the high-level architecture of the human affective system. Damasio also makes an important distinction between emotions and feelings, but only touches on the role that cognition plays in processing and responding to emotions. Furthermore, Damasio's models are descriptive, without the accompanying precision of computational implementations. Gratch & Marsella (2004) on the other hand, focus on a computational model of the interaction of affect with cognition and the social level, including how feelings are represented in a general cognitive architecture, Soar (Newell 1990), and how coping is used to respond to emotions.

Our framework refines a subset of the Gratch model, adding components to model the concepts of Damasio. Moreover, our model, which is an instantiation of the framework, involves *architectural changes* to Soar, so that affective reasoning is directly incorporated into Soar, in contrast to the Gratch implementation where affective reasoning was implemented as knowledge (encoded as rules) in the cognitive system. The research is not mature enough to make precise predictions concerning the generation of specific emotions and feelings in humans. Instead, our goal is to demonstrate that the framework captures important interactions between physiology, cognition, and social interaction. We do this by embedding the model in agents that play a simple cooperative game where emotions can arise. We introduce "lesions" in the agents that disable different aspects of the framework, and confirm that each aspect of our framework contributes to the system's overall behavior.

Prior Work

Gratch's work is notable because it is a model of how emotions and feelings impact cognition and how cognition responds to feelings. Gratch's system, called EMA, uses appraisal theory[1] to describe how an agent evaluates a situation, and how the resulting feeling is represented to the agent. Thus, emotional responses are not hardwired into inputs from the perceptual system, but instead are based on the agent's interpretation (appraisal) of the current situation, which can include arbitrary cognitive processing or even the recall of prior situations from memory. He also describes various coping mechanisms that involve cognitive processing. These mechanisms are strategies (once again, not hardwired responses) that the agent can selectively employ to respond to feelings and include both problem-focused coping strategies, in which an agent takes action in

[1]Appraisal theory is widely studied and many versions exist; for a survey, see Ellsworth & Scherer (2003). In this work we are primarily concerned with the version in EMA.

the world, and emotion-focused strategies, in which an agent reevaluates a situation cognitively.

EMA consists of two basic steps that cycle repeatedly. The agent appraises the current situation, which results in an "emotion." The agent then copes with the "emotion," either by altering its appraisals (emotion-focused) or by altering its situation (problem-focused).

EMA defines several appraisal variables, including Desirability, Likelihood, and Attribution. Desirability is a measure of how much the agent wants to be in the situation. Likelihood, which is the probability of a situation being true, is useful when an agent is considering situations that may have been or might be (either now or in the future). Attribution is the thing that caused the situation, although there may not always be an object for attribution. Combinations of these variables in different ranges give rise to different feelings, as shown in Table 1. Furthermore, the intensity of the feeling is the absolute value of the Desirability multiplied by the Likelihood.

Table 1: "Emotions" (feelings) defined by EMA.

"Emotion" (Feeling)	Aggregated Appraisals
Joy	Desirability > 0 Likelihood = 1
Hope	Desirability > 0 Likelihood < 1
Fear	Desirability < 0 Likelihood < 1
Dismay	Desirability < 0 Likelihood = 1
Anger	Desirability < 0 Object

An example of an appraisal might involve the agent getting dirty because another agent spilled coffee on it. The situation gets a negative desirability, and the likelihood is 1.0 since it actually happened. If the agent views it as an accident, the feeling will come out as dismay. However, if he blames the other agent, he may feel angry. In fact, if he merely suspects that the other agent did it on purpose, then he may feel fear.

A problem-focused coping strategy may involve getting cleaned up. An emotion-focused coping strategy reappraises the situation in a different light. For example, an emotion-focused coping in this situation may involve thoughts along the lines of, "It was going to get dirty eventually anyway."

EMA has some drawbacks for modeling emotions and feelings. Its model of emotion is implemented as knowledge about how to feel in light of various appraisals instead of having an *independent* emotion system that *interacts* with knowledge processing. Furthermore, EMA does not incorporate any notion of the non-cognitive effects the body may have on the agent's affective state. Even if the system were to include the physiological state of the body, this would only influence the agent's affective state through cognitive appraisal.

Damasio, on the other hand, describes the role of the body in determining affect. He also provides the distinction between emotions and feelings, as well as a high-level description of how bodily notions of emotions and feelings interact with cognition. Figure 1 gives a high-level overview. An emotionally competent stimulus affects the body. For our purposes, an emotionally competent stimulus is a stimulus that impacts the emotions of the agent. This stimulus is appraised, both cognitively and bodily. This appraisal causes a change in the state of the body (i.e. the physiology) – this is what Damasio calls the emotion. It is the perception of this emotion that is the feeling. This feeling impacts the decisions made by the agent, which then leads to actions that can affect the environment (and hence the perceived stimulus). Damasio also reports that humans can have non-cognitive emotional responses, even when they are unaware of the emotionally competent stimulus.

There is not complete agreement between Damasio's model and Gratch's. First, there is a vocabulary difference. Gratch uses emotions and feelings interchangeably, whereas Damasio provides a clear distinction. In particular, Damasio distinguishes between the physical state of the body (the emotion) and the perception of that emotion (the feeling). This distinction is not useful for Gratch since EMA does not include a model of physiology. Even so, what Gratch calls an "emotion", Damasio would consider feeling. EMA deals with cognitive interaction, and in order to do this the "emotion" must be represented in working memory – that is, the agent must be aware of it. This awareness is a perception of the emotion, and hence, by Damasio's terminology (which we adopt), it is a feeling. Furthermore, in EMA, feelings come first, and then the emotions. That is, once an agent determines that it feels bad, it generates the appropriate body language to display that – Damasio claims the order is actually reversed, based on several brain imaging and case studies he reviews. Additionally, EMA generates feelings via rule firings that are part of the cognitive system and could be open to the influence of other knowledge as well as cognitive learning, as opposed to a more fixed and architectural mechanism involving the body as suggested by Damasio.

Our framework combines these models to produce one that takes into account both physiological and cognitive effects on emotions and feelings. We use Damasio's theory regarding the overall system architecture and physiological

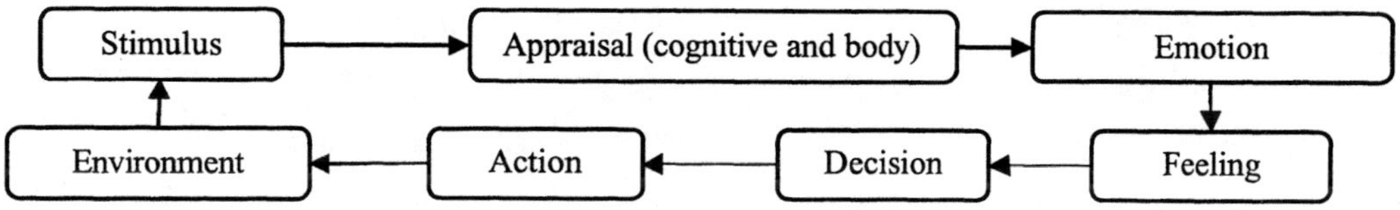

Figure 1: Basic order of events in Damasio's model.

details, but rely on Gratch's approach for modeling appraisal and coping. The challenge is to determine how the body and cognition interact to produce emotions and feelings. Furthermore, we need to reverse the direction of causality in EMA for feelings and physiological effects.

A Simple Cooperative Game

To demonstrate our framework, we have implemented a simple simulated water balloon toss game between two players. In this game, two players toss water balloons back and forth, attempting to score points by successfully catching the balloons (this is cooperative and not competitive); however, things can go awry so that a balloon is missed and one player gets wet or one player might have to dive into bushes (and experience pain) to catch the water balloon. These interactions with the environment and the agents' perception of the body state of each other can give rise to emotions and feelings that impact the game play and lead the agents to remark about each other's performance.

In the game, the players stand in a field of grass and bushes (this is not a full 3D simulation, but a simple abstract simulation of these two possible situations), with one of them the thrower and the other the catcher. The thrower tosses the balloon to the catcher, who then tries to catch it. Following the attempted catch, the thrower gets to remark on the situation. Once the thrower has remarked, the catcher gets to remark. The thrower then gets to consider the catcher's remark, but does not get to take action. Then the next round begins, with the thrower now being the catcher and vice versa.

When the thrower tosses the balloon, he can vary two parameters: the speed (slow or fast) and the aim (near the catcher or far away). Similarly, the catcher can vary two parameters: whether he runs (true or false) and whether he makes an attempt to catch the balloon (true or false). A player can only move when he is the catcher, and then only to the balloon's landing location. There are also two parameters for the remarks: subject (you or me) and type (supportive or critical). A player can also choose silence.

While the thrower may intend to throw a particular way, the result may come out differently (i.e. he may throw fast and far when he meant to throw slow and near). The catcher only sees the actual throw. If the balloon is caught, the players get a point. Otherwise, the catcher may get wet. They have an infinite supply of balloons, so the game can continue even if the balloon is not caught.

Soar-Emote

Our framework has a simplified version of the basic appraisal model used in EMA, but instead of a rule-based implementation of the emotion generation, we introduce architectural mechanisms that are fixed and are used for all tasks. The main differences on the cognitive side consist of an architectural mechanism called the Appraisal Summarizer that translates the appraisals into the cognitive contribution that is then sent to the body. This cognitive contribution is not an emotion, but instead represents the influence of cognition on the emotion. At the same time, the body's Internal Physiology creates an appraisal of the current body state. The Emotion System combines this information from Internal Physiology and the Appraisal Summarizer to create the actual emotion of the system. The agent can have non-cognitive responses to the emotion, such as automatic changes to body language. The perception of this emotion is the feeling the agent gets, which it can then cope with, as per EMA. Like Gratch, we use Soar for the basis for the cognitive system.

Figure 2 shows the overall architecture of a single agent with the cognitive system being on right, and the physical (i.e. non-cognitive) system on the left. The remainder of this section describes the components of the framework and their interactions within the context of the water balloon toss game as told from the point of view of the catcher. Each component is numbered according the descriptions below. We assume that just prior to the beginning of the example, the thrower tossed the balloon far and fast, and the catcher ran and tried to get it and missed. Then the thrower made a critical remark about the catcher. It is now the catcher's turn to remark. There are no specific critical remarks in the program – the agents can just detect whether a remark is critical or not; however, to bring the example to life, you might imagine that the thrower just said, "Nice try Crisco hands!"

(1) External stimuli from the environment arrive at the agent's body. Some stimuli are picked up by perception. In this example, the agent sees that the thrower looks angry, and also hears a critical remark about himself.

(2) External stimuli can also directly affect the agent's physiology, which in turn can be perceived. For example, changes in body temperature that result from the external temperature can be perceived. In this example, the agent is not feeling pain but he is tired since he just ran.

(3) Perception processes stimuli from both external and internal sources and sends them to other systems, namely Working Memory and possibly the Motor System (in the case of reflexes).

(4) In some cases, a stimulus may lead directly to the creation of a motor command. For this simulation, however, reflexive actions were disabled.

(5) Percepts from Perception enter Working Memory. These percepts include processed forms of external stimuli and internal stimuli. Internal stimuli include perceptions of the internal physiology and the current emotion (the perception of which is a feeling). In this example, the current feeling is one of dismay (this stemmed from the immediately prior situation in which there was a bad throw). The agent is also tired and hot but not in pain. The agent becomes aware of these and other details about the situation.

(6) Working Memory contains the objects the agent is currently "thinking" about. These include objects that originated external from the agent's mind (i.e. from the body or the environment) as well as objects it has created internally (i.e. objects it uses in the process of solving problems or otherwise behaving).

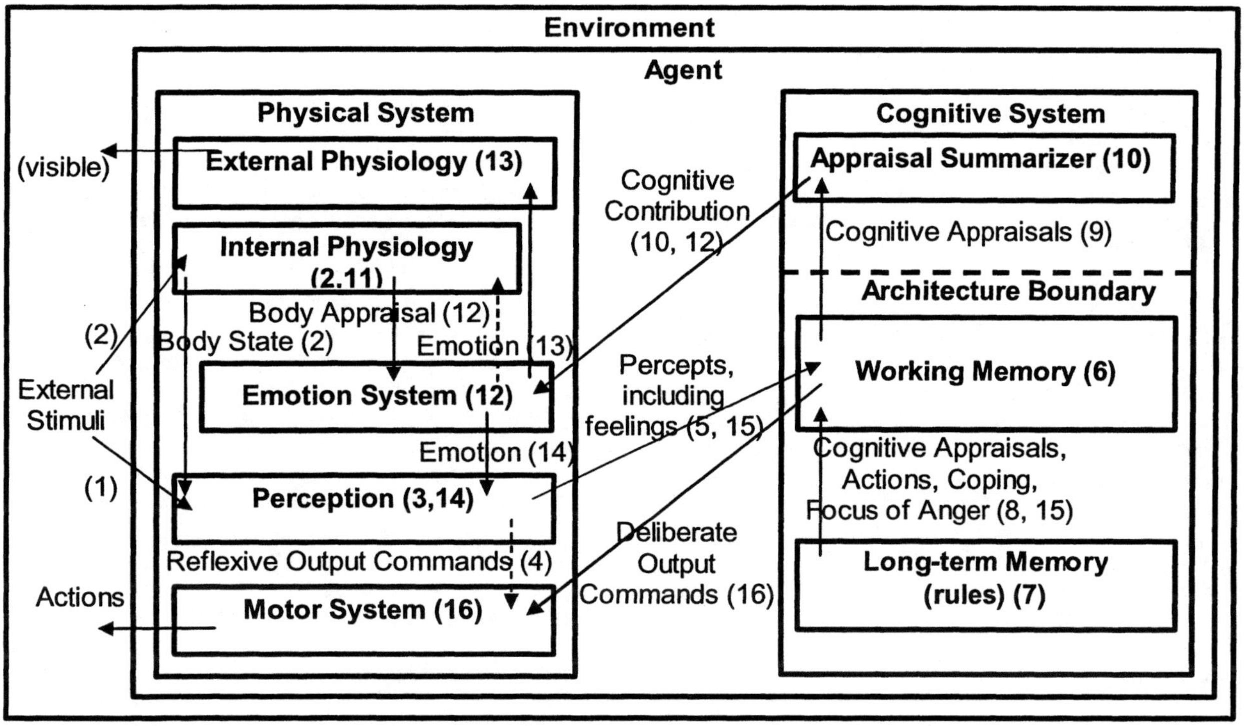

Figure 2: Overview of the Soar-Emote model.

(7) Long-term memory contains rules that encode the agent's knowledge. Thus, objects in working memory may trigger a rule. The kinds of rules present include rules about how to perform actions, such as throwing the balloon and how to evaluate a current situation. Rules to evaluate the situation may be triggered by numerous emotionally competent stimuli; in this case, the evaluations include the current score, the failed catch, the other agent's body language, and the remark.

(8) When rules match and fire, they make changes to Working Memory. In this example, rules fire that create negative appraisals for the agent's body temperature, exertion, failure to catch the balloon, the other agent's angry body language, and the negative remark. In fact, the body language contributes to two appraisals: it gets an appraisal in and of itself, and it also provides context for the remark. Thus, the agent realizes that this is not constructive criticism. Furthermore, the agent attributes some of these, such as the failure to catch the balloon and the negative remark, to the other player.

(9) Appraisals are automatically transferred into the Appraisal Summarizer, which in this case is an extension of the Soar architecture.

(10) The Appraisal Summarizer summarizes these cognitive appraisals and sends the result, which is the cognitive contribution to the next emotion. The intensity of a cognitive contribution is the product of the average desirability and average likelihood. This approach is inspired by EMA, although there are several differences. In EMA, this summary is the next "emotion," whereas in Soar-Emote it is just a contribution. Also, in Soar-Emote, all of the numeric appraisal variables of a particular type are averaged together, and the resulting values are used to generate the cognitive contribution. Instead of this averaging process to produce a single "emotion," EMA allows degrees of multiple emotions at once. The primary reason for this difference at this point is simplicity. The purpose of the framework at this stage is to show how the physical and cognitive systems interact to alter feelings and behavior; thus, the full complexity of EMA is unnecessary.

In this example, the overall desirability is negative and the intensity of appraisals with attributions outweighs those without attributions, so anger is generated for the cognitive contribution.

(11) While all of this activity in the Cognitive System is going on, Internal Physiology generates an appraisal of the current body state based on the agent's body temperature, exertion, pain, and noise. In this example, the agent's temperature is high and he is tired, but he is not in pain. Thus, the body appraisal is negative; however in this case we assume that the added noise is positive, lessening the negative body appraisal.

(12) The contribution from the Cognitive System and the body appraisal from Internal Physiology are combined by the Emotion System into an emotion that the agent will experience. The generation is biased towards the cognitive contribution. In this example, the agent's emotion is still one of anger, but combination with the body appraisal decreases its intensity.

(13) The emotion generated from the Emotion System is sent to External Physiology, which determines the agent's body language. Body language is an abstraction that encompasses externally perceivable aspects of the agent's emotion, such as facial expression and tone of voice. In the current framework, the agent does not have any direct control over his own body language. In our example, the agent is feeling angry. However, it is not very intense, so the body language becomes neutral.

(14) The emotion generated from the body's emotion system also feeds into perception where it becomes the agent's feeling (recall that feelings are the agent's perception of its emotions).

(15) The feeling, along with the latest percepts, are transferred into Working Memory. The agent can appraise the situation again and it can cope with its feelings. It can also explore why it is feeling a certain way; for example, to determine the primary source of anger. All of these are done via the application of knowledge from Long-Term Memory. In Soar-Emote, the agent may reappraise a situation repeatedly until his feeling stabilizes.

In this example, the agent first engages in emotion-focused coping by positively reinterpreting the fact that he failed to catch the balloon. Thus, the cycle of generating a new feeling must repeat. While this reduces the intensity of his anger, he is still angry. He identifies the thrower as the source of his anger, and then he decides to use problem-focused coping by making a critical remark about the thrower (in an attempt, perhaps, to let the thrower know he is angry with the situation).

(16) Output commands are sent to the Motor System, which generates actions in the world, which in this case is consists of a negative remark being made by the agent.

Table 2: Systems involved at each level.

Level	Systems
Biological	Internal and External Physiology, Emotion System
Cognitive	Appraisal Rules, Appraisal Summarizer, Emotion-focused coping
Social	Problem-focused coping (remarks), Perception of External Physiology of others

Influences at Each Level

It is easy to lose the forest for the trees in the previous description of Soar-Emote. Recall that one of the goals for our framework is to be comprehensive and cover the impact of emotion and feelings at three levels: biological, cognitive, and social. Table 2 summarizes how the framework influences these areas.

Results

Our aim is to answer the questions: Do each of the high-level aspects of the system (biological, cognitive, social) contribute to the agent's emotions and feelings enough to change its behavior? What are these differences? It is premature to compare the model's performance to human data; however, we can examine whether the framework achieves our goal of incorporating influences from each of the levels. The agent has a default action for each phase that does not require a feeling (i.e. non-emotional behaviors); for throwing this is the near/slow throw, for catching it is the run/attempt catch, and for remarking it is silence. Other action choices are motivated by emotions and feelings. To demonstrate that the components of Soar-Emote influences behavior across all levels (biological, cognitive, social), we created five types of agents with various "lesions", where groups of components were disabled. Table 2 shows that components correspond to each of the levels we modeled; the non versions of these agents had most of these components disabled. The exception was the Emotion System – for the Non-Biological agent, the agent still experienced feelings, but they were based entirely on the cognitive contribution. Additionally, we tested an Affective agent that had all systems enabled, and a Non-Affective agent that had all of these systems disabled. The Non-Affective agent demonstrates what a Soar agent without any of these augmentations would do. The test consisted of pairing each agent with an agent of the same type for 100 games of 20 rounds each.

In general, agents with different combinations of active components showed different variations in behaviors, stemming both directly and indirectly from differences in the emotions and feelings the agents experienced. We show the results for the remarking and catching behaviors because they had the most variation. Throwing behaviors showed much less variance primarily because the agents do not remember history between rounds and emotions and feelings in our system do not currently have any momentum (they exist only as long as the stimuli that cause them).

The Non-Biological agent does the run/attempt catching behavior significantly more often than the fully-affective agent, and it never choose the attempt only catching action (Figure 4). This suggests that the Non-Biological agent feels better than the fully affective agent most of the time, since not trying the run/attempt catch is indicative of negative feelings (for example, the agent may be angry and refuse to catch the balloon).

The Non-Cognitive agent chooses the silence remark significantly less than the fully affective agent (Figure 3). It also chooses the critical/me remark much more, but never chooses critical/you. This is because without the cognitive influence, the agent has no way of assigning blame, and hence never feels angry at others. However, he can feel dismay, and hence remark critically about himself.

The Non-Social agent always chooses silence, which is a significant departure from the fully affective agent (Figure 3). The Non-Social agent also always chooses the default throw (not shown). This is because, in the extra time between the catch and the next throw, nothing happens, that gives the emotions time to settle down.

In general, the throws (not shown) vary little. This is because the agent lacks history – the remarks happen at the end of the round, and so are immediately forgotten when the

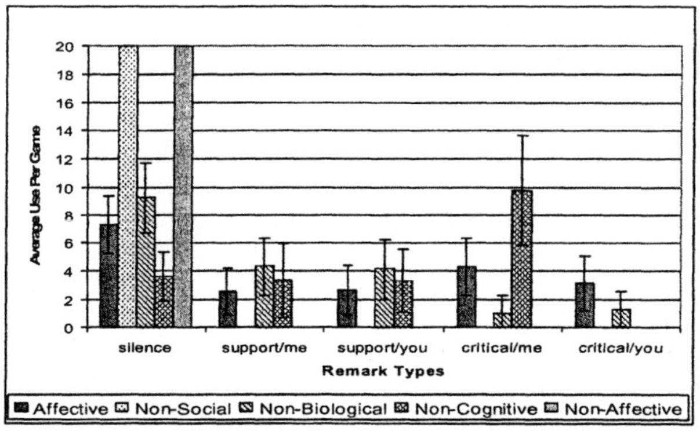

Figure 3: Remarking behaviors by various agent types.

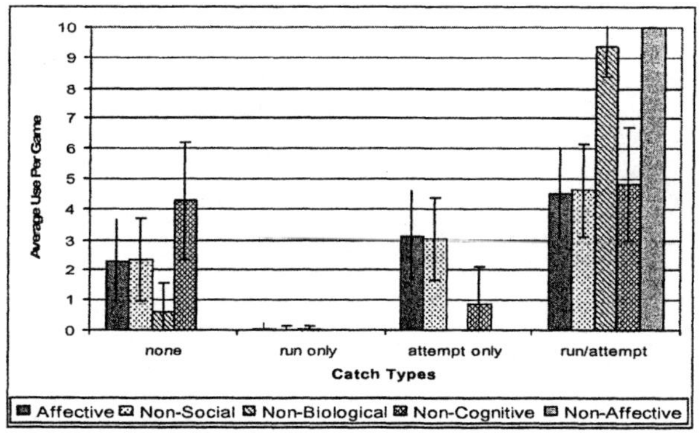

Figure 4: Catching behaviors by various agent types.

next round starts and hence have very limited impact. To rectify this, we added basic history capabilities to the model, and the variability in the throwing behaviors increased significantly, especially for the Non-Social agent.

Overall, we see that the biological parts of the system has a fairly negative impact on the emotions and feelings, whereas the cognitive parts has a more positive impact which helped mediate the biological impact. In general this is probably not true in humans, but given the task and the parts of the agent we are modeling, it is plausible. The social impact was primarily seen in the remarking. The reason the social impact was so limited everywhere else is partially because of the lack of history – when we added history, the non-social agent experienced the most change in the throwing round. The other issue is that the agent's social abilities are very limited, such as a lack of knowledge about what is socially acceptable behavior.

Conclusion

Our primary goal has been to strive towards a comprehensive cognitive framework for emotions and feelings by including biological, cognitive and social influences. We modified the Soar architecture to include ideas from Gratch, and then unified that system with biological systems inspired by Damasio. The full framework includes direct biological and cognitive influences on emotions and feelings. Social influences come in the form decisions to make various remarks, and the appraisals of those remarks and the body language of other agents, and our results show that the influences of these areas had an impact on the agents' behavior.

While we do address several issues, there are many phenomena that the framework does not address adequately or at all. Soar-Emote is a starting point we plan to build on and expand. Future work with Soar-Emote's includes work at all of the levels. For example, biologically, Soar-Emote lacks emotional momentum and non-cognitive modification of the perception of emotionally competent stimuli (i.e. as in the experience of sympathy). Cognitively, Soar-Emote lacks the ability to cognitively moderate automatic emotional responses (Fellous, 2004). We also need to explore the impact of emotions and feelings on the cognitive architecture's processing, such as influencing long-term memory retrieval (Hudlicka, 1997). Socially, our results indicate that history will play an important role. Other changes that impact multiple levels include a more complete theory of appraisal (Smith, 2004), and an explanation of individual differences that goes beyond differences in long-term knowledge (Hudlicka, 1997).

We expect that changes to the framework to incorporate these and other phenomena will involve the creation of new and more complex subsystems, modification of the Soar architecture, and new kinds of rules to take proper advantage of these changes.

References

Biddle, E. S., Henninger, A., Franceschini, R. & Jones, R. M. (2003). Emotion Modeling to Enhance Behavior Representation: A Survey of Approaches. *Interservice/Industry Training, Simulation, and Education Conference*.

Damasio, A. (1994). Descartes' Error: Emotion, Reason and the Human Brain. New York, NY: Avon Books.

Damasio, A. (2003). *Looking for Spinoza: Joy, Sorrow, and the Feeling Brain*. Orlando, FL: Harcourt, Inc.

Ellsworth, P. & Scherer, K. (2003). Appraisal Processes in Emotion. In Davidson, R., Scherer, K. & Goldsmith, H (Eds.), *Handbook of Affective Sciences*. Oxford: Oxford University Press.

Fellous, J. (2004). From Human Emotions to Robot Emotions. *AAAI 2004 Spring Symposium on Architectures for Modeling Emotion: Cross-Disciplinary Foundations*.

Gratch, J. & Marsella, S. (to appear). A domain-independent framework for modeling emotion. *The Journal of Cognitive Systems Research*.

Hudlicka, E. (1997). *Modeling Behavior Moderators In Military Human Performance Models*. Technical Report 9716. Arlington, MA: Psychometrix Associates, Inc.

Newell, A. (1990). *Unified Theories of Cognition*. Cambridge, MA: Harvard University Press.

Smith, C. A. A (2004). Functional Perspective on Emotion Elicitation. *AAAI 2004 Spring Symposium on Architectures for Modeling Emotion: Cross-Disciplinary Foundations*.

Learning to Make Decisions in Dynamic Environments: ACT-R Plays the Beer Game

Michael K. Martin (mkmartin@andrew.cmu.edu)
Dynamic Decision Making Laboratory
Department of Social and Decision Sciences, 5000 Forbes Avenue
Pittsburgh, PA 15213 USA

Cleotilde Gonzalez (conzalez@andrew.cmu.edu)
Dynamic Decision Making Laboratory
Department of Social and Decision Sciences, 5000 Forbes Avenue
Pittsburgh, PA 15213 USA

Christian Lebiere (clebiere@maad.com)
Micro Analysis and Design
Boulder, CO, USA

Abstract

Sterman (1989) proposed that decision makers misperceive the feedback provided by dynamically complex environments, and questioned whether people can learn to make effective decisions in such environments. We provide empirical evidence of learning in a well-known dynamic environment called the beer game. We then describe a preliminary version of an instance-based, dynamic decision making model built using the ACT-R cognitive architecture. The model mimics the general patterns of human behavior observed for aggregate performance across trials and local performance within trials. Implications for research on dynamic decision making are summarized.

Introduction

Dynamic Decision Making (DDM) requires a series of interdependent decisions in an environment whose state evolves over time (see Brehmer, 1992, for a review of DDM). Dynamic decisions often involve choosing control inputs for a dynamic system in a manner that achieves or maintains a desired system state (e.g., a state of equilibrium).

The beer game is a dynamic system used extensively to study the way decision makers perform when confronted by dynamic complexity. Thousands of people from all over the world, ranging from high school students to chief executive officers and government officials, have played the beer game to learn the basic concepts of operations management (Sterman, 2004).

The beer game is not really about beer, and it is not really a game. It is a learning environment of the type called management flight simulators (Sterman, 2004). It provides players an interactive experience that demonstrates the impact of time delays and feedback loops on supply-chain management, and more generally, on coordination among levels in an organization.

In particular this game has been used to demonstrate the *bullwhip* effect, a costly real world phenomenon in which orders oscillate, in increasing amplitude, as one moves farther up the supply chain (Croson and Donohue, 2002). Sterman (1989, 2004) has demonstrated the bullwhip effect in multiple beer game experiments, and has concluded that individuals do not learn to control the system because they misperceive the feedback provided by dynamic systems. Similar results and misperception-of-feedback explanations can be found in other studies (see Croson & Donohue, 2002, for a review of beer game experiments).

We contend that participants in previous experiments performed poorly simply because they did not have enough practice with the system, giving them little opportunity to learn. Proficient DDM typically requires *extended practice* with a system, presumably because it gives decision makers a chance to learn the system dynamics important for control (Kerstholt and Raaijmakers, 1997).

This paper contributes to the current state of affairs in two ways. First, it provides evidence that people learn to adequately control the supply chain when given extended practice. Second, it offers an explanation as to *how* people learn to control the system by providing an ACT-R cognitive model of the learning process.

In the next section we describe the beer game and bullwhip effect in more detail. We then present our study on the effect of extended practice. Next we present the ACT-R cognitive model and comparisons between the model and human. Finally we conclude and present future directions for research.

The Beer Game

The beer game represents a simplified supply chain consisting of a single retailer who supplies beer to consumers (simulated as an external demand function), a single wholesaler who supplies beer to the retailer, a distributor who supplies the wholesaler, and a factory that brews the beer (it obtains it from an inexhaustible external supply) and supplies the distributor.

Individuals play the game in groups of four, with each participant playing the role of one of the four facilities. Their goal is to minimize the cost for the entire supply chain. Each player contributes to this goal by ordering beer from their respective supplier in a manner that maintains enough beer in their respective inventory to meet the demand from their respective customer (i.e., the facility they supply, or the consumer in the case of the retailer).

Costs accrue as follows. Each week, each player is charged a 50¢ holding fee for each case of beer in their inventory. If inventory is too small to meet demand, the shortage is backlogged to be filled as soon as possible. Players are charged a weekly $1 shortage fee for each case of backordered beer. The basic strategy, therefore, is to minimize inventory while avoiding backorders.

The dynamics of the beer game make successful performance difficult. Each week, each player receives an order from their customer, starting with the retailer and working upstream in the supply chain toward the factory. The customer's order is filled with available inventory, and then the player orders more beer from their supplier to replenish the loss from their inventory.

Difficulties arise because players must anticipate demand, as there is a one week delay between when an order is placed and when the supplier receives the order. Assuming that the supplier has enough inventory, there is an additional two week transportation delay before the player receives the ordered beer. If the supplier's inventory is too small to fill the order, additional delays will occur.

The Bullwhip Effect and Experimental Economics

Researchers have identified several causes for the bullwhip effect (Croson & Donohue, 2002). Rational decision makers must use current demand to forecast future demand in an effort to control the impact of order delays, transport delays, production delays, etc. on inventory. Forecasts based on simple ordering formulae (e.g., moving averages) lead to the bullwhip effect. Ordering in batches (e.g., monthly instead of daily) can also create the bullwhip effect. Other causes include fluctuating prices which lead to forward buying, and rationing where suppliers divide limited inventory among customers who then inflate their orders to get a bigger share.

The beer game is much simpler than real world supply chains. Players have no incentive for forward buying because prices are fixed. Order batching is less likely because the frequency with which orders are placed is fixed at one per week. Rationing is not possible because each facility in the supply chain has only one customer. Finally, in the standard scenario, external consumer demand starts at a constant of 4 cases of beer per week and then jumps to a constant of 8 cases per week at the fifth week and remains there for the remainder of what is typically a 52 week scenario.

Sterman (1989) demonstrated that the bullwhip effect emerges even though the beer game presents participants with a nearly ideal supply chain; participants' orders oscillated, and grew in amplitude as orders propagated upstream. This produced oscillations in each participant's net inventory (i.e., inventory – backorders), which also grew in amplitude the farther the facility was from the external consumer. The end result was a supply chain whose operations costs exceeded "optimal" costs by almost 10-fold.

Based on this finding, along with similar findings from experiments with simulations of other supply chains, Sterman (1989) concluded that people misperceive the feedback provided by dynamic systems. According to the misperception of feedback hypothesis, people lack the cognitive machinery to comprehend the dynamic complexity produced by the causal and temporal relationships among system variables. Dynamic complexity is created by delays in a system's response (e.g., transport and order delays), feedback loops, stocks and flows, and nonlinear relationships among system variables. All are commonly found in dynamic systems, and all are present in the beer game.

Extended Practice Experiment

In its strongest form, the misperception of feedback hypothesis implies that people simply cannot learn to control dynamically complex systems. Indeed, researchers often demonstrate that individuals cannot understand the 'basic building blocks' of systems thinking such as the concept of stocks and flows (e.g., Jensen & Brehmer, 2003; Sweeney & Sterman, 2000). This position however, cannot explain how experts in the real world can perform effectively in highly complex dynamic systems such as air traffic control.

A possibility we address here is that although people may not understand the building blocks of dynamic systems, extended practice may help individuals learn to control a dynamic system because it gives them the opportunity to learn the relationships between control inputs and system outputs, and how to anticipate common situations (Kerstholt and Raaijmakers, 1997).

Our experiment required playing the beer game for 20 trials, where each trial used the standard 52-week scenario (described above). The experiment, therefore, required a total of 1,040 ordering decisions in contrast to the typical single-trial experiment that requires a one-time run of 52 weeks and thus 52 ordering decisions.

This experiment simplified game play in two ways. First, participants played alone rather than in teams. Participants played the role of the distributor and the computer played the remaining roles. Second, the computerized players simply ordered the demand. Thus, variability was not added to the external customer demand as it propagated upstream through the supply chain.

Method

Participants. Thirteen Carnegie Mellon University students participated for payment. Participants were paid a base rate of $10, plus performance bonuses of up to $16 (see below).

Procedure. We developed a computerized version of the beer game that presents information in the same way as the in the Systems Dynamics Group www site (http://beergame.mit.edu/). A screenshot of this simulation is presented in Figure 1.

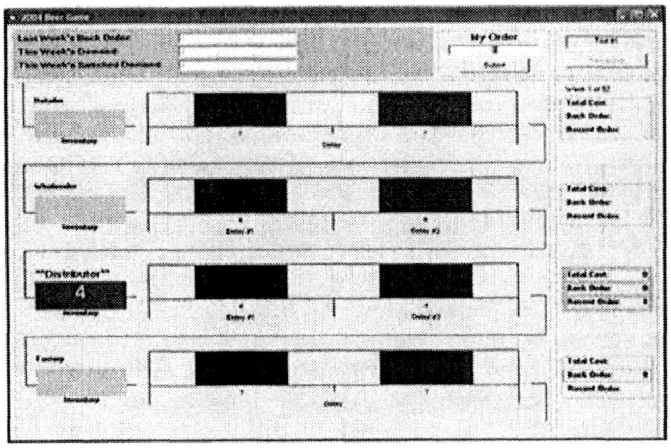

Figure 1: Screenshot of the Beer Game Simulation

The simulation provided information only about the inventory and supply line of the role played by the participant (distributor). Also, only the participant's cumulative cost was displayed. As in the www simulation, the last week's back order, and this week's demand and satisfied demands were displayed.

Participants played the 52-week scenario 20 times. They were instructed to minimize their total cost by ordering beer each week in a manner that allowed them to meet their customer's demand (i.e., the wholesaler's weekly orders). They were told about the cumulative weekly charges, the one week ordering delay, the two week transportation delay, and the possibility that if their supplier (i.e., the factory) could not fill their order, the transportation delay would be longer because of the time it takes the factory to transport raw materials..

The bonus pay schedule was then described. Trials were divided into four blocks of five. A $4 bonus was given for each block of trials in which the designated performance target was achieved at least once. Performance targets (total costs), based on 11 pilot study participants, grew more stringent over the time course of the experiment. The performance targets for blocks 1-4 were total costs of 750, 650, 550, and 450, respectively. (The minimum total cost possible was 396; there were no practical limitations on maximum total cost possible.)

To familiarize participants with the system they played a short 10-week scenario with random external demand. Questions were addressed during this time. Afterward, they played the standard scenario 20 times.

Results

One participant did not complete the 20 trials, so their data set was not considered subsequently. The data set of a second participant was removed after an outlier analysis.

Figure 2 shows the mean cost per trial. A one-way repeated-measures ANOVA using total cost as a dependent variable indicates that performance improved with practice, $F(19,190) = 3.4$, $p < .05$. Helmert contrasts (e.g., Judd & McClelland, 1989) indicate that performance gradually improved until about the ninth trial.

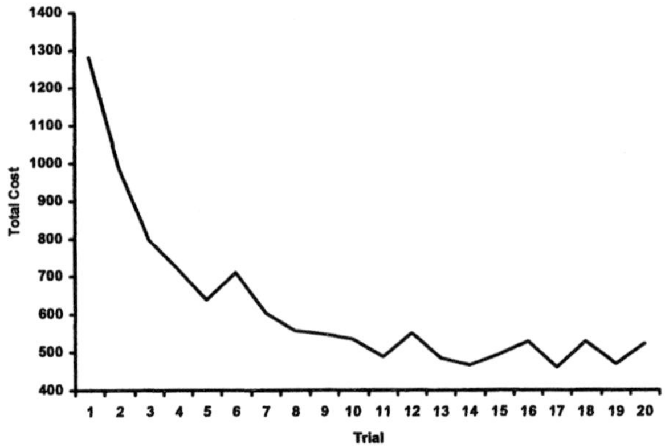

Figure 2: Cumulative Cost as a Function of Practice

Figures 3, 4 and 5 depict performance within trials 1, 9, and 20 respectively. Each shows net inventory (inventory – backorders) across the time course of the 52-week scenario. A net inventory of 0 is ideal.

As Figure 3 shows, our participants exhibited the same behavior as that reported in previous studies. The net inventory oscillates around the ideal of 0. The large deviations from 0, in turn, produce high total costs.

The 3-week delay between placing and receiving orders inevitably leads to back-orders when external consumer demand jumps from 4 to 8 cases per week. (The distributor sees the jump at week 7.) This sudden increase in demand creates a shortage which must be corrected by ordering more beer than indicated by current demand. Too much beer is ordered, creating a slight overshoot in ideal inventory as indicated by the second cycle of positive net inventory. To correct for the overshoot, orders are cut back below current demand, creating yet another cycle of inventory shortages.

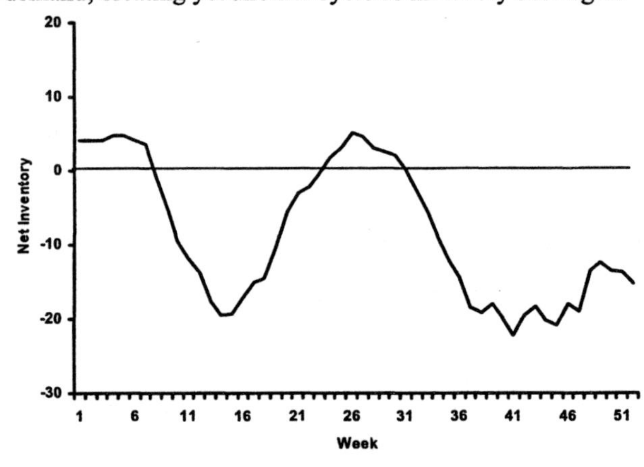

Figure 3: Net Inventory per Week in Trial 1

As with the control of any system with response delays, the only way to avoid oscillations in net inventory is to anticipate demand. Figure 4 shows that by Trial 9 the oscillations in net inventory are still present but participants have learned to dampen them. As can be seen, they anticipate the step increase in external consumer demand and build inventory prior to the increase in demand. The build-up, however, is not yet sufficient, which leads to back-orders and negative net inventory. They continue to overcorrect for back-orders, as indicated by the second cycle of positive net inventory.

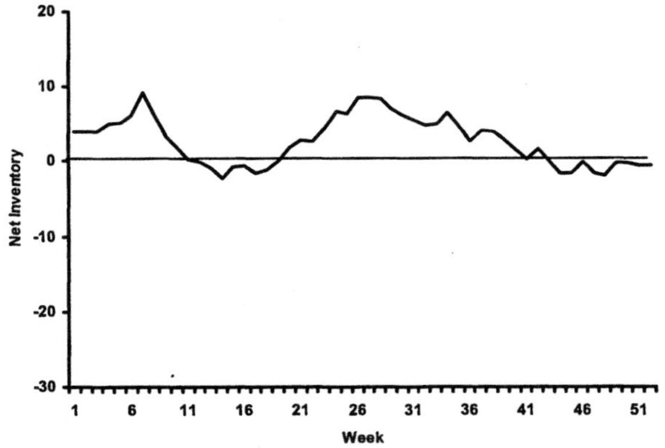

Figure 4: Net Inventory per Week in Trial 9

Figure 5 shows that participants have learned to mostly avoid oscillations in net inventory by Trial 20. The dampening of oscillations between Trials 9 and 20 seems to appear because participants have learned how to correct for back-orders without overshooting the desired net inventory of 0.

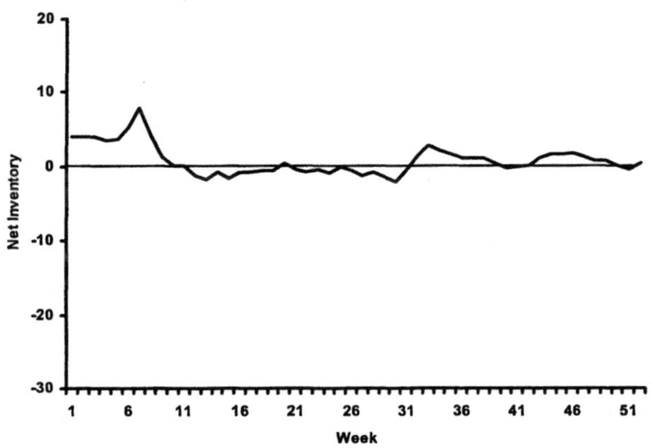

Figure 5: Net Inventory per Week in Trial 20

ACT-R Plays the Beer Game

Our participants learned to play the beer game. But what did they learn, and how did they do it? Gonzalez, Lerch, and Lebiere (2003) proposed Instance-Based Learning Theory (IBLT) to account for DDM performance and concurrent learning processes. IBLT has been successfully applied to multiple dynamic tasks including the Sugar Production Factory and the Tansportation task among others (see Gonzalez and Lebiere, in press).

The gist of IBLT is that dynamic decisions are made by comparing current situations with previously experienced situations. If a similar situation is recalled, the decision associated with that situation is used as an anchor that is adjusted to fit the current situation. Learning occurs as decision makers gradually shift from using simple decision making heuristics to the instance-based anchoring and adjustment process.

IBLT, as implemented in ACT-R, provides a simple explanation of the observed dissociation between verbalizable knowledge and DDM performance (e.g., Berry & Broadbent, 1984). According to IBLT each judgment of an alternative creates an instance, which is represented as a chunk in declarative memory in ACT-R. The slots in the chunks represent the situation, the decision made, and the expected utility of that decision. As declarative knowledge, each instance can be verbalized. However, the subsymbolic parameters that control the retrieval and application of instances (e.g., base-level activation, similarity among chunks, and strengths of association) are not consciously accessible. These subsymbolic parameters represent implicit knowledge of the system, and underlie DDM performance. The implication is that DDM tasks can be learned without explicitly encoding structural and temporal relationships among system variables.

In accordance with IBLT, we enforced the following constraints for modeling beer game performance in ACT-R. First, we represented information only if it was directly available to participants. Second, we represented information only if participants paid attention to it – as indicated by think-aloud protocols from two additional beer game participants. Third, we avoided clever engineering by using only those cognitive mechanisms inherent in ACT-R. This includes using recommended default values for all parameters.

We have also imposed two additional constraints on our modeling efforts to date. The declarative chunks described by Gonzalez et al. (2003) contained slots that represented expected utility. In that model, feedback mechanisms were used to adjust expected utilities. Subsequent application of those instances then depended on their expected utility. We do not include slots for expected utility in the beer game model because of the complications arising from delayed feedback, and the difficulties associated with determining utility. The second additional constraint is that the model reported here uses partial matching only. Base-level learning and blending mechanisms, as used in Gonzalez et al. (2003), have not been used so far.

Because the model operates in a task where contextual attributes vary continuously (e.g., the number of cases of beer in inventory, back-order, etc.), exact matches between context and relevant instances are rare. Partial matching provides a mechanism for retrieving chunks with attribute values that are similar to the current context. Thus, relevant chunks can be retrieved even though they do not exactly match the retrieval cues provided by the current context (i.e., the values of the slots in the goal buffer).

Specifically, the chunk with the highest match score will be retrieved if its activation is higher than the retrieval threshold (-1.0 in our case), where match score M_{ip} is a function of the activation of chunk i in production p (including transient activation noise, .25 in our case) and its degree of mismatch to the desired values:

$$M_{ip} = A_i - MP\sum_{v,d}(1 - Sim(v,d))$$

In the partial matching equation above, *MP* is a mismatch penalty constant (1.5 in our case), while *Sim(v,d)* represents the similarity between the desired value *v* in the goal and the actual value *d* in the retrieved chunk. We used a negatively accelerated similarity function.

The Model

Based on performance, it appears that participants learned: (1) to anticipate the increase in demand and (2) to adjust the size of their orders so that the amplitude of oscillations in net inventory progressively decrease. For our model, we started with the simple heuristic of ordering the demand to replace inventory losses. Verbal protocols indicated that participants frequently examined back-orders and/or inventory immediately after placing an order – even though the change due to that order would not occur until at least 3 weeks later. This observation prompted the addition of slots that represented the changes in back-order and inventory. We then added several more simple heuristics that increase or decrease the base order (i.e., order the demand) according to changes in back-order and/or inventory. These heuristics form the core of the model, and are engaged in the creation of all instances.

At the beginning of each ordering cycle, the model assesses changes in inventory and back-orders, and then attempts to retrieve a relevant instance from declarative memory. The retrieval cue is constructed by projecting the current state of the system onto the next state. That is, current inventory is multiplied by the inventory change that occurred upon entering the current state to produce an expected inventory. An expected back-order is constructed similarly. Expected inventory and expected back-order are then used as retrieval cues.

If the retrieval fails, the heuristics described above are applied to the current demand. If the retrieval is successful, three pieces of information from the projected state are used to construct the current order. First, the demand slot from the projected state indicates the expected demand. The expected demand becomes the current base order. (Notice that this is similar to the first heuristic we created, if it is recognized that expected demand equals current demand in unfamiliar situations.) Retrieval of expected demand thus provides a mechanism by which the model can learn to anticipate the increase in demand.

The next two pieces of information correspond to the changes in inventory and back-orders that produced the projected state. These may be thought of as the size of the adjustments that lead into the projected state, and thus the size of the adjustment that should be made to the current base order.

Results

The results reported herein use the mean of 11 simulated subjects based on the model described above, each playing the beer game 20 times in the standard scenario as human participants did.

The model's mean learning curve approximates the humans' mean learning curve in terms of Total Cost, $r^2 = .875$ (see Figure 6). The model does not perform quite as well as humans but it appears to learn more quickly than humans do. The addition of blending might be expected to help with both of these defects.

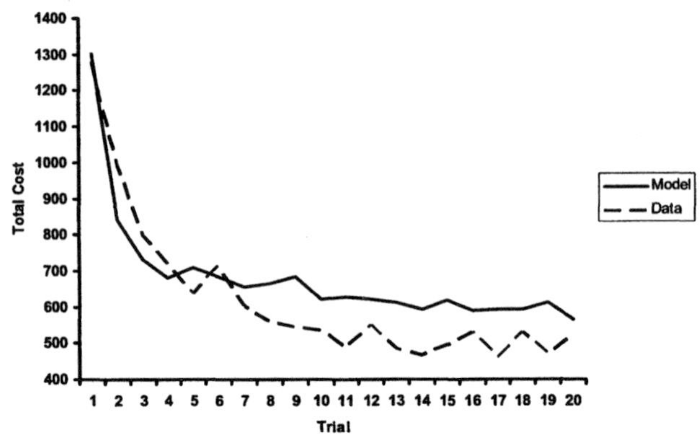

Figure 6: Practice Effect for Model and Humans

Building an ACT-R model that exhibits a learning curve for an aggregate performance measure (i.e., total cost) is fairly straightforward. It is more important for our current efforts that the model learns to control inventory in a manner consistent with that demonstrated by our participants. We can assess this by examining how the patterns of net inventory over weeks in the scenario match those produced by humans.

Figures 7, 8 and 9 depict the model's mean performance in terms of net inventory for trials 1, 9, and 20 respectively. The pattern of the model's performance in trial 1 (see Figure 7) closely mimics that produced by humans. It exhibits the large oscillations in net inventory, along with the overcorrections demonstrated by humans. One difference in the pattern is that the model's cycles of net inventory oscillations have greater amplitude than those of humans. The model also appears to be already learning to dampen the oscillations in net inventory, whereas humans demonstrated a second cycle that was roughly of the same amplitude as their first.

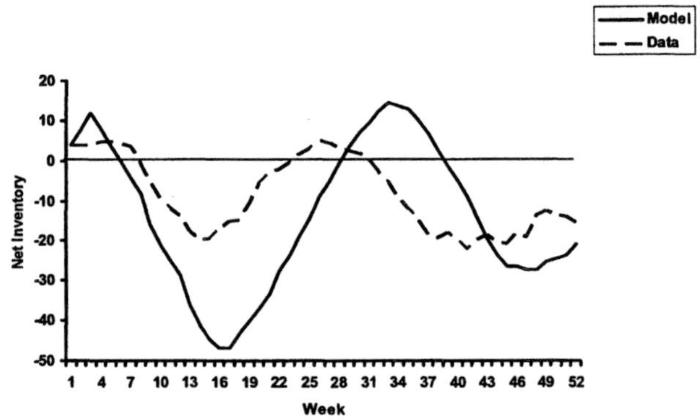

Figure 7: Model's Net Inventory per Week for Trial 1.

By trial 9 the model, like the humans, has learned to partially anticipate the increase in demand, and has learned how to decrease the amplitude of the oscillations in net inventory (see Figure 8). Overall, the pattern of the model's performance is similar to that of humans. One difference is that humans tended to be biased toward a positive inventory, whereas the model appears to be biased toward a negative inventory. This is probably due to the fact that the model, at this point, does not take into account the difference in costs associated with inventory versus back-orders.

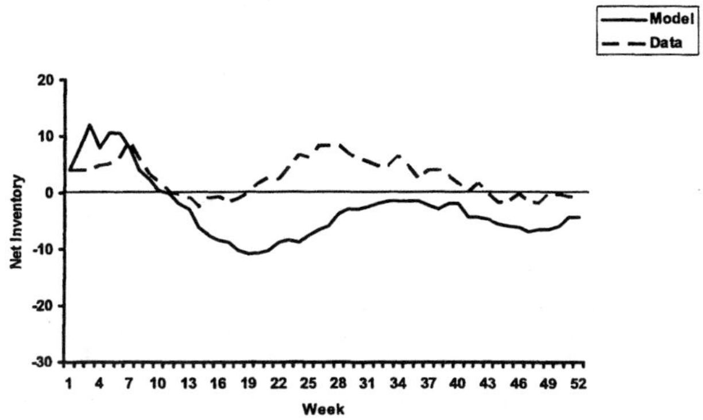

Figure 8: Model's Net Inventory per Week for Trial 9.

By Trial 20 the model's performance indicates further dampening of net inventory oscillations (see Figure 9).

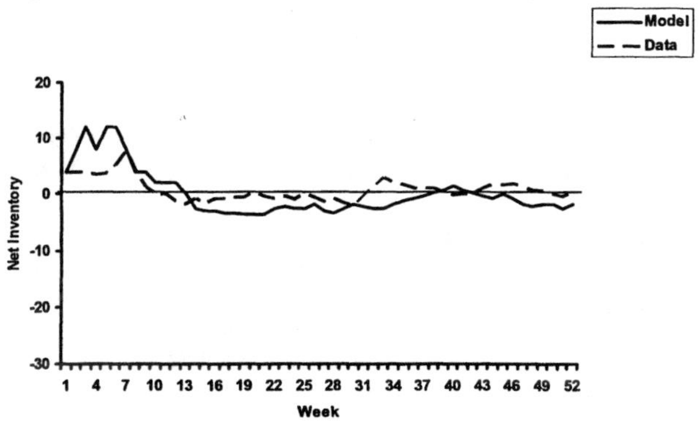

Figure 9: Model's Net Inventory per Week for Trial 20

Conclusions

Learning in dynamic environments is particularly challenging due to the complexity of dynamic problems and cognitive limitations, but our behavioral data showed considerable performance improvements with extended practice in a dynamic task. Our simplifications to the beer game removed the uncertainty in demand created by other players, raising a question of whether it is dynamic complexity or uncertainty that hinder learning.

The cognitive model and the closeness to human data have demonstrated that IBLT implemented on top of a cognitive architecture provides a constrained and reasonably accurate model of the learning process dynamic tasks. The results from the cognitive model support the prediction from IBLT that decision making in dynamic environments is a learning rather than an optimizing process. Humans learn to make better decisions by noticing the changes in an environment, storing examples of each situation experienced, and predicting future situations based on past experience.

Although encouraging, the results presented in this paper are however, far from conclusive. An interesting avenue for future research concerns the robustness of instance-based learning. If people primarily learn the input-output relationships in a dynamic environment rather than more abstract characteristics of dynamic systems, questions arise as to whether and how this type of learning transfers to varying environmental conditions. Our current experimental research is examining this, and is providing preliminary evidence of transfer of knowledge.

Acknowledgments

This research was supported by training grant 5-T32-MH19983 from the National Institute of Mental Health, and the Advanced Decision Architectures Collaborative Technology Alliance sponsored by the U.S. Army Research Laboratory (DAAD19-01-2-0009).

References

Berry, D.C. & Broadbent, D.E. (1984). On the relationship between task performance and associated verbalized knowledge. *Quarterly Journal of Experimental Psychology, 36,* 209-231.

Brehmer, B. (1992). Dynamic decision making: Human control of complex systems. *Acta Psychologica, 81,* 211-241.

Croson, R. & Donohue, K. (2002). Experimental economics and supply chain management. *Interfaces, 32,* 74-82.

Gonzalez, C. & Lebiere, C. (in press). Instance-based cognitive models of decision making. To appear in Zizzo, D. and Courakis, A. (Eds.). *Transfer of knowledge in economic decision making.* McMillan.

Gonzalez, C., Lerch, J.F., & Lebiere, C. (2003). Instance-based learning in dynamic decision making. *Cognitive Science, 27,* 591-635.

Jensen, E., & Brehmer, B. (2003). Understanding and control of a simple dynamic system. *System Dynamics Review, 19,* 119-137.

Judd, C.M. & McClelland, G.H. (1989). *Data analysis: A model comparison approach.* Orlando, FL: Harcourt Brace Jovanovich.

Kerstholt, J.H. & Raaijmakers J.G.W. (1997). Decision making in dynamic task environments. In R. Ranyard, W.R. Crozier, & O. Svenson (Eds.), *Decision making: Cognitive models and explanations.* Ablex: Norwood, NJ.

Sterman, J. (1989). Misperceptions of feedback in dynamic decision making. *Organizational Behavior and Human Decision Processes, 43(3),* 301-335.

Sterman, J.D. (2004). Teaching takes off: Flight simulators for management education. Retrieved April 7, 2004, Massachusetts Institute of Technology, Sloan School of Management website: http://web.mit.edu/jsterman/www/SDG/beergame.html.

Sweeney, L.B., & Sterman, J.D. (2000). Bathtub dynamics: Initial results of a systems thinking inventory. *System Dynamics Review, 16,* 249-286.

Anticipatory Eye Movements in Interleaving Templates of Human Behavior

Michael Matessa (mmatessa@arc.nasa.gov)
NASA Ames Research Center, Mail Stop 262-4
Moffett Field, CA 94035 USA

Abstract

Performance modeling has been made easier by architectures which package psychological theory for reuse at useful levels of abstraction. CPM-GOMS uses templates of behavior to package at a task level (e.g., mouse move-click, typing) predictions of lower-level cognitive, perceptual, and motor resource use. CPM-GOMS also has a theory for interleaving resource use between templates. One example of interleaving is anticipatory eye movements. This paper describes the use of ACT-Stitch, a framework for translating CPM-GOMS templates and interleaving theory into ACT-R, to model anticipatory eye movements in skilled behavior. The anticipatory eye movements explain performance in a well-practiced perceptual/motor task, and the interleaving theory is supported with results from an eye-tracking experiment.

Introduction

Predicting skilled human performance by means of computer modeling is a valuable but difficult process. One easy way for modelers to describe performance would be a series of task-level (e.g., mouse move-click, typing) templates of behavior, laid end-to-end. But skilled performers do not complete all subcomponents of a task before going on to the next task. Instead, some subcomponents of the next task are interleaved into the earlier task. One example of this interleaving is anticipatory eye movements. It has been found that the eyes can move in anticipation of upcoming tasks in domains such as driving (Land & Lee, 1994), tea making (Land & Hayhoe, 2001), and hand-washing (Pelz & Canosa, 2001). In the hand-washing task example, while people perform the subtask of first getting their hands wet they interleave a look to the soap dispenser before performing the motor actions in the subtask of soaping their hands. So an easy-to-use but detailed modeling framework needs both task-level templates of human behavior and a theory of interleaving the lower-level perceptual, cognitive, and motor operators which make up the templates (Matessa et al., 2002). CPM-GOMS (John, 1988; 1990) is an example of such a framework, but it is only recently that the templates and interleaving theory of CPM-GOMS have been automated (John et al., 2002). Ongoing research is developing more templates and investigating the interleaving theory in computational systems. ACT-Stitch (Matessa, 2004) is a framework for automating the templates and interleaving theory of CPM-GOMS in the cognitive architecture ACT-R (Anderson & Lebiere, 1998). This paper will show how the interleaving theory of ACT-Stitch produces interleaved anticipatory eye movements which explain performance in a well-practiced perceptual/motor task. Then empirical support for the interleaving theory is given by results from an eye-tracking experiment.

CPM-GOMS

CPM-GOMS (John, 1988; 1990) uses templates of behavior to package at a task level (e.g., mouse move-click, typing) predictions of lower-level cognitive, perceptual, and motor resource use. Even behavior as simple as a mouse move and click requires coordination of the use of cognitive, perceptual, and motor resources. In order to carefully click on a target, it is necessary to find the target location, move the eyes to that location and perceive the target, verify the target location, move the cursor to the target location, and click the mouse button. CPM-GOMS templates are interleaved to reflect the ability of skilled people to perform parts of one task in parallel with another.

ACT-R

ACT-R (Anderson & Lebiere, 1998) is a computational theory of human cognition incorporating both declarative knowledge (e.g., addition facts) and procedural knowledge (e.g., the process of solving a multi-column addition problem) into a production system where procedural rules act on declarative chunks. Chunks are made up of slots containing information, and production rules which match the information in chunk slots are able to execute. The goal chunk represents the current intentions. Production rules have the ability to perceive objects and make motor movements through perceptual and motor buffers.

ACT-R does not have a built-in theory of multi-tasking which would interleave tasks, although some work has been done in modeling multi-tasking in the ACT-R architecture (Byrne & Anderson, 2001; Lee & Taatgen, 2002; Salvucci, 2002).

ACT-Stitch

ACT-Stitch (Matessa, 2004) uses a process of macro-compilation to translate CPM-GOMS templates of human behavior into ACT-R productions. More specifically, cognitive operators are translated into productions with ACT-R perceptual-motor commands that represent CPM-GOMS perceptual-motor operators. Productions also contain a control structure that allows ACT-R to implement CPM-GOMS interleaving and have productions from one template execute during the execution of productions from another template. This differs from the ACT-Simple system (Salvucci & Lee, 2003) that compiled a sequence of KLM-GOMS tasks into a series of productions which were controlled by an incrementing state counter.

Productions created from macro-compilation must ensure proper sequencing of motor actions, ensure the ability to allow the correct productions in future templates to interleave during the execution of productions in the current template, and ensure the ability to block the incorrect productions in future templates from interleaving with productions in the current template.

These three requirements are accomplished in productions by using information in the current goal as well as perceptual-motor buffers. Slots in the goal are created for the vision and hand resources for both the intended action and target making use of the resource. This makes four slots in the goal: vision action, vision target, hand action, and hand target. To ensure proper sequencing, the action slots in productions of the current template are filled with an intended action appended with the unique number of the current template. Also, the target slots are filled with an intended target. The intended action cannot be used alone since without the template number no sequence information would be stored. The template number cannot be used alone since there may be multiple actions in the same template using the same resource (e.g., mouse move and click). The intended target cannot be used alone since sequence information would be lost if a target appears twice in a sequence (e.g., clicking the same number twice). The intended target cannot be ignored since the same action could be used in a template for two targets (e.g., verify target and verify cursor).

To ensure the ability to interleave productions, separate action slots are used for each resource (vision and hand). This allows, for example, a procedure to initiate a vision action from a future template before a procedure initiates a hand action from the current template. To ensure the ability to block productions from future templates, the action slots are filled with intended actions appended with the current template number. This prevents, for example, moving to the next target while the hand resource is free between moving to the current target and clicking on the current target. The template number cannot be contained in a separate goal slot because that would not allow productions from the next template to execute before the productions of the current template have finished.

Perceptual-motor buffers are also used in sequencing. Productions that interact with the perceptual-motor buffers can fill or empty the buffers and can check the status of the buffers before using them.

These goal slots and buffers could be extended to include resources such as a left hand and buffers such as memory retrieval in future template development.

Empirical Validation

ATM Task

ACT-Stitch was applied to the automated teller machine task used by John et al. (2002) to test their automation of CPM-GOMS. The task was to make an $80 withdraw from a checking account on a simulation of an automated teller machine. Users interacted with the ATM by using a mouse to click on simulated keys or slots. The users were instructed to follow the following steps:

Insert card (click on the card slot)
Enter PIN (click on the 4, 9, 0, and 1 keys in turn)
Press OK (click on the OK button)
Select transaction type (click on the withdraw button)
Select account (click on the checking button)
Enter amount (click on the 8 and 0 keys)
Select correct/not correct (click on the correct button)
Take cash (click on the cash slot)
Select another transaction (click on the No button)
Take card (click on the card slot)
Take receipt (click on the cash slot)

This task was repeated 200 times by the users, and results were analyzed using the means of trials 51-100. This level of practice is comparable to that used by both Card, Moran, and Newell (1983) in a text editing task and Baskin and John (1998) in a CAD drawing task when they explored the effects of extensive practice on match to various GOMS models. As in John et al. (2002), Slow-Move-Click templates were used for clicking on targets that were difficult to select because of size and distance (e.g., the thin card slot) and Fast-Move-Click templates were used for easier targets (e.g., keypad keys). These templates were originally developed for the simple task of clicking on lit circles by Gray & Boehm-Davis (2000) and were successfully reused by John et al. to explain subject performance. The Fast-Move-Click template is made up of operators which find the target location, move the eyes to that location and perceive the target, verify the target location, move the cursor to the target location, and click the mouse button. The Slow-Move-Click template contains the same operators as the Fast-Move-Click template but in addition has operators to perceive the cursor and verify it is at the target. In order to determine eye movement durations in ACT-Stitch, the EMMA (Salvucci, 2000) extension to ACT-R was used.

To get an idea of what a template looks like after being compiled into ACT-R productions, the following shows pseudo-code for the Fast-Move-Click template. Each instance of a template in the task sequence list would have its own set of productions labeled by the position of the template in the list (x).

```
Tx-Init-Move-Cursor
IF
        right hand action goal is to move the cursor in this template
        right hand target goal is this template's object
        motor preparations have completed
THEN
        move cursor
        empty right hand target goal
        set right hand action goal to click the mouse in this template

Tx-Attend-Targ
IF
        vision action goal is to attend target in this template
        vision target goal is this template's object
        visual location and object buffers are empty
        vision is available
THEN
        fill visual location buffer with location where
          this template's object should be
```

```
Tx-Init-Eye-Move
IF
        vision action goal is to attend target in this template
        vision target goal is this template's object
        visual object buffer is empty
        visual location buffer holds object location
THEN
        fill visual object buffer with object at location
        empty visual location buffer

Tx-Verify-Targ-Pos
IF
        vision action goal is to attend target in this template
        vision target goal is this template's object
        right hand target goal is empty
        visual object buffer holds object at location y
        location y is the expected location of this template's object
THEN
        empty visual object buffer
        set visual action goal to attend in the next template
        set visual target goal to next template's object
        set right hand target goal to this template's object

Tx-Init-Click
IF
        right hand action goal is to click the mouse in this template
        right hand target goal is this template's object
        motor preparations have completed
THEN
        click mouse
        set right hand action goal to move the cursor in next template
        set right hand target goal to next template's object
```

Productions that initiate motor movements (Init-Move-Cursor and Init-Click) first check that the motor preparations from previous motor movements have completed. Since motor preparations can happen in parallel with motor executions and finishes in ACT-R, this means that preparations can start during previous executions and finishes. Productions could be written to wait for the previous executions and finishes to complete before starting preparations, but they would not be as efficient.

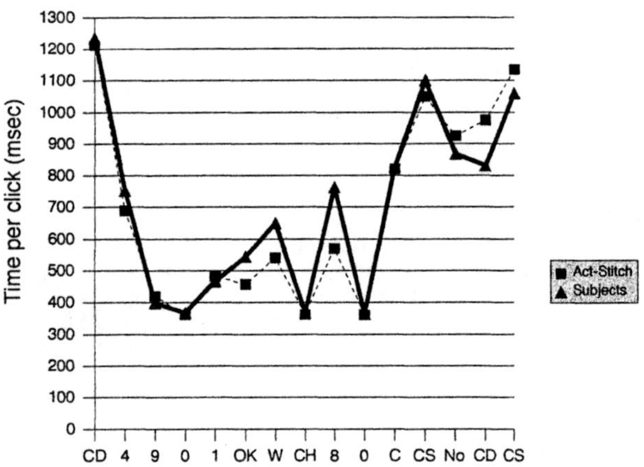

Figure 1: Average subject performance compared to ACT-Stitch predictions

Figure 1 compares ACT-Stitch predictions of mouse click times to average subject mouse click times of trials 51-100. The results are highly correlated (r=.96) with a low average absolute difference of 57msec.

The effect of interleaving on resource use is shown in PERT chart form in Figure 2. This output is from the Sherpa visualization tool developed by John et al. (2002) in their work to automate CPM-GOMS. The top row shows vision execution, the second shows vision preparation, the third cognition, the fourth shows motor preparation, and the bottom shows motor execution and finishing. Resource use is indicated with colored boxes, and instances of resource use in the same template are shown with the same color.

The figure is centered on the template for performing a Fast-Move-Click on the zero key (the lightest colored boxes), which is one of the fastest behaviors in the task for the subjects. ACT-Stitch explains this speed with an anticipatory eye movement to the zero key before the preceding nine key is clicked.

Sequential Response Task

To test the anticipatory eye movement prediction of the interleaving theory in ACT-Stitch, the sequential response task used by Wu & Remington (2004) was modeled. In this task, subjects viewed a series of five letters and responded to each individually. Subjects made sequential fixations to each of the five stimulus characters randomly drawn from the set T, D, and Z, and made choice responses mapped to three response keys (V, B, and N) on a PC keyboard and assigned to the first three digits of the right hand. Eye movements and key presses were recorded, and the stimulus letters were small enough and separated enough so that identification of stimulus letters required separate saccades and fixations. In Experiment 1 of Wu & Remington (2004), the effect of brightness of stimuli was investigated with dim and bright stimuli conditions, but no statistically significant differences were found. The predictions of the ACT-Stitch model will be compared to the results from the bright condition. Subjects were given 24 practice trials with the bright condition, then 120 trials divided into two blocks, one for each brightness condition.

The task was modeled in ACT-Stitch by creating a template for responding to a letter with an appropriate key press and applying this template to each stimulus letter.. The response template consisted of operators for finding the location of a letter, moving the eyes to that location and perceiving the letter, deciding a response, and pressing a key. As with the ATM task, the EMMA (Salvucci, 2000) extension was used to determine eye movement durations. The interleaving theory of ACT-Stitch predicts anticipatory eye movements where productions representing vision initiation operators from a future template can execute during the execution of productions representing operators in the current template.

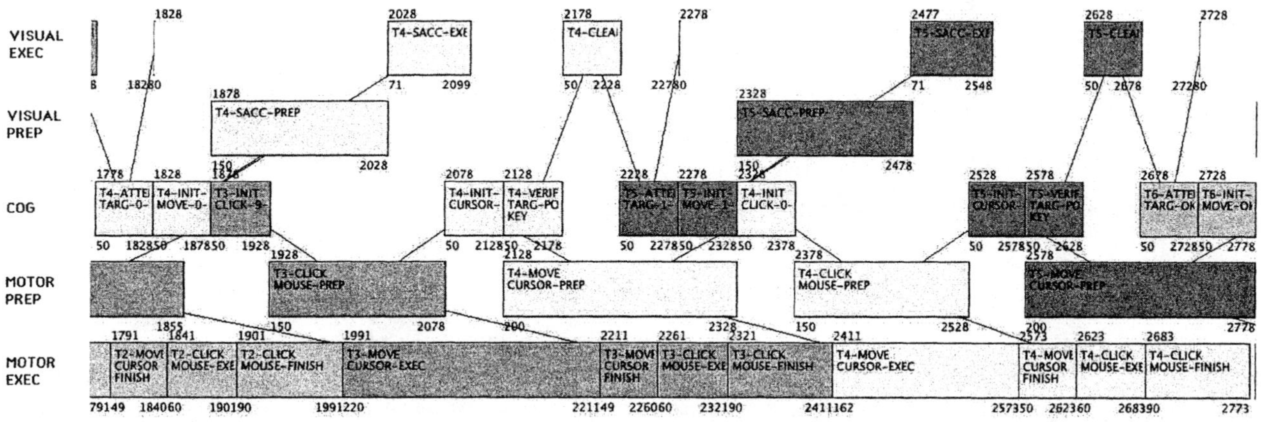

Figure 2: PERT chart of ACT-Stitch interleaving perceptual execution, perceptual preparation, cognitive, motor preparation, and motor execution and finishing resources in the ATM task

Figure 3 shows the time line of ACT-Stitch predictions for eye fixations and typing responses for the first four stimulus items (only four stimuli are presented because subjects had various strategies for where to fixate their eyes during the last stimulus response). Each horizontal bar represents the duration of response to a single stimulus, beginning with the eye fixation on a stimulus and ending with the typed response. The dark area represents eye fixation time ("fixation time") and the light area represents the time between moving the eye fixation to the next stimulus and the response to the current stimulus ("typing time"). Since the typed response for a particular stimulus occurs at the same time as fixations for the next stimulus, the figure shows that ACT-Stitch predicts anticipatory eye movements.

Figure 4 shows the time line of subject performance for eye fixations and typing responses in the bright condition of Experiment 1 of Wu & Remington (2004). The figure shows the anticipatory eye movements that the model predicts. The zero-parameter timing predictions are relatively close, with an average difference of fixation time of 126msec and an average difference of typing time of 64msec. The fixation and typing times of the model are consistently less than those for subjects, perhaps indicating that subjects are doing some processing that is not accounted for by the model. As can be seen in Figure 5 (a PERT chart representation of the model's performance), the fixation time is directly influenced by the time to decide on a mapping between letter and key. A post-hoc change of the decision time from 50msec to 150msec would reduce the average difference of fixation time between model and subjects to 26msec, while keeping the average difference of typing time at 64msec. Another feature of the data not accounted for by the model is the decreasing typing time over subsequent stimuli. These limitations of the model will be discussed later.

General Discussion

The interleaving theory of ACT-Stitch produces anticipatory eye movements that give a good account for data from two tasks, one showing quick motor response of measured mouse clicks and one showing anticipatory eye movements of measured eye fixations.

There is room for improvement, especially in the sequential response task. The fixation and typing times of the model are consistently less than subjects, perhaps indicating that subjects are doing some processing that is not accounted for by the model. Further work with the sequential response task done by Wu, Remington, and Pashler (2004) shows that fixation times on a specific stimulus can be lengthened depending on the response of the previous stimulus, suggesting the processing of the previous stimulus is still occurring after the fixation is started. The ability of ACT-Stitch to interleave productions from different tasks will be useful in trying to develop models to explain this result. Wu et al. also replicate finding of the decrease in time between end of fixation and typed response. It is difficult to explain this result with identical templates that do not make reference to the number of stimuli remaining to be processed because the timing of perceptual and motor processing is linked by cognition. One possible solution may involve perceptual and motor processing of different durations that are decoupled from cognition by means of the motor buffers (derived from EPIC) or visual buffers (derived from EMMA). Since preparation can occur in parallel with execution in these buffers, preparations can put visual or motor actions in a queue that could decouple the start of the execution from the start of the cognitive initiation.

This paper offers only a first step of a template and interleaving theory in ACT-R. Many more templates are needed to test the robustness of the representations used for the interleaving theory. But this work is a direction for easier modeling and multi-tasking in ACT-R.

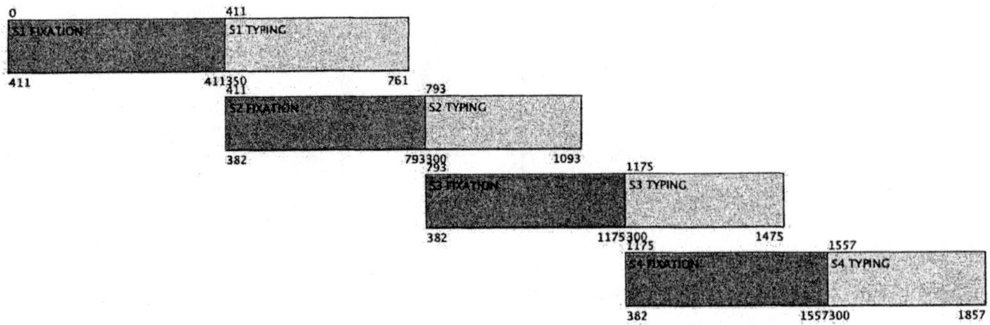

Figure 3: Time line of predictions for eye fixations and typing responses from the ACT-Stitch model

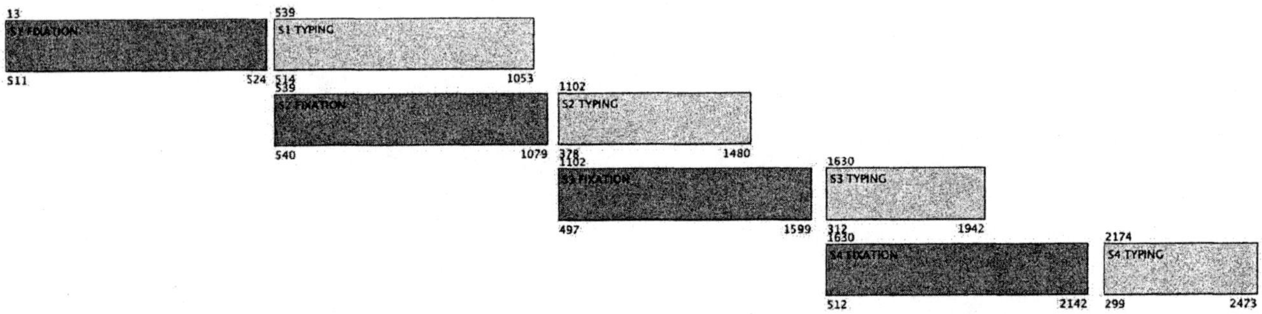

Figure 4: Time line of eye fixations and typing responses from subjects in Wu & Remington (2004)

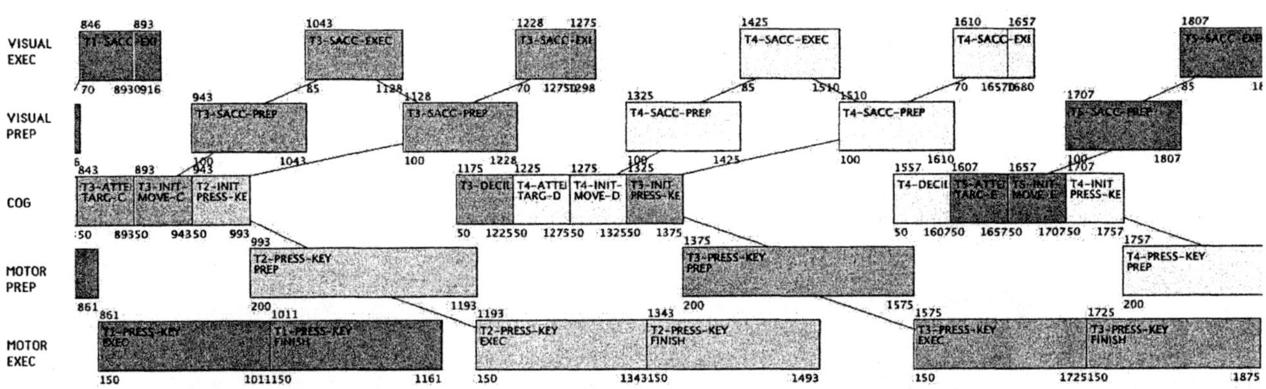

Figure 5: PERT chart of ACT-Stitch interleaving perceptual execution, perceptual preparation, cognitive, motor preparation, and motor execution and finishing resources in the sequential response task

Acknowledgments

This work was supported by Office of Naval Research grant N0001404IP2002 and by funds from the Airspace Operations System project of NASA's Airspace System program.

References

Anderson, J. R., & Lebiere, C. (1998). The atomic components of thought. Hillsdale, NJ: Erlbaum.

Baskin, J. D., and John, B. E. (1998). Comparison of GOMS Analysis Methods. Proceedings of ACM CHI 98 Conference on Human Factors in Computing Systems (Summary) 1998 v.2 p.261-262.

Byrne, M. D., & Anderson, J. R. (2001). Serial modules in parallel: The psychological refractory period and perfect time-sharing. Psychological Review, 108, 847-869.

Card, S. K., Moran, T.P. & Newell, A. (1983). The Psychology of Human-Computer Interaction. Hillsdale, NJ: Lawrence Erlbaum Associates.

Freed, M., Matessa, M., Remington, R. and Vera, A. (2003) How Apex automates CPM-GOMS. Proceedings of the Fifth International Conference on Cognitive Modeling, pp. 93-98. Bamberg, Germany:Universitats-Verlag.

Gray, W. D., & Boehm-Davis, D. A. (2000). Milliseconds matter: An introduction to microstrategies and to their use in describing and predicting interactive behavior. Journal of Experimental Psychology: Applied, 6(4), 322-335.

Gray, W. D., John, B. E. & Atwood, M. E. (1993) Project Ernestine: Validating a GOMS Analysis for Predicting and Explaining Real-World Task Performance, Human-Computer Interaction, 8 (3), pp. 237-309.

John, B. E. (1988) Contributions to Engineering Models of human-computer interaction. Ph.D. Thesis. Carnegie Mellon University.

John, B. E. (1990) Extensions of GOMS analyses to expert performance requiring perception of dynamic visual and auditory information. In proceedings of CHI, 1990 (Seattle, Washington, April 30-May 4, 1990) ACM, New York, 107-115.

John, B. E. & Kieras, D. E. (1996). The GOMS family of user interface analysis techniques: Comparison and Contrast, ACM Transactions on Computer-Human Interaction, 3 (4), pp. 320-351.

John, B. E., Vera, A. H., Matessa, M., Freed, M., and Remington, R. (2002) Automating CPM-GOMS. In Proceedings of CHI'02: Conference on Human Factors in Computing Systems. ACM, New York.

Land, M. F., & Hayhoe, M. (2001). In what ways do eye movements contribute to everyday activities? Vision Research, 41, 3559-3565.

Land, M. F., & Lee, D. N. 1994. Where we look when we steer. Nature, 369, 742-744.

Lee, F.J. & Taatgen, N.A. (2002). Multi-tasking as Skill Acquisition. Proceedings of the twenty-fourth annual conference of the cognitive science society (pp. 572-577). Mahwah, NJ: Erlbaum.

Matessa, M. (2004). An ACT-R framework for modeling interleaving templates of human behavior. In Proceedings of the Twenty-sixth Annual Conference of the Cognitive Science Society.

Matessa, M., Vera, A., John, B., Remington, R., & Freed, M. (2002). Reusable Templates in Human Performance Modeling. In Proceedings of the Twenty-fourth Annual Conference of the Cognitive Science Society.

Pelz, J. B. and Canosa, R. (2001). Oculomotor Behavior and Perceptual Strategies in Complex Tasks. Vision Research, 41, 3587-3596.

Salvucci, D. D. (2000). A model of eye movements and visual attention. In Proceedings of the Third International Conference on Cognitive Modeling (pp. 252-259). Veenendaal, The Netherlands: Universal Press.

Salvucci, D. D. (2002). Modeling driver distraction from cognitive tasks. In Proceedings of the 24th Annual Conference of the Cognitive Science Society.

Salvucci, D. D., & Lee., F. J. (2003). Simple cognitive modeling in a complex cognitive architecture. Human Factors in Computing Systems: CHI 2003 Conference Proceedings. New York: ACM Press.

Wu, S.-C., & Remington, R. W. (2004). Coordination of component mental operations in a multiple-response task. Paper to appear in S.N. Spencer (Ed.), Proceedings of the Eye Tracking Research and Applications Symposium 2004. New York: ACM SIGGRAPH.

Wu, S.-C., Remington, R. W., & Pashler, H. (2004). Coordination of component mental operations in sequences of discrete responses. In Proceedings of the Twenty-sixth Annual Conference of the Cognitive Science Society.

Exploratory Approach for Modeling Human Category Learning

Toshihiko Matsuka (matsuka@psychology.rutgers.edu)
RUMBA, Rutgers University – Newark

Abstract

One problem in evaluating recent computational models of human category learning is that there is no standardized method for systematically comparing the models' assumptions or hypotheses. In the present study, a flexible general model (called GECLE) is introduced that can be used as a framework to systematically manipulate and compare the effects of a limited number of assumptions at a time. Two simulation studies are presented to show how the GECLE framework can be useful in the field of human high-order cognition research.

Introduction

The past fifteen years have seen significant advances in adaptive network models of categorization. In particular, three models of human category learning have attracted much attention, namely ALCOVE (Kruschke, 1992), RASHNL (Kruschke & Johansen, 1999), and SUSTAIN (Love & Medin, 1998). These models share many properties, because they can be considered special cases of Generalized Radial Basis Functions (Poggio & Girosi, 1990), as discussed by several authors (Kruschke, 1993; Matsuka & Corter 2003a; Rosseel, 1996). First, all three models are multilayer adaptive network models, with "reference points" ("basis units" in RBF terminology) in their memory (specifically, in the hidden layer). The models all use similarities between the reference points (RPs) and the input stimuli for calculating activations of RPs. Then, the weighted RP activations are fed forward to output nodes, whose activations are used to categorize the input stimuli. All three models scale feature dimensions independently in calculating these input-to-RP similarities, and this scaling process is interpreted as reflecting dimensional attention processes (Kruschke, 1993). In addition, all models incorporate as their basic learning method a form of gradient descent for incremental adjustments of both association weights and dimension-specific attention parameters. Finally, all three models may be considered as confirmatory models, because they are based on specific a priori assumptions about how humans process information in categorization (e.g. how stimuli are internally represented & how humans pay attention to stimuli). In order to test these specific assumptions, they should be varied systematically, preferably one at a time. The assumptions could then be tested by comparing the fit to empirical data of the resulting models. This is the major theme of the present work.

There are several differences among the three models as well. First, the assumptions about how stimuli are internally represented are different. ALCOVE and RASHNL are exemplar models, in the sense that each stimulus in the training set is allocated as a RP in the "hidden" layer of the networks, and the RPs reside in fixed locations. In contrast, SUSTAIN is a prototype model that uses a reduced number of movable RPs in its hidden layer, corresponding to potential generalizations. In addition, SUSTAIN dynamically allocates new prototypes, thus it may use multiple prototype nodes for each category explicitly defined by the training feedback. Second, how RP activations are utilized in making category predictions and in adjusting parameter estimates during learning are different for the models. SUSTAIN utilizes only the single most activated RP for categorization and learning, whereas ALCOVE and RASHNL utilize the activations of all RPs. Third, the assumptions about attention processes are different. RASHNL assumes limited attention capacity and rapid shifts in attention processes, whereas ALCOVE and SUSTAIN do not. Fourth, the functions for computing similarity measures and RP activations are different.

There have been several studies comparing computational models of categorization, including but not limited to ALCOVE, RASHNL, and SUSTAIN (e.g., Matsuka, Corter, & Markman, 2003; Nosofsky, Gluck, Palmeri, McKinley, & Glauthier, 1994). Although these comparative studies provided information on the models' capabilities for reproducing human-like categorization learning, they did not necessarily provide information that can lead to specific understanding of the nature of human category learning. That is because model-to-model comparisons are not informative for testing the plausibility of each specific assumption, rather such model comparisons are essentially omnibus tests collectively comparing all variations in assumptions at once.

Since it has been difficult to use the results of these previous comparative studies to understand which specific assumptions are supported by the data, it seems desirable to have a general framework for modeling human category learning that allows us to manipulate and test one or a limited number of model assumptions at a time. The framework should be general and flexible, so that we can conduct standardized exploratory modeling of various types of human cognitive processes associated with categorization.

New Model of Human Category Learning

Qualitative Descriptions: The GECLE (for Generalized Exploratory models of Category LEarning) is a general and flexible exploratory modeling approach for human category learning, that is capable of modeling human category learning with many variants using different model assumptions. This general model allows model assumptions to be manipulated separately and independently. For example, one can manipulate assumptions about how stimuli are internally represented (e.g. exemplars vs.

prototypes), or about how people selectively pay attention to input feature dimensions (e.g., paying attention to dimensions independently or not).

The GECLE model uses the Mahalanobis distances (in quadratic form) between the internally represented reference points (corresponding to either exemplars or prototypes) and the input stimuli as the measure of similarity between them. Thus, unlike other NN models of category learning, the GECLE does not necessarily assume that attention is allocated independently dimension-by-dimension. Rather, it assumes that humans in some cases do pay attention to correlations among feature dimensions. This allows the GECLE to model processes interpretable as dimensionality reduction or mental rotation in the perception and learning of stimuli. Such processes may enhance the interpretability of stimuli for categorization task for humans. Another motivation for the use of Mahalanobis distance is that the capability for paying attention to correlations among feature dimensions may be needed for classification tasks defined on integral stimuli. In the GECLE framework, the attention parameters (which are the diagonal and off-diagonal elements of the covariance matrices) can be considered as *shape* and *orientation* parameters for receptive fields or attention coverage areas of the reference points. It should be noted, however, that one can constrain GECLE to incorporate the "dimensional attention processes" assumption (i.e., attention is allocated independently on a dimension-by-dimension basis) by forcing the off-diagonal entries in the covariance matrices to be equal to zero.

Another unique feature of GECLE's attention mechanism is that it allows each reference point to have uniquely shaped and oriented attention coverage area (Figures 1D, 1E, and 1F). I call this "local attention coverage structure". Again, one can impose a restriction on the model's attention mechanism by fixing all covariance matrices to be the same, which I refer as "global attention coverage structure" (Figures 1A, 1B, and 1C). The local attention coverage structure model is complex, but may plausibly model attention processes in human category learning. For example, it allows models to be sensitive to one particular feature dimension when the input stimulus is compared with a particular reference point that is highly associated with category X, while the same feature dimension receives little or no attention when compared with another reference point associated with category Y. Thus the local attention coverage structure causes models to learn and be sensitive to within-cluster or within-category feature configurations, while the global attention coverage structure essentially stretches or shrinks input feature dimensions in a consistent manner for all RP receptive fields and all categories.

Another way of interpreting GECLE's capabilities for paying attention to correlations among feature dimensions and having local attention coverage structures is that the model learns to define what the feature dimensions are for each RP and to allocate attention to those dimensions. In contrast, for almost all previous adaptive models of category learning, the definition of the feature dimensions is static and supplied by individuals who use the models.

The other notable characteristic of the GECLE's attention mechanism is that the user can manipulate characteristics of the distributions assumed for the activations of the reference points. For example, one can have RP activation distributions with thicker tails to obtain more competition among the RPs. In its natural form, the GECLE may be considered as a model using prototype internal representation, because it tries to learn to locate its reference points at centers of each category cluster. However, with a proper user-defined parameter setting, it can behave like a model with an exemplar-based internal representation.

Quantitative Descriptions (Algorithm): The feedforward and learning algorithms of the GECLE are typical for implementation of the Generalized Radial Basis Function (Haykin, 1999; Poggio & Girosi, 1990). GECLE uses the following function to calculate the distances or similarity between internally represented reference points (e.g., prototypes or exemplars) and input stimuli at time n:

$$D_j^n(x^n, r_j) = (x^n - r_j)^T \Sigma_j^{-1} (x^n - r_j) \tag{1}$$

where x^n is an I-ruple vector representing an input stimulus consisted of I feature dimensions presented at time n, r_j, also an I-ruple vector, that corresponds to the centroids of reference point j, expressing its characteristics, and Σ_j^{-1} is the inverse of the covariance matrix, which defines the shape and orientation of the attention coverage area of reference point j. For a model with global attention coverage structure, there is only one global Σ^{-1} for all reference points. The entries (s_{im}) in Σ_j must satisfy the following conditions: $s_{ii} \geq 0$ & $|s_{im}| \leq \mathrm{MIN}(s_{ii}, s_{mm})$.

The psychological similarity measures $D_j(x,r)$ cause some activations in internal "hidden" units or reference points (i.e., exemplars or prototypes). The activation of "hidden" basis unit j, or h_j, is obtained by any differentiable nonlinear activation transfer function (ATF), or

$$h_j = G(D_j(x,r)) \tag{2}$$

given that its first derivative $G'(\cdot)$ exists. An exponential function, $\exp(-cD_j(x,r))$, is an example of an ATF. The ATF must be a differentiable function, because GECLE uses a gradient method for learning, where the partial derivatives are used for updating the learnable parameters. However, it is possible to eliminate this restriction by incorporating a form of derivative-free learning algorithm such as stochastic learning (Matsuka & Corter, 2004).

The activations of hidden units are then fed forward to output nodes. The activation of the kth output node, O_k, is calculated by summing the weighted activations of all hidden units connected to the output node, or

$$O_k = \sum_{j=1}^{J} w_{kj} h_j \tag{3}$$

where w_{kj} is the association weight between output node k and reference point j. The probability that a particular stimulus is classified as category C_k, denoted as $P(C_k)$, is assumed equal to the activity of category k relative to the summed activations of all categories, where the activations are first transformed by the exponential function (Kruschke, 1992)

$$P(C_k) = \frac{\exp(\phi O_c)}{\sum_k \exp(\phi O_k)} \tag{4}$$

ϕ is a real-value mapping constant that controls the "decisiveness" of classification responses.

GECLE uses the gradient method to update parameters. The error function is defined as the sum of squared differences between targeted and predicted output values (i.e., L_2 norm), or

$$E(w,r,\Sigma^{-1}) = \frac{1}{2}\sum_{k=1}^{K} e_k^2 = \frac{1}{2}\sum_{k=1}^{K}(d_k - O_k)^2 \quad (5)$$

Then the following functions are used to update parameters.

$$\Delta w_{jk} = \frac{\partial E}{\partial w_{jk}} = -\eta^w e_k h_j \quad (6)$$

where η^w is the learning rate for the association weights.

$$\Delta r_j = \frac{\partial E}{\partial r_j} = -\eta^r \sum_{k=1}^{K} e_k w_{jk} G'(D_j(x,r))\Sigma_j^{-1}(x - r_j) \quad (7)$$

where $G'(\cdot)$ is a derivative of $G(\cdot)$. Equation 7 can be considered as a function that locates or defines prototypes of stimuli. For the exemplar-based modeling η^r must be set to zero to maintain the static nature of reference points.

$$\Delta\Sigma_j^{-1} = \frac{\partial E}{\partial \Sigma_j} = \eta^{\Sigma} \sum_{k=1}^{K} e_k w_{jk} G'(D_j(x,r))(x - r_j)(x - r_j)^T \quad (8)$$

For models with global attention coverage structure, Equation 8 should be summed over both k and j.

Varieties of Activation Transfer Function: The ATF in the GECLE can be any function as long as it is differentiable. This allows one to manipulate and compare the effects of specific characteristics of the population attention structure (e.g. fatter tail vs. thinner tail). This capability was motivated by the fact that the population attention structure can determine the effectiveness of model predictions. For example, Hanson and Gluck (1991) compared RBFs with Gaussian and Cauchy activation functions, and showed that increased competition by the Cauchy's fatter tails resulted in better fit to the empirical data. Since there is not enough evidence indicating the "true" or best activation transfer function, and to enhance the flexibility of GECLE, ATF is deliberately made user-definable.

Hierarchy of Constraints on Attention Parameters (Σ): There is a hierarchy of constraints that one can impose on the attention parameters Σ to manipulate GECLE's attention mechanisms. There are two levels of uniqueness of Σ (global and local attention coverage structure), in each of which there are three levels of constraints on entries in Σ. The following is a list of six possible levels of restriction, and Figure 1 shows examples of the corresponding attention coverage structures. Note that regardless of the types of restriction, the entries (s_{im}) in Σ_j must satisfy the following conditions: $s_{ii} \geq 0$ & $|s_{im}| \leq$ MIN(s_{ii}, s_{mm}).

Global Attention Coverage Structures

A. Global Pure Radial (GPR): Constraints on Σ_j: $s_{ii} = s$, for all i: $s_{im} = 0$, for all $i \neq m$; $\Sigma_j = \Sigma$ for all reference points j.

B. Global Uncorrelated Non-radial (GUN): Constraints on Σ_j: $s_{im} = 0$, for all $i \neq m$; $\Sigma_j = \Sigma$ for all reference points j.

C. Global Correlated Non-radial (GCN): Constraints on Σ_j: $\Sigma_j = \Sigma$ for all reference points j.

Local Attention Coverage Structures

D. Local Pure Radial (LPR): Constraints on Σ_j: $s_{ii} = s$, for all i; $s_{im} = 0$, for all $i \neq m$.

E. Local Uncorrelated Non-radial (LUN): Constraints on Σ_j: $s_{im} = 0$, for all $i \neq m$.

F. Local Correlated Non-radial (LCN): Constraints on Σ_j: none.

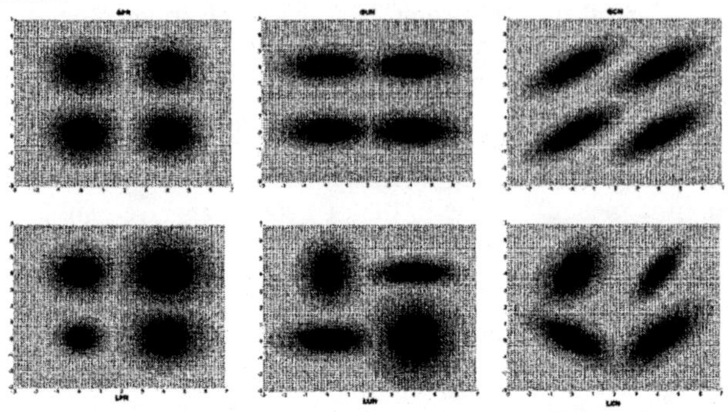

Figure 1. Six types of attention structures in the GECLE framework. Clockwise from top left panel, GPR, GUN, GCN LCN, LUN, and LPR.

Application s of GECLE

Comparing Internal Representation assumptions: There has been an increasing number of studies investigating and debating how stimuli are internally represented in human cognition during the last several years (e.g., Minda & Smith 2002; Nosofsky & Zaki 2002). Most of these debates have been based on quantitative models of categorization, and only a few have considered representational aspects of adaptive or network models of category learning. One limitation of the models of categorization is that, as Shanks (1991) pointed out, its model fitting process is post hoc and thus does not generate predictions on learning processes.

Several studies (Matsuka, 2002; Matsuka et al, 2003) have compared exemplar-based (EB) and prototype-based (PB) adaptive network models of category learning, but there have been no systematic comparison of specific assumptions in EB and PB modeling. For example, the EB and PB models compared in those studies assume different attention processes and utilize reference points differently for categorizing and learning. Thus, differences in the accuracy of reproducing learning curves may not be attributed solely to the plausibility of the EB versus PB representations, but possibly to multiple factors including the models' attention mechanisms. With GECLE, one can systematically compare the plausibility of the EB and PB representations by holding the attention mechanisms constant for both models (i.e., using the same activation transfer function and the same constraints on Σ^{-1}).

Comparing Selective Attention Mechanisms: Selective attention processes have been suggested to play a very important role in human category learning (Shepard, Hovland, Jenkins, 1962). However, only limited numbers of selective attention mechanisms have been modeled and tested. For example, virtually all recent network models of categorization assume dimensional attention processes (i.e., no attention to correlations) and global attention coverage structure (i.e., all reference points have exactly the same shape of receptive field, with independent attention allocation to dimensions).

Again, a general framework like GECLE allows systematic manipulation of models' attention mechanism, and or exploration of various types of attention mechanism that has not be tested, such as distribution of attention to correlated features dimensions. In addition, comparisons on different activation transfer functions $G(\cdot)$ can be informative for understanding how human categorize stimuli.

Investigating Interactions Between Internal Representation & Attention Mechanisms. As a final point, it may be possible that a model with a particular internal representation system (e.g., exemplar-based) performs better with a particular attention mechanism, that does not work as well for models with other representation system (e.g., prototypes). In other words, the effectiveness of internal representation system and attention mechanism may interact with each other in the sense that the effectiveness of model's internal representation systems may depend on its attention mechanism, or the effectiveness of the models' attention mechanism may depend on its internal representation system.

I am not aware of any single study that systematically and simultaneously manipulates models' internal representation system and attention mechanism to investigate possible interaction effects between them. GECLE provides a way to systematically manipulate, by a factorial design, both internal representation system and attention mechanism to tackle this issue of interactivity of internal representation and attention mechanisms.

Simulations

In this section, simulation studies are conducted as examples showing how GECLE can be informative in the field of human category learning. In particular, two simulation studies, investigating possible interactive effects of the models' internal representation and selective attention mechanisms are reported.

Simulation 1: XOR problem

In Simulation 1, a simple exclusive-or (XOR) learning task is simulated with both prototype- and exemplar-based GECLE models. An XOR is one of the simplest stimuli structures with which one can expect interactions between representation system and attention mechanism.

There are eight different models involved in this simulation, namely, E1: an exemplar-based (EB) model with GUN attention mechanism; E2: EB with GCN; E3: EB with LUN; E4: EB with LCN; P1: a prototype-based (PB) model with GUN; P2: PB with GCN; P3: PB with LUN; and P4: PB with LCN. All EB models had four reference points, while all PB models had two. For all eight models, the following one-parameter exponential activation transfer function was used:

$$h_j = \exp\left(-c \cdot D_j(x,r)\right)$$

The models were run in a simulated training procedure to learn the correct classification responses for the stimulus set. The models were run for 250 blocks of training, where each block consisted of a complete set of the training instances. The user-defined parameters (e.g. learning rates) were selected arbitrary.

Results: Tables 1a and 1b show the results of Simulation 1. All exemplar-based models were able to learn to categorize XOR stimuli by utilizing four exemplars (i.e., all unique stimulus configurations) in their "memory", and thus complex attention mechanisms were shown to be ineffective or unnecessary for EB modeling. In fact, when the fit measure was adjusted for model complexity (i.e., number of parameters), the EB model with the simplest attention mechanism resulted in the best (relative) fit. In contrast, for prototype-based modeling, only LCN (i.e., P4) was able to learn to categorize the XOR stimulus set, suggesting that the complex attention mechanism plays an important role for the PB modeling (Figure 2 shows activation areas produced by P4). Note that P4 also resulted in the best (relative) fit among all eight models after controlling for its complexity.

In sum, the results of the present simulation suggest that it is very likely that effectiveness of the model's attention mechanism depends on how the stimuli are internally represented by the model or vice versa; a simple GUN attention mechanism seems sufficient for EB modeling, while a complex LCN is required for PB modeling.

Table 1a: Results of prototype-based GECLE

Model	P1	P2	P3	P4
Attention structure	GUN	GCN	LUN	LCN
No. prototypes	2	2	2	2
No. Learnable parameters	10	11	12	14
SSE	1.328	1.147	1.322	$\varepsilon^{\dagger}$

$\varepsilon^{\dagger} < 10e\text{-}20$.

Table 1b: Results of exemplar-based GECLE

Model	E1	E2	E3	E4
Attention structure	GUN	GCN	LUN	LCN
No. Exemplars	4	4	4	4
No. Learnable parameters	18*	19*	24*	28*
SSE	$\varepsilon^{\dagger}$	$\varepsilon^{\dagger}$	$\varepsilon^{\dagger}$	$\varepsilon^{\dagger}$

$\varepsilon^{\dagger} < 10e\text{-}20$.

* Location parameters for exemplars were static & not subject to error-minimization learning, but it is assumed that optimized locations are learned when the exemplars are created.

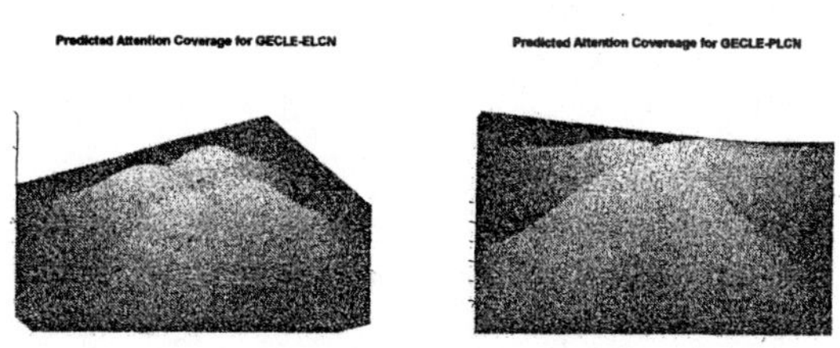

Figure 2: Activation areas & strength of E4 and P4

Simulation 2: Filtration vs. Condensation

Kruschke (1993) claimed that selective dimensional attention processes (i.e., paying attention to each dimension independently) is one of three key principles for models of category learning. His claim was based partly on the evidence that humans learn much better in "filtration" tasks, in which information from only one dimension is required for perfect categorization, than in "condensation" tasks, in which information from two (or more) dimensions is required (Gottwald & Garner, 1975; Kruschke 1993). Thus, a model paying attention to correlations or having diagonal attention coverage may not be able to show the filtration advantage, implying that any model with a GCN or LCN attention mechanism may not be able to replicate such an advantage. If the claim valid, then it is evidence against P4, namely the prototype-based model with diagonal localized attention coverage, as a descriptive model of human cognition. The present simulation study tests if PB model with LCN attention mechanism can replicate the filtration advantage observed in human category learning.

Method: In simulation 2, I revisited Kruschke's claim regarding dimensional attention processes by simulating category learning on both filtration and condensation stimuli using the prototype-based model with LCN (and EB-LCN for a illustrative comparison). The stimulus set presented in Kruschke (1993) is used in this simulation. Table 2 shows the schematic representation of the stimulus set. For the filtration stimulus set, information from only Dimension 1 is required for a perfect categorization (category = A, if D1 < 2; category = B, otherwise) while information on both Dimensions 1 and 2 were required for the condensation set. The same one-parameter exponential ATF used in Simulation 1 is used in the present simulation study. The user-defined parameters were optimized using a simulated annealing method (Ingber, 1989; Matsuka et al. 2003) to reproduce observed empirical learning curves reported in Kruschke (1993).

It should be noted that Kruschke (1993) showed that ALCOVE (i.e., an EB-model with GUN) was able to reproduce the filtration advantage.

Table 2: Stimulus sets used in Simulation 2

Stimulus feature		Category	
Dim 1	Dim 2	Filtration	Condensation
0	1	A	A
0	2	A	A
1	0	A	A
1	3	A	B
2	0	B	A
2	3	B	B
3	1	B	B
4	2	B	B

Results: Figure 3 shows the results of Simulation 2. The prototype-based LCN model was able to show the filtration advantage even when it paid attention to the correlation between the two input dimensions. In contrast, the exemplar-based LCN model showed no filtration advantage.

One possible reason why PB-LCN showed the filtration advantages was that PB-LCN might have been able to locate or define prototypes more easily in the filtration task than in the condensation task. This is because while the condensation stimuli require synchronization of the "correct" movements of centroids of prototypes and the "correct" psychological scaling of the two feature dimensions (i.e., attention processes), the filtration stimuli require "correct" movements and scaling in only one dimension. Thus, synchronization of prototype-movement and scaling was more difficult for the condensation stimuli than in the filtration task for models using prototypes. In other words, category learning by any prototype-based network models is strongly affected by how successfully or how fast the models can find "proper" prototypes and how well psychological scaling of dimensions is synchronized with it. For E4, this was not a problem as it had exemplars in the correct locations from the beginning, resulting in no filtration advantage. These results indicate that correlated (i.e., diagonally-oriented) attention coverage may be more often a required assumption for PB modeling, compared to EB modeling.

As in Simulation 1, the results of the present simulation suggest that it is highly possible that stimuli's internal representation and selective attention mechanisms interact with each other. Furthermore, our simulation studies suggest that human may allocate attention not only to individual dimensions, but also to correlations among dimensions. At the very least, the evidence of a filtration advantage observed in human subjects does not rule out the possibility that humans pay attention to correlations among feature dimensions.

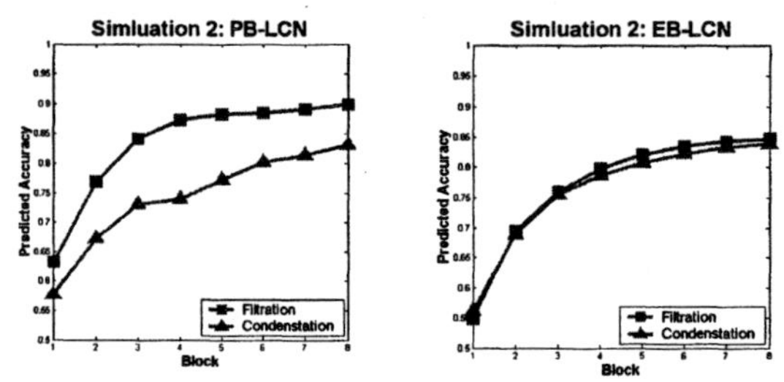

Figure 3: The results of Simulation 2

Discussion and Conclusion

Individual Differences. The results of some simulation studies (e.g., Matsuka, 2002, Matsuka & Corter, 2003b) suggest that NN models of categorization with gradient learning methods are successful in reproducing group learning curves, but tend to underpredict variability in individual-level data. Since GECLE utilizes a gradient method for learning, it too is expected to underpredict individual differences. However, this exploratory model is introduced to compare how well models with different representational and processing assumptions can replicate general tendencies in human category learning. Many of these general tendencies may be best described in terms of such aggregated data. Nonetheless, to account for individual differences, GECLE could be easily modified to incorporate Matsuka & Corter's (2004) stochastic learning algorithm for attention processes, which is shown to be more successful in reproducing individual differences in learning.

Conclusion

One of the most critical problems in evaluating recent computational models of categorization is that there is no standardized method for comparing the models' assumptions systematically. Thus, previous studies involving model comparisons have sometimes been unable to answer which element, assumption, or structure of each model was responsible for successful or unsuccessful replication of observed tendencies in human category learning. In the present study, a flexible general model is introduced, that can be used as a framework to systematically compare a limited number of assumptions at a time.

Two simulation studies are described to show how the GECLE framework can be useful in exploring issues in the field of categorization research. The results of Simulation 1 showed that a pure prototype-based category learning model (i.e., the number of prototypes is equally to that of category) was capable of learning an XOR problem only if it incorporated a very complex attention mechanism, while the exemplar-based model was capable of learning the stimuli with a simple attention mechanism. In Simulation 2, the filtration advantage, which has been used as an argument or evidence for dimensional attention processes (i.e., paying attention to dimension independently with no attention to correlations among feature dimensions), was successfully replicated by the prototype model with a complex attention mechanism capable of paying attention to correlation. This casts some doubt on the claim that the filtration advantage shows that people pay attention to dimensions independently, without attending to correlations among dimensions.

The results of these simulations are argued to provide new insights regarding human category learning, namely that 1) it is very likely that there are interactions between internal mental representation and attention mechanisms, and 2) people may pay attention to correlations among feature dimensions.

References

Gottwald, R. L. & Garner, W. R. (1975). Filtering and condensation tasks with integral and separable dimensions. *Perception & Psychophysics, 2*, 50-55.

Hanson, S. J., & Gluck, M. A. (1991). Spherical units as dynamic consequential regions: Implications for attention and cue-competition in categorization. *Advances in Neural Information Processing Systems #3*. San Mateo, CA: Morgan Kaufman, 656-665.

Haykin, S. (1999). *Neural Networks: A Comprehensive Foundation (2nd ed.).* Upper Saddle River, NJ: Prentice Hall.

Ingber, L. (1998). Very fast simulated annealing. *Journal of Mathematical Modelling, 12:* 967-973.

Kruschke, J. E. (1992). ALCOVE: An exemplar-based connectionist model of category learning, *Psychological Review, 99*. 22-44.

Kruschke, J. E. (1993). Three principals for models of category learning. In G. V. Nakamura, R. Taraban, & D. L. Medin (Eds.), *Categorization by human and machines: The psychology of learning and motivation* (Vol. 29, pp. 57-90). San Diego, CA: Academic Press.

Kruschke, J.K., & Johansen, M. K. (1999). A model of probabilistic category learning. *Journal of Experimental Psychology: Learning, Memory, and Cognition, 25*, 1083-1119.

Love, B.C. & Medin, D.L. (1998). SUSTAIN: A model of human category learning. *Proceeding of the Fifteenth National Conference on AI (AAAI-98)*, 671-676.

Matsuka, T (2002). Attention processes in computational models of categorization. Unpublished Doctoral Dissertation. Columbia University, NY.

Matsuka, T. & Corter, J. E. (2003a). Neural network modeling of category learning using Generalized Radial Basis Functions. Paper presented at 36th Annual Meeting of the Society of Mathematical Psychology. Ogden, UT.

Matsuka, T. & Corter, J. E. (2003b). Empirical studies on attention processes in category learning. Poster presented at 44th Annual Meeting of the Psychonomic Society. Vancouver, BC, Canada.

Matsuka, T. & Corter, J.E (2004). Modeling category learning with stochastic optimization methods. In *Proceeding of International Conference on Cognitive Modelling*. Pittsburgh, PA

Matsuka, T., Corter, J. E. & Markman, A. B. (2003). Allocation of attention in neural network models of categorization. Under review.

Minda, J.P. & Smith, J.D. (2002). Comparing prototype-based and exemplar-based accounts of category learning and attentional allocation. *Journal of Experimental Psychology: Learning, Memory, and Cognition, 28*, 275-292.

Nosofsky, R.M. & Zaki, S. R. (2002). Exemplar and prototype models revisited: Response strategies, selective attention, and stimulus generalization. *Journal of Experimental Psychology: Learning, Memory, and Cognition, 28*, 924-940.

Nosofsky, R.M., Gluck, M.A., Palmeri, T.J., McKinley, S.C., & Glauthier, P. (1994). Comparing models of rule-based classification learning: A replication and extension of Shepard, Hovland, and Jenkins

Poggio, T. & Girosi, F. (1990). Regularization algorithms for learning that are equivalent to multilayer networks. *Science, 247*, 978-982.

Rosseel, Y. (1996). Connectionist models of categorization: A statistical interpretation. *Psychologica Belgica, 36*, 93-112

Shanks, D.R. (1991). Categorization by a connectionist Network. *Journal of Experimental Psychology: Learning, Memory, and Cognition, 17*, 433-443.

Shepard, R.N., Hovland, C.L., & Jenkins, H.M. (1961). Learning and memorization of classification. *Psychological Monograph, 75*(13).

Modeling Category Learning with Stochastic Optimization Methods

Toshihiko Matsuka (matsuka@psychology.rutgers.edu)
RUMBA, Rutgers University - Newark

James E. Corter (corter@tc.exchange.columbia.edu)
Department of Human Development, Teachers College, Columbia University

Abstract

Many neural network (NN) models of categorization (e.g., ALCOVE) use a gradient algorithm for learning. These methods have been successful in reproducing group learning curves, but tend to underpredict variability in individual-level data, particularly for attention allocation measures (Matsuka, 2002). In addition, many recent models of categorization have been criticized for not being able to replicate rapid changes in categorization accuracies and attention processes observed in the empirical studies (Macho 1997; Rehder & Hoffman, 2003). In this paper we introduce stochastic learning algorithms for NN models of human category learning and show that use of the algorithms can result in (a) rapid changes in accuracies and attention allocation, and (b) different learning trajectories and more realistic variability in individual-level.

Introduction

Recent NN models of classification learning, for example ALCOVE (Kruschke, 1992), RASHNL (Kruschke & Johansen, 1999), and SUSTAIN (Love & Medin, 1998), share a number of common aspects, including multilayer architectures and learned dimensional attention weights as well as learned association weights between stimulus input nodes and the output layer. One of these common elements is the use of *gradient-based learning algorithms* to adjust both association weights and dimensional attention parameters. In this method, weights are adjusted based on discrepancies between a training signal and the output activations on the current layer, by computing the *gradient* of the error function in the multidimensional parameter space. This is accomplished by taking the partial derivative of the error function with respect to each of the network parameters (weights) in turn. The algorithm then adjusts each of these weights proportionally to its partial derivative. This learning method is an effective means of finding optimal estimates for parameters, as long as the overall error function is not characterized by strong local minima. Thus the algorithm has normative justification (i.e., it models how people "should" learn or process information).

But, is the gradient method plausible *descriptively* (i.e., does it describe how people actually learn)? It seems implausible that people explicitly compute the gradient on each trial while attempting a classification learning task. On the other hand, people's general learning mechanisms might have evolved so as to approximate gradient learning. But in any case, we should first ask if the gradient-based learning algorithms are successful in replicating empirical data in human category learning. Results in the literature demonstrate that these methods have been successful in reproducing group learning curves (e.g., Kruschke, 1992, Kruschke & Johansen, 1999; Love & Medin, 2000). However, recent studies in our lab suggest these models may underpredict variability in individual-level empirical data, particularly differences in attention allocation measures (Matsuka, 2002; Matsuka & Corter, 2003; Matsuka, Corter & Markman, 2003). In addition, some empirical studies suggest that human's attention allocation to individual dimensions can change quite rapidly (Macho, 1997; Rehder & Hoffman, 2003). Most cognitive models based on gradient-based learning mechanisms appear to have difficulty reproducing such rapid changes in attention. In the present research we explore alternative learning algorithms for NN models of classification learning, specifically stochastic learning algorithms based on simulated annealing.

Stochastic Learning

Our proposed algorithm is based on a specific simulated annealing algorithm (Ingber, 1989). In the present algorithm, initial association weights are randomly selected from a uniform distribution centered at 0, and initial dimension attention strengths are equally distributed across all dimensions. This equal attention allocation in the early stages of learning is motivated by the results of empirical studies (Matsuka, 2002; Rahder & Hoffman, 2003) that showed many subjects tended to evenly allocate attention to the feature dimensions initially. In our algorithm, at the beginning of each training epoch, a hypothetical "move" in the parameter space is computed by adjusting each parameter by an independently sampled term. These adjustment terms are drawn from a prespecified Cauchy distribution. The move (i.e., the set of new parameter values) is then accepted or rejected, based on the computed relative fit of the new values. Specifically, if the new parameter values result in better fit, they are accepted. If they result in worse fit, they are accepted with some probability *P*. This probability is a function of a parameter called the "temperature", which decreases across blocks according to the annealing schedule. This particular annealing algorithm is relatively efficient, in that the adjustment in the network parameters is very rapid initially, and gradually decreases over learning blocks.

Key Assumptions

This learning algorithm can be applied to any feed-forward NN model of category learning. However, we assume that there is no (back) propagation of classification errors in the models. Rather, we propose a very simple operation (i.e., comparison of two values) along with the operation of stochastic processes as the key mechanisms in category learning.

I. Initial network association weights (**w**) are set to small random values, and initial dimension attention weights ($\boldsymbol{\alpha}$) are set equal across dimensions.
II. In learning, the attention strengths and association weights are updated with a random move in the parameter space, based on a prespecified distribution (e.g., the Cauchy distribution).
III. If the new $\boldsymbol{\alpha}$ and **w** result in better categorization accuracies (based on a "mini-simulation" using the network model), then the current move is accepted, and the new attention strengths and association weights replace the old values. In the case of a *decrease* in categorization accuracy due to the move, the move is accepted with some probability P ($0<P<1$) that is functional of both magnitude of error and moment of training (i.e., temperature).
IV. P is relatively large in the early stages of learning, but it decreases as learning progresses. This decrease is associated with a decrease in a parameter called the "temperature", by analogy with the physical process that occurs as a metal cools.

Thus, the present model does not assume that learning is associated with monotonic increases in accuracy (and attention) or continuous search for better categorization processes by human. Rather, it models random fluctuations or "errors" in people's memory and learning processes, and how people utilize and "misutilize" such errors.

As a test of these ideas, we have embedded the present learning algorithm into the ALCOVE model (Kruschke, 1992).

ALCOVE

ALCOVE (Kruschke, 1992), for Attention Learning COVEring map, is an exemplar-based multi-layer adaptive network model of categorization based in part on the Generalized Context Model or GCM (Nosofsky, 1986). The first layer of ALCOVE is a stimulus input layer. Each dimension has an attention strength (α_i) associated with it. The next layer in the network is the exemplar layer. Each node in this layer corresponds to an exemplar, described by its position in the multidimensional stimulus space, and receives input from the input layer. The activation of each exemplar node is calculated based on its similarity to the presented stimulus:

$$h_j = \exp\left[-c\left(\sum_i \alpha_i \mid \psi_{ji} - x_i \mid\right)\right] \qquad \text{(A1)}$$

where ψ_{ji} is the value of exemplar node j on dimension i, x_i is the activation of input feature dimension i, c is a constant called the *specificity* that controls overall attention, and α_i is the attention strength for dimension i. In ALCOVE, the attention strengths essentially stretch and shrink dimensions.

The activity of the exemplar nodes is fed forward to the third layer, the category layer, whose nodes correspond to the categories being learned. The strength of association between category node k and exemplar node j is denoted by w_{kj}. The activation of category node k is then computed as the sum of weighted activations of all exemplars, or

$$y_k = \sum_j w_{kj} h_j \qquad \text{(A2)}$$

The probability that a particular stimulus is classified as category k, denoted as $P(K)$, is assumed equal to the activity of category k relative to the summed activations of all categories, where the activations are first transformed by the exponential function (Kruschke, 1992):

$$P(K) = \frac{\exp(\phi y_k)}{\sum_k \exp(\phi y_k)} \qquad \text{(A3)}$$

where ϕ is a real-value mapping constant that controls decisiveness of classification responses.

The standard version of ALCOVE (Kruschke, 1992) uses a form of gradient descent for updating weights. The error term is defined as the sum of squared differences between the desired and the predicted outputs:

$$E = \tfrac{1}{2}\sum_k (t_k - y_k)^2 \qquad \text{(A4)}$$

Partial derivatives of the error function with respect to the association weights w_{kj} and the attention strengths α_i are used to compute the weight update:

$$\Delta w_{kj} = \frac{\partial E}{\partial w_{kj}} = \lambda_w (t_k - y_k) h_j \qquad \text{(A5)}$$

$$\Delta \alpha_i = \frac{\partial E}{\partial \alpha_i} = -\lambda_\alpha \sum_j \left[\sum_k (t_k - y_k) w_{kj}\right] h_j \cdot c \mid \psi_{ji} - x_i \mid \qquad \text{(A6)}$$

where λ_w and λ_α are the learning rates for the association weights and attention strengths, respectively. It is this gradient-based learning method that we propose to replace with the stochastic learning method.

Stochastic Learning Algorithms

Here, we have evaluated two applications of stochastic learning to ALCOVE: one version in which we implement stochastic learning for adjusting both dimensional attention weights and the network association weights (ALCOVE-CSL, for "completely stochastic learning"), and one in which stochastic learning is used to adjust only the dimension attention weights in ALCOVE (ALCOVE-SAL, for "stochastic attention learning ").

ALCOVE-CSL algorithm

STEP 0: Initialize:
Problem specific parameters (T^0,υ)
T^0 : initial temperature.
υ : temperature decreasing rate

Association weights w_{kj},
$w_{kj} \sim U(MIN_w, MAX_w)$. where MIN_w and MAX_w are minimum and maximum values for **w**.

Attention strengths α_i,
$\alpha_i = 1/I^*(MAX_\alpha - MIN_\alpha)+MIN_\alpha$, for all $i = 1\ldots I$, where I is the number of feature dimensions.

Exemplar ψ_{ji}

$\psi_{ji} = x_{*i}$, where subscript * indicates unique patterns.

STEP 1: Calculate output activations

$$O_k = \sum_J w_{kj} \exp\left[-c(\sum_I \alpha_i | \psi_{ji} - x_i |)\right] \qquad \text{(S1)}$$

STEP 2: Calculate fit index for one training block:

$$f(w^\tau, \alpha^\tau) = \sum_{n=1}^{N}\sum_{k=1}^{K}(d_k - O_k)^2 \qquad \text{(S2)}$$

where K = # categories, N = # input in one block, d_k is a desired output for category node k at time τ.

STEP 3: Accept all weight and attention parameters ($\boldsymbol{\alpha}$ & $\boldsymbol{w}$) at the probability of:

$$P(w^\tau, \alpha^\tau | w^s, \alpha^s) = \left\{1 + \exp\left(\frac{f(w^\tau, \alpha^\tau) - f(w^s, \alpha^s)}{T^\tau}\right)\right\}^{-1} \qquad \text{(S3)}$$

if $f(w^\tau, \alpha^\tau) > f(w^s, \alpha^s)$, or 1 otherwise, where $f(w^s, \alpha^s)$ is the fit index for the previously accepted parameter set.

STEP 4: Reduce temperature:

$$T^t = T^o \exp(-\upsilon \cdot t) \qquad \text{(S4)}$$

STEP 5: Generate new $\boldsymbol{w}$ and $\boldsymbol{\alpha}$

$$w_{kj}^\tau = w_{kj}^s + y(MAX(w) - MIN(w)) \qquad \text{(S5)}$$

$$\alpha_i^\tau = \alpha_i^s + y(MAX(\alpha) - MIN(\alpha)) \qquad \text{(S6)}$$

where

$$y = \text{sgn}(u - 0.5)T\left[(1 + 1/T)^{|2u-1|} - 1\right] \qquad \text{(S7)}$$

Here, u indicates a random number drawn from the Uniform distribution. S7 draws random numbers from the Cauchy distribution (Ingber, 1989).

REPEAT STEPS 1~5 until stopping criterion is met

ALCOVE with SAL

Stochastic Attention Learning (SAL) incorporates both gradient and stochastic methods for learning. In particular, SAL updates its association weights using a gradient method (Equation A5), and attention strengths by the stochastic learning method (Equations S6 & S7).

Since SAL incorporates gradient learning for its association weights, the badness-of-fit index at time t is often less than that at time t-1, even with an "inappropriate" random movement in attention allocation. In other words, the present algorithm as described in the previous section would accept many useless moves for attention distribution, particularly in the early stages of learning. However, this seems both unnecessary and inefficient. Thus, we modified SAL to include a threshold parameter ζ, which controls for the probability of accepting new attention weight values, to make the model accept only moves that satisfy a prespecified criterion (i.e., above the threshold) for categorization accuracy.

Thus for SAL, Equation S3 becomes

$$P(\alpha^\tau | \alpha^s) = \left\{1 + \exp\left(\frac{f(\alpha^\tau) + \zeta - f(\alpha^s)}{T^\tau}\right)\right\}^{-1} \qquad \text{(S3')}$$

if $\{f(\alpha^\tau) + \zeta\} > f(\alpha^s)$, or 1 otherwise, where

$$f(\alpha^\tau) = \sum_{n=1}^{N}\sum_{k=1}^{K}(d_k - O_k)^2 .$$

Simulations

In order to evaluate the abilities of the stochastic learning algorithms to account for human data on classification learning, we conducted three simulation studies. In Simulation 1, we tested if the stochastic learning can replicate rapid changes in attention allocation in category learning for a single simulated subject. In Simulation 2, we simulated the results of a recent empirical study on classification learning (Matsuka, 2002; Matsuka & Corter, 2003) to see if the algorithm can reproduce individual differences in attention processes. In Simulation 3, we examined if the algorithms accurately reproduce aggregated learning curves.

A total of four different ALCOVE-based models are involved in the simulations reported below. The main comparison we are interested in is to compare the performance of standard ALCOVE with ALCOVE incorporating stochastic attention learning (ALCOVE-SAL), and ALCOVE incorporating completely stochastic learning (ALCOVE-CSL). However, for Simulation 2, we also investigate if individual differences could be otherwise accounted for within standard gradient-based ALCOVE. To do this, we also tried another way (besides stochastic learning) of handling random individual differences within the ALCOVE model, namely by randomly varying individual learning rates. This version of standard gradient-learning ALCOVE is referred to here as ALCOVE-RLR.

Simulation 1: Rapid attention shifts

In the present simulation study, we examined if stochastic learning algorithms can replicate rapid changes in attention allocation as observed in empirical studies (Macho 1997, Rehder & Hoffman, 2003). Here, we used the simplest stimulus structure (T1) of Shepard, Hovland and Jenkins' stimulus sets (1961). Table 1 shows schematic representation of the stimuli used in the present simulation (i.e., T1).

Table 1: Schematic representation of stimulus set used in Simulations 1 and 3.

	Stimulus			Category					
	D1	D2	D3	T1	T2	T3	T4	T5	T6
1	1	1	1	A	B	A	A	A	A
2	1	1	0	A	B	A	A	A	B
3	1	0	1	A	A	A	A	A	B
4	1	0	0	A	A	B	B	B	A
5	0	1	1	B	A	B	A	B	B
6	0	1	0	B	A	A	B	B	A
7	0	0	1	B	B	B	B	B	A
8	0	0	0	B	B	B	B	A	B

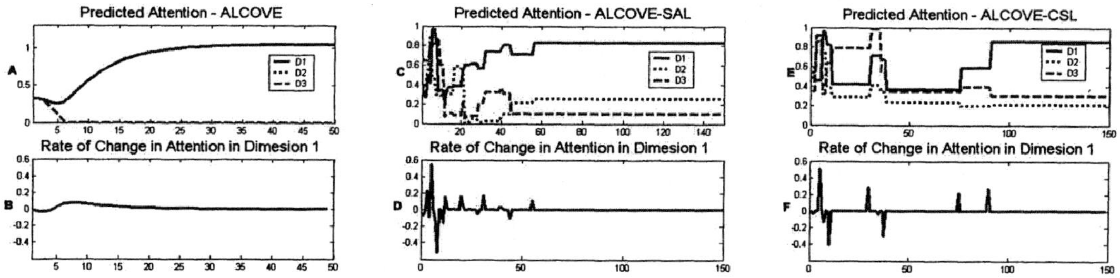

Figure 1: The results of Simulation 1. A: predicted attention allocation to the three feature dimension by standard ALCOVE; B: predicted rate of change in attention allocated to Dimension 1 by ALCOVE; C & E: predictions by ALCOVE-SAL; E & F by ALCOVE-CSL.

Simulation Method: Three ALCOVE-type models of category learning were evaluated in the present simulation studies, namely the standard ALCOVE, ALCOVE-SAL, and ALCOVE-CSL. They were run in a simulated training procedure to learn the correct classification responses. ALCOVE was run for 50 blocks of training, where each block consisted of a complete set of the training instances, while ALCOVE-SAL and ALCOVE-CSL were run for 150 blocks.

For each model, the gradient or rate of change in attention allocated to Dimension 1 was calculated by subtracting the amount of attention allocated to Dimension1 at time *t*-1 from that of time *t*. This measure was used as an index of how rapidly attention distributions changed.

Results: The results of one simulated subject for each model are shown in Figure 1. Note that only accepted attention strengths were graphed for ALCOVE with SAL and CSL. All three models learned to allocate the highest amount of attention to Dimension 1 and learned to ignore or pay less attention to Dimensions 2 and 3. The rate of attention change for ALCOVE was very smooth and its magnitude was smaller than those of ALCOVE-SAL and ALCOVE-CSL. ALCOVE-SAL and ALCOVE-CSL produced oscillating graphs with higher magnitudes of change. These results suggest that our proposed stochastic learning algorithms are capable of replicating rapid changes in attention allocation, while ALCOVE with gradient-based learning is not.

Simulation 2: Individual Differences

In this simulation study, we examined how the models account for individual differences in attention learning. To do this, we simulated the results of an empirical study on classification learning, Study 2 of Matsuka (2002). In this study, there were two perfectly redundant feature dimensions, Dimension 1 & Dimension 2 (see Table 2), and those two dimensions are also perfectly correlated with the category membership. Thus, information from only one of the two correlated dimensions was necessary for perfect categorization performance.

Besides classification accuracy, data on the amount of attention allocated to each feature dimension was collected in the empirical study. The measures of attention used were based on feature viewing time, as measured in a MouseLab-type interface (Bettman, Johnson, Luce & Payne 1993).

To summarize the empirical results that we are trying to simulate, 13 out of 14 subjects were able to categorize the stimuli almost perfectly (Figure 2, left top panel), and on average subjects paid attention to both of the correlated dimensions approximately equally (Figure 2, left middle panel). However, the examination of aggregated data can be misleading. When Matsuka and Corter (2003) analyzed attention data at an individual-level, they found that many subjects tended to pay attention primarily to only one of the two correlated dimensions, particularly in the late learning blocks (Figure 2, left bottom panel). This suggests that the participants utilized only the minimal necessary information for this task.

Table 2: Stimulus structure used in Study 2 of Matsuka (2002)

Category	Dim1	Dim2	Dim3	Dim4
A	1*	1*	3	4
A	1*	1*	4	1
A	1*	1*	1	2
B	2*	2*	2	1
B	2*	2*	3	2
B	2*	2*	4	3
C	3*	3*	1	3
C	3*	3*	2	4
C	3*	3*	3	1
D	4*	4*	4	2
D	4*	4*	2	3
D	4*	4*	1	4

Simulation method: The final parameter values used for each model were chosen by a simulated annealing method (Ingber, 1989; Matsuka, Corter & Markman, 2003) to minimize the objective function (i.e., sum of squared error) in reproducing the classification accuracies by human subjects. The four models were run in a simulated training procedure to learn the correct classification responses for the stimuli of the experiment. ALCOVE and ALCOVE-RLR were run for 48 blocks of training, where each block consisted of a complete set of the training instances, while ALCOVE-SAL and ALCOVE-CSL were run for 480 blocks. For each model, the final results are based on 50 replications

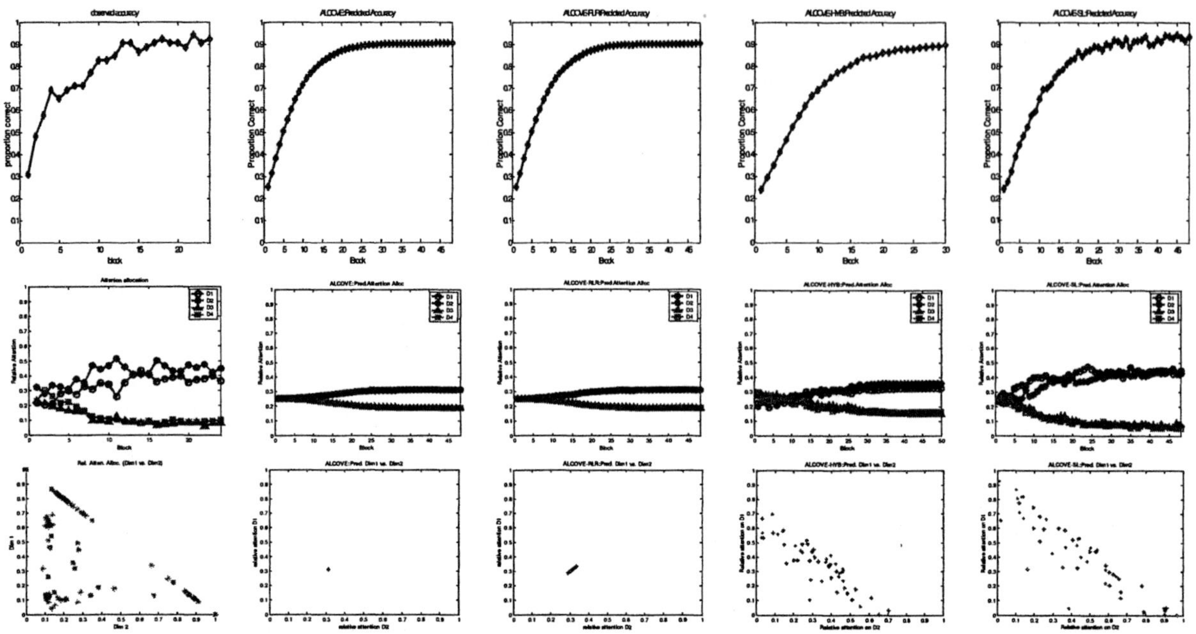

Figure 2: Results of Simulation 2. The left column shows observed learning curve (top panel), observed dimensional attention allocation (middle panel), observed attention allocation to Dimension 1 (Y-axis) and 2 (X-axis) in the last half of the training block (bottom panel). The predictions by ALCOVE, ALCOVE-RLR, ALCOVE-SAL, and ALCOVE-CSL are shown in the 2nd, 3rd, 4th and 5th column, respectively.

Results: Figure 2 summarizes the findings from the simulation study. The left column of this figure shows the empirical data from Study 2 of Matsuka (2002), including the learning curve for overall classification accuracy (top panel), the attention learning curves (middle panel), and attention allocated to Dimensions 1 and 2 (the redundant diagnostic dimensions) in the late training blocks (bottom panel).

Results for each of the models are shown in the remaining rows. Standard ALCOVE learns the classification task quickly, and allocates attention equally to the two diagnostic dimensions, but shows almost no inter-individual variability in attention, no intra-individual variability in the amount of attention allocated to Dimension 1 and 2 (Figure 2, Second row). ALCOVE-RLR, using random learning rate for attention strengths, but still using the standard gradient learning algorithm, shows some inter-individual variability, but virtually no intra-individual variability. ALCOVE-SAL, the version of ALCOVE modified to incorporate stochastic learning of attention weights, showed a less steep learning curve (like the human subjects), and more variability among subjects in attention allocation (again more closely resembling the empirical data). ALCOVE-CSL gave similar results, but exhibited some minor differences from ALCOVE-SAL (including more separation of the diagnostic and non-diagnostic dimensions). Note that the two versions of the stochastic learning models are able to replicate the observed learning curves.

Simulation 3. Replication of Nosofsky et al. 1994

So far, we have shown that our proposed stochastic learning algorithms are successful for reproducing individual-level data (i.e. rapid change & individual differences in attention processes). However, we merely tested the algorithms' capabilities of reproducing aggregated data. In the present simulation study, we simulated a classical study of categorization (Shepard et al. 1961) which is often used as a benchmarking stimulus set (e.g. Nosofsky, Gluck, Palmeri, McKinley & Glauthier, 1994). The stimulus structures are shown in Table 1. The results of previous empirical studies showed that Type 1 (T1) was the easiest to learn to classify, followed by T2, T3, T4, T5, and T6 being the most difficult. More precisely, Nosofsky et al. (1994) showed that the numbers of error made thus difficulties for those stimulus structures were significant except T3, T4, and T5.

Simulation method: In the present study, we tested only ALCOVE-CSL and ALCOVE-SAL, as the standard ALCOVE has previously been shown to be able to replicate the observed learning curves (Nosofsky et al. 1994). The two models were run in a simulated training procedure to learn the correct classification responses for the stimuli. ALCOVE-CSL was run for 250 blocks of training, where each block consisted of a complete set of the training instances, while ALCOVE-SAL was run for 150 blocks. For each model, the final results are based on 500 replications.

Results: Figure 3 summarizes the findings from the Simulation 3. Both ALCOVE-CSL and SAL were able to reproduce the order of difficulty successfully. That is ALCOVE-CSL and SAL find T1 to be the easiest, followed by T2, T3, T4, T5, and T6.

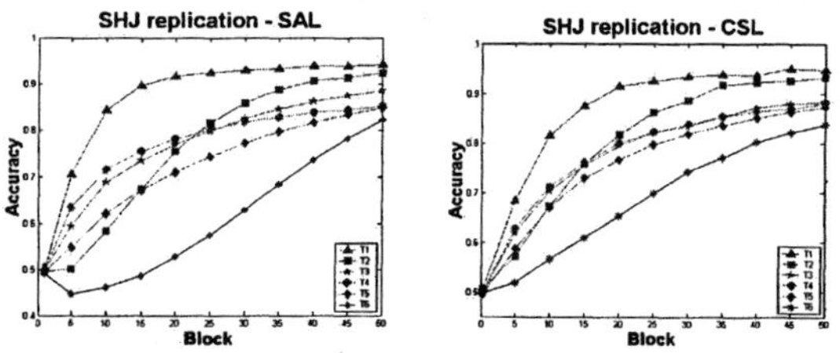

Figure 3. Results of Simulation 3. Both ALCOVE-SAL and ALCOVE-CSL were able to reproduce the order of difficulty successfully

Discussion and Conclusion

We have investigated the possibility of using stochastic learning rather than gradient-based methods in neural network models of human classification learning. In the present simulations we have explored the effectiveness of this method in several variants of the ALCOVE model (Kruschke, 1992). Our main goals were to see if stochastic learning algorithms 1) were able to replicate rapid change in attention processes, and 2) offered a better account of individual differences in final distribution of attention, particularly distribution of attention to the two perfectly correlated dimensions. The simulation studies showed that the new algorithms are satisfactory in these regards.

Stochastic learning algorithms have other desirable properties as well. It could be argued that stochastic learning may be more psychologically plausible than gradient-based methods, which require more mental effort and assume that optimal adjustments are made to the vector of parameters on each trial. One caveat to these results is that the stochastic learning algorithms learn more slowly than the standard gradient methods in categorization tasks with relatively small (both number of stimulus feature dimensions and number of unique exemplars) and well-defined stimulus sets that are usually used in laboratory experiments. However, for more realistic category learning involving complex category structures and/or stimuli with many feature dimensions, stochastic learning may be able to learn faster than ordinary gradient type learning.

Distribution of random numbers: In the present research, the random moves in parameter space were drawn from the Cauchy distribution, mainly because its fatter tails are more likely than the Gaussian distribution to produce rapid and or large shift in attention allocation, which has been reported by some empirical studies. However, with a proper experimenter-defined parameter setting (e.g., initial temperature & temperature decreasing rate), similar results in attentional shifts might have been achieved with some other distributions, including normal, rectangular, skewed, or multi-modal distribution. In addition, we assumed the "temperature" decreases as a function of time or the number of training blocks. But, it may not capture realistic learning mechanisms. Rather, there may be more descriptively valid factor(s) for the annealing schedule, such as classification accuracy. Further simulation and empirical studies seem useful for investigating descriptive validity of stochastic optimization methods as a model of human category learning.

Acknowledgments

This study is partially supported by The National Science Foundation (the award EIA-0205178 granted to Stephan Jose Hanson).

References

Bettman, J.R., Johnson, E.J., Luce, M.F., Payne, J.W. (1993). Correlation, conflict, and Choice. Journal of Experimental Psychology: Learning, Memory, and Cognition, 19, 931-951.

Ingber, L. (1998). Very fast simulated annealing. Journal of Mathematical Modelling, 12: 967-973.

Kruschke, J. E. (1992). ALCOVE: An exemplar-based connectionist model of category learning, Psychological Review, 99. 22-44.

Kruschke, J.K., & Johansen, M. K. (1999). A model of probabilistic category learning. Journal of Experimental Psychology: Learning, Memory, and Cognition, 25, 1083-1119.

Love, B.C. & Medin, D.L. (1998). SUSTAIN: A model of human category learning. Proceeding of the Fifteenth National Conference on AI (AAAI-98), 671-676.

Macho, S. (1997). Effect of relevance shifts in category acquisition: A test of neural networks. Journal of Experimental Psychology: Learning, Memory, and Cognition, 23, 30-53.

Matsuka, T. (2002). Attention processes in computational models of category learning. Unpublished doctoral dissertation. Columbia University, New York, NY.

Matsuka, T. & Corter, J. E. (2003). Empirical studies on attention processes in category learning. Poster presented at 44th Annual Meeting of the Psychonomic Society. Vancouver, BC, Canada.

Matsuka, T., Corter, J. E. & Markman, A. B. (2003). Allocation of attention in neural network models of categorization. Under review

Nosofsky, R.M. (1986). Attention, similarity and the identification – categorization relationship. Journal of Experimental Psychology: General, 115, 39-57

Nosofsky, R.M., Gluck, M.A., Palmeri, T.J., McKinley, S.C., & Glauthier, P. (1994). Comparing models of rule-based classification learning: A replication and extension of Shepard, Hovland, and Jenkins (1961). Memory and Cognition, 22, 352-369.

Rehder, B. & Hoffman, A. B. (2003). Eyetracking and selective attention in category learning. Proceedings of the 25th Annual Meeting of the Cognitive Science Society, Boston, 2003.

Shepard, R.N., Hovland, C.L., & Jenkins, H.M. (1961). Learning and memorization of classification. Psychological Monograph, 75 (13).

A Computational Analysis Model for Complex Open-ended Analogical Retrieval

Junya Morita (morita@cog.human.nagoya-u.ac.jp)
Graduate School of Human Informatics
Nagoya University, Furo-cho, Chigusa-ku, Nagoya City, Japan

Kazuhisa Miwa (miwa@cog.human.nagoya-u.ac.jp)
Graduate School of Information Science
Nagoya University, Furo-cho, Chigusa-ku, Nagoya City, Japan

Abstract

Although many computational models have successfully simulated the results of controlled psychological experiments, few researchers have attempted to apply their models to complex, realistic phenomena. In this study, MAC/FAC ("many are called, but few are chosen"), which models two stages of analogical reasoning (Forbus, Gentner, & Law, 1995), was applied to our experimental data. In our experiment, subjects were presented a cue story and asked to retrieve cases learned from everyday life. Next they rated the inferential soundness (goodness as an analogy) of each retrieved case. For each retrieved case, we used the algorithms of the MAC/FAC to compute two kinds of similarity scores: content vectors and structural evaluation scores. As a result, the computed content vectors explained the overall retrieval of cases well, whereas the structural evaluation scores had a strong relation to the rated scores. These results support the MAC/FAC's theoretical assumption – different similarities are involved in the two stages of analogical reasoning.

Introduction

In the past, many cognitive psychologists have conducted controlled experiments that have lead to scientific understanding of analogical reasoning. The impetus behind these studies might have been the development of computational models (Falkenhainer, Forbus, & Gentner, 1989; Forbus, Gentner, & Law, 1994). Such models have been used to simulate the results of psychological experiments, and they have also guided subsequent studies.

However, we think the above studies are limited because most of the simulated psychological data has been obtained in closed laboratory situations. Thus it is unclear whether the models are suitable for simulating realistic analogical reasoning. Our aim here is to apply a computational model of analogy to more complex and realistic psychological data. Prior to presenting our study, we review previous studies on analogical reasoning.

Framework for analogy research

Analogical reasoning involves two representations: the base and the target. The base is a past case that one is familiar with. The target is a novel case that one is usually less familiar with. The process of analogical reasoning is comprised of two main components: the retrieval of the base and the mapping from the base to the target; the analogy process is guided by similarities between the base and the target. Using propositional representations (predicate-argument formalism), Gentner (1983) distinguished three types of correspondence between the base and the target.

Correspondence of *attributes*: e.g., The sun is round and yellow → The orange is round and yellow [sun (round) sun (yellow) → orange (round) orange (yellow)].

Correspondence of *first-order relations*: e.g., The planets revolve around the sun. → The electrons revolve around the atom [revolve-around (planet, solar) → revolve-around (electron, atom)].

Correspondence of *higher-order relations*: e.g., Because the sun attracts the planets, the planets revolve around the sun. → Because the atom attracts the electrons, the electrons revolve around the atom [cause (attract (solar, planet), revolve-around (planet, solar)) → cause (attract (atom, electron), revolve-around (electron, atom))].

The above discrimination was based on the types of predicates. The attribute is a predicate type that takes a single argument. On the other hand, the first-order and higher-order relations are predicate types that take multiple arguments. There is no depth in the former, but there is in the latter (Gentner, 1983). Therefore, in the studies on analogy, the former is often called surface similarity, while the latter is called structural similarity.

Computational Model of Analogy

Forbus, Gentner, & Law (1995) developed a computational model called MAC/FAC ("many are called, but few are chosen") that aims to simulate the two stages of analogical reasoning: retrieval and evaluation.

In the first stage of the MAC/FAC model called the MAC stage, several candidate bases are recalled from the memory pool. For quick access to the large number of cases in memory, the MAC stage implements a computationally cheap process. Using content vectors (CVectors), which are simple lists of predicates, each memory item is matched with the target. The outputs of the MAC stage are memory items that have a high dot product of CVectors.

CVectors indicate which predicates are contained and how frequently those predicates appear in the representation. Because CVectors do not distinguish between attribute and relation, the dot product is influenced by the commonalities of the attribute as well as by the commonalities of the relation. Also, since the CVectors have nonstructural features, the dot products tend to overesti-

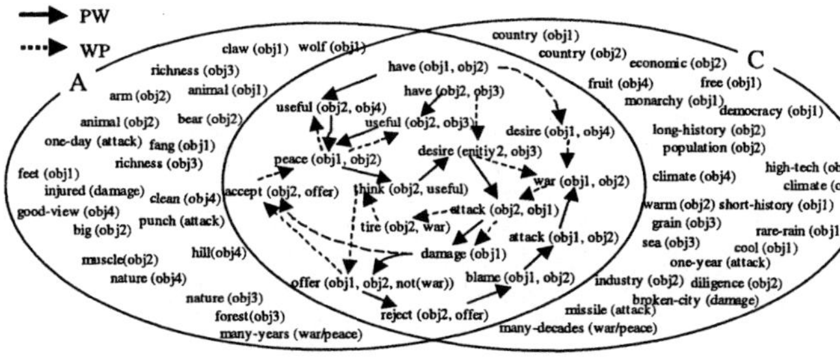

Figure 1: Propositions contained in the target stories.

Table 1: Predicate types shared by target stories.

	A/PW	A/WP	C/PW	C/WP
A/PW	OA+FOR+HOR	OA+FOR	FOR+HOR	FOR
A/WP		OA+FOR+HOR	FOR	FOR+HOR
C/PW			OA+FOR+HOR	OA+FOR
C/WP				OA+FOR+HOR

Note. OA=object attributes; FOR=first-order relation; HOR=higher-order relation.

Table 2: CVectors/SESs between target stories.

	A/PW	A/WP	C/PW	C/WP
A/PW	1.00/28.8	1.00/12.2	0.39/28.8	0.39/12.2
A/WP		1.00/27.9	0.39/12.2	0.39/27.9
C/PW			1.00/28.8	1.00/12.2
C/WP				1.00/27.9

mate structurally dissimilar representations. For example, two propositions ["cause (attract, revolve-around)" and "cause (revolve-around, attract)"] have the same CVector [cause = 1, attract = 1, revolve-around = 1]. Also, the proposition "war (dog1, dog2)" [war = 1, animal = 2] is more similar to a proposition "war (wolf1, wolf2, wolf3)" [war = 1, animal =3] than "war (wolf1, wolf2)" [war = 1, animal =2].

Thus, the candidate bases are further evaluated at the next step called the FAC stage. The FAC stage is modeled by using SME, the *structure-mapping engine* (Falkenhainer, Forbus, & Gentner, 1989). The SME includes the following two sub processes.

Construction of *a local match*, which is a pair of predicates shared by the base and the target. A numerical weight is assigned to each local match according to its relational consistency.

Construction of a *global match*, which is an overall mapping from the base to the target. The global match is constructed by connecting structurally consistent combinations of local matches, satisfying the constraint of *one to one mapping*, which means that the same elements in the base can not match multiple items in the target, or vice versa. For the constructed global match, the structural evaluation score (SES) was computed by summing the numerical weights of the local matches that are included in the global match. The outputs of the FAC stage are the cases that have a high SES.

To summarize, the MAC/FAC assumed a distinction between the two stages of analogical reasoning guided by the different similarities.

Psychological study of analogy

Gentner, Ratterman, & Forbus (1993) provided psychological evidence for the MAC/FAC's assumption. They used story sets manipulated by types of similarity (experiment 2). Each story set was comprised of a base story and four target stories. The target stories shared various predicates with the base story: a surface similarity (SS) that shared attribute and first-order relations, an analogy (AN) that shared first-order and higher-order relations, a literal similarity (LS) that shared all types of predicates, and a story that shared the first-order relations (FOR).

In their experiment, subjects first learned the base story. About a week later, they were presented with the target stories as retrieval cues to help them recall the base stories. Finally they rated the inferential soundness (goodness as an analogy) for each pair of stories. The results implied that different similarities are involved in the process of analogical reasoning. The subjects more often retrieved the base story when they read surface similar stories (LS, SS), rather than when they read structurally similar stories (AN, FOR). However, when evaluating soundness, they rated structurally similar stories (LS, AN) higher than surface similar stories (SS, FOR).

Gentner, Ratterman, & Forbus applied their computational model to the above result. First, they input the base and the targets into the SME and computed the

SES between stories. The results showed that the SME is a good predictor of subjective soundness, suggesting that the story pairs that shared higher-order relations have a higher SES than the other story pairs that did not share the higher-order relations (LS, AN > SS, FOR). In addition, they constructed a memory pool that contained the four target stories (LS, AN, SS, and FOR) and input the base story into the MAC/FAC model as a retrieval cue. As a result, the order of the retrieval rate by the MAC/FAC became LS>SS>AN>FOR. This pattern of retrieval is consistent with the result of human retrieval.

The results of their simulation support the assumed algorithms of the MAC/FAC. However, we think there are limitations to their investigation because they have only studied cases created by the researchers themselves. In real-world situations, individuals make analogy from everyday experience. To apply the model to realistic problems, it is necessary to investigate analogical reasoning using cases that the subjects experienced in everyday life. Therefore, we tested the validity of the algorithms of the MAC/FAC model by applying the model to cases that subjects learned from their everyday life.

Experiment

Materials

Unlike Gentner, Ratterman, & Forbus's study, we did not prepare base stories. Without learning any stories, subjects were presented with a target story as a retrieval cue. The subjects were asked to report remembered cases that came to mind while reading the target story, which consisted of about 600 Japanese characters.

The surface and structural features of the target stories were manipulated. As surface features, a set of attributes related to *animals* (A) and a set of attributes related to *countries* (C) were chosen. As structural features, a story whose plot involved a transition from *peace to war* (PW) and another whose plot was a transition from *war to peace* (WP) were created. Combining the surface and structural features, four types of target stories were prepared: A/PW, A/WP, C/PW, and C/WP.

Figure 1 illustrates the propositions converted from the texts in the target stories. Each of them was included in either set A or set C. Two complements $[(A \cap \bar{C})$ and $(\bar{A} \cap C)]$ contain object attributes, and an intersection $(A \cap C)$ includes first-order relations of two objects. Each of the first-order relations was connected by two types of higher-order relations (PW/WP), represented by two types of arrows (solid/dotted). That is, as the materials in Gentner, Ratterman, & Forbus's study, the four target stories shared first-order relations, varying the attributes and the higher-order relations. Table 1 summarizes the interrelations between the target stories.

Additionally, the similarity scores between the target stories, presented in Table 2, were calculated based on the algorithms of the MAC/FAC. The CVectors were computed as a dot product of the predicate lists, which includes the attributes and the first-order relations as components[1]. The SES was computed by using the SME model with an analogy rule that constructs structurally consistent mapping without matching attributes[2]. The scores shown in Table 2 are consistent with the manipulation of the target stories. The CVectors of the pairs sharing attributes are greater than those of the pairs sharing no attributes, and the SESs of the pairs sharing higher-order relations are greater than those of the pairs sharing no higher-order relations.

Subjects' descriptions	Converted propositions
The story about two tigers. In a forest, two tigers lived. Each of them has a turf. And they battled each other for the turf. One day, an animal that lived in the forest persuaded one of the tigers to stop fighting. After this persuasion, the relationship between the two tigers became peaceful.	((animal tiger1) :name prop3) ((animal tiger1) :name animal2) ((animal animal1) :name animal3) ((have tiger1 turf1) :name have1) ((have tiger2 turf2) :name have2) ((desire tiger1 turf2) :name desire1) ((desire tiger2 turf1) :name desire2) ((war tiger1 tiger2) :name war1) ((and desire1 desire2) :name and1) ((cause and1 war1) :name cause1) ((not war1) :name not1) ((offer animal1 tiger1 not1) :name off) ((accept tiger1 off) :name accept1) ((cause offer1 accept1) :name cause2) ((cause war1 off) :name cause3) ((peace tiger1 tiger2) :name peace1) ((cause accept1 peace1) :name cause4) ((many-tree turf1) :name prop1) ((many-tree turf2) :name prop2)

Figure 2: Examples of coding.

Furthermore, we conducted a pilot experiment to test whether the above SESs are consistent with human feelings of soundness. After the subjects (n = 8) received the four target stories, they rated the inferential soundness of each pair on a 1 ("low") - 5 ("high") scale. As in Gentner, Rattermann, & Forbus's study, soundness was explained as "the degree to which inferences from one story would hold for the other." The results supported the validity of the SME's evaluation scores. Seven of eight subjects judged the structurally similar pairs (A/PW vs. C/PW and A/WP vs. C/WP) as having higher inferential soundness than the other pairs (A/PW vs. A/WP, CPW vs. C/WP, A/PW vs. C/WP, and A/WP vs. CPW).

Participants

Thirty-three undergraduate and graduate students participated in the experiment. They were divided into four groups varying target stories: a group presented with A/PW (n = 8), a group with A/WP (n = 9), a group with C/PW (n = 8), and a group with C/WP (n = 8).

[1]The CVectors were normalized to unit vectors, as Forbus, Gentner, & Law did.

[2]Forbus, Gentner, & Law (1995) computed the SES using a literal-similarity rule that mapped all types of predicates. However, their simulation showed that the CVector predicted results of human retrieval relatively well, and the SME with the analogy rule predicted the soundness evaluation well. Thus, we treated the CVector as a matcher for the initial retrieval, and the SME with the analogy rule as a matcher for the evaluation of soundness.

Procedure

The subjects participated in the experiment individually or in groups of two to four. The experiment was divided into the following three phases.

Retrieval phase: The subjects were presented with one of the four target stories. Then they were told that "while reading the presented story, you should write out any cases from your everyday life that come to mind." After the instructions, they wrote down for twenty minutes any remembered cases.

Evaluation phase: Following completion of the retrieval phase, the subjects rated the soundness of the match between each retrieved case and the presented target story on a 1 – 5 scale.

Explanation phase: Finally, the subjects were asked to explain the retrieved cases in as much detail as possible.

Applying MAC/FAC to the Data

We obtained two types of data from the experiment: cases retrieved by the subject, and the soundness scores rated by the subjects themselves. These two types of data correspond to the two stages of the MAC/FAC model. We directly applied the model to the data to test the correspondence between the data types and the MAC/FAC stages.

First, all of the retrieved cases were converted to propositional representations. The subjects' descriptions were segmented by the appearance of a predicate, and then a coder judged whether each segmented sentence could be represented as a proposition by using predicates contained in the target stories. When possible, a proposition was constructed by complementing proper arguments. While working, the coder used the list that defines the correspondence between the predicates and each word included in the subjects' description. Figure 2 illustrates examples of the coding.

For the converted representations, two scores of similarity (CVector/SES) were computed using the MAC/FAC model. We also computed the two scores of similarity of the retrieved cases with the *target story that was presented to each subject*, as well as the two scores of similarity with the *target story that was not presented to each subject.* The MAC/FAC predicts that the overall average CVector with the presented target story will be higher than the overall average CVector with the target story that was not presented, and the SES with the presented target story will be closely related to the rated scores of soundness, rather than the SES with the target story that was not presented.

Results and Discussion

This section shows the results of applying MAC/FAC to our data[3]. First, we illustrate the overall similarity scores for the experimental groups to test the assumption at the MAC stage (an initial retrieval stage using the

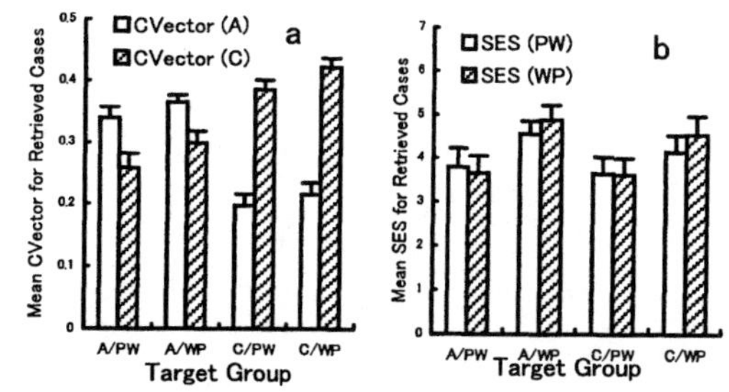

Figure 3: (a) Mean CVector for four groups. (b) Mean SES for four groups. *Note.* Error bars represent one standard error of mean.

CVectors). Next we examine the relationship between the similarity scores and the soundness scores to test the assumption of the FAC stage, which is an evaluation stage using the SME model.

1. Test for the MAC stage

To test the assumption at the MAC stage, we computed four scores of similarity for each retrieved case:

CVector (A): The dot product of CVector of the retrieved case and the CVector of the propositional set A (represented as [animal = 2, wolf =1, bear = 1, forest =1 ... have = 2, desire = 2]).

CVector (C): The dot product of CVector of the retrieved case and the CVector of the propositional set C (represented as [country = 2, democracy = 1, monarchy = 1 ... have = 2, desire = 2]).

SES (PW): The SES computed by inputting the retrieved case and the structural feature PW (solid arrows in Figure 1) into the SME.

SES (WP): The SES computed by inputting the retrieved case and the structural feature WP (dotted arrows in Figure 1) into the SME.

We conducted two analyses of variance to investigate the interaction between the above similarity scores and the experimental groups. If the assumption of the MAC stage is reasonable, the CVector (A) should be higher than the CVector (C) in the subjects who were presented with the surface feature A, and the CVector (C) should be higher than the CVector (A) in the subjects who were presented with the surface feature C. On the other hand, it is predicted that the ANOVA using the SES will not reveal clear difference between the similarity scores.

The CVector as a matching algorithm at the MAC stage Figure 3a shows the mean CVector for each group. A 2 × 2 × 2 surface features of target stories (between) × structural features of target stories (between) × types of CVector (within) ANOVA revealed a significant interaction between the surface features of the target stories and the types of CVector [$F(1, 262) = 206.77, p < .05$]. This indicates that the CVector (A) was higher than the CVector (C) in group A [$F(1, 262) = 30.26, p < .05$], and that the CVector (C) was higher than the CVector (A) in group C

[3]The total number of cases retrieved by the subjects was 266. There was no significant difference on the number of cases among the four experimental groups [$\chi^2(3) = 6.15, ns.$]. Thus, we treated each retrieved case as an individual datum for statistical tests.

Table 3: Types of correspondence to soundness scores.

	OA	FOR	HOR	1-1	r
CVector (presented)	yes	yes	no	no	0.23 ($p < .01$)
CVector (not presented)	no	yes	no	no	0.32 ($p < .01$)
SES (presented)	no	yes	yes	yes	0.44 ($p < .01$)
SES (not presented)	no	yes	no	yes	0.38 ($p < .01$)

Note. OA = object attributes; FOR = first-order relation; HOR = higher-order relation; 1-1 means constraint of one to one mapping.

$[F(1,262) = 220.07, p < .05]$. The retrieved cases were estimated to be more similar to the target story that was presented to the subject than the target story that was not presented to the subject. This result is consistent with the prediction, supporting the assumption of the MAC stage.

The SME as a matching algorithm at the MAC stage Figure 3b shows the mean SES for each group. A 2 × 2 × 2 surface features of target stories (between) × structural features of target stories (between) × types of SES (within) ANOVA revealed a significant interaction between the structural features of the target and the types of SES $[F(1,262) = 8.01, p < .05]$. However, a simple effect was significant only in group WP $[F(1,262) = 7.50, p < .05]$. There was no significant difference of types of SES in group PW $[F(1,262) = 1.60, ns.]$. Those results are not distinctive when compared with the results of the ANOVA using the CVectors. Therefore, it is suggested that the SME, which is assumed as a matching algorithm at the FAC stage, is insufficient to predict the initial retrieval.

2. Test for the FAC stage

To investigate the evaluation of the soundness as the psychological data that corresponds to the FAC stage, we treated the subject groups as counterbalance conditions and reduced the number of factors for the statistical test. Therefore, we computed the following four scores.

CVector (presented) was computed by combining the CVector (A) in group A and the CVector (C) in group C. This score indicates a rough estimate of the overlap between the retrieved cases and the target story that was presented to the subject.

CVector (not presented) was computed by combining the CVector (C) in group A and the CVector (A) in group C. This score indicates a rough estimate of the overlap between the retrieved cases and the target story that was not presented to the subject.

SES (presented) was computed by combining the SES (PW) in the group PW and the SES (WP) in the group WP. This score indicates the depth and breadth of the structural mapping from the retrieved cases to the target story that was presented to the subject.

SES (not presented) was computed by combining the SES (WP) in the group PW and the SES (PW) in the group WP. This score indicates the depth and breadth of the structural mapping from the retrieved cases to the target story that the subject did not receive.

Each score was interpreted as shown in Table 3. The CVector (presented) reflects the commonality of the object attributes (see the second column of Table 3). The SES (presented) reflects the commonality of the higher-order relations (see the fourth column of Table 3). All scores reflect the commonality of the first-order relations (see the third column of Table 3). In addition to the discrimination of predicate types, there is a difference concerning the 1-1 constraint (see the fifth column of Table 3). The CVectors are rough estimations of overlap, which may overestimate the one-to-many and many-to-one mappings. On the other hand, the SES is strictly calculated to satisfy the 1-1 constraint.

The sixth column of Table 3 shows the correlation coefficients between the soundness scores and the similarity scores. There are significant positive correlations in all of the scores, which might indicate that the soundness scores are influenced by the correspondence of the first-order relations, which all of the scores reflect.

To investigate the relation in more detail, we conducted two 5 × 2 ANOVA (soundness scores $1-5$ × presented/not presented), using the CVectors and the SESs as dependent measures.

The CVector as a matching algorithm at the FAC stage Figure 4a shows the mean CVector for each score of soundness $(1-5)$. A 5 × 2 soundness scores (between) × CVector types (within) ANOVA detected a main effect of soundness scores $[F(4,261) = 9.25, p < .05.]$, and a main effect of CVector types $[F(1,261) = 181.64, p < .05]$. There was no significant interaction between the soundness scores and the CVector types $[F(4,261) = 1.59, ns.]$.

The main effect of the CVector types indicates an advantage for the CVector (presented) over the CVector (not presented), regardless of the soundness scores. Therefore, it is suggested that the soundness scores have no relation to the commonality of attributes, which the CVector (presented) reflects (see the second column of Table 3).

Furthermore, multiple comparisons of the soundness scores revealed significant advantages of scores 2, 3, 4, and 5 over score 1, and significant advantages of score 4 over the scores 2 and 3 ($p < .05$), indicating that the CVector increased with the soundness scores. Thus, as shown in the correlation coefficients (see the sixth column of Table 3), the results support a positive relation

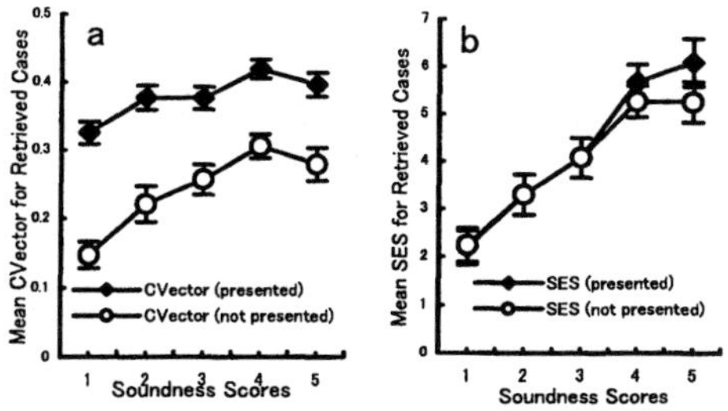

Figure 4: (a) Mean CVector for soundness scores. (b) Mean SES for soundness scores. *Note.* Error bars represent one standard error of mean.

between the soundness scores and the correspondence of the first-order relations (see the third column of Table 3).

The SME as a matching algorithm at the FAC stage Figure 4b shows the mean SES for each score of soundness $(1-5)$. A 5×2 soundness scores (between) $\times$ SES types (within) ANOVA revealed a significant interaction between the soundness scores and the SES types $[F(4,261) = 7.52, p < .05]$.

The simple effects of the soundness scores were significant for both the SES (presented) $[F(4,261) = 15.46, p < .05]$ and the SES (not presented) $[F(4,261) = 11.09, p < .05]$. Furthermore, multiple comparisons for both of the two scores revealed differences between score 1 and scores 3, 4, and 5, between the score 2 and the scores 4 and 5, and between the score 3 and the scores 4 and 5. Compared with the results of the analysis that used the CVectors, there are many significant differences between the soundness scores. Since the SES was calculated to satisfy the 1-1 constraint, unlike the CVector (see the fifth column of Table 3), this result supports the assumption of the FAC stage, which computes accurate structural mapping.

Finally, it was confirmed that the SES (presented) was higher than the SES (not presented) in the cases in which the subjects rated high soundness (scores 4 $[F(1,261) = 8.76, p < .05]$ and 5 $[F(1,261) = 35.73, p < .05]$), suggesting the SES (presented) is more strongly related to soundness than the SES (not presented). Since the difference between the two scores is in the higher-order relations (see the fourth column of Table 3), this suggests that computing a higher-order relation is needed for models of evaluation in analogical reasoning.

General Discussion

In summary, our results support the model's assumption. CVectors were good indicators of the overall difference in the initial retrieval between the experimental groups. On the other hand, the SES reflected a strong relation to the soundness scores. Importantly, the model's algorithm predicted the complex open-ended process of analogical reasoning.

Additionally, we would like to assert the psychological importance of our study. Our results are consistent with those of Gentner, Ratterman, & Forbus's study, which showed the different similarities involved in the two stages of analogical reasoning. However, recently, Blanchette & Dunbar (2000) showed that surface similarity has little effect on analogical retrieval in situations where subjects retrieve the cases that they have learned in everyday life. Their experiment asked subjects to generate analogies to the zero-deficit problem encountered by the Canadian government. As a result, the subjects generated few analogies that had surface features in common with the zero-deficit problem, but they generated many analogies that shared deep structures with the target.

Although there are many differences between our experiment and Blanchette & Dunbar's experiment, the difference in results might be explained by a difference in analysis. Blanchette & Dunbar analyzed generated analogies by categorizing their surface/structural features and comparing the frequencies of the categories. On the other hand, we computed the degree of similarity using a computational model. There is a clear difference of resolution between our analysis and Blanchette & Dunbar's analysis. Our detailed, theory-based analysis might detect the effect of surface similarity on the retrieval.

Thus, this paper defends not only the validity of the MAC/FAC model but also proposes a novel and useful method of investigating complex cognition. In the past, many researchers have analyzed complex, open-ended data using such methods as categorizations. There have been few attempts to apply a computational model directly to psychological data. However, as demonstrated in this paper, using a computational model for analysis could open a new way of firmly validating the connection between a theory and data. Without sufficient models, it is impossible to instantiate complex cognitive conceptual products such as surface similarity or structural similarity.

References

Blanchette, I., & Dunbar, K., (2000). How analogies are generated: The role of structural and superficial similarity. *Memory & Cognition*, *28*, 108–124.

Falkenhainer, B., Forbus, K. D., & Gentner, D., (1989). The Structure-Mapping Engine: Algorithm and Example. *Artificial Intelligence*, *41*, 1–63.

Forbus, K. D., Gentner, D., & Law, K. (1995). MAC/FAC: A model of similarity-based retrieval. *Cognitive Science*, *19*, 141–205.

Gentner, D. (1983). Structure-Mapping: A theoretical framework for analogy. *Cognitive Science*, *7*, 155–170.

Gentner, D., Rattermann, M., & Forbus, K. D. (1993). The role of similarity in transfer: Separating retrievability for inferential soundness. *Cognitive Psychology*, *25*, 524–575.

Thagard, P., Holyoak, K. J., Nelson, G., & Gochfeld, D. (1990). Analog retrieval by constraint satisfaction. *Artificial Intelligence*, *46*, 259-310.

Soar-RL: Integrating Reinforcement Learning with Soar

Shelley Nason (snason@umich.edu)
University of Michigan, 1101 Beal Ave.
Ann Arbor MI 48109-2110 USA

John E. Laird (laird@umich.edu)
University of Michigan, 1101 Beal Ave.
Ann Arbor MI 48109-2110 USA

Abstract

In this paper, we describe an architectural modification to Soar that gives a Soar agent the opportunity to learn statistical information about the past success of its actions and utilize this information when selecting an operator. This mechanism serves the same purpose as production utilities in ACT-R, but the implementation is more directly tied to the standard definition of the reinforcement learning (RL) problem. The paper explains our implementation, gives a rationale for adding an RL capability to Soar, and shows results for Soar-RL agents' performance on two tasks.

Introduction

The Soar architecture has been used extensively, both for developing AI applications and cognitive models. One of its strengths has been the ability to efficiently represent and use large bodies of symbolic knowledge to solve a wide variety of problems using many different methods. It dynamically combines available knowledge for decision-making, and can dynamically create subgoals whenever the knowledge for a decision is incomplete or inconsistent. Soar can also compile the problem solving in subgoals into rules, using a process called *chunking*, so that over time, problem solving in subgoals is replaced by rule-driven decision making. Chunking has proved to be extremely versatile because it stores away whatever problem solving is performed in a subgoal, allowing Soar programs to learn using a wide variety of methods, including explanation-based learning, macro-operator learning, strategy acquisition, learning by instruction, and many others. In general, Soar's processing is symbolic, and although that is sufficient (and necessary) for a wide variety of cognitive activities, it is inadequate (or at the very least extremely inefficient) when encoding probabilities and numeric rewards.

While Soar has strengths in knowledge-rich symbolic reasoning and learning and weaknesses in knowledge-lean, statistical-based learning, the strengths and weaknesses of reinforcement learning (RL) techniques are the reverse. They are successful at capturing statistical regularities related to the expected reward that an agent will receive, but can not encode and effectively use large bodies of symbolic knowledge. In this paper we will present an initial integration of reinforcement learning with Soar, enriching the learning capabilities as well as the representation of knowledge in Soar, while at the same time developing a unique integration of reinforcement learning with symbolic, knowledge-rich reasoning. Specifically, Soar supports dynamic hierarchical task-decomposition, meta-reasoning, and the ability to enrich the state descriptions through internal abstractions. All of these capabilities both complicate and enrich reinforcement learning. This integration requires structural changes to the Soar architecture and we will refer to the unification as Soar-RL.

In the remainder of this paper we first present a simplified description of Soar and the extensions we have made to incorporate reinforcement learning. We then demonstrate the implementation on two simple tasks, highlighting the contributions RL makes to Soar, as well as the capabilities Soar-RL provides beyond standard reinforcement learning. We also compare and contrast Soar-RL to ACT-R, which incorporates a rule-tuning mechanism, comparable to reinforcement learning. We conclude with future directions.

Soar

The structure of Soar's memories is shown in Figure 1. Soar has a declarative working memory that contains its representation of the current situation using labeled graph structures, organized in a hierarchy of states/goals. All long-term procedural knowledge is encoded as production rules. Whenever a rule's conditions match working memory, the rule is fired and its actions performed. Actions may involve adding or removing structures from working memory. They may also create preferences used to select operators.

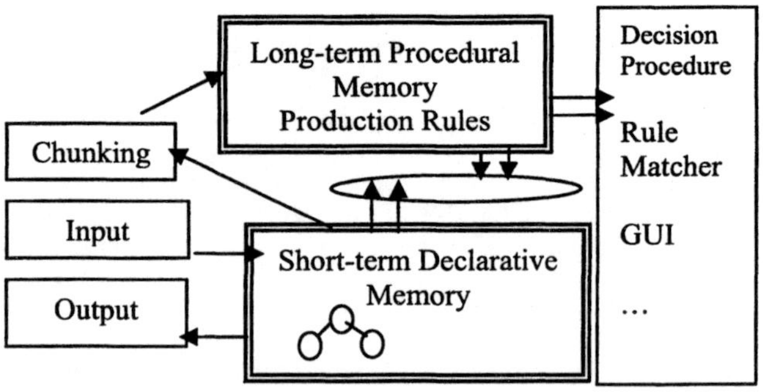

Figure 1: Soar's Structure

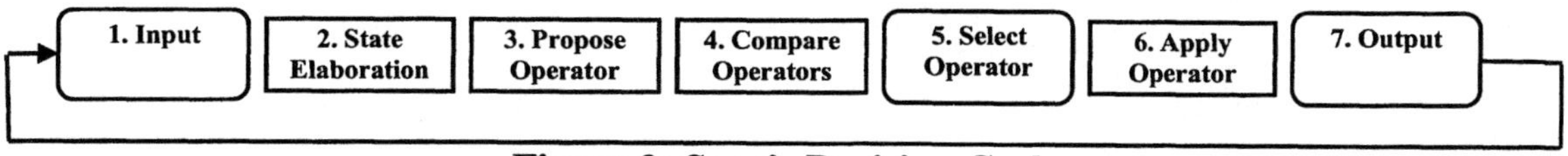

Figure 2: Soar's Decision Cycle

Soar's learning mechanism, chunking, monitors problem solving and automatically creates new rules, which are added to long-term memory during execution.

Soar's basic reasoning cycle is illustrated in Figure 2.

1. Input. Changes to perception are processed and Soar's perceptual buffer in working memory is updated.
2. State elaboration. Rules that newly match are fired in parallel to retrieve relevant information. For example, in a robotic task, a rule might test the distance to the object and the robot's available reach and determine if the object is within reach.
3. Proposing operators. Rules can propose operators by creating *acceptable* preferences for specific operators. In general, the rules' conditions test the situation so that an operator is proposed only when it is relevant.
4. Comparing and evaluating operators. Rules can test the proposed operators and other features of the situation and create preferences, which make assertions about the absolute or relative merit of the operators. Multiple preferences can be generated for a single operator.
5. Selecting the current operator. The preferences are evaluated to select the current operator. If the preferences are insufficient or contradictory, an impasse ensues and Soar creates a substate in which the goal is to resolve that impasse. This provides Soar with meta-reasoning so that it can reflect on its own processing.
6. Applying operator. Rules match against the currently selected operator and the state and make changes to the state, including creating motor commands.
7. Output. All new motor commands are processed.

Stages 2-4 are intermixed during execution. The only decision making in Soar is the selection of operators, which is the sequential bottleneck in Soar, as only a single operator can be selected for a state at a time (rules fire in parallel).

Soar's original set of symbolic preferences included preferences specifying that operators should be rejected, that one operator is better than another, or that an operator can be a default. Recently, we have added numeric preferences, which allow a real number to be associated with an operator (Wray & Laird, 2003). The numeric preferences are considered only if the symbolic preferences are insufficient for determining a single best choice. For each candidate operator O_i, the values of all numeric preferences proposed for O_i are summed into a total score $Sum(O_i)$ for the operator. The winning operator is chosen probabilistically according to the Boltzmann distribution:

$$\frac{e^{Sum(O_i)/Temperature}}{\sum_i e^{Sum(O_i)/Temperature}}.$$

Temperature is a parameter that controls the "peakedness" of the probability distribution. These selection rules provide a means of incorporating domain-specific probabilistic selection knowledge.

Adding Reinforcement Learning

The core problem in reinforcement learning is to learn how to act in the world so as to maximize a reward signal. In most approaches, an agent learns a value function, a mapping from either a state or a state-operator pair to an estimation of the expected sum of future rewards that can be achieved from that state or after applying that operator in that state. In Soar-RL, we use numeric preferences to represent a state-operator value function. We have further extended Soar by adding a reward as one of the inputs from the environment. The value of the reward can also be modified by internal knowledge (additional rules) that generates a reward for subgoal achievement.

One recurring problem in RL is how to represent the value function for a large state or state-operator space. The simplest way to store a value function is as a table with all the possible states or state-operator pairs enumerated with their associated values. However, this representation is both bulky and slow, since it does not allow learned values to be generalized over sets of states, and so is impossible for the sorts of tasks we expect to encode in Soar.

By summing over the values suggested by multiple rules, the recommendation of the numeric preferences shares a characteristic with the recommendation of the symbolic preferences – they are both the result of the combination of different pieces of advice, each of which covers a portion of the state-operator space. Each individual rule can be considered to be a specialized complex feature detector, and the predicted utilities for the operators are the runtime linear combination of the values of these complex features. Learning linear functions over features is one of the best understood methods of value function approximation in RL. Automatically constructing features, however, is not well understood, and this is part of our research (necessary, for an agent with many tasks; helped by the flexibility of Soar's language). Incorporating RL in Soar has two parts, first, adjusting the values of numeric preferences for existing rules, and later, creating new rules that test different feature sets while creating numeric preferences.

Adjusting Numeric Preferences

A numeric preference for an operator is generated by a rule associating a numeric value with a particular set of features in working memory. This set should incorporate features of the operator and the current state, possibly including aspects such as the current goal and surrounding context. Such a rule can be general enough to apply across many different states and operators if it is selective in the features it tests. As an example, a maze-world agent might have the rule[1]:

```
sp {RL*rule1
   (state <s> ^task mazeworld
              ^location C5
              ^destination A1
              ^operator <o>)
   (<o> ^name move
        ^direction east)
  -->
   (<s> ^operator <o> = -.82)}
```

Figure 3a

The RL task is to adjust the value of this rule (currently, -.82) as the agent encounters rewards in the world. The aim of these adjustments is that the summed value of numeric preferences for operator **o** given the working memory state **s** should approximate the action value, notated **Q(s,o)**. The action value is the expected sum of rewards to be received after choosing operator **o** in state **s**, and thereafter following the agent's current policy. Such adjustments, combined with our soft greedy policy for operator selection, move the agent toward actions that maximize its expected reward.

Soar-RL performs updates of numeric preferences immediately - the preferences that fired for the operator selected on the tth decision cycle are updated during the $(t+1)^{st}$ decision cycle – and so manages to store very little history. The updating procedure is a variant of the SARSA algorithm (Rummery & Niranjan, 1994), which translates to the following steps in Soar-RL:

1. On cycle t, operator o_t is selected. This selection may or may not result solely from symbolic preferences, but, in either case, any numeric preference rules for o_t in state s_t are fired. This, possibly empty, set of instantiated numeric preference rules is stored, as is the sum over these preferences, which represents $Q(s_t,o_t)$.
2. The system fires rules to apply the operator o_t, which in turn leads to new rules matching to select the following operator. At this point, the reward r_t for this decision phase is recorded and the next operator o_{t+1} is selected.
3. By this stage, the system has summed the numeric preferences for o_{t+1}, the quantity representing $Q(s_{t+1},o_{t+1})$. The estimate $Q(s_t,o_t)$ will be improved by adjusting it toward $r_t + Q(s_{t+1},o_{t+1})$. More precisely, $Q(s_t,o_t)$ is updated according to:
$$Q(s_t,o_t)=Q(s_t,o_t)+\alpha(r_t+\lambda\times Q(s_{t+1},o_{t+1})-Q(s_t,o_t))$$
where α is a learning rate and λ is the rate at which future rewards are discounted, both currently constant values set by hand.
4. The update is distributed over the saved instantiated preference rules for o_t by dividing the portion $\alpha(r_t+\lambda\times Q(s_{t+1},o_{t+1})-Q(s_t,o_t))$ **(Formula 1)** by the number of preference rules and adding this quantity to the numeric value of each rule.

[1] Note on Soar syntax. The If and Then portions of a rule are separated by -->. The syntax of a condition is (identifier ^attribute value). And anything in < > brackets is a variable.

For instance, a maze-world agent chooses a **go east** operator that has a summed value –5.82 computed from the preferences in Figure 3a and Figure 3b:

```
sp {RL*rule2
   (state <s> ^task mazeworld
              ^has_monster east
              ^operator <o>)
   (<o> ^name move
        ^direction east)
-->
   (<s> ^operator <o> = -5)}
```

Figure 3b

If the total update (**Formula 1**) computed on the next step is –1, then these two rules would be updated to have the actions (<s> ^operator <o> = -1.32) and (<s> ^operator <o> = -5.5) respectively.

Currently, the Soar-RL agent will do no reinforcement learning unless the programmer provides numeric preference rules for it to update. These rules can be written in the form of variabilized prototypes such as

```
sp {RL*prototype
 (state <s> ^task mazeworld
            ^location <c>
            ^destination <d>
            ^operator <o>)
 (<o> ^name move
      ^direction <direction>)
-->
 (<s> ^operator <o> = 0)}
```

When this rule fires in a specific situation, it will build a new version of itself with the variables <c>, <d>, and <direction> replaced by constants, as in Figure 3a. At that point and thereafter, this new rule will have its value updated by the procedure described above. Using this prototype allows the agent to associate values with all combinations of location, destination, and direction without hand-writing a rule for each – plus, it only has to build a specific rule for combinations that it actually encounters. Combined with monotonic inferences, the prototypes can also support relational tests. For instance, to test that *location* was north of *destination* would require one production to elaborate working memory whenever the north relationship existed, and one numeric preference prototype to match against this elaboration.

Results

What is detailed in the previous section is what has been implemented in Soar, and is available to any domain that is encoded in Soar. We tested the system in two domains. In both cases, we wrote Soar agents with little or no operator selection knowledge, and so relatively random initial behavior. We looked for improvement in their performance, relative to their reward functions, as they gained experience in the world. Only reinforcement learning was used – there was no chunking.

Missionaries & Cannibals: Our first test domain was the puzzle-problem Missionaries & Cannibals. To turn this into an RL problem, we gave the agent rewards of +1 for moving into a success state (all persons on right bank), -1 for moving into a failure state (majority cannibals on some bank), and 0 for other state transitions. In this world, working memory always contained a description of the current state in terms of the number of missionaries and cannibals on each bank and the location of the boat. Thus, the state was directly observable, and, since the state space was small, we could use the representation of one numeric preference rule for each state-operator pair. Below are the results of 500 training runs, in which the agent had no control knowledge other than its improving numeric preferences. In the graphs, black vs. gray indicates whether a run ended in success or failure, and the y-value measures the number of steps in the run.

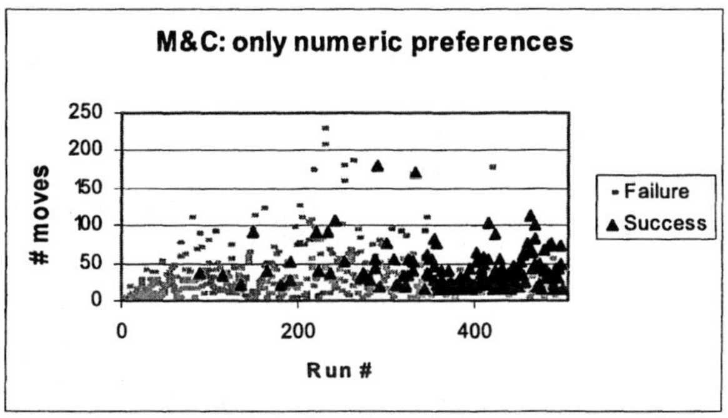

By the end of these runs, the agent is generally successful. Mistakes are attributable to the persistent randomizing effect of a soft greedy policy. Learning can be accelerated considerably by the addition of heuristic symbolic preference rules. For instance, after modifying the Soar-RL agent above to give a symbolic *worst* preference to a move reversing the previous move, the results over 500 training runs were (note different y axis scale: 45 vs. 250):

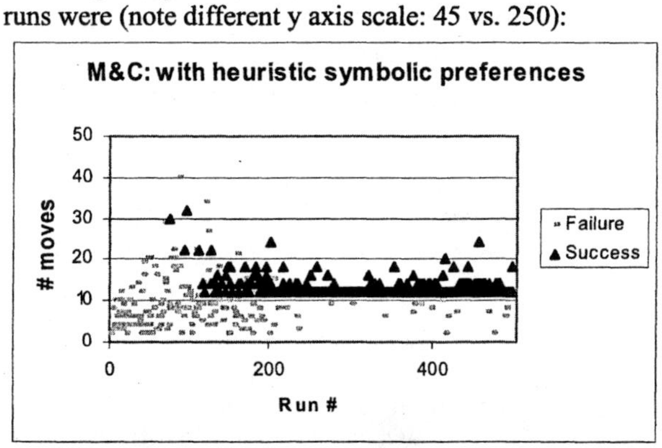

Eaters: Puzzle problems are often used to demonstrate the speed-up effect of Soar's chunking mechanism, and they are useful for evaluation because the desired behavior is known. However, puzzles have a couple of faults as test domains for reward-based learning. First, they tend to offer rewards only as one-time marks of success or failure, rather than as a varied stream of feedback. Second, they frequently demand less interesting, more table-based representations of value functions, because their puzzling nature is often achieved by forcing the agent to consider all features in combination, rather than calculating the implications of each feature independently. Our second test domain attempts to avoid these faults. It is called Eaters and is played by a Pacman-like agent, called an eater, moving around a board.

The Eaters board initially is filled with food of two types (bonusfood and normalfood) as well as some interior walls. An eater receives a reward of +10 for moving onto a bonusfood, +5 for moving onto a normalfood, and 0 for moving into an empty cell. (Contents are consumed when the agent moves onto them.) Additionally, there is a discount factor that favors earlier rewards. The agent can move in the four cardinal directions, if unblocked by a wall, and can sense only the contents of the cells in the 5x5 square centered on itself.

Given the agent's sensing capabilities, there are a variety of features of the state that could be represented – for instance:

1. The contents of one of the four adjacent cells.
2. The contents of all of the four adjacent cells.
3. The contents of the three new cells that could be visited on its second move given that its first move is in direction x.
4. The contents of the twenty-five cells that it can sense.

Of course, to build a numeric preference rule, a feature must be combined with an operator – for instance, the contents of the cell to the east *plus* move north or the contents of an adjacent cell *plus* an operator to move to that cell.

We tested two RL agents in this world. The first agent associated numeric preferences with the contents of the next cell; ultimately, it learned only three numeric preference rules – one each for **move to bonusfood, move to normalfood, move to empty**. The second agent had rules of this type, plus rules of type 3 in the list above; therefore its predicted value for a move was a sum of the immediate value associated with the move and a value associated with the resulting options for its second move.

These two agents were trained for 20 runs of 600 moves each and then their performance was tested. During testing, reinforcement learning was turned off and the parameter controlling randomness in operator choice was reduced, so that the agent had less of a tendency to explore. For comparison purposes, we also tested a random agent and a non-RL agent with symbolic preferences favoring **move to bonusfood** over **move to normalfood** over **move to empty**. The results are graphed below (total reward as a function of move #), where each line represents an average over twenty runs.

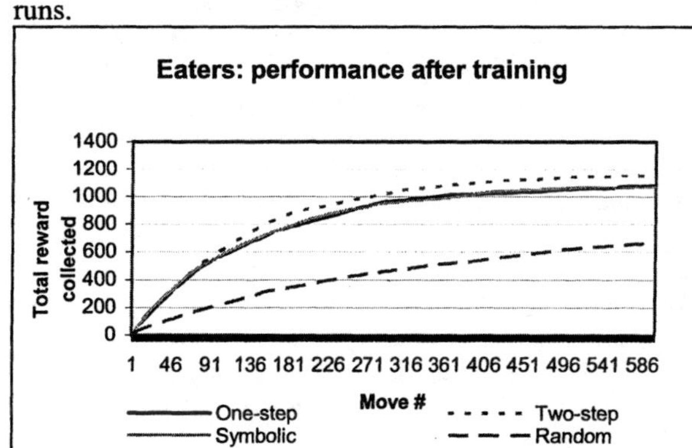

The major points demonstrated in this graph are

- Both trained RL agents have learned well enough to soundly beat the random agent.
- The performance of the one-step look-ahead agent is identical to that of the symbolic agent. The one-step look-ahead RL agent learned the values 50.12, 29.38, and 5.66 for **move to bonusfood**, **normalfood**, and **empty**, respectively, so it would be expected to have the same behavior as the symbolic agent.
- Adding the two-step look-ahead feature yields a relatively small but consistent gain in performance. The nature of this world is such that a move that looks best from the one-step perspective is generally the best move. The two-step look-ahead is most useful in deciding between moves when the one-step cell contents are the same. For instance, in the pictured situation, the eater has a numeric preference of 14.5 for all four **move to empty** operators, but a two-step preference of -6.6 for **(empty, empty, normalfood)** to the east versus -13.8 for **(empty,empty,wall)** to the west. The eater agent learned 56 two-step numeric preference rules vs. 3 one-step rules. These numbers are clearly out of proportion to the gains in performance; much of the needed control knowledge in this world can be captured by the one-step rules. Still, we avoided learning the 3x56 rules that would have been required if our two types of features had not been considered independently.

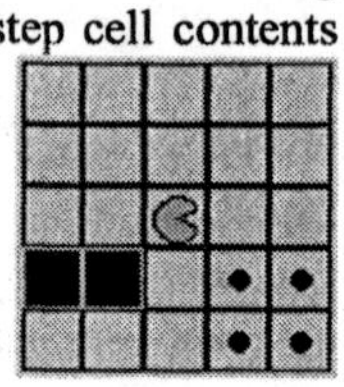

Comparison with ACT-R

ACT-R has a learning component that is comparable to reinforcement learning. The locus of selection in ACT-R is at the level of individual rule selection instead of at the level of operators in Soar. In ACT-R, only a single rule is fired at a time and the selection is done probabilistically based on a utility that is associated with each rule, with the selection distribution similar to the Boltzmann distribution. The utilities are a function of the probability of and expected cost of goal completion given that a particular production is chosen, as well as of the value of the current goal. The value of a goal is set by the programmer, and the other two numbers are learned from experience.

Below is a list of key differences and their possible implications between Soar-RL and ACT-R rule tuning:

1. Soar allows both symbolic and numeric preferences to contribute to the decision. Thus, it is possible to encode control knowledge that one option is better than another. There is no way to directly encode similar knowledge in ACT-R for preferring one rule to another.
2. The utilities captured by the numeric preferences in Soar do not rely solely on the idea of progress toward a goal, but instead are associated with a more flexible reward function. This reward function is capable of representing goal-directed activity, but can also be used for ongoing activities, such as the top-level of an agent, which monitors energy levels or selects goals to be pursued.
3. The utilities for a rule in ACT-R are related to a single goal, whereas in Soar, there can be many different rules for an operator that are sensitive to different goals. Thus, agents developed in ACT-R, may have difficulty learning in environments with shifting or multiple goals.
4. Ultimately, the utility of an action and its likelihood of being selected in ACT-R are based on the conditions of the individual rule. There is no run-time combination of the values as there is when the numeric indifferent preferences are combined in Soar. Thus, Soar distinguishes between the conditions for when an action/operator is valid and when it is desirable, which in turn gives it a richer representation for capturing complex reward/utility functions.
5. Although ACT-R can get the effect of combining multiple pieces of advice by using multiple rules with different conditions for the same actions, its learning mechanism can give credit to only one of these rules. In Soar-RL, all numeric preferences that contribute to the selection of an operator receive credit for the results of selecting that operator.

In this paper we have described a significant extension to the Soar architecture – the addition of reinforcement learning. This provides a new, architectural approach to learning control knowledge based on expected rewards. The implications for Soar as an AI architecture are clear – all Soar programs will now be able to automatically adjust to feedback from their environments in cases where symbolic preferences are inadequate to make decisions. The impact on Soar for cognitive modeling is less clear, but intriguing. Soar-RL, together with recent work in adding activation to Soar (Nuxoll, Laird, & James, 2004), moves Soar much closer to ACT-R in terms of combining numeric and symbolic computation and learning. The underlying mechanisms share many common features, but also differ enough that they might lead to different computational models of human behavior. Even if these models do not surpass current ACT-R theory, they should provide us with new insights into what aspects of the ACT-R models are most important.

Future work

For future work, we have two basic thrusts. One is to build models of human performance, comparing and contrasting these models with existing ACT-R models. The second thrust, is to greatly expand the use of reinforcement learning in Soar to be more complete, which will push our implementation in directions that have not yet been explored in other cognitive architectures.

1. Task independent learning parameters. Reinforcement learning invariably has parameters that are used to tweak learning for a specific problem. In addition, we have a Boltzmann decision procedure that has a temperature parameter that influences the evenness of probabilistic selection. Although there may be truly task-independent values for these parameters, embedding the decision procedure and learning mechanism in a general architecture allows us to explore the possibility that these parameters can be tied to a general feature of the current task that is detectable by the architecture. For example, temperature works best in the decision procedure if its value is high during earlier explorations of a domain so that exploration is preferred to exploitation. As the domain is explored and the knowledge matures, exploitation is usually preferred, so that temperature can be lowered. As demonstrated in adaptive simulated annealing (Ingber 1993), dynamically adjusting the temperature can greatly improve performance. To begin with, we will derive the temperature from the certainty of the numeric preferences. When the variance is high, the temperature will also be high leading to a more uniform distribution of the decision space and encouraging exploration, but when the variance is low, the temperature will be low, biasing the decision to those operators with higher numeric preference.
2. Learning the appropriate features to associate with predicted rewards. The state spaces that Soar works in are structured (involving relations) and huge, so that many standard approaches to representing the association between features and state values are inadequate. We use rules for such associations, but that then begs the question as to where those rules come from. We plan to investigate schemes where rules are dynamically generated by specializing existing rules.
3. Hierarchical Learning. Soar automatically generates subgoals. This gives us the opportunity for the system to learn about not only the overall task it is trying to achieve, but also the component parts of the task: subtasks and meta-reasoning. These components are more likely to be shared across many different tasks, so that learning about subgoals can improve performance on novel tasks with common subtasks. Thus, an important issue is to determine how reinforcement learning should work across subgoals, with the main issue being how to introduce rewards at the end of a subtask. These rewards should be related to subgoal achievement, and may be independent of environmental rewards. This is the approach taken in several hierarchical RL systems, in particular MAXQ (Dietterich, 2000) and options (Sutton, Precup, & Singh 1999), but neither system explains how these subgoal rewards may be generated automatically. For an autonomous agent, the origin of those reward functions is unclear because subgoals may not be preprogrammed but arise from a lack of knowledge.
4. Internal, task-independent reward functions. Our goal is to integrate reinforcement learning into Soar so that it can be used on any task. One key component of reinforcement learning is obtaining a reward from the environment. However, for many tasks, the environmental rewards are extremely sparse, making learning very slow. We want to investigate domain-independent reward functions that augment external reward functions and help steer an autonomous agent to useful parts of its environment – these are almost meta-rewards in that they are based on characteristics of the knowledge that has been learned, emphasizing achieving improvements in the agent's ability to predict its environment (Kaplan 2003). Emotions may be an alternative (or possibly related) source of reward that can direct behavior.

References

Anderson, J. R. & Lebiere, C. (1998). *The Atomic Components of Thought.* Mahwah, NJ: Erlbaum.

Dietterich, T. (2000). Hierarchical Reinforcement Learning with the MAXQ Value Function Decomposition. *Journal of Artificial Intelligence Research*, 13.

Ingber, L. (1993) Simulated Annealing: Practice versus Theory, Mathematical Computer Modeling, 16 (11), 29-57.

Kaplan, F. (2003) Bootstrapping Awareness, In Dautenhan, K. and Nehaniv, C., editor, Proceedings of Second International Symposium on Imitation in Animals and Artifacts. 48-57.

Newell, A. (1990). *Unified Theories of Cognition.* Cambridge, MA: Harvard University Press.

Nuxoll, A., Laird, J., James, M. (2004). Comprehensive Working Memory Activation in Soar..International Conference on Cognitive Modeling.

Rummery, G. A., & Niranjan, M. (1994). On-line Q-Learning using Connectionist Systems. Technical Report CUED/F-INFENG/TR 166. Engineering Department, Cambridge University.

Sutton, R. & Barto, A. (1998). *Reinforcement Learning: An Introduction.* Cambridge, MA: The MIT Press.

Sutton, R., Precup, D., & Singh, S. (1999). Between MDPs and Semi-MDPs: A Framework for Temporal Abstraction in Reinforcement Learning. *Artificial Intelligence*, 112(1-2).

Wray, R., & Laird, J. (2003). Variability in Human Behavior Modeling for Military Simulations. Proceedings of the 2003 Conference on Behavior Representation in Modeling and Simulation. Scottsdale, AZ.

Towards a Theory of Balancing Exploration and Exploitation in Probabilistic Environments

Stefani Nellen (snellen@andrew.cmu.edu)
Marsha C. Lovett (lovett@cmu.edu)
Carnegie Mellon University,
Department of Psychology, 5000 Forbes Ave
Pittsburgh, PA 15213, USA

Abstract

Learning to make good choices in a probabilistic environment requires that the Decision Maker resolves the tension between exploration (learning about all available options) and exploitation (consistently choosing the best option in order to maximize rewards). We present a mathematical learning model that makes selections in a repeated-choice probabilistic task based on the expected payoff associated with each option *and* the *information gain* that will result from choosing that option. This model can be used to analyze the relative impact of exploration and exploitation over time and under different conditions. It predicts the aggregated and individual learning trajectories of participants in various versions of the task sufficiently well to support our basic argument: *Information gain* is a valid and rational criterion underlying human decision making. Future modeling work will be addressing the exact nature of the interaction between exploration and exploitation.

Introduction

Decision makers are often placed in novel situations that offer them a finite variety of choices. They know that each of the choices is associated with some probability of leading to a positive outcome, but they don't know what these probabilities are. They might also know that making a choice will constrain the available choices in the subsequent decision cycle, but, again, they don't quite know in which manner this will happen. All they know is that they will have *n* opportunities to make one choice at a time, and that the long-term goal is to maximize accumulated rewards. Each Decision Maker has to resolve the tension between exploration (learning about the payoffs of the options, which is achieved by selecting them and observing the outcome) and exploitation (consistently choosing the best option). Maximizing rewards depends on accurate payoff estimates and therefore on sufficient information. On the other hand, this exploration must eventually be discarded in favor of consistently choosing the best option if the goal is to be met. This example shows that exploration is a necessary prerequisite of probability learning. It also shows that the benefit of exploration depends on the amount of information that has already been accumulated. Therefore, it will not remain constant over time. We are convinced that it is possible to predict the relative impact of exploration and exploitation under different conditions over time, and that this relative impact varies in interaction with the probabilistic structure provided in the environment. The model we present here provides an implementation of this basic idea.

In the context of decision making research, the representation of probabilistic information/learning has been somewhat understudied, partly because probabilistic representations are often assessed by eliciting one-shot probability estimates from participants instead of observing changes in their actual behavior (see Gigerenzer, 1994, for a critique of the notion of single-event probabilities). Additionally. following a long history of models of probability learning (e.g. Estes, 1964;), recent cognitive models of heuristics or decision making algorithms that rely on probabilistic cues often provide participants with the explicit probabilities from the outset (Broeder, 2000; Payne, Bettman and Johnson, 1993; Rieskamp & Hoffrage, 2000) or assume that the representations have already been formed (models often achieve this during a separate "training phase") and are ready to be used. In ACT-R (Anderson & Lebiere, 1998), the selection of cognitive operators (production rules) is based on their "expected utility", which is partly determined by estimates of the probability of success associated with that operator (the other components being "cost/ effort" and noise). This expected utility can gradually be learned by experience. *Information gain*, i.e. the increase in knowledge about the payoff-structure of the system, does not play an explicit role in utility learning and operator selection in any of these models right now.

The model presented in this report proposes a definition of *information gain* and an explanation of information gain and estimated payoffs interactively influence decision making. The model also explains how the relative impact of information gain and estimated payoffs on decision making changes over time.

Before describing the model and its predictions in more detail, we will provide a description of the task in which it was developed and tested, a repeated-choice probability learning task with immediate and delayed rewards. Afterwards, we will present the fit between some model predictions and behavioral data, compare this fit to that of a version of the model that does not take *information gain* into account, and explore future directions for refining the model.

A Probabilistic Learning Task

A version of the task that was used for developing this model has previously been used by Brown and Lovett (2001) in order to assess people's ability to learn to prefer long-term over short-term benefits. In that context, it has been dubbed a "Single Player Version of the Prisoner's Dilemma (PD)". However, the particular connotations of the PD-game are of less relevance for the present work and might be confusing here. For this reason, we will simple be referring to the "Probabilistic Learning Task (PLT)".

Figure 1 shows a schematic overview of the PLT, including its underlying rewards and constraints structure. Participants go through the task at the computer. They are presented with four options in the form of four closed doors, two red ones and two green ones (in the figure, the dark doors in the top row are both red). In the instructions, participants are given the following pieces of knowledge about the task ahead of them: They know that they are asked to make a series of choices, selecting one door at a time. They need a key to open a door of the same color, and they are given a red key at the beginning of the task. By opening a door, the current key will be given up. Upon opening a door, two things will happen: The participants may or may not receive a reward (5 "points"). Additionally, they will receive a new key that will be either red or green. This key will constrain their available options in the subsequent trial, because red keys open only red doors and green keys open only green doors. Finally, participants are told that the goal in this game is to gain as many points as possible.

Upper Left (UL):	Upper Right (UR):
p(reward)=0.6 Key: Red Door Color: red	p(reward)=0.8 Key: Green Door Color: Red
Bottom Left (BL):	**Bottom Right (BR):**
p(reward)=0.2 Key: Red Door Color: Green	p(reward)=0.4 Key: Green Door Color: Green

Figure 1: Overview of the choices and their outcomes in the PLT. The outcomes are shown for clarification. Participants have to learn them .

It may be noted that green doors have a lower probability of reward than red doors. However, the upper right red door (UR), which has a higher reward probability than the upper left red door (UL) gives a green key. To solve the game, participants have to learn that UL has the highest payoff probability in the long run, even though its immediate rewards are less probable than those of UR. The behavioral manifestation of having solved the game is to choose UL consistently, disregarding the other options.

Before being able to solve the game, participants must learn the reward-probabilities associated with each door and the association between door and keys (the latter connection is deterministic). Participants complete a total of 200 trials in one session with this task.

The "Expected Utility Differences (EUDs)" in Two Versions of the Task

Given the constraint of the keys, there are three sequences of choices that can be done repeatedly: Choosing UL a number of times, choosing first UR and then BL, and Choosing BR repeatedly. These three strategies can be evaluated in terms of the payoffs associated with the involved doors. In the PLT, the expected values of the strategies follow the following general constraint:

$$2*payoff(UL) > payoff(UR)+payoff(BL) > 2*payoff(BR)$$

The payoff probabilities of the doors can be manipulated to characterize different versions of the PLT, which differ in the extent of the "greater than" relation. This relation is called the "Expected Utility Difference" (EUD). Inserting the reward probabilities from fig. 1 in the equation above and multiplying the outcome with five (because of the "points" that form the reward) yields

$$2*(0.6*5) > (0.8*5+0.2*5) > 2*(0.4*5), \text{ or}$$
$$0.6 > 0.5 > 0.4$$

Therefore, the EUD in this scenario is 1 (in [arbitrary] tenth of an expected point units). The other version of the PLT we are interested in has the following probabilities of reward (the mapping of keys to doors is identical): P(reward UL)=0.8, p(reward UR)=0.9, p(reward BL)=0.1, p(reward BR)=0.2. This results in an EUD of 3. Brown and Lovett (in preparation) have found that participants find it very hard to learn how to solve the game when the EUD is 1, while many more participants are able to learn the solution when the EUD is 3. The onset of learning is also much earlier in that condition. Interestingly, this effect is independent of the specific probabilities that are used to form the EUDs (Brown & Lovett, 2001). This justifies our decision to regard the EUD 1 and 3 versions of the task as two truly distinct conditions, and to compare the behavior of the model to human behavior under exactly these two conditions.

Description of the Model

Like participants, our model for learning probabilities and making choices in the PLT starts out with very little knowledge. The only initial constraint on its selection is the required mapping between keys and doors of the same color. Like participants, it will be making 200 choices, and learn about the probabilistic reward structure of the system during the process of making the choices and from the feedback following them.

At each "choice point", there are two available options (two doors matching the color of the key). The model selects the option that has a higher current *evaluation.* The following three factors contribute to the overall evaluation of each option:

(1) The current estimate of the probability of receiving a reward upon opening that door.

(2) The current estimate of the value of the key that will be given by that door, weighed by a parameter relating

the importance of future rewards to that of immediate rewards.
(3) A measure of *information gain*, that expresses how much knowledge of the characteristics of all available options, i.e. of the system as a whole, will increase as a consequence of "opening that door". This is weighed by a parameter that grades the importance of *information gain*.

A formal definition of the model's selection S at time t is:

$$S_t = \max(E_{it}, E_{jt}),$$

where E_{it} and E_{jt} are the evaluations of the two available choices i and j at time t. The evaluation of both options is analogously computed as

$$E_{it} = \frac{successes_{it}}{successes_{it} + failures_{it}} + k * \max(A_t; \frac{B_t + C_t}{2}) + c * N_{it}^{-.5}$$

The first term is the estimate of the probability of success of door i at time t. The second term is the value of the key given by the door, weighed by the parameter k. The value of a key is the estimate of the best future rewards that can be expected from having that key. This is either the current estimate of the repeatable cell whose door matches the current key (denoted by A), or the expected value of alternative sequences that starting with the current key but will give a different key (denoted here by the average of the expected value of the door that can be opened with the current key and will itself yield a different key (B) and the current value of that different key (C). The third term, finally, denotes the *information gain* associated with choosing door i at time t, weighed by the parameter *c*. This measure simply decreases as a power function of N_{it},, the number of times door i has already been chosen at point t. This expresses the assumption that we learn something about the system each time we make a choice (which is true, because we get feedback), but that this information shows marginally decreasing returns. The selection of a power function to represent this effect quantitatively is partly based on the fact that the posterior distributions of the reward probabilities associated with the four doors are beta distributions, the variance of which decreases as a power function of additional observations. This characteristic of the posterior probability distribution of an event again points to the crucial interaction between exploration and exploitation we are trying to capture here: the accuracy of the estimates, both of the reward-probabilities and the value of the keys, critically depends on sufficient exploration, more specifically: a sufficiently large number of observations. However, the impact of exploration does not remain constant, but instead decreases systematically: rapidly in the beginning, marginally later on.
The model presented is a learning model in the sense that all estimates are updated after each choice, as is the information gain measure associated with that choice. It does not use noise; the only sources of variability are changes in the estimates, which in turn are caused by the probabilistic feedback, and changes in the *information gain* measure, which leads the model to abandon familiar options and explore less familiar ones.
The parameters in the model (*k, c*) are basically free parameters that can be adjusted to reflect a different impact of future rewards (*k*) or Information Gain (*c*), either under different circumstances or even between (simulated) participants. However, in all simulations presented here, they have been kept constant throughout, with *k*=3 and *c*=4.

Some Basic Predictions and Mechanisms

As the relative impact of the two competing components of the evaluations changes with time, the learning and behavior of the model can be describes as follows (note that they are not assuming discrete stages but continuous changes, the segmentation in the following paragraph was made to serve clarification):
(1) *Pure exploration*. Choices will be made based on the Information Gain measure, i.e. choices that have not yet been explored will be explored. When *c*=4, the estimates of the actual payoffs (which are initialized to 0.5) are still too small to counteract this during the first few cycles (i.e. until a sufficient number of experiences has been accumulated).
(2) *Early, inaccurate Estimates*. Because the measure of Information Gain decreases relatively rapidly in the beginning, the actual estimates of the reward probabilities and the key values begin to impact the models choices. However, due to the probabilistic feedback, and the fact that they are still based on relatively few observations, the estimates might not reflect the true ranking of the options, particularly in conditions where EUD=1. Two forces counteract these inaccurate representations. First, by choosing the currently best option repeatedly, the model gathers information that corrects its estimate. Now previously lower-ranked options can compete again for selection. Second, by exercising its bias towards the less explored options, it obtains a more accurate estimate of their payoff probabilities as well. Consequently, the model quickly recovers from initial false estimates. Now, the model has established estimates of the reward probabilities that are robust enough to remain unfazed by the occasional "failure"-feedback. The model will then consistently chose the option with the highest estimated long-term payoff
(3) *Familiarity breeds contempt*. Flexibility is maintained for a while beyond this point, because, as one option, or a combination of two options, is chosen repeatedly, its *information gain* measure decreases, while that of the other options remains constant. Thus, there is the chance that the model abandons an option again in favor of another, particularly if this competitor has a similar expected payoff. It is clear how appropriate this behavior is in the task described here, where the EUD between choices are sometimes as close as 1 unit, and the chance to "accidentally" settle on the wrong solution is pronounced.
(4) *Optimal choices with Intermissions*. Eventually, the model will learn to solve the game, because its estimates are becoming more and more accurate. However, because of the dynamics described in the previous section, the model will continue to explore alternative options. The Intervals between these "exploration fits" also follow a Power Function: the number of experiences between "exploration

fits" becomes much longer each time, until they are so widely spaced as to be without any relevance anymore.
We will examine how the model learns and behaves under different EUD-conditions of the task described here in the next section.

Comparison Between Model and Data

The data to which we compare the behavior of our model were collected by Brown and Lovett (in preparation). A total of 80 participants worked on the version of the PLT described in this paper (for related work using a deterministic version, see Brown & Lovett, 2001). 60 Participants worked under the EUD 1 condition (participants were collapsed from different groups that used different sets of probabilities to form an EUD of 1, because the specific probabilities had no effect on behavior, as reported by Brown & Lovett, 2001). 20 Participants worked under a EUD = 3 condition. The reward probabilities for both groups are given in Table 1.

Table 1: Reward Probabilities in the EUD1 and EUD3 conditions.

	UL	*UR*	*BL*	*BR*
EUD=1	p=0.6/0.7/0.8	p=0.8/0.9/0.9	p=0.2/0.3/0.5	p=0.4/0.5/0.6
EUD=3	p=0.8	p=0.9	p=0.1	p=0.2

Participants' choices in both groups were recorded on a trial by trial basis. Choices of UL and BL, which are the choices that yield the "better key" (red) were coded as "1", to indicate subjects' attention to the keys, as opposed to the immediate rewards (which would favor UR and BR, respectively, choosing these options was recorded as "0"). Based on the binary raw data, the proportion of "choosing left" was computed for each of the 200 trials. Note that only a repeated choice of UL can lead to a proportion > 50%. An increase above that level is thus indicative of choosing the solution, UL, more often than not. Finally, the proportions were averaged over ten-trial blocks, resulting in 20 ten-trial blocks that depict the aggregated learning trajectory for all participants. Additionally, response latencies were obtained for all trials.
It was our goal to have the model produce data of the same format as the empirical data. To this effect, we implemented it in a spreadsheet and recorded each of the 200 choices it made per run, coding them the same way the human data were coded. We produced 3*20 model runs under the EUD1 condition, using the probabilities given in table 4, and 60 model runs using the EUD3 condition, also using the same probabilities. The results of these model runs were aggregated in the same manner as the human data. The model (obviously) was constrained to open doors with the appropriate key in the same manner as humans were. Its choices were based on the evaluations elaborated in the preceding section.
We will first present and comment on the correspondence between model and data for both EUD conditions on the aggregate level. However, the quality of a model can also be assessed by determining how well its individual runs resemble the individual learning trajectories of participants, particularly when behavior is very variable, as is the case here. Therefore, we present some comparisons between individual participants and individual model runs that show that the model can produce a range of behavior consistent with that shown across the sample of participants. A note on the parameters: The values of k=3 and c=4 were chosen to obtain a good the fit to the data in EUD1. They were kept constant for all other comparisons, including the ones on the individual level.

EUD1 and EUD 3: aggregate learning trajectories

It is easy to see how the behavior of the model differs under different EUD conditions. When the EUD is 1, i.e. very low, the model needs more trials to arrive at estimates that are accurate enough to warrant exploitation of one option. However, the small advantage associated with the winner, combined with the model's bias towards exploration, limits the stability of behavior under this condition. While the model will eventually converge to solving the game even under this condition, it is, like humans, often unable to do so within the 200 trials allotted in the experimental task described here. However, if the EUD is as high as 3, learning is considerably sped up: the accuracy of the estimates increases faster, because the variability in binary feedback is lower as the probability of success (or 1 vs. 0) becomes more extreme. and the gap between the two options' estimates increases faster. Exploration continues to be beneficial under this condition, but is more rare, as the differences between the estimates are pronounced enough to lead to appropriate exploitation and to counteract the impact of the *information gain* measure.
Figure 2 shows the (aggregated) learning trajectories of data and model under conditions 1 and 3. The difference between the groups is captured nicely by the model, especially the logarithmically shaped learning curve in the EUD 3 condition. Note that neither humans nor model arrive at 100% exploitation of UL under EUD 3. This reflects (in the model) the response to the probabilistic feedback and the ensuing recurrent, brief "exploration bursts". Note particularly the downward dip in the final four 10-trial-blocks that is shared by empirical and model data. In the model, this is the consequence of having chosen the solution, UL, a considerable number of times. As a consequence, its *information gain* Measure has decreased sufficiently to allow the competitors a few more explorations: familiarity breeds contempt. Neither humans nor model show much learning under the EUD 1 condition, for the reasons outlined above.
An even more powerful test of the validity of our model, in particular our claim that a component of *information gain* is essential for understanding and predicting human behavior, is a comparison between the data and a version of our model that sets c=0, thereby completely eliminating the component of *information gain*. The results of this comparison are shown in Fig. 3.

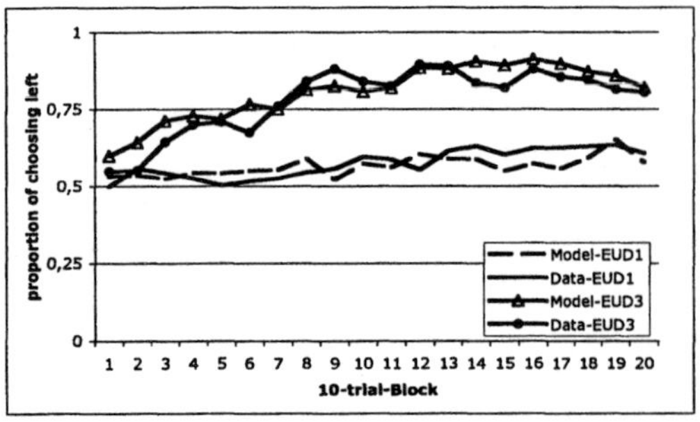

Figure 2: Model and Data Curves under EUD1 and EUD3 conditions.

The "no-info-gain" model does not capture the learning under the EUD3 condition. Levels of "choosing left" remain constant during all trials for this model. Even more importantly, the "no-info-gain" model operates without noise, and therefore does not exhibit any variability over time. The fact that the proportion of choosing left remains below 100% for the no-info-gain-model is an artifact of this: Some model runs *always* choose UL from the beginning, others *always* choose ER-BL throughout all trials. No changes occur, because none of these options has a bad enough payoff to jolt the model out of its inertia. The same inflexible behavior is true for the EUD1 version of the no-info-gain-model.

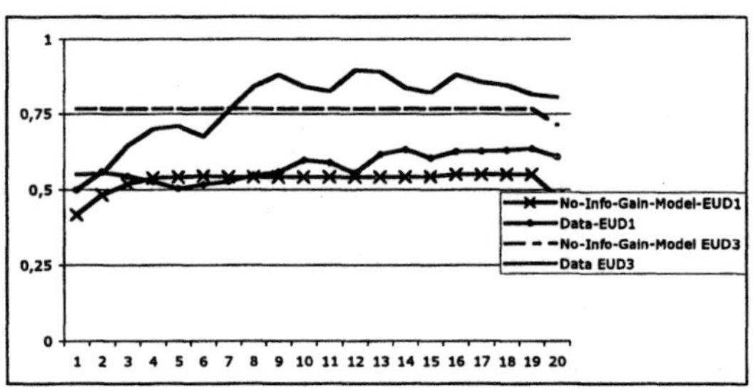

Figure 3: Predictions of a model without *information gain.*

This begs the question whether the addition of noise to the evaluation mechanism wouldn't have the same effect as the notion of explicit information gain. This is still an issue for future exploration, especially since there are two kinds of noise that can play a role here: The estimates themselves can be noisy, or the selection process that operates on them can involve noise, the subtle difference between these two concepts of noise, and possible integrations with the present model are the objective of future work. Here, the following argument can be made against the use of noise and in favor of the notion of *information gain* proposed here. Cognitive models of probability matching within the ACT-R framework have to assume an extremely high level of noise in order to capture the observation that there is still variability in participants behavior after a large number of experiences. We will show examples of this kind of behavior in the current task in the next section, and show how our model can reproduce this without assuming any noise. We will address this issue again in the discussion.

Individual Participants and Individual Model Runs

Another measure of a model's quality that goes beyond the comparison of average curves involves the inspection of individual model runs with individual subjects. Especially in tasks that cause a high variability in behavior, this is interesting, because it enables us to inspect the flexibility of both model and humans. Therefore, we inspected whether we could identify individual model runs, in the set that fed the average curves shown in Fig. 2, that match the learning trajectories of individual subjects.

One characteristic of the model, both in the EUD 1 and the EUD3 condition, is that it can exhibit relatively "gradual" learning. Essentially, the model settles on a "current best" based on the current estimates, and is drawn away from it again by increasingly correct estimates (which perhaps reveal that the option it has settled on is not that superior after all, as well as the decrease in that option's *information gain* relative to the that of the other options. It gradually converges towards an increased choice of the true best, in this case UL. Figure 4 shows an example of this, the model being matched to participant 126.

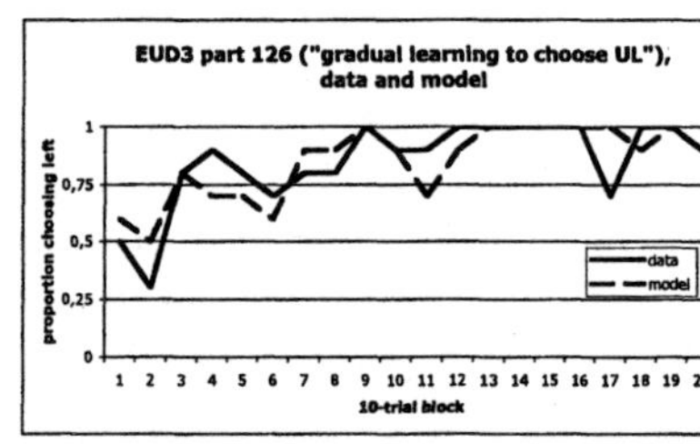

Figure 4

We can see similar patterns of learning in individual model runs under the EUD1 condition, even though the model does, on average, hardly learn. One striking and frequent pattern under the EUD1-condition, and one that we believe is hard to capture by a model that uses only noise as variability-inducer, is the complete abandonment and later re-uptake of the option UL. This pattern, which we call spikes (one of many examples is shown in Fig. 5) is due to two factors: Firstly, the quality of UL as "best" choice is less clear in EUD1 that in EUD2, so its estimate will remain closer to that of its competitors, occasionally falling below them. Secondly, the advantage that UL might have over the other options in terms of expected payoff is not big enough to counter the fact that its *information gain* measure will decrease as it is chosen more often, falling below that of the competitors: Familiarity breeds contempt. These two factors taken together model the fact that, under EUD1, the model can't establish sufficient "trust" in an option in order to exploit it: As it repeatedly chooses UL, its estimate remains mediocre, at the same time, the other options begin to seem more attractive again. If there is one option that has been explored particularly rarely, this option will promise a high *information gain and* will be chosen for a couple of trials, until it has been established that its current estimate is below the UL and its *information gain* has decreased. The result is a "spike", as the one seen below. Fig. 6 shows one of the many examples of this pattern that can be found in the data and in the set of model runs. Here, one model run is matched to participant 293.

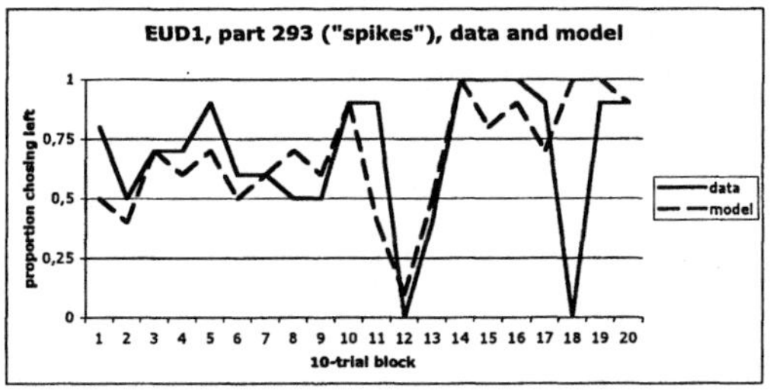

Figure 6.

Discussion

It has been our goal to demonstrate that Information gain, not just actual payoffs, can drive Decision Making in a probabilistic environment. To this end we have created a model that learns to make choices in a probability learning task, choosing among options based on its estimates of their actual payoffs and the *information gain* associated with selecting that option. Upon detecting that this model fits human behavior much better than a model that ignores *information gain,* we ask ourselves two questions: Is the behavior of such a model rational? And: Is the model correct? The answer to the first question is, in our opinion, a clear "yes". Exploration is appropriate in probabilistic environments, because it increases the accuracy of the probabilistic representations. The model, like humans, is able to adjust its amount of exploration to the structure of the probabilities, abandoning exploration early when they are easier to discriminate. Its need for exploration prevents the model from being stuck with the wrong choices, but this need is also systematically related to the structure of the environment, such that its impact will be smaller the fewer data are needed to arrive at reliable estimates.

But: Is the model as it stands now correct? This is unlikely. It has only recently been formulated and has only been tested with the datasets reported here. We regard the present version of the model as a skeleton containing the elements we believe are essential for explaining human choices and learning in a probabilistic situation. However, as the word "skeleton" suggests. augmentation is clearly called for. For instance, it is likely that the value of exploration, i.e. of information gain, varies from situation to situation. It is even more likely that Humans themselves can adapt its importance to different situational demands. Essentially, this calls for modeling work regarding systematic changes in the c parameter. Another open question concerns the precise definition of Information Gain. Right now it is a "raw", content independent indicator of how much we have learned by making a choice, and it always decreases according to the same function. This is a strong assumption, which must be tested empirically. It remains to be seen whether the monotonous decrease of Information Gain remains adequate to model behavior in situations with, e.g., non-stationary probabilities. Human adjustment to this additional complexity will certainly pose another challenge.

Finally, the view put forth in this paper is that choices in probabilistic environments can be influenced by the explicit, active wish to explore. This notion is partially at odds with models that only assume a noisy estimation, or a noisy choice process in order to account for variability in behavior. A comparison between these two approaches can be resolved on two different levels: Experiments can be designed in which the two models make clearly distinct predictions, and formal analyses of both model can be conducted to reveal how the two models might "fit" different situations, and under which circumstances their choices converge. These efforts might lead toward a more complete theory of how these two drives of exploration and exploitation might be interacting in driving human behavior, an endeavor of which the model reported here merely scratches the surface.

Acknowledgments

S.N. would like to thank Niels Taatgen (CMU, Department of Psychology) and Howard Seltman (CMU, Department of Statistics) for their extremely helpful comments on this work.

References

Anderson, J.R. & Lebiere, C. (1998). *The Atomic Components of Thought.* Mahwah, New Jersey: Erlbaum.

Broeder, A. (2000). Assessing the Empirical Validity of the "Take The Best" heuristic as a model of human probabilistic inference. *Journal of Experimental Psychology: Learning, Memory and Cognition, 26(5),* 1332-1346.

Brown, J.C. & Lovett, M.C. (2001). The effect of reducing information in a modified Prisoner's Dilemma Game. In J.D. Moore & K. Stenning (Eds.), Proceedings of the 23rd Annual Meeting of the Cognitive Science Society (pp. 162-167). Mahwah, New Jersey: Erlbaum.

Brown, J.C. & Lovett, M. C. (in preparation). Learning to Choose in a Non-Deterministic, Single-Player Version of the Prisoner's Dilemma Game. Carnegie Mellon University, Pittsburgh, PA.

Estes, W.K. (1964). Probability Learning. In A. W. Melton (Ed.), *Categories of human learning.* New York: Academic Press.

Gigerenzer, G. (1994). Why the distinction between single-event probabilities and frequencies is relevant for psychology (and vice versa). In G. Wright & P. Ayton (Eds.), *Subjective Probability.* New York: Wiley.

Payne, J.W., Bettman, J.R. & Johnson, E.J. (1993). *The Adaptive Decision Maker.* New York: Cambridge University Press.

Rieskamp, J. & Hoffrage, U. (2000). *When do people use simple heuristics and how can we tell.* In G. Gigerenzer, P.M. Todd & the ABC Research Group, *Simple Heuristics that Make Us Smart.* New York: Oxford University Press.

A Cognitive Model of Episodic Memory Integrated With a General Cognitive Architecture

Andrew Nuxoll (anuxoll@umich.edu)
University of Michigan, 1101 Beal Ave.
Ann Arbor MI 48109-2110 USA

John E. Laird (laird@umich.edu)
University of Michigan, 1101 Beal Ave.
Ann Arbor MI 48109-2110 USA

Abstract

Episodic memory provides a mechanism for accessing past experiences and has been relatively ignored in computational models of cognition. In this paper, we present a framework for describing the functional stages for computational models of episodic memory: encoding, storage, retrieval and use of the retrieved memories. We present two implementations of a computational model of episodic memory in Soar. We demonstrate all four stages of the model for a simple interactive task.

Introduction

Episodic memory, as described by Tulving (1983, 2002), captures a history of what happens to an entity. In contrast to semantic learning, episodic learning remembers events and history that are embedded in experience, while semantic learning attempts to extract out facts from their experiential context. Thus a memory of looking out over the grand canyon during your last family vacation is an episodic memory; however, if someone asks what state the Grand Canyon is in, we would typically use semantic memory to answer (unless the answer relies on the recall of a specific episode where someone is telling you where the Grand Canyon is).

In humans, episodic memory has several characteristics that define it and may have an impact on its implementation. Below are some of the most distinguishing aspects of episodic learning:

1. Architectural: The mechanism is used for all tasks and does not compete with knowledge-based reasoning. Reasoning can impact episodic memory by affecting the determination of what is stored such as through deliberate rehearsal.
2. Automatic: Memories are created without a deliberate decision.
3. Autonoetic: A retrieved memory is distinguished from current sensing.
4. Autobiographical: The rememberer remembers the episode from his or her own perspective.
5. Variable Duration: The time period spanned by memory is not fixed.
6. Temporally Indexed: The rememberer has a sense of the time when the episode occurred.

The obvious value of episodic learning is that it gives an entity a personal history. The recorded history can later be recalled to answer questions about the past, to aid in decision making through predicting the outcome of possible courses of action, possibly by creating an internal model of the entity and its environment, or to help keep track of progress on long-term goals. The history can also be used for deliberate reflection about past events that can improve behavior through other types of learning, such as reinforcement learning (such as the classic "I can't believe I was so stupid! I'll never do that again."). The movie *Memento* (Nolan 2001) gives a taste for how crippled humans would be without it.

There are many challenges in creating computational episodic memory, but the one that stands out is the design of the episodic memory storage and retrieval system. By its very nature, the possible cues for retrieving an episode are not known when it is stored – will your memory of dancing at your high school prom be useful for remembering the name of your date, your favorite song that was playing in the background, or how uncomfortable you felt in that pair of shoes? A further challenge is that, the number of memories continually increases as new episodes are stored without significantly degrading the time it takes to retrieve specific memories.

The goal of our research is to explore models of episodic learning that are useful both for creating AI entities and for modeling human behavior. From the AI side, current agents have no general automatic mechanisms for remembering and accessing their past. From the cognitive modeling side, current models are restricted to single tasks where prior behavior is either forgotten or must be deliberately remembered. Although in the future it may be worthwhile to deviate from some of the details of human episodic learning when building artificial systems, there is such a dearth of understanding of both human and artificial episodic ıing, it is worthwhile to use whatever sources of vledge and research available to advance our rstanding of both.

Prior Work

Episodic memory was first described by Endel Tulving in the early 1970s and later formalized in his book *Elements of Episodic Memory* (Tulving 1983). Research on episodic memory has concentrated either on the phenomenological aspects of the behavior or studies to find a neurological basis for it (Baddeley et al. 2002). Unfortunately there have not been any comprehensive computational models of episodic memory in the psychology literature and AI has paid scant attention to episodic learning. The closest work in AI has been research on case-based reasoning (Kolodner 1993). Episodic memory can be thought of as the mother-of-all case-based reasoning problems – how to store and retrieve cases about everything relevant in an entity's existence. Most case-based reasoning research has avoided these issues. One exception is a Continuous Case-Based Reasoning agent created for a robot navigation domain (Ram & Santamaría 1997). The robot recorded its sensory input at each time increment along with the current distance from its goal. By comparing a series of previous actions and their results to the most recent actions it had taken, it was able to improve its navigation. Two key differences are that their agent relies upon the cases being represented as purely quantitive data and that it performed a match for retrieval using multiple temporally contiguous cases (episodes). To date their work has not been applied outside of the original navigation domain.

There are few instances in which AI systems have an episodic memory (Vere and Bickmore 1990; Rhodes 1997). Vere Basic Agent maintained a limited episodic memory of events. However, the memory did not address issues of how to search and retrieve episodes effectively and efficiently. Rhodes created a wearable computer that continually recorded incoming information and provided the ability to search through the recorded history, but the memory was not integrated in a reasoning architecture and had a preset class of cues that could be used for retrieval. In the cognitive science community, Altmann & John (1999) built an episodic memory model using the Soar architecture that was based upon the recorded behavior of a computer programmer. However, the model was built specifically for that task and did not include a general architectural episodic learning mechanism (it was neither architectural nor automatic and required deliberate processing that could compete with the task at hand). ACT-R's (Anderson & Lebiere, 1998) mechanism for creating new chunks has some characteristics of episodic memory (it stores away partial contents of working memory for subsequent retrieval), but it does not store complete descriptions of working memory, nor are there any methods for retrieving temporally related items, and once a memory is retrieved it is not distinguished as a memory of a prior event.

Methodology

Our research can be viewed as a search through the design space of episodic learning systems, where different choices for the components (encoding, storage, retrieval) lead to different systems with different capabilities and properties. Our plan is to explore the space, using constraints from what is generally known about human episodic memory, with additional constraints coming from the need to create a computational model embedded in a general cognitive architecture. We have adopted an iterative design methodology, developing models that address some, but not all of the constraints, and then refining them. For example, our models do not encode memories that are temporally indexed or have variable duration.

Our models of episodic learning are *architectural*, meaning that episodic learning will be part of the underlying cognitive architecture and thus available for all tasks. In addition, an architectural learning mechanism does not interfere with other reasoning tasks. Although the underlying implementation is architectural and task independent, what is learned and retrieved will be influenced by task knowledge – both the situation when memories are created and stored, and the situation when memories are retrieved.

Our models of episodic memory are integrated in the Soar cognitive architecture (Newell 1990). Soar is a general cognitive architecture that has been used to model a wide variety of phenomena. It shares many of the features of other architectures such as ACT-R and EPIC (Kieras & Meyer 1997), having a short-term declarative memory and a long-term procedural knowledge encoded as rules. The version of Soar we are using has been extended to include activation of working memory elements (Nuxoll, et. al 2004),

Framework for Episodic Memory

Episodic learning can be decomposed into the following major phases: encoding – how a memory is captured and stored; storage – how a memory is maintained; retrieval – how a memory is retrieved; and use – how the memory is used to improve behavior. Some of the most basic design decisions, such as the structure of a memory, have impact across all phases.

Encoding

- Encoding initiation: when an episode is recorded. Initiation must be automatic, but there are many possible events that could trigger encoding.
- Episode determination: what information is stored in an episode.
- Feature selection: what features in an episode will be available for retrieval.

Storage

- Episode structure: how the encoded episode is stored.
- Episode dynamics: how stored episodes change over time (such as possible decay and removal).

Retrieval

- Retrieval initiation: how retrieval is triggered – automatically and/or deliberately.
- Cue determination: which data are used to cue the retrieval of an episode.
- Retrieval: which episode is retrieved given the current cue.
- Retrieved episode representation: how the retrieved episode is represented.
- Retrieval meta-data: what meta-information about the retrieved episode is available.

Use

- Once the episode is retrieved, how it is used to aid reasoning. This is not a part of the design of the episodic learning system, but depends on the capabilities of the embedding architecture, general methods, and task knowledge.

Pilot Implementation

Our first implementation was an attempt to confront and explore the issues that arise when developing a comprehensive, architectural episodic learning mechanism, without worrying about getting all the details right. To speed the implementation, we did what was easiest to do within Soar. Thus, episodes were based on the contents of working memory and episodes were stored as rules, with a subset of the current situation being designated as features used for retrieval, while other aspects of the situation were designated as the content of the retrieved episode.

Originally, the conditions and actions of the rules were based on a hand-selected subset of working memory that we knew was optimal for our domain. We experimented with using only the most highly activated working memory elements, but then faced the expected problem that if too few features were selected, the retrieval was too general in situations that were not appropriate, whereas if too many features were selected, the retrieval was overly specific and did not provide significant transfer to new situations. We were unable to develop a task-independent method to make sure that just the right number was recorded.

Although the use of rules for memories was limiting, we completed an end-to-end episodic memory system. Memories were automatically recorded whenever there was a significant change in the highest activated working memory elements. Memories were retrieved when a special symbol was created in working memory to trigger retrieval (all episodic memory rules tested for the existence of this symbol). The retrieved memories were created in an agent-specified area of working memory so that they could be distinguished from the rest of working memory.

There could be many uses of the episodes, but in the agents we developed, the episodes include sufficient information for the agent to determine what feedback it would get from the environment if it did the same action i the same situation in the future. The agents can use th episodes in one of two ways. One way aids decision makin by using retrieved episodes to predict the results of the actions and then choosing the action with the best predicted result. In the other case, episodes were retrieved as the entity deliberately reflected about the task, generating possible situations and recalling the results of its actions. In both cases, Soar's chunking mechanism learned control rules to speed future decision making.

The strength of this implementation is that it demonstrated that it is possible to create an architectural mechanism that automatically creates episodic memories and that those memories can be recalled in the future to improve problem solving without confusing the agent (partially fulfilling the first four properties of episodic memories). Its major failing arose from encoding the episodes as rules that required complete matches in order to retrieve the memories, which forced us to handpick the features encoded in the memory.

Current Implementation

The most significant change in the current implementation is that we have created a separate episodic memory in Soar, one where complete episodes are stored (similar to cases in case-based reasoning). Retrieval of the episodes is patterned on the use of buffers in ACT-R, where a retrieval is made when a cue is placed in a special location in working memory. The retrieval is based on a partial matching algorithm. This design removes all domain-dependent aspects of the previous system. The new episodic memory system fits within our episodic memory framework as follows:

Encoding:

- Encoding initiation. A new memory is encoded each time the agent takes an action in the world.
- Episode determination. All working memory elements, including their activations, are included in the episode.
- Feature selection. All features in the episode can participate in retrieval, so no feature selection is made.

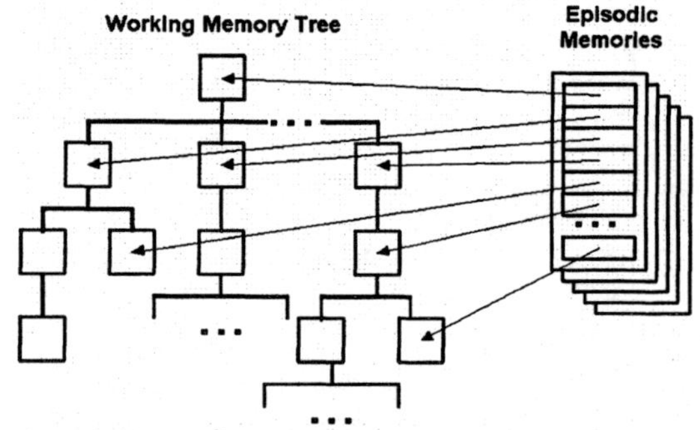

Figure 1: Episodic Memory Structure

Storage

Episode structure. As shown in Figure 1, our episodic memory structure attempts to minimize the overall storage requirements by building up a single structure (the working memory tree) that holds a single instance of all elements that have ever been in working memory. Each episode consists of a list of pointers to each of the working memory elements it contains, in a canonical order. In addition, all elements in the tree have pointers to the memories that contain them. (This is not depicted in the diagram.) The activation level of each element at the time of storage is also recorded.

- Episode dynamics. Currently, episodic memories do not change over time.

Retrieval

- Retrieval initiation. Retrieval is initiated deliberately by placing a cue in a special location in working memory.
- Cue determination. A cue is deliberately constructed by the rest of the cognitive system (using rules) in a reserved location in working memory. The cue can have any number of working memory elements.
- Retrieval. The cue is compared to all stored episodes, selecting the episode that "best matches" the cue. The match is determined by totaling the number of working memory elements that are shared between the cue and the episode. The match uses the following algorithm:
  1. The system traverses the tree containing all known working memory elements (see **Figure 1**) finding all of the elements in the cue.
  2. Each element that is matched in the tree contains a list of references to every episode that contains it. (These references are not depicted in the figure.) A list of all episodic memories that contain at least one element from the cue is created. by merging the references from each matched part of the cue.
  3. The complete cue is then compared to each episodic memory in the episodic memory list and the one that best matches the cue is selected.

  Because the number of episodic memories in the system continues to grow and any memory may potentially match a given cue, the retrieval phase may slow as more episodes are stored. However, this above approach was straightforward to implement and allowed us to create a working episodic memory system rather quickly.

  Once an episode has been retrieved, the system can also retrieve a series of episodes in temporal order via a special "next episode" command.
- Retrieved episode representation. The complete episode is retrieved in a labeled area of working memory to avoid confusion with the current state of the agent.
- Retrieval meta-data. Currently there is no meta-data retrieved with the episode.

Use

- Initially our implementation is used to support the same use as in the pilot implementation: evaluation of alternative actions. The following section describes this in detail for a simple interactive task.

The Eaters Environment

To test the system, we created episodic memory agents with different episodic memory variants in a simple game called Eaters. An eater is a Pacman-like agent that moves around a 16x16 grid world. Each cell in the grid is either empty or contains a wall, normal food (● = 5pts), or bonus food (■ = 10 pts). The eater is able to move in each of the four cardinal directions unless there is a wall in its way. Each time it moves into a cell containing food; it eats the food (receiving the appropriate score). When an eater leaves a cell it becomes an empty cell. The eater's goal is to get the highest score it can, as fast as it can. The eater's sensory input includes the contents of nearby cells, its current score, its color, and number of moves taken so far. The figure below depicts the input available to an eater (although it is represented completely symbolically for the eater).

Figure 2: An eater's sensory input

To test the episodic memory system, we created an eater to run in Soar-EM (Soar version 8.5 extended to include the episodic memory system). This eater does not know the relative value of the objects it senses, nor does it have a model of how its actions move it through the world. Our goal is for it to use its episodic memory in place of that knowledge to aid in selecting which direction it should move. For each of the possible directions, it creates a memory cue composed of its current sensory input and the proposed direction of travel. Once it has retrieved a memory of prior situations, it then retrieves the next memory (in temporal order). This new memory includes the score it received after taking the proposed action in the retrieved situation. The change in score between the two memories provides a quantitative evaluation for the proposed action. This evaluation is used to compare the proposed action to the other possible actions (whose evaluations are computed in the same manner). The agent selects the action with the highest evaluation.

There is no guarantee that an appropriate prior episode will be retrieved – that will depend on the set of recorded episodes and the algorithm used for retrieving episodes from memory. If no prior memory is found the agent must still assign a default evaluation to the action so that it can compare the action to the other possible actions. If the default value is greater than bonus food, the eater will be [illegible]d to explore new actions, which in turn would build up [illegible]pisodic memory. If a low value is used, the eater will [illegible]d the unknown. Although we tested the eaters with a variety of values, the differences in behavior were slight and the results we report are for the eater that favored exploration.

Results

To evaluate the effectiveness of episodic memory, we created two baseline eaters (without episodic memory) for comparison: a random eater and a greedy eater. The random eater has no knowledge of the value of cells and selects its next move randomly. The greedy eater always selects the move that will bring it the highest immediate score. The random eater serves as a lower bound on performance while the greedy eater provides an approximation to an upper bound on performance.

Our first episodic memory system implementation ranked the episodes by the number of features matched against the cue, with the episodes having the highest number of matching features picked. Ties were broken randomly. In this task, there are a total of 31-36 features per state; however, the correct decision is determined by the specific values of only two features: the direction being moved and the content of the cell being moved into. To explore the impact of having the episodic memory filled with different numbers of episodes, the experiment had five iterations where we measure the score of the eater for each iteration. For each iteration, the eater made 1500 internal decisions, which resulted in an average of 130 actions in the world. (Multiple internal decisions are required to evaluate the alternative actions and finally select the external action.) After each iteration, the contents of eater's world map were reset (randomly rearranged and refilled with food) and the score was reset to 0. The eaters were allowed to retain their episodic memories between iterations. The experiment was repeated five times and the results were averaged together.

The first set of bars in **Figure 3** show the episodic memory eaters' average scores after their 100th move in each iteration. The dashed lines show the average scores for the greedy and random eaters.

These results suggest that the unbiased eaters (darker columns on left of each pair) benefit from using their episodic memory to select actions. However, this benefit is

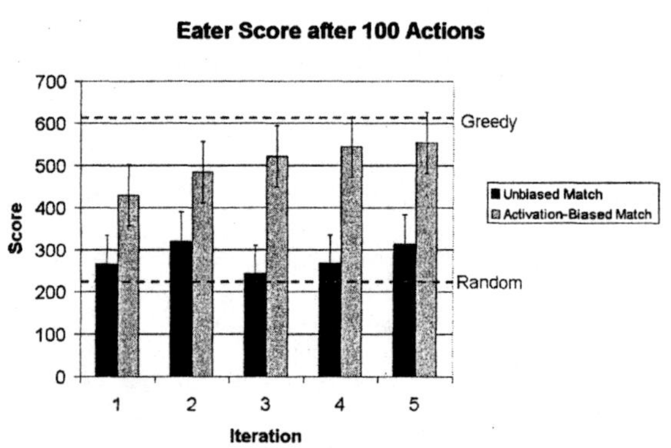

Figure 3: Average score after 100 actions

small as its score is only moderately better than it woul receive for random actions. Also, the eaters' performanc did not improve significantly on subsequent iterations despite having more memories available.

Another intriguing question is why the eater requires relatively few memories to achieve better than random performance but fails to improve beyond this level. This turns out to be of the result of an interaction between features of the task and our specific implementation. Recall that the eater creates a cue to retrieve a memory for each direction that it is considering. Since these cues differ by only a single feature (the proposed direction of travel), it is often the case that this difference is overwhelmed by a good match with other features of the cue. As a result, the eater recalls the same episode for each direction. Thus, the eater will often behave randomly, particularly when it has fewer memories in its episodic store. The eater is able to perform better than random as it gains more memories because it eventually has a large enough memory store that the single difference between the cue affects the match.

Given that memories are never lost, the expectation that an eater using an unbiased match will eventually achieve greedy behavior is probably true. However, by rough estimation, there are over 1 quadrillion different possible inputs in even this simple domain. Since only two of the features determine the correct match, the agent would have to have a *lot* of memories before its unbiased best match became a reliable predictor.

Clearly, the key problem with an unbiased partial match is that the likelihood of retrieving a truly relevant episode is small because of the large number of irrelevant features that have equal say in whether an episode is retrieved. What is needed is some mechanism that biases the selection toward relevant features. However, the relative value of matching different working memory elements is highly task dependent. A general purpose episodic memory must have a way of ranking features in a domain independent manner. To resolve this issue, we hypothesized that working memory elements which were highly active when the episode was created would be more likely to be relevant.

The activation mechanism used to bias the match is a recent addition to Soar (Nuxoll, Laird & James, 2004) with the following properties.

- Working memory elements receive an initial activation based upon the activation level of existing elements which were tested by the production that created the new element.
- An element receives an activation boost whenever a production fires that has tested it.
- Activation levels decay over time using a formula similar to that used by ACT-R (Anderson and Lebiere 1998).

For our second trial, we modified the retrieval mechanism to be biased in favor of episodes that had highly activated features which matched the cue. For each match, the activation values of all matching elements were summed. These sums were compared and the highest value was selected as the best match.

The previous experiment was repeated with the new ation-based matching scheme. The results are shown as econd set of bars in **Figure 3**. The episodic memory with an activation-based match begins by performing significantly better than its unbiased match predecessor. Furthermore, as it gains more memories the eater's performance continues to improve until it performs comparably to the greedy eater.

An alternative view of the results is in **Figure 4**. This graph depicts the fraction of correct evaluations that the agent has made over the course of an entire run (five iterations of 1500 cycles). As before, this data is an average of five runs. Both agents made over 2200 actions during the run. This graph shows that the eater with activation-biased match is improving its ability to assess its situation at a consistently faster rate than the eater with unbiased match.

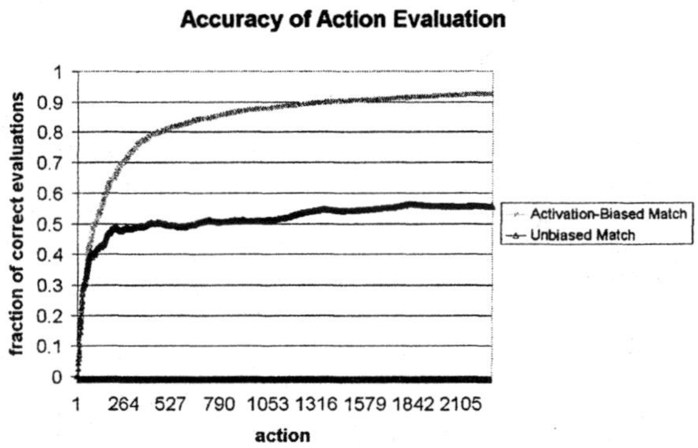

Figure 4 Comparison of eater action evaluation

Issues and Future Work

We obviously need to apply this model to more complex tasks and start comparing to human data, although it is difficult to find experiments that single out only episodic memory. Beyond that, the episodic memory framework we've established provides a good basis for examining issues and shortcomings in the current model and providing an agenda for future research.

Encoding

- Encoding initiation: As an implementation convenience we abandoned the pilot approach of using significant changes in activation to trigger an encoding. It is unclear what is the "right" trigger for initiation and we need to explore our original approach as well as others.
- Episode determination: We currently store and retrieve the complete state of working memory. As part of our attempts to improve efficiency, we may need to store and/or retrieve only subsets of working memory.

Storage

- Episode structure: We may need to modify the structure to improve retrieval efficiency. We also need to find ways of encoding temporal information in the episodes, which we expect to be more difficult than just recording the time of storage. The temporal qualities of retrieved episodes appear to be more subtle than time tags.
- Episode dynamics: Human memory is rarely eidetic How is it that some parts of memories are lost an others retained?

Retrieval

- Retrieval initiation: Initiation is always deliberate based on a specific cue. Another possibility is automatic cuing based on the contents of working memory.
- Retrieval: In the worst case retrieval is linear with the number of stored episodes, which will greatly limit the usefulness of episodic memory. We need to develop algorithms that approach constant time, although this may be possible only with a parallel implementation.

Use

- There are many avenues for using episodic memory for improving agent behavior. Some of our future efforts will include:
  - creating episodic agents in more environments where the relationships between cues and episodes is much more complex;
  - exploring the relationship between episodic and semantic memory;
  - using of episodic memory in reflective learning;
  - exploring the influence of affect on episodic memory.

References

Altmann, E. M. & John, B.E. (1999) Episodic indexing: A model of memory for attention events. Cognitive Science, 23, 117-156.

Anderson, J. R. & Lebiere C (1998) The Atomic Components of Thought. Mahwah, NJ: Lawrence Erlbaum Associates.

Baddeley, A., Conway, M. & Aggleton, J. (2002) Episodic Memory: New directions in research. Oxford University Press, Oxford.

Kieras, D. & Meyer, D.E. (1997). An overview of the EPIC architecture for cognition and performance with application to human-computer interaction. Human-Computer Interaction, 12, 391-438.

Kolodner, J. (1993) Case-Based Reasoning, Morgan Kaufmann Publishers, San Mateo, CA.

Newell, A. (1990) Unified Theories of Cognition. Harvard University Press, Cambridge, Mass.

Nolan, C. (Director) (2000) Memento [Film]. Newmarket Films. http://us.imdb.com/Title?0209144

Nuxoll, A., Laird, J., James, M. (2004). Comprehensive Working Memory Activation in Soar. International Conference on Cognitive Modeling.

Ram, A. and Santamaría, J.C. (1997) Continuous Case-Based Reasoning. Artificial Intelligence, (90)1-2:25--77.

Rhodes, B.J. (1997) The Wearable Remembrance Agent: A System for Augmented Memory. Personal Technologies.

Tulving, E. (2002). Episodic Memory: From mind to brain. *Annual Review of Psychology*, 53, 1-25.

Tulving, E. (1983). *Elements of Episodic Memory*. Oxford: Clarendon Press.

Vere, S. and Bickmore, T. (1990). A Basic Agent. *mputational Intelligence*. 6, 41-60.

Comprehensive Working Memory Activation in Soar

Andrew Nuxoll (anuxoll@umich.edu)
University of Michigan, 1101 Beal Ave.
Ann Arbor MI 48109-2110 USA

John E. Laird (laird@umich.edu)
University of Michigan, 1101 Beal Ave.
Ann Arbor MI 48109-2110 USA

Michael R. James (mrjames@umich.edu)
University of Michigan, 1101 Beal Ave.
Ann Arbor MI 48109-2110 USA

Abstract

Memory activation has been modeled in symbolic architectures in the past, but usually at the level of individual chunks or productions in long term memory. Recent research (Chong 2003) has demonstrated activation at the level of individual elements of working memory. In this paper, we present a comprehensive implementation of working memory activation in Soar that takes advantage of the unique characteristic of Soar's working memory structure, namely persistence. We also explore modifications to activation so that the activation of new working memory elements is not a fixed level, but is based on the activation of the working memory elements tested in its creation. We demonstrate our model in terms of how it aids the selection of features relevant to learning.

Introduction

Since its inception, the Soar architecture (Newell 1990) has focused on symbolic reasoning to the exclusion of numeric processing. Even the learning mechanism, chunking, collected conditions and actions for new production rules through an analytic method that did not require any numeric processing. One can think of these earlier versions of Soar as experiments in the sufficiency of purely symbolic reasoning.

In recent years, we have looked to expand the types of knowledge that Soar can learn that has involved moving beyond a single architectural learning mechanism, to incorporate an episodic memory (Nuxoll & Laird, 2004) in Soar. A critical part of episodic memory is determining which stored episode is the best match for the current situation. Using a purely symbolic match is inadequate because of the large number of irrelevant features in the episode. Thus, we needed some way of biasing the match to the most relevant features. One possible bias is the "activation" of the features. Activation of the contents of working memory was pioneered by the ACT family of architectures (Anderson & Labiere, 1998) and was initiall implemented in Soar by Ron Chong (2003) to suppoı forgetting in working memory.

In this paper we describe extensions to that original implementation that include a more comprehensive implementation (Chong's implementation supported activation only in a subpart of working memory), a more efficient implementation, and an implementation that takes advantage of the unique structure of Soar's working memory and distinguishes it from ACT-R's activation scheme. Furthermore, we present results of using this activation-based memory to improve the retrieval of episodic memories.

Overview of the Soar Architecture

Soar is a production rule-based cognitive architecture. Like most other architectures of this type, Soar has two types of knowledge: *working memory* (short term, declarative) and *production rules* (long-term, procedural). Working memory consists of a collection of attribute-value pairs. The agent's current state, including both external sensing and internal inferences, is stored in working memory. Production rules consist of actions and conditions. If the conditions of a production match the contents of working memory then that production *fires* its actions which create (or remove) one or more elements in working memory. These changes may cause the agent to take an action in its environment, which can result in additional changes to working memory via its sensory input. These changes to working memory may in turn trigger the firing (or retraction) of additional productions so that this "match-fire" cycle repeats indefinitely.

The Soar architecture has several unique characteristics. We describe here only the characteristics that are relevant to this research:

- Simultaneous Production Firing: Many production systems allow only one production to fire at a time. Soar allows all productions whose conditions match working memory to fire in parallel.
- Operators: To avoid the conflicting behavior that might result from simultaneous production firing, Soar supports a special knowledge structure called an

operator. A Soar production may *propose* an operator by creating it in working memory. Multiple operators may be proposed in a given situation but only one can be *selected*. Selection is controlled by the creation of *preferences*: structures created by production rules that test for proposed operators and details of the current situation. The selection of an operator can trigger additional productions to fire *apply* the operator, making changes to working memory.

- Decision Cycles: Soar operators extend the traditional match-fire cycle of Soar to a three-phase *decision cycle*: propose, select and apply.
- Persistent working memory: Soar distinguishes between two types of persistence in working memory. Working memory elements (WMEs) that are created as part of an application of an operator persist indefinitely; they remain in working memory until they are explicitly removed. These are called *o-supported* WMEs. The remaining WMEs are removed as soon as the *instantiation* of the production that created them ceases to match working memory. (An instantiation is the collection of WMEs that match a production). These are known as *i-supported* WMEs. The advantage of i-support is that it automatically removes WMEs that are no longer relevant to the current situation. For example, an operator is proposed based on specific features of the situation, when one of these features change, the operator is automatically retracted. The same is true for elaborations of the state, such as the calculation that a block is clear based on the fact that no blocks are above it. This is a simple inference that is automatically retracted when a block is placed on top of it, so that it is no longer clear.

Original Implementation of Activation in Soar

The original addition of working memory activation to Soar by Chong (2003) allowed persistent working memory elements (o-supported WMEs) to decay from working memory over time in a manner very similar to the decay of chunks in ACT-R. The implementation reserved a section of Soar's working memory for activated WMEs. The activation level of these WMEs changed as follows:

- WMEs received an initial, fixed activation when they were first created.
- Any time that a WME was tested by a production that fired, it received an activation boost.
- Any time an action would attempt to add an existing WME, the existing WME would receive an activation boost.
- WME activation levels decayed over time using an exponential decay formula identical to that used by Chong:

$$A_i = \beta + \ln\left(\sum_{j=1}^{n} t_j^{-d}\right) + \varepsilon$$

A_i is the activation of a WME at time i. β is a base leve constant. t_j is the number of cycles since the WME wa referenced for the jth time. d is a learning rate parameter which we set at the same default as in previous research (0.8). ε is a noise component. We set this value to zero for our experiments.

Throughout this paper we will use the term "reference" to refer to an incident that led to an activation boost (i.e., a WME is tested or recreated).

Extensions to Chong's Activation Implementation

Although Chong's work established the possibility of using activation in Soar, several extensions are possible.

- Extend activation to all of working memory, not just a subset.
- Improve the efficiency. Calculating activation values can be computationally expensive. We improved the efficiency by not updating the activations every cycle, and only calculating the activation values when they were used. Other optimizations included using a bounded approximation of each WME's complete history (an optimization also used by Chong) and pre-calculating commonly used activation values. To evaluate these improvements, we ran ten iterations of a moderately complex task in Soar with three types of memory activation: a) no memory activation (i.e., the original, unmodified Soar) b) a memory activation mechanism without the efficiency enhancements and c) the enhanced memory activation system. Although the unoptimized activation implementation is significantly slower than Soar without activation, the optimized implementation is not (see Figure **1**).

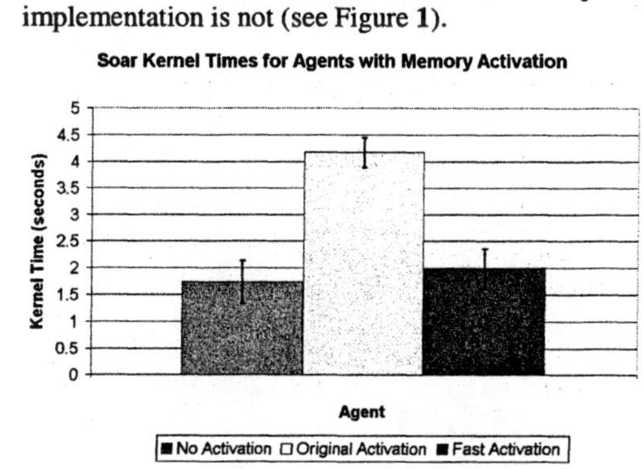

Figure 1: Demonstration of Efficiency Improvements

- Apply activation differentially to i-supported and o-supported working memory elements. This is the major conceptual change and is described in detail below.
- Base the activation of newly created WMEs on the activation of the WMEs matched in the conditions of the rules. This is a departure from both the ACT and Chong approach to activation, where newly created WMEs get fixed boost. The impact of this change is demonstrated n the results section.

Memory Activation Implementation

Like Chong, our original implementation ignored i-supported WMEs because even without activation, i-supported WMEs are automatically removed when the WMEs they are based on are removed. Thus, the o-supported WMEs are the persistent structures in working memory that the i-supported structures are entailments of. If we correctly maintain the o-supported structures, the i-supported structures will take care of themselves. The activation of an o-supported WME is based on a decay function of all prior references to that WME – each of these references are decayed individually, with the activation being the sum of all of the current values of those references. Therefore, in our implementation, all WMEs have an associated history of prior references (see Figure **2**).

The conceptually simplest approach to maintaining activations is to update the activation values of every WME every cycle. However, the majority of WMEs are not accessed on a given cycle, so that the calculated activations are never used. An alternative approach is to only update the activation when a WME is accessed or when it needs to be removed. But how do we know when a WME must be removed? We can "predict" which cycle a WME should be removed by using the reference history and decay function, and then store that cycle on a "timelist." Then on each cycle we check the timelist to see which elements need to be removed. If a WME is referenced on a cycle, its removal time will be updated on the timelist. As shown in Figure **2**, the timelist is an array indexed by removal time and treated as a large circular queue. Each entry in the array is a future cycle that contains a list of WMEs that will be removed on that cycle if they are not referenced again before that cycle.

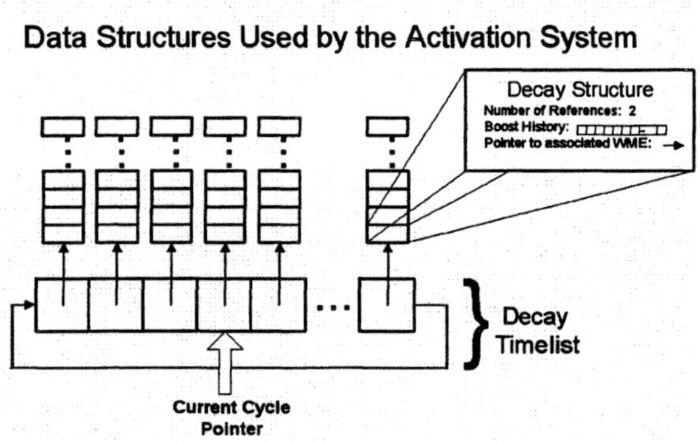

Figure 2: Activation Data Structures

The decay data structure does *not* maintain an actual numerical activation level for the WME. This value is calculated only when the WME is referenced and it is necessary to determine the WME's new position in the decay timelist

The "activation life cycle" of a WME has three stages:

Creation: When a new WME has been created.,a decay data structure is created and attached to the WME. The WME is assigned an initial reference history.

Activation: At the conclusion of each decision cycle all referenced WMEs have their boost histories updated and their position in the decay timelist adjusted appropriately. This step is by far the most computationally intensive. As discussed previously, we were able to reduce the time required by making some efficiency improvements. As a result, the time required for this step is now linear in the number of WMEs that were referenced during the cycle.

Removal: Once all the WMEs in the decay timelist have been adjusted, any WMEs that remain at the current position in the timelist are removed from working memory.

Evaluation of Activation Implementation

Although one use of activation is to decay and possibly remove WMEs, it is difficult to evaluate the direct impact of the activation scheme on behavior. Our initial implementations of activation and decay schemes closely model both ACT-R and Chong's Soar implementation, so there is little to be gained by running any new simulations of decay and removal. Instead, we have chosen to focus on the role activation can play in aiding learning, specifically in selecting relevant features for retrieving past experiences from an episodic memory we are implementing in Soar (Nuxoll & Laird, 2004).

Our initial research on episodic memory is being carried out in a simple interactive domain called Eaters, which consists of a 16x16 gridworld. An agent (i.e., eater) in the world can sense the contents of the cells around it and must make decisions about what direction to move in that world. To aid in its decision, the eater has approximately 30 sensory inputs from the outside world. Most of these inputs consist of the contents of the cells that surround the eater. Cells can contain normal food (●), bonus food (■) or a wall. If an eater has already visited a cell it will be empty. Figure **3** depicts the sensory input of an eater.

Figure 3: The Eaters World

An important feature of the Eaters world is that different sensory inputs have varying levels of significance. If an eater is considering moving west, the content of the cell just west of it can have a dramatic impact on the results of its ision whereas the content of cells in other directions are important.

Initial Results

For our experiments, an eater was created that used an episodic memory to record its past situations and recall them to aid in future decisions. The eater would evaluation each action it was considering by attempting to recall a similar situation in its past. The results of that past action were used to evaluate the action the eater was currently considering. As a result, the quality of the eater's action depended heavily upon its ability to select the memory that best matched the current situation and proposed action.

We began initially with an eater that selected the best episodic memory via an unbiased partial match. We then added an activation-bias to the partial matching algorithm. When episodes were recorded, the activation level of the WMEs in the episode were also recorded. Rather than calculate the exact activation level of each WME (a time intensive process), we used its position in the decay timelist as a discretized measure of activation. When a partial match was being performed, WMEs with a higher recorded activation added more weight to the match than those with lower activation. (i.e., The match score was equal to the sum of the activation levels of the matching WMEs.)

We ran both eaters for five iterations of 7500 decision cycles and averaged the results. During a single iteration the eater took more than 2200 actions in the world. (The episodic retrievals and comparisons require multiple decision cycles.) After every 1500 decision cycles the contents of the gridworld would be randomly refilled and rearranged so as to present a variety of situations to the eater rather than a sparse grid.

Figure **4** shows a comparison of these two eaters. The ordinal is a measure of the fraction of retrievals the eater made which were correct (i.e., led to an accurate evaluation of the proposed action). A perfect eater would create a horizontal line across the top of the graph. In this graph, the eater with an activation-biased match is showing about a 30% improvement over the unbiased eater.

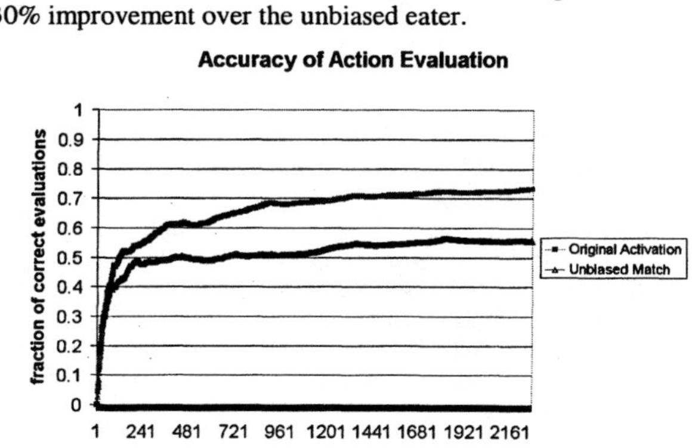

Figure 4: Effectiveness of Activation-Bias

Modifications to the Activation System

During the course of this research, we made two observations about the design of activation system that led us to believe we could improve its ability to predict salient features of an agent's environment. As mentioned above, the original implementation does not activate i-supported WMEs..However, the i-supported WMEs can "block" the activation of the underlying o-supported WMEs responsible for the creation of the i-supported WMES. Consider the case where a production tests an i-supported WME. That WME does not receive any activation boost since it is not activated. However, the o-supported WME(s) that were tested in order to create that WME *also* do not receive a boost. Thus i-supported WMEs are masking o-supported WMEs from activation.

This phenomenon is shown in Figure **5**, where six o-supported WMEs (A-F) are depicted along with five production rule instantiations that create i-supported WMEs (1-5). As mentioned earlier, i-supported WMEs do not directly receive activation, but instead exist "at the pleasure" of the o-supported elements they are based on. Thus, if D is removed, that will cause 4 to be retracted, which in turn will lead to 5 being removed.

In terms of activation, if WME 5 is tested, it will not receive any activation boost (because it is i-supported). Moreover, WMEs E and F (and indirectly C and D) also do not receive any activation boost although they are responsible for maintaining 5. Once they decay, WME 5 will be removed despite the fact that it has been referenced.

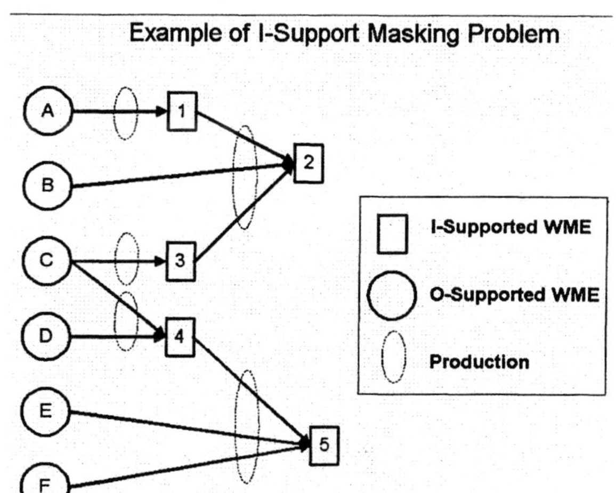

Figure 5: Example of the I-Support Masking Problem

In response to this problem we added a "pay it backward" approach for passing references from i-supported WMEs to o-supported ones. The activation system now calculates the *ıf o-support* for any referenced i-supported WME and boosts the activation of WMEs in the o-support set. In example, the set of o-support for WME 5 includes

WMEs E and F, which are directly tested by the rule that creates 5, as well as C and D, which are indirectly tested by the WMEs that create 5 (via WME 4).

The second observation we made was in regard to initial activation for newly created WMEs. In the original implementation, these WMEs receive a fixed initial boost equivalent to a single reference. (The same approach is used in ACT-R.) In terms of decay and removal, this may be sufficient because newly created WMEs will usually immediately lead to additional rule firing (and receive a boost in activation) or if they are not relevant to the situation, will not be tested and be removed after a short time. However, when using memory activation as a measure of the importance of features of the agent's state, this flat level of initial activation can be misleading. An agent might test multiple WMEs with high activation and create one new WME that would have much lower activation. Thus, at creation time, when this feature will most likely be very important as a cue for future retrieval, its activation is much lower than the activation of the features that led to its creation. This may be a new manifestation of an old problem. Chong noted that in both his model and some ACT-R models, newly created WMEs/chunks can decay rapidly and never have a chance to participate in reasoning.

To ameliorate this problem, the second modification we implemented was a "pay it forward" approach for setting the activation level of new WMEs. Thus, the activation of a new WME is based on the activation levels of the set of o-support WMEs tested in its conditions, which includes all o-supported WMEs tested in the conditions, as well as o-supported WMEs that were precursors to i-supported WMEs tested in the conditions. Although it is convenient to think of each WME as having activation, the system never directly represents a specific activation value. Instead, it stores a history of all of the activation boosts each WME has received. Thus, we base the activation of a newly created WME on the activation levels of the set of o-support WMES, by averaging the reference histories of those WMEs and assigning that average history to the new WME.

Final Results

The two improvements described in the previous section were incrementally added to the activation system and the experiment was repeated. The results can be seen in Figure **6** along with the results from the previous experiment. Modifying the system so that the o-support sets received boosts in activation raised the accuracy because it was more likely that the correct features (which are o-supported) would contribute to retrieving an episode. Modifying the boosting of newly created WMEs had an even bigger impact because one of the most important features (the proposed direction of movement) is created just before an episode is stored. In the original versions, this feature would have activation that was significantly below others, making it a marginal feature for biasing retrieval. The modificatio boosted the proposed direction of movement feature – it creation was dependent on highly active features – an made a marked improvement in the agent's ability to retrieve appropriate episodes and select the correct action.

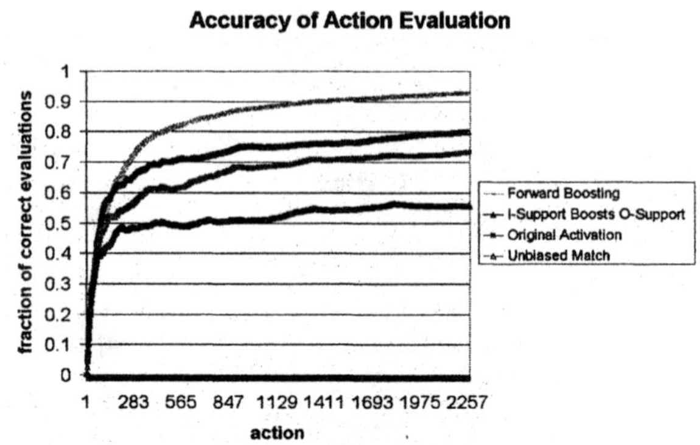

Figure 6: Result of Activation System Modifications

Future Work

We currently see three areas where this work can be expanded. First, this research demonstrates that a comprehensive activation system can be an effective technique for selecting the salient features of an agent's working memory. It would be useful to test the effectiveness of the activation-biased match in other domains and with other learning mechanisms (e.g., reinforcement learning).

Second, the changes we have made to the activation system will likely have an impact on the effectiveness of that system as a memory decay mechanism.

Finally, the use of memory activation as a search control mechanism (i.e., selecting which matching production will be selected to fire) has a long tradition in the ACT-R architecture. What impact might this approach to memory activation have if it was used to influence operator selection in Soar?

References

Anderson, J. R. & Lebiere C (1998) The Atomic Components of Thought. Mahwah, NJ: Lawrence Erlbaum Associates.

Chong, R. (2003) The addition of an activation and decay mechanism to the Soar architecture. Proceedings of the 5th International Conference on Cognitive Modeling. Bamberg, Germany. April, 2003.

Laird, J. E., Newell, A., and Rosenbloom, P. S., (1987) Soar: An architecture for general intelligence. Artificial Intelligence, 33(3), 1-64.

Newell, A. (1990) Unified Theories of Cognition. Harvard University Press, Cambridge, Mass.

Nuxoll, A. & Laird, J. (2004). A Cognitive Model of Episodic Memory Integrated With a General Cognitive Architecture. International Conference on Cognitive odeling 2004.

Modelling performance in the Sustained Attention to Response Task

David Peebles (D.Peebles@hud.ac.uk)
Department of Behavioural Sciences, University of Huddersfield,
Queensgate, Huddersfield, HD1 3DH, UK

Daniel Bothell (db30@andrew.cmu.edu)
Psychology Department, Carnegie Mellon University,
Pittsburgh, PA 15213, USA

Abstract

We present a computational model of human performance on the Sustained Attention to Response Task, a computer-based task in which people must withhold responses to infrequent and unpredictable stimuli during a period of rapid and rhythmic responding to frequent stimuli. The model, formulated within the ACT-R cognitive architecture, accounts for human performance in terms of two competing strategies and the dynamic modification of priorities given competing task demands to minimise both response time and error. The model suggests that such strategic factors may be responsible for the observed speed-accuracy trade-off rather than the alternative proposal based on sustained attention.

Introduction

The Sustained Attention to Response Task (SART; Robertson, Manley, Andrade, Baddeley, and Yiend, 1997) is a computer-based task designed to measure a person's ability to withhold responses to infrequent and unpredictable stimuli during a period of rapid and rhythmic responding to frequent stimuli. As its name implies, the creators of the task interpret performance as reflecting the ability to sustain attention, which they define as "the ability to self-sustain mindful, conscious processing of stimuli whose repetitive, non-arousing qualities would otherwise lead to habituation and distraction to other stimuli" (Robertson et al., 1997, p. 747).

In the standard version of the task, participants are presented with the digits 1 to 9 in random order at a rate of one every 1.15 s. Each digit is presented for 250 ms followed by a 900 ms mask and participants are required to respond to each digit as rapidly as possible by clicking the mouse, apart from when they see the number 3 when they must withhold the response. The task consists of a total of 225 trials (25 of each of the 9 digits) and lasts approximately 4.3 min. Participants are instructed to respond as quickly as possible while making as few errors (i.e. incorrectly responding to a 3) as possible.

The central idea behind this paradigm is that the continuous performance over 225 trials together with the long and unpredictable intervals between targets encourages the development of an automatic response to non-target 'go' trials and that vigilant monitoring is required to withhold this response on the infrequent 'no-go' target trials. This contrasts with previous perceptual detection paradigms used to test sustained attention which have typically required participants to respond to the infrequent target (e.g., Loken, Thornton, Otto, & Long, 1995; Parasuraman, Mutter, & Molloy, 1991; Whyte, Polansky, Fleming, Coslett, & Calvallucci, 1995). Robertson et al. (1997) argue that in these paradigms, it is responses to the target stimuli that can become automatised which tends to increase the level of performance and makes the detection of attention lapses harder.

The main focus of research with the SART has been to investigate the performance of patients with traumatic brain injury (TBI) affecting the frontal lobes, a region previously associated with the ability to sustain attention (e.g., Rueckert & Grafman, 1996; Wilkins, Shallice & McCarthy, 1987). In several studies, Manly, Robertson, and their colleagues have shown that performance on the SART is diminished following TBI to the frontal lobes and that SART performance correlates with other measures of sustained attention (e.g., Manly, Robertson, Galloway, & Hawkins, 1999; Manly, Davison, Heutink, Galloway, & Robertson, 2000; Manly et al., 2003; Robertson et al., 1997).

The relationship between response time and error in the SART

In several studies, Manly and Robertson have found that individuals' rates of responding were significantly predictive of the number of errors they made (Robertson et al., 1997; Manly et al., 1999). In particular, the mean RT for the four trials prior to an error of commission (i.e. incorrectly responding to a 3) was found to be significantly faster than that for the four trials prior to a correct response to a no-go trial. This suggests that performance may be determined to a large extent by an individual's particular emphasis when trying to satisfy the competing task instructions to minimise both response time and the number of erroneous responses.

To study the relationship between RT and error in more detail, Manly et al. (2000) pooled data from 109 neurologically healthy participants who had carried out the SART in a number of different studies. They found that the mean RT for responses to a go trial was 375 ms (*SD* 65) and that subjects made on average 6.36 (*SD* 4.36) errors of commission and 1.06 (*SD* 3.41) errors of omission (incorrectly withholding a response on a go trial). They also found a significant negative correlation ($r = -0.49$) between the mean RT for go trials and the number of errors of commission made, indicating that,

Table 1: Mean Number of Responses and Mean Response Time (RT in ms) for Trials Immediately Before and After Correct and Erroneous Responses to the Target Digit 3 in the SART (Manly et al., 2000).

	Pre-correct		Pre-error		Post-correct		Post-error	
	Response	RT	Response	RT	Response	RT	Response	RT
Mean	17.6	384	6.0	333	17.30	351	6.07	364
SD	4.34	68.4	4.03	63.7	4.32	60.6	3.95	95.17

as in the previously reported studies, participants who responded more rapidly were more likely to make more errors. This relationship is shown in Figure 1 which plots the number of errors of commission against mean RT to non-targets for each participant.

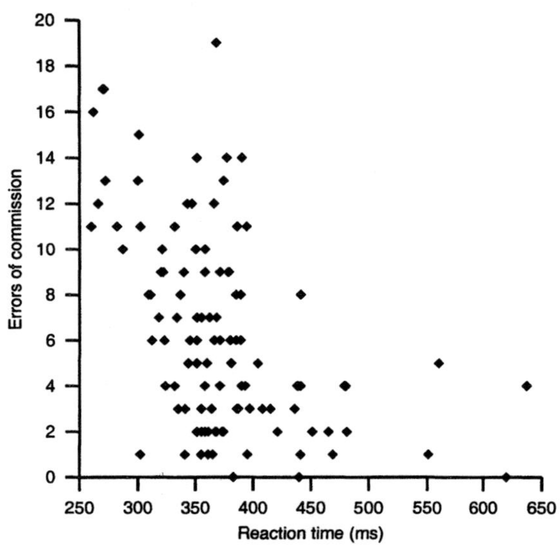

Figure 1: Plot of errors of commission against reaction time to non-targets in the SART (reproduced from Manly et al., 2000).

Manly et al. also analysed the within-subjects pattern of RTs relative to no-go trial responses. To do this they identified four categories of go trial: (a) those immediately prior to a correctly withheld response to the target digit (pre-correct); (b) those immediately prior to an incorrect response to the target digit (pre-error); (c) those immediately after a correctly withheld response to the target digit (post-correct); (d) those immediately after an incorrect response to the target digit (post-error). Their findings are summarised in Table 1. The mean RT for pre-error trials (333 ms) was significantly faster than that for pre-correct trials (384 ms). In addition, RTs for post-error trials (364 ms) were significantly slower than those for pre-error trials (333 ms), indicating that after participants had detected an error, they made a conscious decision to slow down. Finally, the data also revealed that the mean RT for post-correct trials was significantly faster than that for the pre-correct trials, indicating that in general, participants tended to speed up after correctly withholding a response to 3.

In an attempt to reduce the strength of the relationship between RT and error rates found in their study, Manly et al. (2000) produced a modified 'response-locked' version of the SART in which subjects were required to respond in time with an auditory signal presented 100 ms after stimulus onset. This modification was designed to reduce the variability in RTs both between and within subjects by setting a target RT close to that found in the previous study to be associated with correct responses to no-go trials (approximately 380 ms). Manly et al. argued that the modified task should be able to differentiate between the alternative accounts of the speed-accuracy trade-off in RTs. For example, if the main determinant of performance is the ability to sustain attention, then the number of errors of commission in the response-locked SART should be the same as in the standard version, although the amount of variation within an individual participant's RTs and also between participants RTs should be reduced. However, if participants' balancing of priorities in relation to the conflicting task demands is the primary determinant of performance, then setting the target RT to a value that has previously been associated with correct responses to no-go trials should reduce the number of errors of commission as well as the variability in RTs.

In an experiment in which 30 normal participants carried out both response-locked and standard versions of the SART, Manly et al. (2000) showed that the auditory signal significantly reduced the variability in individual participant's RTs and also between participants' RTs in the response-locked condition. They also found that the modification reduced the correlation between RT and error in the response-locked condition ($r = -0.19$). The comparative results were somewhat less conclusive. In the previous study, the proportion of erroneous responses to target trials was 25.44%. In the standard experiment condition, that proportion increased to 27.67% but in the response-locked condition it decreased to 22.57%. It seems, therefore, that introducing an auditory signal in the response-locked condition reduces the number of errors of commission as well as the variability in RTs, supporting the view that participants' balancing of priorities is a strong factor in performance. The difference between the two experiment conditions was found to be only approaching statistical significance, $F(1, 29) = 3.16$, $p = .09$. However the standard deviations of the errors for both the standard SART ($M = 8.30$, $SD = 5.11$) and response-locked SART ($M = 6.77$, $SD = 4.80$) are relatively large so it is possible that this issue may only be made less equivocal by a further study using a larger sample size.

A process account of the SART

The principal relationship between RT and error rate may, therefore, be characterised as a speed-accuracy trade-off, but as Manly et al. (2000) argue, this merely describes the relationship but does not provide an account of it. Manly et al. suggest that this pattern results from the waning of sustained attention that occurs during the course of the task. An alternative interpretation, however, is that observed performance is the result of *strategic* factors, in particular the attempt by subjects to find an appropriate strategy that allows them to satisfy the conflicting demands to minimise both RT and error. It is an open and interesting question, therefore, whether a strategy-based interpretation can account for performance in the SART. The purpose of the modelling work reported here is to address this question by producing and analysing a process model of the task.

The ACT-R cognitive architecture

ACT-R 5.0 (Anderson et al., submitted) is the most recent version of the ACT-R cognitive architecture (Anderson & Lebiere, 1998) that adds perceptual and motor modules to the existing cognitive module. The perceptual and motor modules provide ACT-R with rudimentary speech and audition capabilities, visual attention and processing mechanisms, and elements of motor control to simulate interaction with a computer keyboard and mouse.

ACT-R's cognitive system consists of a module for maintaining information about goals and two memories: a procedural memory in the form of a set of production rules and a declarative memory which is a network of chunks. The interface between the production rules and the other modules is through a set of *buffers*. The buffers hold information about such things as the current goal, the item of declarative knowledge that is currently available to the system, and the current state of the perceptual and motor modules. Each buffer may contain only one chunk of information, and each new request for information replaces the current contents of the buffer. Productions are rules of the form "IF <conditions> THEN <actions>," the conditions specifying a pattern of information that must be present in the buffers for the rule to apply and the action specifying the actions to be taken should this occur. The actions can be direct modifications of the buffers' contents or requests to other modules to change the buffers, for instance a request of the memory system to retrieve a chunk or a request to the visual system to encode a particular stimulus.

ACT-R is able to interact with a computer through its visual and motor modules. The visual system provides the model with the ability to detect changes in a computer's display and to attend to and encode the items that are there. The manual system allows the model to click a button on a simulated mouse or press a key on the keyboard. The timing parameters for those systems are built in and based on previous research. In particular the motor system is heavily influenced by the EPIC architecture (Kieras & Meyer, 1997).

Currently, ACT-R has one explicit attention mechanism, the primary purpose of which is to determine how activation from the current goal is used to retrieve knowledge in declarative memory. Processing in ACT-R is driven by the current goal (which is also represented as a chunk in declarative memory) and elements of the goal are viewed as the main focus of attention. ACT-R has a limited quantity of *source* activation (ACT-R's W parameter) which is shared equally between the elements in the goal and subsequently spread to associated chunks in declarative memory when retrieval requests are made. However, the SART is a very rapid, stimulus-driven respond-don't-respond paradigm which can be modelled in ACT-R with the assumption that participants are required to retrieve little, if anything, from declarative memory (the model assumes that the target number 3 is an element of the goal). As a consequence, this attention mechanism is not utilised in the model reported here.

A description of the model

An ACT-R model was constructed that was able to interact with the SART through the same interface as the human participants used—text presented on a computer screen and a mouse to enter the response. The model consists of 11 production rules, illustrated in Figure 2.

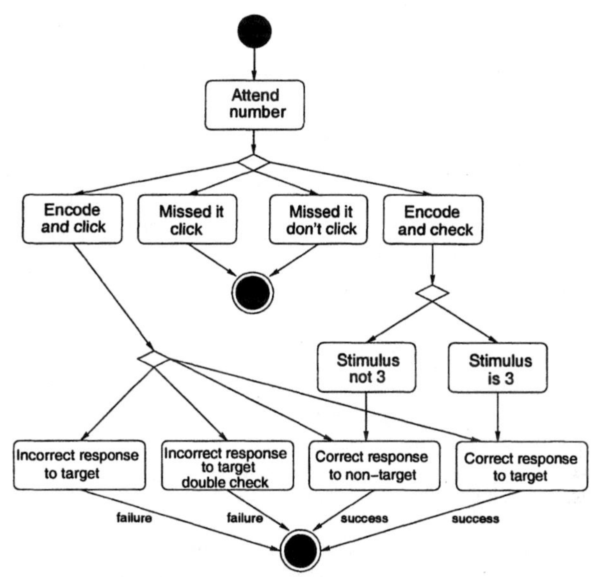

Figure 2: Flow of processing in the ACT-R SART model. Each state in the chart corresponds to one production rule in the model.

The model comprises two competing strategies, each represented by one production rule. The first is a faster, but less accurate method in which the model simply clicks the mouse after detecting the stimulus (encode-and-click). The second strategy is more accurate but slower. In this case, the model first checks the stimulus to ensure that it should click the mouse (encode-and-check), and then does so when appropriate.

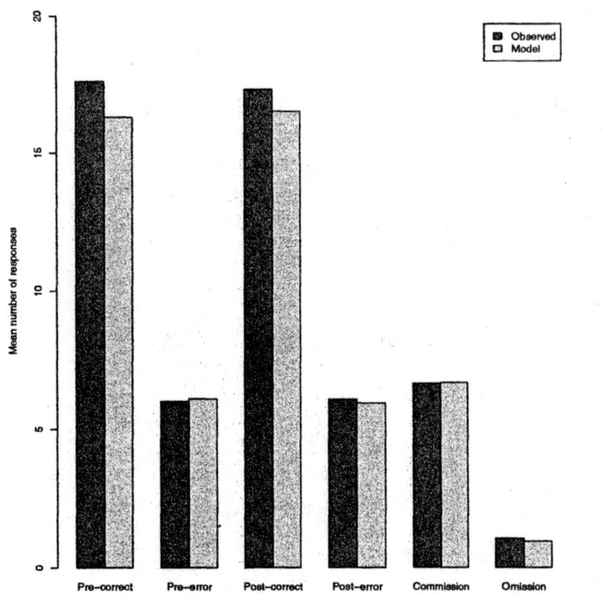

Figure 3: Observed and model mean number of responses for trials immediately before and after correct and erroneous responses to no-go trials in the SART and mean errors of commission and omission.

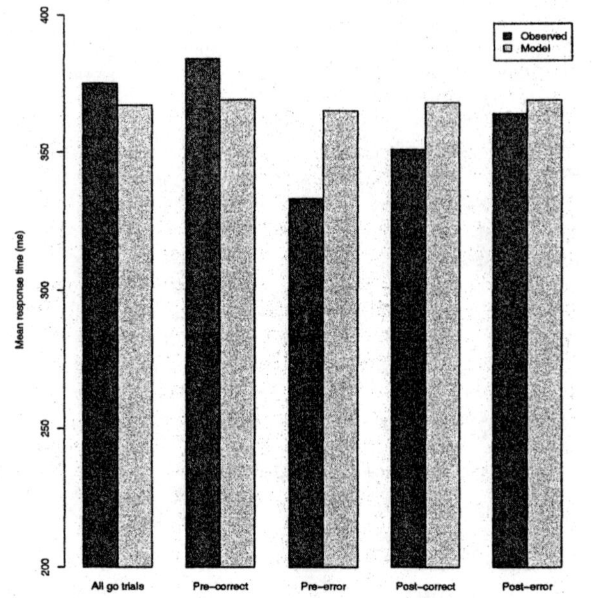

Figure 4: Observed and model mean response times for all go trials and trials immediately before and after correct and erroneous responses to no-go trials in the SART.

The choice between the two strategies is controlled by a quantity called *utility* which ACT-R uses to determine which production will fire next. The utility, U_i of a production i is defined as

$$U_i = P_iG - C_i + \sigma \quad (1)$$

where P_i is the probability of successfully achieving the goal if production i fires and reflects the history of successes and failures for the production, G is a global parameter that represents the cost (measured in time) of the current goal, and C_i is the measure of cost (also measured in units of time) associated with the use of production i until the goal has been achieved. When choosing from a number of candidate productions to fire, ACT-R's conflict resolution mechanism selects the production with the highest utility. To introduce an element of stochasticity to this process, however, a noise value, σ is added to each production's utility.

Initially, the utility values of the two productions that instantiate the strategies are equal. During the course of the task, the history of success and failure for each production, along with the time taken to produce those successes or failures, are used by ACT-R's utility learning mechanism to tune the P and C values of both productions. As a consequence, the model's preference for one strategy or the other changes from trial to trial.

In addition to the subsymbolic utility learning mechanism, the model also incorporates an explicit symbolic process for modifying its strategy. On each trial, after applying a strategy, the model verifies that it responded appropriately. In cases where the model verifies a correct response (the correct-response-to-target and correct-response-to-non-target productions) it simply waits for the next trial. If the model detects that it incorrectly clicked for a target, however, it can do one of two things; either just wait for the next trial (incorrect-response-to-target) or make an explicit choice to use the more accurate strategy on the next trial (incorrect-response-to-target-double-check). The production implementing this second option simply creates an additional element in the new goal to ensure that the encode-and-check production fires on the next trial, representing what is essentially an overt decision to be more careful in subsequent trials, (as evidenced by the significant increase in RTs for the post-error trials relative to pre-error trials observed by Manly et al., 2000).

There is one other aspect of the model can affect its performance. Like the EPIC architecture (Kieras & Meyer, 1997), completion times for ACT-R's perceptual-motor operations can be stochastic and one consequence of this is that the model sometimes fails to encode the stimulus during the 250 ms presentation period. On the relatively rare occasion that this happens, the model must decide whether to click the mouse (missed-it-click) or not (missed-it-don't-click). The structure of the SART implies that clicking the mouse is the more frequent response so a small preference for this option is implemented by giving the missed-it-click production a slightly higher initial utility value.

Applying the model

The ACT-R model was run 150 times to produce a data set from 150 simulated participants that could be compared with the RT and response data from the human subjects in the Manly et al. (2000) study. To fit the model to the data, two parameters which control the learning of production utilities were adjusted. One is the s parameter that adjusts the amount of the variance in the noise added to the calculations. The noise parameter was set to a low value (.01) to allow the model to be sensitive to the changes in the utility values of each production from trial to trial while still having some variability in its performance. The other is the G parameter representing the value of the goal in the utility calculation. The definition of G is based on the expected time to complete the goal and, as trial durations in the SART are very short (1.15 seconds) so is the expected goal completion time. Hence, the value of G was lowered from its default value of 20 seconds to 0.45 seconds to reflect this.

The results of the simulation are shown in Figures 3 and 4. The model provides a very close fit to the observed pattern of responses, accounting for 99% of the variance in the observed data (R^2 = .998, *RMSE* = .756). The model makes on average 6.67 errors of commission and 0.95 errors of omission (the latter occurring on occasion when the model fails to encode the stimulus). The RTs produced by the model are reasonably close to the observed data (R^2 = .665, *RMSE* = 19.933) but do not reproduce the pattern of variation between trial categories. However the mean RT for all go trials produced by the model (367 ms) is very close to the observed (375 ms).

The discrepancy between the observed and model RTs may be explicated somewhat by looking at the distribution of observed RTs in Figure 1, where it can be seen that the large majority of participants fall within a range of 340–400 ms. At either end of this range is a smaller group of participants who either respond very rapidly (between 250 and 340 ms) and make more errors or respond relatively slowly (between 400 and 650 ms) and make fewer errors. It is possible that the difference between the mean pre-correct and pre-error RTs is being magnified by the responses of these participants at the extreme ends of the range. The performance of these two groups of participants may reflect a strong preference for satisfying one of the task requirements over the other. In terms of the current model, this would be represented as a prior preference for a particular strategy, implemented by increasing the initial utility value for one of the two relevant productions.

To illustrate the range of behaviour covered by the model, Figure 5 plots the number of errors of commission against mean RT to non-targets for the 150 runs in the simulation. Although the range of RTs produced by the model is smaller than the observed, the model does produce a similar relationship between error and RT with faster means correlating with more errors. The correlation between mean RT on go trials and the number of errors of commission by the model is significant ($r = -0.788$, $p < .01$).

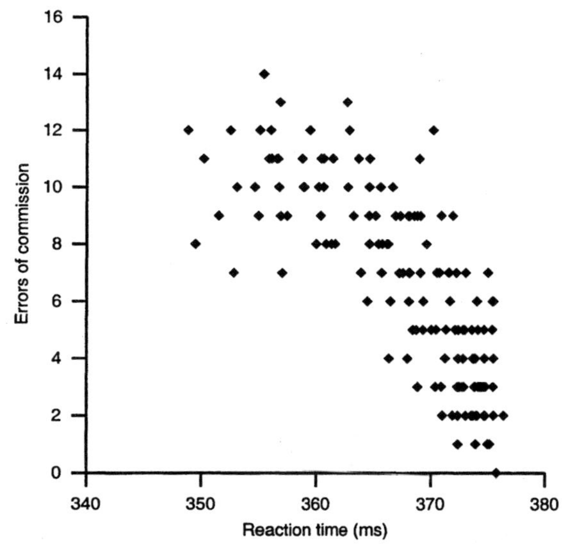

Figure 5: Plot of errors of commission against RT to non-targets for 150 runs of the ACT-R SART model.

Discussion

Although there is little doubt that the ability to sustain attention is an important factor affecting performance in many tasks, it remains an open question to what extent performance in the SART is determined by sustained attention or by the subject's attempt to satisfy conflicting task demands.

The model presented here questions the explanatory role of sustained attention in the SART by providing a plausible alternative account of performance in terms of strategy choice. The main process underlying this account is ACT-R's utility learning mechanism which increases the likelihood of a strategy being used if it has been successfully used before. As both strategies lead to a majority of successful trials, if the model starts to prefer one over the other, it will continue with that preference, but this choice of strategy will effect the number of errors and the mean RT produced by the model. The additional factor in the model is the explicit symbolic process for modifying its strategy after detecting that it has incorrectly clicked for a target. This has the effect of biasing the model in the direction of the encode-and-check strategy after an error of commission, resulting in the greater number of model runs producing slower RTs but fewer errors in the bottom right corner of Figure 5.

The RT and error data produced by the model provide a close fit to the observed data and, although the range of RTs currently captured by the model is limited to a range within the main section of the distribution, it does reproduce the error profile and the relationship between error and RT observed in Manly et al. (2000), which was the primary aim of the current effort. One proposal for future work under consideration is to use the current version of the model as a basis for analysing the performance of individuals at the extreme ends of the RT and error scales, possibly due to preferences in response selection (i.e. ACT-R's G parameter), small variations in strategy, or overall response speed (e.g., related to the speed parameters of ACT-R's motor module).

The phenomenon of speed-accuracy trade-off is commonly observed in human behaviour. In ACT-R, speed-accuracy trade-off is typically addressed by the conflict resolution mechanism for selecting the next production rule to fire—more specifically, the G parameter in Equation 1 (Anderson & Lebiere, 1998, p. 60). Changing the value of G affects the balance between speed and accuracy; a lower value of G places greater emphasis on the time cost of a production, C, resulting in a model that will be more likely to choose the rapid but less accurate option over the more successful, but slower strategy. Conversely, a higher G will increase the emphasis on successfully achieving the goal, even though this might take longer. This mechanism enables ACT-R models to account for a range of RT and error data along the trade-off spectrum. This model, however, is able to account for a range of behaviour in the SART with a fixed value of G. In addition, by incorporating an explicit symbolic level strategy to correct for detected errors in the model's responses, that range of behaviour was found to correspond to that of the human participants. The use of the utility learning mechanism to govern both the speed and accuracy of the model's behaviour is, to our knowledge, an original feature of this model.

It may be the case that Manly et al's (2000) first study can be completely accounted for in terms of the dynamic modification of priorities in relation to the conflicting demands of the task, without any reference to attention. However, an alternative conception of the model is also possible, one that views the differences in the amount of effort required to process a trial between the two strategies in terms of attention to the task. In this sense, the encode-and-click production represents the inattentive automatic response while the encode-and-check production embodies the additional 'attention' required to check the identity of the stimulus to successfully withhold the response to 3. This attentional interpretation of production rules has recently been adopted by other researchers attempting to model attention at the symbolic rule level (Wang, Fan, & Johnson, 2004), although it remains to be seen whether—and how far—this interpretation can be utilised in accounting for attention processes in executive control, particularly as production rules have traditionally been interpreted as representing unconscious, automatic, procedural knowledge. In terms of the new modular ACT-R 5.0 architecture, it is unclear whether a simple interpretation of attention at the production rule level will be adequate or whether an additional attention module and buffer may be necessary. It is clear, however, that the issue of how to understand and model attentional phenomena in cognitive architectures such as ACT-R is a pressing one that must be addressed.

Acknowledgements

The first author would like to thank Tom Manly for his kind help in providing information about the SART.

References

Anderson, J. R., & Lebiere, C. (1998), *The atomic components of thought.* Hillsdale, NJ: Lawrence Erlbaum Associates.

Anderson, J. R., Bothell, D., Byrne, M. D., Douglass, S., Lebiere, C., & Qin, Y. (submitted). An integrated theory of the mind. *Psychological Review.*

Kieras, D. E., & Meyer, D. E. (1997). An overview of the EPIC architecture for cognition and performance with application to human-computer interaction. *Human-Computer Interaction, 12*, 391–182.

Loken, W. L., Thornton, A. E., Otto, R. L., & Long, C. L. (1995). Sustained attention after closed head injury. *Neuropsychology, 9*, 592–598.

Manly T, Robertson I. H., Galloway, M., Hawkins, K. (1999). The absent mind: Further investigations into sustained attention to response. *Neuropsychologia, 37*, 661–670.

Manly, T., Davison, B., Heutink, J., Galloway, M., & Robertson, I. H. (2000). Not enough time or not enough attention?: Speed, error and self-maintained control in the Sustained Attention to Response Task (SART). *Clinical Neuropsychological Assessment, 3*, 167–177.

Manly, T., Owen, A. M., McAvinue, L., Datta, A., Lewis, G. A., Scott, S. K., Rorden, C., Pickard, J. & Robertson, I. H. (2003). Enhancing the sensitivity of a sustained attention task to frontal damage. Convergent clinical and functional imaging evidence. *Neurocase, 9*, 340–349.

Parasuraman, R., Mutter, S. A., & Molloy, R. (1991). Sustained attention following mild closed-head injury. *Journal of Clinical and Experimental Neuropsychology, 13*, 789–811.

Robertson I. H., Manly, T., Andrade, J., Baddeley, B. T., & Yiend, J. (1997). Oops!: Performance correlates of everyday attentional failures in traumatic brain injured and normal subjects. *Neuropsychologia, 35*, 747–758.

Rueckert, L., & Grafman, J. (1996). Sustained attention deficits in patients with right frontal lesions. *Neuropsychologia, 34*, 953–963.

Wang, H., Fan, J., & Johnson, T. R. (2004). A symbolic model of human attentional networks. *Cognitive Systems Research, 5*, 119–134.

Whyte, J., Polansky, M., Fleming, M., Coslett, H. B., & Calvallucci, C. (1995). Sustained arousal and attention after traumatic brain injury. *Neuropsychologia, 33*, 797–813.

Wilkins, A. J., Shallice, T., & McCarthy, R. (1987). Frontal lesions and sustained attention. *Neuropsychologia, 25*, 359–365.

A Pandemonium Can Have Goals

Giovanni Pezzulo (pezzulo@ip.rm.cnr.it)
Institute of Cognitive Science and Technology – CNR. Viale Marx, 15 – 00137 Roma, Italy
and Università degli Studi di Roma "La Sapienza", Piazzale Aldo Moro, 9 – 00185 Roma, Italy

Gianguglielmo Calvi (calvi@noze.it)
Institute of Cognitive Science and Technology - CNR
Viale Marx, 15 – 00137 Roma – Italy

Abstract

AKIRA is an open-source framework for agent-based cognitive and socio-cognitive modeling and simulation. In this introductory paper we explain the underlying theoretical assumptions that lead to its central features, e.g. hybridism of the components, access to a common pool of resources, homeostasis of the system. In order to show the potentiality of AKIRA for cognitive modeling, we provide and example of a Goal Directed Agent. However, we are not focused on a single agent model, mechanism or computational tool; we use AKIRA as an "experimental laboratory" for modeling and implementing many cognitive functions (e.g. belief and goal dynamics, epistemic actions, anticipation, attention), exploring how higher-order cognition emerges from the interplay of many specialized agents and coalitions that compete, coöperate and learn how to exploit each other.

Introduction

The open source project AKIRA is a multi agent framework for cognitive and socio-cognitive modeling and simulations. The framework is inspired by the work of many cognitive scientists and philosophers (Minsky, 1986; Dennett, 1991), sharing some features with related computational models (Kokinov, 1994; Franklin, 1995; Sloman, 1999; Hofstadter, 1995), but retaining some unique ones.

Although our first interest is in high-order cognition, mainly in BDI-like (Rao & Georgeff, 1995) style, it is possible to use AKIRA for implementing a range of agents and architectures having different levels of complexity: reactive, logic-based, layered and autonomous agents; constraint satisfaction and connectionist networks. AKIRA is suitable for socio-cognitive, situated simulation, where cognitively rich agents are required in order to model e.g. trust and reputation dynamics (Falcone, 2004; Conte, 2001).

In order to explain the basic structure of AKIRA, we assume the Pandemonium metaphor (Jackson, 1987), describing its agents as Daemons that cooperate and compete, form Coalitions and communicate trough a Blackboard. In order to show how Daemons and Coalitions can be exploited for cognitive modeling we explore a range of built-in features, such as the hybridism of the agents and the homeostasis of the system (Cannon, 1939).

Assuming *goal-directness* (Castelfranchi, 1995) as the basic functioning of an autonomous agent, we describe the facilities for implementing e.g. goal and belief dynamics. We also show the interplay between epistemic structures, motivations and activity in the autonomous agents, adopting a constructivist and situated perspective (Piaget, 1975).

The Framework

The multi-agent framework AKIRA (http://akira-project.org/) is a run-time C++ multithreading environment for building and executing Agents and a web/system development platform to model their behavior and their interaction, as well as for interacting with the environment. AKIRA is implemented using state-of-the-art tools and design: this allows to build applications that are scalable and solid.

AKIRA provides a MACRO language and many templates for building Agents of different complexity (e.g. reactive, BDI-like). The whole system is written in C++ and integrates many different open source libraries. A number of soft computing technologies are included as basic features, e.g. Fuzzy Logic and Fuzzy Cognitive Maps (Kosko 1986).

BDI constructs are among the facilities embedded within the framework and reliable for agent programming. A strong multithread model ensures parallel, distributed computation.

Fig. 1. sketches the dynamics of the agents in AKIRA. The agents pass energy through an energetic network and communicate via a blackboard. Their computational resources (priority and memory space) can change during execution. Resources are an index of *contextual relevance*: more relevant agents have more resources for their operations and introduce a stronger *pressure* over the system, e.g. activating or inhibiting other agents.

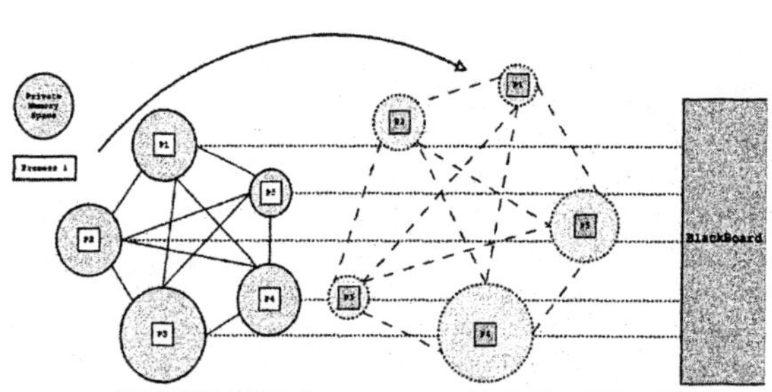

Fig. 1: The Daemons and their dynamics at run time.

AKIRA follows the Pandemonium metaphor (Jackson, 1987); its components are: the **Pandemonium** (kernel), the **Daemons** (micro agents) and the **Coalitions** (Daemon sets);

the **Blackboard** (XML stream); the **Energy Pool** (an abstraction for the computational resources).

The Pandemonium

The Pandemonium is the system kernel, the main process that instances the threads that are necessary to execute the Daemons (agents) and that executes all the monitoring and control operations over the single components. It is the "father process" that during start-up identifies the agents to load (accordingly to the constraints in the initial configuration) and during execution monitors the content of the XML Stream and of the single agents.

Demons

The **Daemons** (micro agents) are the minimal computational elements, each carrying its code, that are implemented as single threads. They are instantiated and executed by the Pandemonium during the system lifetime. Daemons have some features: a **priority** (set by the programmer), that gives a measure of absolute relevance; a **current activation level** (updated at each cycle), that gives a measure of contingent, contextual relevance; a **tap power**, that sets the access to the concurrent resources; a **symbolic operation**, that is the functional body; a **labeled link list**, that points to some other daemons and is the medium for spreading activation with a mechanism inspired by the Slipnet (Hofstadter 1995): the links are strengthened if the concepts associated to the labels are more contextually relevant.

Daemons are hybrid (Kokinov, 1994): they have both a symbolic and a connectionist side. With respect to their **symbolic side**, they can: *execute* the symbolic action they carry on, that can be performed if the contextual conditions are met (like productions) and if the energy is sufficient; *shout*, notifying their current activity and status to the other Daemons via the Blackboard. With respect to their **connectionist side**, they can *tap* energy from the Energy Pool; *spread*, giving it to linked Daemons; *release* it to the Energy Pool; *join* other agents in order to form complex structures called Coalitions.

Coalitions

The **Coalitions** are communities of cooperating daemons that can be created on-the-fly. Their purpose is to solve together complex, non atomic tasks. For example, in a composite pattern matching problem, each Daemon carries the code for matching a part (e.g. the subject, the sender and the address of an email). Differently from pure connectionist dynamics, Coalitions can cooperate and coordinate exchanging messages (e.g. for interactive tasks requiring explicit coordination). Coalitions arise and die in a dynamic way during computation as Daemons shift between them.

Coalitions can be nested and they can result both from bottom-up (*Bands*) and top-down (*Hordes*) pressures.

Bands. Bands arise when some Demons start to Shout in order to find help for a non atomic problem, in which its single symbolic operation is a non sufficient part; other Demons can respond to the Shout and start to Join. When two or more Demons Join in this way, a Band arises.

A Band is the resultant of auto-organization in a bottom-up fashion; its semantic is mainly driven from similarity and proximity (via the Link List) and concomitant activation (only active Daemons can join Coalitions). The topological structure of a Band carries semantic information about the Role that each Daemon assumes into it (as explained later).

Hordes. Hordes arise in a more top-down way and have a more structured, hierarchical shape. Normally special purpose Demons called **Archons**, start to Shout in order to recruit Demons. What's new is that they carry a non-atomic *Structure*, in which specific *Roles* for single Demons have to be fit. While in Bands the aggregation starts to build a structure in a "blind" way, in Hordes the prototypal skeleton of the structure is carried by Archons. In order to execute its symbolic operation, Archons try to recruit other Daemons, spreading them some energy, and taking advantage of their joining or of their symbolic operations.

Archons can be seen as active focuses recruiting Demons for their goal oriented operations; normally they are fueled by higher-level Daemons, e.g. representing internal drives.

The Blackboard

The communication medium is a shared data structure (Blackboard), divided into blocks containing XML packets, where the messages are concurrently written and read. Daemons can address their messages to specific Daemons, or to classes of Daemons, or to everything. Daemons actions and activation levels are notified to the Blackboard, too.

The Energy Pool

Energy is not a component; however, its dynamics are central in explaining the functionalities of the system. It exist in AKIRA as a global variable, *energy*, shared by all the agents, that gives a measure of the available computational resources. So all the daemons reside in an intrinsically concurrent environment with limited resources.

For each agent, more energy means more resources (e.g. more computational time). The total energetic level of the system, summing up all the priorities for all the agents, their tap power and the Pandemonium energy, gives an upper bound to the possible activation sequences of the threads during their lifetime. Energy can migrate between the agents, driving the sequences of activation of the threads.

Constraints for Cognitive Modeling

A main requirement for the AKIRA platform is to allow high-order cognitive modeling, mainly exploiting BDI-like constructs. AKIRA allows to implement *goal-directness* exploiting both the top-down control structures (e.g. from goals to plans, subgoaling) and the dynamics of parallel and concurrent computation (e.g. reactivity, behavioral modules, emergence and self organization).

The general AKIRA paradigm is MAS, but the level of complexity of an agent can be set at different levels. While for many kind of simple agents (e.g. reactive) it is sufficient [illegible]e daemon, a cognitive agent is conceived as a *macro-* [illegible] composed by many cooperating and concurrent [illegible]*agents* (Daemons), each representing e.g. a goal, or a belief, etc. The environment, with its logic and dynamics, can be modeled as an agent, too, interacting with the other agents within the framework.

Here we describe the underlying theoretical assumptions.

Hybridism and Locality Principle

AKIRA agents have both a *symbolic* and a *connectionist* component. The **symbolic component** involves the set of operations an agent can perform. The symbolic operations can range from simple operations to very complex ones (e.g. involving reasoning; AKIRA furnishes to each agent a set of modules and facilities). However, accordingly to the underlying distributed, MAS approach, complex tasks are more likely performed by complex-agents or by Coalitions of cooperating agents. The **connectionist component** involves the activation level of the agent as well as the energy exchanges between the agents. A mayor difference exist with many neural-like architectures: the whole system is *homeostatic* and there are limited resources shared by all the agents, stored into the Energy Pool; the agents can tap or release energy. Spreading energy to another agent means losing it (like in Behavior Networks; Maes, 1990).

We also rely upon a **Locality Principle**: every interaction between the agents, both connectionist and symbolic, is implemented locally; the agents interact globally only with the Energy Pool. The medium for all the operations is the Blackboard; however, it is only a functional abstraction.

Hybridism is a central property of the system; it allows to perform a continuum of computation styles, ranging from centralized, hierarchical control to distributed, emergent computation. The *connectionist side* of AKIRA endorses the emergent and distributed properties of cognitive and socio-cognitive phenomena: the MAS perspective allows complex patterns of actions using autonomous and specialized computational units, i.e. Daemons. The *symbolic side* allows both to introduce top-down drives and structure and to manage operations requiring semantic compositionality.

Two Metaphors

The agents dynamic follows an **Energetic Metaphor** (Kokinov, 1994): greater activation corresponds to a greater computational power, i.e. speed. Each agent has an amount of computational resources (energy) that is proportional to its activation level (and is a measure of its relevance, both absolute and contextual, in the current computation). More active agents have a priority in their symbolic operations and their energetic exchange, and more frequent access to the Energetic Pool. This mechanism allows to model a range of cognitive phenomena such as context and priming effects, as explored mainly by Kokinov (1994).

At the same time, the system implements also a **Physical Work Metaphor**: performing symbolic operations has a cost in energy, that is paid by the performing agent to the Energetic Pool: this keeps the system conservative.

Energetic and Physical Work Metaphors let Demons compete for energy and for the access to the resources (e.g. access to the effectors). They also allow to characterize the interesting concept of *Temperature* (as introduced by Hofstadter, 1995) as an emergent property of the system. Temperature of the system is represented by the currently used energy; it can increase and decrease over time and it is proportional to how far it is from a "solution", if we make the assumption that many Daemons and Coalitions can compete as concurrent "hypothesis" in order to fit the data.

As in Copycat (Hofstadter, 1995) a *hot system* is far from stabilization and performs quick-and-dirty computation, with a rapid hypothesis shift; a *cold system* means stability and successful solutions, computing in a more accurate and conservative way. We can also give a meaning to some local dynamics; e.g. interesting, unsolved problems lead to hot Coalitions, with many Daemons joining it (see Baars, 1988).

Resources, Urgency and Pressures

A Demon or Coalition can act (i.e. *execute*) only if it has sufficient energy, because each symbolic operation has a cost. The cost can be seen as inverse to *urgency* of the behavior: less cost means more easily activated. So, urgent behaviors, like stimulus-response behaviors (as well as *alarms*, Sloman, 1999), can be represented with very low-cost operations; more complex cognitive operations are slower: they need to recruit a lot of energy, or exploit operations by other Daemons, or wait for one or more *join*.

As a consequence of system dynamics, each Daemon introduces a *pressure* over the computation in virtue of its presence. The system shows also implicit, contextual forces and pressures (e.g. set points) that may lead to Coalitions.

Daemons and Coalitions introduce contextual pressures in many ways: perceptual, goal-driven, cultural, conceptual, memory contexts are among those possible. As an effect of the Archons work, Demons that are somewhat related to the contexts are able to recruit more energy; this is true even if they are not able to join the Horde. But this is also true of Bands, where the pre-existing link topology is charged to embed in an implicit way a "similarity" semantic.

Learning and Coalitions. Daemons can exploit connectionist learning (involving the Link List) as well as symbolic learning. For example, a new Archon can be created by to bottom-up pressures (e.g. if a Band shows persistence) or by top-down motivational pressures (e.g. via analogy). The prototypical structure of an Archon can be modified interacting with new exemplars of situations, as in Case Based Reasoning. These processes have the potential to model *assimilation* and *accommodation* (Piaget, 1975).

AKIRA Agents and Agent Societies

All the features hence described represent what we call the *constraints* given by the framework to any possible agent implementation: they are built-in and they cannot be violated, setting the "expressive power" of the framework. AKIRA allows to implement a MAS that is *homeostatic* and endorses *intrinsic concurrence between the agents*. Which kind of Agents and Agent Societies can be so implemented?

We define an *AKIRA Agent Society* as a set of agents working under a common Energetic Pool and competing for its limited resources; an *AKIRA Agent* is the unity which has e, concurrent access to the Pool, proportionally to its ion/pertinence (both absolute and contextual). All the energetic exchanges are local between the agents; all the symbolic operations have a cost in energy. So the system is conservative and it allows emergent phenomena. It is also possible to have many kernels implementing multiple societies that interact; to interface external components (e.g. agents or objects); and to implement agents persistence.

Although it is out of the boundaries of the present paper to argument the cognitive plausibility of the constraints embedded in AKIRA, they are not accidental but rather an essential part of the cognitive model: the expressive power of the each architecture is bounded by its implementation.

The first aim in designing AKIRA constraints and peculiarities is making it well suited for modeling a large set of cognitive and socio-cognitive phenomena. In this phase we do not commit to a specific architecture or model; we plan to use AKIRA as a framework for implementing and testing a number of functions and mechanisms, and trying to find interesting ways to let them interact and cooperate; this path to high-order cognition is inspired by the Society of Mind (Minsky, 1986). As an example of AKIRA's peculiarities for cognitive modeling, here we describe a Goal Directed Agent integrating BDI-like features.

Goal Directed Systems

According to Castelfranchi (1995), an autonomous, goal-directed agent is: able to generate its own goals, to select between multiple alternative goals to pursue, and to decide to adopt goals from others (e.g. for its own purposes).

We use AKIRA as a "cognitive modeling laboratory": we have included many modeling tools (including BDI) as libraries, making it possible to implement and test many cognitive agent architectures. Moreover, AKIRA allows to situate the agents, building environments that constrain their actions and representations. We provide an example of an agent model, showing AKIRA's potential and expressive power. We sketch its motivational apparatus (involving Desires and Goals) and some of its epistemic features; we discuss its goal dynamics and belief building capabilities.

Desires and Goals. Desires are internal, top-level drives and the source of activation for Goals: an unique feature of self-motivated systems. Desires are *maintain* conditions (e.g. fuzzy variables to be kept into a given interval); they are stronger and they spread more activation if they are farer from the condition. Under certain conditions they can also build new Goals. Sometimes Desires intervention ca be seen as normative ("adverbial"), in the sense that they constrain how the goals are performed (e.g. behave *politely*). So, they are well suited for implementing a (poor) version of Norms.

Goals have a condition to satisfy (e.g. in the form of a fuzzy variable); they become stronger the more they are close to the satisfying conditions; this accounts for a form of implicit commitment. Goal dynamics can be implemented by Archons; their concurrence is a built-in feature and there is no need of an external interpreter. Archons structure can represent several features: for example a temporal sequence or a control scheme (useful for Plans); or they can assume a role into complex structures. Goals activate Plans whose effects are positive for them and inhibit Plans whose effects are negative. A Goal can be activated by Desires, as well as by bottom-up pressures, by Plans and even by Beliefs.

Planning and Subgoaling. Plans are control structures with Preconditions and Effects. They activate Goals that correspond to Preconditions to met; in this way an automatic subgoaling mechanism is achieved. Plans can activate and inhibit other Plans as well as Goals.

Plans do not need to be fully represented in the system; they can either be a part of an Archon, or only partially pre-determined: in this case, as in Behavior Networks (Maes, 1990), planning and control result from the dynamics of the system. Pre-planning everything is not always an advantage.

A Bridge between Knowledge and Action

Explicit Plans are action schemes, intrinsically calling procedures (*heuristics*) for dealing with the current situation as represented in the Archon structures. Such heuristics can work at the structural, descriptive level rather tan the semantic one. Depending from the situation, given the same Plan structure, many strategies are suitable, such as: serial vs. parallel subgoal processing; direct goal exploitation vs. facilitation of side conditions, etc. In order to explain how to bridge Plans activity with representations, we introduce the epistemic role of the Coalitions and their structures, using the *Description & Situations* formalism (Gangemi, 2003).

Descriptions and Situations

Coalitions are not simply aggregates of Demons: they can have *structures*. We call these structures **Descriptions**; they represent a prototype of a concept, a situation, a theory, the abstract form of a solved problem. They contain also slots for *Roles* to be filled in e.g. by Daemons joining the Coalition. Descriptions carry on also the functional counterpart of the problem itself, i.e. *heuristics*: they are prototypical operations for dealing with them. Heuristics can not be directly applied to Descriptions (that are abstract entities), but to their concrete counterpart: a reified Description (e.g. in a Horde) represents a **Situation**.

Descriptions can be e.g. *hypothesis* that compete for explanation of phenomena. Following Activity Theory, representational activity is "organizing for use" but also *constraining* it: an active structure, once instantiated, constraints the way stimuli can be successively perceived and dealt with. This process is not only bottom-up, stimuli-driven: a cognitive agent has its internal drives. i.e. the axiological and operational counterpart is driven by the motivational apparatus, which embodies the desires of the system that compete in order to activate their functions.

Situations pair intelligibility with action possibility. Problem-solving capabilities are embodied in situations as their functional-effective counterpart. For instance, some Demons can be specialized in building or dealing with sequences, regardless of the objects that they are putting in the sequence, thus performing operations on the structures themselves. The constructive operation "put-in-sequence" builds a situation that is associated to some "plans-for-cing" and some "plans-to-deal-with-objects-in-a-ce". There is a functional link between situations and objects that can take a *role* into, as well as there are constraints for the (structural) *heuristics* the system can apply to them. Therefore, constructive operations contribute to build a "cognitive map" of the environment, fully grounded in the internal, functional structure of the system.

Epistemic activity. Starting from perception, a stimulus is an "open problem" for a cognitive system: many Demons try to interact with it, singularly or in Coalitions (for more complex operations). A Demon that interacts with a stimulus "cuts it from the noise"; this constructive operation results in applying a *description* and building a *situation*. For example a Demon that carries a pattern matching operation only matches a certain pattern -and it is constrained to describe data in that way. Representing is not mirroring the environment, but the constructive operation of fitting stimuli into a schematic, functional structure: the constructive operation of perceiving a stimulus fits it into a description and constrains how the system can deal with it.

Since they carry little structural information, active Bands are conceived as a (mainly reactive) auto-organization in response to the stimuli of their environment, as unorganized or proto-organized data. At the contrary, Hordes can be seen as more structured attempts to extract (or impose) an interpretation, e.g. formulating an hypothesis. In both cases the epistemic apparatus is constrained by action, either in a reactive fashion or retaining the goal-oriented perspective of the top-down pressures. Due to the functional counterparts of the structures, knowledge and action are both endorsed by the same structure; this also results in a connection between procedural and declarative knowledge.

A Pandemonium Without Satan. Since all the information (e.g. about Daemons activity) is notified to the Blackboard, it can be exploited as data by the other agents, allowing *introspection*: some Daemons can be specialized in monitoring, interpreting and applying their functions to patterns of activity of other Daemons. It has to pointed out that introspective agents have no special access to the private memory and resources of the other agents: they only interpret the results of their actions as meaningful data (they only have to know what the agents do, not how they do it).

Meta-reasoning is not a distinct module, but it is implemented by structures interpreting other structures as data, in a constructive way. The approach is **non-modular**: many active agents and coalitions, e.g. different *points of view* or conflicting Descriptions, as well as specialized cognitive functions, can be active (or partially active) at the same time, communicating, blending, exploiting and interrupting each other. Conflict management can be left to the energetic dynamics or solved by specialized agents, too.

Goals Dynamics: the Watchdog Example

Here we present an example of goal dynamics describing an agent having a set of Goals and a Norm.

The Watchdog agent patrols a house; it has a Norm: *stay always close to the house*, and some active Goals: #1 *walk around the house*; #2 *bark if* you *see an intruder*; #3 *chase and follow the intruders*; etc. Goals inhibit each other, too.

In order to fulfill goal #1 respecting the Norm, the Watchdog will follow circular trajectories around the house, standing always *close* (in fuzzy terms) to it. When an intruder arrives, in order to fulfill goal #2, the Watchdog will bark; if the intruder tries to fly out, in order to fulfill goal #3 the Watchdog has a pressure to follow it. In this case the Norm and Goal #3 are two contrasting pressures: the first to stay close to the house, the second to leave the house. The Watchdog trajectory results from a mix of those factors. Moreover, the internal dynamic of the system will follow some built-in rules for Goals and Norms: the goal becomes stronger the closer it is from its realization; the norm becomes stronger the farer it is from its realization; both become stronger as the Watchdog follows the intruder.

Assuming a slightly higher priority for the Norm, the Watchdog will follow the intruder, until: either the dog reaches him, or it goes too far from the house and the pressure of the Norm becomes stronger. The behavior of the Watchdog simply results from diverging pressures: the trajectory as well as the exact point where it comes back home are not pre-calculated. However, the effect can be amplified by a symbolic operation; e.g. after a while (when its clause is far from realization) the Norm can activate another goal: #4 *come back to the house*. An explicitly planned activity can also intervene: a Goal can activate a Plan involving a rigid "sentinel routine" e.g. follow a certain trajectory that includes each corner, bark each minute, etc.

The Watchdog behavior thus emerges from the *interplay between top-down and bottom-up components and pressures*. It can start as a reactive, stimuli-driven action, be modulated by contextual pressures, activate a Goal and shift to a proactive, top-down control sequence regulated by a Plan; all is done without a central interpreter.

The Watchdog behavior can also follow epistemic drives, e.g. proactively performing *epistemic actions* (Kirsh, 1994).

A Case Study: Constructive Belief Building

Let's sophisticate the Watchdog example, furnishing it the ability to perform *epistemic actions*. These are actions explicitly aimed at acquiring information (e.g. look at the world, waiting for) and building or revising a belief.

Building a Belief is a constructive, goal-driven action: normally a Belief is built because a Goal requires it (e.g. as a precondition). Rather than only collecting input data, the agent can proactively "ask questions to" (the environment, a set of data, another agent); questions always embeds the "point of view" and the intention of the asker, i.e. the (goal driven) Descriptions, as well as some contextual pressures.

In a cognitive perspective, Beliefs are *explicit* epistemic atoms (mainly declarative[1]). In a specific implementation, building a belief can mean e.g. assigning a value to a fuzzy variable: this operation carries on the schematic structure of the goal (e.g. the Description it carries and the metric for that specific fuzzy variable). It also introduces a number of top-down constraints derived by Daemons dynamics, e.g. selective attention (driven by the activity), priming and contextual effects, interference with other Daemons. Using

ybrid formalism allows also to model *implicit knowledge* as emic pressure, distributed and not explicitly represented.

fuzzy logic a Belief, like any predicate, can express *degrees* (i.e. this place is *quite* far). A belief has also an associated *strength*, a degree of epistemic certainty (i.e. I am *rather sure* that this place is quite far). According to Castelfranchi (1995), the strength of a belief is a function of its sustaining sources: how many sources I have queried; how much accurate and reliable they are. Thus, building a belief consists in a set of epistemic actions towards a number of sources; since the sources can be of different kinds, this can lead to different pragmatic actions: e.g. consider perceptual or stored data; ask a question to another agent. This activity is similar to a "detective" that formulate, confirms and falsifies hypothesis; however it has to be stressed that the main drive of the epistemic action is action-oriented. All the information has to be unified into a single Belief: it is a complex cognitive activity that exploits a set of heuristics (e.g. for mixing converging and diverging sources, for managing contradictions) and that subdues to many bias and contextual influences (Castelfranchi, 1995). As a consequence of the constructive procedure, a *degree* and a *strength* are always associated to any Belief. In the same way, exploiting more abstract Descriptions, we can build complex epistemic structures (e.g. causal explanations), in which beliefs take a Roles (e.g. "the core of the theory")[2].

This belief building procedure is broad enough to comprehend even perception: it is both data-driven and proactive, depending on the current hypothesis of the agent.

Giving the Watchdog the possibility to perform epistemic actions and to proactively "ask questions" allows it to enrich its epistemic apparatus (formulating and comparing hypothesis, e.g. "this corner is now secure"). It can also perform new behaviors (e.g. search and follow the footprints of the intruder). We use the belief building mechanism for modeling anticipatory, proactive Watchdogs, exploiting constructive perception: they build expectations and "asks questions" in order to confirm or falsify them. They do not simply react to input data, but they "look at the world" with a set of hypothesis, thus reacting to the discrepancy between expectations and observations, i.e. to *surprise*.

Conclusions

We have presented AKIRA, a framework for cognitive and socio-cognitive modeling. We have described some of its features and assumptions, including hybridism, energetic and physical work metaphor, allowing to model and implement a range of agent architectures. We have described how to model Goal Directed Agents in AKIRA.

Currently we are using AKIRA for implementing a set of cognitive models and functions (mainly developed at ISTC-CNR) including: plans for delegation, monitoring and control; a quantification of the strength of the beliefs as a function of the trust in their sources (Falcone, 2004); expectations and epistemic actions (Pezzulo, 2004); uncertainty and belief revision (Pezzulo, 2004). They will be included into the framework as a set of functions for agent modeling; allowing to explore how their interplay realizes high-order cognition. We plan to include the functionality of simulators (Barsalou, 1999); to interface the foundational ontology DOLCE (Gangemi, 2003); and to implement some architectural features such as data analysis tools, programming interfaces and data exchange protocols.

Even if our models and functions are not jet mature for empirical verification, in order to validate and refine them we are performing a set of experiments involving human subjects: e.g. how do they mix different, possibly discordant information; how do they build and revise trust; how do they manage uncertainty. Allowing empirical testing will be crucial in order to exploit AKIRA for cognitive modeling.

2 In computational terms another important question is: when does an agent consider a belief "solid enough", i.e. when does it stops asking the sources? For this dimension, we are investigating the *certainty* parameter (Pezzulo, 2004): it involves *ignorance* (how much things I do know that I don't know), *perceived contradiction* (the degree of contradiction in my data), and *uncertainty* (that involves a comparison with competing hypothesis).

References

Baars B. J.. *A Cognitive Theory of Consciousness*. Cambridge University Press, Cambridge, MA, (1988).

Barsalou L.W., Perceptual symbol systems, *Behavioral and Brain Sciences*, 22, 577-609, 1999.

Cannon W. B.. *The wisdom of the body*. Norton, New York, 1939.

Castelfranchi C.. Guarantees for autonomy in cognitive agent architecture. In M. Wooldridge and N. R. Jennings, editors, *Intelligent Agents: Theories, Architectures, and Languages*, LNAI 890, pages 56–70. Springer-Verlag, 1995

Conte, R. Emergent (Info)Institutions, *Cognitive Systems Research*, 2001, Volume 2, Issue 2

Dennett D. *Consciousness Explained*, Boston: Little, Brown, 1991

Falcone R., Pezzulo G., Castelfranchi C., Calvi G.. Why a cognitive trustier performs better: Simulating trust-based Contract Nets. *Proceedings of AAMAS 2004*.

Franklin S., *Artificial Minds*. Cambridge, MA: The MIT Press, 1995

Gangemi A, Mika P. Understanding the Semantic Web through Descriptions and Situation. *LNCS 2519*, Springer Verlag, 2003.

Hofstadter D. R. and FARG, *Fluid Concepts and Creative Analogies: Computer Models of the Fundamental Mechanisms of Thought*, Basic Books, New York, (1995).

Jackson J. V., Idea for a Mind. *Siggart Newsettler*, 181, (1987).

Kirsh, D. & Maglio, P. (1994). On distinguishing epistemic from pragmatic action. *Cognitive Science*, *18*, 513–549.

Kokinov B. N., The context-sensitive cognitive architecture DUAL, in *Proceedings of the Sixteenth Annual Conference of the Cognitive Science Society*, Lawrence Erlbaum Associates, (1994).

Kosko, B. (1986). Fuzzy Cognitive Maps. *International Journal Man-Machine Studies*, vol. 24, pp.65-75.

Maes P., Situated Agents Can Have Goals. *Robotics and Autonomous Systems*, 6 (1990).

Minsky M. *The Society of Mind*. Simon and Schuster, NY, 1986.

Pezzulo G., Lorini E.. Calvi G. How do I Know how much I don't Know? A cognitive approach about Uncertainty and Ignorance. *Proceedings of COGSCI 2004*.

Pezzulo, G., Calvi G.. AKIRA: a Framework for MABS. *Proceedings of MAMABS 2004*

Piaget J.. *L'équilibration des structures cognitives: problème central du développement*. Pr. Univ. de France, Paris (1975).

Rao A., Georgeff M., BDI Agents from Theory to Practice, *Technical Note 56, AAII*, April 1995.

A. *What sort of architecture is required for a human-like* *t*? In: M. Wooldridge. A. Rao (eds). Foundations of onal Agency, Kluwer Academic Publishers, 1999.

PsychSim: Agent-based modeling of social interactions and influence

Stacy C. Marsella and David V. Pynadath
Information Sciences Institute
University of Southern California
4676 Admiralty Way, Marina del Rey, CA 90292 USA
{marsella,pynadath}@isi.edu

Stephen J. Read
Department of Psychology
University of Southern California
Los Angeles, CA 90089-1061 USA
read@usc.edu

Abstract

Agent-based modeling of human social behavior is an increasingly important research area. For example, it is critical to designing virtual humans, human-like autonomous agents that interact with people in virtual worlds. A key factor in human social interaction is our beliefs about others, in particular a theory of mind. Whether we believe a message depends not only on its content but also on our model of the communicator. The actions we take are influenced by how we believe others will react. However, theory of mind is usually ignored in computational models of social interaction. In this paper, we present PsychSim, an implemented multiagent-based simulation tool for modeling interactions and influence among groups or individuals. Each agent has its own decision-theoretic model of the world, including beliefs about its environment and recursive models of other agents. Having thus given the agents a theory of mind, PsychSim also provides them with a psychologically motivated mechanism for updating their beliefs in response to actions and messages of other agents. We discuss PsychSim and present an example of its operation.

Introduction

Human social interaction is complex. Rarely are our interactions independent, nor do they fall into narrow categories of full cooperation or competition. People may share some goals but not others, may cooperate at some times and compete at others. To navigate this complexity, we rely on forming beliefs about the goals and behavior of others, what is called a *theory of mind* [Whiten, 1991]. These beliefs inform not only our decisions about what to do, but also what to believe.

There are a range of applications where the computational modeling of such social interactions and influence in general [Prietula and Gasser, 1998, Liebrand and Hegselmann, 1998, Gilbert and Troitzsch, 1999], and theory of mind in particular, is important. Theory of mind, for example, plays a role in childhood aggression whereby one child's aggression may be rooted in misattribution of another child's intent or outcome expectancies on how people will react to the violence [Schwartz, 2000]. To develop a better understanding of the causes and remedies of school bullying, we could use agent models of the students that incorporate a theory of mind to simulate and study classroom social interactions. Models of social interaction have also been used to create social training environments where the learner explores high-stress social interactions in the safety of a virtual world (e.g., [Marsella et al., 2000, Swartout et al., 2001]). In particular, one might imagine an environment where a young teacher could train for handling unruly students by playing the role of a teacher in a virtual classroom populated with virtual students (e.g., see www.victec.org). The teacher could explore in the safety of a virtual world the same sorts of situations and dilemmas he is likely to face in the real world, with interaction between human and virtual characters dynamically unfolding based on choices they make in their various roles.

To facilitate such explorations and applications, we have developed a social simulation tool, called PsychSim, designed to explore how individuals and groups interact and how those interactions can be influenced. PsychSim allows an end-user to quickly construct a social scenario, where a diverse set of entities, either groups or individuals, interact and communicate among themselves. Each entity has its own goals, relationships (e.g., friendship, hostility, authority) with other entities, private beliefs and mental models about other entities. The simulation tool generates the behavior for these entities and provides explanations of the result in terms of each entity's goals and beliefs. A user can play different roles by specifying actions or messages for any entity to perform. Alternatively, the simulation itself can perturb the scenario to provide a range of possible behaviors that can identify critical sensitivities of the behavior to deviations (e.g., modified goals, relationships, or mental models).

A central aspect of the PsychSim design is that agents have fully specified decision-theoretic models of others. Such quantitative recursive models give PsychSim a powerful mechanism to model a range of factors in a principled way. For instance, we exploit this recursive modeling to allow agents to form complex attributions about others, enrich the messages between agents to include the beliefs and goals of other agents, model the impact such recursive models have on an agent's own behavior, model the influence observations of another's behavior have on the agent's model of that other, and enrich the explanations provided to the user. The decision-theoretic models in particular give our agents the ability to judge degree of credibility of a message in a subjective fashion that factors in a range of influences that sway such judgments in humans. In this paper, we present PsychSim, discuss key aspects of its approach to modeling social interaction, and illustrate its operation on a school bullying example.

PsychSim Overview

PsychSim is shown in Figure 1. The user sets up a simulation by selecting generic agent models that will play the roles of the various groups that will be simulated and spe-

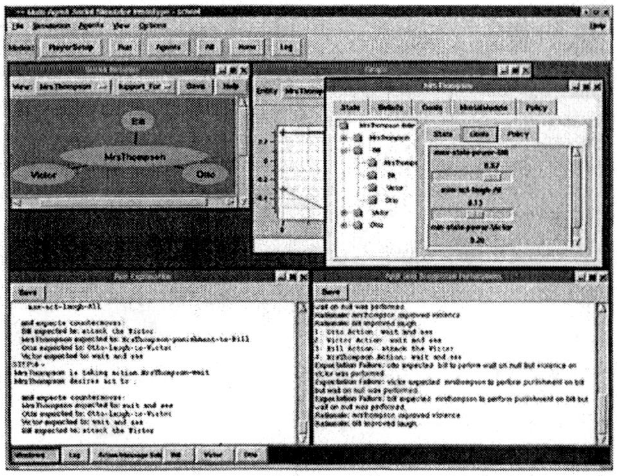

Figure 1: Screenshot of PsychSim interface

cializing those models as needed. To facilitate setup, PsychSim uses an automated fitting algorithm. For example, if the user wants the bully to initially attack a victim and wants the teacher to threaten the bully with punishment, then the user specifies those behaviors and the model parameters are fitted accordingly [Pynadath and Marsella, 2004]. This degree of automation has the potential to significantly simplify simulation setup.

Execution of the simulation incorporates a variety of visualization techniques along with analyses of the simulations. PsychSim allows one to explore multiple tactics for dealing with a social issue and to see potential consequences of those tactics. How might a bully respond to admonishments, appeals to kindness or punishment? How might other groups react in turn? What are the predictions or unintended side-effects?

Finally, there is an analysis/perturbation capability that supports the iterative refinement of the simulation. The intermediate results of the simulation (e.g., the reasoning of the agents in their decision-making, their expectations about other agents) are all placed into a database. Inference rules analyze this database to explain the results to the user in terms of the agents' motivations, including how their beliefs and expectations about other agents influenced their own behavior and whether those expectations were violated. Based on this analysis, the system also reports sensitivities in the results, as well as potentially interesting perturbations to the scenario.

The rest of this paper describes PsychSim's underlying architecture in more detail, using a school bully scenario for illustration. The agents represent different people and groups in the school setting. The user can analyze the simulated behavior of the students to explore the causes and cures for school violence. One agent represents a bully, and another represents the student who is the target of the bully's violence (for young boys, the norm would be physical violence, while young girls tend to employ verbal abuse and ostracizing). A third agent represents the group of onlookers, who encourage the bully's exploits by, for example, laughing at the victim as he is beaten up. A final agent represents the class's teacher trying to maintain control of the classroom, for example by doling out punishment in response to the violence.

The Agent Models

We embed PsychSim's agents within a decision-theoretic framework for quantitative modeling of multiple agents. Each agent maintains its independent beliefs about the world, has its own goals and it owns policies for achieving those goals. The PsychSim framework is an extension to the Com-MTDP model [Pynadath and Tambe, 2002].

State: Each agent model includes several features representing its "true" state. This state consists of objective facts about the world, some of which may be hidden from the agent itself. For our example bully domain, we included such state features as `power(agent)`, to represent the strength of an agent, though the agent may have its own subjective view of its own power. It is impacted by acts of violence, conditional on the relative powers of the interactants. `trust(truster,trustee)` represents the degree of trust that the agent `truster` has in another agent `trustee`'s messages. `support(supporter,supportee)` is the strength of support that an agent `supporter` has for another agent `supportee`. We represent the state as a vector, $\vec{s}^t$, where each component corresponds to one of these state features and has a value in the range $[-1, 1]$.

Actions: Agents have a set of actions that they can choose to perform in order to change the world. An action consists of an action type (e.g., `punish`), an agent performing the action (i.e., the actor), and possibly another agent who is the object of the action. For example, the action `laugh(onlooker, victim)` represents the laughter of the onlooker directed at the victim. Actions affect the state of the world, via a transition function, $T(\vec{s}^t, a)$, that returns a modified state vector, $\vec{s}^{t+1}$, that captures the effects of the performed action, a.

Goals: An agent's goals represent its incentives (and disincentives) for behavior. In PsychSim's decision-theoretic framework, we represent goals as a reward function that maps the current state of the world into a real-valued evaluation of benefit for the agent. A goal of **Minimize/maximize** `feature(agent)` corresponds to a negative/positive reward proportional to the value of the given state feature. For example, an agent can have the goal of maximizing its own power. A goal of **Minimize/maximize** `action(actor, object)` corresponds to a negative/positive reward proportional to the number of matching actions performed. For example, the teacher may have the goal of minimizing the number of times any student teases any other.

We can represent the overall goals of an agent, as well as the relative priority among them, as a vector of weights, $\vec{g}$, so that the product, $\vec{g} \cdot \vec{s}^t$, quantifies the degree of satisfaction that the agent receives from the world, as represented by the state vector, $\vec{s}^t$. For example, in the school violence simulation, the bully's reward function consists of goals of maximizing `power(bully)`, minimizing `power(victim)`, minimizing `power(teacher)`, and maximizing `laugh(onlookers, victim)`. We can model a sadistic bully with a high weight on the goal of minimizing `power(victim)` and an attention-seeking bully with a high weight on maximizing `laugh(onlookers, victim)`. In other words, by modifying the weights on the different goals, we can alter the motivation of the agent and, thus, its behavior in the simulation.

Beliefs: The simulation agents have only a *subjective* view of the world, where they form beliefs, denoted by the vector $\vec{b}^t$, about what they *think* is the state of the world, $\vec{s}^t$. The structure of an agent's beliefs is identical to the representation of the objective world state. In other words, each agent's beliefs consist of models of all of the agents (including itself), representing their state, beliefs, goals, and policy of behavior. For example, an agent's beliefs may include its subjective view on states of the world: "The bully believes that the teacher is weak", "The onlookers believe that the teacher supports the victim", or "The bully believes that he/she is powerful." These beliefs may also include its subjective view on beliefs of other agents: "The teacher believes that the bully believes the teacher to be weak." An agent may also have a subjective view of the *goals* of other agents: "Teacher believes that the bully has a goal to increase his power."

Policies: Each agent's policy is a function, $\pi(\vec{b}, \vec{g})$, that represents the process by which it selects an action or message based on its beliefs and goals. An agent's policy allows us to model critical psychological distinctions such as reactive vs. deliberative behavior. PsychSim models this process as a table of "Condition⇒Action" rules. The left-hand side conditions may trigger on an *observation* of some action or a *belief* of some agent (e.g., such as the bully believing himself as powerful). The right-hand side takes on one of the following forms: action, wait, or bounded lookahead. An action on the right-hand side causes the agent to execute the specified action. A *wait* right-hand side means that the agent does nothing.

We model each agent's real policy table as including a bounded lookahead policy rule that seeks to best achieve the agent's goals given its beliefs. To do so, the policy considers all of the possible actions/messages it has to choose from and measures the results by simulating the behavior of the other agents and the dynamics of the world in response to the selected action/message. They compute a quantitative value, $V_a(\vec{b}^t)$, of each possible action, a, given their beliefs, $\vec{b}^t$.

$$V_a(\vec{b}^t) = \vec{g} \cdot T(\vec{b}^t, a) + V(T(\vec{b}^t, a)) \quad (1)$$

$$V(\vec{b}^{t+1}) = \sum_{\tau=1}^{N} \vec{g} \cdot \vec{b}^{t+\tau} \quad (2)$$

Thus, an agent first uses the transition function, T, to project the immediate effect of the action, a, and then projects another N steps into the future, weighing each state along the path against its goals, $\vec{g}$. Thus, the agent is seeking to maximize the expected value of its behavior, along the lines of decision policies in decision theory and decision theory. However, PsychSim's agents are only boundedly rational, given that they are constrained, both by the finite horizon, N, of their lookahead and the possible error in their belief state, $\vec{b}$.

In addition, we give PsychSim's agents erroneous beliefs about the policies of other agents. In particular, the agents *believe* that the other agents do not do much lookahead, even though, in reality, they use lookahead exclusively when choosing their own actions. This achieves two desirable results. First, from a human modeling perspective, the agents perform a shallower reasoning when thinking about other agents, which provides an accurate model of the real-world entities they represent. Second, from a computational perspective, the direct action rules are cheap to execute, so the agents gain significant efficiency in their reasoning by avoiding expensive lookahead.

Messages: Messages are attempts by one agent (or the user) to influence the beliefs of recipients. Messages have five components: a source, recipients, a message subject, content and overhearers. For example, the teacher (source) could send a message to the bully (recipient) that the principal (subject of the message) will punish acts of violence by the bully (content). Finally, overhearers are agents who hear the message even though they are not one of the intended recipients. Messages can refer to beliefs, goals, policies, or any other aspect of other agents. Thus, a message may make a claim about a state feature of the message subject ("the principal is powerful"), the beliefs of the message subject ("the principal believes that he is powerful"), the goals of the message subject ("the bully wants to increase his power"), or the policy of the message subject ("if the bully thinks the victim is weak, he will pick on him").

Mental Models: An agent's beliefs about another agent are realized as a fully specified agent model of the other agent, including goals, beliefs and policies. To simplify the setup of the system, these mental models are realized as stereotypes. For our simulation model of a bullying scenario, we have implemented mental models corresponding to *selfishness*, *altruism*, *dominance-seeking*, etc. For example, a model of a selfish agent specifies a goal of increasing self-wealth as paramount, while a model of an altruistic agent specifies a goal of helping the weak. Similarly, a model of an agent seeking dominance specifies a goal of having more power than its competitors.

Modeling Influence and Belief Change

A challenge in creating a social simulation is addressing how groups or individuals influence each other, how they update their beliefs and alter behavior based on observations of, as well as messages from, others. Although many psychological results and theories must inform the modeling of such influence (e.g., [Cialdini, 2001, Abelson et al., 1968, Petty and Cacioppo, 1986]), they often suffer from two shortcomings from a computational perspective. First, they identify factors that affect influence but do not operationalize those factors. Second, they are rarely comprehensive and do not address the details of how various factors relate to each other or can be composed. To provide a sufficient basis for our computational models, our approach has been to distill key psychological factors and map those factors into our simulation framework. Here, our decision-theoretic models are helpful in quantifying the impact of factors and in such a way that they can be composed.

Specifically, a survey of the social psychology literature identified the following key factors:

Consistency: People expect, prefer and are driven to maintain consistency, and avoid cognitive dissonance, between beliefs and behaviors. This includes consistency between their old and new information, between beliefs and behavior, as well as consistency with the norms of their social group.

Self-interest: Self-interest impacts how information influences us in numerous ways. It impacts how we interpret appeals to one's self-interest, values and promises of reward or punishment. The inferences we draw are biased by self-interest (e.g., motivated inference) and how deeply we analyze information in general is biased by self-interest. Self-interest may be in respect to satisfying specific goals like "making money" or more abstract goals such as psychological reactance, the tendency for people to react to potential restrictions on freedom such as their freedom of choice (e.g., the child who is sleepy but refuses to go to bed when ordered by a parent.)

Speaker's Self-interest: If the sender of a message benefits greatly if the recipient believes it, there is often a tendency to be more critical and for influence to fail.

Trust, Likability, Affinity: The relation to the source of the message, whether we trust, like or have some group affinity for him, all impact whether we are influenced by the message.

Providing each agent with a model of itself and, more importantly, fully specified models of other agents gave us a powerful mechanism to model this range of factors in a principled way. We can model these factors by a few simple mechanisms in the simulation: *consistency*, *self-interest*, and *bias*. We can render each as a quantitative function on beliefs that allows an agent to compare alternate candidate belief states (e.g., an agent's original $\vec{b}$ vs. the $\vec{b}'$ implied by a message).

Consistency is an evaluation of whether the content of a message or an observation was consistent with prior observations. In effect, the agent asks itself, "If this message is true, would it better explain the past better than my current beliefs?". An agent assesses the quality of the competing explanations by a re-simulation of the past history. In other words, it starts at time 0 with the two worlds implied by the two candidate sets of beliefs, projects each world forward up to the current point of time, and compares the projected behavior against the behavior it actually observed. In particular, the consistency of a sequence of observed actions, $\omega^0, \omega^1, \ldots$, with a given belief state, $\vec{b}$, corresponds to:

$$\text{consistency}(\vec{b}^t, [\omega^0, \omega^1, \ldots, \omega^{t-1}]) = \sum_{\tau=0}^{t-1} V_{\omega^\tau}(\vec{b}^\tau) \quad (3)$$

Note that the value function, V, computed is with respect to the agent performing the action at time τ. In other words, we are summing the value of the observed action to the acting agent, given the set of beliefs under consideration. The higher the value, the more likely that agent is to have chosen the observed action, and, thus, the higher the degree of consistency.

Self-interest is similar to consistency, in that the agent compares two sets of beliefs, one which accepts the message and one which rejects it. However, while consistency requires evaluation of the past, we compute self-interest by evaluating the future using Equation 2. In other words, if $V(\vec{b}) > V(\vec{b}')$ (all else equal), then an agent prefers believing $\vec{b}$. An agent can perform an analogous computation using its beliefs about the sender's reward to compute the sender's self-interest in sending the message.

Bias factors act as tie-breakers when consistency and self-interest fail to decide acceptance/rejection. We treat support (or affinity) and trust as such a bias on message acceptance. Agents compute their support and trust levels as a running history of their past interactions. In particular, one agent increases (decreases) its trust in another, when the second sends a message that the first decides to accept (reject). Similarly, an agent increases (decreases) its support for another, when the second selects an action that has a high (low) reward, with respect to the goals of the first. In other words, if an agent selects an action a, then the other agents modify their support level for that agent by a value proportional to $\vec{g} \cdot T(\vec{b}, a)$, where $\vec{g}$ corresponds to the goals of the agent modifying its support.

Upon receiving any information (whether message or observation), an agent must consider all of these various factors in deciding whether to accept it and how to alter its beliefs (including its mental models of the other agents). For a message, the agent determines acceptance using a weighted sum of the five components: consistency, self-interest, speaker self-interest, trust and support.

We see the computation of these factors as a toolkit for the user to explore the system's behavior under existing theories, which we can encode in PsychSim. For example, the elaboration likelihood model (ELM)[Petty and Cacioppo, 1986] is a two-process model that argues that the way messages are processed differs according to the relevance of the message to the receiver. High relevance or importance would lead to a deeper assessment of the message, which is consistent with the self-interest calculations our model performs. For less relevant messages, more peripheral processing of perceptual cues such as "liking for" the speaker would dominate. PsychSim's linear combination of factors is roughly in keeping with ELM because self-interest values of high magnitude would tend to dominate.

Example Scenario in Operation

The research literature on childhood bullying and aggression provides interesting insight into the role that theory of mind plays in human behavior. Although a number of factors are related to bullying, two social cognitive variables have been shown to play a central role. One variable discussed is a hostile attributional style [Nasby et al., 1979], wherein typical playground behaviors are interpreted as having a hostile intent. Children who tend to see other children as intending to hurt them are more likely to display angry, retaliatory aggression. A second variable is outcome expectancies for the effectiveness of aggression. Children develop outcome expectancies for the effectiveness of aggression depending on whether in the past they have been rewarded for its use or found it to be ineffective or punished for it.

Investigations of bullying and victimization [Schwartz, 2000] have identified four distinct types of children: those who characteristically display reactive aggression (aggressive victims), those who display proactive aggression (nonvictimized aggressors), those who are typically victimized (nonaggressive victims), and normal children. Nonaggressive victims display a hostile attributional style and have negative outcome expectancies for aggression. Aggressive victims tend to have a hostile attributional style, but neither positive nor negative outcome expectancies for aggression. Nonvictimized aggressors have positive outcome expectancies for aggression, but do not have a hostile attributional style.

We have begun to use use PsychSim to explore psychological theories by demonstrating how PsychSim can represent both *attributional style* and *outcome expectancies* in a simulation of school violence. The user can manipulate each factor to generate a space of possible student behaviors for use in simulation and experimentation. For example, an agent's attributional style corresponds to the way in which it updates its beliefs about others to explain their behavior. A hostile attributional style corresponds to an agent who tends to adopt negative mental models of other agents. In our example scenario, agents with a hostile attributional style mentally model another student as having the goal of hurting them (i.e., minimizing their power).

Our agents already compute the second factor of interest, outcome expectancies, as the expected value of actions (V_a from Equation 1). Thus, when considering possible aggression, the agents consider the immediate effect of an act of violence, as well as the possible consequences, including the change in the beliefs of the other agents. In our example scenario, a bully has two incentives to perform an act of aggression: (1) to change the power dynamic in the class (i.e., weaken his victim and make himself stronger), and (2) to earn the approval of his peers (as demonstrated by their response of laughter at the victim). Our bully agent models the first incentive as a goal of maximizing `power(bully)` and minimizing `power(victim)`, as well as a belief that an act of aggression will increase the former and decrease the latter. The second incentive must also consider the actions that the other agents may take in response. The agents' theory of mind is crucial here, because it allows our bully agent to predict these responses, albeit limited by its subjective view.

For example, a bully motivated by the approval of his classmates would use his mental model of them to predict whether they would enjoy his act of aggression and laugh along with him. Similarly, the bully would use his mental model of the teacher to predict whether he will be punished or not. The agent will weigh the effect of these subjective predictions along with the immediate effect of the act of aggression itself to determine an overall expected outcome. Thus, the agents' ability to perform bounded lookahead easily supports a model for proactive aggression.

We explored the impact of different types of proactive aggression by varying the priority of the two goals (increasing power and gaining popularity) within our bully agent. When we ran PsychSim using an agent model where the bully cares about each goal equally, then the threat of punishment is insufficient to change the bully's behavior, because he expects to still come out ahead in terms of his popularity with his peers. On the other hand, a threat against the whole class in response to the bully's violence is effective, because the bully then believes that an act of violence will *decrease* his popularity among his peers. If we instead use an agent model where the bully favors the first goal, then even this threat against the whole class is ineffective, because the bully no longer cares about his popularity in the class.

Of course, this example illustrates one outcome, where we do not change any of the other variables (e.g., bully's power relative to victim, teacher's credibility of threats). PsychSim's full range of variables provide a rich space of possible class makeups that we can systematically explore to understand the social behavior that arises out of different configurations of student psychologies. We have also begun developing algorithms that are capable of finding the configuration that best matches a real-world class dynamic, allowing us to find an underlying psychological explanation for a specific instance of behavior [Pynadath and Marsella, 2004]. Furthermore, as illustrated, we can try out different interventions in the simulation to understand their impact under varying student models. As we have seen, alternate scenarios will have different results, but by systematically varying the scenario, we can draw general conclusions about the effectiveness of these different intervention methods. Finally, although this section uses a specific taxonomy of student behavior to illustrate PsychSim's operation, the methodology itself is general enough to support the exploration of many such taxonomies.

Related Work

PsychSim agents model each other's beliefs, goals, policies, etc. and are able to reason with it. This is essentially a form of recursive agent modeling [Gmytrasiewicz and Durfee, 1995], but specifically organized to model psychological factors that play a role in influence, human communication about theory of mind and human social interaction in general. Other agent theories have also addressed the modeling of social interaction, including many based on the Beliefs-Desires-Intentions (BDI) framework [Rao and Georgeff, 1995]. Existing BDI models are based in first-order logic, which can often lead to ambiguous conclusions (e.g., multiple beliefs are consistent with the observations). PsychSim's decision-theoretic foundation supports the quantitative evaluation of beliefs (e.g., computing *degrees* of consistency), as well as the development of novel algorithms for automatic model construction [Pynadath and Marsella, 2004].

Typical work on multi-agent based social simulation has used often thousands or more agents used to study complex systems (e.g. [Terna, 1998]). However, the focus of such work is often the study of the emergent properties of the interactions of simpler agents. For example, Mosler and Tobias use a simulation of 10,000 agents to study the emergence of collective action. Although the agents employed in the work are rather sophisticated by the standards of such simulations, they are nevertheless far simpler than the agents in PsychSim. The agents' decision-making is parametrized by exogenous variables, rather than controlled by deep recursive modeling of others as in PsychSim. Of course, in a simulation with 10,000 agents exploring aggregate properties that emerge from simple behaviors, it would probably not be desirable to give the agents theories of mind and complex lookahead reasoning. Thus, the complexity is in the system, not the individual agents.

Directions and Conclusion

We have presented PsychSim, an environment for multi-agent simulation of human social interaction that employs a formal decision-theoretic approach using recursive models. This approach allows us to model phenomena rarely if at all addressed in simulated worlds. For example, our agents can reason about the behavior and beliefs of other agents. In psychological terms, they have a theory of mind. This allows agents to communicate beliefs about other agent's beliefs, goals and intentions and be motivated to use communication to influence other agents' beliefs about agents. Within PsychSim, we have developed a range of technology to simplify the task of setting up the models, exploring the simulation and analyzing results. This includes new algorithms for fitting multi-agent simulations. Within PsychSim, there is also an ontology for modeling communications about theory of mind. We have exploited the recursive models to provide a psychologically motivated computational model of how agents influence each other's beliefs.

The rich belief structure of PsychSim's agents offers avenues for future exploration. For example, our current reward structures can represent goals on the true state of the world, but it is straightforward to extend it to encompass goals regarding *beliefs* about the world. For example, the bully may have a goal of maximizing the perception that he is powerful, rather than his actual power. We can easily implement such a goal as being proportional to the power that the victim believes the bully to have. We can define many more such goals by applying reward functions to any aspect of the agents' beliefs. These beliefs also include the agents' subjective view of the world dynamics. In our current implementation, the agents all have correct beliefs about the world dynamics. In other words, the bully's beliefs about how beating up the victim affects his power is identical to the true effect. We could easily open up such beliefs about dynamics to the same errors and biases that are possible within the other beliefs. We can also enrich the message space. Certainly, as we expand the belief space to more possibilities for deviation from reality, the more possibility there is for interesting communication between entities with differing perspectives. Furthermore, it is reasonable to enrich the types of performatives supported. Our current messages are intended as statements of fact, but we could easily imagine request messages as well. For example, an agent who has uncertainty in some areas of belief could ask a trusted second agent about its belief in those areas. We believe PsychSim has a range of innovative applications, including computational social science and the model of social training environments. Our current goals are to expand the exploration already begun in the school violence scenario and begin evaluating the application of PsychSim there and in these other areas.

Acknowledgments We would like to thank Dr. Kelton Rhoads for his help in developing the critical factors in PsychSim's model of influence.

References

[Abelson et al., 1968] Abelson, R., Aronson, E., McGuire, W., Newcomb, T., Rosenberg, M., and Tannenbaum, P., editors (1968). *Theories of Cognitive Consistency: A Sourcebook.* Rand McNally, Chicago, IL.

[Cialdini, 2001] Cialdini, R. (2001). *Influence: Science and Practice.* Allyn and Bacon, Boston, MA.

[Gilbert and Troitzsch, 1999] Gilbert, N. and Troitzsch, K., editors (1999). *Simulation for the Social Scientist.* Open University, Buckingham, UK.

[Gmytrasiewicz and Durfee, 1995] Gmytrasiewicz, P. and Durfee, E. (1995). A rigorous, operational formalization of recursive modeling. In *Proceedings of the International Conference on Multiagent Systems*, pp. 125–132.

[Liebrand and Hegselmann, 1998] W. Liebrand, A. N. and Hegselmann, R., editors (1998). *Computer Modeling of Social Processes.* Sage, London, UK.

[Marsella et al., 2000] Marsella, S. C., Johnson, W. L., and LaBore, C. (2000). Interactive pedagogical drama. In *Proceedings of the International Conference on Autonomous Agents*, pp. 301–308.

[Nasby et al., 1979] Nasby, W., Hayden, B., and DePaulo, B. (1979). Attributional biases among aggressive boys to interpret unambiguous social stimuli as displays of hostility. *Journal of Abnormal Psychology*, 89:459–468.

[Petty and Cacioppo, 1986] Petty, R. and Cacioppo, J. (1986). *Communication and persuasion: Central and peripheral routes to attitude change.* Springer, New York, NY.

[Prietula and Gasser, 1998] M. Prietula, K, C. and Gasser, L., editors (1998). *Simulating Organizations: Computational Models of Institutions and Groups.* AAAI Press, Menlo Park, CA.

[Pynadath and Marsella, 2004] Pynadath, D. V. and Marsella, S. C. (2004). Fitting and compilation of multiagent models through piecewise linear functions. *Proceedings of the International Conference on Autonomous Agents and Multiagent Systems*, to appear.

[Pynadath and Tambe, 2002] Pynadath, D. V. and Tambe, M. (2002). Multiagent teamwork: Analyzing the optimality and complexity of key theories and models. In *Proceedings of the International Joint Conference on Autonomous Agents and Multi-Agent Systems*, pp. 873–880.

[Rao and Georgeff, 1995] Rao, A. S. and Georgeff, M. P. (1995). BDI-agents: from theory to practice. In *Proceedings of the International Conference on Multiagent Systems*, pp. 312–319.

[Schwartz, 2000] Schwartz, D. (2000). Subtypes of victims and aggressors in children's peer groups. *Journal of Abnormal Child Psychology*, 28:181–192.

[Swartout et al., 2001] Swartout, W., Hill, R., Gratch, J., Johnson, W., Kyriakakis, C., LaBore, C., Lindheim, R., Marsella, S., Miraglia, D., Moore, B., Morie, J., Rickel, J., Thiébaux, M., Tuch, L., Whitney, R., and Douglas, J. (2001). Toward the holodeck: Integrating graphics, sound, character and story. In *Proceedings of the International Conference on Autonomous Agents*, pp. 409–416.

[Terna, 1998] Terna, P. (1998). Simulation tools for social scientists: Building agent based models with SWARM. *Journal of Artificial Societies and Social Simulation*, 1(2).

[Whiten, 1991] Whiten, A., editor (1991). *Natural Theories of Mind.* Basil Blackwell, Oxford, UK.

Letter Spirit: A Model of Visual Creativity

John Rehling (rehling@andrew.cmu.edu)
Carnegie Mellon University, 5000 Forbes Avenue
Pittsburgh, PA 15213 USA

Douglas R. Hofstadter (dughof@cogsci.indiana.edu)
Center for Research on Concepts and Cognition
Indiana University
510 North Fess Street
Bloomington, IN 47408 USA

Abstract

The Letter Spirit program is a model of human creativity in the domain of typeface design. Creativity is involved in most human activities, but it is by its nonquantifiable nature at odds with many modeling approaches. The Letter Spirit program models the inherently creative task of typeface design with a two-level approach. The top level loop coordinates three modules that create letterforms, evaluate them aesthetically, and decide which letterforms need to be re-designed, until all 26 lowercase letters have been rendered suitably. On the lower level, each module carries out its task by modeling the way that subcognitive pressures combine and conflict to produce emergent behavior on the cognitive level. While certain types of reasoning can be modeled with the manipulation of symbols whose meaning is considered to be absolutely fixed, this work helps to illustrate why it is difficult to demonstrate creative behavior with systems of that kind.

Introduction

The Letter Spirit domain is based upon a grid consisting of 56 quanta, each of which is a horizontal, vertical or diagonal line segment connecting adjacent points in a 3x7 grid. Some sample *gridletters* are displayed later in Figure 1.

The task

The task of Letter Spirit is to accept as input a few gridletters from a human designer, with those *seed* letters intended to represent the same style. Using the seeds as a beginning, Letter Spirit creates versions of the remaining lowercase letters of the roman alphabet until it has produced an entire *gridfont* of 26 stylistically consistent gridletters. Letter Spirit can also begin a new gridfont from scratch, beginning without any human-created seeds and those gridfonts are usually stylistically coherent and are sometimes stylistically interesting

Implementation

Letter Spirit consists of three modules, each being a relatively complex program solving a vital subtask of gridfont design, plus the top-level loop that coordinates the three modules into a single strategy of design called *review-and-revision.* In order to carry this strategy out, the program has one module, the Drafter, that, given a particular letter category (e.g., 'b') and a particular style (usually exemplified for the program by means of several examples of different letter categories in that style) creates an attempt at a gridletter that is a good representative both of that letter category and that style.

Superficially, it may seem sufficient to create but one such attempt per letter category until the gridfont is complete. Creating one attempt according to a prescribed method of creation is the approach of many computational models of creativity, but when the output of such a method is of low quality (as if often the case), the program consequently shows the brittleness that is characteristic of many efforts in AI. To get around this problem, Letter Spirit hands the Drafter's output to the other modules, called the Examiner and the Adjudicator, and they rate each Drafter-created gridletter for how well it fits its intended letter category and the intended style, respectively.

The FARG approach

Foundations

While the three modules differ significantly in their implementation (that is inevitable, given their very different subtasks), they are all based upon an architecture used in other cognitive models that have been implemented by the Fluid Analogies Research Group. As the name implies, those efforts were primarily concerned with modeling how people create analogies by means of concepts that are fluid, defined in terms of norms that may or may not be upheld rather than rigid, strictly-enforced rules. Most of those projects, including Copycat (Mitchell, 1993), Tabletop (French, 1992), and Metacat (Marshall, 1999) aimed at implementing increasingly refined models of analogy. One of the Letter Spirit modules, the Examiner (McGraw, 1995) pursues a similar approach, although its task is gridletter categorization rather than analogy.

The FARG programs all have some common architectural elements, loosely inspired by the Hearsay-II speech recognition architecture of Erman, et al. (1980).

The Coderack: Fine-grained processing

In the FARG architecture, the counterpart to what would be, in a traditional computer program, a procedure call is a *codelet*, a relatively short routine that performs some small operation. No individual codelet does very much of the program's work. They are intended to correspond meaningfully to small-scale cognitive events. The Coderack is the repository of codelets, and execution begins with the nondeterministic selection of one codelet from the Coderack. The selection process is nondeterministic, weighted by the individual *urgency* value attached to each codelet when it is posted to the Coderack. The selected codelet is removed from the Coderack and then executed. This activity is repeated until either the Coderack is empty or some other condition for halting is reached. A run begins with the placing of some codelets on the Coderack. As codelets are run, the number of codelets remaining on the coderack generally decreases, but codelets may, as part of their work, add more codelets to the Coderack. The length and composition (in terms of what types of codelets are present) of the Coderack both change over the course of a run

With each individual codelet performing only a very small portion of the overall task, FARG models progress towards producing an output only through the cumulative action of many codelets. As a spectator, one may infer implicit high-level goals in the activity of the Coderack although such goals are not explicit anywhere in the Coderack, and instead emerge as trends in the activity carried out one codelet at a time on a lower level. Multiple high-level activities can chronologically overlap as their associated codelets run in an interleaved order. That leads to implicit parallelism and, consequently, behavior that allows multiple high-level pressures to compete or cooperate as the program progresses towards a final answer.

Systems whose behavior emerges from the combined action of many small agents have been inspired by and likened to the work of an anthill as carried out by many ants, or the activity in a cell, with countless proteins each performing a tiny part. The benefits can be understood by means of an analogy with integral calculus. One of the insights underlying calculus is that the area under a curve can be approximated by fitting rectangles onto the Cartesian plane, between the curve and the x-axis, and calculating the sum of their areas. The thinner the rectangles, the better the approximation to the actual area bounded by the curve. In a Coderack architecture, likewise, the small grain size of processing by codelets offers the potential for a tight fit to the mental processes that the architecture models.

Long-term and short-term memory

Each FARG model has a short-term memory called the *Workspace* that consists of structured representations that relate how the symbols contained therein relate to one another. Further details vary widely from model to model.

Many of the models also have a localist connectionist network that allows (among other operations) a scheme of spreading activation to mediate, throughout the course of a run, how important concepts (represented by nodes in the network) appear to be to processing at any given point in time. In the case of Copycat, Tabletop, and Metacat, the network is called a *Slipnet*, which is imbued with some sophisticated characteristics that are not implemented in any of the Letter Spirit modules, and will not be described more in this paper. Letter Spirit's Examiner module has a simpler *Conceptual Network* that is a conventional localist connectionist network (McClelland and Rumelhart, 1981) with nodes on one level for whole letters and nodes on another level for letter parts; during a run, activation can spread between nodes so that when the relevance of one concept has been detected or suspected, the program can suspect the presence of related concepts. For example, when 'i' is activated, activation spreads to the 'dot' node as well, and future processing is then more likely to see any appropriate structure that is present *as* a dot.

Long-term memory is also important for each model, with the repository of necessary domain knowledge included in a store called the *Conceptual Memory*. The details of this also vary greatly from model to model.

Parallel terraced scan

The activity of a FARG model can be seen as the investigation of multiple competing possible answers to the same problem or subproblem, frequently with more than one possible answer being considered at the same time. A goal of the FARG architecture is to create models that direct effort preferentially towards answers that are apparently more promising. The *parallel terraced scan* describes the way in which a program can explore many possible answers at once (thus, in parallel) but devotes more time to pursuing those answers that seem more worthwhile (thus, terraced).

The result of a well-implemented parallel terraced scan is a program that finds good answers without having to commit equal amounts of computational resources to all possible answers. In general, a Coderack program implementing the parallel terraced scan carries out search (if its activity is to be described in that way) very differently from either a traditional search algorithm or a generate-and-test algorithm that is applied exhaustively to the set of possible answers. In a FARG model, good answers (not necessarily perfect answers) bubble up to the top naturally. The effect of the parallel terraced scan is to makes models of cognition more efficient while also raising the quality of their output. Decisions regarding how much computational effort should be directed towards each answer are made implicitly by the program during runtime, sensitive to the conditions of the particular run, rather than being made by the programmer, during programming time, and insensitive to the conditions of the future run. The parallel terraced scan is concerned with making a program that is intelligent not only with regard to its eventual output, but also in regard to how it goes about producing that output.

The details of how a FARG model can incorporate the parallel terraced scan vary according to the other details of the model. The process of imbuing a model with the parallel terraced scan (or increasing the *extent* to which it utilizes the parallel terraced scan) is documented in our account of how the Examiner was optimized from its original implementation (Rehling and Hofstadter, 1997).

Representing letters

People do not conceptualize letters as exact graphical shapes. If they did, deviations from the definition would be seen as inherently flawed. In reality, there is no one "right" font, although some certainly seem more normal than others.

Work on the Letter Spirit project has been guided by the belief that letters are represented by roles – abstract concepts on a level beneath that of letters. The constituent strokes from which children learn to draw letters approximately indicate the level of roles. Letters are defined by one or more role-sets, which specify which roles make up the letter and how they ought to touch or otherwise relate spatially to one another. An actual physical rendering of a letter – a *letterform* – can be decomposed into physical *role-fillers*, which satisfy (to some degree) the definitions of their corresponding roles. It is important to note that role-fillers are exact shapes, but roles are not. Roles are concepts, flexible enough that a great variety of shapes may serve as role-fillers for them. The role hypothesis and support for it is explained in greater detail in (McGraw, 1995).

Representing style

Stylistic properties

The work on Letter Spirit aims to capture a reasonable range of styles by implementing (with routines allowing for perception as well as creation) a number of stylistic properties, which can define many kinds of gridfont, though not all of the ones that people can invent. This implementation of Letter Spirit realizes three kinds of stylistic property, and diverse styles are represented as combinations of these.

One type of stylistic property is *motifs*. A motif is a particular shape that recurs in numerous gridletters. "Shape" can be thought of in a number of levels of literalness. Most literal would be a set of contiguous quanta with precise location on the grid specified. Less literal versions allow translated, reflected, or rotated versions to count as the same shape. Letter Spirit allows for different levels of literality in motifs, with an emphasis on detecting and using those motifs that are inherently more noteworthy. The larger and more literal a motif is, the more noteworthy it for it to occur.

Abstract rules are properties that may be present in an individual gridletter. These are "Thou shalt not" rules that forbid quanta of certain orientation, angles of certain measure, quanta within certain zones of the grid, and collinear stretches of segments of various lengths.

The third kind of stylistic property is that which occurs when role-fillers deviate significantly from their corresponding roles; these conflicts are called *norm violations*. If a role-filler lacks a particular property expected for its abstract role, that does not destroy its membership in the category, although it makes it a less prototypical member of the category.

The Thematic Focus

The style of the gridfont that Letter Spirit is in the process of designing is represented in a structure called the *Thematic Focus*. The Thematic Focus consists of a handful of levels (six) with different levels indicating the degree to which the SPs contained in that row are important to the style of the gridfont. Being located on a higher level indicates that a stylistic property is more important. When a stylistic property has been found in one gridletter, it is placed on the lowest level of the Thematic Focus. Promotion (and demotion) of stylistic properties to higher (and lower) levels throughout a run is probabilistic, so that a stylistic property's frequency of occurrence in the gridfont roughly correlates with how high in the Thematic Focus it is located. The higher in the Thematic Focus that a stylistic property is found, the greater its influence in the subsequent calculation of style ratings as well as in decisions pertaining to the creation of new gridletters.

At the beginning of each run, the Thematic Focus is empty, and it is gradually filled and modified as more examples of the gridfont are used to form the basis of the goal style. It tends to form a pyramidal shape, with a style being defined in terms of a very large number of stylistic properties that are relatively unimportant to the style and a few stylistic properties that are considered important to the style. A filled Thematic Focus may have well over one hundred stylistic properties in it, although typically somewhere between zero and two of them are on the highest level. A style is thus typically a very complex thing, although a brief list of a couple of traits can sometimes suffice as an approximate description of it. We believe that this kind of representation of style is consistent with the observation that people can readily recognize a style when they see it, but cannot easily define the style explicitly.

The Library

The style of a gridfont is also represented with the help of the *Library*, a repository of all the gridletters and role-fillers that Letter Spirit has found among the seeds and among its own creations that have been used to define the style. The shapes in the Library can be useful in defining a style, both in creating new gridletters and in calculating ratings of style membership. This will be discussed in greater detail below.

Examiner

The Examiner determines the letter category of an input gridletter. It was implemented by Gary McGraw as a stand-alone program, and can be discussed as such, although its intended and current purpose as to serve as a module within Letter Spirit. In addition to categorizing the input as an 'a', or a 'b', etc., it also segments the input into constituent parts and identifying for which abstract roles those parts are meant to fill. Finally, it returns a numeric rating of how well the input serves a member of its perceived letter category.

Processing in the Examiner follows an opportunistic course which is initially bottom-up, but seeks to exploit any information gained to the maximal benefit in guiding future processing. A run that accepted a relatively plain 'b' as input might proceed as follows: First, a Gestalt codelet would activate nodes in the Conceptual Memory letter categories based upon the coarse shape properties of the whole letterform; for a 'b', the Gestalt codelet would typically activate 'b'. This would lead to an attempt to segment the whole letterform into parts that are appropriate for 'b', namely a post on the left side distinguished from a bowl on the right side. Activation in Conceptual Memory would spread from 'b' to the nodes for 'left-post' and 'right-bowl'. After the program has attached descriptive labels (such as 'tall', 'thin') to each part, a Sparker codelet would attempt to bind a part to a corresponding role in memory, based upon matches between the part's labels and the role's permanent definition, but also favoring those roles that are more highly activated. Typically, a very simple letterform will be recognized in just a few dozen codelets. When early attempts have failed, codelets will very likely run that attempt to glue parts together or break bigger parts in two, seeking a segmentation that leads to successful recognition. After hundreds of codelets have run, the standards by which a part is expected to match a role are gradually loosened, allowing more and more exotic parts to be seen as members of less obvious roles. Atypical gridletters are thus often recognized after more time has gone by, when the right combination of segmentation and, if necessary, role loosening, has taken place.

The Examiner has received more development effort than the other portions of Letter Spirit combined. Logically, its performance is most crucial to the program as a whole (it is not possible to know what a fancy 'g' is without knowing what a 'g' is). It is has been tested as a stand-alone gridletter recognizer, and the latest version of the Examiner achieves 93.5% correct categorization on a large test set for which literate human subjects without special expertise in typefaces were only 84.0% correct.

Adjudicator

The Adjudicator performs three tasks. First, it fills the Workspace with a representation, in terms of stylistic properties, of the style of a given gridletter. Second, and concurrent with the first task, it finds ways in which the Workspace would modify the Thematic Focus, should the gridletter be accepted as a member of the gridfont and worthy of being part of the definition of the goal style. Third, it uses the degree of fit between the Workspace and the Thematic Focus to calculate a style goodness rating for the gridletter.

The Adjudicator's codelet-based implementation differs from that of the Examiner in that the Examiner's processing often involves emergent loops, as one gridletter segmentation is attempted, and then rejected if it does not lead to successful recognition. The Adjudicator, in contrast, runs more like a straight-line program. Although the choice of its codelets, which detect a variety of stylistic properties, is still nondeterministic, the course of a run is not open-ended. A fixed number of style-detecting codelets runs in nondeterministic sequence, building up a letterform-style structure that may differ somewhat from one run to another. The Adjudicator serves as a style evaluator as well as a style builder, producing a numeric rating of how well a single gridletter's style matches that of a gridfont as a whole. It can build up a style from many gridletters by accumulating properties from the set, promoting more frequently occurring ones to levels of greater significance in the Thematic Focus.

The Adjudicator depends crucially upon the Examiner, and expects as input the kind of output that the Examiner provides, not just a mere gridletter, but the gridletter annotated, as it were, in terms of letter category, parts, and roles. Thus, when the Adjudicator judges the style of a gridletter, it judges it *as* a 'b', or an 'x', or a 't'. Key aspects of the Examiner's design reflect that it is not a mere Optical Character Recognition program, but a servant to the Adjudicator.

Drafter

The Drafter's task is to create a gridletter; the interaction of abstract roles and the concrete grid lead to a three-level hierarchy of activity to consider:

(I) Drafting a gridletter.
(II) Drafting a role-filler.
(III) Drafting a single quantum.

For the Drafter to render a gridletter, it must act on the top level. There are two ways to go about this. One is to create the gridletter in a single step, by borrowing and adapting a letterform from the Library. For example, an attempt at 'd' can be rendered by taking the mirror image of the 'b' (if there is one) that is part of that style. It should be stressed that this is not an infallible method of creating a good 'd'; for some styles, the mirror image of 'b' is **not** a good 'd' in that style.

The alternative to that is to create the gridletter one role-filler at a time. At that level, the Drafter may borrow a role-filler wholesale from the Library, or it may create the role-filler quantum by quantum. The drafting of a quantum is carried out by a run of the Coderack, which selects the next quantum for the role-filler out of all the eligible candidates. In this mode of the Drafter, role-fillers are drawn one quantum at a time, via successive point-to-point moves, as though an ink pen were doing the drawing. A run of the Drafter begins with the pen in some location, with various amounts of information on where and why it should draw next, and the output of one run of the Drafter is to decide which quantum, if any, should be drafted next. These coderack runs consult the Thematic Focus, as built up previously by the Adjudicator, and also the platonic definition of the intended letter category. The selection of each quantum can be thought of as a democratic process, with each codelet voting for the penstroke that most suits its cause, which might be for one codelet the continuation of a motif that is common to the typeface; for another, making tall a part that whose role calls for tallness; for another, making the part shorter because the style calls for shortness. Codelets can and do conflict with one another. The eventual output must agreeably balance all constraints. The Drafter is in this way a good example of how the FARG architecture can avoid the brittleness that is so often a feature of AI.

The Drafter's output is not uniformly good. In fact, many of its attempts at a gridletter are poor versions of the intended letter category, or style, or both. Some attempts are even completely unrecognizable. This does not weaken the Letter Spirit program because no gridletter is accepted as a final member of the gridfont without the approval of both the Examiner and the Adjudicator.

The top-level loop

The Examiner, Adjudicator, and Drafter are each programs of considerable sophistication. The top-level control of Letter Spirit ties them together into a relatively detailed model of human creativity in the domain of gridfont design. Its task is to carry out "the central feedback loop of creativity", in which everything the program creates is inspected for quality before it is deemed worthy of inclusion in the final output. A second objective that occurs simultaneously (in some versions of the program) is the evolution of a new style, in which new additions to the output can change the goal style that governs future output.

The Letter Spirit top-level program consists of two phases:

The first phase uses the seed letters given to the program to create a representation of the style that they have in common. The second phase of the program is a loop, in which a letter category is selected (nondeterministically, but favoring those categories that do not yet have a good version of them in the gridfont), and then the Drafter renders a gridletter that, ideally, incorporates the goal style as well as that letter. The Drafter's attempt is run past the Examiner and the Adjudicator, and if the attempt is the best version thus far for that category, as determined by the scores that the Examiner and the Adjudicator generate, then it is kept as the current version of that category in the gridfont. This loop runs many times, and as it does so, the quality of the gridfont should incrementally increase.

The top-level program of Letter Spirit lacks most of the characteristics of the FARG architecture, and is, in fact, a relatively simple program. The simplicity serves a number of causes. First, this makes it a relatively clean test of the strategy of the central feedback loop of creativity. Second, as a model of higher-level activity, it requires less in the way of fine-grained detail to model humanlike activity. And, of course, it is easier to begin with a simple implementation and add features and functionality in future work. A thorough exploration of Letter Spirit as it stands now is a valuable precursor to any future work that will increase the sophistication of the top-level program.

Results

Letter Spirit and its constituent modules have each been tested in a variety of circumstances. For the sake of brevity, the only results that will be reported here are those of the whole Letter Spirit program. These runs each began with the program receiving five gridletters (in each case, 'b', 'c', 'e', 'f', and 'g') taken from one of five different human-designed gridfonts. Then the program strove to finish each gridfont by designing the remaining gridletters. In these runs, the program used up to 300 Drafter runs to generate multiple (a mean of 14) attempts for the 21 non-seed categories, keeping the attempt it rated best in each case. The resulting gridfonts appear in Figure 1.

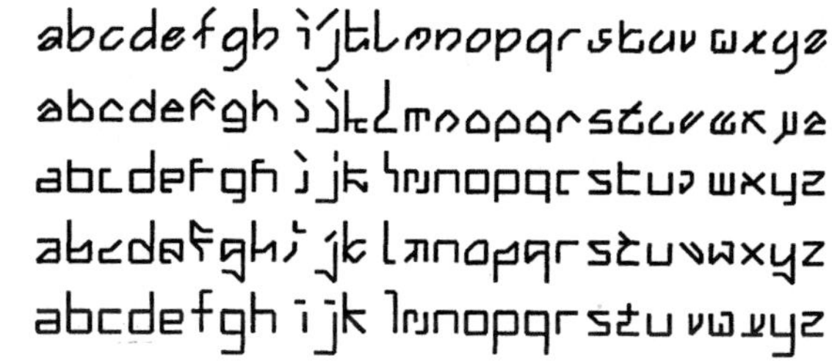

Figure 1: Five Letter Spirit-designed gridfonts.

Observations

The gridfonts are both relatively legible and relatively coherent stylistically. Out of the non-seed gridletters (105 in all), all but perhaps five are easily recognized as members of their intended letter category. By scanning across the figure, one can easily see the general stylistic trend of each gridfont, although each gridfont has a few members that show little, if any, of the intended style. Approximately 30% of the gridletters are the exact ones used by the human designer in the original gridfont. Many of the remaining ones are only slightly different, or reflect the intended style despite going about the matter in a manner different than that of the original designer. In other cases, such as the tall 'd' in the third gridfont of Figure 1, which otherwise features short ascenders and descenders, or the squared-off 'w' of the first gridfont, one sees inconsistency in the program; it is not the case that Letter Spirit always does good work, but that it is capable of doing so. Fine examples of incorporation of letter with style are seen in the 'z' of the first gridfont, the 'v' and the 'x' of the second gridfont. Letter Spirit output has been exhibited in art shows (Rehling and Hofstadter, 2000) and, subjectively speaking, is of moderate, although not excellent, quality.

Essential aspects of this approach

Letter Spirit has commonalities with other approaches in cognitive modeling, but diverges from them in important ways. For example, the Coderack has certain computational similarities to the way that production rules are implemented in architectures like Soar and ACT-R. Also, the review-and-revision strategy is a species of the widespread and more general generate-and-test approach that is ubiquitous in computer programming of all sorts.

However, it is important to note key aspects of the Letter Spirit program without which these mechanisms would be incapable of producing aesthetically plausible output. First, the nondeterministic production style of the Drafter is essential to the success of the review-and-revision approach. Simply put, it is of no benefit to produce multiple attempts at a product if all of those many attempts are identical. Second, the subcognitive level of codelets in all three Letter Spirit modules crucially distinguishes this work from cognitive modeling in the symbolic tradition. In order to display purposeful variability in behavior at one level of processing, a model must concern itself with a lower level. In the case of typeface design, there can be no rules at the role-filler level that instruct the model in definitive terms how to make tradeoffs between letter and style. Instead, the model must allow these pressures to compete with one another, allowing a range of possible results in the output, because it is not predictable a priori which influences should prevail in any given situation.

While it is usually the case that mature work in cognitive modeling includes a quantitative comparison between human and model data for the sake of partial validation of the approach, we have no intention of undertaking such an analysis here. It is *possible* in the case of models of aesthetic creation to perform strained versions of the comparison-with-subjects exercise but it is, we feel, for numerous reasons not only beside the point but also misleading. The goal of this work was not to implement most or all of the complexity of human-level creativity but rather to identify mechanisms that are sufficient to enable a semblance of creative behavior and to provide a case study that allows one to begin to consider which of those mechanisms are necessary. Any effort to quantitatively assess progress would derive more from the statistical methods used than it did from the model. It is certainly not a coincidence that artistic creativity is an endeavor that is hard or undesirable to evaluate quantitatively and that it has been less the subject of modeling work than human behavior that *is* easily captured quantitatively. Members of the artistic community itself have argued vehemently, if not always with logic, that quantitative expressions of aesthetic quality are misguided (Rehling, 2000), and we would argue that adhering to the usual standards of empirical validation is actually a hindrance to the enterprise of modeling creative behavior.

Analysis

Letter Spirit produces moderately impressive output thanks to several positive aspects in its implementation, described in greater detail than is possible here in (McGraw, 1995) and (Rehling, 2000). One strength is the cognitive plausibility of its modules, particularly that of the Examiner. Another is the review-and-revision strategy of creativity, an approach that is probably hard to beat in terms of performance with one-pass approaches that do not review their own output and hope to create high-quality output on their first and only try.

Letter Spirit also has some clear shortcomings that could be addressed in future work. First, the Drafter completely finishes drafting an attempt of a gridletter before that gridletter is evaluated for quality (by the other modules). This allows the program to draft blindly and leads to Drafter output being clearly inferior to human-designed gridletters. A second problem is that gridletters that are rated poorly by the Examiner and Adjudicator are simply discarded, so that the Drafter must begin its next attempt at that category from scratch. A human designer can modify flawed products, correcting their flaws to make a good final product. Future work should allow Letter Spirit to review-and-revise on the role-filler level, as well as on the letter level. A third problem is that style needs to be represented in richer and more flexible ways. It is hoped that this work's strengths will motivate future work on this project and projects applying the same architecture to other domains.

References

Erman, L., Hayes-Roth, F., Lesser, V., and Raj Reddy, D. (1980). The Hearsay-II Speech-understanding System: Integrating knowledge to resolve uncertainty. *Computing Surveys*, 12(2):213-253.

French, R. (1992). *Tabletop: An emergent stochastic computer model of analogy-making*. PhD thesis, University of Michigan, Ann Arbor, Michigan.

Grebert, I., Stork, D., Keesing, R., and Mims, S. (1992). Connectionist generalization for production: An example from GridFont. *Neural Networks*, 5.

Hofstadter, D., and the members of FARG (1995). *Fluid Concepts and Creative Analogies: Computer Models of the Fundamental Mechanisms of Thought.* New York: Basic Books.

McClelland, J., and Rumelhart, D. (1981). An interactive-activation model of context effects in letter perception: Part 1, an account of basic findings. *Psychological Review*, 88(5):375-407.

McGraw, G. (1995). *Letter Spirit (part one): Emergent high-level perception of letters using fluid concepts*. PhD thesis, Indiana University, Bloomington, Indiana.

Marshall, J. (1999). *Metacar: A self-watching cognitive architecture for analogy-making and high-level perception*. PhD thesis, Indiana University, Bloomington, Indiana.

Mitchell, M. (1993). *Analogy-making as Perception.* Cambridge, Massachusetts: MIT Press/Bradford Books.

Rehling, J. (2000). *Letter Spirit (part two): Modeling creativity in a visual domain.* PhD thesis, Indiana University, Bloomington, Indiana.

Rehling, J. and Hofstadter, D. (1997). The parallel terraced scan: An optimization for an agent-oriented architecture. *Proceedings, IEEE First International Conference on Intelligent Processing Systems.*

Rehling, J. and Hofstadter, D. (2000). Letter Spirit: A model of creativity in a visual domain. *Art show at the XVI Congress of the International Association of Empirical Aesthetics.*

Smolensky, P. (1988). On the proper treatment of connectionism. *Behavioral and Brain Sciences*, 11:1--23.

Cognitive modeling versus game theory: Why cognition matters

Matthew F. Rutledge-Taylor (mrtaylo2@connect.carleton.ca)
Institute of Cognitive Science, Carleton University, 1125 Colonel By Drive
Ottawa, Ontario, K1S 5B6 Canada

Robert L. West (robert_west@carleton.ca)
Institute of Cognitive Science, Department of Psychology, Carleton University, 1125 Colonel By Drive
Ottawa, Ontario, K1S 5B6 Canada

Abstract

We call into question game theory, as a account of how people play two player zero-sum games. Evidence from a modified version of the game Paper, Rock, Scissors suggests that people do not play randomly, and not according to certain play probabilities. We investigated the relationship between game theory predictions and a cognitive model of game playing based on the detection of sequential dependencies. Previous research has shown that the sequential dependency model can account for a number of empirical findings that game theory cannot. The sequential dependency model has been implemented using both simple neural networks and ACT-R. In this paper we used simple neural networks (a description of how our findings relate to the ACT-R model is included in the Conclusion section). For simple games, such as Paper, Rock, Scissors, game theory has been able to correctly predict aggregate move probabilities. In this paper we show that this is an artifact of the symmetry of the payoffs, and that for asymmetrical payoffs the game theory solution does not predict human behavior. Furthermore, we show that the model of game playing that underlies game theory cannot be used to predict the results no matter what move probabilities are used. Finally, we show that the results can be accounted for by augmenting the network sequential dependency model so that the reward system is related to the game payoffs.

Introduction

Game theory (VonNeumann & Morgenstern, 1944) was not created as a cognitive model, that is, it was not intended to account for mechanisms underlying the manner in which humans play games. Rather it was intended to be a formal mathematical system for understanding game playing from a rational perspective. However, game theory has had an enormous influence on theories and ideas about how people play games. Because empirical studies have shown that game theory is poor at predicting how humans play games, most researchers do not believe that game theory is a good model of how humans process information during games (Pool, 1995). Nevertheless, the game theory model of a player is still extremely influential in terms of how we understand human game playing.

Game theory is a large collection of mathematical principles that are used in accounts of a variety of human interactions. It is used by a variety of researchers including economists, mathematicians, and behavioral psychologists to name a few. The focus of this paper is specific to the application of game theory as an account how people play a relatively constrained set of non-zero sum games. Henceforth, the use of the term 'game theory', will be used to refer to a subset of principles within game theory commonly used to describe this type of game.

In game theory, a game is defined as a situation in which two players each select one move from a discrete set of choices, the combination of which determines a payoff for each player. There are two varieties of games. In zero-sum games the payoffs for the players must sum to zero, and there is no option for cooperation (i.e., one player's gain is always equal to the other player's loss). In non-zero-sum games this constraint does not exist and cooperation can be an option. In either case, to apply game theory it is necessary to assume that players can make random selections from their move choices according to specific probabilities assigned to each move. This amounts to two assumptions about the cognitive abilities of players: (1) they have some way of calculating or learning move probabilities, and (2) they are able to select moves at random according to these probabilities. In what follows, we will refer to these assumptions concerning players' abilities as *the game theory player model.* Assuming that a player matches the game theory player model, it is generally possible to use game theory to calculate the optimal set of move probabilities. We will refer to the outcome of game theory calculations as *the game theory solution*.

Most research on human game playing has focused on non-zero-sum games, where it has been shown that people generally do not follow the game theory solution. One reason for this is that people are often influenced by factors that are extraneous to the game theory solution, such as social norms concerning cooperation (Poundstone, 1992; Samuelson, 1997). However, these results show only that people do not play according to the game theory solution. The game theory player model is still viable as it is possible to explain the results in terms of people playing according to the game theory player model (i.e., probabilistically), but with move probabilities inconsistent with the calculated game theory solution. In zero-sum games there is no option to cooperate so there should be less extraneous influences. However, there is very little direct research on human zero-sum game playing (Poundstone, 1992). The general view of human zero-sum game playing is based mainly on experimental comparisons between humans and the game theory player model. With regard to this, psychological studies have shown that people are very poor at the two essential skills required by the game theory player model: (1) learning optimal move probabilities (e.g., Gazzaniga, 1998), and (2) behaving randomly (see Wagenaar, 1972 for a review).

Based on this, a common view of individual humans as game players is that we are poor game theory players. That is, we play in a way consistent with the game theory player model but we are not good at learning the optimal move probabilities, we are poor at being random, and we are influenced by considerations extraneous to the game theory framework (e.g., norms about cooperation). Attempts have been made to adapt the game theory player model to better capture human behavior. The focus in this area has been on introducing a learning component to explain how humans acquire move probabilities. Game theorists have studied the effects of using learning algorithms to acquire move probabilities (Fudenburg & Levine, 1998), and there have been some attempts to build cognitively plausible versions of the game theory player model using cognitive architectures (e.g., see Ritter & Wallach, 1998 for examples using ACT-R and SOAR). However, all of this work is based on the implicit acceptance of the game theory player model as an appropriate framework for understanding human game playing behavior.

An alternative player model

A psychologically plausible alternative to the game theory player model is that instead of trying to learn advantageous move probabilities, people try to detect sequential dependencies in their opponents' play and use this to predict their opponents' moves (Lebiere & West, 1999; West, 1998; West & Lebiere, 2001). That is, players learn recognize patterns of consecutive moves in their opponents' play. This model is consistent with a large amount of psychological research showing that when sequential dependencies exist, people can often detect and exploit them (e.g., Estes, 1972; Restle, 1966; Rose & Vitz, 1966; Vitz & Todd, 1967). It also explains why people tend to do poorly on tasks that are truly random - they persist in trying to predict the outcomes even though doing so results in sub-optimal results (e.g., Gazzaniga, 1998; Ward, 1973; Ward, Livingston, & Li, 1988).

West and Lebiere (2001) used neural networks to examine the possibility that people play games by attempting to detect and exploit sequential dependencies in their opponent's play. The networks were designed to detect sequential dependencies in the game of Paper, Rock, Scissors (hence forth PRS). PRS was chosen because it is familiar to most people and because it is very easy to play. It is also a zero-sum game and therefore does not involve the complications associated with the option to cooperate. The players were modeled using very simple two layer neural networks rewarded by adding 1 and punished by subtracting 1 from the connection weights (all of which started with a weight of 0). The inputs to the network were the opponent's previous moves (referred to as lags), and the outputs were the moves the player would make on the current play. The goal in creating these networks was to use the simplest possible model of sequential dependency detection.

The simulations revealed that processing more lags is an advantage. That is, a network that processed the last two lags (a lag 2 network) would reliably win against a network that processed only the last lag (a lag 1 network). Also a network that treated ties as losses (an aggressive network) could reliably win against a network that was neither punished nor rewarded for ties (a passive network). Furthermore, these effects were additive and approximately equal in magnitude. Another important finding was that the interaction between the networks produced a chaos-like behavior that made them appear to be playing randomly. Subsequent to examining the play of the neural network models, West & Lebiere (2001) investigated the play of humans against the models. They found that humans could reliably beat both the aggressive lag 1 network and the passive lag 2 network. This suggested that humans play similarly to the aggressive lag 2 network. Although there was a small but statistically significant tendency for people to lose against the aggressive lag 2 model rather than tie, this was attributed to the humans being unable to play as consistently as the network model. This interpretation was supported by the fact that subjects reported getting frustrated when playing the aggressive lag 2 network (i.e., playing hundreds of trials is only fun if you are winning). Both the network model and the game theory solution predicted that, on average, people would play each of the three play options with equal frequency. However, the game theory model (i.e., the combination of the game theory player model and the game theory solution) predicted that people would tie against the networks, which was not the case.

Aggregate behavior

The West & Lebiere (2001) results show that the sequential dependency player model can account for results in simple zero-sum games that game theory model cannot. In addition, the sequential dependency player model is consistent with the empirical facts. Specifically, people are poor at being random and poor at learning optimal move probabilities because they are instead trying to detect and exploit sequential dependencies. However, the game theory solution did correctly predict that, on average, humans played paper, rock, and scissors with approximately equal frequency. This raises the question of whether game theory can still be considered viable for predicting aggregate move probabilities for this type of game. That is, regardless of the details of how people play, does game theory capture certain higher-level stochastic properties of game playing behavior? After all, even if people do not process game information in the manner suggested by the game theory player model, it may still be the case that across time and across individuals, human game playing can legitimately be viewed as (pseudo) randomly emitting moves according to certain probabilities. To test this possibility and to probe deeper into the relationship between game theory and the sequential dependency player model we tested a variant of PRS.

Rock=2

An aspect of PRS that makes it a very simple game is that each of the three play options is functionally identical to the other two. That is, each move beats one of the other two moves and loses to the remaining move. In addition, a win is worth the same amount for each move. Thus, it is not surprising that the game theory solution for playing PRS is to play the three options with equal probabilities. The reason this is somewhat problematic is that the agreement between the game theory solution and human behavior for this game may be an artifact of the simplicity and symmetry of the game. To clarify this issue, a modified version of PRS was developed. The new game was identical to the original PRS game except that a win using rock counted for 2 points while a win with scissors or paper counted for only 1 point. In this way, each of the three choices were unique. Rock could win two points and lose only one, scissors could lose two points and win only one, and paper could win or lose only one point. The game theory solution to this modified version of Paper, Rock, Scissors differs depending on whether a zero-sum or non-zero-sum interpretation of the game is adopted. That is, whether a player is trying to maximize the difference in points between himself and the other player, or whether each player is attempting only to maximize the total number of points for themselves. However, since we instructed our subjects to try to maximize the points difference we will focus on the zero-sum interpretation. In this case, the game theory solution is to play paper 50% of the time, rock 25% of the time, and scissors 25% of the time, for the expected outcome of a tie.

The simulated opponents

For this study we used the same simple network models as West & Lebiere (2001) and created three simulated opponents for our human subjects to play against. The first two opponents were taken directly from the West & Lebiere (2001) study. They were, the aggressive lag 2 model and the aggressive lag 1 model. We did this to test the hypothesis that people simply try to maximize wins in this type of game. If this were the case then the results against these two models would replicate the results of West & Lebiere (2001) as neither the humans nor the models would be influenced by rock wins being worth more points. To create the third simulated opponent we adapted the aggressive lag 1 model so that it rewarded the relevant connection weights by 2 instead of 1 when it won with rock. This model was created to pit the human players against a model that might better take advantage of winning with rock, but still had some weaknesses that humans could exploit (i.e., it was only a lag 1 network and it was not set to avoid losing with scissors thereby allowing the opponent to win with rock). The reason for this was that games where humans win are much more informative than games where humans lose, as the loss can be attributed to extraneous factors such as lack of effort, boredom, or frustration. To distinguish this model we will refer to it as the rock=2 lag 1 model.

Method

Ten human subjects played against each of the three network models. They were instructed to try to maximize the points difference in their favor by as much as possible. This goal corresponded to the zero-sum interpretation of the game. The subjects were also told that the network models did not play randomly and that they could be beaten. Additionally, the subjects were instructed to play naturally, not to play too slowly, nor to think too much about their play. The order in which the subjects played the network models was random. Subjects were required to play one game (300 trials) against each of the three different network models. This process took from between 30 to 45 minutes in total depending on the speed at which the subject played.

Results and Discussion

The points differences between each of the human players and each of the network models were calculated for each trial by subtracting the network score from the human score. Thus a positive score indicated that the human was ahead and a negative score indicated that the network was ahead. The mean total points difference at the end of each game (see Table 1) revealed that the humans were able to win against all of the network models. To test the significance of this we ran a regression on the group points difference data for each different type of opponent across trials. The regression coefficients thus corresponded to the average rate of points accumulation (i.e., points difference/trials) for the humans against each network opponent (the intercept was forced through zero). 95% confidence intervals for the coefficient values revealed that all of them were significantly above zero. That is, against each network models, there was a significant tendency for the humans to win.

The fact that people could beat the aggressive lag 2 model under these conditions, whereas they tended to lose in West & Lebiere (2001), where all three varieties of wins were of equal value, indicates that they were able to exploit the fact they knew that wins using rock were rewarded with 2 points. Thus, the hypothesis that people simply try to maximize the number of wins regardless of the number of points awarded for wins, was refuted. That is, the profiles of the humans' play suggested wins with rock were preferred to wins with paper or scissors.

Given that people were sensitive to the payoff information the next question was whether, as in West & Lebiere (2001), the game theory solution predicted the move probabilities for the human players. Figure 1 displays the probabilities for playing paper, rock and scissors for the human subjects, for each of the opponents they faced, with 95% confidence intervals. Figure 1 also displays the predicted probabilities from the game theory solution. As can be seen, the game theory solution was significantly different from the human probabilities.

Table 1: The game results of the human versus the neural network models.

Network Model	Play ratios (P,R,S)	Mean points difference	Expected points difference	Strategy Index
Lag 1	28.8%, 39.6%, 31.5%	16.5	4.86	11.64
Lag 2	32.3%, 38.1%, 29.5%	5.7	11.38	-5.68
Rock=2 lag 1	33.5%, 39.7%, 26.8%	25.6	5.76	19.84

Table 2: Regression analysis on the performance of human subjects against the three network models.

Network Model	Regression coefficient	R Squared	Confidence Intervals Lower 95%	Upper 95%
Lag 1	0.0886	0.0208	0.0835	0.0938
Lag 2	0.0212	-0.0163	0.0160	0.0264
Rock=2 lag 1	0.0539	0.0251	0.0488	0.0590

We also examined whether the human results could be explained by using the game theory player model without the optimal game theory solution. That is, did the human subjects win by using move probabilities that exploited non-optimal move probabilities produced by the network models? Table 1 shows the average difference in score along with what the difference in score would be if it were determined solely by the overall move probabilities of the two players. The strategy index number is the difference between these two. A score of zero on the strategy index would indicate that the score difference could be accounted for entirely by move probabilities. In all cases the strategy index was significantly different from zero ($P<0.05$, determined by confidence intervals). Given the move probabilities, against the two lag 1 models the humans played significantly better than expected, while against the lag 2 model they played significantly worse than expected. This can be interpreted as the humans being better at exploiting sequential dependencies than the lag 1 models, but not as not as good as the lag 2 model. This agrees with the results of West & Lebiere (2001) and suggests that humans were able to narrowly beat the lag 2 because of their knowledge of the payoffs.

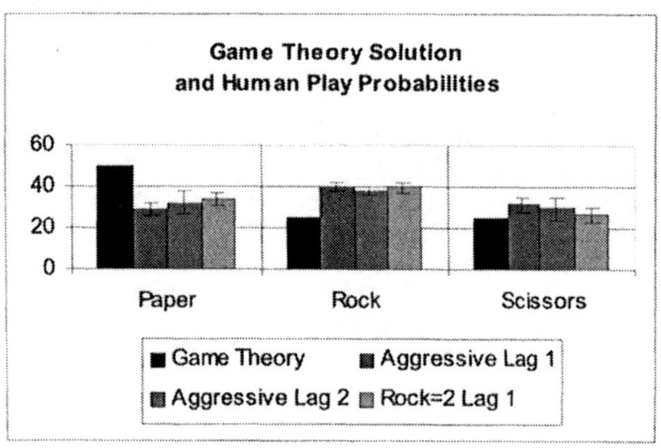

Figure 1: Game Theory Solution and Human Play Probabilities.

Modeling the Human results

Our next step was to construct a neural network model of how the humans played. For the model we assumed that people detect sequential dependencies in a way similar to a lag2 network. Although the results of this paper and West & Lebiere (2001) show that in games against the lag 2 network, humans seem to be slightly worse at detecting sequential dependencies, we again assumed this was due to humans finding the lag 2 less fun to play against because it is a stronger opponent. In both studies the advantage for the lag 2 network was relatively small. Additionally, West & Lebiere (2001) found that they could account quite well for the results of the other games by modeling humans as aggressive lag 2 networks. To account for the findings in this study we modeled people as aggressive lag 2 networks with the ability to adjust their rewards and punishments so as to best take advantage of the payoffs in the game.

To get an idea of how people could be adjusting the rewards and punishments we obtained self-reports on the strategies used by several of our more successful subjects. These reports generalized to favoring rock wins to paper wins, and paper wins to scissors wins. That is, they were focused first on getting rock wins and second on blocking the opponent from getting rock wins. With this in mind we ran a genetic algorithm to find a system of rewards and punishments for the neural network model that would match the human point difference results. The result was the following: rock wins = 3, paper wins = 2 scissors wins = 0; rock tie = -1, paper tie = -1, scissors tie = 0; and -3 for all losses. Note that not rewarding scissors wins makes sense as winning frequently with scissors would be associated with the opponent playing paper more often and that would block rock.

The model was played against each of the three networks that the humans faced. Each simulation consisted of 1000 games of 300 trials each. For each game, the net points difference between the models, the probabilities by which the human model selected each of the three play choices, and the strategy index for the human model was recorded. These three measures were used to determine how well the model fit the human data.

Results

The results showed that the model matched the point differences well (see Figure 2). The model also reproduced the human move probabilities against each opponent with a high degree of accuracy. The correlation between the model move probabilities and the human move probabilities was 0.964 ($p<0.0005$). Also, the model provided a good overall fit to the human strategy index data (see Figure 2). Although the model did not match the data as well when the opponent was the lag 2 network, this is actually consistent with our position that humans do not detect sequential dependencies as well against the lag 2 network due to confounding factors, which would not apply to the network model of human behavior. Also, the fact that the model was able to match two sets of results that it was not explicitly designed to match (the move probabilities and the strategy index values) suggests that the results were reasonably robust.

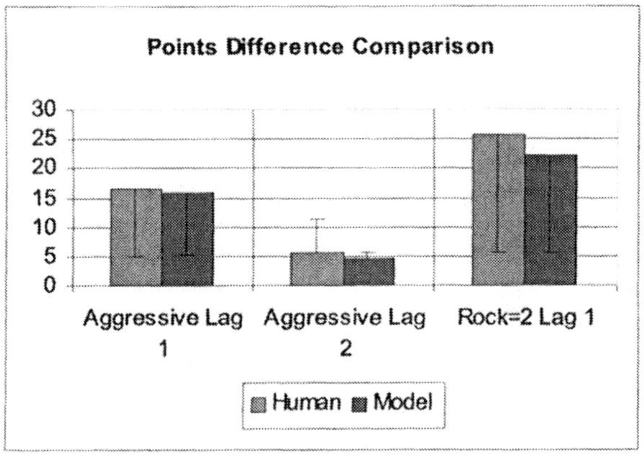

Figure 2: Points Difference Comparison. The shaded bars indicate the mean final points difference. The heads of the T-bars represent the expected final points difference; and, the length of the T-bars indicate the magnitude of the strategy index values.

Conclusions

The results of this study replicate the West & Lebiere (2001) findings that the commonly used probabilistic game theory model (as defined in this paper) cannot account for the game results when humans play against agents programmed to play by exploiting sequential dependencies. We also demonstrated that when the game payoffs are not all equal, the game theory solution does not predict the aggregate move probabilities. We further demonstrated, using the actual move probabilities, that the results could not be accounted for by the game theory player model. That is, the actual move probabilities did not predict the final points differences. These results show that the game theory player model, with or without the game theory solution, is fundamentally different from the way people process information in this type of situation.

In terms of modeling, we replicated the West & Lebiere (2001) result that this type of human game playing can be accurately modeled using simple lag 2 networks. Furthermore, we extended the original model by showing that people are sensitive to different game payoffs and that this can be modeled by adjusting the rewards and punishments associated with different play outcomes. These results are also consistent with a number of ACT-R studies showing that people play a variety of games using the lag 2 strategy (PRS: Lebiere & West, 1999: non-zero-sum games: Lebiere, Wallach, & West, 2000; baseball: Lebiere, Gray, Salvucci, & West, 2003). The ACT-R model works by using the ACT-R declarative memory system as a neural network for detecting sequential dependencies, and produces results similar to the simple networks we used (Lebiere & West, 1999). An ACT-R model equivalent to the one in this paper could be created by "popping" the "chunks" representing sequential dependency patterns a different number of times for different outcomes. Likewise, a genetic algorithm could be used to fit the model. However, this approach would sidestep the next important issue, which is modeling how humans adjust their reward structure in response to the game payoffs.

References

Estes, W. K. (1972). Research and theory on the learning of probabilities. *Journal of the American Statistical Association*, 67, 81-102.

Fudenburg, D., & Levine, D. K. (1998). *The theory of Learning in Games*. Cambridge, MA: The MIT Press.

Gazzaniga, M. S. (1998). The split brain revisited. Scientific American, July, 50-55.

Lebiere, C., Gray, R., Salvucci, D., & West, R. L. (2003). Choice and Learning under Uncertainty: A Case Study in Baseball Batting. *Proceedings of Cognitive Science.*

Lebiere, C., Wallach, D., & West, R. L. (2000). A Memory-based Account of the Prisoner's Dilemma and Other 2x2 Games. *The proceedings of the Third International Conference on Cognitive Modeling.*

Lebiere, C., & West, R. L. (1999). Using ACT-R to model the dynamic properties of simple games. *Proceedings of Cognitive Science.*

Pool, R. (1995). *Putting game theory to the test.* Science, 267, 1591-1593.

Restle, F. (1966). Run structure and probability learning: Disproof of Restle's model. *Journal of Experimental Psychology*, 72, 382-389.

Ritter, F. E., & Wallach, D. P. (1998). Models of two-person games in ACT-R and Soar. *Proceedings of the Second European Conference on Cognitive Modelling.* 202-203. Thrumpton: Nottingham University Press.

Rose, R. M., & Vitz, P. C. (1966). The role of runs of events in probability learning. *Journal of Experimental Psychology*, 72, 751-760.

Samuelson, L. (1997). *Evolutionary games and equilibrium selection.* Cambridge, Mass: MIT Press.

Poundstone, W. (1992). *Prisoner's dilemma.* NY: Doubleday.

Vitz, P. C., & Todd, T. C. (1967). A model of learning for simple repeating binary patterns. *Journal of Experimental Psychology*, 75, 108-117.

VonNeumann, J., & Morgenstern, O. (1944) *Theory of games and economic behaviour*. Princton, N. J.: Princton University Press.

Ward, L. M. (1973). Use of markov-encoded sequential information in numerical signal detection. *Perception and Psychophysics*, 14, 337-342.

Ward, L. M., Livingston, J. W., & Li, J. (1988). On probabilistic categorization: The markovian observer. *Perception and Psychophysics*, 43, 125-136.

Wagenaar, W. A. (1972). Generation of random sequences by human subjects: A critical survey of the literature. *Psychological Bulletin*, 77, 65-72.

West, R. L. (1998). Zero Sum Games as Distributed Cognitive Systems. *Proceedings of the Complex Games Workshop*. Tsukuba, Japan: Electrotechnical Laboratory Machine Inference Group.

West, R. L., & Lebiere, C. (2001). Simple games as dynamic, coupled systems: Randomness and other emergent properties. *Cognitive Systems Research*, 1(4), 221-239.

Dimension-Wide vs. Exemplar-Specific Attention in Category Learning and Recognition

Yasuaki Sakamoto (yasu@psy.utexas.edu)
Department of Psychology, The University of Texas at Austin
Austin, TX 78712 USA
Toshihiko Matsuka (matsuka@psychology.rutgers.edu)
RUMBA, Department of Psychology, Rutgers University
Newark, NJ 07102 USA
Bradley C. Love (love@psy.utexas.edu)
Department of Psychology, The University of Texas at Austin
Austin, TX 78712 USA

Abstract

Items that violate a category rule are remembered better than items that follow the rule. This finding cannot be predicted by exemplar models when all exemplars share the same attention along a dimension. With dimension-wide attention, violating and rule-following items are treated equally. When each exemplar selects which dimensions to attend to, exemplar models can predict the memory advantage for violating items. With exemplar-specific attention, attention is distributed uniformly for exemplars encoding violating items but is allocated to the rule dimension of exemplars encoding rule-following items. This differential attention makes violating items distinctive in memory. In addition to exemplar-specific attention, exemplar models need the ability to distinguish important errors from negligible ones to predict better memory for items that violate a stronger than a weaker rule.

Humans are confronted with more information than they can process. Consequently, the ability to selectively attend to salient information is fundamental to our cognitive behavior. Many category learning models utilize the same attention at all locations along a dimension in the representational space (e.g., Kruschke, 1992; Love, Medin, & Gureckis, 2004; Nosofsky, 1986). The dimension-wide attention is well suited for many artificial category learning studies, in which categories are symmetric and category members are differentiated by the values on the same dimensions. For example, category A members may be large on the size dimension, whereas category B members may be small. In natural categories, however, there are inconsistent items that do not follow the structure. For example, penguins do not fly but are members of category birds, whereas bats fly but belong to category mammals. Dimension-wide attention may not fare as well when categories contain inconsistent members.

Some laboratory work does suggest that attention is specific to the region along a dimension in the representational space (e.g., Aha & Goldstone, 1992; Barsalou & Medin, 1986; Lewandowsky, Kalish, & Ngang, 2002). Humans attend to different dimensions of an item depending on the context the item is in. For example, humans may attend to the color dimension when shopping for clothing but not as much when shopping for a computer.

Exemplar models have a long history of explaining key psychological phenomena in the category learning research (Kruschke, 1992; Medin & Schaffer, 1978; Nosofsky, 1986). However, one finding that is problematic for exemplar models is that people better remember items that violate a structure (e.g., rule) than structure-consistent (e.g., rule-following) items (Palmeri & Nosofsky, 1995; Sakamoto & Love, in press). By storing every studied item as a separate trace and using the dimension-wide attention, current exemplar models treat consistent and inconsistent items in the same fashion and cannot predict the memory advantage for inconsistent items.

In this paper, we show that an exemplar model with exemplar-specific attention (see Kruschke, 2001 for a related model with exemplar-specific "specificity") can differentiate violating items from rule-following items and predict the memory advantage for violating items. While exemplars encoding rule-following items result in attention allocated to the rule dimension, exemplars encoding violating items result in attention distributed to the non-rule dimensions. This differential attention makes violating items distinctive in memory.

We further show that in addition to the exemplar-specific attention, exemplar models need a mechanism that accentuates larger errors and minimizes the impact of smaller ones to predict a better memory for items that violate a stronger rule than items that violate a weaker rule (e.g., Sakamoto & Love, in press). In the remainder of the paper, we review previous category learning work that examines recognition memory for violating items, introduce the models, present model fits to previous findings, and discuss our modeling results.

Memory Advantage for Exceptions

Palmeri and Nosofsky (1995) found that humans remember exceptions to the category rule better than items that follow the rule. In their study, subjects learned to classify geometric stimuli into two contrasting categories. An imperfect rule successfully classified the majority of study items (e.g., most small items were in category A, whereas most large items were in category B), but two exceptions violated the rule (e.g., a large item that was a member of category A). Following learning, subjects completed a recognition test consisting of studied items and novel items that served as foils. The basic finding was that people recognized the exceptions better than the studied rule-following items. The special status of violating items is also suggested by the schema (e.g., Rojahn & Pettigrew, 1992) research and is not specific to category learning studies.

Exemplar models with dimension-wide attention, such as ALCOVE (Kruschke, 1992) and the context model (Medin & Schaffer, 1978), cannot predict the memory advantage of exceptions. Exemplar models store every studied item as a separate trace. The likelihood of recognizing an item is determined by the item's absolute similarity to all exemplars from both categories A and B. Exemplar models cannot predict the enhanced memory for exceptions because the exceptions share the same similarity relations with other items in memory as rule-following items do. Exceptions are distinguished from rule-following items because the exceptions' category assignment runs counter to the rule. According to exemplar models, this reversal is not germane to recognition.

In contrast, a rule-based model, such as RULEX (Nosofsky, Palmeri, & McKinley, 1994), can predict the memory advantage for exceptions. RULEX constructs rules and stores exceptions to the rules. Rule-following items are not individually stored but rather are captured by the rule. Information about exceptions is explicitly stored. The likelihood of recognizing an item is determined by the items in the exception store. The separate storage of exception information allows RULEX to predict the memory advantage of exceptions.

However, RULEX cannot predict the entire pattern of Palmeri and Nosofsky's results. In addition to the memory advantage for exceptions, they found better recognition for studied rule-following over novel items. RULEX cannot predict the memory advantage for rule-following over novel items, which the context model can predict, because rules encode very little information about rule-following items. Thus, Palmeri and Nosofsky combined RULEX and the context model (see Erikson & Kruschke, 1998 for a similar approach involving knowledge gating). The combined model was able to predict the entire pattern of their recognition data.

The inability of RULEX to predict the memory advantage of rule-following over novel items suggests that humans encode more than just rules to represent rule-following items. In support of this idea, Allen and Brooks (1991) found that even when a rule is explicitly applied to a novel item, humans are still somewhat sensitive to the similarity between the novel item and previously encountered examples. This line of work argues that humans have both rule and exemplar systems for learning and recognition.

An alternative approach is clustering as in SUSTAIN (Love, Medin, & Gureckis, 2004). SUSTAIN has aspects of both rule violation and exemplar memory and can predict better memory for exceptions than studied rule-following items as well as better memory for studied rule-following items than novel items. We predict that introducing exemplar-specific attention to exemplar models will allow those models to be sensitive to rule violation.

Exemplar-Specific Attention

An exemplar model, such as ALCOVE (Kruschke, 1992), treats exceptions and rule-following items in the same manner because all exemplars share the same attention along a dimension. Consequently, ALCOVE cannot predict the memory advantage of exceptions. In contrast, ALCOVE with exemplar-specific attention (ES-ALCOVE) should be able to predict the memory advantage for exceptions. ES-ALCOVE shifts attention to the rule dimension. To classify an exception in the correct category, ES-ALCOVE will distribute attention to the non-rule dimensions so that the exception is distinguished from the rule-following items from the opposing category. While attention will be distributed to the non-rule dimensions for the exceptions, the rule-following items will receive attention on the rule dimension. ES-ALCOVE could predict a memory advantage for exceptions because the exceptions will be "differentiated" from the rule-following items. We simulate ES-ALCOVE to Palmeri and Nosofsky's results to test this intuition. In the next section, we formalize ES-ALCOVE.

Formalism

ES-ALCOVE stores every training item as a separate trace. The probability that a stimulus x_i is classified into category A is determined by:

$$P(A|x_i) = \frac{exp[\phi \cdot O_A(x_i)]}{exp[\phi \cdot O_A(x_i)] + exp[\phi \cdot O_B(x_i)]} \quad (1)$$

where the parameter ϕ controls the decisiveness of classification response. $O_A(x_i)$ is the category A output activation given a stimulus x_i defined as:

$$O_A(x_i) = \sum_{j=1} \omega_{Aj} \cdot S(x_i, y_j) \quad (2)$$

where ω_{Aj} is the strength of association between category A and exemplar j. $S(x_i, y_j)$ is the similarity between a stimulus x_i and a stored exemplar y_j given by:

$$S(x_i, y_j) = exp[-c \cdot D(x_i, y_j)] \quad (3)$$

where the free parameter c scales the strength of overall similarity. $D(x_i, y_j)$ indicates the distance between x_i and y_j defined by:

$$D(x_i, y_j) = \sum_{k=1} \alpha_{jk} \cdot |x_{ik} - y_{jk}| \quad (4)$$

Table 1: Recognition ratings observed in Palmeri and Nosofsky (1995) and predicted by ES-ALCOVE. The fits of ALCOVE are also included for a comparison. Item types are exceptions (Exc), studied rule-following items (Rul), and novel items (Nov).

Item	Observed	ES ALCOVE	ALCOVE
Exc	6.92	6.58	6.39
Rul	5.74	5.94	6.39
Nov	5.30	5.20	5.13

where k is the number of dimensions and α_{jk} is the attentional weight for the kth dimension of an exemplar y_j. Unlike dimension-wide attention, in which all stored exemplars share the same attention along a dimension, each stored exemplar selects which of its dimensions will receive attention.

The familiarity $F(x_i)$ of a stimulus x_i is determined by the summed similarity of the stimulus item to the stored exemplars of both categories A and B:

$$F(x_i) = \sum_{j=1} S(x_i, y_j) \quad (5)$$

where $S(x_i, y_j)$ is defined in Equation 3.

Simulation 1

ES-ALCOVE was fit to the mean recognition ratings provided by human subjects in Palmeri and Nosofsky (1995). Prior to recognition, the models learned to classify the members of Categories A and B. Most of the members followed the imperfect rule, but each category contained an exception that violated the rule. The models updated the association (ω) and attention weights (α) after the presentation of each training item by gradient descent (see Kruschke, 1992 for derivations) that reduced the sum squared differences between the target and the predicted output values (i.e., Equation 2).

After training, ES-ALCOVE generated recognition ratings. A linear relationship was assumed between the human recognition ratings and ES-ALCOVE's familiarity (i.e., Equation 5). The basic finding was that exceptions received the highest recognition ratings, followed by rule-following items, followed by novel items.

Results

As shown in Table 1, ES-ALCOVE was able to capture the observed pattern. For a comparison, Table 1 displays that ALCOVE cannot predict the observed results by using the same attention along a dimension for exemplars representing exceptions and rule-following items. Table 2 shows that as predicted,

Table 2: The attention weights obtained by ES-ALCOVE for the rule (D_{rule}) and the non-rule dimensions (sum for $D_{non-rule}$) of exemplars encoding exceptions (Exc) and rule-following items (Rul).

Exemplar	D_{rule}	$D_{non-rule}$
Exc	.07	.80
Rul	.22	.64

ES-ALCOVE distributed more attention to the non-rule dimensions of the exemplars encoding exceptions than to the non-rule dimensions of the exemplars encoding rule-following items. The opposite pattern was observed for the rule dimension.

The differential attention for exceptions made the exceptions distinctive in memory. Attention scales the distance when the dimension values mismatch. The greater attention for the non-rule dimensions of exemplars encoding exceptions suggests that items with mismatching values on the non-rule dimensions become highly dissimilar. As in RULEX, exceptions are distinguished from rule-following items and remembered better. Unlike RULEX, ES-ALCOVE also correctly predicts a memory advantage for rule-following items over novel items because it retains information about non-rule dimensions.

Further Test of Exception Memory

ES-ALCOVE was able to predict a memory advantage for exceptions by differentiating the exceptions from rule-following items in terms of attention allocated to exemplars encoding those items. A related finding from the category learning research that examines memory for exceptions is that the memory advantage for exceptions is greater when the violated rule is stronger (Sakamoto & Love, in press).

In Sakamoto and Love, as in Palmeri and Nosofsky, most of the members of categories A and B followed an imperfect rule, and each category contained an exception. Rule strength was manipulated by varying the frequency of rule-following items. Category A contained eight rule-following items, whereas category B contained only four. The classification learning procedure encouraged subjects to entertain the rules "If value 1 on the first dimension, then category A" and "If value 2 on the first dimension, then category B." Category B's exception violated category A's rule, whereas category A's exception violated category B's rule. After training, these exceptions were remembered better than the rule-following items, replicating Palmeri and Nosofsky. Furthermore, as in the schema research (e.g., Rojahn & Pettigrew, 1992), memory for the category B exception, which violated more frequent category A's rule, was enhanced.

As in Palmeri and Nosofsky, exemplar models, such as ALCOVE and the context model, were unable to account for the enhanced recognition of the category B exception because they did not provide a role for knowledge structures in encoding exceptions and rule-following items. RULEX captured rule-governed behavior with actual rules and was insensitive to the rule frequency manipulation (cf., Pinker, 1991; Smith, Langston, & Nisbett, 1992). Because the context model and RULEX had trouble with different aspects, the combined model was also unable to predict the results.

Although ES-ALCOVE was able to account for the memory advantage for exceptions, it was unable to predict the greater memory advantage for the category B exception. This failure arises because the attention shift in ES-ALCOVE does not distinguish the two exceptions. Because there are more category A rule-following items, it is harder to learn the category B exception than the category A exception. The category B exception results in larger discrepancies between target and predicted output values. ES-ALCOVE treats large and small discrepancies in the same manner and cannot differentiate the two exceptions. As a result, ES-ALCOVE is unable to predict the rule frequency effect observed in Sakamoto and Love.

We created another version of ALCOVE called ESSW-ALCOVE for Exemplar-Specific Squeaky Wheel ALCOVE. In addition to the exemplar-specific attention, ESSW-ALCOVE has a mechanism that emphasizes larger errors and minimizes the impact of smaller ones. ESSW-ALCOVE should be able to predict Sakamoto and Love's results by distributing more attention to the non-rule dimensions of the category B exception, which results in larger errors, than to the non-rule dimensions of the category A exception. ESSW-ALCOVE was fit to Sakamoto and Love's results to test this intuition.

Simulation 2

In ESSW-ALCOVE, the sum squared error between target and predicted output values minimized by ALCOVE (and ES-ALCOVE) was exaggerated to the 10th power. This causes larger errors to remain relatively large but smaller errors to become extremely small. As a result, relative to ES-ALCOVE, attention learning is accentuated when an item leads to a larger error, and the exemplar representing the error-prone item will receive more uniform attention (hence squeaky wheel). ESSW-ALCOVE can predict Palmeri and Nosofsky's results.

ESSW-ALCOVE learned to classify the training items. Category A contained eight rule-following items, whereas category B contained only four. Each category contained one exception item. After training, ESSW-ALCOVE made two-alternative forced choice judgments. The pairs matched on the rule dimension. The basic finding was that the category B exception, which violated more rule-following items, was remembered better than the category A exception. The probability of choosing the studied (old) item was determined by the exponential decision function:

$$P(old) = \frac{exp[\rho \cdot F(old)]}{exp[\rho \cdot F(old)] + exp[\rho \cdot F(new)]} \quad (6)$$

where $F(old)$ is the model's familiarity (see Equation 5) for the studied item, $F(new)$ is the models' familiarity for the novel item, and ρ is the recognition decision parameter.

Results

As shown in Table 3, ESSW-ALCOVE, which accentuated errors, captured the observed pattern. For a comparison, Table 3 shows that ES-ALCOVE cannot predict the observed results despite its exemplar-specific attention. The rank order (1 is most and 4 is least) of ESSW-ALCOVE's attention shown in Table 4 reveals that as in Palmeri and Nosofsky, exemplars encoding exceptions resulted in more attention

Table 3: Recognition performance (with 95% confidence intervals) observed in Sakamoto and Love (in press) and predicted by ESSW-ALCOVE. The fits of ES-ALCOVE are also included for a comparison. The category B exception (Exc B) violated more rule-following items from category A (Rul A). The category A exception (Exc A) violated fewer rule-following items from category B (Rul B).

Item	Observed	ESSW ALCOVE	ES ALCOVE
Exc B	87±3	85	82
Exc A	79±3	80	82
Rul B	69±3	70	70
Rul A	70±3	70	70

Table 4: The rank order of ESSW-ALCOVE's attention (with attention weights) for the rule (D_{rule}) and the non-rule dimensions (sum for $D_{non-rule}$) of the exemplars encoding the category B exception (Exc B), the category A exception (Exc A), the category B rule-following items (Rul B), and the category A rule-following items (Rul A). The category B exception violated more rule-following items from category A, and the category A exception violated fewer rule-following items from category B. Smaller values represent greater attention.

Exemplar	D_{rule}	$D_{non-rule}$
Exc B	4 (0.460)	1 (0.331)
Exc A	3 (0.463)	2 (0.314)
Rul B	1 (0.470)	3 (0.279)
Rul A	2 (0.467)	4 (0.260)

distributed to the non-rule dimensions than exemplars representing rule-following items. Moreover, as predicted more attention was distributed to the non-rule dimensions for the exemplar encoding category B's exception than for the exemplar encoding category A's exception.

There were more errors involving the category B exception than the category A exception because more rule-following items in category A were similar to the category B exception. Such errors were accentuated in ESSW-ALCOVE, and the category B exception resulted in more uniform attention. ESSW-ALCOVE predicts a memory advantage for the category B exception because it is differentiated from many similar rule-following items from category A.

Filtration and Condensation

Incorporating exemplar-specific attention and accentuated errors allowed ALCOVE to predict previously challenging data. It is crucial that a model that incorporates new mechanisms can still account for other basic psychological phenomena. Examining filtration and condensation tasks (Gottwald & Garner, 1975; Kruschke, 1993, Matsuka, in press) is important because these tasks investigate how humans allocate attention.

Humans (and ALCOVE) find it easier to learn filtration tasks, in which information from only one dimension is required for perfect classification, than condensation tasks, in which information from two (or more) dimensions is needed (e.g., Kruschke, 1993). This filtration advantage was predicted by ESSW-ALCOVE (and ES-ALCOVE). In filtration, all exemplars attend to the predictive dimension. This leads to increased psychological distances between category A and category B members. In condensation, some items are closer to the opposing category's exemplars even with exemplar-specific attention, and ESSW-ALCOVE has less "confirmations" from some exemplars. ESSW-ALCOVE finds it easier to learn the filtration task, in which it receives more "evidence" for correct category membership.

Discussion

Humans can flexibly attend to different dimensions of an item depending on the values of the dimensions that are not critical for classification (Aha & Goldstone, 1992). Such flexible attention allowed ES-ALCOVE to differentiate exceptions from rule-following items and predict a memory advantage for exceptions. In ES-ALCOVE, each exemplar selected which dimensions to attend to. ES-ALCOVE attended to the non-rule dimensions of exemplars encoding exceptions but to the rule dimension of exemplars encoding rule-following items. This differential attention made exceptions distinctive in memory.

However, ES-ALCOVE was unable to account for the finding that memory for a violating item is stronger when the violated structure is stronger. This finding was predicted by ESSW-ALCOVE, which accentuated errors. With errors raised to the 10th power, ESSW-ALCOVE distinguished important errors (e.g., miss-classification) from trivial ones (e.g., correct classification with 90% confidence level). ESSW-ALCOVE learned attention more rapidly in response to larger errors and minimized the impact of smaller errors. A similar effect can be obtained by updating the attention weights multiple times on each training trial (e.g., Kruschke, 2001). ESSW-ALCOVE better remembered items that violated a stronger rule because those items were associated with "important" errors.

In addition to the exception memory findings reviewed in this paper, ESSW-ALCOVE was able to predict a filtration advantage. We should note that ALCOVE that accentuated errors without exemplar-specific attention was unable to predict the exception memory findings. Thus, ALCOVE needs both exemplar-specific attention and accentuated errors to account for all of the findings described in this paper. One question to ask is whether these mechanisms are also required by other models.

Clustering Approach

SUSTAIN (Love et al., 2004), which uses dimension-wide attention, can predict the memory advantage for exceptions as well as the greater memory advantage for exceptions that violate a stronger rule. SUSTAIN clusters together similar items and recruits a new cluster in response to a prediction error. SUSTAIN develops rule-following clusters and shifts attention to the rule dimension. All clusters share the same attention along a dimension. When an exception item elicits a prediction error, SUSTAIN recruits an additional cluster to encode the item. While rule-following items tend to cluster with one another, each exception item will be isolated in its own cluster. This differential storage makes exceptions more distinctive in memory.

SUSTAIN also predicts better recognition for the category B exception that violates more rule-following items from category A. A prediction error occurs when SUSTAIN attempts to cluster together highly similar items from competing categories. The exception clusters brought about such errors by attracting rule-following items from the opposing category. Because there were more category A rule-following items, there were more opportunities for such errors involving the category B exception to occur, and a greater number of category A rule-following clusters were recruited. These clusters formed a highly contrastive backdrop for the category B exception and enhanced recognition.

As ESSW-ALCOVE, SUSTAIN treats "important" and "negligible" errors differently. Large discrepancies between target and predicted output values result in prediction errors. SUSTAIN recruits new clusters in response to such errors but "ignores" small discrepancies.

SUSTAIN's success in predicting these results using dimension-wide attention suggests that attention mechanisms interact with internal representations of a model (cf., Matsuka, in press). Clustering allows SUSTAIN to be sensitive to rule violation without exemplar-specific attention. In contrast, by storing every studied item, ALCOVE needs exemplar-specific attention to capture the rule-violating nature of exceptions.

Future Direction

A more flexible model may use dimension-wide or exemplar-specific attention depending on a given task. For example, humans may initially be biased to attend to the same dimensions for all exemplars, analogous to a prior, but over time optimize learning by utilizing a separate attention for certain items. In learning about the rule-plus-exception category structures, humans may initially attend to the rule dimension for all the items. When an exception appears and a prediction error occurs, a separate attention may be used for the exception. Such processes suggest that humans have simplicity bias (cf., Matsuka, 2004). For example, filtration tasks may result in dimension-wide attention, whereas condensation tasks may lead to exemplar-specific attention. Future work should examine when exemplar-specific or dimension-wide attention is more appropriate.

Acknowledgments

This work was supported by AFOSR Grant FA9550-04-1-0226 to B.C. Love, and by the James McDonnell Foundation and the National Science Foundation (EIA-0205178).

References

Aha, D. W., & Goldstone, R. L. (1992). Concept learning and flexible weighting. In *Proceedings of the 14th Annual Conference of the Cognitive Science Society* (pp. 534–539). Hillsdale, New Jersey: Lawrence Erlbaum Associates.

Allen, S. W., & Brooks, L. R. (1991). Specializing the operation of an explicit rule. *Journal of Experimental Psychology: General, 120*, 3–19.

Barsalou, L. W., & Medin, D. L. (1986). Concepts: Static dimensions or context-dependent representations? *Cahiers de Psychologie Cognitive, 6*, 187–202.

Erikson, M. A., & Kruschke, J, K. (1998). Rules and exemplars in category learning. *Journal of Experimental Psychology: General, 127*, 107–140.

Gottwald, R. L., & Garner, W. R. (1975). Filtering and condensation tasks with integral and separable dimensions. *Perception & Psychophysics, 2*, 50–55.

Kruschke, J. K. (1992). ALCOVE: An exemplar-based connectionist model of category learning. *Psychological Review, 99*, 22–44.

Kruschke, J. K. (1993). Three principles for models of category learning. In G. V. Nakamura, R. Taraban, & D. L. Medin (Eds.) *Categorization by human and machines: The psychology of learning and motivation* (Vol. 29, pp. 57–90). San Diego, CA: Academic Press.

Kruschke, J. K. (2001). Toward a unified model of attention in associative learning. *Journal of Mathematical Psychology, 45*, 812–863.

Lewandowsky, S., Kalish, M., & Ngang, S. K. (2002). Simplified learning in complex situations: Knowledge partitioning in function learning. *Journal of Experimental Psychology: General, 131*, 163–193.

Love, B. C., Medin, D. L., & Gureckis, T. M. (2004). SUSTAIN: A Network Model of Human Category Learning. *Psychological Review.* In press.

Matsuka, T. (in press). Generalized exploratory model of category learning. International Journal of Computational Intelligence.

Matsuka, T. (2004). Biased stochastic learning in computational model of category learning. In *Proceedings of the 26th Annual Conference of the Cognitive Science Society.*

Medin, D. L., & Schaffer, M. M. (1978). Context theory of classification learning. *Psychological Review, 85*, 207–238.

Nosofsky, R. M. (1986). Attention, similarity, and the identification-categorization relationship. *Journal of Experimental Psychology: General, 115*, 39–57.

Nosofsky, R. M., Palmeri, T. J., & McKinley, S. C. (1994). Rule-plus-exception model of classification learning. *Psychological Review, 101(1)*, 53–79.

Palmeri, T. J., & Nosofsky, R. M. (1995). Recognition memory for exceptions to the category rule. *Journal of Experimental Psychology: Learning, Memory, & Cognition, 21*, 548–568.

Pinker, S. (1991). Rules of language. *Science, 253*, 530–535.

Rojahn, K., & Pettigrew, T. F. (1992). Memory for schema-relevant information: A meta-analytic resolution. *British Journal of Social Psychology, 31(2)*, 81–109.

Sakamoto, Y., & Love, B. C. (in press). Schematic influences on category learning and recognition memory. *Journal of Experimental Psychology: General.*

Smith, E. E., & Langston, C., & Nisbett, R. E. (1992). The case for rules in reasoning. *Cognitive Science, 16*, 1–40.

Toward an ACT-R General Executive for Human Multitasking

Dario D. Salvucci (salvucci@cs.drexel.edu)
Yelena Kushleyeva (yk45@drexel.edu)
Frank J. Lee (fjl@cs.drexel.edu)
Department of Computer Science, Drexel University
3141 Chestnut St., Philadelphia, PA 19104

Abstract

While cognitive models of complex tasks have begun to incorporate increasingly sophisticated models of human multitasking, most models have utilized customized executives (Kieras et al., 2000) that fine-tune specific multitasking mechanisms for particular applications. This paper proposes a general executive for multitasking that facilitates the integration of separate task models and subsequent prediction of the effects of multitasking and task interaction. Developed in the ACT-R cognitive architecture, the general executive is specified as a new goal module that orders goals by urgency and their own requested running times. The paper demonstrates the predictive power of the general executive in the driving domain by integrating separate models of control and monitoring and predicting drivers' gaze distributions across various regions of the driving environment.

Introduction

Whereas much of traditional cognitive modeling has focused primarily on cognitive processes in small-scale laboratory tasks, modeling efforts have been rapidly maturing to address increasingly complex task domains and phenomena. Some of the most successful models for such complex tasks have arisen in work using unified cognitive architectures such as ACT-R (Anderson et al., in press), EPIC (Meyer & Kieras, 1997), and Soar (Newell, 1990) — for example, models of driving (e.g., Aasman, 1995; Salvucci, Boer, & Liu, 2001), piloting combat aircrafts (Jones et al., 1999), and air-traffic control (e.g., Lee & Anderson, 2001). As such efforts evolve, the modeling community is facing several major new challenges in the study and modeling of complex tasks, and among the most critical of these is the study of human multitasking — the ability to integrate, interleave, and perform multiple tasks and/or component subtasks of a larger complex task.

Recent explorations of multitasking in cognitive architectures have examined the integration of two tasks, where the tasks are either discrete (i.e., short tasks of roughly 1-10 sec) or continuous (i.e., extended tasks). For instance, researchers have investigated how architectures can account for switching costs in successive discrete tasks (e.g., Sohn & Anderson, 2001), psychological refractory period (PRP) effects in concurrent discrete tasks (e.g., Byrne & Anderson, 2001; Meyer & Kieras, 1997b), and error effects in elementary continuous tasks (see Kieras et al., 2000). In addition, the models for the complex tasks listed above (driving, piloting, etc.) all incorporate some multitasking in having to perform various subtasks at various intervals (e.g., occasionally checking wind direction in the air-traffic control task). These models, however, use *customized executives* for multitasking (Kieras et al., 2000) that have been fine-tuned to the particular domain, resulting in domain-specific models of multitasking that can be difficult to generalize.

In this paper, we propose an initial formulation of a *general executive* for human multitasking in the ACT-R cognitive architecture. The general executive manages a set of current goals and dictates when each goal may proceed given ordering constraints based on desired initiation times for each goal. The proposed general executive attempts to balance the individual goals' desires for unhindered progress in each task with the overall system's need for fair resource allocation across tasks as well as achievement of higher-level goals. The executive also encourages modularity and model re-use by allowing for the integration of separate models, possibly developed in isolation of one another, into an integrated task model that interleaves multiple subtasks.

We first describe the general executive mechanism as developed in the framework of the ACT-R architecture, including a discussion of the many considerations that arise in specifying such a mechanism. We then demonstrate the predictive power and usefulness of the mechanism in an application to the domain of driving, showing how the general executive can automatically interleave the subtasks of control (steering and speed control) and monitoring (for situation awareness) and accurately model human drivers' allocation of attention to these two subtasks.

The ACT-R General Executive

The ACT-R general executive proposed here provides a mechanism that schedules and interleaves multiple subtasks. The goal of creating a general executive has both scientific and engineering implications. As a scientific endeavor, we desire a psychologically plausible mechanism that fits well theoretically with the existing cognitive architecture and generates sound predictions that fit available human data. As an engineering endeavor, we desire a mechanism that facilitates independent development of task models as well as model re-use. Of course, taken as a whole, the topic of multitasking spans an enormous array of phenomena and empirical literature. For the purposes of this paper, we focus specifically on the integration of two continuous tasks — that is, two tasks that must be performed continually for a long time (several minutes to hours) and must be interleaved at short intervals (hundreds of milliseconds to seconds), like the driver control and monitoring subtasks analyzed in a subsequent section. Of note, we are not currently addressing explicit discrete task switching or the one-shot behavior that arises in psychological refractory period (PRP) tasks; instead, we focus on continuous tasks in which people (and models) must schedule and interleave multiple tasks for an extended period of time, sharing cognitive and other resources to maintain execution of all tasks in a fair manner.

ACT-R, Buffers, and Goals

The ACT-R cognitive architecture (Anderson et al., in press; see also Anderson & Lebiere, 1998) is a production-system architecture that posits two types of knowledge: declarative knowledge comprising factual chunks and procedural knowledge comprising condition-action production rules. The most recent version of the architecture (5.0: Anderson et al., in press) centers on "buffers" that pass information between central cognition and various modules, such as the retrieval module for memory recall and the visual and aural modules for perceptual input. To use a module and its associated buffer, a production rule typically passes a request to the module and when the module has obliged the request, another production rule "harvests" the result by examining the buffer; for example, if a production rule requests that the visual module look at and encode an object at a specific location, the visual module performs this task and places the result in the visual buffer and a subsequent rule can examine this buffer and use the information therein.

One of the buffers posited in the architecture is the goal buffer, which stores the current goal (itself a declarative chunk) and directs the system to perform a specific task. One natural way to interpret multitasking in ACT-R is that multitasking represents the scheduling and management of what goal is currently in the goal buffer. Interestingly, although ACT-R has a goal buffer, it currently has no full-fledged module associated with this buffer as it does for other modules — production rules simply set the goal buffer explicitly without making requests in the same manner as the other buffers. Older versions of ACT-R (Anderson & Lebiere, 1998) used a goal stack that allowed for pushing onto and popping from the stack which in a sense provided an instantiation of a possible goal module, but the stack-based instantiation was difficult to justify from the standpoint of clear psychological plausibility (Altmann & Trafton, 2002). We view our ACT-R general executive as a specification for a new goal module, one that maintains a set of active goals and manages the setting of goals through the goal buffer.

Specification of the General Executive

The proposed general executive centers on the maintenance and scheduling of a *goal set* that contains all currently active goals. Maintenance of the goal set is accomplished by allowing production rules to add and remove goals to and from the goal set. To add a new goal, a rule specifies the new goal chunk and also notes the goal's *delay time* — that is, when the goal should run, specified as a time offset from the current time. The delay time, noted as a slot in the goal chunk, models the fact that some goals may not need to run immediately but rather can be scheduled for execution at a later time — for instance, the periodic monitoring of some aspect of a display, such as checking wind direction in the air-traffic control task (Lee & Anderson, 2001). For convenience, we also allow for a constant 'now' as special a value for the delay time that mandates that the goal execute immediately. For removal of goals, a rule simply notes the termination of the current goal and thus its removal from the goal set.

The implementation of this mechanism in ACT-R is fairly parsimonious. In the current ACT-R syntax, the command "+goal" sets the current goal to a new goal; we simply alter the command's semantics such that this same command *adds* a goal to the current goal set — in essence, specifying the desire that a certain goal be performed, much like requests to the retrieval or visual modules. For convenience and also to maintain consistency with current syntax, the first such command also removes the current goal from the goal set. Whenever a goal is added or removed, the goal set is reordered according to the scheduling criteria described previously, and the most urgent goal is placed in the goal buffer for execution.

Given the current set of goals, we require some method of ordering and scheduling the goals to ensure fairness across tasks. All goals on the set are ordered based on their *urgency* as indicated by their desired start time (i.e., creation time plus delay time): the executive simply orders the goals from most urgent to least urgent, that is, from earliest to latest start time. However, we also expect some amount of noise in this ordering scheme, just as many mechanisms in the ACT-R architecture and the human system itself exhibit variability in behavior. For this purpose, and to maintain consistency with the architecture, we utilize a logistic noise distribution on start times with variance $\pi^2 = \pi^2 s^2 / 3$ as dictated by the noise parameter s; this same distribution is used for noise throughout the architecture including noise on expected gain of potential production-rule instantiations.

Figure 1 illustrates an example of how two task goals might be scheduled. In the first frame, Task 1 has a delay time of 400 ms while Task 2 is scheduled to execute right away. Even with noise added to these times, it is most likely that Task 2 will be allowed to execute. In the second frame, Task 2 (specifically a new goal that continues the task) has a new start time of 250 ms, and again this most likely wins out over Task 1. In the third frame, Task 2 has a later start time than Task 1, and thus the first task most likely proceeds; note that when the goals come closer together in time, as in the third frame, the noise increases the chance of a later goal superceding an earlier goal.

Theoretical and Practical Considerations

Kieras et al. (2000) have reviewed a number of theoretical issues and considerations for the development of a general executive within a cognitive architecture, many of which are related to basic concepts used in today's computer operating systems. One of the most critical issues is that of task

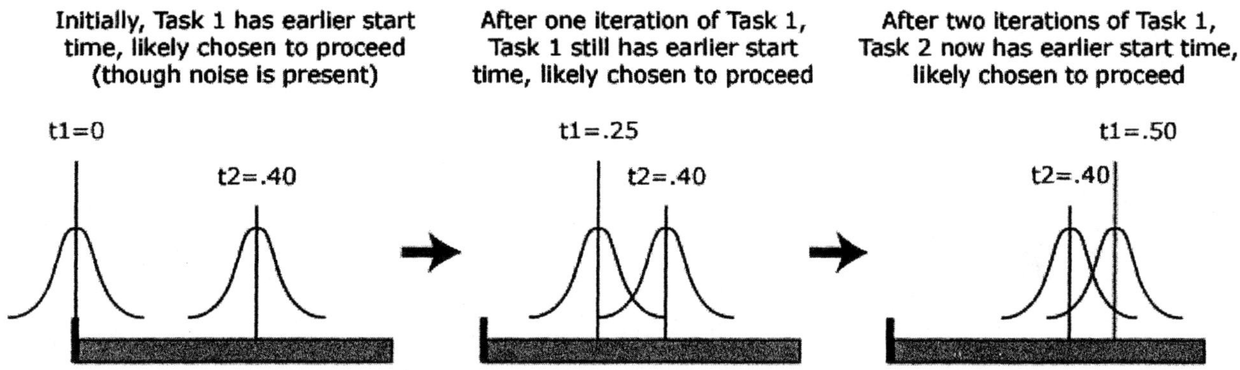

Figure 1: Example of scheduling and execution of two goals in the *goal set* as a function of time.

scheduling. Our proposed general executive uses a variant of a "first come, first served" scheduler that also appears in operating systems. However, the executive is more flexible than a strict first-come, first-served algorithm in two ways: first, it allows for time delays in the start time of processes (i.e., the goal delay times); and second, it incorporates noise into start times to enable randomness in task scheduling.

As a related point, another critical issue is that of conflict resolution, where the executive must avoid deadlocks within processes and ensure that no process experiences "starvation" with respect to needed resources. In a broader context, the executive must strike a balance between allowing processes to utilize necessary resources as much as possible but at the same time interrupting processes to allow others to proceed. Our executive interrupts processes *between* goals — corresponding roughly to the level of unit tasks — without interrupting *within* goals. We believe that this approach provides a good balance of fairness between tasks as desired by a robust general executive. It does, however, rely critically on the structure of declarative knowledge in that different declarative goal structures (e.g., as might arise from different instructions) may greatly affect when and how people perform multiple tasks. However, this reliance on declarative structure is not necessarily a disadvantage; in fact, a recent empirical study has demonstrated such a dependence (Vera, personal communication) and thus might be best modeled with an executive like that proposed here.

Along with such theoretical considerations, a general executive can have practical considerations in providing a robust workable mechanism that facilitates modeling in the cognitive architecture. One advantage of our proposed executive is that the task models can be developed independently of one another and later merged. Such modularity greatly facilitates model re-use and, in turn, increases the predictive power of newly integrated models (see, e.g., Salvucci, 2001). The executive also does not affect the behavior of existing ACT-R models, since existing models' explicit setting of goals is equivalent to a goal set with exactly one goal throughout execution.

Case Study: Driver Control and Monitoring

To illustrate how the proposed general executive facilitates the modeling of human multitasking, we now describe an application to the domain of driving. As a complex yet ubiquitous task, the task of driving is actually an integration of several subtasks performed fluidly by the driver to successfully navigate the environment. Of these various subtasks, the two (arguably) most critical subtasks are control and monitoring: control of steering and acceleration that maintains the car's safe position, speed, and heading; and monitoring that maintains a consistent, up-to-date awareness of the current situation as is critical for normal maneuvers (e.g., lane changes) or emergency maneuvers (e.g., swerving to avoid an obstacle). We now explore how the general executive facilitates the integration of basic models of driver control and monitoring into a single multitasking model that switches between these tasks. Of course, the evaluation of any general mechanism requires application not in a single domain but in many domains; nevertheless, we consider the driving application as an initial foray into evaluating and testing the benefits and limitations of our general executive in a complex, dynamic, real-world domain.

Component Models

To build a model of driver behavior in the spirit of the general executive, we started with models for the component subtasks involved in driving and then integrated these using the general executive. We extracted the component models from an existing ACT-R model of driver behavior (Salvucci, Boer, & Liu, 2001) that navigates a highway environment. The component models studied here are:

- **Control model**: The control model updates the positions of the steering wheel and the accelerator and brake pedals. This update is based on observed positions and changes in salient visual points in the driving environment (e.g., the vanishing point of the road or the position of a lead vehicle).

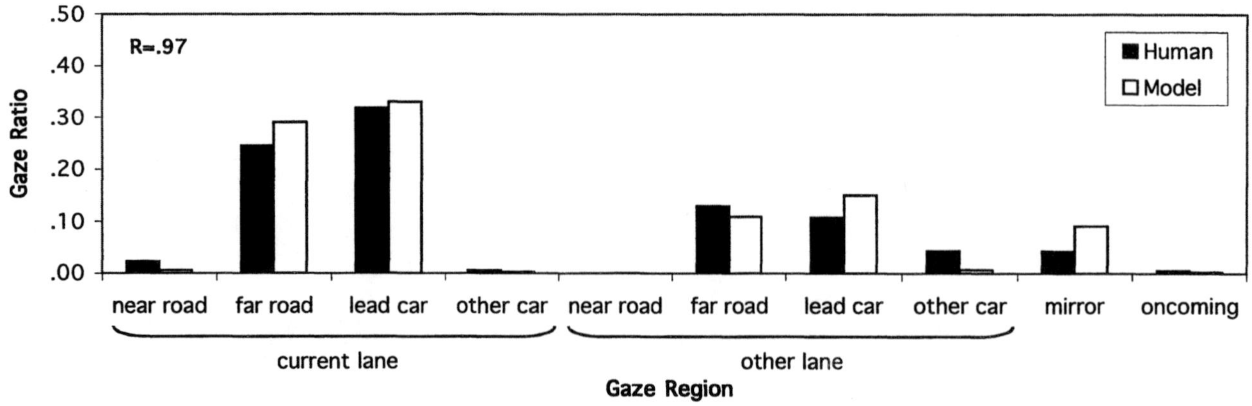

Figure 2: Ratio of gaze dwell time in various regions of the driving environment.

- **Monitoring model**: The monitoring model randomly samples the environment — with equal probability, it chooses a lane (left or right in a two-lane highway) and direction (forward or backward in the mirror) and looks there for the presence of a vehicle; if a vehicle is present, the model notes its position in declarative memory.

The original driver model has a third component model for decision making that decides whether and when to change lanes based on its current situation awareness; due to space constraints in this paper, we ignore decision making here and focus solely on control and monitoring. Readers can find further details about the component models in the full description of the model (Salvucci, Boer, & Liu, 2001).

The component models have associated goals *control* and *monitor* that perform the respective subtasks. It is critical to note that both of these goals perform incremental work of approximately 100-300 ms and then pass control to a new goal of the same type: one *control* goal will update steering and acceleration and then create a new *control* goal to perform the next iteration, and one *monitor* goal will attempt one glance in a chosen lane/direction and then create a new *monitor* goal to continue this task. As a result, both models operate independently (in fact, they could each run alone without the other) and continually refresh their respective subtasks with a new goal.

Model Integration

The creation of the original integrated model required a customized executive that passed control back and forth between control and monitoring. Now, with the general executive, model integration happens easily: we simply add one production rule that adds both *control* and *monitor* to the goal set, and the general executive handles the rest. As each incremental goal finishes and creates a new goal, the general executive reschedules the current goals in the set and allows the most urgent goal (as computed using start times with noise) to proceed.

The default integration above treats both subtasks equally, with one iteration of control and one of monitoring alternating one after another during execution. While this integration indeed successfully navigates a realistic environment, the resultant task switching does not necessarily match people's tolerance for when to switch between tasks. Thus, we explored two possibilities for changing the default scheduling:

- *Control stability*: The original driver model contains a stability threshold that switches away from control only when the external environment is "stable," where stability measures the lateral displacement from the lane center and the lateral velocity of the vehicle. We maintained this scheme in the new integrated model: if the control goal finds that the environment is *not* stable, it dictates that the next control goal should be done immediately by means of a *now* value for the *when* slot. In addition, we define a parameter F_{stable} that scales the threshold to allow us to estimate drivers' degree of stability acceptance.
- *Control delay time*: The default integrated model specifies a delay (*when*) time of 0 for new control goals, indicating that they should execute as soon as possible. However, given a stable environment, there is the possibility that control need not execute for a given period of time. Therefore, we define a second parameter $D_{control}$ that specifies the delay time for a new control goal assuming a stable environment.

We also have a third parameter dictated by the general executive, namely the s parameter that controls the amount of noise in the scheduling times; note that eventually as the general executive extends to more domains, this parameter should be estimated across all domains like the other noise parameters in the ACT-R architecture.

Parameter estimation was done first informally to find reasonable settings for parameter values, then more systematically to improve the fits to the results below. The resulting values for the parameters were as follows: F_{stable} = .71, $D_{control}$ = 500 ms, s = .075.

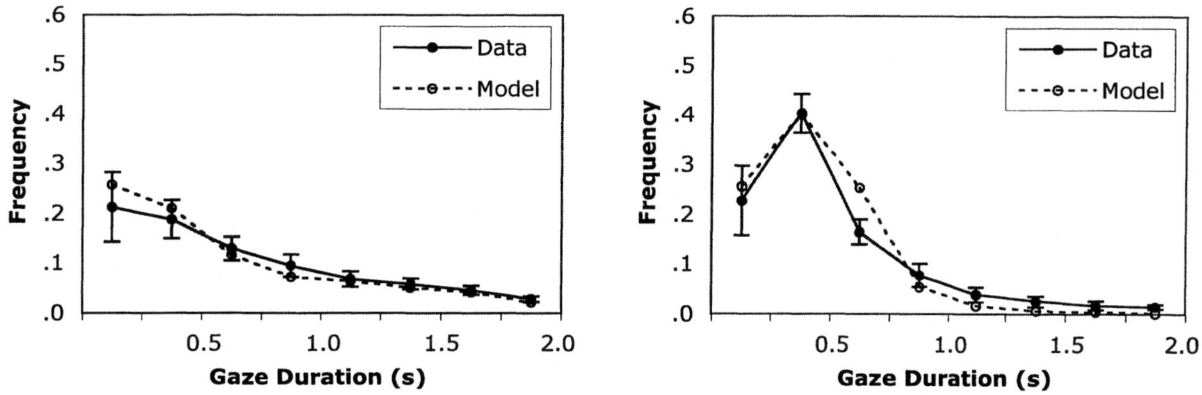

(a) (b)

Figure 3: Frequency of gaze durations in various regions of the driving environment for (a) control and (b) monitoring. Data points represent frequencies of durations within a .25 s range; bars indicate standard deviations.

Data Collection

To validate the integrated model, we would like to compare the model's behavioral data to that of human drivers. The human driver data comes from the original study (Salvucci, Boer, & Liu, 2001) in which 11 drivers navigated a multi-lane highway for a total of 311 minutes (548 km) of driving data. The model data comes from running three 10-minute simulations in the same conditions as the human data; because the model drives in the same environment as did human drivers in the driving simulator (minus the rendering of complex graphics), a model simulation generates a completely analogous protocol that can be analyzed in an identical manner as the human data.

Results

To analyze drivers' switching between control and monitoring, we examine and compare the eye movements produced by the human drivers and the integrated driver model (through its perceptual mechanisms and EMMA model of eye movements: Salvucci, 2001b). Figure 2 shows the ratio of gaze dwell time in various regions of the driving environment: the near road, far road, lead car, and other cars in the current lane; the same regions in the other lane (given the two-lane highway); the rear-view mirror; and oncoming cars (in the opposite two lanes). For both the human drivers and the model, the majority of the gaze is directed at the far road and lead car of the current lane — both critical regions for controlling the vehicle (both laterally and longitudinally). The far road and lead car of the other lane also garner some attention, as does the rear-view mirror, as indicators of occasional monitoring glances to these areas. Also, both humans and the model rarely look at the near road, at other cars in the two lanes, and at oncoming cars. Thus, the model nicely captures the overall distribution of gaze across the visual regions, R=.97.

While the overall distribution provides some sense of drivers switching between tasks, we can also analyze more revealing finer-grain data that indicate exactly when drivers switched between these tasks. As a first step in exploring this idea, we classified the gaze regions as either control or monitoring regions based on their function in the environment: the control regions included the far and near road regions as well as the lead car, and the monitoring regions included the lead and other car in the other lane as well as the rear-view mirror. Using this classification, we extracted all single gazes at one region (consecutive eye-movement samples ignoring single-sample "blips") and computed histograms of the frequencies of gaze durations, using 1/4-second increments to alleviate data noise.

Figure 3(a) shows the human and model gaze-duration histograms for control gazes, with points representing frequencies of gaze durations within a .25 s range. The graph of the human data (solid line) clearly indicates a drop over time, with the highest frequency (and highest standard deviation) occurring in the first 1/4 second and steadily decreasing out to 2 seconds. The pattern for the model data (dotted line) arises primarily from its stability threshold for switching: often the model can switch away after controlling up to 1/4 second, but on occasions when the vehicle incurs some instability (such as steering into and out of a curve), the model must stay with control until stability is again achieved. The model thus produces a very similar histogram to that of human drivers, R=.99.

Figure 3(b) shows the analogous histograms for monitoring gazes. For the human data, we see a very different pattern compared to the control gazes: the peak of the distribution lies around 1/4-1/2 s, and drops steeply thereafter, with very few gazes lasting more than 1 s. The model nicely captures the human data, the critical aspect being the control delay time incorporated into the model. In essence, the control goal requests a new goal only after the control delay time (500 ms), thus giving monitoring time to perform several iterations before switching back to control. At the same time, after the delay time threshold has passed, the model's urgency to switch back becomes increasingly larger, thus producing very few gaze durations greater than 1 second and a good overall fit to the human drivers' gaze pattern, R=.97.

General Discussion

We have described a general executive for multitasking in the ACT-R architecture that schedules multiple tasks and interleaves them to execute reasonable multitasking behavior. One of the most significant implications for this work is the ability to develop task models *independently* and later integrate them into a fuller model. While we have demonstrated such an integration for driver control and monitoring, there are of course many real-world applications for such integration. In the driving domain itself, there is a clear application with great potential — namely, to secondary device use and driver distraction. While recent work has shown the predictive power of integrated ACT-R models to account for driver distraction (Salvucci, 2001; Salvucci & Macuga, 2002), these models have customized executives that specify how the models should interleave driving and the secondary task. Now, the new general executive offers the potential to make *a priori* predictions about when and how interleaving takes place, allowing a modeler to develop secondary-task models independently and then easily integrate them with the driver model for predictions of their distraction potential.

This general executive nicely handles the type of multitasking in driving — and we would argue in similar complex dynamic tasks — that interleave multiple tasks at roughly the level of hundreds of milliseconds to a few seconds. However, there are clearly other levels of multitasking that are not yet addressed by this general executive. On one end of the spectrum, people can multitask at the level of several minutes to hours — e.g., word processing and checking email. On the other end, people can perform discrete rapid tasks together in a highly optimized manner to produce PRP effects (Byrne & Anderson, 2001; Meyer & Kieras, 1997b). At this time, we have focused our general executive on a level in between, but we hope that in the long term, this mechanism can generalize to account for multitasking at other such levels.

Another critical aspect involves the modeling of improvement in multitasking over time (Chong, 1998), or put another way, transitions between the "stages of multitasking skill acquisition" (Kieras et al., 2000). Learning in multitasking of course involves learning of the individual tasks, but even when people are skilled at particular individual tasks, the integration adds another layer of complexity (e.g., driving and talking on a phone). Again, we view the proposed general executive as currently occupying one point on this continuum: the multitasking being represented should be sensible but not necessarily highly optimized. One of our most important long-term goals involves extending the general executive to handle learning and optimization for particular sets of tasks — for instance, using ACT-R's production compilation mechanism (Taatgen & Lee, 2003) to create new "compiled" production rules for both tasks and integrate these tasks in an increasingly automated manner.

Acknowledgments

This work was supported by ONR grant N00014-03-1-0036 to the first author. We thank John Anderson, Christian Lebiere, Niels Taatgen, and Alex Chavez for their helpful comments and suggestions regarding this work.

References

Aasman, J. (1995). *Modelling driver behaviour in Soar.* Leidschendam, The Netherlands: KPN Research.

Altmann, E.M., & Trafton, J.G. (2002). Memory for goals: An activation-based model. *Cognitive Science*, 26, 39-83.

Anderson, J.R., Bothell, D., Byrne, M.D., Douglass, S., Lebiere, C., & Qin, Y. (in press). An integrated theory of the mind. *Psychological Review*.

Anderson, J.R., & Lebiere, C. (1998). *The atomic components of thought*. Hillsdale, NJ: Erlbaum.

Byrne, M.D., & Anderson, J.R. (2001). Serial modules in parallel: The psychological refractory period and perfect time-sharing. *Psychological Review, 108*, 847-869.

Chong, R.S. (1998). Modeling dual-task performance improvement: Casting executive process knowledge acquisition as strategy refinement. Doctoral Dissertation, University of Michigan.

Jones, R.M., Laird, J.E., Nielsen P.E., Coulter, K., Kenny, P., & Koss, F. (1999). Automated intelligent pilots for combat flight simulation. *AI Magazine*, *20*, 27-42.

Kieras, D.E., Meyer, D.E., Ballas, J.A., & Lauber, E.J. (2000). Modern computational perspectives on executive mental processes and cognitive control: Where to from here?. In S. Monsell & J. Driver (Eds.), *Control of Cognitive Processes: Attention and Performance XVIII* (pp. 681-712). Cambridge, MA: MIT Press.

Lee, F.J., & Anderson, J.R. (2001). Does learning of a complex task have to be complex? A study in learning decomposition. *Cognitive Psychology*, *42*, 267-316.

Meyer, D.E., & Kieras, D.E. (1997). A computational theory of executive cognitive processes and multiple-task performance: Part 1. Basic mechanisms. *Psychological Review, 104*, 3-65.

Meyer, D.E., & Kieras, D.E. (1997b). A computational theory of executive cognitive processes and multiple-task performance: Part 2. Accounts of psychological refractory period phenomena. *Psychological Review, 104*, 749-791.

Newell, A. (1990). *Unified theories of cognition*. Cambridge, MA: Harvard University Press.

Salvucci, D.D. (2001). Predicting the effects of in-car interface use on driver performance: An integrated model approach. *International Journal of Human-Computer Studies*, *55*, 85-107.

Salvucci, D.D. (2001b). An integrated model of eye movements and visual encoding. *Cognitive Systems Research*, *1*, 201-220.

Salvucci, D.D., Boer, E.R., & Liu, A. (2001). Toward an integrated model of driver behavior in a cognitive architecture. *Transportation Research Record*, 1779, .

Salvucci, D.D., & Macuga, K.L. (2002). Predicting the effects of cellular-phone dialing on driver performance. *Cognitive Systems Research*, *3*, 95-102.

Sohn, M.-H., & Anderson, J.R. (2001). Task preparation and task repetition: Two-component model of task switching. *Journal of Experimental Psychology: General*, 130, 764-778.

Taatgen, N.A., & Lee, F.J. (2003). Production compilation: A simple mechanism to model complex skill acquisition. *Human Factors*, *45*, 61-76.

A computational account of latency impairments in problem solving by Parkinson's patients

Patrick Simen (psimen@math.princeton.edu)
Applied and Computational Mathematics / Center for the Study of Brain, Mind and Behavior; Princeton University

Thad Polk (tpolk@umich.edu)
Department of Psychology; University of Michigan, Ann Arbor

Rick Lewis (rickl@umich.edu)
Department of Psychology; University of Michigan, Ann Arbor

Eric Freedman (freedman@umich.edu)
Department of Psychology; University of Michigan, Flint

Abstract

Cognitive slowing is a feature of problem solving performance by Parkinson's patients. Here we present a computational model of the Tower of London problem solving task that provides a straightforward explanation of this latency impairment. The model is a neural network consisting of a set of idealized neurons whose activation levels are continuously varying quantities ranging between 0 and 1, organized hierarchically into a collection of structured, columnar assemblies. The model differs from many neural network models in that it intentionally approximates the discrete stages of processing and rule-based computation of production systems. It differs from many symbolic models in that it does not assume an instantaneous transition between different stages of processing and does not rely on a central, system-wide clock signal. As a result, the model faces timing problems that can prevent it from approximating an idealized discrete-time system closely enough to perform symbolic computation. The columnar assembly provides a solution to these problems by imposing a direction on the flow of activation between units and by controlling the rate of that flow. Columns are proposed to play the role of distributed timers implemented in cortical columns and frontostriatal loop circuits in the brain. Dopamine depletion in Parkinson's is proposed to reduce the rate of information transmission through a frontostriatal timer circuit critical for the generation of subgoal representations in prefrontal cortex. The model fits latency impairments in problem-solving by Parkinson's patients relative to controls and predicts that only problems that require the generation of subgoals will produce a significant latency impairment in Parkinson's patients.

Contribution of prefrontal cortex and basal ganglia to problem solving

Many researchers have noted the similarity between cognitive impairments in patients with Parkinson's disease (PD) and patients with lesions of the prefrontal cortex (PFC). Both groups show impairments related to problem solving, planning and set-shifting (Owen *et al.*, 1992; Owen *et al.*, 1995). Both groups are impaired on the Tower of London problem solving task depicted in Fig. 1 (which is the task that we model). Prefrontal patients have difficulty achieving an optimal solution to the puzzle (a solution trajectory that involves a minimal number of moves), but only when the required number of moves to solution exceeds three (Shallice, 1982). The same pattern of impairments has been seen in severe cases of PD (Owen *et al.*, 1995) (although another study showed no such accuracy impairment (Owen *et al.*, 1992)). Here we propose a computational account of the function of brain structures affected in both conditions that may explain the similarities and differences of these cognitive deficits.

PD involves the death of dopamine-producing neurons in the substantia nigra that project to the striatum, and it is this anatomical fact that suggests a connection between the effects of PD and prefrontal lesions. Prefrontal areas of cortex are connected back to themselves by multi-synaptic 'frontostriatal loops' whose first segments consist of excitatory axonal projections from deep cortical layers to the striatum in the basal ganglia. These striatal neurons send inhibitory projections to neurons in the globus pallidus and subthalamic nucleus (also considered structures of the basal ganglia). These in turn inhibit neurons in the thalamus that send excitatory projections back to the areas of cortex near the starting points of the loops. At each stage, a notable degree of topographical organization is seen, such that the loops can be considered to be, to some degree, segregated and parallel (Alexander, DeLong & Strick, 1986). Functionally, signals from cortex to the basal ganglia have the effect of disinhibiting the thalamus and exciting cortex in a form of positive feedback.

In the striatum, dopamine appears to potentiate the excitatory effect of cortical glutamate received by medium spiny neurons, as well as to modulate synaptic plasticity (Alexander *et al.*, 1986). A natural hypothesis, therefore, is that the disruption of frontostriatal loops (which are one means by which PFC communicates with itself) in PD mimics the effects of prefrontal lesions, and that this is why the two patient groups exhibit similar cognitive deficits. However, while similarities exist between prefrontal and Parkinsonian cognitive impairments, there are salient differences. In all levels of PD severity, impairment in problem solving latency (the time to begin problem solving once a problem has been presented) is seen (see Fig. 2). Interestingly, this impairment again appears only on problems requiring more than three moves. But a similar latency effect is not seen in prefrontal patients (Owen *et al.*, 1992; Owen *et al.*, 1995).

We have previously argued that the primary role of dorsolateral PFC in problem solving is to represent subgoals that guide the selection of actions (Polk *et al.*, 2002), thereby providing a potential explanation of suboptimal problem solving in prefrontals as resulting from a failure to select actions consistent with subgoals. In this paper, we argue that latency impairments observed in PD arise as a result of slowed generation of subgoals after action selection impasses occur.

Tower of London task

The Tower of London (TOL) task, is shown in Fig. 1. This task has been used extensively to assess planning impairments and is thought to depend crucially on goal management (Shallice, 1982). It is a variant of the Tower of Hanoi problem and involves moving colored balls on pegs from an initial configuration until they match a goal configuration. Unlike the Tower of Hanoi problem, there are no constraints specifying which balls can be placed on which others, but the pegs differ in how many balls they can hold at one time (the first peg can hold one ball, the second peg can hold two, and the third peg can hold three). There is typically one red, one green and one blue ball. Participants are often asked to try to figure out how to achieve the goal in the minimum number of moves and are sometimes asked to plan out the entire sequence of moves before they begin (Shallice, 1982). Normal, prefrontal and Parkinson's performance from Owen *et al* (1995) is shown in Fig. 2.

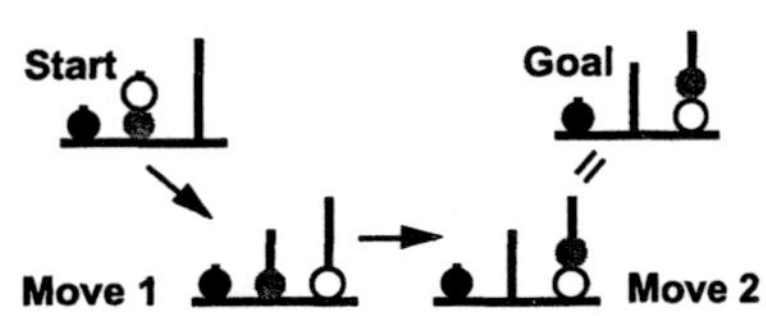

Figure 1: The Tower of London task.

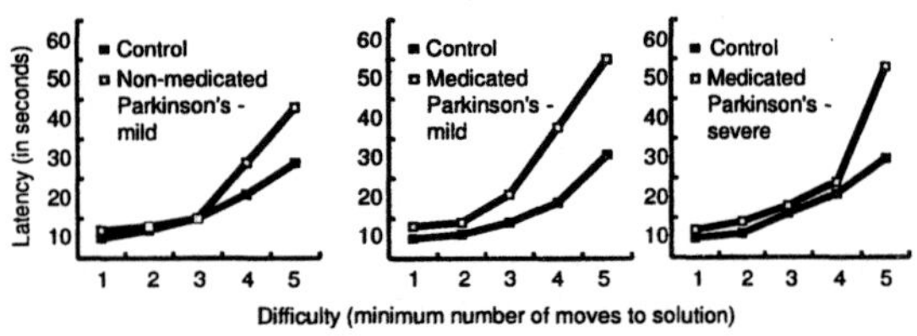

Figure 2: PD patients display latency impairments (slower planning prior to first move) in the Tower of London task. Increased latency relative to control subjects is shown by patients at two different stages of the disease, with and without medication in the early stage.

Components of a neural cognitive architecture

For the most part, the model we present works by implementing in neural networks some of the computational assumptions of symbolic production systems. We have argued previously that there is a natural mapping from such symbolic systems onto the brain (Polk *et al.*, 2002). Production systems typically assume a production matching cycle that occurs at regular intervals governed by a central clock, and that the matching and firing of a production is an all-or-none affair. In the brain, however, there is no evidence for a single, system-wide clock signal capable of sequencing events at the level of simple cognitive operations. For this reason, the architecture we present makes a commitment to distributed timing, in which components at the lowest level time their own operations. Also, since cortical neurons appear to encode information in some form of analog, rather than binary, code, the architecture makes a further commitment to analog representations of the state of a system. These commitments require an approach to duration encoding and sequential processing that is different than the one taken in almost all digital systems. This approach also sets the architecture apart from previous connectionist implementations of production systems such as DCPS (Touretzky & Hinton, 1988).

The architecture we present consists of modular winner-take-all networks composed hierarchically. Each module is layered, with distinct input and output layers, and an intermediate layer that delays propagation from input to output by a controllable amount. In fact, the modules are themselves constructed out of a more basic primitive, a structured column of neural units. Columns are arranged in parallel with lateral inhibitory connections to form modules. Individual units are taken to model the spatiotemporal average firing rate of a population of spiking neurons.

Model neurons

A unit's activation at any moment is represented by a number between 0 and 1. The activation of unit i, $V_i \in [0,1]$, is determined by a standard nonlinear differential equation that is taken to model the firing rate of a population of neurons, possibly averaged over time so that more recent firing contributes more to the average than firing that occurred longer ago:

$$\frac{dV_i}{dt} = -V_i + \frac{1}{1+e^{-\lambda(NetIn_i - \beta_i)}} \qquad (1)$$

where $NetIn_i = \sum_{j=1}^{n} w_{ij}V_j$, and w_{ij} is the synaptic weight on the connection from unit j to unit i (Cohen & Grossberg, 1983). A small random noise term is also often added to V_i. The sigmoid function $f(NetIn_i) = \frac{1}{1+e^{-\lambda(NetIn_i-\beta_i)}}$ forms a curve defining equilibrium values of activation for any given (constant) level of net input. Curves of this type are depicted in Fig. 3.

Self-excitation

Positive feedback within individual units plays an important role in what follows. We define a *self-exciting unit* to be a unit i whose output value V_i is weighted by a nonzero synaptic strength w_{ii} and added to the $NetIn_i$ term of its own activation function. Non-self-exciting units have $w_{ii} = 0$.

We state without proof that the dynamics of self-exciting units receiving constant inputs from other units are completely characterized by the 'self-excitation diagrams' of Fig. 3 (proofs can be found in Simen, (2004)). In plot A, two different activation curves are shown.

Each corresponds to a different level of net excitation received from other units: net excitation results in a leftward shift of the activation function by precisely the amount of the excitation, and inhibition results in a similar rightward shift. The horizontal axis reflects only input to a unit from itself. A stair step trajectory formed between the shifted activation function and the straight line with slope $1/w$ (hereafter referred to as the 'reference line'), where w is the self-excitatory connection strength, determines the equilibrium level of activation ultimately achieved by the unit (assuming no change in external inputs), and the size of the stair steps determines the speed of approach to the equilibrium value (larger steps imply faster approach). As a result, strong and weak self-excitation results in qualitatively different behavior. With weak self-excitation, activation that ramps up or down at a controllable and relatively constant rate is possible. With strong self-excitation, depicted in plot B, approximately all-or-none activation levels and memory in the form of reverberating activation are possible.

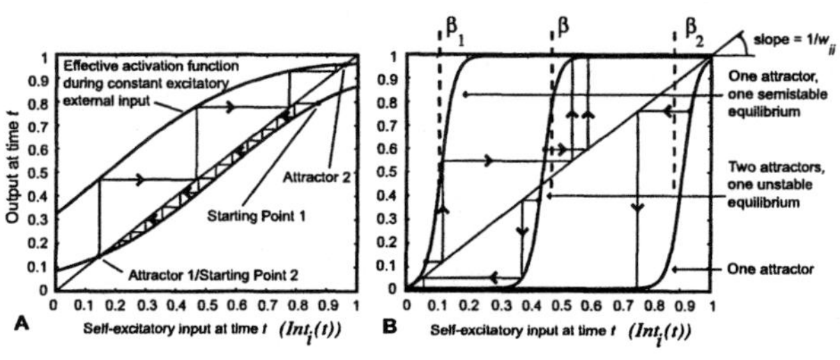

Figure 3: Self-excitation diagrams for units with weak (A) and strong (B) recurrent connections to themselves. The rate at which a unit's activation approaches the nearest attractor is determined by the height of the stair steps depicted.

Decision-making/memory modules

The basic structural primitive of the architecture is a column consisting of a small number of units, but the architecture is easier to approach by beginning with a simple, modular network structure that does not involve columns. This structure, henceforth called simply a 'module', is a fully connected recurrent network in which each unit is connected with a bidirectional connection to every other unit, and in which each unit also excites itself with a strength of 0 or more. The typical mode of operation of a module is as a winner-take-all network, in which one unit becomes highly active and all others become inactive as a result of external input, or 'votes'. Thus modules collect preferences for various decisions and amplify the preference for the most preferred outcome at the expense of those less preferred.

Self-excitation diagrams are useful for module design because they allow a programmer to determine the inter-module connection strengths necessary to create modules that act like AND and OR gates and inverters in digital logic, as well as memory components like latches and flip-flops. Most importantly, they allow the setting of connection strengths that cause modules to implement an analogue of production matching and conflict resolution in production systems. Simen (2004) describes this strength-setting recipe in detail.

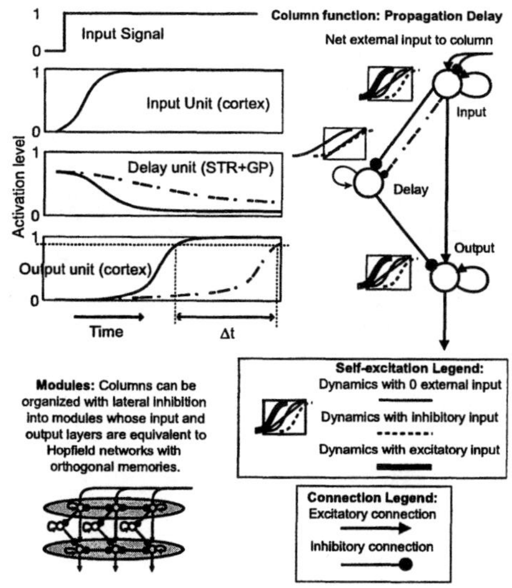

Figure 4: Activation levels in a single column show the effect of weakening the inhibitory strength of the connection from Input to Delay: an increase in propagation delay of size Δt results from slower inactivation of the Delay unit. Modules organized from laterally connected columns are shown at the lower left. Input and Output units model deep and superficial layers of cortex, respectively, and Delay units model the sequence of connections from neurons in the striatum (STR) to the globus pallidus (GP) to the thalamus.

Model cortical columns

A physical aspect of electronic digital logic components that can have major effects on the performance of a circuit is the internal delay that signals face when propagating through them. In order for high-level descriptions of circuits to work, their components must satisfy a set of timing constraints. The same is true for neural circuits, and a means of adjusting the internal propagation delay within a module to allow timing constraint satisfaction is now presented.

In fact, though, the need for control over propagation is more extensive here than in digital circuit design, because these internal delays will be the sole source of time measurement in the architecture, and timing will be largely asynchronous (without reference to a single, system-wide clock). For this purpose, the architecture elaborates on the module concept with a column structure.

The columnar version of a module is schematically depicted in Fig. 4. Instead of a single, fully connected recurrent network, a module now consists of two identical copies of such a network. One copy functions as the input interface to the module, and the other functions as the output interface. Each input layer unit, Input, in addition to the lateral inhibition it sends and receives from other input layer units, sends a feedforward excitatory connection to its counterpart Output in the output layer. The Output unit is inhibited by a self-exciting unit, Delay, that also receives inhibitory input from Input. The Delay unit serves to prevent rapid transmission to Output of large jumps in the activation of Input. The rate of transmission from Input to Output is determined by how strongly Input inhibits Delay.

Variable delay characteristics derive from the fact that strong inhibition from Input shifts the Delay unit's effective activation sigmoid to the right. Typically the Delay unit's activation curve sits far enough to the left that it has a single equilibrium value near 1. When shifted to the right, the Delay unit's activation approaches a new equilibrium value near 0. If the rightward shift places the activation curve near to the reference line, this approach will be slow. A sequence of columns imposing a propagation delay on a signal propagating through it can therefore be used as a timer. The effect of weakened Input → Delay inhibition is shown in Fig. 4. This effect is at the heart of the explanation of latency impairments provided here.

Neural model of the Tower of London task

In the model solver, illustrated in Fig. 5, a set of Sensory modules, one for each position of the gameboard, is initialized to patterns encoding the color of a ball at that position, if any, and these representations then persist until reinitialized by changes in the environment. They excite the representations of legal moves in a separate Move module devoted to action representations, and inhibit illegal ones. Attractor dynamics within the Move module results in the selection of a single action for execution, completing the simulation of a simple production of the form: 'if the red ball is in position X, then place it in position Y'. In this respect, the model is similar to the neural network Tower of London model of Dehaene & Changeux (1997). That model, however, did not explicitly represent subgoals, and that aspect of the current model is critical to its ability to explain latency impairments.

Goals and subgoals

In the Tower of London solver, one set of winner-take-all modules is dedicated to the representation of externally defined goals and another to internally generated subgoals. Activation in the goal modules biases the competition taking place in the Move module, favoring one column over the others. This biasing is just another form of production, but the if-condition is semantically special: it represents a desired state of the environment. Further, the biasing strength of such a production is insufficient to activate its then-condition without support from some other module, as in the case of the Sensory module just discussed. Technically, this Goal → Move

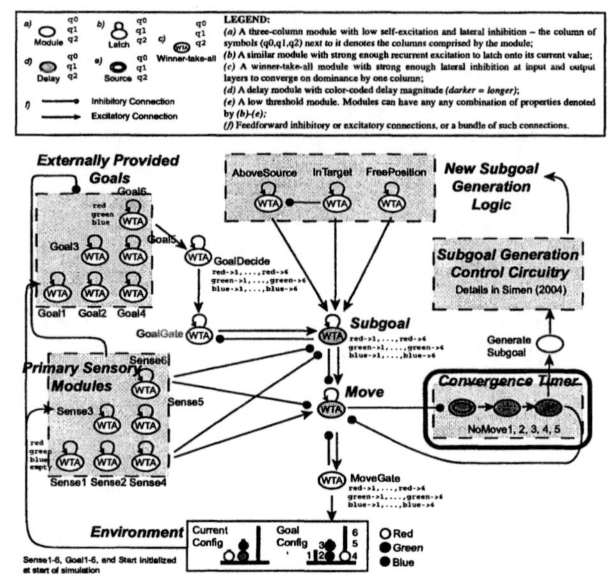

Figure 5: Tower of London model schematic. Each oval is a multi-layer module as depicted in Fig. 4. The highlighted convergence timer component is the only component to make significant use of internal propagation delay. Slowing in this component resulting from weak Input → Delay inhibition is the source of the model's latency impairment.

excitation should be considered only a component of a production of the form: 'if Goal is X and Percept is Y, then Do Z'.

Convergence timing

The scheme so far described for building problem-solving neural networks addresses critical issues, but it is not obvious that a system of continuous-time winner-take-all modules composed with between-module connections that form closed loops will work properly. The prefrontal/normal model of Polk *et al.* (2002), for example, is actually a hybrid neural/symbolic system. It requires a non-neural component to read the output of a feedforward composition of modules with no cycles. This output emerges in the Move module representing the current action. If a single, clear winner emerges in this network during voting by Sensory and Goal modules, the system takes the prescribed action and then reinitializes several modules in the network to new values. It thereby closes the loop that feeds output information back into the system. However, in the smoothly continuous systems proposed here, such a process cannot be instantaneous: new module values will have a nonzero rise time, and old values will have a nonzero decay time. Thus a timing mechanism is required for ensuring proper 'setup' of inputs for the next cycle of computation.

In many situations, however, an action should never emerge, because no atomic action can achieve the current goal. In such a case, the Move module remains near base-

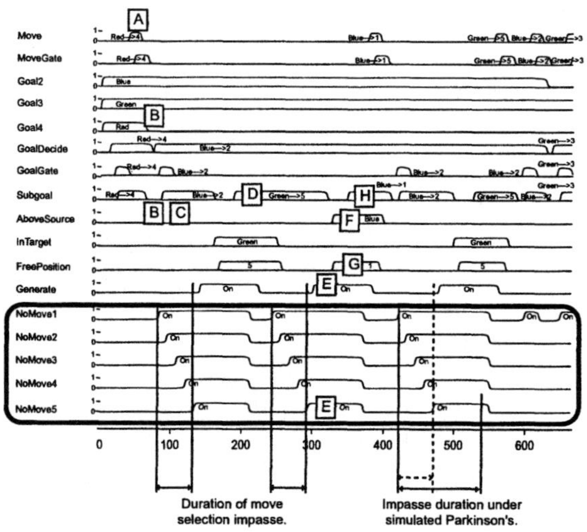

Figure 6: Time course of Output unit activation in most modules of the model during the solution of a five-move-minimum problem, depicted in the Environment panel at the bottom of Fig. 5. Delay between onset of activation in NoMove1 and NoMove5 defines the time window in which a move can be selected before a subgoal is generated. This delay increases as Input → Delay inhibition is weakened, producing the model's latency impairment.

line activation, thereby signaling that a subgoal ought to be generated in order to produce environmental conditions suitable for taking actions to achieve the parent goal. This raises a second question: for how long should an election be allowed to continue? Convergence times within attractor networks (of which winner-take-all networks are a special case) are difficult if not impossible to predict in many cases.

The simple solution presented here is simply to wait for some period of time after one pattern of activation has remained approximately unchanged to declare either a winner or convergence to baseline: the waiting period can be determined by the cost of waiting too long to get a true winner relative to the cost of responding too quickly with a false winner. If we can sample the activation of a module only after a 'safety period' has elapsed, we can reduce the chance of picking a false winner. If we are able to construct a timer which is triggered whenever module activation fails to satisfy the criteria for representing a winner, and which is inactivated by the emergence of a winner, we can produce a convergence-to-baseline detector that is triggered when the timer reaches a threshold duration without being inactivated. This is the function of the Convergence Timer mechanism in Fig. 5.

Model performance

Timecourses of activation in key components of the model are shown in Fig. 6. The problem, shown at the bottom of Fig. 5, requires five moves for solution and therefore requires that some balls be moved to positions other than their final, goal positions. Thus it requires the internal generation of subgoals for efficient solution. The Sense modules, like the Goal modules, are initialized at the beginning of the simulation and excite potentially legal moves. A winner, 'Red to 4' is selected at time point A, and the corresponding unit in MoveGate is caused to rise to threshold, achieving the move and wiping out the move-generating command in Move. At this point, the simulated environment causes an update of the Sense modules, which in turn extinguish any goal or subgoal activation pattern in the Goal system or Subgoal which represent goals to create the current environmental configuration (point B). This allows the next most preferred goal to be retrieved and worked on, as can be seen in Subgoal at point C. At no point is the clock circuit involved.

Now the next goal, 'Blue to 2', which is unachievable, has been selected, and this in turn generates a subgoal to remove an obstacle. Once a subgoal is selected ('Green to 5', since Green is in the target position of the blue ball, at time point D), the first element of the NoMove timer sequence begins to ramp up, and finally maximal activation reaches the last timer in the sequence at time E (this also happens for the previous goal). This activates the Generate module for generating a subgoal. Finally, the subgoal generation logic computes that the ball above the green source ball is blue, at time F, and that the lowest position on a peg which is neither the source nor the target of the goal is position 1 at time G, and Subgoal responds to this voting at time H. The model continues on in this way until eventually solving the problem in 5 moves, as is shown in the sequence of moves selected by the model.

Effect of simulated Parkinson's disease

PD is simulated by reducing the inhibitory effect of Input connections to Delay units within columns. By weakening connections, the propagation delay inherent in any column will be lengthened. For columns in which no propagation delay is required, the effect of this dopamine depletion is expected to be minimal. For columns in which weak Input → Delay connections are needed to model slow propagation delay, the effect is expected to be pronounced. The convergence timer that determines the maximum interval during which computation of a new move can occur is therefore susceptible to simulated dopamine depletion.

In problems which do not require the generation of a subgoal, because all balls can be moved directly to their goal positions, the convergence timer never needs to expire: a move will always emerge well before the time limit. In problems which do require generation of subgoals (some 3 move problems, and by definition, all 4 and 5 move problems), the timer will expire once for each subgoal generated in the simulated dorsolateral PFC.

Fig. 7 shows the interactive effect of simulated dopamine depletion and problem difficulty on the total time to solution of the Tower of London model. Since the model does not store plans, the comparison to the latency data of Owen *et al.*, (1995) is based on the notion that the complete solution of a problem by the model is equivalent to the complete generation of a plan prior to execution by subjects (which is assumed to precede the first move made by the subject – latency in Owen *et al.* (1995) is therefore the time from presentation of the problem to execution of the first move). The model was run on all possible problems requiring five moves or less, except for problems which would cause difficulty for the simple algorithm implemented by the model. Including modules implementing the heuristics necessary to solve the remaining problems would not be expected to change reaction times significantly, because the modules implementing this logic would not require internal propagation delay (assuming, that is, that these heuristics did not produce a much larger number of moves, but this was not what happened in the hybrid model of Polk *et al.* (2002)).

Simulated dopamine depletion slows the performance of the model in difficult problems. This effect is most pronounced at the five-move-minimum level, and almost nonexistent at the two-move-minimum level. This supralinear increase in latency impairment with increasing problem difficulty is also seen in the performance of PD patients vs. controls. Results therefore support the notion that demands for subgoal representation and maintenance are at the heart of problem solving impairments in prefrontals and PD patients in the Tower of London task.

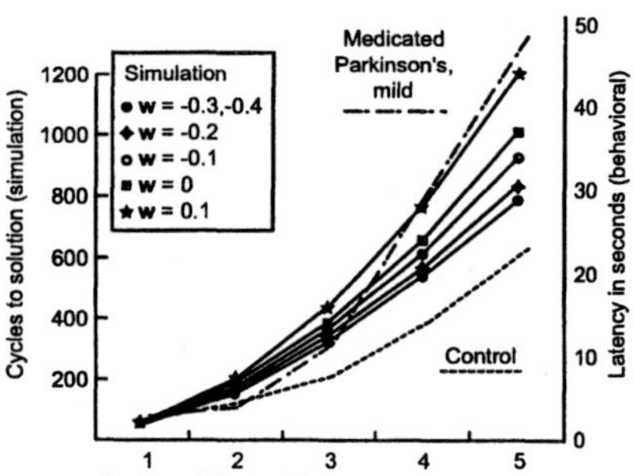

Figure 7: Comparison of average simulation reaction time in solid lines to empirical results in dashed lines (control subjects) and dotted-dashed lines (medicated patients with mild PD). More positive values of **w**, the connection strength from Input to Delay units, corresponds to increasing severity of PD symptoms.

Discussion

The model presented here provides a potential explanation for latency impairments in problem solving by PD patients, in accordance with a previous model that potentially explains prefrontal accuracy impairments (Polk *et al.*, 2002). The structure of the model also hints at the power of a form of neural network modeling that borrows key computational aspects from production systems while adhering to parallel distributed, analog representations of time.

Since dopamine depletion of sufficient severity should also impact the ability of subgoal representations to guide action selection according to this model, accuracy impairments should be expected in severe PD, and this impairment seems supported by the data (Owen *et al.*, 1995). More modeling work is necessary before this expectation can be confidently called a prediction of the model, however. Conversely, since timer circuits are proposed to contain both a cortical component and a component residing in the basal ganglia, the lack of latency impairments in patients with dorsolateral prefrontal lesions remains unexplained. Two possible explanations that do not necessarily conflict with the model are: 1) that timing operations are distributed across a larger area of cortex than just dorsolateral PFC, and thus are not disrupted by focal lesions; and 2) since the model only attempts to capture the planning rather than the execution phase of problem solving, it may be that prefrontals show no latency impairment in initiating problem solving despite disrupted internal timing because they impulsively begin execution before a plan is fully developed. Thus they trade off a latency impairment for an accuracy impairment. Future work in which plans are stored, evaluated and finally executed will better draw out the implications of the model presented here.

References

Alexander, G., DeLong, M. & Strick, P. (1986). Parallel organization of functionally segregated circuits linking basal ganglia and cortex. *Annual Review of Neuroscience, 9*, 357–381.

Cohen, M. & Grossberg, S. (1983) Absolute stability of global pattern formation and parallel memory storage by competitive neural networks. *IEEE Transactions on Systems, Man and Cybernetics, 13*, 815–826.

Dehaene, S. & Changeux, J. (1997) A hierarchical neuronal network for planning behavior. *Proceedings of the National Academy of Sciences, USA, 94*, 13293–13298.

Owen, A., James, M., Leigh, P., Summers, B., Marsden, C., Quinn, N., Lange, K., & Robbins, T. (1992). Fronto-striatal cognitive deficits at different stages of Parkinson's disease. *Brain, 115*, 1727–1751.

Owen, A., Sahakian, B., Hodges, J., Summers, B., Polkey, C., & Robbins, T. (1995). Dopamine-dependent frontostriatal planning deficits in early Parkinson's disease. *Neuropsychology, 9*, 126–140.

Polk, T., Simen, P., Lewis, R. & Freedman, E. (2002). A computational approach to control in complex cognition. *Cognitive Brain Research, 15*, 71–83.

Shallice, T. (1982). Specific impairments in planning. *Phil. Trans. Royal Soc. London, B, 298*, 199–209.

Simen, P. (2004). *Neural mechanisms for control in complex cognition.* PhD Thesis, University of Michigan, Ann Arbor.

Touretzky, D. & Hinton, G. (1988) A distributed connectionist production system. *Cognitive Science, 12(3)*, 423–466.

Episodic versus Semantic Memory: An Exploration of Models of Memory Decay in the Serial Attention Paradigm

Chris R. Sims (simsc@rpi.edu)
Cognitive Science Department
Rensselaer Polytechnic Institute

Wayne D. Gray (grayw@rpi.edu)
Cognitive Science Department
Rensselaer Polytechnic Institute

Abstract

We investigate the implications of an episodic versus semantic memory representation in a model of serial task switching. Altmann and Gray (2002; 2004) have developed a model in which the rate of memory activation, decay, and interference are tightly coupled in an explanation of learning, forgetting, and performance. A central assumption of their model is the use of an episodic memory trace for task cues. The current research explores the centrality of this assumption by exploring the predictions made by three mathematical formulations of interference and decay.

Introduction

It is clear that we can have memories for individual events such as the birth of a child or buying our first new car. It is also clear that we have memories for common facts such as the name of the President of the United States or the number of states in the Union. Tulving (1972) advocated a distinction between these types of memory, calling the former *episodic* and the latter *semantic* memories. Typically, episodic memory is viewed as an instance-based representation: each perceptual experience results in the encoding of a new instance of memory. In contrast, semantic memory is typically viewed as a non-instance based representation. Although we have all encountered the fact that 2+2=4 hundreds of times in our lives, we might only have one memory representation of this fact.[1] Although memory and memory representation are popular topics to study, it is rare to find a phenomenon in which the rate of memory activation, decay, and interference are tightly coupled in an explanation of learning, forgetting, and performance. It is even rarer to find a phenomenon where both the speed of performance and number of errors are tightly predicted. It is rarer still to find a precise mathematical formulation of such a phenomenon that depends, as its most basic assumption, on the idea that the memory it is studying is episodic, not semantic.

The Altmann and Gray (2002, 2004) theory of *serial attention* provides this rare combination of constraints. In the next section we introduce the paradigm they used and their results, as well as the *functional decay* model they propose to account for these findings. In the following section we examine three alterations to the original model. The first holds all but the assumption of episodic memory constant, while the second two compare the predictions of two different equations of activation and decay. We intend our discussion to shed light on the functional differences between episodic and semantic memory representations and the implications of these differences for cognitive modeling. We also believe that our close comparison of three popular mathematical formulations of interference and decay sheds light on the strengths and weaknesses of each. Other modelers who depend on one or more of these formulations should find our comparisons to be at least interesting and, perhaps, disturbing.

Task Switching and the Serial Attention Paradigm

The study of task switching examines the costs and mechanisms required to rapidly switch from one task to another. In the basic task switching experiment, participants either perform one task repeatedly or switch between different tasks. In some experiments (e.g., Logan & Bundesen, 2003) each trial is preceded by a task cue, indicating which task to perform, whereas other experiments use task cues that are implicit in the trial stimuli themselves (e.g., Allport & Wylie, 2000). In general, participants who are required to switch tasks perform more slowly as compared to those performing consecutive blocks of the same task, a phenomenon generally labeled *switch cost*. Allport, Styles, and Hseih (1994) showed that the magnitude of switch cost can be reduced, but not eliminated, by increasing the interval or rest period between tasks. A variety of related phenomena have also been discovered, including restart cost, residual switch cost, and within-run slowing (Altmann & Gray, 2002).

Unfortunately, the wealth of data available in this research domain does not converge on an explanation. The earliest findings in this field were reported by Jersild (1927), and since that time dozens of different theories have been proposed. The explanations range from postulating executive control mechanisms responsible for reconfiguring cognition and selecting tasks (e.g., Monsell, 2003; Rogers & Monsell, 1995; Kieras, Meyer, Ballas & Lauber, 2000), to attempting to account for task switching phenomena as a result of some combination of well-

[1] A notable exception to this classification is Logan's (2002) instance theory of memory, in which both semantic and episodic memories are represented as instances.

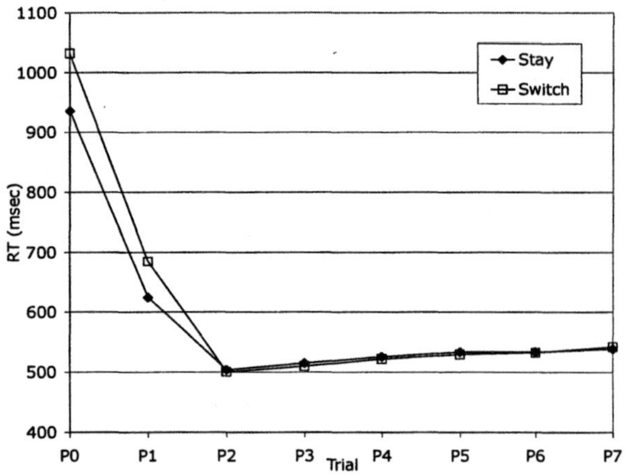

Figure 1: Mean response time by task continuity. Data taken from Altmann and Gray (2004).

established mechanisms (e.g., Altmann, 2003; Altmann & Gray, 2004). (A more classic way of phrasing the difference is as between explanations of task switching that postulate new *hypothetical constructs* versus those constructed out of established *intervening variables*, see MacCorquodale & Meehl, 1948).

Altmann and Gray (2004) favor the intervening construct approach and have constructed a model of task switching that requires no special executive control processes. Instead, their model predicts complex behavior almost entirely from basic memory processes such as interference and decay. The next section will provide an overview of their paradigm and experimental results. Following this, their functional decay model and its underlying assumptions will be examined in detail.

Paradigm and Results

In their experiment 1A, Altmann and Gray (2004) presented participants with one of two tasks on each trial. The two tasks were a Groupsize task or a Position task for a string of repeated letters (for example, 'AAAA'). The Groupsize task was to determine whether a string contained greater or fewer than five characters, while the Position task was to determine whether the letters occurred at the beginning or end of the alphabet. The task trials were grouped into runs, such that participants would perform one task for the duration of the run. Since the trial stimuli were ambiguous as to task, before the start of each run an instructional cue would indicate whether the Groupsize or the Position task was to be performed. This cue would remain on the screen until dismissed by a keypress. Run lengths were chosen randomly according to a uniform distribution in the range $7 \leq n \leq 13$, constrained so that each successive pair of runs would sum to 20 trials. Each successive pair defined a task block.

Figure 1 shows the mean response latency for the second run of each block, separated by continuity; that is, whether the task *switches* or *stays* the same. P0 represents the presentation of the task cue, while P1 through P7 are the first seven trials of that task in the run. This figure demonstrates several phenomena relevant to task switching. First, switch cost is observed as the difference between P0 Switch and P0 Stay. The data also shows residual switch cost, as the lingering latency difference into the first trial of the run. Finally, and most relevant to this research, within-run slowing is observed as the gradual but steady decline in performance over trials 2 through 7. In their experiment 3, Altmann and Gray varied the task runlength between participants. Using runlengths of 10 and 20, they found that the slope of within-run slowing is sensitive to the runlength of the task, such that longer runlengths result in a more gradual decay than shorter runlengths.

Functional Decay Model

To explain these various phenomena, Altmann and Gray assume that a task cue for the current task is retained in memory. On each trial, participants must successfully retrieve this cue in order to know which of the two tasks to perform. Additionally, Altmann and Gray assume an episodic, instance-based representation of task cues. Each time a task cue is encountered at the beginning of a run, it is encoded in memory as a new trace, regardless of whether the current cue was identical to the task cue from the previous run. According to their functional decay model, this process of encoding corresponds to rapidly rehearsing the task cue until it reaches a threshold level of activation.

In the empirical data in Figure 1, the process of encoding can account for the relatively slow response time for P0 as compared to P2 through P7. Residual encoding is also used to explain the lingering latency on trial P1. The functional decay model predicts that the activation of a task cue decreases from its peak encoding value. The reasoning behind this is that if task cues did not decay in memory, each successive encoding would have to reach a higher threshold than the previous, resulting in an escalating spiral of encoding. The theory that task cues decay across each run also explains the within-run slowing phenomenon: in the memory model used (discussed in the next section) retrieval time is a function of activation level. Hence, as declarative memory chunks decay, their retrieval times increase as well as becoming more prone to retrieval failures. Thus, across a run both retrieval time and errors should increase. Using this functional decay model, Altmann and Gray were able to obtain close quantitative fits to their data.

Modeling Memory

Each of the three memory functions considered in this paper share the three main assumptions of functional decay theory. First, each assumes an activation account of memory. This account proposes that the encoding of information imparts the memory representation (or trace) with an activation that decays as a function of time. Second, each assumes that interference among memory elements may affect the success of retrieving the correct memory. Third, a key feature of the functional decay model is the notion that as memories decay they become more susceptible to the effects of interference. Although decay and interference have historically been seen as competing accounts of memory, in recent years the use of decay in combination with interference has been justified by a large and growing body of research (see Altmann & Schunn, 2002 for a review).

According to the ACT-R cognitive architecture (Anderson & Lebiere, 1998), memory is stored in declarative and procedural memory subsystems. For the purpose of this paper, declarative memory is the system of interest, as it encompasses both episodic and semantic memory. Knowledge in declarative memory is represented in the form of memory chunks, each of which has a base level of activation that decays according to a power law of forgetting, and increases through rehearsal. Equation 1 shows the ACT-R chunk activation function.

$$a_i = ln(\sum_{j=1}^{n} t_j^{-d}) \quad (1)$$

In this function, a_i is the activation of chunk i, t_j is the time lapse since its jth retrieval, and n is the total number of retrievals during the lifetime of the chunk. d is a parameter governing the rate of decay, typically set to the value 0.5. Assuming that retrievals are evenly spaced across the lifetime of the chunk, the summation in Equation 1 can be replaced by $N \cdot (N/T)^{-d}$, where N is the total number of retrievals, and T is the total lifespan of the chunk. This then reduces to the function given by Equation 2:

$$a_i = ln(\frac{N}{1-d}) - d \cdot ln(T) \quad (2)$$

For computational efficiency reasons, Equation 2 is used by default in the ACT-R implementation, and can be set with the ACT-R *:ol* parameter. However, as will be discussed, the choice of episodic or semantic memory representation has important implications for the justification of using this simplified equation.

Although other classes of equations have been considered (see Wickens, 1998, for a review), power functions have been successfully used to fit a large body of data (Anderson, 2000). Additionally, Anderson & Schooler (1991) show that power functions also find theoretical interpretation as a reflection of the probability of needing information in the natural environment. They argue that it is rational for memory to mirror the need structure of the environment.

In the subsections that follow, we consider each of the two ACT-R decay equations presented above, along with an equation proposed by Pavlik and Anderson (2003) to account for the spacing effect. Each candidate equation will be considered in the context of episodic versus semantic memory representation in a task switching paradigm.

Episodic Assumptions and the Model Used by Altmann & Gray

Altmann and Gray claim that the use of an episodic memory representation in their model is critical to the flexibility of the cognitive system. In this section we question the necessity of this assumption in accounting for the data that their theory currently explains. In addition, we raise the issue as to whether an acknowledged limitation of the functional decay model is inherent in the model or due to their choice of an episodic rather than semantic representation.

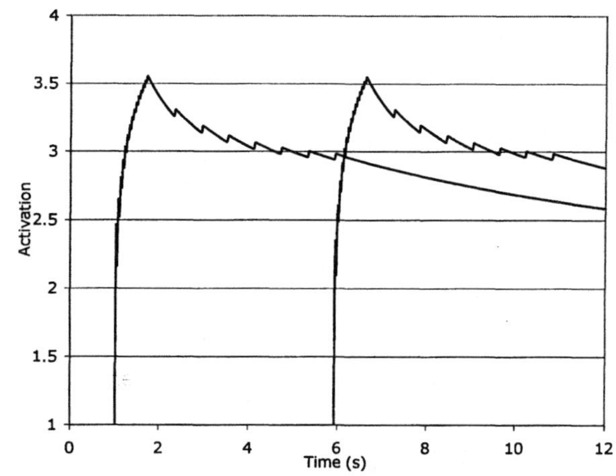

Figure 2: Memory activation trace for two task cues encoded in episodic memory.

Figure 2 shows the activation trace of two episodic task cues for the simplified decay equation used by Altmann and Gray (Equation 2). The initial spike for each task cue represents a period of rapid rehearsal while the task cue is present on the screen. Following this, the task cues are retrieved from memory once per trial. The spacing of these retrievals is insufficient to maintain the peak activation at the time of task encoding, resulting in decay. After a run of seven trials, a new task cue is presented and encoded in episodic memory, regardless of whether the same task is being performed.

In contrast to the episodic assumption, if the same two task cues are reused across runs (a non-instance based, or semantic representation), then activation becomes relatively stable as the number of retrievals increases. Hence, the activation for the memory of the current task will not differ significantly from the activation of the prior task. However, this argument is based on ACT-R's simplified decay equation. While activation does become stable at large values of N and T (see Equation 2), the equation assumes an approximately uniform spacing between retrievals. This assumption does not hold in the case of the functional decay model, since most retrievals of a trace occur in the rapid period of encoding at the beginning of the episodic trace's lifetime. Further, when task switches occur, the task cue for the previous task will not be retrieved for relatively long stretches of time.

In contrast, the complete ACT-R decay equation takes both the number and distribution in time of individual retrievals into account. As we discuss in the next section, using this form of the decay function (see Equation 1), the activation of a single (i.e., non-episodic) trace may fluctuate enough to produce the differences in activation that are key to Altmann and Gray's theory of serial attention.

An acknowledged limit of the functional decay model is its inability to account for the variable slope of within-run slowing. In an experiment in which subjects had either an average run length of 10 or 20 trials, Altmann and Gray found that increasing the run length had the effect of decreasing the rate of slowing by approximately half. Crucially, this between-Ss difference in slowing rate was apparent across all trials of a given run. Altmann and Gray (2004) leave this effect unexplained. Indeed, it would be hard to explain in a purely episodic task cue representation without introducing additional mechanisms.

To account for variable rate within-run slowing, the cognitive system must be differentially modified by short and long runs. This possibility requires that an additional mechanism must persist *across runs* in order for it to be modified. Episodic memory chunks are unlikely to fit this description, as they are never retrieved (except in cases of retrieval error) beyond the duration of one task run. The next section explores the possibility that semantic task cues might serve this purpose in the cognitive system.

Non-episodic Functional Decay

Altmann and Gray's central argument for an episodic memory representation is the cognitive inflexibility of the decay equation over a large number of retrievals. As discussed above, this problem might be more indicative of the unsuitability of the simplified decay equation rather than a theoretical limitation of the architecture. In order to examine this possibility, we modified our MATLAB simulation to use the complete ACT-R decay equation. By conducting the simulation in MATLAB we were able to plot activation values as a continuous function of time, a task that is significantly harder in the ACT-R implementation. In addition, we modified the simulation to re-use old task cues rather than creating new memory chunks at the start of each run.

Figure 3 shows a comparison between the simplified and full ACT-R decay equations using a semantic memory representation of task cues. The sequence in Figure 3 corresponds to a participant performing the same task for four runs in a row (the *stay* condition). At the presentation of each task cue, the rehearsal is performed on an existing memory chunk rather than a newly created chunk. As predicted, the simplified decay equation quickly stabilizes on a value.

The differential effect of the two equations on predicted activation is notable even in the first run. The full curve decays faster and is more sensitive to within-run retrievals (as shown by the sharp jags) than is the simplified curve. Across runs the simplified curve decays less and less making it clear that an episodic representation is needed to support Altmann and Gray's theory of serial attention. However, it is also clear from Figure 3 that the necessity of the episodic assumption may be an artifact of the simplified equation.

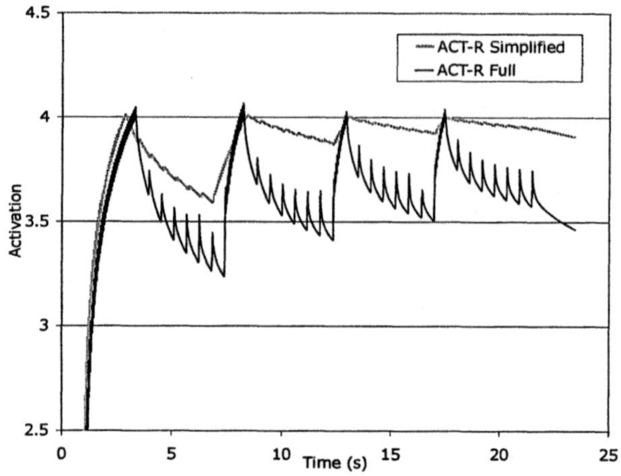

Figure 3: Comparison of the full and simplified ACT-R decay equations in a semantic memory representation.

It may be objected that although the full ACT-R equation demonstrates more flexibility with a semantic trace than does the simplified version, Figure 3 also shows that with each run of trials the activation fluctuates less with each retrieval. However, participants in the serial attention paradigm were not presented with a single task for a large number of consecutive runs. Hence, in this paradigm, the task cues would have an opportunity to decay while the second task intervenes. As far as we can determine, our modified model that assumes the full equation and semantic representation is theoretically consistent with all of the phenomena accounted for by the current functional decay model. Since task cues are re-used, on stay runs the task cue chunk starts from a higher activation. This results in less need for encoding, and therefore a P0 response time difference between switch and stay runs. Like the functional decay model, within-run slowing can be explained by the decay across runs and consequent increases in retrieval time as well as in retrieval failures.

It remains to be seen whether a semantic memory representation solves the problem of variable rate within-run slowing. Since task cue chunks now serve a functional role across runs, it seems both plausible and parsimonious for these chunks themselves to implicitly encode information about the length of a run. This model would eliminate the need for an additional cognitive mechanism that monitors the duration of each task. The question then, is how are chunks differentially modified as a consequence of long versus short run lengths?

An obvious answer is that the chunk activation itself somehow encodes information about the duration of the task. However, this alone cannot be the answer. As Figure 3 demonstrates, the primary determinant of chunk activation is not the quantity, but rather the spacing of memory retrievals. In the serial attention paradigm, trials progressed at the same speed regardless of run length. Because of this, by the end of a run, chunks that are retrieved 10 times following encoding will tend to have similar activation levels to chunks retrieved 20 times, and so activation level alone cannot be used as a predictor for run length. Hence, although a combination of a semantic representation and the full decay equation (Equation 1) leads to the same predictions as the functional decay model, neither model seem capable of accounting for variable rate within-run slowing.

The Pavlik & Anderson Decay Function

An alternative answer may be that activation duration, rather than amplitude, leads to the observed effect of variable within-run slowing. Since longer runs result in more spaced rehearsals of a task cue, it seems plausible that these chunks might decay more gradually, and therefore lead to a decreased rate of slowing. This theory is consistent with a widely studied phenomenon in experimental psychology, known as the spacing effect. The general finding is that increasing the interval between practice episodes has the effect of increasing the probability that the studied information can be correctly recalled at a later time. Conversely, the more closely two practice sessions occur, the more likely the practiced item is forgotten.

Pavlik and Anderson (2003) proposed a model of the spacing effect that utilizes a modified form of the decay function shown in Equation 1. This model assumes that the decay rate of a chunk is controlled not only by the frequency of retrieval, but also by the activation of the chunk at the time of its retrieval. Equations 3 and 4 give the Pavlik and Anderson decay function.

$$d_j = ce^{a_j} + b \tag{3}$$

$$a = ln(\sum_{j=1}^{n} t_j^{-d_j}) \tag{4}$$

Equation 4 is identical to the full ACT-R decay function, except the decay constant d is replaced by the function d_j. In Equation 3, a_j is the activation of chunk a at the time of its jth retrieval, while b and c are scaling parameters. These equations imply that retrievals of a chunk with a high level of activation have little impact on the long-term behavior of that chunk. At the same time, infrequent but steady retrievals will cause the memory chunk to decay more slowly.

Using Equation 4 in a non-episodic version of Altmann and Gray's functional decay model should have two consequences. First, the high density of retrievals during the rehearsal phase should have relatively little impact on the long-term activation of the chunk, since these occur at a high activation level compared to the rest of the run. Secondly, increasing the run length of a task has the effect of increasing the number of spaced retrievals of that task cue. Therefore, longer tasks should result in memory representations that decay more slowly, and consequently demonstrate a lower rate of within-run slowing.

Figure 4 shows the activation traces of two task cues overlaid. The first task uses a run length of 20 while the second task uses a run length of 10. The parameters used in this plot were [$c = 0.232$, $b = 0.5$], although the results obtained are not sensitive to the particular values chosen. Unfortunately, the effects expected of Pavlik and Anderson's equation were not observed. The expected effect was a smaller rate of decay during disuse for the longer run-length trace. This result would have helped explain the variable rate of performance decline observed by Altmann and Gray. However, as Figure 4 shows, during the period of disuse (to the right of the dotted line), both memory traces decay at the same rate.

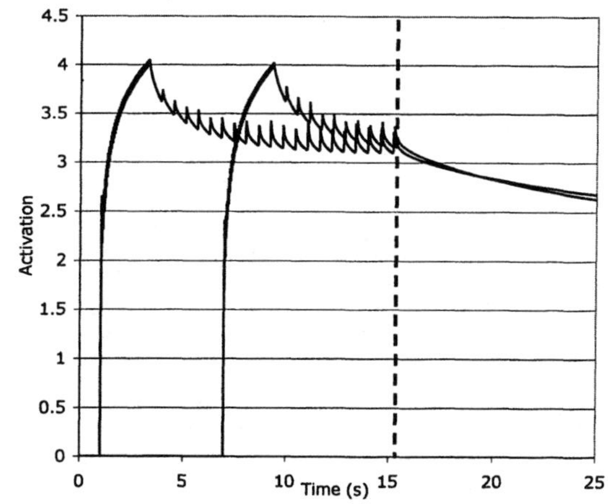

Figure 4: Comparison of decay of task cues at run lengths 20 and 10 using Pavlik & Anderson's (2003) decay equation. The dotted line indicates the disuse phase of both traces.

We hypothesize that these unexpected results are due to the high number of retrievals performed during the encoding phase of each task. Although the longer run-length task receives more spaced practice, the relative proportion of spaced practice is only slightly higher than in the shorter run-length condition. Additionally, Pavlik and Anderson's equation was developed to account for spacing differences on the order of seconds or minutes. The equation may not apply to the much shorter duration of continuous rehearsal proposed by the functional decay model.

Summary and Conclusions

This paper has examined issues regarding the use of an episodic versus semantic memory representation in cognitive modeling. It is not our intention to suggest that all memory processes must be purely semantic or purely episodic. Indeed, Tulving (1972) has suggested that any non-trivial task necessitates the use of both forms of memory. Rather, we have focused on the ways in which the choice of a memory model seems to dictate the choice of representation in the functional decay model. The exploration was grounded in the context of a well-developed model of serial task switching by Altmann and Gray (2004). A critical assumption of this model is the use of an episodic memory representation for current task cues. Although by using this approach

Altmann and Gray were able to explain a number of phenomena, some evidence suggests that their assumption is not fully justified.

By simulating the ACT-R memory activation equations we uncovered a potentially disturbing limitation of the simplified function. If memory chunks are not retrieved uniformly or at random, the decay delivered by the simplified equation will have a more shallow slope and will asymptote at a higher level than that predicted by the ACT-R theory. Since the simplified equation is used by default, many ACT-R modelers may be unaware of the consequences of equation choice. This issue is especially relevant to models using a non-instance based memory representation, as in these cases retrievals are likely to exhibit distinct patterns of use and disuse.

After simulating the full ACT-R equation, we found that contrary to Altmann and Gray, memory traces can maintain flexibility even after a relatively large number of retrievals. Building on this finding, we explored the possibility of using a semantic memory representation in the functional decay model. At a qualitative level, the semantic representation model is able to explain the same phenomena as its episodic counterpart.

We also experimented with an equation that Pavlik and Anderson (2003) developed to account for the spacing effect. By using this equation, we hoped to explain the variable rate within-run slowing phenomenon observed in Altmann and Gray's experiment 3. Although the Pavlik and Anderson equation failed to explain the intended phenomenon, its use did not diminish the performance of the general functional decay model.

Finally, it is not clear to us whether the current set of memory decay functions are applicable to the sort of rapid memory rehearsal proposed by the functional decay model. Indeed, the development and evidence for all three was in the context of data observed across timescales in the range of 10s of seconds to years. An interesting area for future research will be the investigation of how interference and decay alter behavior on the level of milliseconds.

Acknowledgments

We would like to thank Phil Pavlik and Erik Altmann for their feedback during the early stages of this research. The work reported was supported by a grants from the Air Force Office of Scientific Research AFOSR #F49620-03-1-0143, as well as the Office of Naval Research ONR #N000140310046.

References

Allport, A., Styles, E. A., & Hsieh, S. (1994). Shifting intentional set: Exploring the dynamic control of tasks. In C. Umilta & M. Moscovitch (Eds.), *Attention and performance IV* (pp. 421–452). Cambridge, MA: MIT Press.

Allport, A. & Wylie, G. (2000). Task switching, stimulus-response bindings, and negative priming. In S. Monsell & J. Driver (Eds.), *Attention and Performance XVIII: Control of cognitive processes* (pp. 35-70). Cambridge, MA: MIT Press.

Altmann, E. M. (2003). Task switching and the pied homunculus: where are we being led? *Trends in Cognitive Sciences*, 7(8), 340–341.

Altmann, E. M., & Gray, W. D. (2002). Forgetting to remember: The functional relationship of decay and interference. *Psychological Science*, 13(1), 27-33.

Altmann, E. M., & Gray, W. D. (2004). Beyond switch cost: Modeling memory processes in task switching. *Manuscript submitted for publication.*

Altmann, E. M., & Schunn, C. D. (2002). Integrating Decay and Interference: A New Look at an Old Interaction. *Proceedings of the 24th Annual Conference of the Cognitive Science Society.* (pp.65-70). Mahwah, NJ: Erlbaum

Anderson, J. R. (2000). *Learning and Memory* (2nd ed.). New York: Wiley

Anderson, J. R. & Lebiere, C. (1998). *The Atomic Components of Thought.* Mahwah, NJ: Erlbaum.

Anderson, J. R. & Schooler, L. J. (1991). Reflections of the environment in memory. *Psychological Science*, 2, 396–408.

Jersild, A.T. (1927). Mental set and shift. *Archives of Psychology*, Whole No. 89.

Logan, G. D., (2002). An instance theory of attention and memory. *Psychological Review*, 2, 376–400.

Logan, G. D. & Bundesen. (2003) Clever homunculus: Is there an endogenous act of control in the explicit task-cuing procedure? *Journal of Experimental Psychology: Human Perception and Performance*, 29(3), 575–999.

MacCorquodale, K., & Meehl, P. E. (1948). On a distinction between hypothetical constructs and intervening variables. *Psychological Review*, 55, 95–107.

Monsell, S. (2003). Task switching. *Trends in Cognitive Sciences*, 7(3), 134–140.

Monsell, S. (2003). Task-set reconfiguration processes do not imply a control homunuculus: Reply to Altmann. *Trends in Cognitive Sciences*, 7(8), 341–342.

Pavlik Jr., P. I., & Anderson, J. R. (2003). An ACT-R model of the spacing effect. In F. Detje, D. Dorner & H. Schaub (Eds.), *Proceedings of the Fifth International Conference on Cognitive Modeling.* (pp. 177–182). Germany: Universitats-Verlag Bamberg.

Rogers, R. D., & Monsell, S. (1995). Costs of a predictable switch between simple cognitive tasks. *Journal of Experimental Psychology: General*, 124(2), 207–231.

Tulving, E. (1972). Episodic and semantic memory. In E.Tulving & W. Donaldson (Eds.), *Organization of memory.* New York: Academic Press

Wickens, T.D. (1998). On the form of the retention function: Comment on Rubin and Wenzel (1996): A quantitative description of retention. *Psychological Review*, 105, 379–386.

Simple Agents Learning to Add Useful Structures to the World

Terry Stewart (tcstewar@connect.carleton.ca)
Sanjay Chandrasekharan (schandra@sce.carleton.ca)
Institute of Cognitive Science, Carleton University
Ottawa, Canada, K1S 5B6

Abstract

We provide a computationally tractable model of how organisms can learn to add structures to the world to reduce the complexity of their tasks. This model is then implemented using two techniques: first using a genetic algorithm, and then using the Q-learning algorithm. The results clearly show that organisms with only reactive behavior can learn to systematically add structures to the world to reduce their cognitive load. We show that such learning can happen in both evolutionary time and within an agent's lifetime. An extension of this model (currently being implemented) is then illustrated, where organisms with just reactive behavior learn to systematically generate and use internal structures akin to representations.

Many organisms generate stable structures in the world to reduce their cognitive load. Wood mice distribute small objects, such as leaves or twigs, as points of reference while foraging. Under laboratory conditions, they will spontaneously make use of plastic discs for this purpose. Stopka & MacDonald (2003) show that this 'way-marking' diminishes the likelihood of losing interesting locations during foraging. Red foxes use urine to mark food caches they have emptied. This marking acts as a memory aid and helps them avoid unnecessary search (Henry, 1977, reported in Stopka & MacDonald, 2003). The male bower bird builds colorful bowers (nest-like structures), which are used by females to make mating decisions (Zahavi & Zahavi, 1997). Ants drop pheromones to trace a path to a food source. Many mammals mark up their territories.

At the most basic level, cells in the immune system use antibodies that bind to attacking microbes, thereby 'marking' them. Macrophages use this 'marking' to identify and destroy invading microbes. Bacterial colonies use a strategy called 'quorum sensing' to know that they have reached critical mass (to attack, to emit light, etc.). This strategy involves individual bacteria secreting molecules known as auto-inducers into the environment. These accumulate in the environment, and when it reaches a threshold, the colony moves into action (Silberman, 2003).

Given that this 'doping' of the world is so common in these simpler creatures, it is somewhat surprising that there has been relatively little investigation into the use of this technique by *homo sapiens*. More than any other species, humans generate these external structures to reduce the amount of physical and cognitive effort required to perform tasks in their daily lives. Examples include markers, color-codes, page numbers, credit-ratings, badges, shelf-talkers, speed bugs, road signs, post-it notes – an almost endless list.

Epistemic Structures

The pervasiveness of such structures across species indicates that adding structure to the world is a fundamental cognitive strategy (Kirsh, 1996). Note that these structures serve to make tasks easier for agents. Some of these structures have referential properties, but they do not exist for the purpose of reference. We use the term *epistemic structures* to refer to these, in deference to Kirsh's (1994) distinction between epistemic and pragmatic action.

Kirsh's (1996) model of "changing the world instead of oneself", postulates that such generation of structures involve task-external actions, and these structures work by deforming the state space, so that paths in a task environment are shortened. Such structures also allow new paths to be formed in the task environment. However, Kirsh only explicitly addresses the generation of *tools*, rather than the direct modification of the world to reduce cognitive load.

Extending his idea to develop a full computational model of how organisms generate such structures, we make two reasonable assumptions. One, organisms sometimes randomly generate structures in the environment (pheromones, urine, leaf piles) as part of their everyday activity. Two, organisms can track their physical or cognitive effort (i.e. they get 'tired'), and they have a bias to reduce tiredness.

Given these assumptions, some of the randomly generated structures are encountered while executing tasks like foraging and cache retrieval. In some random cases, these structures make the task easier for the organisms (following pheromones reduces travel time, avoiding urine makes cache retrieval faster, avoiding leaf-piles reduce foraging effort). In other words, they shorten paths in the task environment. Given the postulated bias to avoid tiredness, these paths get preference, and they are reinforced. Since more structure generation leads to more of these paths, structure generation behavior is also reinforced.

The Simulation

To test and investigate the above model of epistemic structure generation, we have developed a computational model, where simple agents in a simple world, given only feedback in terms of their 'tiredness' (i.e. the effort required to perform their task), learn to systematically add structures to their environment.

The task we have chosen is analogous to foraging behavior, i.e. navigating from a home location to a distant target location and back again. Our environment consists of a 30x30 toroidal grid-world, with one 3x3 square patch representing the agent's home, and another representing the target. This 'target' can be thought of as a food source, to fit with our analogy to foraging behavior.

Agent Actions

At any given time, an agent can do one of five possible actions. The first and most basic of these is 'moving randomly'. This consists of going straight forward, or turning to the left or right by 45 degrees and then going forward. The agent does not pick which of these three possibilities occurs (there is a 1/3 chance of each).

In deciding the actions available to the agent, we needed to postulate some basic facilities within each agent. In our case, we felt it was reasonable to assume that the agents could distinguish between their home and their target. To do this, we added two more actions to the agents' repertoire. These are exactly like the first action, but instead of moving randomly, the agent would move to whichever square is sensed to be the most 'home-like' (or the most 'target-like'). Initially, the only things in the environment that are 'home-like' or 'target-like' are the home and the target themselves.

One way to think about these actions is to consider the pheromone-following ability of ants. Common models of ant foraging (e.g. Bonabeau et al, 1999) consist of the automatic release of two pheromones: a 'home' pheromone and a 'food' pheromone. The ants go towards the 'home' pheromone when they are searching for their home, and they go towards the 'food' pheromone when foraging for food. This exactly matches these two actions in our agents. The 'home' pheromone would be an example of a 'home-like' structure in the ant environment.

The fourth and fifth possible actions provide for the ability to generate these 'home-like' and 'target-like' structures. In the standard ant models, this could be thought of as the releasing of pheromones. However, our simulation has an important and very key distinction. Here, this ability to modify the environment is something the agents can do *instead* of moving around. That is, this generation process requires time and effort. The best way to envisage this is to think of an action that a creature might do which inadvertently modifies its environment in some way. Examples include standing in one spot and perspiring, or urinating, or rubbing up against a tree. These are all actions which modify the environment in ways that might have some future effect, but do not provide any sort of immediate reward for the agent. Kirsh (1996) terms these 'task-external actions'.

It must be stressed here that we are not presuming any sort of long-term planning on the part of the agents. We are simply specifying a collection of actions available to them, and they will choose these actions in a purely reactive manner (i.e. based entirely on their current sensory state). It may also be noted that our 'actions' are considered at a slightly higher level than is common in agent models. Our agents are not reacting by 'turning left' or 'going forward'; they are reacting by 'following target-like things' or 'moving randomly'. Furthermore, they do not initially have any sort of association between the action of making 'home-like' structures and the action of moving towards 'home-like' things. Any such association must be learned (either via evolution, or via some other learning rule).

Also, our agents are not designed to form structures automatically as they wander around (as is the case in standard ant models). In our simulation, a creature must expend extra effort to systematically generate these structures in the world. An agent that does this will be efficient only if the effort spent in generating these structures is more than compensated for by the effort saved in having them. Moreover, these are not permanent structures. The agents' world is dynamic and the structures do not persist forever. The 'home-likeness' or 'target-likeness' of the grid squares decrease exponentially over time. Furthermore, these structures also spread out over time. A 'home-like' square will make its neighboring squares slightly more 'home-like'. This can be considered similar to ant pheromones dispersing and evaporating, or leaf/twig piles being knocked over and blown around by wind or other passing creatures.

Table 1: The five actions available to the agent.

Agent Actions
Move randomly
Move toward 'home-like' structures
Move toward 'target-like' structures
Make 'home-like' structure
Make 'target-like' structure

Agent Sensing

Since our agents are reactive creatures and thus do no long-term planning, they require a reasonably rich set of sensors. We have given them four sensors, two external and two internal, to detect their current situation. The two external sensors sense how 'home-like' and how 'target-like' the current location is (digitized to 4 different levels). The internal sensors are two simple bits of memory. One indicates whether the agent has been to the target yet, and the other indicates how long it has been since the agent generated a structure in its environment (up to a maximum of 5 time units). This is all that the agents can use to determine which action to perform. This configuration gives each agent 192 (4 x 4 x 6 x 2) different sensory states.

Table 2: The four sensory values available to the agent.

Agent Sensory State
Similarity to 'home' of current location (0-3)
Similarity to 'target' of current location (0-3)
Amount of time since the agent last made a change to the environment (0-5)
Currently carrying anything (0-1)

Alternative State/Action Representations

The particular choices we made for the set of actions and sensory states the agent has available to it (as shown in tables 1 and 2) may be a bit non-intuitive. They are not the first representations that we tried. Initially, instead of the 'move randomly', 'follow home-like', and 'follow target-like' actions, we used the more traditional actions of 'move forward', 'turn left', and 'turn right'. Using these actions requires a more complex sensory state; we would have had to add sensor values to detect how home-like and target-like the squares ahead of, to the left of, and to the right of the agent.

However, when we used this approach of having a more complex sensory state and a less complex set of actions, the agents were unable to learn to create structures. For this reason, we used the state/action representations shown in tables 1 and 2 for the results given in this paper. We also found that the 'time since a change was made' sensor was also needed for the agents make use of their abilities to change the world.

The Learning Rules

Given our representations, we needed some way of determining which action the agent will perform in each of these 192 states. Note that by having this sort of mapping, we are implying a purely reactive agent. We investigated two different methods for matching sensory states to actions: a Genetic Algorithm, and Q-Learning.

Stage 1: The Genetic Algorithm

For our first model, we used a genetic algorithm to determine which action to take in each situation. The genome consisted of a simple list of actions, one to perform in each state. To evaluate a particular genome, we started 10 agents in the home location and ran the simulation for 1000 time steps. The evolutionary fitness was the agents' average tiredness (i.e. how long it took each agent to make it back home from the target).

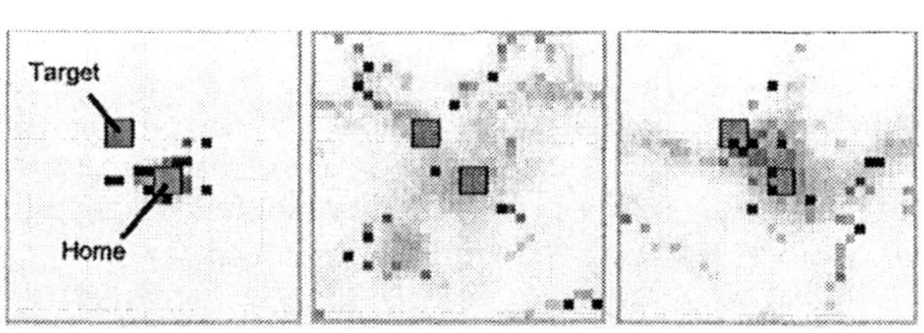

Figure 1: The computer model at 10, 100, and 300 time steps. Black dots are the agents. The shading is darker the more 'home-like' or 'target-like' a particular square is. This run shows typical agent behavior after 300 generations.

Result: Initially, the agents behaved randomly. Starting at the 'home', they would wander about and might, by chance, find the target and then, if they were very lucky, their home. Indeed, most agents did not find the target and make it back within the 1000 time steps. On average, we found that each agent was completing 0.07 foraging trips every 100 time steps. After a few hundred generations, the agents were soon completing an average of 1.9 trips in that same period of time. In other words, the agents were able to, on an evolutionary time scale, learn to make use of their ability to sense and generate structures in the world. Furthermore, this ability provided a very large advantage over completely random behavior.

This result confirmed that it is possible for agents to learn to systematically generate and use structures in the world in an evolutionary time scale. It also showed that we had not chosen an impossible task for the agents to learn. However, for our purposes, we were much more interested in an individual agent learning to generate epistemic structures within that agent's lifetime. To investigate this, we turned to the Q-Learning algorithm.

Stage 2: Q-Learning

The heart of our investigation was to determine whether a simple, general learning algorithm would allow our agents to discover and make use of the strategy of systematically adding structures to the world. In keeping with our 'tiredness' theory, the only feedback the learning mechanism had was an indication of the exertion or effort. The delayed-reinforcement learning rule known as Q-Learning (Watkins, 1989) seemed best suited for this task. (Sarsa and other TD-Learning algorithms would also be suitable). The Q-Learning algorithm[1] develops an estimate of the eventual outcome of performing a given action in a given situation. The agent then performs the action with the highest expected payoff.

Using the Q-Learning algorithm, we again ran 10 agents for 1000 time steps. To indicate 'tiredness', we gave them a reinforcement value of -1 all the time (indicating a constant 'punishment' for expending any effort). When they returned home after finding the target, they were given a reinforcement of 0, and they were then sent back out again for another trip. Each agent independently used the Q-Learning algorithm, and there was no communication between the agents.

Result: The dark line in figure 2 shows the results averaged over 100 separate trials. We can clearly see that the agents are improving over time (i.e. they are spending less time to perform their foraging task).

Stage 3: Confirmation

Although we have observed improvement over time, we still need to show that it is the agents' ability to systematically

[1] The estimated reward for performing action a in state s is $Q(s,a)$. This is increased by $\alpha(r+\gamma\max(Q(s',b))-Q(s,a))$, where r is the immediate reward/punishment, s' is the resulting state, γ is the future discounting rate (set to 0.5), and α in the learning rate (0.2). We used ε-greedy action selection with ε set to 0.1, so the agents choose the action with the highest expected reward 90% of the time, and the other 10% they perform an action at random.

add structures to the world that is causing this effect. To prove this, we re-ran the experiment, this time removing the agents' ability to generate structures in the world. No other changes were made.

Result: We found that when the agents were unable to generate structures in the world, Q-Learning did not provide as much improvement[2]. This result is shown in the lighter line in Figure 2. There is still a small improvement given by Q-Learning, but we can conclude that the significant improvement seen in the previous experiment (the dark line in Figure 2) is due to the agents' ability to modify their environment.

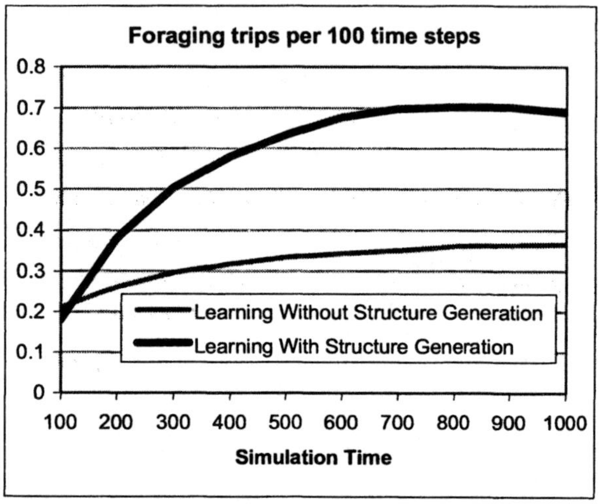

Figure 2: The effect of epistemic structure generation. The foraging rate is measured in trips per 100 time steps. A foraging rate of 0.5 means that trips require an average of 200 time steps to complete.

We can also see from Figure 2 that having these extra actions available does incur some cost in the early stages. Initially, the agents perform slightly worse. However, the advantage of being able to form epistemic structures quickly improves the agents' performance. By the end of the simulation, agents require only around 150 time steps to make a complete trip (a foraging rate of 0.66 trips in 100 time steps). This is twice as quick as agents without the structure-forming ability.

When we analyzed the actions of the agents, we found that they actually spent 58% of their time generating structures. This is striking, since time spent generating these structures means less time for wandering around trying to find the target or their home. Table 3 gives the breakdown of how time was allocated to different actions. The data indicates that epistemic structure generation allowed the agents to go from spending 300 time steps down to 150 time steps to complete their foraging task, even though over half of those 150 time steps are spent standing still. There is clearly a large efficiency advantage to making use of these structures.

Table 3: Time spent performing various actions over 1000 time steps.

Action	With Structure Generation	Without Structure Generation
Move randomly	10%	32%
Toward 'home-like'	19%	36%
Toward 'target-like'	13%	32%
Make 'home-like'	35%	
Make 'target-like'	23%	

Model Capabilities

The Q-Learning system is a concrete implementation of our model: a simple learning mechanism that allows agents with purely reactive behavior to systematically add structures to the world to lower search.

The Q-Learning model implemented in this simulation can explain the generation of structure that is used both by the agent generating the structure, and by the other agents in the environment. The agents ended up forming structures that were useful for everyone, even though they were just concerned about reducing their own tiredness. This was only possible because the agents were similar to each other. This is similar to how paths are formed in fields: one person cuts across the field to reduce his physical effort, others, sharing the same system and wanting to reduce their effort, find the route optimal. As more people follow the route, a stable path is formed.

Other Models

It is worth noting that our model presents a novel simulation of ant behavior. The closest existing models are those in (Bonabeau et al, 1999) which use the 'home-pheromone' and the 'food-pheromone'. This is in contrast to such models as (Nakamura & Kurumatani, 1996), where a land-based and an airborne pheromone are used, or any models of the Cataglyphis species of ant, which uses a complex landmark-navigation scheme which allows it to return directly to the nest (Miller & Wehner, 1988).

That said, all of these other models assume both that pheromones are continually being released while the ant forages, and that there is no learning happening during the foraging behavior. Our model does not make either of these assumptions.

We were unable to find references indicating that real ants might, in fact, learn to use pheromones, or any research that indicates that the effort required to produce these pheromones might interfere with the foraging behavior. This indicates our model may not be a good one for understanding ants. However, the fact that our agents are able to learn to reflexively generate these cognitively

[2] Q-Learning also did not provide significant improvement if the agents were only able to generate one type of structure, or if any of the agent's sensors were removed.

beneficial structures in the absence of any immediate feedback to their benefit, indicates a simpler way to model more complex creatures that exhibit such behavior.

Conclusions

The model presented here shows that a simple agent using Q-Learning can learn to modify its environment in such a way as to reduce the amount of effort required to perform a task. This ability to change the environment is one that is common in simple creatures, but has not been the focus of attention of computational modeling. This ability to change the world is known to be fundamental for a broad range of human activity. This result indicates a new domain of investigation for more complex learning agents in more dynamic and realistic environments.

Future Work

Interestingly, this same model could explain generation and tracking of *internal* structures in organisms. The actions which generated structure in our simulation were actions that affected the environment. But this does not have to be the case. Just as we had both internal and external sensors, we can have actions which affect either the state of the world *or* the state of the agent itself. In other words, we can use this model to investigate the generation of *internal* structure (i.e. representations).

As an example, consider foraging bees. Suppose that, just as our agents left traces in the world of their activity via their structure-generating actions, we have the bees leave a sequence of internal memory traces corresponding to landmarks (say a tall tree, a lake, a garden) as a result of their everyday foraging activity. In some foraging trips of some bees, the trace sequences match to some degree the external structures they perceive. Such trips involve less search, because they lead to food more directly, i.e. they form shorter paths in the task environment. Over time, using the exact same learning mechanisms that apply in the external case, the bias against tiredness leads to such paths being used more, and so they are reinforced. This could lead to landmark-based navigation, which does, in fact, exist in bees (Gould, 1990). As in the case of external structures, the generation of such memory traces is reinforced because more traces lead to more such shorter paths in the task environment. We are currently working on a computational model of this example. Interestingly, recent research shows a similar use of landmarks by homing pigeons, which follow highways, railways and rivers to reach their destination with less cognitive effort (Guilford, 2004).

This idea presents a situated cognition model of how memory structures come to be used as task-specific structures, and why such internal structures are systematically generated. If such task-specific memory structures are considered to be representations (that is, they stand for something specific in the world), then the model explains, in a computationally tractable manner, how organisms with just reactive behavior can learn to generate and use representations.

The model also explains what such 'primitive' representations are: they are internal traces of the world that allow the agent to shorten paths in a task environment. Roughly, they are computation-reducing structures (and equivalently, energy-saving structures). They are internal 'stepping stones' that allow organisms to efficiently negotiate the ocean of stimuli they encounter. This means the traditional cognitive science view, that thinking is computations happening over representations, presents a secondary process. In the stepping stone view, representations are crucial for organisms, but they are just useful, incidental entities, not fundamental entities by themselves.

All source code for the simulations can be found at: http://www.carleton.ca/iis/TechReports/code/2004-01/

References

Alcock, J. (1998). *Animal Behavior: An evolutionary approach*, Massachusetts, Sinauer Associates.

Bogen, J. (1995). Teleological explanation. In Honderich (Ed.), *The Oxford Companion to Philosophy*. Oxford University Press, New York.

Bonabeau E., Dorigo M. and Theraulaz G. (1999) *Swarm intelligence: From natural to artificial systems*. Santa Fe Institute studies in the sciences of complexity. Oxford University Press, New York.

Clark, A. (1997). *Being There: putting brain, body, and world together again*, Cambridge, Mass., MIT Press.

Gould, J.L. (1990). Honey bee cognition. *Cognition*, *37*, 83-103.

Guilford, T., Roberts, S. & Biro, D. (2004). Positional entropy during pigeon homing II: navigational interpretation of Bayesian latent state models. *Journal of Theoretical Biology*.

Henry, J.D. (1977). The use of urine marking in the scavenging behaviour of the red fox (Vulpes vulpes). *Behaviour*, *62*, 82-105.

Kirsh, D., & Maglio, P. (1994). On distinguishing epistemic from pragmatic action. *Cognitive Science*, *18*, 513-549.

Kirsh, D. (1996). Adapting the environment instead of oneself. *Adaptive Behavior*, *4:3/4*, 415-452.

Miller, M., & R. Wehner (1988). Path integration in desert ants, Cataglyphis fortis. *Proceedings of the National Academy of Sciences, 85*: 5287-5290.

Nakamura, M., & Kurumatani, K. (1996). Formation mechanism of pheromone pattern and control of foraging behavior in an ant colony model. *Proceedings of the Fifth International Conference on Artificial Life*, 67-74.

Silberman, S. (2003), The Bacteria Whisperer. *Wired*, Issue 11.04, April 2003.

Stopka, P. & Macdonald, D. W. (2003) Way-marking behavior: an aid to spatial navigation in the wood mouse (Apodemus sylvaticus). *BMC Ecology, 3:3.*

Watkins, C. (1989). *Learning From Delayed Rewards*, Doctoral dissertation, Department of Psychology, University of Cambridge, Cambridge, UK.

Zahavi, A., & Zahavi, A. (1997). *The Handicap Principle: A missing piece of Darwin's puzzle*. Oxford: University Press, Oxford.

A Dynamic, Multi-Agent Model of Peer Group Formation

Terrence C. Stewart (tcstewar@connect.carleton.ca)
Institute of Cognitive Science, Carleton University
1125 Colonel By Drive, Ottawa, ON, K1S 5B6, Canada

Robert L. West (robert_west@carleton.ca)
Institute of Cognitive Science; Department of Psychology, Carleton University
1125 Colonel By Drive, Ottawa, ON, K1S 5B6, Canada

Robert Coplan (robert_coplan@carleton.ca)
Department of Psychology, Carleton University
1125 Colonel By Drive, Ottawa, ON, K1S 5B6, Canada

Abstract

A computational model of the formation of peer groups in children is presented. We used standard sociometric measurements (the CDC classification of Popular, Rejected, Neglected, and Controversial individuals) to compare the model to empirical data. The model fit this data well in terms of category distributions and stability, even without introducing individual differences. When individual differences were added, the model went on to accurately predict a separate set of empirical results. Furthermore, patterns arose in the results which suggested an underlying relationship between certain types of individual differences.

Introduction

In recent years, we have started to see an increasing willingness among researchers to examine high-level models of human group behavior. Interesting results have been found regarding wealth distribution (Bouchaud & Mézard, 2000), human rioting (Granovetter, 1978), crowd panic (Helbing, Farkas, & Vicsek, 2000), and friendship networks (Watts & Strogatz, 1998), to name just a few. We have also seen the development of larger frameworks for investigating social cooperation to achieve goals in general (for example, Sun, 2001). These models provide new insight into these fields, along with new predictions that can be tested.

In this paper, we present a computational model of the formation of friendship groups among peers. In the model, individuals meet and interact over time, becoming friends (and possibly enemies) with each other. This process has been empirically investigated by developmental psychologists, who have developed standardized measures of popularity (see Cillessen & Bukowski, 2000, for a review). In this paper we will evaluate our model's results by comparison to the CDC measure (named after its creators' initials: Coie, Dodge, & Coppotelli, 1982), which is the most commonly used system.

Psychologists do not consider 'popularity' to be an intrinsic aspect of an individual. Rather, 'popularity' is a dynamic construct formed among a group of individuals. That is, an individual's popularity is determined by how others feel about them. However, explanations as to why a person is popular or unpopular focus on individual characteristics, particularly personality factors. Individual factors do influence popularity; a child could be unpopular because they are shy, aggressive, and/or socially incompetent (Newcomb, Bukowski, & Pattee, 1993). However, Rubin, Hymel, & Mills (1989) reviewed the findings in this area and found that only a small percentage of the variance could be accounted for by individual differences.

While it is possible that the unaccounted for variance is simply due to an inability to measure popularity and/or personality factors well enough, another possibility is that it can be attributed to the dynamics of the system. That is, people's popularity may have more to do with the dynamics of their peer group interactions than with their own personality. If so, this would mark a radical departure from the common understanding of popularity as a product of personality. We investigated this possibility through the use of multi-agent modeling.

The CDC Measure

The goal of the CDC measure is to classify people into one of five categories: *Popular*, *Rejected*, *Controversial*, *Neglected*, or *Average*. Using interviews or questionnaires, each person is asked to name three people in their peer group that they like, and three people that they dislike. The simplicity of this measurement is important for measuring popularity in young age groups. Using the survey results, each individual is given an *Acceptance* score (the total number of times that person is listed by other people as someone they like) and a *Rejection* score (the number of times they appear on the 'dislike' lists). A *Preference* value (Acceptance minus Rejection) and an *Impact* value (Acceptance plus Rejection) are also created; where preference refers to whether you are more liked or disliked and impact refers to how much people pay attention to you. Individuals are then classified into the five categories according to the rules shown in Table 1.

Table 1: The decision rules for classifying with CDC. All values are normalized to a mean of 0 and a standard deviation of 1.

Category	Rule
Popular	Preference>1 Acceptance>0 Rejection<0
Rejected	Preference<-1 Acceptance<0 Rejection>0
Neglected	Impact<-1 Never appears on anyone's 'like' list
Controversial	Impact>1 Acceptance>0 Rejection>0
Average	None of the Above

Comparison Data

Newcomb, Bukowski, & Pattee (1993) provide a considerable amount of data to test our model against. To begin with, they give complete results from nine different studies that used CDC categorization on a total of 2,571 students, ranging from kindergarten to grade 9. These nine studies give the following 95% confidence intervals for the percentage of people in each category.

Table 2: The 95% confidence intervals indicating the percentage of people in each CDC category.

Popular	Rejected	Neglected	Contro.	Average
7%-32%	12%-26%	0-28%	1.6-16%	5.9-69%

This data set gives us our first basis of comparison between the model and reality. In particular, if we find that a model gives results outside of this range, then we can conclude that it is not a suitable model.

The next set of data that we can use for comparison is the stability of these measurements over time. Cillessen, Bukowski, & Haselager (2000) give the results of a meta-study which collected the change in CDC categorization over periods of time ranging from one month to four years. The results give the percentage of people who were in the same category at the beginning and at the end of the study period. The following figure gives the 95% Confidence Interval for this data.

Table 3: The 95% confidence intervals for the stability of each CDC category.

Popular	Rejected	Neglected	Contro.	Average
33-44%	39-49%	20%-30%	24%-36%	51%-69%

In addition, the results showed no significant change related to the amount of time in the study period. In other words, there is no significant difference between the stability over four months and stability over four years. We should thus expect a good computational model to not only match the stability data shown in table 3, but also to give this result independent of the number of interactions simulated.

Creating a Benchmark

To further evaluate our model, we created a benchmark for evaluating CDC results. Specifically, we wanted to create a null condition result, representing "no effect" for the CDC measure. To do this used a completely random response scheme. That is, we determined what would happen if the 'like most' and 'like least' responses were generated by having each person nominate three others completely at random. This data was then subjected to the standard CDC analysis, as described. We chose a group size of 30 to be consistent with real-life situations.

After generating 1000 groups, we collected the CDC classification data and found the following distribution.

Table 4: The distribution of individuals in the benchmark.

Popular	Rejected	Neglected	Contro.	Average
12%	12%	7%	2%	67%

Surprisingly, the null condition results did not fall outside the confidence intervals for the categorization data in table 2. Thus the categorization distribution results cannot be considered to be different from chance. When we turn to the stability data, however, we do find a statistically significant difference. Given the simplicity of this random model, since an individual has a 12% chance of being classified as Popular, when we re-run the simulation that same person will still have only a 12% chance of being Popular this time. This means that the stability data will be identical to the previous table

Table 5: The stability of categorization in the benchmark.

Popular	Rejected	Neglected	Contro.	Average
12%	12%	7%	2%	67%

These results were well outside the confidence intervals presented in table 3 ($p<0.001$). The null condition clearly does not capture important aspects of the process. The results for the stability data show that the CDC findings are meaningful. However, the results for the categorization data show that the apparent differences in the sizes of the categories can be viewed as an artifact of the way the CDC system works.

Given this baseline, we can see that predicting the category distribution data (Table 2) would only provide weak support for a model. However, if a model was able to predict both the distribution and the stability data, then we could be confident that it is matching the real-world data well.

Our Model

In developing our model, we tried to determine the simplest system possible that had the following characteristics:

1. Each person should remember how much they like each other person

2. People should use this memory to determine how they will 'interact' with others
3. People should use the results of this 'interaction' to change how much they like the person they just interacted with

This led us to the following algorithm:

1. Let a[i,j] be the amount person i likes person j
2. For all pairs of individuals i,j:
   a. Have i use a[i,j] to decide how to behave towards j
   b. Have j use a[j,i] to decide how to behave towards i
   c. Update a[i,j] based on how j behaves towards i
   d. Update a[j,i] based on how i behaves towards j
3. Repeat (2) for as many iterations as desired

To implement this, we needed to define an algorithm for each agent to use to decide how to behave towards another, based on how much they like each other (steps 2a and 2b). We chose a simple method: take the value of a[i,j], add a random value from a Gaussian distribution, and use the resulting value to represent how 'good' individual i is going to behave towards j.

$$b[i,j]=a[i,j]+N(0,1) \qquad \text{(formula 1)}$$

The random variable has a deviation of 1. Changing this deviation does not affect the behavior of the model.

Similarly, we needed to update how much i likes j, based on j's actions. One idea would be to simply add b[j,i] (how 'nicely' j has behaved toward i) to a[i,j]. However, this approach can be easily shown to cause a positive feedback loop which would mean that if i currently likes j and j currently likes i, they will keep increasing how much they like each other to astronomical values. This is clearly does not capture the ebb and flow of real human relations. Instead, we use a slightly more complex formula which acknowledges the role that a[i,j] (the amount person i likes person j) could play in determining how person i would react to the behavior of person j.

$$\Delta a[i,j] = w_{lr}(b[j,i]-a[i,j]) \qquad \text{(formula 2)}$$

The result of this was that individuals in the model evaluated the actions of others according to the expectations they had for those others. This was based on the insight that we expect a friend to be friendly and an enemy to be unfriendly, and we are not surprised when this happens. However, if a friend were to be unfriendly we could be quite hurt. Likewise, if an enemy were to be kind it would surprise us and could cause us to reevaluate our feelings toward them.

Since repeated iterations of this model will cause a[i,j] and a[j,i] to interact with one another, it is instructive to view what typically happens to these variables over time. Figure 1 illustrates the effect. The results show that the model can potentially capture the ups and downs of real human relationships.

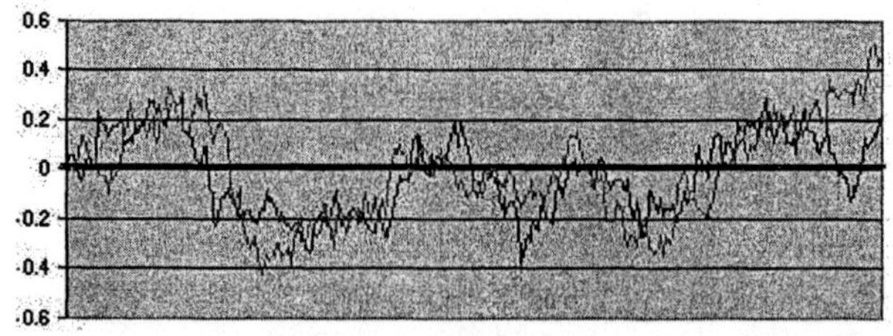

Figure 1: Two individuals interacting over 300 iterations of the model. Shown is a[i,j] and a[j,i]. w_{lr} is 0.1.

It should be noted that in doing this we have introduced the first parameter into this model: w_{lr}. This is the 'learning rate' and controls how quickly an individual changes its opinions of others. It can be considered to be similar to the learning rate parameters found in a wide variety of other computational models. This is a very interesting parameter because creates a potential link between cognitive abilities and popularity, something that has not been previously explored. Essentially, the model examines the role of memory, learning and expectations in human relationships.

Category Distribution Results

Interestingly, it turns out to be very easy to show that this new model is exactly as good as the random model at predicting the distributions of categories. We first note that, given the preceding algorithm, the values a[i,j] and a[j,i] change in a manner that is independent of anything else in the simulation. This means that when the CDC nominations occur (i.e. when we check each individual to see which three others they like and dislike the most), the chance of a particular person j nominating another person i is independent of person k also nominating person i. This in turn means that, for any given CDC evaluation, the number of nominations an individual receives (both positive and negative) will have the same sort of distribution as the initial completely random model. This predicts that we will continue to have exactly the same distributions of categories in this new model as under the random model. We ran the simulations anyway, and found that we did, in fact, get the same distributions (for all values of w_{lr}).

Table 5: The distribution of individuals in the model.

Popular	Rejected	Neglected	Contro.	Average
12%	12%	7%	2%	67%

Category Stability Results

We can now determine how well this model predicts the category stability data. In order to do this, we first need to choose a value for w_{lr} (the learning rate which controls the speed of adaptation of an individual's 'liking' of another). Possible values range from 0 to 1. To deal with this, we can run the simulation for multiple values of w_{lr}.

We also need to determine how many iterations our model should be run for. Given the real-world data from Cillessen, Bukowski, & Haselager (2000), we know that a good model will result in asymptotic stabilities. That is, over time the stability of the Popular category should approach a value between 33% and 44%. Similar expectations exist for the other four categories.

Figure 2 shows our model's stability results for each of the five CDC categories. Each line in each graph represents a different setting for w_{lr}. The x-axis indicates the number of iterations (i.e. time) between the two category measurements. The far right gives our results after 20 time steps. Overlaid on each of these graphs is the 95% confidence interval from the real-life stability data.

Analyzing this data, we can see that the precise value of w_{lr} does not significantly affect our results. That is, we can set our learning rate parameter anywhere within its possible range and still get very similar results. Our model is thus effectively a zero-parameter model.

The other promising result is that each of the five stability measurements end up in or near the 95% Confidence Interval. This is a very encouraging finding, since no parameter tweaking was required.

Individual differences

Although our model matches the data well without introducing individual differences, it is still the case that these differences have been found to have some influence on popularity. To investigate this factor, we systematically introduced different sorts of variations in our model to investigate four well-known effects. In all of these simulations, we allowed a particular aspect of each individual in the model to vary across a normal distribution. The size of this distribution was made as large as possible while still matching to the real-life data.

To begin with, we looked at the Hostile Attribution Bias. This refers to the fact that certain individuals tend to interpret the actions of others in a more negative light than is intended. Research has shown that rejected children (particularly those who are aggressive) are more likely to assume malevolent intent when they are faced with ambiguous social cues (Crick & Dodge, 1994; Dodge, Lansford, Burks, Bates, Pettit, Fontaine, & Price, 2003). To model this we inserted this individual difference into our model by adjusting formula (2) in the following way:

$$\Delta a[i,j] = w_{lr}(b[j,i] + B[i] - a[i,j])$$

In this new formula, B is an individual interpretation bias. Note that in our simulation, agents could be biased to be either overly negative or overly positive in how they interpreted the actions of others.

We then ran the simulation over 1000 groups of 30 agents each. Each agent had a value of B chosen from a normal distribution with a deviation as large as possible while still matching the aforementioned results. After 50 simulation iterations, CDC classification was performed. We then measured the effect size. This involved determining how many standard deviations above or below the mean the values of B were within each category. This method was used for all individual differences investigated.

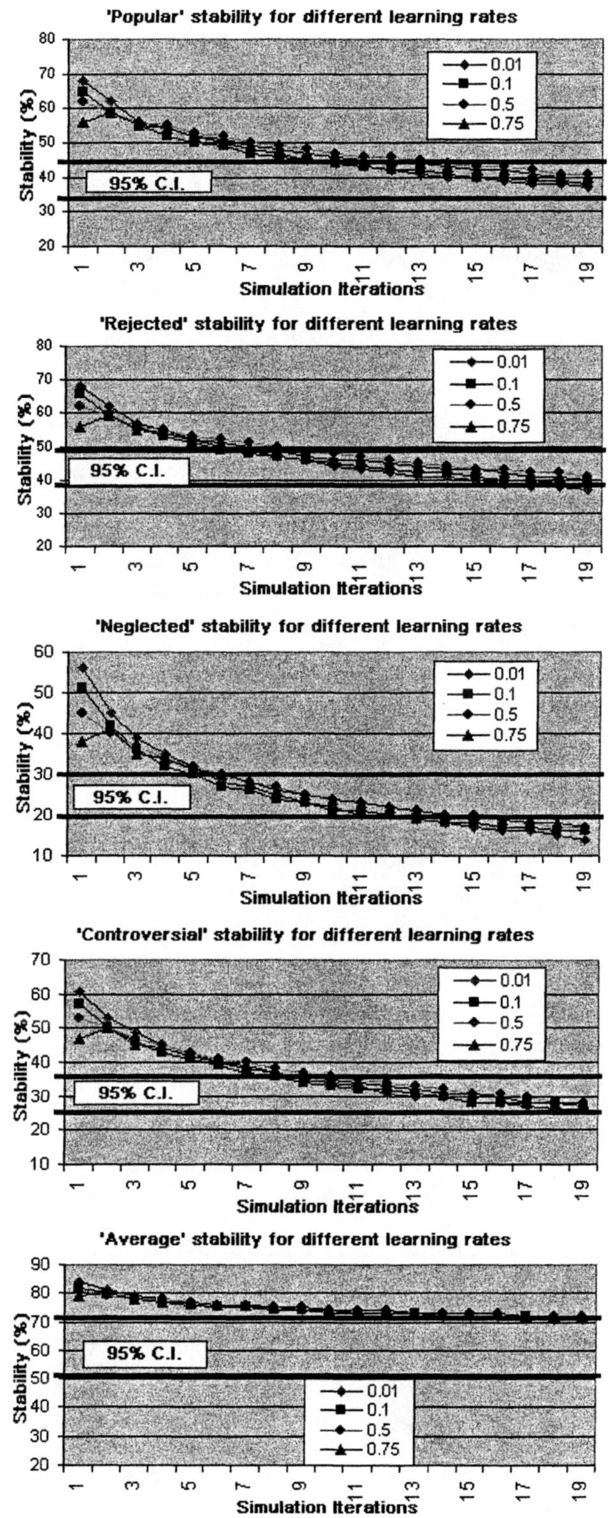

Figure 2: Simulation stability results compared to real-world data from (Cillessen, Bukowski, & Haselager, 2000), compared over a range of settings for w_{lr}.

The results in Figure 3 show that, as predicted, the Rejected agents tended to be overly negative in interpreting the behaviors of others, while the Popular agents tended to be overly positive. It was also interesting that the Neglected agents also tended to be somewhat negative.

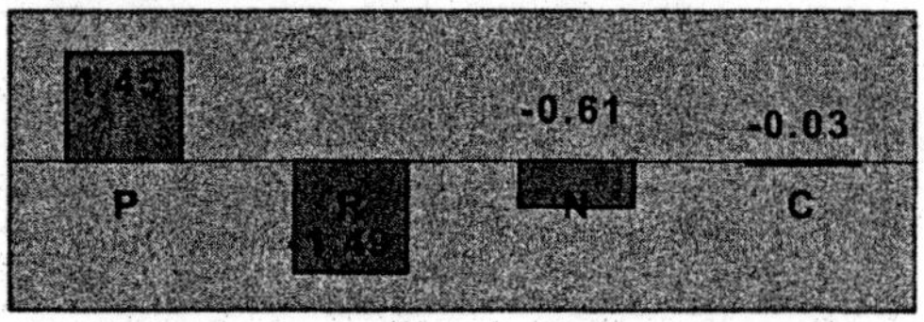

Figure 3: Effect size of varying interpretation bias (the average number of deviations a group is from the mean)

Another well-known finding is that Popular children tend to have good social skills whereas rejected children tend to have poor social skills. (e.g., Coie, Dodge, & Kupersmidt, 1990; Newcomb, Bukowski, & Pattee, 1993; Parkhurst & Hopmeyer, 1998). To reflect this in the model we biased the mean of the Gaussian distribution used in Formula 1. Specifically, a positive value created a bias toward behaving nicely and a negative value created a bias toward behaving badly. Figure 4 displays the results and shows that the manipulation had the intended effect. Interestingly, the neglected agents were again shown to be somewhat similar to the rejected agents.

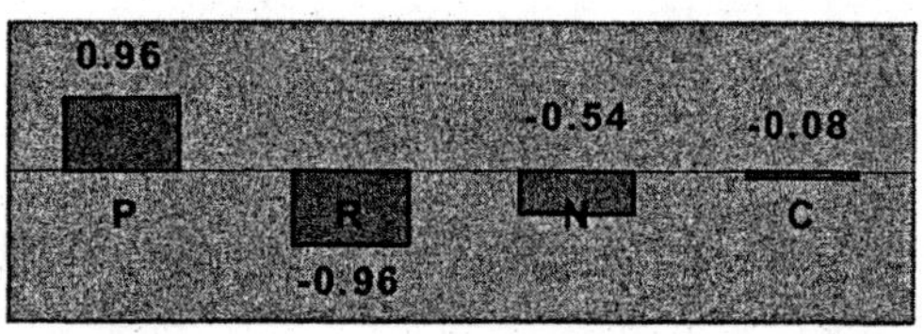

Figure 4: Effect size of varying behavior bias (the average number of deviations a group is from the mean)

The third effect we investigated was that neglected children have been shown to interact with their peers less frequently than average children (Dodge, Coie, & Brakke, 1982; Coie & Dodge, 1988). To reflect this in our model, we added an interaction probability for each agent. The percentage chance for two agents interacting was determined by multiplying their interaction probabilities together. As illustrated in Figure 5, this manipulation was successful in capturing the effect. However, it also produced the unexpected effect that a high level of interaction was associated with being controversial.

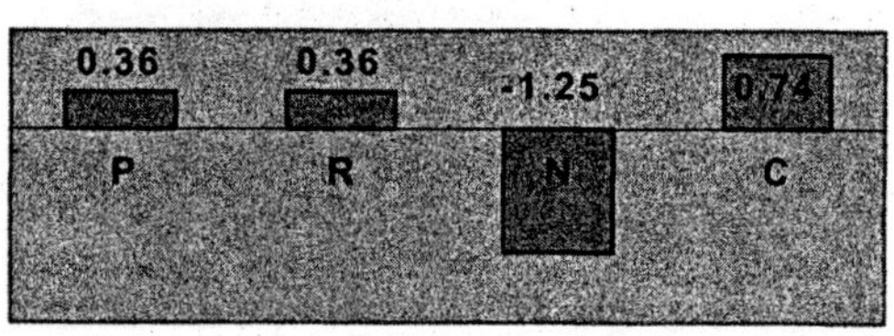

Figure 5: Effect size of varying interaction probability (the average number of deviations a group is from the mean)

Finally, we examined the finding that controversial children tend to display a combination of positive and negative social behaviors (e.g., Coie & Dodge, 1988). We modeled this lack of consistency by varying the standard deviation of the Gaussian distribution in Formula 1. The results, displayed in Figure 6 supported our interpretation that Controversials tend to be highly variable in their behavior. The results also revealed an unexpected effect in which neglected individuals tended to be more reliable in their behavior (lower variability).

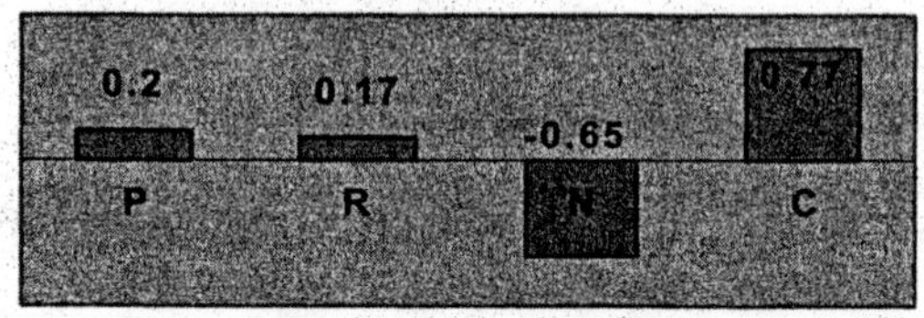

Figure 6: Effect size of varying behavior deviation (the average number of deviations a group is from the mean)

Insights and Predictions

The model also gave us the opportunity to explore other possible individual differences. The first that we looked at was the initial value for a[i,j]. For this simulation some agents started off more predisposed to liking everyone (a high value of a[i,j] for all j), and others more predisposed to disliking everyone (a low value of a[i,j] for all j). This was meant to represent the effect of previous experience, before entering the group (e.g., family experiences or other peer group experiences). The results of this simulation are displayed in Figure 7. Note that the results of this simulation were very similar to the results of the hostile attribution simulation and the social skills simulation. All of these showed an expected association between strong positive biases and being popular, and strong negative biases and being rejected. But they also all showed a somewhat unexpected association between moderate negative biases and being neglected. Taken together, these results show that the model is very robust in producing this pattern of results in response to factors that bias an agent to be more positive or more negative.

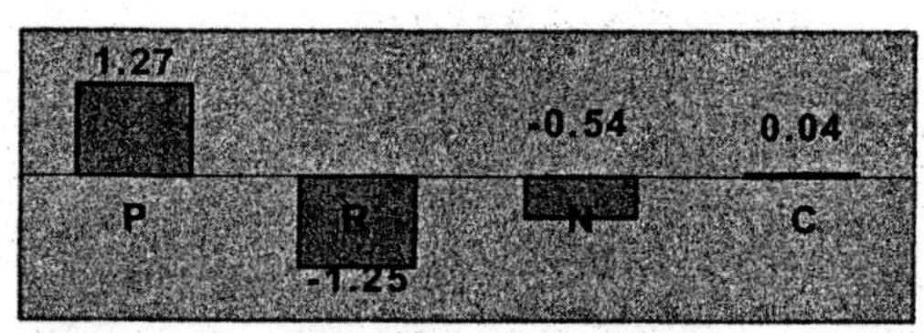

Figure 7: Effect size of varying the initial value of a[i,j] (the average number of deviations a group is from the mean)

We also analyzed the effect of individual differences on w_{lr}. That is, how does having a different learning rate affect one's eventual CDC categorization. The results, displayed in Figure 8, were interesting, in that they were very similar to the results of the interaction probability simulation and the behavior consistency simulation. The most straightforward interpretation of this is that having a high learning rate or having many interactions can have the effect of making one appear to be more variable in behavior. This result is also important since we earlier (see Figure 2) showed that varying the overall learning rate among all the individuals had no effect.

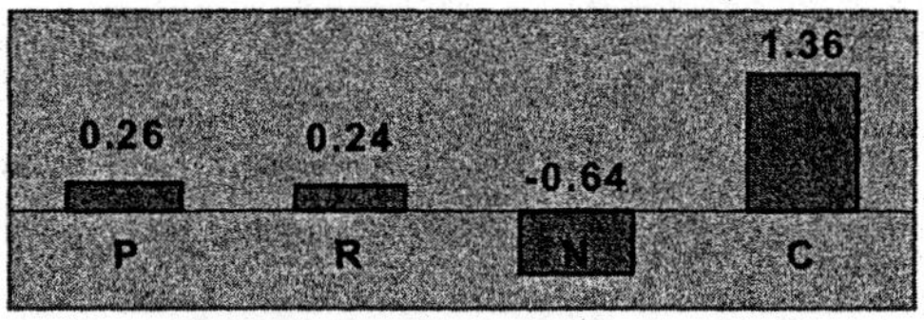

Figure 3: Classification effect size of varying w_{lr} (the average number of deviations a group is from the mean)

Conclusions

This model has presented us with a number of intriguing results. We have shown that the distribution of individuals into the five CDC categories is an artifact of the measurement system itself. The model we have presented predicts both the distribution data and the long-term stability data of these categories very well (the model predicts Neglected individuals have a stability which is 5% lower than the real-life data, but for all other categories it is within the 95% confidence interval). Our model can do this without recourse to any individual differences. Instead, these categorizations arise from the dynamics of the interpersonal interactions. These results hold over all settings for the model's one parameter, which means it is effectively a zero-parameter model.

When we introduced individual differences into the model, it predicted a number of standard effects of such differences in real children. Furthermore, these effects indicated underlying similarities in the processes involved. For example, we have consistently seen that changes which correlate with an individual being Rejected tend to also have a smaller effect associated with being Neglected. Also, we showed a striking similarity between the results of varying the learning rate, the behavioral variability, and the interaction probability. These patterns suggest new ways of looking at current research in popularity.

References

Bouchaud, J-P. & Mézard, M. (2000). Wealth condensation in a simple model of economy, *Physica A*, *282*, 536.

Cassidy, J., Scolton, K.L., Kirsh, S.J., & Parke, R.D. (1996). Attachment and representations of peer relationships. *Developmental Psychology, 32*, 892-904.

Cillessen, A., Bukowski, W., & Haselager, G. (2000). Stability of Sociometric Categories. *New Directions for Child and Adolescent Development, 88*, 75-93.

Coie, J.D. & Dodge, K.A., (1988). Multiple sources of data on social behavior and social status. *Child Development, 59*, 815-829.

Coie, J.D., Dodge, K.A., & Kupersmidt, J.B. (1990). Peer group behavior and social status. In S.R. Asher & J.D. Coie (Eds.), *Peer rejection in childhood.* New York: Cambridge University Press.

Cohn, D.A. (1990). Child-mother attachment of six-year-olds and social competence at school. *Child Development, 61*, 152-162.

Crick, N. R., & Dodge, K. A. (1994). A review and reformulation of social information-processing mechanisms in children's social adjustment. *Psychological Bulletin, 115*, 74-101.

Dodge, K.A., Coie, J.D., & Brakke, N.P. (1982). Behavioral patterns of socially rejected and neglected preadolescents: The roles of social approach and aggression. *Journal of Abnormal Child Psychology, 10*, 389-409.

Dodge, K. A., Lansford, J., Burks, V., Bates, J.E., Pettit, G., Fontaine, R., & Price J.M. (2003). Peer rejection and social information-processing factors in the development of aggressive behavior problems in children. *Child Development, 74*, 374-393

Elicker, J., Englund, M., & Sroufe, L.A. (1992). Predicting peer competence and peer relationships in childhood from early parent-child relationships. In R. Parke & G. Ladd (Eds.), *Family-peer relationships: Models of linkage.* Hillsdale, NJ: Erlbaum.

Granovetter, M. (1978). Threshold models of collective behavior. *American Journal of Sociology. 83:6*, 1420-1443.

Helbing, D., Farkas, I., & Vicsek, T. (2000). Simulating Dynamical Features of Escape Panic. *Nature 407*, 487-490.

Newcomb, A.F., & Bukowski, W.M., & Pattee, L. (1993). Children's peer relations: A meta-analyic review of popular, rejected, neglected, controversial, and average sociometric status. *Psychological Bulletin, 113*, 99-128.

Parkhurst, J.T., & Hopmeyer, A. (1998). Sociometric popularity and peer-perceived popularity: Two distinct dimensions of peer status. *Journal of Early Adolescence, 18*, 125-144.

Rubin, K.H., Hymel, S., & Mills, R.S. (1989). Sociability and social withdrawal in childhood: Stability and outcomes. *Journal of Personality, 57*, 237-255.

Sun, R. (2001). Cognitive science meets multi-agent systems: a prolegomenon. *Philosophical Psychology, 14:1*, 5-28.

Watts, D.J. & Strogatz, S.H. (1998). Collective dynamics of 'small-world' networks. *Nature, 393*, 440-442.

Time Perception: Beyond Simple Interval Estimation

Niels Taatgen[12] (taatgen@cmu.edu), Hedderik van Rijn[12] (hedderik@ai.rug.nl) and John Anderson[1] (ja@cmu.edu)

[1]Carnegie Mellon University, Department of Psychology, 5000 Forbes Av.
Pittsburgh, PA 15213 USA

[2]University of Groningen, Department of Artificial Intelligence, Grote Kruisstraat 2/1
NL-9712 TS Groningen, Netherlands

Abstract

Reasoning about and with time has never been studied extensively in the context of cognitive architectures. We present a temporal reasoning system that can make accurate predictions for two classic interval estimation experiments by Rakitin et al. (1998) and Penney et al. (2000). The system is implemented as an additional module for ACT-R. In combination with ACT-R's other mechanisms, it makes accurate predictions about experiments that go beyond just estimating time. We demonstrate this by modeling data from a choice-reaction task by Grosjean et al. (2001).

Temporal Reasoning

Despite the fact that most cognitive architectures make predictions about time, they do not support models that reason about time itself. People on the other hand usually have a good sense of the passage of time, and use this capability, often implicitly, to handle time intervals in their reasoning. The classical example is approaching a traffic light that turns from green to yellow. The decision drivers are faced with is whether to brake or drive on, which depends on their (earlier established) sense of time about when the light will turn red. A sense of time may also be necessary in the coordination of multi-tasking. For example, when driving a car and keying a number on a cell-phone at the same time, it is necessary to keep track of elapsed time between consecutive checks of the road condition. In interaction with computers, it is not uncommon that some time elapses between the user's input and the computer's response, especially when it concerns the internet. It is important for the user to be able to assess whether the waiting time is reasonable, because an interval that is too long might indicate a problem (either in the software or the user's understanding of the interface).

When one wants to model the estimation of time intervals in a cognitive architecture, the important question to ask is whether or not interval timing is an architectural feature or something that can be modeled with an appropriate set of knowledge. The only way to keep track of time in the context of the ACT-R architecture[1] (Anderson & Lebiere, 1998) is explicit counting, making it almost impossible to account for tasks in which an implicit sense of time partly determines behavior. Even more, experimental results indicate that humans can assess time intervals when explicit counting is discouraged or forbidden. There are also biological and neuropsychological reasons to favor an architectural solution. It is well-known from the behaviorist literature that animals can learn time intervals. For example, if rats or pigeons receive a reward when they press a bar, they quickly learn to anticipate this. However, when this reward is only given every 30 seconds, the rats will increase their bar-pressing when the 30 second deadline approaches, clearly showing a sense of the duration of the time interval (Gibbon, 1977).

More recent neuropsychological research suggests that there are particular areas in the brain related to the estimation of time intervals. Based on several studies, Matell and Meck (2000) have constructed a model of interval timing, which we summarize in Figure 1a. The general idea is that certain stimuli can synchronize neurons in a certain region of the cortex, effectively acting as a starting sign. As each of the neurons produces its own particular pattern of activation in time, each moment in time is associated with an unique pattern of activation. These patterns are recognized by striatal spiny neurons, whose activations are then integrated by certain basal ganglia output nuclei (globus pallidus, entopeduncular nucleus and substantia nigra reticulata), and then relayed to the thalamus for behavioral expression.

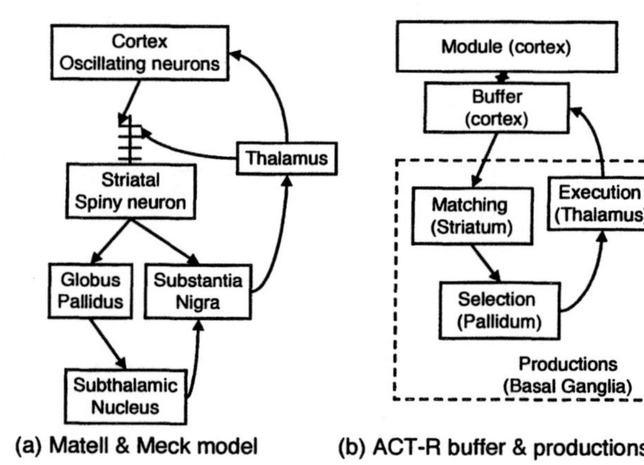

Figure 1. Matell and Meck (2000) model and the general ACT-R production matching schema

The neural mechanisms proposed by Matell and Meck show a close resemblance to the mappings from ACT-R onto the brain, shown in Figure 1b. ACT-R assumes several cortical buffers corresponding to the goal, retrieval,

[1] Note that time is used for utility calculations. However, ACT-R has no direct access to subsymbolic quantities.

perceptual and motor systems. The contents of these buffers are matched in the striatum, after which one production rule is selected in the pallidum. The thalamus then takes care of the execution of the production rule, which can for example reset the internal time. We therefore propose to model time estimation by adding an extra module to ACT-R that implements the perception of time. Mapping the M&M model onto ACT-R produces a model in which most systems correspond to general architectural areas, with the exception of the special-purpose temporal buffer and module.

The Temporal Module

The general idea, based on Matell and Beck (2000), is that an internal timer can be started explicitly to time the interval between two events. Matell and Beck hypothesized that a start event resets the cortical neurons. Given that different neurons have different associated firing patterns, each point in time is characterized by a different subset of firing neurons. Therefore, at any given point in time, the state of the neurons is a representation of the time passed since the resetting of the clock. Since some neurons have a greater periodicity than others, similar durations will be associated with similar patterns. As coarse time-grained neurons are the main determiners of the time elapsed for longer intervals, timing becomes less precise for longer durations.

To approximate this process in the simulation we represent the current neural pattern by an integer. A reset event sets this integer to zero, after which it is increased as time progresses. The idea is that the temporal module acts like a metronome, but one that starts ticking slower and slower as time progresses. The interval estimate is based on the number of ticks the metronome has produced. More precisely, the duration of the first tick is set to some start value:

$$t_0 = \text{starttick}$$

Each tick is separated from the previous tick by an interval that is a times the interval between the previous two ticks. Each interval has some noise drawn from a logistic distribution added to it. The distribution of this noise is determined by the current tick-length.

$$t_{n+1} = at_n + \text{noise}(mean = 0, sd = b \cdot at_n)$$

We have estimated values for the parameters in these equations on the basis of an optimal fit to the first experiment, interval estimation. There we found 11 ms for start tick, 1.1 for a, and 0.015 for b. These values also provided excellent fits to the other experiments discussed in this paper.

The starting of the timer is triggered by a production action. This timer generates incremental timestamps that virtually represent firing neurons. The current state of the timer can be matched in the condition of a production. This match can take two forms. The first form is reading out the current value of the timer. This occurs when the variable being matched is not yet bound to a value. In that case the variable is instantiated to contain the current value of the timer. If we store this time pattern in the goal or in declarative memory, it can be saved for future use. This future use is also the second form of using the patterns: by matching the temporal buffer to a specific value. The match only succeeds if the buffer (approximately) has the value we try to match, thereby making it possible to have production rules "wait" until their time to fire.

Example: Estimate and Reproduce a Time Interval

Suppose we want to reproduce a time interval, as represented in the first horizontal bar in Figure 2, that is defined as the time between the start of a trial and the moment a light comes on.

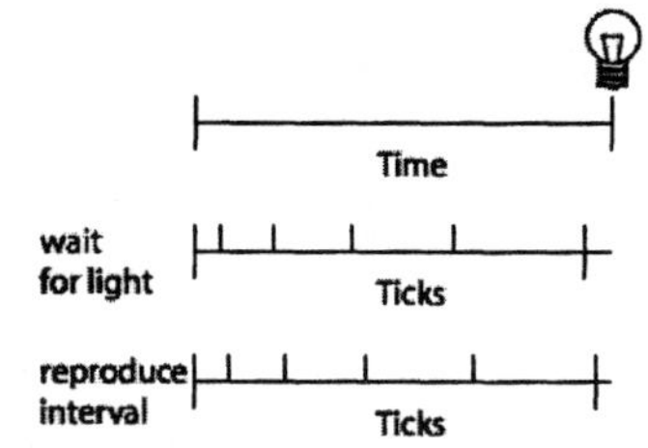

Figure 2. Illustration of the temporal module

A first production will start the timer:

(p start-learn =goal> isa reproduce-interval state begin	IF the goal is to reproduce an interval
==> +temporal> isa time =goal> state wait)	THEN initiate the time module AND wait

The "+temporal>" on the right hand side initiates the time module by resetting the internal timer. After the timer has been reset, it starts recording the increasing ticks as shown in the second bar of Figure 2. As soon as the light comes on, a second production rule fires and reads out the current tick value:

(p respond-to-light =goal> isa reproduce-interval state wait =visual-location> isa visual-location =temporal> isa time ticker =time	IF the goal is to reproduce a pattern and we are waiting AND a visual stimulus has been found AND the current time is *time*
==> =goal> time =time state wait-again +temporal> isa time)	THEN store *time* in the goal AND restart the timer

As the temporal module's state is in the interval between tick 5 and 6, the value read from the buffer will be 5. Now that we have a pattern, a new production rule can initiate the key-press after approximately the same amount of time has elapsed. Note that in the above production rule, the time module is reset again, so a new series of ticks, represented in the bottom bar of Figure 2, is initiated. As soon as the temporal state resembles the stored value, in the example in Figure 2 slightly later due to noise, the production rule below initiates a key press.

(p respond-to-predicted-light =goal> isa reproduce-interval state wait-again time =time =temporal> isa time ticker =time ==> +manual> isa press-key =goal> state done)	IF the goal is to reproduce an interval and we are waiting for interval *time* to pass AND the temporal buffer has reached value *time* THEN press the key

The estimation process here seems to under predict the interval because the real time is rounded down to the lower tick (integer) value. In practice this effect is much smaller than the effect of the noise.

Model of Interval Estimation

In interval estimation experiments, participants are exposed to a certain time interval a number of times, and are then asked to reproduce it. We modeled Experiment 3 from Rakitin et al. (1998). In this experiment participants were first trained on a certain time interval (8, 12 or 21s). Training consisted of 10 trials in which a blue rectangle appeared on the screen, which changed to magenta when the time interval had elapsed. In the 80 test trials they had to predict the interval by pressing a key when they expected the rectangle to change color. In 25% of the test trials the rectangle changed color when the interval had elapsed. The results are based on the remaining 75%, in which the rectangle stayed blue. Participants were forbidden to count.

Figure 3-5 show the scaled density plots of the responses (triangles). The peak of the distribution is at the appropriate time, the variance grows with the length of the interval, and the distribution is slightly skewed. The distributions satisfy the so-called scalar property of time estimate, in that the variance in the estimates grows linearly with the duration of the interval.

The model of this experiment closely resembles the example model outlined in the previous section. The learning phase is used to estimate the number of ticks in the interval (the model takes the average of the ten presentations), and during the testing phase the model waits until the appropriate number of ticks has passed. We estimated the three model parameters (multiplication factor, initial tick and noise) to obtain the best (least-squares) fit (shown by the solid lines).

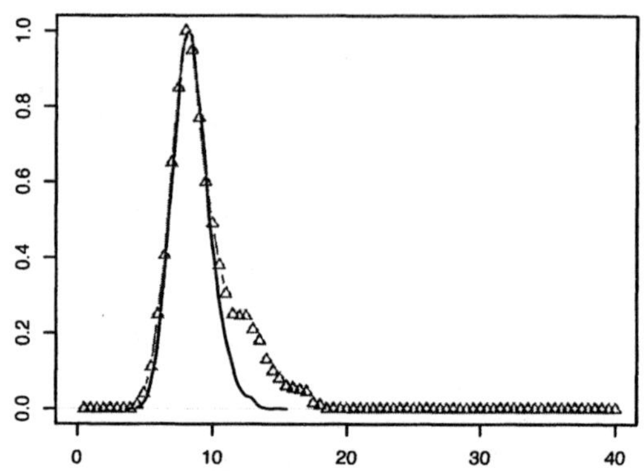

Figure 3. Data (triangles) and model (line) for 8 seconds

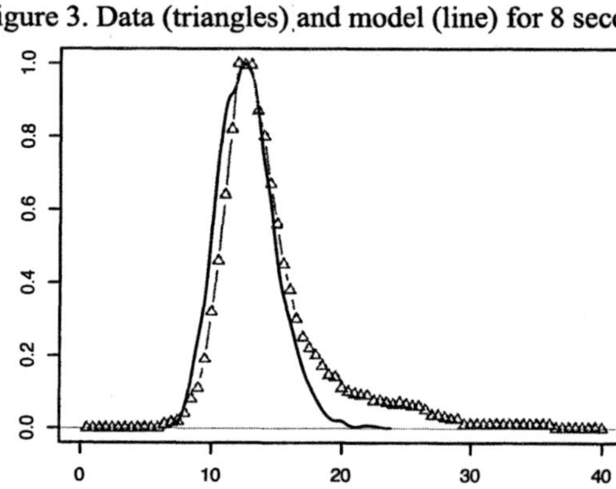

Figure 4. Data (triangles) and model (line) for 12 seconds

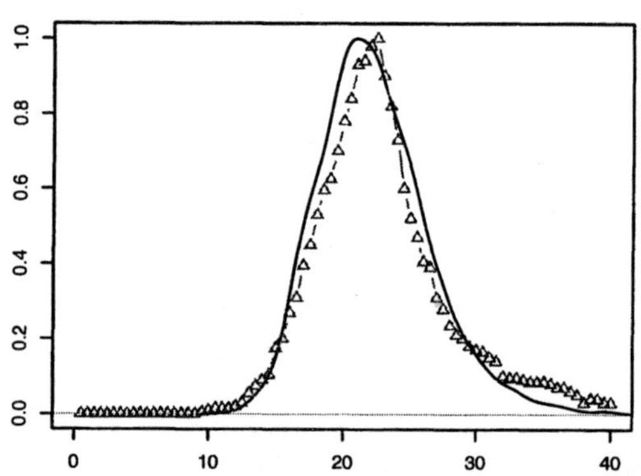

Figure 5. Data (triangles) and model (line) for 21 seconds

The fit between model and data is overall very good. The only aspect of the data that the model does not predict is the tail of the distribution for the 8 and 12 second conditions. In Experiment 1 of Rakitin et al. (1998), which is similar to Experiment 3, the tails of the distributions are much shorter. We therefore decided to leave the temporal mechanisms as simple as possible and to let further experience with the temporal buffer determine if an extension is necessary.

Model of Bisection Experiments

Another paradigm in time perception are so-called bisection experiments. In these experiments, participants are first trained on two time intervals, one short interval and one long interval. After this learning phase, they are exposed to new time intervals that are either equal to the short or the long interval, or somewhere in-between. Participants are then asked to judge whether the presented interval is closer to the short or to the long interval. We will model Experiment 2 from Penney, Gibbon and Meck (2000). In this experiment, three short-long pairs of intervals were used: 3 and 6 seconds, 2 and 8 seconds, and 4 and 12 seconds. In the training phase 10 tones of either the short or the long duration were presented to the participant. After that participants were tested for 100 trials, 30% of which were anchor point intervals (short or long), and 70% were tones of intermediate duration.

The model uses the training phase to determine the timing for both the short and the long interval. During testing, it times the presented interval, and then compares the value to both anchor intervals. If the value is closer to that of the short interval, it chooses that, if it is closer to the long interval, it decides that it is long. The parameters for the model are identical to that of the earlier fitted interval estimation experiment, that is, they have not been estimated anew for these data.

Figures 6-8 show the results of the experiment and the model. A property of both participants' behavior and the model's predictions is the fact that the interval that is judged long 50% of the times is shorter than the mean of the short and long interval. For example, in the 2-8 second version of the task (Figure 7), the 50% point is at 4 seconds instead of 5. The model explains this by the fact that its "ticks" increase in duration: there are approximately the same number of ticks between 2 and 4 seconds as between 4 and 8 seconds.

Role of timing in choice-reaction tasks

The previous two models show the validity of the temporal module, but they do not go beyond estimation and judgment of time intervals. To show that the temporal module has more than marginal value for an integrated cognitive architecture, we will now examine an experiment in which timing is important, but where it is not the focus of the experiment. It concerns a choice-reaction time experiment by Grosjean, Rosenbaum and Elsinger (2001, Experiment 1).

The experiment itself is straight-forward: on each trial a "+"-like figure appeared on the screen, of which the vertical line was always in the same place, but the horizontal either had two-thirds of its length to the left or to the right of the vertical line. Participants were instructed to press the "d" key if the horizontal line would be more towards the left, or the "k" key when it would be more towards the right.

The manipulation in this experiment was the inter-stimulus interval (ISI), here defined as the elapsed time between the key press of the previous trial and the

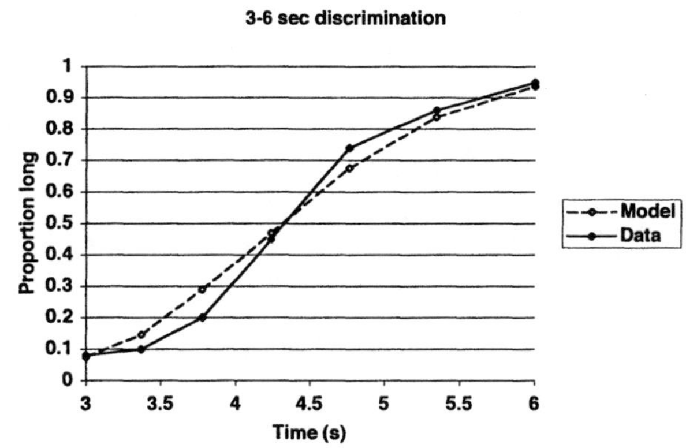

Figure 6. Proportion of "long" responses for intervals between 3 and 6 seconds

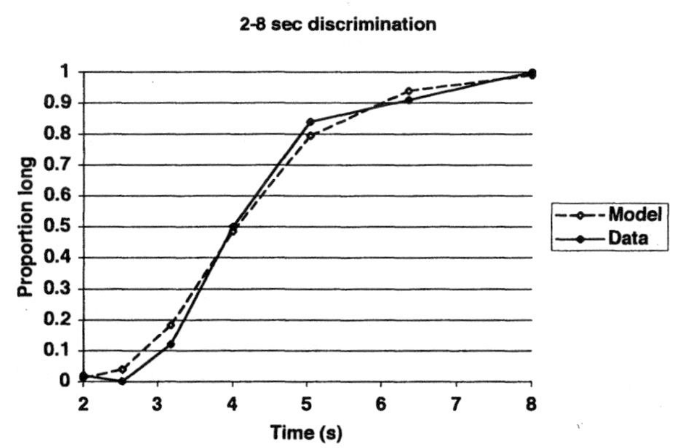

Figure 7. Proportion of "long" responses for intervals between 2 and 8 seconds

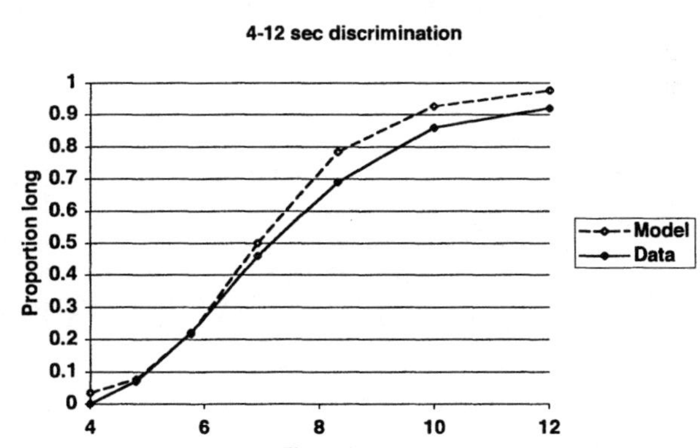

Figure 8. Proportion of "long" responses for intervals between 4 and 12 seconds

presentation of the next trial. Each block consisted of 16 trials. This ISI was kept constant for the first 15 trials, but was changed on trial 16. More specifically, there were three conditions: **Shortened:** the ISI on trial 1-15 was 700 ms, while the ISI on trial 16 was 467 ms, **Control**: the ISI in all the trials was 467 ms, and **Lengthened**: the ISI on trial 1-15 was 350 ms, while the ISI on trial 16 was 467 ms. Figure 9 shows the reaction times for these conditions per trial.

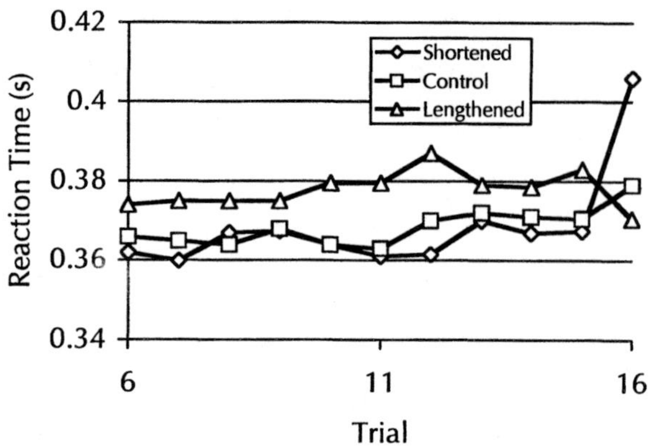

Figure 9. Results of the Grosjean et al. (2001) experiment

The strongest effect in the results is the increased reaction time on trial 16 in the shortened condition. Apparently, there is an advantage of knowing when the stimulus will arrive, and when it arrives early this advantage is cancelled. In our model we assume that the advantage of knowing when and where the stimulus will arrive allows one to start attending the stimulus at just the right time. In ACT-R, attending a stimulus normally takes two steps. First, it is detected that there is something on the screen (a visual-location). Then a production rule has to fire to direct attention to this location. Firing a production rule takes 50 ms, so this is the amount of time we can save if we can direct attention to the right location at the right time. Let us examine the model in more detail.

First, a rule starts the timer, and waits for the stimulus:

(p start-timer =goal> isa crt status start ==> +temporal> isa time =goal> status waiting)	IF the goal is a choice-reaction-task THEN start the timer and wait

When a stimulus appears on the screen, the model starts attending it, but at the same time it stores the time and location on which it appears in the goal:

(p found-visual-location =goal> isa crt status waiting	IF the goal is a choice-reaction-task and we are waiting
=visual-location> isa visual-location =temporal> isa time ticks =time ==> +visual> isa visual-object screen-pos =visual-location =goal> time =time loc =visual-location status wait-visual	AND there is a visual stimulus AND time *time* has elapsed THEN attend the visual stimulus AND store the time and the location of the visual stimulus in the goal

Other rules then handle the visual stimulus and make the appropriate key presses. On the next trial, the model uses the stored time and location to estimate when the stimulus will appear, and already starts the attention process:

(p expect-visual-location =goal> isa crt status waiting time =time loc =loc =temporal> isa time ticks =time ==> +visual> isa visual-object screen-pos =loc =goal> status wait-visual)	IF the goal is a choice-reaction-task and we are waiting, and the previous stimulus came up at time *time* and location *loc* AND time *time* has elapsed THEN attend location *loc*

This rule can fire before the stimulus comes up, and can therefore, in principle, save up to 50 ms. Let us examine in some more detail how the model manages to do this. If the *expect-visual-location* rule is to be of any benefit, it has to fire before *found-visual-location*. If the time estimate would be completely accurate, this would never happen: both rules would have their conditions satisfied at the same time. However, timing is inaccurate. Suppose the value in the timing buffer is 20 after the first stimulus presentation. For the next trial, the *expect-visual-location* can fire when the temporal buffer becomes 20. If this happens before the stimulus comes up (because the timer is a bit faster than the previous time), then *expect-visual-location* will fire, and will produce some benefit on the reaction time. However, if the timer is slow on this trial, *found-visual-location* will fire, because the stimulus appears before the timer runs out. In that case, the value of the timer is lower at the moment the stimulus appears, for example 19. The *found-visual-location* rule stores this new value in the goal. Therefore there is a downward pressure on the value of the time estimate until this value consistently causes *expect-visual-location* to fire before *found-visual-location*.

For the model results we have used the same temporal module parameters as in the previous models, and used the default ACT-R parameters for all other modules. Figure 10 shows the results of the model. There is a 25 ms difference between all the model's predictions and the data. Although we could have easily mended this by lowering the visual-attention parameter from 85 ms to 60 ms, defending this by the fact that this 85 ms normally also includes a here unnecessary eye-movement, we chose to leave the results as they first came out of the model. The model reproduces the main effect in the data, that is, the slower reaction time on trial 16 when the ISI is shortened, but also shows two other effects present in the data. A first effect is that in the lengthened condition the reaction time on trial 16 is slightly faster than the reaction time for the control condition. This can be explained by the fact that if the predicted time on trial 16 is shorter than the ISI, the model will always save the full 50 ms, while in the control condition it can be less.

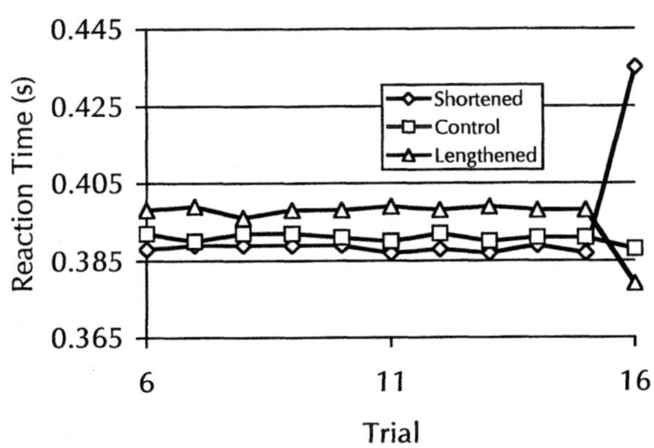

Figure 10. Results of the model for Grosjean et al. (2001)

A second effect that the model produced was the small difference in reaction times on trials 6 to 15. It appears that the longer the ISI, the shorter the reaction time, an effect normally attributed to the need to prepare for the next trial. As the model needs no preparation, but also shows the effect, this model offers a different explanation for this stereotypical effect: with the shorter ISI's, the model is more accurate in estimating the interval. However, in this case being accurate is counter-productive. Inaccuracies in the time estimate will cause the model to lower its value of the time estimate. The lower this value, the higher the gain in reaction time (up to 50 ms).

Discussion

The temporal module we propose is a simplification of the neural model that Matell and Meck (2000) outline. Instead of complex neural patterns ordinary integers are used. Nevertheless it matches the behavior shown in experiments. The integers in the temporal buffer can be considered to act isomorphically to the more complex patterns in a neural implementation. It is not clear to what extend the neural representation allows all the operations that can be done on integers (comparison, subtraction, etc.). We therefore are careful in specifying what can and cannot be done with the values in the temporal buffer, currently limiting the actions to simple comparisons.

The temporal buffer is not the answer to all issues regarding time. It can be used for intervals up to a minute, but certainly not for longer intervals. Its performance can however be augmented by the appropriate cognitive strategies. For example, by allowing a model to count, it can become much more precise in its estimations, mirroring human performance.

The model of the Grosjean et al. data produces some surprising explanations, and poses some new questions. One assumption is that it is no problem to attempt to attend a location early. Why don't people do that all the time? Maybe the attention process is effortful, and people try to avoid expending too much effort. Another explanation might be that by already attending a certain location, other new information may be missed. This is irrelevant in this particular experiment, but may be important in general. Indeed, ACT-R will normally automatically attend new stimuli, provided that the visual system is available. This means that timing the onset of a stimulus is a good strategy because other stimuli can be attended to in the period before the stimulus, and visual attention can be directed to the right location at exactly the right time.

Acknowledgements

This research is supported by NASA grant NCC2-1226 and ONR grant N00014-96-01491. N.T. would like to thank Stefani Nellen for her comments on an earlier draft.

References

Anderson, J.R. & Lebiere, C. (1998). *The Atomic Components of Thought*, Mahwah, NJ: Erlbaum.

Gibbon, J. (1977). Scalar expectancy theory and Weber's Law in animal timing. *Psychological Review, 84*, 279-335.

Grosjean, M., Rosenbaum, D.A. & Elsinger, C. (2001). Timing and Reaction Time. *Journal of Experimental Psychology: General, 2*, 256-272.

Matell, M.S. & Meck, W.H. (2000). Neuropsychological mechanisms of interval timing behavior. *BioEssays, 22*, 94-103.

Penney, T.B., Gibbon, J., & Meck, W.H. (2000). Differential effects of auditory and visual signals on clock speed and temporal memory. *Journal of Experimental Psychology: Human Perception and Performance, 26*, 1770-1787.

Rakitin, B.C., Gibbon, J., Penney, T.B., Malapani, C., Hinton, S.C. and Meck, W.H. (1998). Scalar Expectancy Theory and Peak-Interval Timing in Humans. *Journal of Experimental Psychology: Animal Behavior Processes, 24*, 15-33.

Categorical Data Displays Generated from Three Architectures Illustrate Their Behavior

Kevin Tor (tor@cse.psu.edu)
Steven R. Haynes (shaynes@ist.psu.edu)
Frank E. Ritter (frank.ritter@psu.edu)
School of Information Sciences and Technology, The Pennsylvania State University
University Park, PA 16802

Mark A. Cohen (mcohen@lhup.edu)
Business Administration, Computer Science, and Information Technology, Lock Haven University
Lock Haven, PA 17745

Abstract

The following introduces CaDaDis, an extension to the architecture-independent VISTA toolkit. CaDaDis creates visible accounts of cognitive model behavior through a set of Categorical Data Displays. It includes a standard Pert Chart showing tasks by category (or resource), a Nonstandard Pert Chart that shows the temporal dependencies, and a Gantt chart that helps show occurrences of agent events along a time line. Perhaps most usefully, it can display categorical and numeric data generated by models as they run. Structured messages are sent on behalf of the cognitive model by several example script-based interfaces. This paper presents a series of examples of CaDaDis with three different cognitive architectures. The displays, for example, show that the sample models included in the Soar distribution have different structures.

Introduction

Textual traces from cognitive models are often considered unhelpful by observers trying to understand them, and these traces are essentially unintelligible to non-programmers. Modelers and subject matter experts want to know the structure of models as well as how they work (Avraamides & Ritter, 2002; Councill, Haynes, & Ritter, 2003). One approach that has been relatively well received is to provide a graphic representation of a model's internal processing and behavior. Where this has been done, observers have at least thought they understood the models more, and in some cases have seen and learned new things about their models (Ritter, Jones, & Baxter, 1998; Ritter & Larkin, 1994).

We present a general tool for displaying categorical data generated by cognitive models. It can be and has been used by multiple cognitive architectures to display their internal processing. It is based on a no-cost toolkit, implemented in a widely used programming language, and should help many models to be understood. The display tool is designed to support the reuse that Newell (1990) referred to in his *Unified theories of cognition*, in this case by helping models be understood and by being used itself in multiple applications.

It appears to take an expert programmer about an afternoon to start to use CaDaDis with an architecture that they know, and about an hour for a typical graduate student to use it with an new model within an existing supported architecture. In the first case, the user must adapt the architecture to pass information that they want displayed to CaDaDis through a socket. For ACT-R, for example, this meant creating a socket and modifying an existing function in the architecture to pass information when a rule is fired. In the second case, the user has to download CaDaDis, and load into the ACT-R Tcl interface a file, and load a file into the ACT-R Lisp system. In each case, a user has to get permission to use Vista and download it through a web site.

This is an update on a tool that we have previously introduced to a different community (Tor, Ritter, Haynes, & Cohen, 2004). We report here several new features and new analyses and this is directed at cognitive modelers, not the simulation and synthetic forces community.

We start by describing several displays that have inspired us and provide lessons for our design. We provide example displays created quickly to work with Soar, ACT-R, and the Jess agent architecture. These displays helped us understand the models we did not write ourselves, and show the types of knowledge that can be gleaned from models using categorical displays of their internal processing. We hope that readers are inspired to create such displays for their models, and that they use our system, CaDaDis.

Previous Systems

A wide range of graphical displays have been used by cognitive modelers. Not every model and not every cognitive architecture has had one, but the displays that have been available appear to help explain models. The following is a brief summary of the previous work in graphical displays for cognitive models. A slightly more comprehensive review of these systems is available (Tor, et. al., 2004).

Gray, John, and Atwood (1993) implemented, by hand, some task analysis for a new and old telephone workstation in CPM-GOMS. This study gave us the groundwork for our nonstandard Pert Chart, where resources appear on the left of the display. The main problem with this was the lack of automation.

Automation was introduced with APEX (John, Vera, Matessa, Freed, & Remington, 2002). Pert charts were created with a CPM-GOMS approach with such success that it was later added to the ACT-R architecture.

The Developmental Soar Interface (DSI) and the Tcl/Tk Soar Interface (TSI) provide graphical traces of running Soar models automatically (Ritter, Jones, & Baxter, 1998; Ritter & Larkin, 1994). The DSI showed that cyclical behavior can be found in some models and the TSI provided multiple views of model's behavior.

The connectionist model community initially had rather poor interfaces that would print out all of the nodes each epoch (e.g., the early PDP toolkits). But the users could at least see the processing of their model. Later versions showed it helps to be able to visualize model behavior.

Among the desired functionality motivating CaDaDis development are searching for cyclical model behavior and providing multiple views, two tasks shown to be important in previous work on cognitive model data displays and assumed to be important for any cognitive architecture. These visualizations will aid novice to experienced programmers, and perhaps prove useful for even non-programming model users.

CaDaDis

We have created a general tool for displaying categorical data generated by cognitive models. The system is implemented in Java and VISTA (the latter we explain below). We provide a general rational for the design first, and then explain in order of increasing complexity the displays that we and others have created.

The Use of Java and VISTA

We chose to create these displays in Java because it is portable and increasingly used to create displays for cognitive models (e.g., ACT-R's new interface, and JACK's eye and hand: Norling & Ritter, 2001). We used the Visualization Toolkit for Agents toolkit (VISTA) (Taylor, Jones, Goldstein, & Frederiksen, 2002) because it supports creating displays and a variety of communication channels between the display and cognitive models. VISTA is supported and available under a non-commercial license from Soar Technology. As part of this project we also created and used a VISTA tutorial web site (acs.ist.psu.edu/vista) to learn VISTA.

VISTA facilitates the creation of model visualization applications. This toolkit provides an infrastructure for communication between models and VISTA enabled applications. Using this communication channel, models can convey changes in their internal attributes to a listening VISTA enabled application, which then updates its display to reflect these changes. The VISTA toolkit also provides the ability to record and playback model activity (Soar Technologies, Inc., 2002). Importantly VISTA includes useful Java objects and methods for creating displays of cognitive models (including the ability to parse and deparse objects communicated), as well as a series of examples that can be modified and reused. Using VISTA as the infrastructure for communication between a cognitive architecture and a visualization tool eliminates a significant amount of development time and effort. VISTA proves to be well designed, easy to use, and stable.

VISTA also provides classes to represent some of the major categories of objects based on an analysis of cognitive models, including goals, events, time, and so on, along with corresponding sample display methods. These classes are not directly available in Swing or other Java packages. Also, VISTA's connection manager (via SoarComm) offers some remote discovery capabilities, communication utilities, and usability additions that have often been built to support model/interface communications.

CaDaDis Design

A VISTA window displays the objects traced in the simple (modified-PERT), Pert, or Gantt view. All views are generated in parallel and a view menu allows the user to toggle between views. Advanced options will be added as this project continues.

Each view consists of a scrollable canvas for graphical objects. Objects represent actions of the model and are specified as labeled rectangles. Each object consists of a unique id, an associated code (operator name, etc.), start and end times for execution duration, and an optional list of constraints drawn as lines. Currently, a default value of 50 ms is being used for Soar objects in the Gantt view. This is our assessed time for a decision cycle in the model's run. Constraints are displayed as lines between rectangles. They represent dependencies for the target action. If constraints are not specified, a default constraint line can be drawn from the previous action's rectangle.

A useful design guideline when developing distributed systems is to keep the interface between the client and the server simple. It is important to realize that exposing the visualization client's objects and their methods can tightly couple the client and the cognitive architecture. For complex displays, this level of dependency can be necessary and beneficial, and VISTA supports such complex displays. However, if the designer is not careful strong dependencies can make the system difficult to maintain, and, for simple displays, provide no added benefit. As a result, when designing CaDaDis the classes exposed to the cognitive architecture were minimized.

Our only public datatype is a CaDaDis Object (CDDObject). It can be used to represent any event in the execution of the model whether it is the firing of a rule, or the entrance into a new problem space, state or operator. The generality afforded by this implementation provides flexibility to the users of this tool.

Simple Charts

A simple version of the categorical data from a model can be plotted. It is sometimes called a modified PERT. It can be quite helpful, for example, to sequentially chart Soar operators in a model. Figure 1 shows a CaDaDis display of operator firings for the Waterjug demo included with the Soar 8 distribution. The model has four operators (seen on the left as codes) and each point on the right side represents the firing of the corresponding operator. Creating this display required building a CaDaDis and a Soar-Tcl interface file, now included in CaDaDis. This file contains generic Soar rules used to generate data for CaDaDis such as states and operators from any Soar model. Figure 1 shows that this task had interesting combinations of operators and no clear cycles.

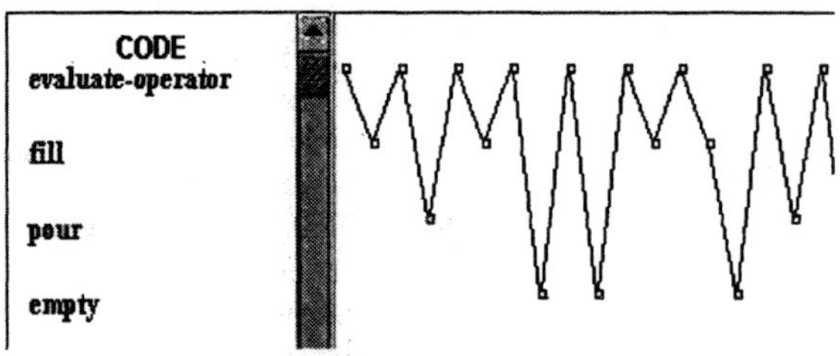

Figure 1: A simple display of a Soar model (Waterjug) with a categorical display of the operators as they are executed.

The order of the codes is initially arbitrary in all of these displays. They can be set by the user to more accurately reflect their meaning. The order can be modified dynamically by moving them with the mouse, by loading a file, or, by default, they are put in order of first use.

Pert Charts

A Pert chart is a display used to do task analysis. It is used to estimate the time to completion for a particular task (Harris, 1996). Based on dependencies of actions, subtasks can be started earlier if they are independent. Therefore, the end result of the Pert chart is the realization of the critical path of a task—what is the sequence of actions to complete a task in the least amount of time?

There are two variations on the Pert chart in CaDaDis. The first is the Standard Pert chart implementation. The Standard Pert chart is a simple drawing canvas that contains a set of rectangles representing tasks and arrows representing dependencies between tasks. Once the model has completed, the view can adjust to show the critical path. This is available in CaDaDis but no models have been yet instrumented to use it. Doing so would require modifying an architecture to pass the relevant information of start and stop times, and the actions' dependencies.

The other is based on John's CPM/GOMS Pert chart, a Nonstandard Pert chart, where codes are located on the left-hand side of the window. Objects are drawn in the row of corresponding to their category.

Gantt Charts

A Gantt chart uses a time line to show execution time for a particular action. The codes are on the left side as tasks in the Nonstandard Pert chart and rectangles are on the right canvas. The size of the rectangle is based on time of execution with respect to the time line. This is available in CaDaDis but no models have been yet instrumented to use it.

A new modification we are working on, called the Modified Gantt chart, will examine the life of an operator from proposal to completion with respect to the time line. A dashed line will represent an operator that is proposed. When the operator is selected, the dashed line will become a solid line.

Interface Features

These displays provide enough functionality that users will want to take advantage of their features, but will need support to do so. Table 1 lists several of the manipulations users can perform to their displays. Features in the table with an asterisk (*) are in the process of being implemented.

Example Uses

We present some examples to illustrate CaDaDis and its use. The first example displays operator applications. It uses an example Soar model in the Nonstandard Pert View, which is easy to create — the operator names are placed in the left pane and large or small boxes are drawn along the line corresponding to the operator that fires in the right pane. All the figures here use the Nonstandard Pert. As a default, dependencies are specified as the box representing the last operator or production that fired. A portion of this display is shown in Figure 2.

Soar provides a set of demos with its TSI distribution. We used these demos as a test set for CaDaDis. Each demo was run in CaDaDis to see if any interesting behavior could be found. (Figure 1 showed the demo model for the Waterjug problem.) At this stage of development, CaDaDis receives only operators from the Soar models to display. However, even with operators, the demo models can be categorized.

Table 1: CaDaDis Features.

- Rescale x and y, and the objects
- Choose fields to display (i.e., hide some fields to create abstract display, or highlight remaining fields)
- Print the figure to a printer*
- Save the figure to file, both as a jpeg and semantically
- Change the type of display and provide a best-effort display under the new type *
- Provide information about CaDaDis and the type of display
- Provide help functions *
- Set display options, such as autoscroll and absorb new codes (or generate an error)
- Gantt timeline in milliseconds or decision cycles *
- Reorder categories in nonstandard Pert by mouse or by loading a file
- Logging and log replay support for Soar and ACT-R
- Other architectures including: JACK, and CAST *

The next demo examines the 8-puzzle problem. This consists of 8 tiles numbered one through eight on a 3x3 board. There is a space to allow movement. A series of movements is performed to achieve the proper configuration (numerical order) of the tiles. The demo model attempts to solve the puzzle. The model uses two operators: evaluate-operator and move-tile. The first operator examines the current board configuration to determine the next move and the second, predictably, moves one of the tiles. The initial run showed the model examining the board, then moving the tile or tiles, before repeating this cycle (Figure 2), and sometimes not using an operator.

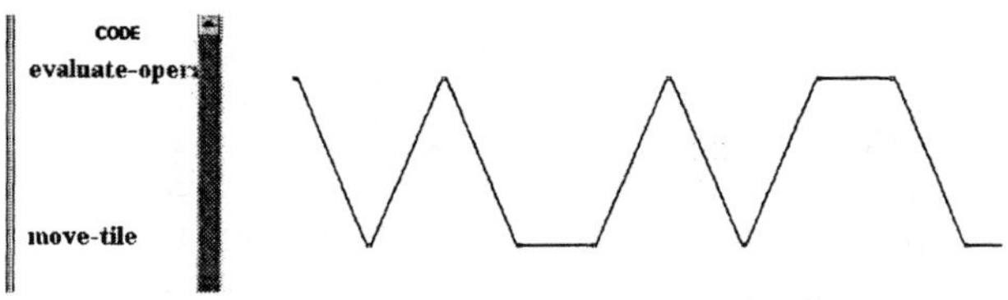

Figure 2: Operator trace of the Soar 8-puzzle model.

The Tower of Hanoi is a famous task where a number of disks in ascending size are found on one peg. The Soar model consists of one operator, move-disk. A run of CaDaDis does not provide an interesting picture, as shown in Figure 3, but it does display the model's behavior. This result suggests that operator-based traces require more detailed attribute displays to expose meaningful information about model behavior. These has since been added; displays using this feature are described below.

Some models employ a Wait operator. No decision needs to be made because the world does not require (or allow) a decision. The Blocks World model shows this operator in use (Figure 4).

Another example of a model using the wait operator is found in a run of dTank (acs.ist.psu.edu/dTank). dTank is an environment that allows models to play in a simple tank game. In this example, CaDaDis displays the operator applications in a simple Soar dTank model. Once again, dependencies are specified as the previously entered state.

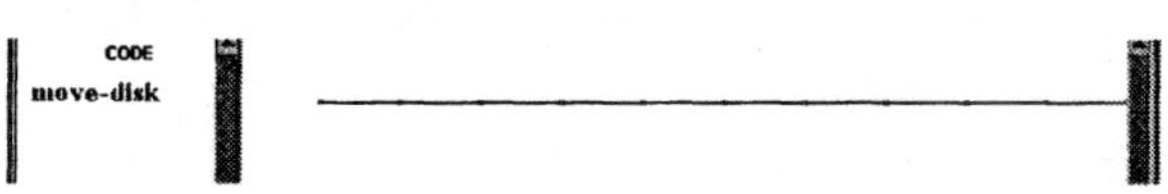

Figure 3: Operator trace of the Soar Tower of Hanoi model.

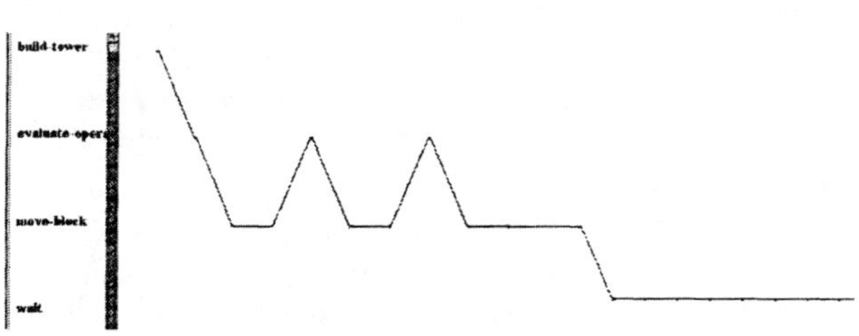

Figure 4: Blocks World example of wait operator.

Our dTank trace (Figure 5) comes from a basic tank agent. This display contains the portion of the run where the model has just found an opponent to attack. The horizontal portion of the trace represents the idle period (of Wait operators) of the tank waiting for something to come into view. The first set of spikes show the recognition of the opponent. The waiting period is waiting for more perceptions to arrive. The decision to attack is made and shown with the subsequent spikes in the trace. This is a nice illustration of agent behavior in the dTank environment.

In revisiting the 8-puzzle demo, we implemented an extension to the model that would give more information about the model. Similar to the Tower of Hanoi model, most of the action concerns the operator attributes. Our extension pulled information about the attributes out of the model and sent them to CaDaDis as categorical data. Figure 6 presents the 8-puzzle model run with an initial configuration that is unsolvable. We added one Soar rule and a Tcl function to the model to pass attributes from the operator of the tiles as they moved. Cyclical behavior emerging from the model as Soar attempts to solve the puzzle in now visible. Also, with some inspection, one can see that the blank tile is labeled as Tile 3 because this tile is not used.

Finally, because of the VISTA toolkit's architecture independence, CaDaDis can be used by a number of different cognitive architectures. The final examples are from models written in ACT-R and Jess. Production names are passed to CaDaDis from the model as they fire. The instance of a production firing is represented by a dot in the

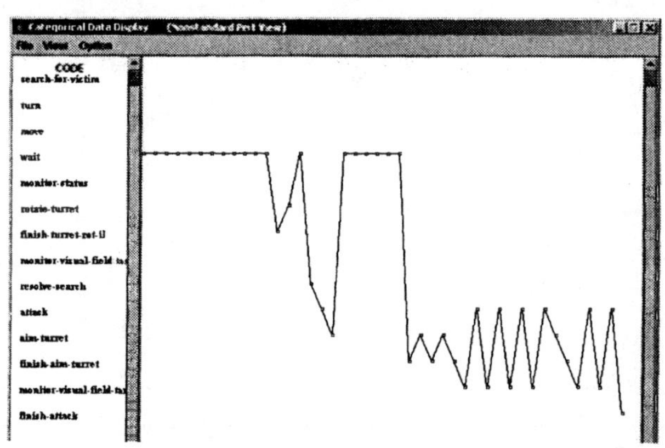

Figure 5: A Soar dTank model trace of tank behavior.

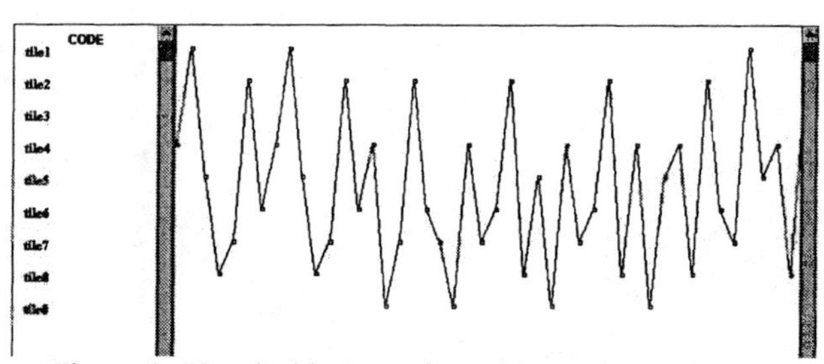

Figure 6: Unsolvable 8-puzzle problem using an attribute display to show individual tile movements.

graph. The ACT-R model traced in Figure 7 is a Serial Subtraction model being revised (Ritter, Avraamides, & Councill, 2002; Ritter, Reifers, Klein, Quigley, & Schoelles, in press). The model starts with a four-digit number and continually subtracts seven. This behavior is naturally cyclical as the figure shows. The productions consist of checking for the need to borrow, the act of borrowing, and simple subtraction. This portion of the trace illustrates some cycles where the subtraction was simple but ends with a need to borrow from the hundreds position, hence the downward spike. This display uses CaDaDis and a Lisp file addition to ACT-R (now included with CaDaDis) that passes information to CaDaDis when a rule fires.

Recently, CaDaDis has been successfully used with Jess (Friedman-Hill, 2003). A Jess model for dTank used in a programming languages course was adapted to send strings to CaDaDis. The project took three hours but included generating a reusable interface for Jess and CaDaDis. A display of a Jess model in Figure 8 shows that it has similar behavior to the Soar dTank model. The model waits and scans for enemies. Then, it aims and fires until the enemy is out of sight or dead. The display shows that Jess works on productions, and that this model is simpler than the Soar model.

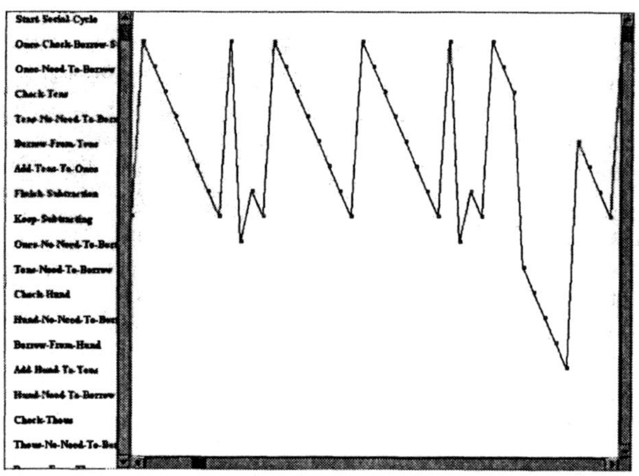

Figure 7: Serial Subtraction Model in ACT-R.

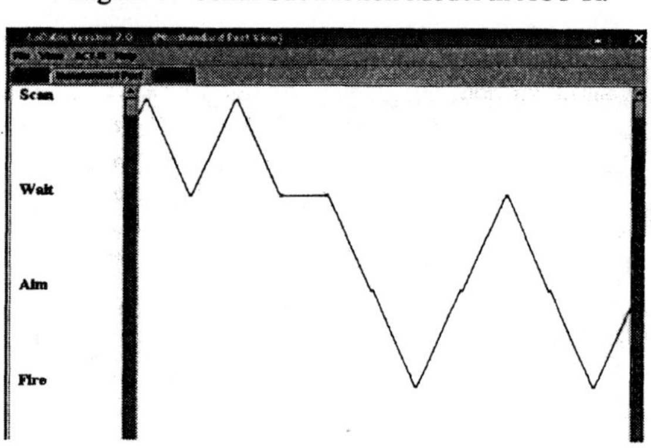

Figure 8: A Jess dTank model trace of tank behavior.

Conclusions

In addition to providing a useful tool for understanding cognitive models and providing a documented and reusable display in VISTA, creating CaDaDis provided several lessons for cognitive modeling and behavior representation. These lessons can be grouped into lessons about our CaDaDis system and the models examined, about VISTA as a tool, and for this process. We take them up in turn.

Lessons about CaDaDis

Preliminary results show that CaDaDis is successful in showing model behavior. It can create unique displays showing information with more clarity than textual traces. It provides nice displays of model activity in two different cognitive architectures and an agent architecture. Furthermore, it can prove useful in debugging cognitive models by analyzing rule usage, whether certain operators fire, and so on.

The CaDaDis displays included here have become easy to create because creating new displays can be done by modifying previous examples. The displays work with multiple architectures and provide multiple views. They need to be expanded to include more of the functionality in Table 1.

To realize their potential, these displays need to be reused by others. To this end we have put these displays on the VISTA tutorial web site (acs.ist.psu.edu/vista). The displays here created in CaDaDis suggest that the ability to help models explain their behavior can be significantly enhanced with CaDaDis and the VISTA toolkit.

For many applications of VISTA, CaDaDis can serve as a library, providing much of the additional graphics needed. For systems that need more complex graphics, or that need to modify the displays we have created, the source code for the graphic displays that make up CaDaDis have been designed to be extended and are documented (acs.ist.psu.edu/CaDaDis). We thus find that we will need libraries of graphical displays for cognitive models as much as we need libraries of model components. It may be possible, however, that interfaces are more reusable.

Lessons about Vista

Developing the interface to CaDaDis required extending the VISTA toolkit. The VISTA source code and manual are both available facilitating changes such as those we required. VISTA provides a base library for 2d-drawing that can be taken advantage of. However, in the example application included in the VISTA distribution, there were no classes related to the specific categorical display objects required for CaDaDis.

The creation of CaDaDis also provides a chance to reflect on the VISTA toolkit as it is used externally. We found that VISTA requires some overhead to learn and to create displays, but provides a worthwhile framework for creating these displays and provides a productive approach to reuse.

This level of understanding is not required unless the user wishes to modify the or extend these displays.

In summary, VISTA is a well-designed, easy to use, and powerful tool. Its use can reduce the development time of cognitive model visualization tools and allow the developer to concentrate on the domain specifics of the application—as opposed to the communication infrastructure.

Reuse in Cognitive Models

The reuse of displays with ACT-R, Soar, and Jess suggests that the first major reuse of cognitive modeling behavior may be in interfaces and not of knowledge. This might not be that surprising, given that the interface code looks more like the code that currently gets reused. Interfaces make up about 50% of most systems (Myers & Rossen, 1992). If this is true, which we believe it can be for cognitive models and agents using CaDaDis and VISTA, this resuse is a very worthwhile result.

Acknowledgments

This work was supported by the Office of Naval Research through a subcontract from Soar Technology, contract VISTA03-1. Doug Hogan, Andrew Reifers, Glenn Taylor, and Al Wallace provided useful suggestions about the design and implementation of these displays and about this presentation. Isaac Council and Andrew Reifers have provided extensions to CaDaDis. Doug Hogan and Andrew Reifers generated displays for this paper. Geoff Morgan and three anonymous reviewers provided useful comments.

References

Avraamides, M., & Ritter, F. E. (2002) Using multidisciplinary expert evaluations to test and improve cognitive model interfaces. In *Proceedings of the 11th CGF Conference*. 553-562, 02-CGF-100. Orlando, FL: U. of Central Florida.

Councill, I. G., Haynes, S. R., & Ritter, F. E. (2003). Explaining Soar: Analysis of existing tools and user information requirements. In F. Detje, D. Doerner, & H. Schaub (Eds.), *Proceedings of the Fifth International Conference on Cognitive Modeling*. 63-68. Bamberg, Germany: Universitats-Verlag Bamberg.

Friedman-Hill, E. (2003) *Jess in Action*. Greenwich, CT: Manning Publications, Co.

Gray, W. D., John, B. E., & Atwood, M. E. (1993) Project Ernestine: Validating a GOMS analysis for predicting and explaining real-world task performance. *Human-Computer Interaction, 8*(3), 237-309.

Harris R. L. (1996) *Information graphics: A Comprehensive illustrated reference*. Atlanta, GA: Management Graphics.

John, B., Vera, A., Matessa, M., Freed, M., & Remington, R. (2002). Automating CPM-GOMS. In *Proceedings of the CHI'02 Conference on Human Factors in Computer Systems*. 147-154. New York, NY: ACM.

Myers, B. A., & Rossen, M. B. (1992) Survey on user interface programming. In *Proceedings of CHI'92*. 195-202. New York, NY: ACM Press.

Newell, A. (1990) *Unified theories of cognition*. Cambridge, MA: Harvard University Press.

Norling, E., & Ritter, F. E. (2001). Embodying the JACK agent architecture. In M. Stumptner, D. Corbett, & M. Brooks (Eds.), *AI 2001: Advances in Artificial Intelligence. Proceedings of the 14th Australian Joint Conference on Artificial Intelligence*. 368-366. Berlin: Springer.

Ritter, F. E., Avraamides, M., & Councill, I. G. (2002) An approach for accurately modeling the effects of behavior moderators. In *Proceedings of the 11th Computer Generated Forces Conference*. 29-40, 02-CGF-002. Orlando, FL: U. of Central Florida.

Ritter, F. E., & Larkin, J. H. (1994). Using process models to summarize sequences of human actions. *Human-Computer Interaction, 9*(3), 345-383.

Ritter, F. E., Jones, R. M., & Baxter, G. D. (1998) Reusable models and graphical interfaces: Realising the potential of a unified theory of cognition. In U. Schmid, J. Krems, & F. Wysotzki (Eds.), *Mind modeling - A cognitive science approach to reasoning, learning and discovery*. 83-109. Lengerich, Germany: Pabst Scientific Publishing.

Ritter, F. E., Reifers, A., Klein, L. C., Quigley, K., & Schoelles, M. (in press). Using cognitive modeling to study behavior moderators: Pre-task appraisal and anxiety. In *Proceedings of the Human Factors and Ergonomics Society*.

Soar Technology, Inc. (2002). VISTA Developer's Handbook. Soar Technology. www.soartech.com/research.downloads.vista.php

Taylor, G., Jones, R. M., Goldstein, M., & Frederiksen, R. (2002) VISTA: A generic toolkit for visualizing agent behavior. In *Proceedings of the 11th Computer Generated Forces Conference*. 29-40, 02-CGF-002. Orlando, FL: U. of Central Florida.

Tor, K., Ritter, F. E., Haynes, S. R., & Cohen, M. A. (2004) CaDaDis: A Tool for Displaying the Behavior of Cognitive Models and Agents. In *Proceedings of the 13th Conference on Behavior Representation in Modeling and Simulation*. acs.ist.psu.edu/papers/torRHC04.pdf

Using a Genetic Algorithm to Optimize the Fit of Cognitive Models

Kevin Tor (tor@cse.psu.edu) and Frank E. Ritter (fer2@psu.edu)
School of Information Sciences and Technology and Department of Computer Science and Engineering
The Pennsylvania State University
University Park, PA 16801

Abstract

This paper analyzes the use of a genetic algorithm (GA) for auto-fitting a non-linear, multivariable model to data. The example model and task used to illustrate the effectiveness of a GA is the Tower of Nottingham task. Empirical data collected from adults completing the task were used to create an initial model. This model's variables were adjusted with a GA to simulate the performance of a child completing the task. The optimized fit was better than the previous fit generated by hand. We note the steps required to duplicate the process with other models.

Introduction

Achieving the perfect picture on a television can be a painstaking problem. Settings for hue, color, and other variables interact. As the people on the screen turn red, then purple, the user frenetically switches back and forth from tint to brightness making slight adjustments.

Adjusting the fit of a cognitive model to data, like adjusting a television, can depend on many parameters. Changing only one may not be enough to achieve the desired outcome. Hand optimization of a model can take an extraordinary amount of time in situations involving the adjustment of multiple parameters to understand how they influence problem solving. This occurs, for example, when modifying parameters to represent development or the effect of behavior moderators on cognition.

Complications also come from the size of the search space. The search space for a model of multiple parameters grows exponentially with the number of parameters. Quick solutions for these problems, which are attained by hand, can be simply the result of luck and cannot be relied on for consistency. For this reason, it is better to pursue an option involving automatic search (Ritter, 1991). These problems appear to have local minima and certainly have noisy evaluation functions, so we chose to explore the use of Genetic Algorithms (Davis & Ritter, 1987). Other approaches should also be tried, such as gradient descent.

Overview of Example Problem

To provide a basis for testing the effectiveness of a GA in optimizing the fitting of a model, an example problem must be used. Here, we use a cognitive model simulating a child putting together a puzzle. Empirical data was taken from both children and adults who constructed the puzzle. This data was used as a comparison for the model simulation.

Puzzle Construction. The puzzle is a pyramid consisting of five levels and a top. It is known as the Tower of Nottingham puzzle or, simply, the Tower Task (Wood & Middleton, 1975). The puzzle is constructed one level at a time and each level is made up of four pieces. The bottom level is constructed first followed by the next largest. Typically, two levels are not constructed simultaneously. Thus, the problem solver is concerned with finding and manipulating at most four pieces of the puzzle at a time.

Model construction and fit to the data. Adults provided the initial empirical data for the task. A model was created and fit to the adult's data by adjusting a set of parameters used by various prominent developmental psychologists in their theories of 'what develops' (Jones, 1998; Jones, Ritter, & Wood, 2000). These parameters range from memory capacity to understanding whether the puzzle's edges are flush. Adults have an accepted set of values for the parameters and a model was created to fit the adult data. Therefore, using this model and data collected from five seven-year-old children, the parameters were adjusted to fit the behavior of children. The problems lay in determining which parameters need adjustments, and what the appropriate ranges are for these parameters, and how much of an adjustment they need.

Use of the genetic algorithm. Once the desired parameters have been chosen and a set of ranges determined, the GA can commence fitting the model. In this case, a randomly generated bit string represents an individual or child. In the GA metaphor, this is a genotype. The bit string is parsed in a programmer-defined way to become the parameters of the genotype's runs. (The model that realizes the parameters is the phenotype). Here, a gene might represent processing speed and another whether block edges were checked.

After each genotype completes a series of constructions (the number of times the genotype puts the puzzle together), a fitness function provides a value based on a statistical measure of performance over those runs for how close the model's performance fits the data. This fitness measure is then multiplied in our GA by the fraction of successful constructions. In some cases, a phenotype does not complete the puzzle or does not complete it properly, resulting in a penalty on average performance.

This process occurs for all phenotypes in the population. Once the entire population has completed its construction sets, a best phenotype is chosen. If the model deems the best phenotype close enough to the children's performance, the GA terminates and the model is considered fit. Otherwise, a new generation is created by selecting some genotypes (bit strings) from the phenotypes (successful models) that performed well and creating new genotypes through combination and mutation.

Preview of Results. The GA finds a good fit to the data. With this particular GA, a run was conducted of 10 genotypes per population each constructing the model three times. In the first full run, the model ran for 77 generations. Possible solutions based on this population set were found (Tor, 2004). Further runs with larger populations and more constructions per genotype will be needed to achieve complete optimization of the model.

Example Problem

To illustrate the application of GA to this problem, an example model will be optimized. Jones (1998) created a model of child development with multiple parameters. This is an ideal example for running our experiment. The values for the parameters can be incorporated in a GA. Due to the number of parameters being monitored, a substantial number of generations may be needed, showing the effectiveness of the GA in a large search space.

Developmental Theories Examined

Psychologists argue over which changes in which mechanisms explain the changes in thought. There are several theories about what mechanisms change. We worked with a subsample of the set of theories Jones (1998) used. Table 1 shows how the theories were implemented in the model, including the ranges for each parameter. This set of ranges was checked with Lebiere (2003).

Pascual-Leone's (1987) M-Power theory explains that what develops is the number of working memory elements that could be managed at one time. His theory states that as children grow, their memory capacity increases allowing them to have access to more information at a given moment.

Siegler believes that development stems from strategy choice. Children have less ability in choosing an appropriate strategy than adults (Siegler & Shipley, 1995). There are many ways to make strategy choices less rational: increasing noise to decisions, decreasing the, or decreasing the accuracy in checking for obstructions of fit.

Kail (1996) argues that processing speed in humans increases with age. Essentially, a child takes longer to perform a cognitive task than an adult.

Piaget (1952) had a stage theory about how humans developed. Each stage represents a step in complexity of thinking that can be performed and thus understanding the environment around them. Children represented by this model are 7 years-old and, according to Piaget, are at the end of the pre-operational stage and about to enter the concrete operational stage. This means that they have trouble hypothesizing mental manipulations to objects found in the outside world. In this case, the objects would be blocks or block constructions, and it has been operationalized in the model by modifying the accuracy of checking flush edges in the puzzle.

Source of Data

The empirical data was taken from work done in Jones' (1999) thesis. For the adults, he filmed five undergraduate students completing the Tower task. The students had no a priori knowledge of the task and were given five attempts at the puzzle. Throughout the attempts the students were asked to give verbal accounts of their behavior and choices. The exhibited behavior was noted and subsequently coded. This became the basis for the original model.

The childrens' data was taken from observations on videotapes of children performing the task (Reichgelt, et al., 1993). Jones analyzed the children's videotapes the same way the adult data was analyzed.

Tower of Nottingham Model

Jones wrote the model for the Tower of Nottingham Task using the ACT-R 3 cognitive architecture (Anderson, 1993). Jones included a graphical representation of the Tower using Garnet (Myers, et. al., 1990), a toolkit for building user interfaces in Common Lisp.

Jones looked at the developmental theories and wrote a separate model to incorporate each one. He ran each model and adjusted their parameters by hand. Although each model provided favorable results, Jones' best results came from the strategy choice model.

Cornwell (2001) in his thesis furthered the exploration of what develops. He took Jones' models and created a single model combining the theories. This model had multiple parameters to adjust. Hand adjustment was too difficult so a GA was added. Cornwell left the project with a usable model but the graphics implementation made running the GA quite slow. His GA runs were thus limited in number of genotypes and number of generations. We started our project based on his GA and optimized its performance.

The genetic algorithm used on this model is called GAL. It is open-source code downloadable from Carnegie-Mellon (Spears, 1991). The implementation of this particular GA maintains diversity amongst the population with low mutation rate, distinct genotypes across generations, and by limiting the spread of the best individuals. All that needs to be added to the GA is the fitness function, a mapping from binary genotypes to the model parameters, and the termination function, all of which we took from Cornwell (2001).

Optimization of Model Runs

The initial system (Cornwell, 2001) was created in a Macintosh environment and was nearly complete. We ported the program to Red Hat Linux, putting it in a Unix environment. Because Unix often offers access to faster machines and offers a more advanced operating system for testing GA implementations, this port was essential.

Table 1: The incorporation of the developmental theories in the ToN model and GA.

Theorist	Theory as it Relates to ToN Model	Theory Implementation in ToN Model	GA Implementation (genes)
Piaget	7-year-olds fall at the end of the pre-operational stage—they may not be able to manipulate internal representations of objects and events in the external world.	Conditionally removed mental check to determine if two blocks will have flush outer edges when fit together	Conditionally loaded different production rules for mental fitting operations.
Siegler	Children should be slower and less accurate at both choosing and implementing an effective strategy than adults.	Added noise to rule firings with ACT-R's expected gain noise parameter(:egn)	:egn parameter set to integer value between 2 and 9
		Varied accuracy of check to determine if the outer edges of two blocks would be flush if fit together	probability of correctly checking for flush edges set between 1 and 0.5 (chance)
		Varied accuracy of check to determine if obtruding features would obstruct the fit of two blocks	probability of correctly checking for obstructing features set between 1 and 0.5 (chance)
Pascual-Leone	The number of knowledge structures that can be simultaneously held in memory increases with age.	Limited the number of DMEs that could be active at once with ACT-R's retrieval threshold parameter (:rt)	:rt parameter set to integer value between 0 and 7 [between -2 and 7]
Kail	Processing speed increases with development	Varied the default time allotted to rule firings with the ACT-R's default action time parameter (:dat)	:dat parameter set between 65 and 100 ms [between 40 and 100 ms]
		Varied the time allotted to hand movements	:effort parameter for productions involving hand movements set between 700 and 1,050 ms
		Varied the time allotted to eye movements	:effort parameter for productions involving eye movement speed set between 65 and 100 ms

A fitness function for the genetic algorithm focuses on 24 measures of comparison between the 7 year-old empirical data and the simulated data as determined by Jones (1998). These measures include layer completion times, total completion time, number of construction attempts for a layer, number of errors, and number of blocks examined.

Statistical t-tests are used to measure differences between the real and simulated data. The sum of the 24 t-tests is the major component of the fitness function. The sum is inverted to create a value that increases as the sum of the values of the t-tests decreases. That is, as fit increases, the fitness increases. (This is an example fitness function. It should not be inferred that we wish to accept the null hypothesis this way, merely that the numbers provide a useful and intuitive measure of data fit. Many others, quite possibly better ones, are possible.) Finally, a penalty is assessed by a successful construction ratio consisting of the number of successful constructions over the number of total constructions. A successful construction consists of the complete construction of the puzzle without skipping a level or constructing a level incorrectly. Therefore, the number of successful constructions over the genotype's total constructions is multiplied against the fitness value.

There were a few problems. The state of the model left by Cornwell appeared to have a memory leak. Jones' model was written to run for one construction and was not scalable to accommodate multiple runs. For the purposes of the GA, the model needed to run continually. Attempts to reset the graphics for the next puzzle construction were unsuccessful.

To alleviate these problems, many options were explored. Ultimately, we spawned Unix processes that each executed a model constructing the puzzle. Once a set number of constructions are completed, a fitness value is attained for that genotype and the next genotype begins. Using system processes works with a model that was originally written to run once.

Results

To illustrate how the GA works, a sample, small run is explained. The following is the preliminary set of results for the first full run, consisting of 77 generations of ten genotypes, each evaluated three times. Each evaluation takes about 5 minutes on standard PC, so this represents about 8 days of processing.

The results presented are the best three genotypes found during the first full 77 x 10 x 3 run. These genotypes are then rerun more times to validate the soundness of the achieved fitness values. Then, they are compared to the previous best of Jones.

The model was started with a randomly-generated, initial population of 10 genotypes, each made up of 24-bit strings as specified by the GA. The bit string is parsed and converted into the parameters that are used in the puzzle construction simulation. At this point, the GA runs the resulting model and collects the results of the fitness function for each of the genotypes. It takes the results and manipulates the better genotypes to create the next generation. We also saved the best genotypes.

The best three genotypes found are displayed in Table 2, including their fitness and t-test values. There is some commonality between the three in terms of expected gain noise and the motor and perceptual speeds. The fitness value and the first t-test sum were attained from the initial GA run. The t-test sum is the sum of the t-test values for the 24 measures. The order of the genotypes in the chart corresponds to the order that they were generated.

As can be seen, the GA was finding higher fitness values with time. Genotype 3 performed better than Genotypes 1 or 2 but did not check for flush edges during construction. It also had a longer processing time that the Adult meaning that the child model takes more to apply one of its rules.

Logically, Genotype 3 would not appear to be a good set of parameters. It does not check for flush edges, causing the puzzle to create more incorrect constructions. Also, its retrieval threshold is extremely high, allowing very few pieces of knowledge to appear in memory. Finally, it is drastically different from Genotypes 1 and 2. It may represent a local minimum and provide evidence that a GA is an appropriate optimization technique.

For further analysis these three genotypes were run 10 more times. Genotypes were run until 10 perfect construction sets of 3 runs each were completed. Construction sets where any of the three constructions failed were excluded with no penalty. The 10 fitness values were averaged to get the average t-test sum shown in the table. The max t-test measure, shown in Table 2, is the measure of the genotype that performed worst. As a group, the genotypes appear to perform poorly on timing measures.

The average t-test analysis showed that Genotypes 1 and 2 were initially performing correctly as they did not vary much from the initial t-test sum. They did experience regression to the mean as the average t-test sum did increase slightly over the initial t-test score. However, Genotype 3 had a unique, initial performance for those parameters leading to the belief that three constructions per genotype may not be enough to attain stable measures. In order for the GA to improve the genotypes, it would need to run with more constructions and more genotypes per generation.

The GA's performance should be considered successful based on these preliminary results. The initial fitness values of the three best genotypes did increase as the generations progressed. A termination function, created as specified by the GA, was not used in this run. This was done purposely to see how close a fit could be reached

Comparison to Jones' Results

The results attained in this study should be compared against other data previously found in this area, that is, Jones' hand optimization models. The easy method is to take the best model from his research and compare it to the three best genotypes discussed earlier. Jones found that the best performance came in the implementation of the Siegler model (Jones et al., 2000). This model affects the strategy choice by adding noise to the decision process. The model with these parameters was run until 10 successful puzzle construction sets were found in the same manner as with the three best genotypes. This is illustrated in Table 2

The average t-test sum for the Siegler model turned out to perform better than two of the top-performing genotypes. However, Table 2 shows that Jones' Siegler model took 67 attempts to get 10 puzzle construction sets that finished. This shows that his Siegler parameter set often cannot solve the Tower task. In fact, the parameters find the solution only about 1 in 7 times whereas the GA-computed parameters solve the task at a rate of about 1 in 1.5 to 2. The problem is that there was no penalty assessed for poor performance in Jones' analysis. Forcing the Siegler model to finish 10 perfect sets did give a good average t-test sum but may not be a good measure of its fit to the children's data. It also suggests that we need to go back and examine how often Wood's children completed the task on their own. As this task was designed to study contingent tutoring, the model predicts that many children cannot solve the task without assistance.

These results take a small step towards answering the question of what develops. The results in Table 2 suggest that multiple aspects develop. The best fitting model modified the children's model in several dimensions, providing support for several theories. The fit (and direction of parameter modifications) is not clear enough and good enough to suggest that we have a complete understanding of this task and data. There remains what appears to be some noise in the fit, and there may be multiple parameter sets that fit the data well. We will have to run the GA longer to get more stable and clear results.

Conclusions

This work shows that a genetic algorithm can be used to optimize models' fit to data where the fit consists of a large search space due to multiple model parameters. Hopefully, genetic algorithms will be used to optimize other complex, nonlinear models. The following is a discussion of possible improvements to this approach and how implementing a GA to optimize other should be approached.

Table 2: Comparison of GA found models with Siegler Model as implemented by Jones.

	Genotype 1	Genotype 2	Genotype 3	Siegler (Jones)	Adult
Check flush edges	T	T	Nil	T	T
Expected gain noise	2	2	2	6	0.4
Flush edge accuracy (%)	90	95	100	100	100
Obstruction detection accuracy (%)	100	100	75	100	100
Retrieval Threshold	3	3	6	0	0
Processing Speed (s)	0.065	0.065	0.08	0.05	0.05
Eye Speed (s)	0.095	0.095	0.085	0.2	0.2
Hand Speed (s)	0.95	0.95	0.8	0.55	0.55
Average fitness (x 10 sets)	46.7	47.4	55.1	46.9	
Number of runs to get 10 correct	16	19	18	67	10
Maximum t-test measure	3.45 (Size 4 Layer Time)	3.31 (Time Taken To Complete)	6.28 (Time Taken To Complete)	2.66 (# of Incorrect Disassembled)	

Improvements

The GA needs some measure to compare its performance against. This measure comes from the empirical data set. Given this statistical measure for comparison, the GA can optimize the model. Next is the idea of how close a fit is sufficient. Determining a measure of an appropriate fit is arguably the most complicated part of the GA application.

In this example, the GA did not perform as well as we would have liked; the largest problem appears to be a matter of having enough puzzle constructions per genotype to get stable evaluations. Three constructions may not be enough to get acceptable results where 5, 10, or even 20 may. Running the model on a larger scale will cause the t-tests to give results closer to the averages seen in Table 2 and remove the possibility of the unique performance seen in Genotype 3, where a higher fitness value was achieved initially but the average was significantly more.

Another plan is to adjust the GA to keep the best performing genotypes in the current population. the GA we used prefers diversity and uses the best performing genotypes to create the next generation but they, themselves, may not appear in the next generation. This would allow the fitness value for a particular genotype to be updated keeping performance realistic.

The successful construction ratio used to penalize a poor run's fitness can be squared to increase the penalty. The bit string could be lengthened to 48 bits providing more granularity for the parameters, creating more options in the range.

Applying the Optimization to Other Models

Given a model that needs to be fit and empirical data to fit the model to, optimization can be accomplished by following a set of guidelines. The model should have multiple parameters needing adjustment and fine granularity in the range of acceptable values for those parameters. This provides a large and expensive search space that makes a GA a useful choice.

The genotypes need a way to be represented in the GA. In this project bit strings were used because the GA operated on bit strings. Arrays of the parameter values can also be used depending on the GA implementation. Then, simple functions to parse the strings or arrays can be written to incorporate the values into the model. Theoretically, these representations are equivalent; the bit string is easier for the GA to manipulate, and the symbolic representations are easier for modellers.

A set of measures need to be created for use in the comparison of the simulation run data and the empirical data. T-tests or other statistical measures, such as correlations, should be used to provide a fitness value for the model's correspondence to the data.

Graphics are a nice addition to a running simulation model but can be problematic when optimizing fits. A decision should be made as to whether the model being optimized needs graphics. If graphics are added, they should be as independent of the model as possible and capable of being shut off. Graphics slow models down.

GAs need to run for many generations, which can take time. The run of a genotype's solution should take the smallest amount of time possible. This will help minimize the overall run time of the GA and allow the search space to be better explored. A summary of these lessons are shown in Table 3.

Table 3: List of Using a GA with another Model.

1. Multiple parameter model
2. Empirical data and measures of comparison
3. Candidate strings and parsing function
4. Fitness function that increases as fit improves
5. Graphics that can be "reset" by model
6. Minimized genotype run time
7. Loads of time

Further Work

Using this methodology to fully understand what develops in children will not be complete until a larger population of models has been examined over further generations. We have arranged supercomputer support from the U.S. Army Research Laboratory in Maryland to perform this analysis. The processing power found at their facilities will allow the runs to be completed on the order of 100x faster.

Acknowledgments

The previous work by Gary Jones and Jason Cornwell provided the foundation for this project. Thanks to the ACT-R community and Christian Lebiere for providing their input. Ian Schenck and three anonymous reviewers provided useful comments.

References

Anderson, J.R. (1993). *Rules of the mind.* Hillsdale, NJ: Lawrence Erlbaum.

Cornwell, J. B. (2001). *Using genetic algorithms to create and optimize models of development.* Unpublished undergraduate honor's thesis, Pennsylvania State University, University Park, PA.

Davis, L. W., & Ritter, F. (1987). Schedule optimization with probabilistic search. *Proceedings of the Third Conference on Artificial Intelligence Applications.* IEEE Computer Society. 231-236.

Jones, G. (1998). *Testing mechanisms of development within a computational framework.* Unpublished doctoral dissertation, University of Nottingham, Nottingham, England.

Jones, G., Ritter, F. E., and Wood, D.J. (2000). Using a cognitive architecture to examine what develops. *Psychological Science, 11*, 1-8.

Kail, R. (1996). Nature and consequences of developmental change in speed of processing. *Swiss Journal of Psychology*, 55, 133-138

Myers, B. A., Guise, D. A., Dannenberg, R. B., Vander Zanden, V., Kosbie, D. S., Pervin, E., Mickish, A., & Marchal, P. (1990). Garnet: Comprehensive support for graphical, highly-interactive user interfaces. *IEEE Computer*, *23*, 71-85.

Newell, A. (1990). *Unified theories of cognition.* Cambridge, MA: Harvard University Press.

Pascual-Leone, J. (1987). Organismic processes for neo-Piagetian theories: A dialectical causal account of cognitive development. In A. Demetriou (Ed.). *The neo-Piagetian theories of cognitive development: Towards an integration.* Amsterdam: North-Holland.

Piaget, J. (1952). *The origins of intelligence in children.* New York: International Universities Press.

Reichgelt, H., Shadbolt, N., Paskiewicz, T., Wood, D. and Wood, H. (1993) EXPLAIN: On implementing more effective tutoring systems. In A. Sloman, D. Hogg, G. Humphreys, D. Partridge, & A. Ramsey (Eds.), *Prospects for artificial intelligence* (pp. 239-249). Amsterdam: IOS Press.

Ritter, F. E. (1991). Towards fair comparisons of connectionist algorithms through automatically generated parameter sets. In Proceedings of the Thirteenth Annual Conference of the Cognitive Science Society. 877-881. Hillsdale, NJ: Lawrence Erlbaum.

Siegler, R. S., & Shipley, C. (1995). Variation, selection and cognitive change. In T. Simon & G. S. Halford (Eds.). *Developing cognitive competence: New approaches to process modeling*, 31-76. Hillsdale, NJ: Lawrence Erlbaum.

Spears, W. M. (1991). [Computer Software]. GAL. U. S. Navy Center for Applied Research in Artificial Intelligence, USA.

Tor, K. (2004). *Using a genetic algorithm to optimize the fit of cognitive models.* Unpublished Master's thesis. Pennsylvania State University, University Park, PA.

Wood, D., & Middleton, D. (1975). A study of assisted problem solving. *British Journal of Psychology, 66*, 181-191.

Modeling Behavioral and Brain Imaging Phenomena in Transcription Typing with Queuing Networks and Reinforcement Learning Algorithms

Changxu Wu (changxuw@umich.edu)
Department of Industrial & Operations Engineering, University of Michigan
1205 Beal Avenue, Ann Arbor, MI, 48109 USA

Yili Liu (yililiu@umich.edu)
Department of Industrial & Operations Engineering, University of Michigan
1205 Beal Avenue, Ann Arbor, MI, 48109 USA

Abstract

Transcription typing is one of the basic visual-motor control tasks with practical importance. Brain imaging studies (fMRI and PET) have discovered 2 phenomena related to transcription typing, involving different activation patterns of brain areas. In behavioral studies, Salthouse (1986) summarized 29 quantitative behavioral phenomena of transcription typing, 19 of which were modeled by TYPIST (John, 1996)—the most comprehensive computational model of transcription typing. This paper first proposes a queuing network model that integrates queuing networks with reinforcement learning algorithms, and then describes how the model simulated the two brain imaging phenomena and three behavioral phenomena in learning process. Overall, the queuing network model successfully modeled not only all the phenomena modeled by TYPIST, but also 11 additional behavioral phenomena and the 2 brain imaging phenomena. All the phenomena emerged as outcomes of the natural operations of the human information processing queuing network, with no need to draw situation-specific scheduling charts that are required by TYPIST.

Introduction

Transcription typing involves intricate and complex interactions of perceptual, cognitive, and motoric processes and it is one of the major activities in human-computer interaction. Numerous studies in experimental psychology (Salthouse, 1984a), brain imagining (Gordon et al., 1995), and cognitive modeling (John, 1996) have been conducted to explore and quantify transcription typing phenomena in the past 70 years.

Behavioral Phenomena

Salthouse (1986) reviewed the major behavioral experiment results in transcription typing and summarized 29 phenomena in this area. John (1988) summarized 2 additional behavioral phenomena found by Gentner (1983) and Salthouse & Saults (1987). These 31 behavioral phenomena include 12 basic phenomena, 5 error phenomena, 6 phenomena in typing units, and 8 skill learning phenomena in transcription typing. We have developed a queuing network model that successfully modeled 32 behavioral phenomena in transcription typing including 3 newly discovered eye movement phenomena and 29 of these 31 behavioral phenomena except 2 phenomena related to reading and comprehension–whose modeling requires significant extensions of our model to include production systems and is a topic of our on-going research (Wu & Liu 2004a; Table 2 in this paper). In this paper we focus on modeling the learning aspect of the behavioral phenomena and brain imaging phenomena. Due to space constraint, the simulation results of 6 phenomena in typing units and 5 of the 8 skill learning phenomena are summarized in Table 2 of this paper. The other 3 skill learning phenomena are described below and discussed in the paper:

First, interkey time of transcription typing decreases with practice following the power law of practice (Gentner, 1983). Typing speed of an unskilled typist (about 30 words per minute, John 1996) can be improved to that of a skilled typist (about 68 words per minute, Salthouse 1986).

Second, variability of interkey interval decreases with the increased skill of the typist. The interquartile range of interkey intervals correlates significantly with net interkey interval ($p<.05$, Salthouse 1984a).

Third, rate of repetitive tapping is greater among more skilled typists. The correlation between tapping speed with same finger and net typing speed was significant ($p<.05$, Salthouse 1984a).

Brain Imaging Phenomena

Recently, brain imaging studies (fMRI and PET) have discovered two phenomena related to transcription typing.

First, at the beginning stage of learning a visuomotor control task including transcription typing, mainly the dorsal lateral prefrontal cortex (DLPFC), the basal ganglia and the pre-SMA are activated (Jueptner et al., 1998; Sakai et al., 1998). After practice, activation of the DLPFC disappeared and strong activation was observed in the supplementary motor area (SMA), the basal ganglia and the primary motor cortex (M1) and with small activation at the somatosensory cortex (S1) (Gordon et al., 1995).

Second, in the well-learned stage (skilled typist in Gordon et al.'s study, 1995), when stimuli to be typed are repetitive letters (e.g. AAA…), the M1 was strongly activated; when stimuli to be typed are multiple-digit sentence (e.g. JACK AND…), the M1 was also strongly activated with stronger activation was observed in the SMA, the basal ganglia, and the S1.

Previous Model of Transcription Typing

TYPIST is the most comprehensive cognitive models in simulating the phenomena of transcription typing (John,

1988, 1996). It quantified 19 of the 29 Salthouse phenomena and 2 additional phenomena. However, TYPIST quantified transcription typing only along the time dimension with scheduling charts; it can simulate neither different brain areas' activation patterns nor decreased performance variability with higher typing skill.

Is it possible to quantify the skill learning and brain imaging aspects of transcription typing with a unified model? The current study examines this question by integrating queuing networks and reinforcement learning algorithms.

A Queuing Network Model with Reinforcement Learning Algorithms

The Static Part of the Queuing Network Model: A Queuing Network of the Major Brain Areas

Queuing Networks in Modeling Human Behavior A queuing network is a network of servers and queues that allow two or more servers to act serially, in parallel, or in any network arrangement (Liu, 1996, 1997). Computational models based on queuing networks have successfully integrated a large number of mathematical models in response time (Liu, 1996) and in multitask performance (Liu, 1997). A queuing network modeling architecture called the Queuing Network – Model Human Processor (QN-MHP) has been developed and used to generate behavior in real time (Liu, Feyen & Tsimhoni, 2004), including simple and choice reaction time (Feyen & Liu, 2001) and driver performance (Tsimhoni & Liu, 2003). The model in this paper extends QN-MHP by integrating reinforcement learning algorithms and strengthening its long-term memory and motor sub-network.

Basic Assumptions Numerous PET and fMRI studies in brain mapping indicated that the major brain areas are localized in the brain cortex with the neuron pathway connected to each other (Roland, 1993; Smith & Jonides, 1998; Bear & Connor, 2001). In addition, each major brain area may have certain information processing capacity and time (see Table 1). Therefore, it is assumed that the major brain areas form a queuing network with brain areas as the servers, information processed as entities, and neuron pathways as routes (see Figure 1). Within this general information processing structure, the major brain areas activated in transcription typing task were identified by the fMRI and PET studies (Jueptner et al., 1998; Sakai et al., 1998; Gordon et al., 1995, see Figure 1).

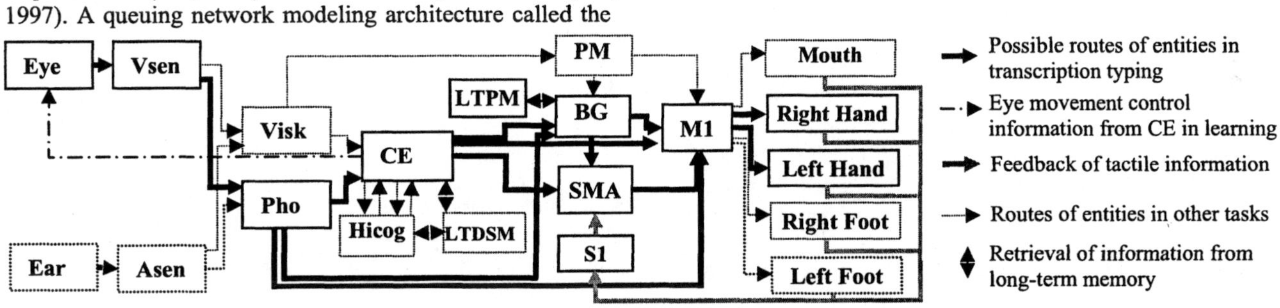

Figure 1: The general structure of the queuing network model (QN-MHP) with routes and servers involved in transcription typing tasks highlighted (server names, capacity, brain structures, and processing logic and time are shown in Table 1)

Table 1: Server name, major function and brain structure (Capacity is based on Card et al., 1986; Liu, 1997. Processing logic and time is based on Rudell & Hu, 2001; Feyen & Liu, 2001; Romero et al., 2000)

Server (Capacity)	Brain Structure	Major function (Processing logic)	Processing Time
Eye (∞)	Eye, LGN, SC, visual pathway	Visual sampling & signal transmission	[a] Eye movement: 50ms
VSen (17 letters)	Distributed parallel area & superior frontal sulcus, dorsal and ventral system	Visual sensory memory & perception	Normal distribution: M=263ms, SD=11ms
Pho (3 chunks)	Left posterior parietal cortex, inferior parietal lobe	Phonological loop to store auditoria & textual information	Included in CE server
CE (3 chunks)	Dorsal lateral prefrontal cortex & ACC	Mental process & response inhibition and selection	Exponential distribution: Mean=70ms, Min=25ms
BG (1 chunk)	Basal ganglia	Motor program retrieval	[b] $A_i+B_i \operatorname{Exp}(-\alpha_i N_i)$
LTPM (∞)	Striatal and cerebellar systems	Long term procedural knowledge storage	Included in BG server
SMA (2 letters)	Supplementary motor area and pre-SMA	Motor program assembly, error detection and bimanual coordination	188 ms
M1 (2 letters)	Primary motor cortex	Addressing spinal motorneurons	70 ms
S1 (2 letters)	Somatosensory cortex	Sending the sensory information to other areas	80 ms
Hand (1 letter each)	-	Execution of motor movement	[c] $I_m \log_2(D/S+0.5)$

a. Eye fixation time depends on the perception time and motor processing time of the previous word

b. See the part "learning process of individual server" in this paper. c. Fitt's Law: I_m=26.3,D=1.3cm, S: movement distance.

As shown in Figure 1, upon completing service at the Pho server, entities have numerous possible routes to follow to traverse the network: 1) At the Pho server, the entities can choose one of 3 three routes to depart Pho: from Pho to CE, BG, or M1. 2) At the CE server, they can choose from CE to BG, SMA, or M1. 3) At the BG server, they can select routes from BG to SMA or M1. Therefore, there are a total of 3*3*2=18 possible routes for the entities to be processed in the network in transcription typing.

An important question is, therefore, how the entities choose among these routes that activate (utilize) different brain areas (servers) in different learning stages or with different stimuli at well-learned stage? This question could be answered by the dynamic part of the model.

The Dynamic Part of the Queuing Network Model: Self-organization of the Queuing Network with Reinforcement Learning Algorithms

Ungerleider et al. (2002) found evidence of reorganization of brain areas with practice, which indicates that individual brain areas may change their information processing speeds in learning. Moreover, some brain areas may have error detection functions but others may not (see Table 1). Because the routes of the queuing network are composed of different brain areas (servers), different routes chosen by the entities may lead to different information processing speeds or errors. If the entities try to maximize their performance, they may choose an optimal route with maximum speed and minimum error. Therefore, in different situations, different routes may be chosen by the entities, activating different brain areas (servers), which forms the dynamic, self-organization aspect of the queuing network.

Consequently, there are two levels of learning within the queuing network: learning processes of the individual servers and self-organization of the queuing network depending on the stages of learning or the type of stimuli presented.

Learning Processes of the Individual Servers In the motor learning process, the basal ganglia, striatal and cerebellar systems (BG and LTPM servers) play a major role in procedural knowledge acquisition (Bear & Connor, 2001). Therefore, the current model focuses on the BG and LTPM servers in quantifying the learning process of the individual server. It is assumed that the time for the BG server to retrieve a motor program from LTPM decreases exponentially as a function of the number of practices (see Equation 1) because exponential function fits learning process of memory search, motor learning, visual search, and mathematic operation tasks better than power law (Heathcote et al., 2000) and it also has been applied in modeling long term memory retrieval (Anderson & Lebiere, 1998).

$$1/\mu_{BG}= A_{BG}+B_{BG}\,\mathrm{Exp}(-\alpha_{BG}N_{BG}) \qquad (1)$$

$1/\mu_{BG}$: motor program retrieving time; A_{BG}: the minimal of processing time of BG server after practice (314 ms, Rektor et al., 2003); B_{BG}: the change of expected value of processing time from the beginning to the end of practice (2*314=628 ms, assumed). α_{BG}: the learning rate of server BG (0.00142, Heathcote et al., 2000); N_{BG}: number of digraphs (letter pairs excluding the space key) processed by server BG, which is implemented as a matrix of diagraph frequency recorded in LTPM server.

Self-organization of the Queuing Network If the entities try to maximize their information processing speed and minimize error, it is appropriate to apply the reinforcement learning algorithms to quantify this dynamic process because reinforcement learning algorithm is a computational approach to quantify how an agent tries to maximize the total amount of reward it receives in interacting with a complex, uncertain environment (Sutton et al., 1998) and it has been applied in modeling motor learning in neuroscience (Nakahara et al., 2001). To integrate the reinforcement learning algorithms with queuing network, it is necessary to define the state, transition and reward in reinforcement learning with the concepts of queuing network.

1) State: the status that an entity is in server i.

2) Transition: An entity routed from server i to j.

3) Time-saving reward (r_t'): $r_t'=(1/w_q)+\mu_{j,t}$ (2)

w_q: time the entity spent waiting in the queuing of the server; $\mu_{j,t}$: processing speed of the entity at that server.

4) Error-saving reward (r_t''): $r_t''=1/(\mathrm{Nerror}_{j,t}+1)$ (3)

$\mathrm{Nerror}_{j,t}$: number of action errors of the previous entities made in the next server j at t th transition.

Mechanism of Choosing Different Routes Q online learning algorithms in reinforcement learning are used to quantify the process through which entities choose different routes based on the rewards.

1) Q online Learning Algorithm of Time-Saving Reward

$$Q_T^{t+1}(i,j)\leftarrow Q_T^{t}(i,j)+\varepsilon\{r_t'+\gamma \max_k[Q_T^{t}(j,k)]-Q_T^{t}(i,j)\} \qquad (4)$$

ε: learning rate of Q online learning ($0<\varepsilon<1$, $\varepsilon=.99$)

γ: discount parameter of routing to next server ($0<\gamma<1$, $\gamma=.3$)

$Q_T^{t+1}(i,j)$: online Q value if entity routes from server i to server j in t+1 th transition based on time-saving reward

$\max_k[Q_T^{t}(j,k)]$: maximum Q value routing from server j to the next k server(s) ($k>=1$)

Equation 4 updates a Q value of a backup choice of routes ($Q_T^{t}(i,j)$) based on the Q value which maximizes over all those routes possible in the next state ($\max_k[Q_T^{t}(j,k)]$). In each transition, entities will choose the next server according to the updated $Q_T^{t}(i,j)$.

2) Q online learning Algorithm of Error-Saving Reward

$$Q_E^{t+1}(i,j)\leftarrow Q_E^{t}(i,j)+\varepsilon\{r_t''+\gamma \max_k[Q_E^{t}(j,k)]-Q_E^{t}(i,j)\} \qquad (5)$$

3) Trade-off of the Two Q Values

The choice of route is determined by the trade-off between the two Q values. Currently, it is assumed that $Q_E^{t+1}(i,j)$ of error-saving reward has the higher priority than the $Q_T^{t+1}(i,j)$ of time-saving reward: if $Q_E^{t+1}(i,j)> Q_E^{t+1}(i,k)$, the entity will choose the next server j whatever the value of

Q_T^{t+1}(i,j); if Q_E^{t+1}(i,j)=Q_E^{t+1}(i,k), entity will choose the next server with greater Q_T^{t+1}; if Q_E^{t+1}(i,j)=Q_E^{t+1}(i,k) and Q_T^{t+1}(i,j)= Q_T^{t+1}(i,k), entity will choose next server randomly.

Model Predictions of 3 Skill Learning Phenomena and 2 Brain Imaging Phenomena

The 3 skill learning phenomena and the 2 brain imaging phenomena of transcription typing described earlier in this paper can be predicted by the queuing network model with reinforcement learning.

Prediction of the 3 Skill Learning Phenomena Because the processing times of CE, BG, SMA server follow the exponential distribution (see Table 1), if Y_1.. Y_k are k independent exponential random variables, their sum X has an Erlang distribution. Based on the feature of Erlang distribution, we have:

$$X = \sum_{i=1}^{k} Yi \quad E[X] = E[\sum_{i=1}^{k} Yi] = \sum_{i=1}^{k} E[Yi] = k\frac{1}{\lambda}$$

$$Var[X] = Var[\sum_{i=1}^{k} Yi] = \sum_{i=1}^{k} Var[Yi] = k\frac{1}{\lambda^2}$$

These mathematical results can be used to predict the skill learning phenomena, together with the prediction described below that entities may learn to skip certain server(s). First, because E[X]=k (1/ λ), if k'<k, then E[X']<E[X]. This may be one of the reasons that the skipping of server(s) can explain reduction in interkey time in typing normal text (the 1st skill learning phenomenon in this paper) and repetitive letters (the 3rd skill learning phenomenon). Second, skipping some of the servers will decrease the variance of the Erlang distribution because if k'<k, then *Var*[X']<*Var*[X]. This is one possible reason that skipping of server(s) can account for the reduction in the variability of interkey time in the learning process (the 2nd skill learning phenomenon).

Prediction of the First Brain Imaging Phenomenon a) At server Pho, at the beginning of the learning, entities can go through CE server for eye movement control to locate the specific position o,f a target key on the keyboard (Deseilligny et al., 2003, see Figure 1) and response selection and inhibition (Smith & Jonides, 1998). Entities can also choose the route from Pho to BG, but it takes longer time than going through CE because BG may not work effectively in retrieval of the motor program from LTPM (Bear & Connor, 2001) and its Q value of time-saving reward is smaller than that of CE. Entities can also choose the route from Pho→M1 directly. However, the occurrence of typing error will decrease the Q value of error-saving reward from Pho→M1. As the number of practice increases, the route Pho→BG is selected by the majority of the entities because the functions of CE were gradually replaced by the BG with less process time based on the parallel cortico-basal ganglia mechanisms (Nakahara et al., 2001).

b) At CE server, entities can one of routes from CE to BG, SMA or M1. If entities select the first route, the correct motor program will be retrieved without decreasing the Q_E^{t+1}(i,j). If the second or the third route is chosen, its Q_E^{t+1}(i,j) will decrease because no correct motor program is retrieved.

c) At server BG, because the stimuli keeps changing in typing multi-digit sentences, if the entity goes from the BG directly to M1 skipping SMA whose function is motor program assembling (Roland, 1993) and movement accuracy ensuring for error detection (Gordon et al., 1995), Q_E^{t+1}(i,j) will deceases in route BG→M1.

In sum, at the beginning of the learning process, entities will go through: Pho→CE→BG→SMA→M1. After learning, the majority of entity will travel: Pho→BG→ SMA→M1.

Prediction of the second brain imaging phenomon If the stimuli change from repetitive letters to regular words in the same task, the entity will change the route Pho→M1 to Pho→BG→SMA→M1 because error-saving reward decreases in route Pho→M1 without the motor program function of BG and sequencing function of SMA.

Simulation of the 3 Skill Learning Phenomena and the 2 Brain Imaging Phenomena

The First and the Second Skill Learning Phenomena Simulation results showed that the simulated interkey interval in the learning process followed the power law of practice (R square=0.8, p<.001). The simulated interkey interval improved from 385 ms to 180 ms, which was consistent with existing experimental data about performance change from the unskilled typist (interkey time 400 ms) with estimation error 3.75% (estimation error=|Y-X|/X*100%, Y: simulation result; X: experiment result) to the skilled typist (177 ms interkey time) with estimation error 1.69% (see Figure 2).

As shown in Figure 2, the change of the quartile range (75% quartile-25% quartile) is significantly correlated with the change of the simulated interkey interval (the Pearson correlation coefficient is 0.727, p<.05), which is consistent with the experimental result of Salthouse (1984a). This was one of the phenomena not covered by TYPIST (John, 1996).

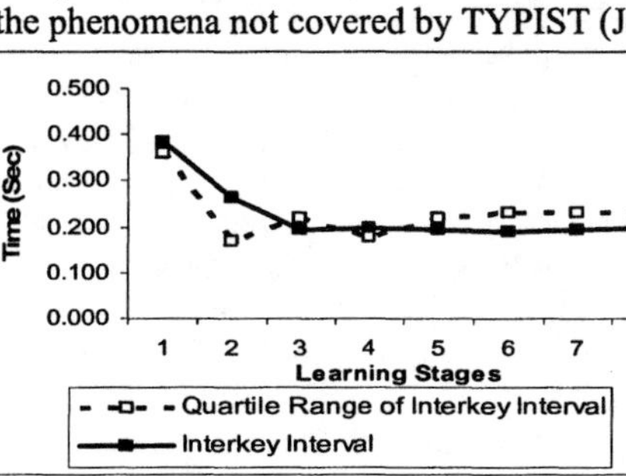

Figure 2: Simulated variability of interkey interval and interkey interval in the learning process. Each stage represents 352,125 keystrokes

The Third Skill Learning Phenomena The simulated tapping rate (interkey interval in typing repetitive letters) and typing speed of text (interkey interval in typing multi-digit sentence) during the learning process are strongly correlated (Pearson correlation efficient equals .968, p<.01), which is consistent with the experimental result of Salthouse (1984a) who found the significant correlation between the two variables (p<.01).

The First Brain Imaging Phenomena As shown in Figure 3, at the beginning of practice, CE (including DLPFC) and BG server are the highly utilized servers with a small percentage of utilization at SMA server (including pre-SMA) (3%) and M1 & 2 hand servers (15%). After 352,125*8 trials of practice, CE server (DLPFC) decreased its utilization greatly to 0%. Percentage of utilization of SMA server is increased by 47%. M1& 2 hand servers and S1 are also increased their percentage of utilization during the learning process by 85% and 22% respectively. These simulation results are consistent with the experiment results in PET and fMRI studies (Jueptner et al., 1998; Sakai et al., 1998; Gordon et al., 1995).

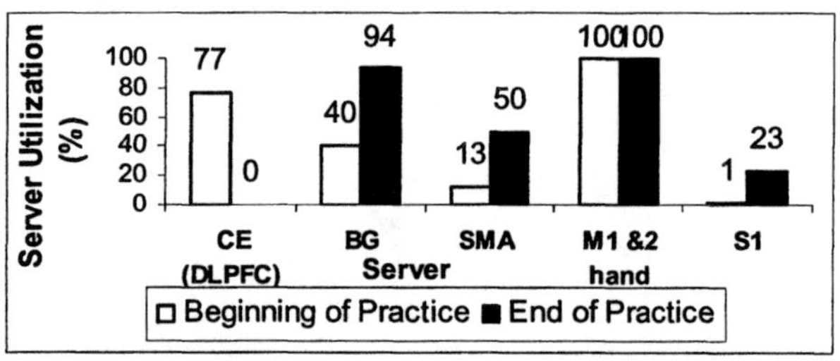

Figure 3: Server utilization at the beginning and end of practice in learning to type multi-digit sentence

The Second Brain Imaging Phenomena After the model finished its learning process, it can simulate the second brain imaging phenomenon of skilled typist in typing different stimuli. The 1600 letters to be typed by the model changed following this pattern: 1st-800th letters: repetitive letters; 801st-1600 letters: multi-digit sentence.

Figure 4 shows the percentage of utilization of the major servers in the different stimulus conditions. When the model is typing repetitive letters, mainly M1 & two hand servers is utilized. When the stimuli changed from repetitive letters to multi-digit sentence, the utilization of SMA, BG and S1 is increased by 49%, 90%, and 22% respectively. It demonstrated that fewer entities travel from Pho to M1 directly when the stimuli presented change from repetitive letters to multi-digit sentence. These results are consistent with the fMRI study of Gorden et al. (1995).

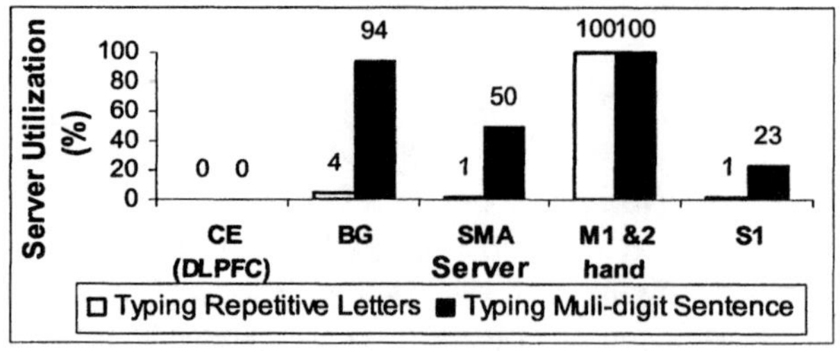

Figure 4: Server utilization when stimuli presented changed in the well-learned transcription typing situation

Simulation Results of a Sample of Additional Behavioral Phenomena

Table 2 summarizes a sample of 11 additional behavioral phenomena simulated by QN-MHP in typing units and skill effects of transcription typing. Our queuing network model has successfully modeled 32 behavioral phenomena thus far, most of which are not described in this paper due to space constraint.

Table 2: Experiment and simulation results of other 11 behavioral phenomena

Behavioral Phenomena (Salthouse, 1986)	Simulation Results (Estimation Error)
13. Copy span is 7-40 letters	11.0 letters (22.6%)
14. Stopping Span is 1 or 2 letters	2.3 letters (18%)
15. Eye-hand span is 3-8 letters (5 letters)	7.8 letters (0%)
16. Eye-hand span is 1.75 letters for meaningless material	1.82 letters (4%)
17. Replacement span is 3 letters	3.25 letters (12.1%)
Post 29. Detection span is about 8 letters	8.5 letters (6.25%)
§ 23. Two-hand and two-finger diagram (correlates with speed: -2.23) improves faster than one-finger and one-letter diagram (-1.38)	-1.92 (13.9%) -1.54 (11.7%)
26. Eye-hand span correlated with interkey time (0.5-1.2 letters/20 wpm increase in skill)	1.28 letters/20 wpm (6.7%)
27. Replacement span increases with skill (correlates with speed: .80)	.99 (23.75%)
28. Copy span is moderately related to skill (correlates with speed: .35)	.388 (10.8%)
29. Fast typists have larger stopping spans (correlates with speed: .57)	.6 (5.3%)

§ Phenomena not modeled by TYPIST

Discussion

With the task-independent structure of the queuing network and dynamic self-organization within the network by reinforcement learning algorithms, adding the 18 behavior phenomena simulated by the same model (Wu & Liu, 2004a), overall, QN-MHP successfully modeled 34 phenomena in transcription typing including 32 behavioral phenomena (21 of them modeled by TYPIST) and 2 brain imaging phenomena (average estimation error 9.3%) with higher estimation accuracy than TYPIST (average estimation error 21.3%).

In practice, because our queuing network model was built in a general structure of brain areas, it can be easily transformed to model other task situations, e.g. PRP (Wu & Liu, 2004b). Moreover, in contrast to TYPIST, the current model can generate behavioral results by the interaction of the servers without drawing complex scheduling charts. These unique features offer great potential of the model for learning and easy application by researchers in cognitive modeling and human factors.

Acknowledgments

This article is based upon work supported by the National Science Foundation under Grant No. NSF 0308000. However, any opinions, findings and conclusions or recommendations expressed in this article are those of the authors and do not necessarily reflect the views of the National Science Foundation (NSF).

Reference

Anderson, J.R. & Lebiere, C. (1998). *The atomic components of thought.* Mahwah, NJ: Erlbaum

Bear, M. & Connor, B. (2001). *Neuroscience: Exploring the Brain.* Lippincott Williams & Wilkins Publisher

Card, S., Moran, T., & Newell, A. (1986). The Model Human Processor: An engineering model of human performance. In Boff, K. (Eds.) *Handbook of Perception and human performance,* New York: Wiley

Deseilligny, C. & Muri, M. (2003). Cortical control of ocular saccades in humans: a model for motricity. In: C. Prablanc, D. Pelisson etc. (Eds.), *Neural Control of Space Coding and Action Production,* Elsevier. Progress in Brain research, v142: 1-19

Feyen, R. & Liu, Y. (2001). Modeling Task Performance Using the Queuing Network Model Human Processor (QNMHP) *Proceedings of the the 4th International Conference on Cognitive Modeling*, 73-78

Gentner, D.R. (1983). The acquisition of typewriting skill. *Acta Psychologica*, 54, 233-248

Gordon, A.M., & Soechting, J.F. (1995). Use of tactile afferent information in sequential finger movements. *Experimental Brain Research*, 107, 281-292

Heathcote, A., Brown, S., & Mewhort, D.J. (2000). The power law repealed: the case for an exponential law of practice. *Psychological Bulletin,* 7 (2), 185-207

John, B.E. (1988). *Contributions to Engineering Models of Human-Computer Interaction,* Department of Psychology, Carnegie-Mellon University, Ph.D. thesis

John, B.E. (1996). TYPIST: A theory of performance in skilled typing. *Human Computer Interaction,* 11, 321-355

Jueptner, M. & Weiller, C. (1998). A review of differences between basal ganglia and cerebellar control of movements as revealed by functional imaging studies. *Brain*, 121, 1437–1449

Liu, Y. (1996). Queuing network modeling of elementary mental processes, *Psychological Review*, 103, 116-136

Liu, Y. (1997). Queuing network modeling of human performance of concurrent spatial and verbal tasks. *IEEE Transactions on Systems, Man, Cybernetics,* 27, 195-207

Liu, Y., Feyen, R., and Tsimhoni, O. (2004). *Queuing Network – Model Human Processor (QN-MHP): A Computational Architecture for Multitask Performance in Human Machine Systems.* Technical Report 04-05. Department of Industrial and Operations Engineering, University of Michigan

Nakahara, H., Doya, K., & Hikosaka, O. (2001). Parallel Cortico-Basal Ganglia Mechanisms for Acquisition and Execution of Visuomotor Sequences. *Journal of Cognitive Neuroscience,* 13, 5, 626–647

Rektor, I, Kanovsky, P., Bares., M. (2003). A SEEG study of ERP in motor and premotor cortices and in the basal ganglia. *Clinical Neurophysiology,* 114, 463–471

Roland, P.E. (1993). *Brain Activation*, John Wiley &Sons

Romero, D.H, Lacourse, M.G, Lawrence, K.E. (2000). Event-related potentials as a function of movement parameter variations during motor imagery and isometric action. *Behavioral Brain Research,* 117, 83–96

Rudell, A.P. & Hu, B. (2001). Does a warning signal accelerate the processing of sensory information? Evidence from recognition potential responses to high and low frequency words. *International Journal of Psychophysiology,* 41, 31-42

Salthouse, T.A. (1984a). Effects of age and skill in typing. *Journal of Experimental Psychology: General,* 113, 345-371

Salthouse, T.A. (1986). Perceptual, cognitive, and motoric aspects of transcription typing. *Psychology Bulletin*, 99 (3), 303-319

Salthouse, T.A. & Saults, J.S. (1987). Multiple spans in transcription typing. *Journal of Applied Psychology,* 72, 2, 187-196

Smith, E.E. & Jonides, J. (1998). Neuroimaging analyses of human working memory. *Proc. Natl. Acad. Sci. USA*, 95, 12061–12068

Sutton, R.S. & Barto, A.G. (1998). *Reinforcement Learning: An Introduction.* MIT Press: Massachusetts

Tsimhoni, O., & Liu, Y. (2003). Modeling steering with the Queuing Network-Model Human Processor (QN-MHP). *Proceedings of the 47th Annual Conference of the Human Factors and Ergonomics Society*, 81-85

Ungerleider, L.G., Doyon, J., & Karni, A. (2002). Imaging Brain Plasticity during Motor Skill Learning. *Neurobiology of Learning and Memory.* 78, 553–564

Wu, C. & Liu, Y. (2004a). Modeling human transcription typing with queuing network-model human processor. *Proceedings of the 48th Annual Meeting of Human Factors and Ergonomics Society*, New Orleans, Louisiana, USA (In press).

Wu, C. & Liu, Y. (2004b). Modeling Psychological Refractory Period (PRP) and Practice Effect on PRP with Queuing Networks and Reinforcement Learning Algorithms, *Proceedings of the Sixth International Conference on Cognitive Modeling*, Pittsburgh, PA, USA (In press).

Modeling Psychological Refractory Period (PRP) and Practice Effect on PRP with Queuing Networks and Reinforcement Learning Algorithms

Changxu Wu (changxuw@umich.edu)
Department of Industrial & Operations Engineering, University of Michigan
1205 Beal Avenue, Ann Arbor, MI, 48109 USA

Yili Liu (yililiu@umich.edu)
Department of Industrial & Operations Engineering, University of Michigan
1205 Beal Avenue, Ann Arbor, MI, 48109 USA

Abstract

PRP (Psychological Refractory Period) is a basic but important form of human information processing in dual-task situations. This article describes a queuing network model of PRP that successfully modeled PRP without the need of setting up complex lock/unlock performance strategies employed in the EPIC model of PRP or drawing complex scheduling charts employed in the ACT-R/PM model of PRP. Further, by integrating queuing networks with reinforcement learning algorithms, the queuing network model successfully simulated practice effect on PRP, which has not been modeled in existing PRP models. The current research indicates that depending on an individual's degree of practice, cognition can be either serial or parallel at the level of production or response selection. Extensions of queuing network model in modeling other tasks and its easiness in modeling concurrent tasks in practice are also discussed.

Introduction

PRP (Psychological Refractory Period) is one of the most basic and simple forms of dual task situations, and has been studied extensively in the laboratory for half a century (Meyer & Kieras, 1997a). In the basic PRP paradigm, two stimuli are presented to subjects in rapid succession and each requires a quick response. Typically, responses to the first stimuli (Task 1) are unimpaired, but responses to the second stimuli (Task 2) are slowed by 300 ms or more (Ruthruff et al., 2001). In the PRP paradigm of Van Selst's study (1999), in task 1, subjects were asked to discriminate tones into high or low tones by vocal responses (audio-vocal responses); in task 2, subjects watched visually presented characters and performed choice reaction task by manual responses (visual-motor responses). They found that practice dramatically reduced dual-task interference in PRP.

The basic PRP has been modeled by several major computational cognitive models based on production rules, notably EPIC (Meyer & Kieras, 1997a) and ACT-R/PM (Byrne & Anderson, 2001). Based on the major assumption that production rules can fire in parallel, EPIC successfully modeled the basic PRP effect by using complex lock and unlock strategies in central processes to solve the time conflicts between perceptual, cognitive and motor processing (Meyer & Kieras, 1997a). In contrast to the parallel cognitive processing assumption of EPIC, ACT-R/PM assumes serial firing of production rules in cognition. It modeled the basic PRP effect by relying on drawing scheduling charts manually to quantify the quick switching of central cognition in information processing, which shares certain similarities with the lock and unlock strategies in EPIC (Byrne & Anderson, 2001). All of these complex lock/unlock strategies or scheduling charts came from the arrangements by the researchers rather than the natural interactions among the processors or modules. Moreover, because of the complexity of dual tasks in practical situations, it may be difficult for a system designer or human factor specialist to figure out these complex strategies or scheduling charts to arrange the activities of the central cognitive processor.

Furthermore, neither EPIC nor ACT-R/PM modeled the practice effect on PRP (Ruthruff et al., 2001).

The current paper describes a model of PRP that integrates queuing network theory (Liu, 1996, 1997) and reinforcement learning algorithms (Sutton & Barto, 1998) and was developed on the basis of neuroscience findings. Model simulation results were compared with experimental results of both the basic PRP paradigm and the practice effect on PRP (Van Selst et al., 1999). All of the simulated human performance came from the natural interactions among servers and entities in the queuing network without setting up lock and unlock strategies or drawing complex scheduling charts.

Modeling the Basic PRP and the Practice Effect on PRP with Queuing Networks

Modeling Human Performance with Queuing Networks

A queuing network is a network of servers and queues that allow two or more servers to act serially, in parallel, or in any network arrangement. Computational models based on queuing networks have successfully integrated a large number of mathematical models in response time (Liu, 1996) and in multitask performance (Liu, 1997) as special cases of queuing networks. A queuing network modeling architecture called the Queuing Network – Model Human Processor (QN-MHP) has been developed and used to generate behavior in real time (Liu, Feyen & Tsimhoni, 2004), including simple and choice reaction time (Feyen & Liu, 2001), driver performance (Tsimhoni & Liu, 2003) and transcription typing (Wu & Liu, 2004a, 2004b). The model in this paper extends QN-MHP by integrating reinforcement learning algorithms and strengthening its long-term memory. The simulation model is easy to use and is implemented with Promodel®, one of the most popular simulation software programs in industry.

Because major brain areas with certain information processing time and capacity are localized and connected with each other in brain cortex via neural pathways (Bear & Connor, 2001; Smith et al., 1998; Roland, 1993, see Figure 1 and Table 1), it is assumed that they form a queuing network. One or several brain areas are regarded as servers of the network with neural pathways regarded as routes and information to be processed in these brain areas regarded as the entities. Cognitive performance is the outcome of processing the entities in the queuing network.

Modeling the Basic PRP and Practice Effect on PRP with Queuing Networks and Reinforcement Learning Algorithms

Because the PRP effect prior to or at beginning of learning (the basic PRP) is a special case of the PRP effect during the learning process, the two phenomena of PRP (basic and learning) are modeled with the same mechanism in our queuing network model. The experimental tasks and data of Van Selst et al. (1999) were used to test the model.

Brain areas (servers) and their routes related to the two PRP tasks in Van Selst's study were identified within the general queuing network structure based on recent neuroscience findings (Mitz et al., 1991; Fletcher et al., 2001; Bear & Connor 2001, see Figure 1).

Entities of task 1 (audio-vocal responses) cannot bypass the Hicog server because the function of phonological judgment is mainly located at the Hicog server, and thus there is only one possible route for the entities of task 1 (see the dotted thick line in Figure 1). However, the function of movement selection in task 2 (visual-motor responses) is located not only in the Hicog server but also in PM server. Therefore, there are two possible routes for the entities of task 2 starting at Visk server (see the gray and black solid lines in Figure 1).

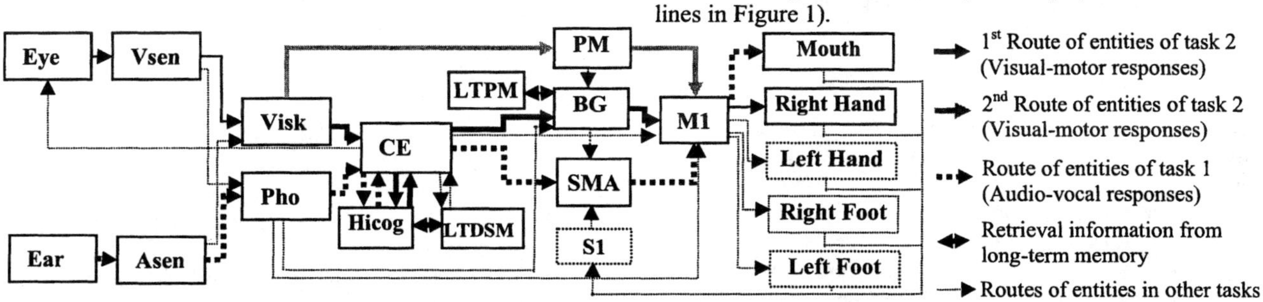

Figure 1: The general structure of the queuing network model (QN-MHP) with servers and routes involved in the PRP task highlighted (server names, brain structures, major functions and processing time are shown in Table 1)

Table 1: Server name, major function and brain structure (Capacity is mainly based on Card et al., 1986; Liu, 1997. Processing time is based on Rudell & Hu, 2001; Feyen, 2001; Romero, 2000; Nakahara et al., 2001)

Server(Capacity)	Brain Structure	Major function (Processing logic)	Processing Time
Eye (∞)	Eye, LGN, SC, visual pathway	Visual sampling & signal transmission	Eye movement: 50ms
VSen (17 letters)	Distributed parallel area, SFS, dorsal & ventral system	Visual sensory memory & perception	Normal distribution: M=263ms, SD=11ms
Visk (3 chunks)	Right posterior parietal cortex	Visuospatial sketchpad to store the graphic information	Included in CE server
Ear (∞)	Ear, auditory pathway	Converts sound waves to neuron signals	10 ms
Asen (5 entities of tones)	Primary auditory cortex & planum temporale	Auditoral sensory memory and perception	50 ms
Pho (3 chunks)	Left posterior parietal cortex, inferior parietal lobe	Phonological loop to store auditoria & textual information	Included in CE server
CE (3 chunks)	Dorsal lateral prefrontal cortex & ACC	Mental response inhibition & selection, improve processing rate of premotor cortex via parallel learning mechanism	Exponential distribution: Mean=70ms, Min=25ms
Hicog (1 chunk)	Left interior partial cortex, IPS and VLFC	Phonological judgment, visuomotor choices by retrieving production rules at LTDSM	[a]A_i+B_i Exp(-$\alpha_i N_i$)
LTDSM (∞)	Hippocampus	Long-term declarative memory (e.g. production rules in judgments and choices) and spatial memory	Included in Hicog
PM (1 chunk)	Premotor cortex (BA 6)	Select movement in learning visuomotor association	[a]A_i+B_i Exp(-$\alpha_i N_i$)
BG (1 chunk)	Basal ganglia	Motor program retrieval	[a]A_i+B_i Exp(-$\alpha_i N_i$)
LTPM (∞)	Striatal and cerebellar systems	Long term procedural knowledge storage	Included in BG server
SMA (2 letters)	Supplementary motor area & pre-SMA	Motor program assembly, error detection	188 ms
M1 (2 letters)	Primary motor cortex	Addressing spinal motorneourons	70 ms
Mouth (1 letter)	-	Converts neuron signal to movement of vocal organ	10 ms
Hand (1 letter)	-	Execution of motor movement	[b]$I_m \log_2(D/S+0.5)$

a. See the part "learning process of individual servers" in this paper. b. Fitt's Law: I_m=26.3,D=1.3cm, S: movement distance.

However, how do the entities of task 2 choose one of the two alterative routes in the network? What is the behavioral impact of this choice on PRP and the practice effect on PRP? This can be answered by integrating queuing networks with reinforcement learning algorithms.

Before exploring the mechanism with which entities of task 2 select from the 2 routes, it is necessary to understand the learning process of individual brain areas. It was discovered that each individual brain area reorganizes itself during the learning process and increases its processing speed (Ungerleider, 2002). For example, for the simplest network with 2 routes (see Figure 2), if servers 2 and 3 change their processing speeds, different routes chosen by an entity (1→3→4 or 1→2→4) will lead to different performance. Without considering the effect of error, entities will choose the optimal route with the shortest processing time if they want to maximize the reward of performance.

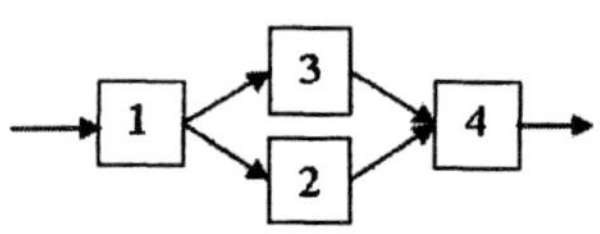

Figure 2: The simplest queuing network with 2 routes

Consequently, to model learning, it is first necessary to quantify the learning process in individual servers. Based on that, the condition under which an entity switches between the 2 routes shown in Figure 2 can be established and proved by integrating queuing network with reinforcement learning. Finally, this quantitative condition of route switching can be applied to the more complex model of 18 servers with 2 routes (see Figure 1) to generate the basic PRP and reduction of PRP during the learning process.

Learning Process in Individual Servers Based on the function of the servers in Table 1, the two long-term memory servers (LTDSM and LTPM) play the major roles in learning phonological judgments (task 1) and choice reaction (task 2) (Bear & Connor, 2001). Because the learning effects of long-term memory are represented as speed of retrieval of production rules and motor programs from the two long-term memory servers at Hicog and BG servers, it is important to quantify the processing time of Hicog and BG servers. In addition, because premotor cortex (PM) server is activated in learning visuomotor association (Mitz et al., 1991), change of the processing speed of PM server is also to be considered in the learning process of the model.

Because exponential function fits the learning processes in memory search, motor learning, visual search, and mathematic operation tasks better than power law (Heathcote et al., 2000), it is applied to model the learning process in the individual servers (Equation 1).

$$1/\mu_i = A_i + B_i \, \mathrm{Exp}(-\alpha_i N_i) \qquad (1)$$

μ_i: processing speed of the server i; $(1/\mu_i)$ is its processing time; A_i: The minimal of processing time of server i after intensive practice; B_i: The change of expected value of processing time of server i from the beginning to the end of practice; α_i: learning rate of server i; N_i: number of customers processed by server i.

For BG server, $1/\mu_{BG}$: motor program retrieving time; A_{BG}: the minimal of processing time of BG server after practice (314 ms, Rektor et al., 2003); B_{BG}: the change of expected value of processing time from the beginning to the end of practice (2*314=628 ms, assumed). α_{BG}: the learning rate of server BG (0.00142, Heathcote et al., 2000); N_{BG}: number of entities processed by server BG which is implemented as a matrix of frequency recorded in LTPM server.

For Hicog and PM servers, to avoid building an ad-hoc model and using the result of the experiment to be simulated directly, nine parameters in Hicog and PM servers were calculated based on previous studies (see Appendix 1).

Learning Process in the Simplest Queuing Network with 2 routes Based on the learning process in individual servers, the condition under which an entity switches between the 2 routes in the simplest form of queuing networks with 2 routes (each capacity equals 1) (from route 1→2→4 to route 1→3→4, see Figure 2) was quantified and proved by the following mathematic deduction.

1) Q Online Learning Equation (Sutton & Barto, 1998).

$$Q^{t+1}(i,j) \leftarrow Q^t(i,j) + \varepsilon \, \{r_t + \gamma \max_k [Q^t(j,k)] - Q^t(i,j)\} \qquad (2)$$

- $Q^{t+1}(i,j)$: online Q value if entity routes from server i to server j in t+1 th transition
- $\max_k[Q(j,k)]$: maximum Q value routing from server j to the next k server(s) (k>=1)
- $r_t = \mu_{j,t}$: reward is the processing speed of the server j if entity enters it at t th transition
- N_{jt}: number of entities goes to server j at t th transition;
- ε: learning rate of Q online learning ($0 < \varepsilon < 1$)
- γ: discount parameter of routing to next server ($0 < \gamma < 1$)
- p: probability of entity routes from server 1 to server 3 do not follow the Q online learning rule if Q (1,3)> Q (1,2). For example, if p=0.1, then 10% of entity will go from server 1 to server 2 even though Q (1,3)> Q (1,2).

State is the status that an entity is in server i; Transition is defined as an entity routed from server i to j. Equation 2 updates a Q value of a backup choice of routes ($Q^{t+1}(i,j)$) based on the Q value which maximizes over all those routes possible in the next state ($\max_k[Q(j,k)]$). In each transition, entities will choose the next server according to the updated $Q^t(i,j)$. If Q (1,3)>Q (1,2), more entity will go from server 1 to server 3 rather than go to server 2.

2) Assumption

- ε is a constant which do not change in the current learning process ($0 < \varepsilon < 1$)
- Processing speed of server 4 (μ_4) is constant

3) Lemma 1. At any transition state t (t≠0), if $1/\mu_{2,t} < 1/\mu_{3,t}$, then $Q^{t+1}(1,2) > Q^{t+1}(1,3)$

Proof of Lemma 1 (see Appendix 2).

Based on Lemma 1 and equation 1, we got Lemma 2:

4) Lemma 2. At any transition state t (t≠0), if $A_2 + B_2 \mathrm{Exp}(-\alpha_2 N_{2t}) < A_3 + B_3 \mathrm{Exp}(-\alpha_3 N_{3t})$, then $Q^{t+1}(1,2) > Q^{t+1}(1,3)$

Prediction of the Basic PRP and the Practice Effect on PRP with the Queuing Network Model Based on equation 1 and Lemma 1 and 2, we can predict the possible simulation result of PRP and PRP practice effect.

For the entities in task 2 (see Figure 1), at the beginning of the practice phrase, because the visual-motor mapping is not ready in PM (Mitz et al., 1991), PM takes a longer time to process the entities than the CE and Hicog. Thus, the Q value from Visk to PM (Q (1,3)) is lower than the Q value from Visk to CE (Q (1,2)). According to Lemma 1, the majority of the entities will go to the CE and Hicog server at the beginning of the learning process in dual tasks. Consequently, because entities from task 1 also go through the CE and Hicog server, a bottleneck at the Hicog server would form at this stage which produces the basic PRP effect.

During the learning process, CE will send dummy entities which increase the processing speed of PM based on the parallel learning mechanism between visual loop (including CE) and motor loop (including PM) (Nakahara et al., 2001, see Table 1). Therefore, when Q value of the 2nd route of task 2 increases, more and more entities of task 2 will travel in the 2nd route, forming an automatic process and two parallel routes in this dual-task situation. However, because the learning rate of PM server (1/16000) is lower than that of Hicog server for the entities in task 2 (1/4000), the majority of the entities will still go through the Hicog server.

Simulation Result

Figure 3 shows the simulation result of the basic PRP effect compared with the experiment data (Van Selst et al., 1999). The linear regression function relating the simulation and experiment results is: Y =1.057X-58 (Y: Experiment Result; X: Simulated Result; R square =.984, p<.001; sig. of constant > .05).

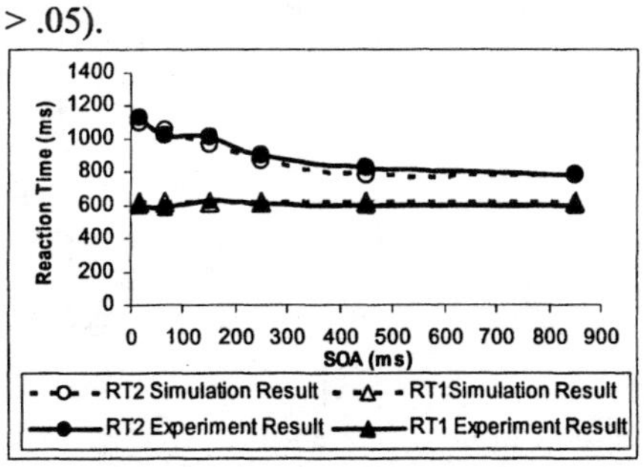

Figure 3: Comparison of simulation and experiment results at the beginning of practice (basic PRP effect)

Figure 4 compares of simulation and experiment result of PRP effect at the end of practice (after 7200's trials). The linear regression function relating the simulated results and experiment results is: Y=1.03X+105 (R square =.891, p<.001; sig. of constant > 0.05).

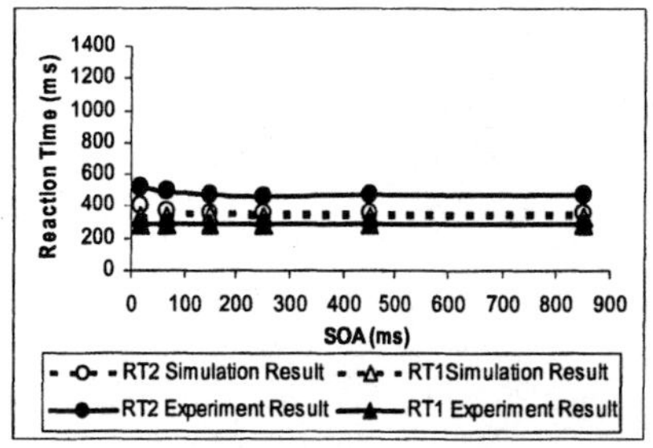

Figure 4: Comparison of the simulation and experiment results at the end of practice

Figure 5 shows the comparison of simulation and experiment results during the practice process (7200 trials). The linear regression function relating the simulated results and experiment results is: Y=0.965X +10 (R square =.781, p<.001; sig. of constant > 0.05). Moreover, it was found that the Q value of the 2nd route of task 2 never exceeded that of the 1st route of task 2 during the practice process and the majority of entities of task 2 went through the 1st route rather than the 2nd route.

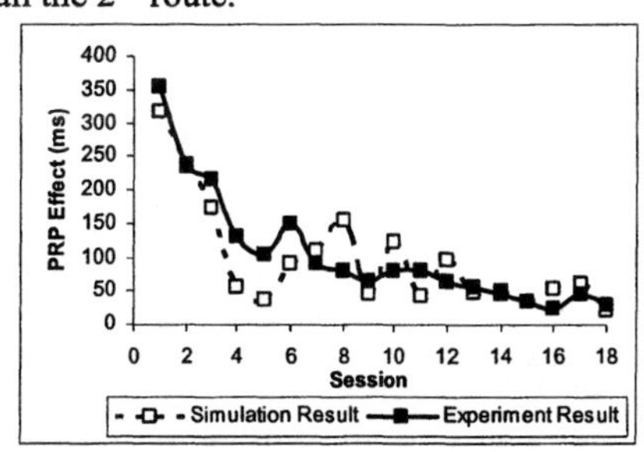

Figure 5: Comparison of simulation and experiment results during the practice process (7200 trials)

Discussion

Integrating queuing networks and reinforcement learning algorithms, the extended version of QN-MHP simulated both basic PRP effect and the practice effect on PRP in Van Selst (1999)'s study without setting up lock and unlock strategy or drawing complex scheduling chart.

With the formation of an automatic process during learning, two parallel routes were formed in the dual-task situation, which partially eliminated the bottleneck at the Hicog server. The PRP effect is reduced greatly with the decrease of the processing time in both the Hicog and the PM server. However, since the majority of the entities of the two tasks still went through the Hicog server, the effect of the automatic process on PRP reduction does not exceed the effect of the reduction of RT 1 on the PRP. This is consistent with the result of Van Selst (1999) that automatic process does grow from weak to strong but only contributes small PRP reduction.

The current model indicates that depending on an individual's degree of practice, cognition can be either serial or parallel at the level of production or response selection. At the beginning of practice, a bottleneck at the Hicog server within the central cognition is critical for the generation of PRP effect, which is consistent with the assumption of ACT-R/PM that cognition must be serial without considering the effect of practice. However, the formation of a parallel process in production selection after practice indicates that there are two response or production selection processors running in parallel in the cognitive system in the well-learned situation, which is different from the basic assumption in ACT-R/PM. This parallel processing in the well-learned situation in the current model is supported by Van Selst (1999)'s study and the review of fMRI study by Collette (2001) who found that in dual task situations not only the brain area of CE server (BA9/46: DLPFC & ACC) but also the area of well-learned visuomotor control (BA 6: premotor area, PM server) were activated simultaneously. In addition, it is also important to note that the formation of two parallel processing routes in the current model is different from the assumption and mechanism of EPIC because the two routes were formed with reinforcement learning naturally without using any unlock and unlock strategies.

Because the queuing network model was built in a general structure of human information processing, it can be easily transformed to model other task situations involving concurrent activities in practical situations, e.g. dual tasks in transcription typing (Wu & Liu, 2004b). In contrast to complex unlock and lock strategies in EPIC or drawing complex scheduling charts in ACT-R/PM, the current model can model concurrent tasks more easily by adding a route for the secondary task in the model, feeding the stimuli of the secondary task to the model, and coding the task description into an easy-to-use Excel sheet. This unique feature offers great potential of the model for easy application and learning by researchers and engineers in practice.

Acknowledgments

This article is based upon work supported by the National Science Foundation under Grant No. NSF 0308000. However, any opinions, findings and conclusions or recommendations expressed in this article are those of the authors and do not necessarily reflect the views of the National Science Foundation (NSF).

Reference

Bear, M., & Connor, B. (2001). *Neuroscience: Exploring the Brain (2nd edition)*. Lippincott Williams& Wilkins Publisher

Byrne, M.D., & Anderson, J.R. (2001). Serial Modules in Parallel: The Psychological Refractory Period and Perfect Time Sharing. *Psychology Review*, 108, 4, 847-869

Card, S., Moran, T., & Newell, A. (1986). The Model Human Processor: An engineering model of human performance. In Boff, K. (Eds.) *Handbook of Perception and human performance*, New York: Wiley

Collette, F. & Linden, M.V. (2002). Brain Imaging of the Central Executive Component of Working Memory. *Neuroscience and Biobehavioral Review*, 26, 105-125

Feyen, R.G. & Liu, Y. (2001). Modeling Task Performance Using the Queuing Network Model Human Processor (QN-MHP). *Proceedings of the Fourth International Conference on Cognitive Modeling*, 73-78

Fletcher.P.C., & Henson, R.N.A. (2001). Frontal Lobes and Human Memory: Insights from Functional Neuroimaging. *Brain*, 124, 849-881

Flynn, B.M. (1943). Pitch discrimination: the form of the psychometric function and simple reaction time to liminal differences. *Archives of Psychology (Columbia University)*, 280, 41

Heathcote, A., Brown, S., & Mewhort, D.J.K (2000). The power law repealed: the case for an exponential law of practice. *Psychological Bulletin*, 7, 2, 185-207

Liu, Y. (1996). Queuing network modeling of elementary mental processes. *Psychological Review*, 103, 1, 116-136

Liu, Y. (1997). Queuing network modeling of human performance of concurrent spatial and verbal tasks. *IEEE Transactions on Systems, Man, Cybernetics*, 27, 195-207

Liu, Y., Feyen, R., and Tsimhoni, O. (2004). *Queuing Network – Model Human Processor (QN-MHP): A Computational Architecture for Multitask Performance in Human Machine Systems*. Technical Report 04-05. Department of Industrial and Operations Engineering, University of Michigan

Meyer, D.E & Kieras, D.E. (1997a). A computational theory of executive cognitive. *Psychology Review*, 104, 1, 3-65

Mitz, A. R., Godshalk, M., & Wise, S.P. (1991). Learning – dependent activity in the premotor cortex: activity during the acquisition of the conditional motor assassinations. *The Journal of Neuroscience*, 11, 6, 1855-1872

Mowbray, G. H., & Rhoades, M.V. (1959). On the reduction of choice reaction time with practice. *Quarterly Journal of Experimental Psychology*, 14, 1-36

Mustovic. H., Scheffler, K., Di Salle, F. (2003). Temporal integration of sequential auditory events: silent period in sound pattern activates human planum temporale. *NeuroImage*, 20, 429–434

Nakahara, H., Doya, K., & Hikosaka, O. (2001). Parallel Cortico-Basal Ganglia Mechanisms for Acquisition and Execution of Visuomotor Sequences-A Computational Approach. *Journal of Cognitive Neuroscience*, 13, 5, 626–647

Rektor, I., Kanovsky, P., Bares., M. (2003). A SEEG study of ERP in motor and premotor cortices and in the basal ganglia. *Clinical Neurophysiology*, 114, 463–471

Roland, P.E. (1993). *Brain Activation*. John Wiley &Sons

Romero, D.H, Lacourse, M.G, Lawrence,K.E. (2000). Event-related potentials as a function of movement parameter variations during motor imagery and isometric action. *Behavioral Brain Research*, 117, 83–96

Rudell, A.P. & Hu, B. (2001). Does a warning signal accelerate the processing of sensory information? Evidence from recognition potential responses to high and low frequency words. *International Journal of Psychophysiology*, 41, 31-42

Ruthruff, E., Johnston, J.C & Selst, M.V. (2001). Why Practice Reduces Dual-Task Interference. *Journal of Experimental Psychology: Human Perception and Performance*, 27, 1, 3-21

Schmidt, R.A. (1988). *Motor control and learning* Human Kinetics Publishers, 82-298

Smith, E.E, & Jonides, J. (1998). Neuroimaging analyses of human working memory. *Proc. Natl. Acad. Sci. USA*, 95, 12061–12068

Sutton, R.S. & Barto, A.G. (1998). *Reinforcement Learning: An Introduction.* MIT Press: Cambridge, Massachusetts.

Tsimhoni,O., & Liu, Y.(2003). Modeling steering with the Queuing Network-Model Human Processor (QN-MHP). *Proceedings of the 47th Annual Conference of the Human Factors and Ergonomics Society*, 81-85

Ungerleider, L.G., Doyon, J., & Karni, A. (2002). Imaging Brain Plasticity during Motor Skill Learning. *Neurobiology of Learning and Memory*, 78, 553–564

Van Selst, M., Ruthruff, E. & Johnston, J.C. (1999). Can Practice Eliminate the Psychological Refractory Period Effect? *Journal of Experimental Psychology: Human* Perception and Performance, 25, 5, 1268-1283

Wu, C. & Liu, Y. (2004a). Modeling behavioral and brain imaging phenomena in transcription typing with queuing networks and reinforcement learning algorithms. *Proceeding of Sixth International Conference on Cognitive Modeling*, Pittsburgh, PA, USA (In press).

Wu, C. & Liu, Y. (2004b). Modeling human transcription typing with queuing network-model human processor. *Proceedings of the 48th Annual Meeting of Human Factors and Ergonomics Society*, New Orleans, Louisiana, USA (In press).

Appendix

1. Parameters Setting at Hicog and PM server

1) $A_{Hicog\text{-}symbol}$: minimal value of the processing time of task 2 entity in Hicog server. Since choice reaction time (RT) of 4 alternatives can be reduced to RT of 2 alternatives with practice (Mowbray et al., 1959), after intensive practice, RT of 8 alternative choices in Van Selst's experiment will reduce to RT of 4 alternatives without intensive practiee. $A_{Hicog\text{-}symbol}$ equals the RT of 4 alternatives (Hick's Law, intercept:150ms, slope:170ms/bit, Schmidt, 1988) minus 1 average perception cycle (100ms), 2 cognitive cycles (2*70 ms) and 1 motor cycle (70ms) (Card et al., 1986). Therefore, $A_{Hicog\text{-}symbol}$=150+170*Log$_2$(4)-100-2*70-70=180 ms.

2) $B_{Hicog\text{-}symbol}$: change of processing time of task2 entity in Hicog server at the beginning and end of practice. At the beginning of the practice in single task 2, RT of the 8 alternatives (Hick's Law, intercept:150ms, slope:170ms/bit, Schmidt, 1988) is composed of 1 perception cycle (100ms), maximum processing time at Hicog ($A_{Hicog\text{-}symbol}$+$B_{Hicog\text{-}symbol}$), and 1 motor cycle (70ms) (Card et al., 1986). Therefore, $B_{Hicog\text{-}symbol}$=150+170*Log$_2$(8)-100-$A_{Hicog\text{-}symbol}$-70=170ms.

3) $\alpha_{Hicog\text{-}symbol}$, $\alpha_{Hicog\text{-}tone}$: learning rate of Hicog server in processing the task 2 and task 1 entities. Based on α=0.001 approximately in Heathcote et al. (2000)'s study, learning difficulty increased 4 times because of the 4 incompatible alternatives. Thus, $\alpha_{Hicog\text{-}symbol} = \alpha_{Hicog\text{-}tone}$=0.001/4=1/4000.

4) $A_{Hicog\text{-}tone}$: minimal value of the processing time of task1 entity in central executive. After intensive practice, the discrimination task of the 2 classes of tones in Van Selst (1999) 's experiment can be simplified into a choice reaction time of two alternatives, requiring the minimum value of 1 cognitive cycle (25ms) (Card et al., 1986).

5) $B_{Hicog\text{-}tone}$: change of processing time of task1 entity in Hicog at the beginning and end of practice. At the beginning of the single task 1, the reaction time to discriminate the 2 classes of tone is 642ms (Flynn, 1943), which is composed of 1 perception cycle (100ms), 2 cognitive cycles (70*2 ms), ($A_{Hicog\text{-}tone}$+$B_{Hicog\text{-}tone}$) and 1 motor cycle (70ms). Therefore, $B_{Hicog\text{-}tone}$=642-100-2*70-$A_{Hicog\text{-}tone}$-70=307ms.

6) $A_{PM\text{-}symbol}$: minimal value of the processing time of task 2 entity in PM. After intensive practice, RT of the 8 alternative choices in Van Selst's experiment will transform to RT of 8 most compatible alternatives (RT=217ms, Schmidt, 1988) which is composed 1 perception cycle and 1 motor cycle. Therefore, $A_{PM\text{-}symbol}$=217-100-70=47ms.

7) $B_{PM\text{-}symbol}$: change of processing time of task2 entity in PM at the beginning and end of practice. At the beginning of practice in single task 2, RT of 8 alternative choice reaction time (Hick's Law: 50ms, slope: 170ms/bit) is composed of 1 average perception cycle (100ms), ($A_{PM\text{-}symbol}$+$B_{PM\text{-}symbol}$), 1 motor cycle (70ms). Thus, $B_{PM\text{-}symbol}$=150+170*Log$_2$(8)-100- $A_{PM\text{-}symbol}$-70=443 ms.

8) $\alpha_{PM\text{-}symbol}$: learning rate of PM in processing the task2 entity The speed of formation of the automatic process in PM is slower than Hicog because it receive the dummy entities from CE server via the indirect parallel learning mechanism with the 4 incompatible alternatives (Nakahara et al., 2001). Thus, $\alpha_{PM\text{-}symbol}$=(0.001/4)/4=1/16000.

2. Proof of Lemma 1

Lemma 1. At any transition state t (t≠0), if $1/\mu_{2,t} < 1/\mu_{3,t}$, then $Q^{t+1}(1, 2) > Q^{t+1}(1, 3)$

Proof. Using mathematic deduction method

i) At t=0: $Q^1(1,3)=Q^1(1,2)=Q^1(2,4)=Q^1(3,4)=0$

ii) At t=1: Using the online Q learning formula:

$Q^2(1,3)=Q^1(1,3) + \varepsilon [r_t + \gamma Q^1(3,4)-Q^1(1,3)] = \varepsilon\mu_{3,1}$

Note: because entity routes to only one server (server 4) $\max_b Q^t(S_t+1,b)=Q(3,4)$, $Q^2(1,2)= \varepsilon\mu_{2,1}$, $Q^2(3,4)= \varepsilon\mu_4$, $Q^2(2,4)= \varepsilon\mu_4$; If $1/\mu_{2,1} < 1/\mu_{3,1}$, then $\varepsilon\mu_{3,1} < \varepsilon \mu_{2,1}$ (given $0< \varepsilon<1$), i.e. $Q^2(1,2) > Q^2(1,3)$ Thus, Lemma is proved at t=1.

iii) According to Mathematic Deduction Method, Lemma 1 is correct: i.e. at transition state t=k: if $1/\mu_{2,k} < 1/\mu_{3,k}$, then $Q^{k+1}(1,2)>Q^{k+1}(1,3)$. We want to prove at transition state k+1, Lemma is still correct: i.e.At transition state t=k+1: if $1/\mu_{2,k+1} < 1/\mu_{3,k+1}$, then $Q^{k+2}(1,2)>Q^{k+2}(1,3)$

At t=k+1:

$Q^{k+2}(1,2)= Q^{k+1}(1,2)+ \varepsilon [\mu_{2,k+1} + \gamma\varepsilon\mu_4-Q^{k+1}(1,2)]$

$Q^{k+2}(1,3)= Q^{k+1}(1,3)+ \varepsilon [\mu_{3,k+1} + \gamma\varepsilon\mu_4 -Q^{k+1}(1,3)]$

$Q^{k+2}(1,2)- Q^{k+2}(1,3)$

$= Q^{k+1}(1,2)+ \varepsilon [\mu_{2,k+1} + \gamma\varepsilon\mu_4-Q^{k+1}(1,2)] -\{Q^{k+1}(1,3)+ \varepsilon [\mu_{3,k+1} + \gamma \varepsilon \mu_4-Q^{k+1}(1,3)]\}$

$= (1- \varepsilon)[Q^{k+1}(1,2)- Q^{k+1}(1,3)]+ (\varepsilon \mu_{2,k+1} - \varepsilon \mu_{3,k+1})$

With equation 3 and $0<\varepsilon<1$, we have:

$(1- \varepsilon)[Q^{k+1}(1,2)- Q^{k+1}(1,3)]>0$

Given $1/\mu_{2,k+1} < 1/\mu_{3,k+1}$ and $0<\varepsilon<1$, then $(\varepsilon \mu_{2,k+1} - \varepsilon \mu_{3,k+1})>0$, i.e. $Q^{k+2}(1,3)- Q^{k+2}(1,2) >0$

Thus, Lemma 1 is correct at t=k+1. Lemma 1 is proved.

Vergence Control in Fixation with Minimal Disparity Information

Weilie Yi (wyi@cs.rochester.edu)
Department of Computer Science, University of Rochester
Rochester, NY 14627 USA

Dana H. Ballard (dana@cs.rochester.edu)
Department of Computer Science, University of Rochester
Rochester, NY 14627 USA

Abstract

Vergence control is an important part of biologically plausible models of active vision, which plays an important role in cognition. Our vergence system is based on the output of three disparity-selective neurons corresponding to zero, near, and far disparities. A neuron's output is defined as the distance between the feature vectors of two points in the left and right images. Vergence control is generated by using these three disparity values, in search for global minimum of disparity. To escape from local minima, the image is subsampled, and gradually expanded to the original scale. Empirical results shows that the method is effective and robust when applied to targets in natural scences.

Introduction

In the human binocular visual system, because of the different viewpoints of the eyes, two retinas receive similar, but slightly different images of the physical world. To ensure the object of interest is in the fovea, where the majority of optical sensors gather (Yarbus, 1967), the brain sends motor control signals to muscles in the eyes and orient the eyeballs, according to the images perceived by the visual cortex. When the intersection of the lines of sights falls on the object of interest, the object is fixated on, and the vergence control is completed.

A central component of any visual system is the ability to perform figure-ground segmentation. Vergence control ability can greatly facilitate this process by isolating the parts of the world that have small disparities near the point of gaze. The small disparity constraint can be as powerful as color and shape cues(Coombs, 1992) in isolating useful objects. Vergence is also special because it provides depth information needed for reaching.

Vergence control has been implemented in various artificial binocular active vision systems (Hansen and Sommer, 1996; Manzotti et al., 2001; Sturzl et al., 2002). The goal of active control is to minimize the distance between the central pixel/patches of the two views. For the sake of simplicity, one camera could be designated as the reference camera, and computation of vergence control parameters only occurs on the other camera.

There are basically two methods to determine the control parameters to verge the second camera. One is to solve the Stereo Correspondence Problem. This problem asks which pixel in the first image corresponds to which pixel in the second one, or, in this case, which pixel in the verging camera's image corresponds to the central pixel of the dominant camera's image. Solving this problem involves heavy computation, because every possible pixel in the vergence view has to be compared with the central pixel in the reference view, and the one with maximal similarity is labeled as the corresponding pixel(Manzotti et al., 2001).

The other method is to use a close loop control module, which takes error as input and uses PID control to generate vergence parameters(Hansen and Sommer, 1996). This method does not guarantee convergence because the error curve could have multiple minima with various values due to the change of image characteristics in an nonlinear and unpredictable fashion(Manzotti et al., 2001).

In a symmetrical binocular vision system, which doesn't have a dominant camera, the same problem persists(Sturzl et al., 2002).

To make the vergence control computationally efficient and biologically more plausible, we incorporated the idea of disparity selective cells(Ohzawa et al., 1996) and minimized the amount of information needed for the vergence control problem. The motivation is that, in a foveated visual system, the majority of information comes from the fovea. So the computation has to focus on the centers of retinal images. In our approach, three neurons are responsible for detecting zero, tuned near and tuned far disparities respectively. According to the relative values of these three neuron's output, control command is generated and the two views are updated, entering the next loop of execution. Inspired by psychopysiological studies, we also applied Scale Space Theory to ensure a better convergence.

The rest of this paper is arranged as follows. The next section introduces the simple cell model and our basic algorithm for vergence control. Section 3 provides a modified version of this algorithm. Experimental results are presented in Section 4. Finally, conclusive remarks are given in Section 5.

Vergence Control Algorithm

In our binocular vision system, only horizontal disparities exist. Further more, since the fovea lies in the center of the image, we only consider the distance between the central point in one image and points on the middle horizontal line of the other image, neglecting vertical varieties. As shown in Figure 1, an error curve has all the information we need to compute the vergence control

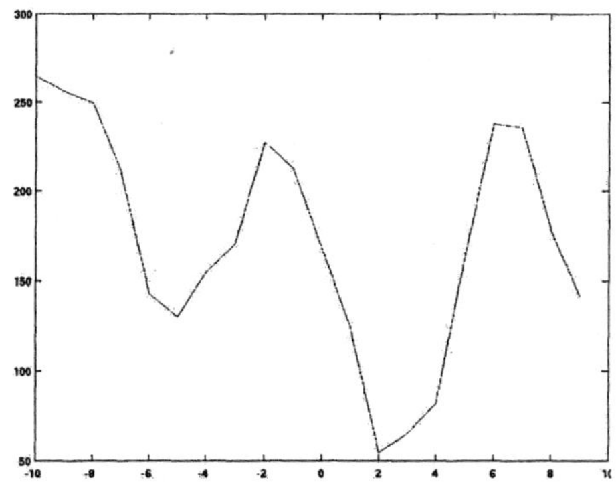

Figure 1: Error curve for the right eye. The X axis is the relative horizontal position from the center of the retinal image. The Y axis is the Euclidean distance, between the image patch at position X in the right image and the one at position -X in the left image. For example, the patches in the center of the two images have an error of about 150, while the patch centered at (4, 0) in the right image and the patch centered at (-4, 0) in the left image have an error of around 80. To fixate on any particular point, the vergence system must reduce its error to zero.

parameters. The minimum of this curve corresponds to the optimal verge angle. In our algorithm, we only use a minimal subset of it to direct the vergence, instead of searching for the minimum in a brute force manner.

Simple Cell Model

On the visual pathway of the brain, visual stimuli are sent from the retina to the primary visual cortex via LGN (lateral geniculate nucleus) (Hubel, 1988). The majority of the orientation sensitive neurons are called *simple cells*, whose receptive fields can be modeled by Gabor filters (Petkov and Kruizinga, 1997). In the classical view, *binocular neurons* in the visual cortex receive inputs from simple cells and they are tuned to specific disparity values (Hubel, 1988) (Qian, 1994) (Qian and Zhu, 1997). Disparity selective binocular neurons can be classified to five categories, according to their tuning curve: tuned zero disparity cells, tuned near/far disparity cells and untuned near/far disparity cells.

Receptive Fields

The base representation, or the receptive fields of the first layer image processing units, could be modeled by Gabor filters (Daugman, 1980). The general form of a 2-D Gabor filter is:

$$F(x,y) = \frac{1}{2\pi\alpha\beta} e^{-\pi[\frac{(x-x_0)^2}{\alpha^2} + \frac{(y-y_0)^2}{\beta^2}]} e^{i(\xi_0 x + \nu_0 y)} \quad (1)$$

We will use Gaussian derivatives to approximate Gabor filters (Rao and Ballard, 1995):

$$G(x,y) = \frac{1}{\sqrt{2\pi}\sigma} e^{-\frac{x^2+y^2}{2\sigma^2}} \quad (2)$$

$$G_1^0 = \frac{\partial}{\partial x} G(x,y) = -2xe^{-(x^2+y^2)} \quad (3)$$

$$G_1^{\pi/2} = \frac{\partial}{\partial y} G(x,y) = -2ye^{-(x^2+y^2)} \quad (4)$$

Basing on Eqs. 3 and 4, orientation selective receptive fields with the same spatial frequency could be synthesized as:

$$G_1^\theta = k_{11}(\theta) G_1^0 + k_{21}(\theta) G_1^{\pi/2} \quad (5)$$

where θ is the preferred orientation and

$$k_{11}(\theta) = cos(\theta)$$
$$k_{21}(\theta) = sin(\theta)$$

Similar steerability exists in other orders. For each pixel on the image, a feature vector $\mathbf{V}$ is computed basing on a set of selective fields, typically 0th to 3th Gaussian derivatives, by doing convolution with discrete filters on the image.

$$V(x,y) = \sum_{|i-x|<R, |j-y|<R} I(i,j) G(i-x, j-y) \quad (6)$$

where $\mathbf{G}$ is a vector of discretized Gaussian derivative receptive fields. A screen shot of the computation of base representation is in Fig. 2.

Minimal Information Model

Of the five categories of disparity selective neurons, we only used one neuron from the first 3 categories (tuned zero, tuned near and tuned far), whose receptive field is easy to implement. The zero disparity neuron's output is defined as the Euclidean distance between the feature vectors of the central pixels of two images.

$$d^{zero} = (V^l(\hat{x},\hat{y}) - V^r(\hat{x},\hat{y})) \cdot (V^l(\hat{x},\hat{y}) - V^r(\hat{x},\hat{y}))^T \quad (7)$$

where $< \hat{x}, \hat{y} >$ is the central pixel of an image, and V^l and V^r stand for the feature vectors of left image and right image respectively. The near and far disparities are defined as

$$d^{near} = (V^l(\hat{x}+1,\hat{y}) - V^r(\hat{x},\hat{y})) \cdot (V^l(\hat{x}+1,\hat{y}) - V^r(\hat{x},\hat{y}))^T \quad (8)$$

$$d^{far} = (V^l(\hat{x}-1,\hat{y}) - V^r(\hat{x},\hat{y})) \cdot (V^l(\hat{x}-1,\hat{y}) - V^r(\hat{x},\hat{y}))^T \quad (9)$$

The vergence control algorithm only uses the output of these three disparity tuned neurons.

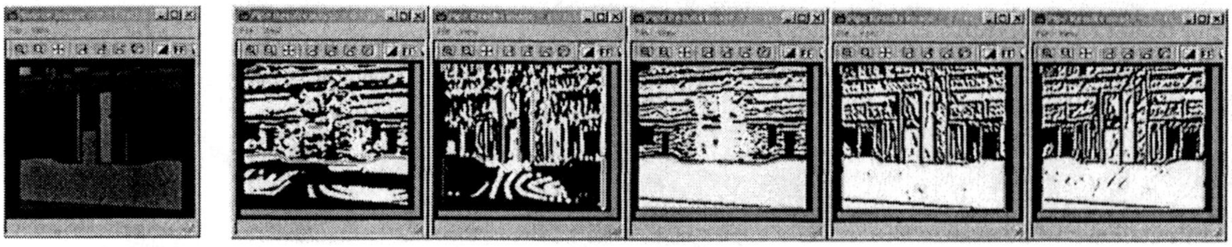

(a) (b)

Figure 2: Filter Outputs. (a) shows the raw image sent over network, and (b) shows five filtered images used in a vergence control operation. These filters, from left to right, are 1st order and 2nd order Gaussian derivatives. The sizes of these filters are all 15 by 15

The Basic Vergence Control Algorithm

Our basic vergence control algorithm, which doesn't work well, is like follows.

Vergence Control Algorithm 1

```
calculate errors;
WHILE (d_near < d_zero OR d_far < d_zero)
{
  IF (d_near < d_zero)
     converge the eyes;
  ELSE
     diverge the eyes;
  re-calculate errors;
}
```

The vergence control here is symmetric: Both cameras adjust their orientation symmetrically basing on the output of the three disparity neurons. In the loop, outputs of the three disparity neurons are compared. We tried to used quadratic function to model the local error curve according to the three values, but failed due to the unpredictable nature of the views. This algorithm terminates when the minimum is found, e.g. the zero disparity value is smaller than both near and far disparities.

A problem with this algorithm is that it is always looking for a minimum which is next to the starting point. The minimum it finds is not guaranteed to be a global one. We will address this problem in the next section.

Saccadic Suppression and Scale Space Theory

The solution of the local minimum problem is motivated by a psychological observation called saccadic suppression and the Scale Space theory in computer vision.

Psychological studies discovered that when the eye is performing a saccadic movement, the resolution of the images sent to the visual cortex is reduced to keep the stability of the view(Ross and Ma-Wyatt, 2004; Thilo et al., 2004). This is mainly because of the prohibitive lateral retinal interconnections(Hubel, 1988). Computationally speaking, retinal images are subsampled before being sent over the optic nerve. Since every saccade is generally followed by a fixation, fixation can take advantage of saccadic suppression by processing the shrunken images.

In our revised algorithm, vergence control begins with a highly shrunken image pair. Suppression is implemented as another loop which gradually decreases the suppression factor so that the images become finer and more details are unveiled. As the original sized images are restored, the accuracy of vergence comes to a maximum. With this suppression loop, the pixel pair with global minimum of distance is always close to the centers of the images.

Closely related to the saccadic suppression observation, Scale Space Theory was introduced to study the properties of images in different scale levels(Lindeberg and Romeny, 1994; Lindeberg, 1994). For any image I, a continues family of images $I(x, y, \sigma)$ is a series of blurred versions of $I(x, y)$ in which $I(x, y, \sigma)$ is the original image $I(x, y)$ when $\sigma = 0$:

$$I(\sigma) = I \otimes G(\sigma) \tag{10}$$

where $G(\sigma)$ is a Gaussian kernel $G(x, y, \sigma) = \frac{1}{\sqrt{2\pi}\sigma} e^{-\frac{x^2+y^2}{2\sigma^2}}$ and $\otimes$ is a convolution operator. The space (x, y, σ) is called a scale space, and σ the scale parameter.

One of the basic properties of the scale space is "non-creation of extrema"(Lindeberg, 1994), which means, as the scale parameter increases, local extrema would merge together, and when σ is above a certain value, only one extremum exists.

The Enhanced Vergence Control Algorithm

Inspired by the theories described above, we revised our algorithm to simulate the suppression process. Intuitively when the scale parameter, or equivalently the suppression factor, increases, local minimum tend to disappear and only the global one persists, at a slightly different location, because of the blurring effect. Our enhanced vergence control algorithm is as follows.

Vergence Control Algorithm 2

```
initialize scale
WHILE(scale > 0)
{
  calculate errors;
  WHILE (d_near < d_zero OR d_far < d_zero)
  {
    IF (d_near < d_zero)
       converge the eyes;
    ELSE
       diverge the eyes;
    re-calculate errors;
  }
  lower scale;
}
```

Note that this algorithm doesn't guarantee convergence to the global minimum of error, which in some extreme cases, e.g. wall paper illusion, is impossible(McKee and Mitchison, 1988).

Experiments

We implemented this algorithm in the Virtual Reality Laboratory at University of Rochester. A virtual human running on an SGI station walks in a town, and the images it sees are sent over network to a PC where the algorithm runs. This PC has a Datacube™ MaxRevolution image processing board, which is dedicated to the computation of image convolutions. Vergence control commands are sent back to the virtual human and the view ports of its two eyes are updated.

The changes of disparity curves are illustrated in Figure 3. Note only the three disparity values, d^{near}, d^{zero} and d^{far}, as shown in Figure 3(b), were actually used in the experiments, while Figure 3(a) gives the complete disparity curves for reference.

Since vergence facilitates depth perception. We used the deictic information, the distance between two eyes, to compute how far the fixated object is from the viewer. In our experiment, the object gradually moves away and so the vergence angle decreases. We calculated the distance by solving the following equation

$$e/2 = d \tan(\alpha/2) \qquad (11)$$

where e is the distance between two eyes, d is the distance between the fixated object and the viewer, and α is the vergence angle. In our experiments, $e = 0.05m$. We used linear regression to extract the relation between the computed distance and the true distance which is available from the simulator program. The result is shown in Figure 4.

Conclusions

We presented a biologically and psychologically plausible model for vergence control. In this single cell based model, minimal disparity was used to determine the vergence command. Inspired by Scale Space Theory and saccadic suppression, a coarse-to-fine loop was introduced to help the process converge.

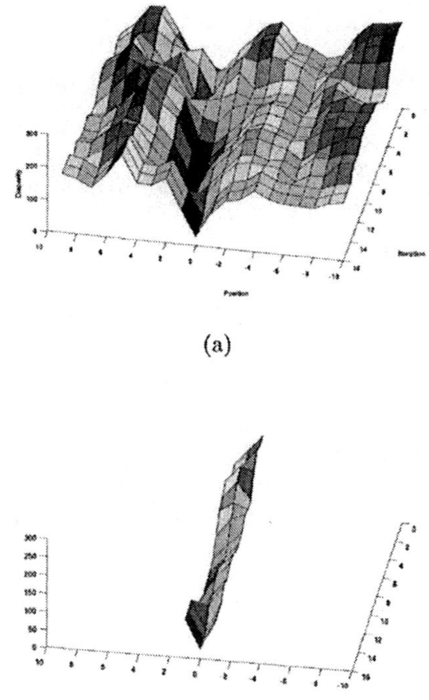

(a)

(b)

Figure 3: Vergence Control and Error Curves. This plot shows how the error curve (the curve on an error-position plane) changes over the iterations. Only the iterations with the starting scale parameter are illustrated. Later ones with scale parameters down to zero are not displayed for the sake of clarity. Subfigure (a) shows the complete error surface, while (b) only shows the output of the three disparity tuned neurons, which correspond to positions -1, 0 and 1. Only these three values are used in our algorithm, which successfully reduce the error at the point x=0 to zero.

We cannot compare this model to others at this time because we do not have a proper criteria. For instance, it is not reasonable to compare the computational complexity of various models without considering the physiological mechanism, which is the object of these models, and which is unknown. Our work suggests, it is likely that the brain can go around the Correspondence Problem and do vergence control with visual information which is very local to the fovea.

Acknowledgments

This work was supported by NIH grants MH60624 and P41 RR 09283.

References

Coombs, D. J. (1992). *Real-Time Gaze Holding in Binocular Robot Vision.* PhD thesis, University of

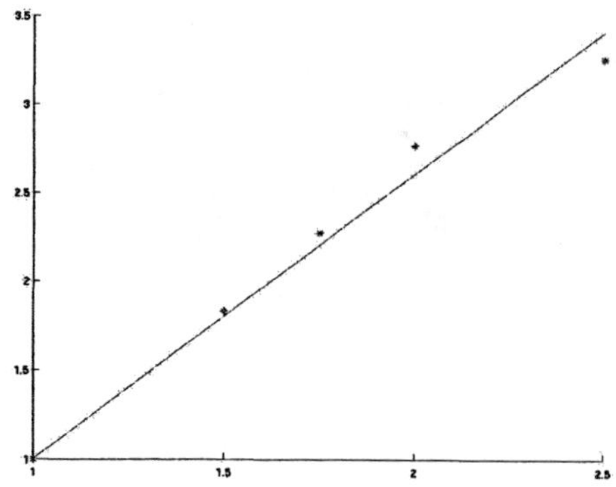

Figure 4: Computation of distance. The X axis is the computed distance between the fixated object and the viewer, while the Y axis is the true distance known from the simulating software.

Rochester.

Daugman, J. G. (1980). Two-dimensional spectral analysis of cortical receptive field profile. *Vision Research*, 20:847–856.

Hansen, M. and Sommer, G. (1996). Active depth estimation with gaze and vergence control using gabor filters. In *Proc., 13th Int. Conf. Pattern Recognition*, volume A, pages 287–291, Vienna, Austria.

Hubel, D. H. (1988). *Eye, brain, and vision.* Scientific American Library.

Lindeberg, T. (1994). *Scale-Space Theory in Computer Vision.* Kluwer Academic Publishers.

Lindeberg, T. and Romeny, B. M. T. H. (1994). *Linear scale-space*, chapter 1-2. Series in Mathematical Imaging and Vision. Kluwer Academic Publishers.

Manzotti, R., Gasteratos, A., and Metta, G. (2001). Disparity estimation in log polar images and vergence control. *Computer Vision and Image Understanding*, 83:97–117.

McKee, S. P. and Mitchison, G. J. (1988). The role of retinal correspondence in stereoscopic matching. *Vision Research*, 8:1001–1012.

Ohzawa, I., DeAngelis, G. C., and Freeman, R. D. (1996). Encoding of binocular disparity by simple cells in the cat's visual cortex. *J. Neurophysiol.*, 75:1779–1805.

Petkov, N. and Kruizinga, P. (1997). Computational models of visual neurons specialised in the detection of periodic and aperiodic oriented visual stimuli: bar and grating cells. *Biological Cybernetics*, 76(2):83–96.

Qian, N. (1994). Computing stereo disparity and motion with known binocular cell properties. *Neural Computation*, 6(3):390–404.

Qian, N. and Zhu, Y. (1997). Physiological computation of binocular disparity. *Vision Research*, 37:1811–1827.

Rao, R. P. and Ballard, D. H. (1995). An active vision architecture based on iconic representations. *Artificial Intelligence*, 78:461–505.

Ross, J. and Ma-Wyatt, A. (2004). Saccades actively maintain perceptual continuity. *Nature Neuroscience*, 7(1):65–69.

Sturzl, W., Hoffmann, U., and Mallot, H. A. (2002). Vergence control and disparity estimation with energy neurons: Theory and implementation. In Dorronsoro, J. R., editor, *Proc., Int'l. Conf. on Artificial Neural Networks*, pages 1255–1260.

Thilo, K. V., Santoro, L., Walsh, V., and Blakemore, C. (2004). The site of saccadic suppression. *Nature Neuroscience*, 7(1):13–14.

Yarbus, A. L. (1967). *Eye movements and vision.* New York, Plenum Press.

The Data Learning Problem Arises also in Cognitive Architectures other than Soar

Richard M Young (r.m.young@herts.ac.uk)
Department of Psychology, University of Hertfordshire
Hatfield, Herts AL10 9AB, UK

Abstract

The *data learning problem* is a phenomenon that arises when an agent employing a cognitive architecture faces the task of acquiring declarative information from an external source, such as the "answer" to a "question". Because the agent has to pay attention to both question and answer in order to learn the association between them, it is problematic for the agent to learn to produce the answer in response to the question alone. This observation helps shape the basic characteristics of human memory. The problem was first reported with the Soar architecture, but it arises also in ACT-R, and this paper argues that it will occur in any cognitive architecture, connectionist as well as symbolic, which is specified in a sufficiently explicit manner to avoid having the theorist act as an implicit homunculus for the agent.

The Data Learning Problem

The *data learning problem* is a phenomenon that arises when an agent faces the task of acquiring declarative information provided by an external source. Paradigmatic tasks are, in the laboratory, paired associates, where in response to a stimulus word such as *shelf,* the agent has to learn to say the digit *eight;* or in everyday life, learning the correct answer to a question, such as that *the capital of France* is *Paris.* In such situations, both the question and the answer (or the "stimulus" and the "response") must be present and attended to by the agent, in order to learn the connection between them. Thus the agent must process both the question "capital of France?" and the answer "Paris" if it is to acquire the relevant fact. It is therefore problematic for the agent to acquire a directed association from the question to the answer, such that it will be able to retrieve the answer in response to the question alone, without the answer also being present.

The data learning problem was first identified in work with the Soar cognitive architecture. However, this paper shows that the problem occurs with ACT-R as well, and argues that it arises in any architecture which is sufficiently theoretically explicit.

The Data Learning Problem in Soar

The data learning problem was first noticed and discussed (and termed the *data chunking problem*) by Rosenbloom, Laird & Newell (1987; Newell, 1990) in early work with Soar, and later substantially re-analysed by others (Vera, Lewis & Lerch, 1993; Young & Lewis, 1999). All these authors trace the problem back to the learning mechanism employed by Soar.

In Soar, all knowledge — both declarative and procedural — is encoded as *if U then* associations in production rules. These production rules implement a search in a state space, by repeatedly identifying an operator to apply to the current state, thereby tranforming it to a new state. Learning can occur only when Soar encounters a situation, an *impasse,* where it cannot proceed with its normal process of choosing operators and applying them to the current state. In the impasse, Soar sets up a new state, called a *substate,* and starts a new search whose purpose is find further information that resolves the impasse and allows the original search to continue. Soar learns a new production rule when it finds information that helps the resolution of the impasse and which it associates with the original state. For that new production rule, the 'then' part simply consists of the new information. The 'if' part consists of the information that existed before the impasse was encountered and to which the new information is connected by the sequence of production rule firings.

That description sounds a bit complicated, but the process is quite simple in concrete cases. It also has some surprising consequences. Consider a Soar agent learning about the capital of France. As analysts outside of the process, we can see that the production rule the agent needs to acquire is of the form:

Q: capital of France? U A: Paris (R1)

However, in the actual learning situtation, the agent needs to be told both the question and the answer, otherwise it will have no way of knowing what the answer is. So the agent encounters an impasse because it doesn't know the answer to the question *What is the capital of France?,* and that impasse is resolved when it is told that the answer is *Paris* because it can then respond "Paris". But because both the question and the answer were consulted in giving that response, the production rule actually learned takes the form not of R1 but of:

Q: capital of France? & A: Paris U A: Paris (R2)

Rule R2 encodes the information "If you are given the question *What is the capital of France?* and the answer *Paris,* then respond 'Paris'". This is obviously not equivalent to the desired rule R1, because it is applicable only when it is given both the question and the answer. It therefore will not fire when given only the question, which is what R1 does and what the agent needs to be able to do if it is to know how to answer the question. How, then, can the agent ever acquire rule R1?

The solution to this problem (Rosenbloom *et al,* 1987) requires us to note that although rule R2 cannot serve to *recall* the answer to the question, it can serve as an *episodic recognition* rule encoding the fact that the question and answer have been encountered together. Rosenbloom *et al* point out that if the agent is able to generate plausible candidates for the answer, then the recognition rule R2 can fire to pick out which one was previously presented as the answer. In this case the agent asked *What is the capital of France?* might generate and consider well-known cities in France, such as *Calais,* or *Lyon,* or *Marseilles,* or *Paris.* For the first three nothing would happen, but for the fourth, rule R2 would fire, identifying Paris as the answer and enabling the agent to respond. Because this time the answer is generated internally, the impasse occurs when only the question is present, not the answer, and so when *Paris* is given as the response the new rule learned is R1. Thereafter the agent can answer the question by recalling the answer, i.e. by firing rule R1, and the generation process does not have to be repeated.

Young & Lewis (1999) re-cast the process as a choice between four methods of answering the question:

1. If there is a production rule that provides the answer to the question, then it fires to deliver the answer.
2. If the answer already exists in working memory, then it gets picked up and delivered as the answer.
3. If possible candidates for the answer can be generated, and if a production rule exists that will recognise the answer, then candidates are generated until the answer is recognised.
4. The answer is sought from external sources.

These methods are listed in order of increasing effort. It is therefore rational for the agent to attempt the methods in that order. On the first pass, when the agent is first asked *What is the capital of France?,* none of the first three methods is applicable: the agent does not yet have a rule like R1; the answer is not available in working memory; and although the agent can try generating plausible candidates, there is no basis for recognising the correct answer. So the agent needs to be told (method 4) that the answer is *Paris.* It then has the answer in working memory, and so can respond (using method 2). The result is the acquisition of rule R2, the recognition rule.

On the second pass, when the agent is asked the question again, once more neither method 1 nor method 2 is applicable. Method 3 however does now work. As just described, the agent proceeds to generate plausible candidates, and when *Paris* is generated rule R2 fires, identifying it as the correct answer. In consequence, the recall rule R1 is acquired.

On the third and subsequent passes, once the agent is asked the question rule R1 fires (method 1) to give the answer. The agent has now clearly learned the geographical fact.

In their discussion, Young & Lewis (1999) stress several characteristics of this process:

- Learning the fact, i.e. acquiring the recall rule R1, requires the agent at one stage to generate the answer from its own internal, cognitive resources (Newell, 1990). This observation connects with wider discussion of the "reconstructive" nature of human memory (e.g., Bahrick, 1970).
- The sequence of learning emerges as a by-product of the agent's rational attempts to answer the question. At no point is there a deliberate attempt to learn, and there is no appeal to any mechanism for "self-programming", i.e. deliberate learning.
- The learning process is inherently multi-pass, not because of any kind of gradual strengthening but because of the inherent informational dependencies in the process. The first pass is needed to acquire the recognition rule, which is used in the second pass to acquire the recall rule. Only in the third and later passes can recall occur.

The learning process just sketched arises as a consequence of the architecture, and is lean in its theoretical assumptions. Young & Lewis see these as theoretically desirable properties of any explanation of learning. In particular, the analysis argues against the need for any homunculus-like mechanism for deciding what should and what should not get learned.

Generality of the Data Learning Problem

Although the data learning problem has to date been documented and analysed only for Soar, our account of its origins suggests that the problem is potentially of wide generality and should arise also for other, or perhaps even all, cognitive architectures. After all, the only assumption needed to carry the argument is that learned memory associations incorporate the information presented — and that would seem to be a reasonable tenet of most architectures and associative theories.

So, is the data learning problem just a quirk of Soar, or is it instead pointing us to a deep truth about cognitive architectures in general? If we strip away the particularities of Soar, the bare bones of the argument look something like this:

- We assume that memory (or at least, part of memory) is encoded in the form of directed associations, *cue* ∪ *response.*
- When learning a new association, the *response* derives from some response previously produced by the agent, and the *cue* reflects the information presented to the agent.
- When learning a new fact, such as a question-answer pair, the agent has to be presented with both the question and the answer, since by supposition this is a new fact that the agent does not already know.
- Therefore, the agent is likely to acquire an association

  cue & response ∪ *response*

  which does not serve to recall the *response* given just the *cue.*

If we apply this argument to the case of an agent learning that the capital of France is Paris, it follows that the agent will be presented with both the question *capital of France?*

and the answer *Paris,* and will learn a new association based on

Q: capital of France? & A: Paris ∪ A: Paris.

This is the data learning problem. The solution to the problem appears to be that:

- By some means or other, the agent has to generate the *response* from its own cognitive resources. (The learned association just above may be of help in this process.)
- The effect of generating the response with only the cue presented will be to learn a new association based on just the cue, e.g.,

  Q: capital of France? ∪ A: Paris

  which is exactly the association wanted.

Stated that way, the argument makes no appeal to features or mechanisms specific to Soar. Indeed, the assumptions would seem to be widely shared by most if not all associative theories and architectures, which means that they too should exhibit the data learning problem.

If the logic of this argument is correct, it raises a number of puzzling questions. Why has the problem been reported only for Soar? Does the problem not in fact arise for other theories and architectures, or has it simply not been noticed? If it does arise, why has it not been identified and discussed before now?

Such questions are perhaps more for historians and philosophers of science to answer than for working scientists, but we can go some way towards resolving the puzzle by considering the operation of the many theories that posit a direct learning of cue-response associations. It appears that the authors of such theories have no hesitation in assigning to the cue and the response different roles in the postulated learning mechanism, even when both are of the same kind (e.g., written words) and are simultaneously presented to the agent. For example, in network models (for more on which, see below), the theorist might treat the cue as the input to a network, while the response is treated as the target output for supervised training of the network. Even as careful a theorist as Estes (e.g., 1994), for example, seems implicitly to assume that in verbal paired-associate learning, subjects have enough deliberate control over their cognitive mechanism that they can treat the *cue* part of a paired associate in one way and the *response* part in another.

The plausibility of this assumption would intuitively seem to depend upon the nature of the items being associated. In the case of classical conditioning, say, the 'response' is definitely a *response,* i.e. something that the subject actually does (such as a dog's salivating: Pavlov, 1927), and is quite distinct in kind from the perceptual stimulus used as the cue (in that case, the ringing of a bell). In such cases, it seems reasonable to suppose that the cue is treated by the cognitive machinery as the input to an associative network, with the response being the output. But at the other end of the spectrum, with verbal materials, the situation is quite different. There, in many studies, the cue and response are exactly the same kind of thing: written words or nonsense syllables, where the desired 'response' is as much an input (presented to the subject) as an output. Indeed, the same item can appear as both a cue and a response, in the same or different lists. How plausible is it then that just because the experimenter tells the subject "When you see *this,* I want you to say *that*", the subject's cognitive system will treat the two items in such contrasting ways?

To take just one concrete illustration from the many possible, when presenting their "stimulus sampling theory", Atkinson & Estes (1963) state that

> "... most contemporary learning theories utilize a common conceptualization of the learning situation in terms of *stimulus, response,* and *reinforcement.* The stimulus term of this triumvirate refers to the environmental situation with respect to which behavior is being observed, the response term to the class of observable behaviors whose measurable properties change in some orderly fashion during learning, ..." (p.123)

However, in describing a paired-associate learning experiment, they give an example of a pair

S	R
xvk	*house*

and say "On a reinforced trial the stimulus and response members were exposed together, as shown." (p.126). So despite its earlier definition in terms of "observable behaviors", the desired 'response' is presented as a perceptual input to the subject just as much as the 'stimulus' is.

The (implicit) assumption that the cue and response can be treated in contrasting ways by the agent's cognitive mechanism makes sense in terms of task analysis, because they do indeed play asymmetric roles in the task: given the cue, the agent is supposed to recall the response. However, the assumption is much less plausible from the viewpoint of the cognitive architecture. It seems unlikely that two structurally similiar perceptual stimuli can be treated by the cognitive mechanism in such structurally contrasting ways. From the perspective of a more complete computational account, such as is given by Soar or other integrated cognitive architectures, it becomes clear that the theorist — albeit unintentionally — is playing the role of a *deus ex machina* in relation to the mechanism postulated by the associative theories, making sure that the cue goes *here* and does *this,* and the response goes *there* and does *that* within the mechanism.

According to the viewpoint adopted in this present paper, this role being taken by the theorist serves to mask the data learning problem, leaving the theorist unaware of it. Thus, the answer to the puzzle is that, in principle, the data learning problem does indeed arise for other learning theories, and becomes evident once they are formulated sufficiently carefully and explicitly. However, in their usual formulations, the theorist plays a homunculus-like role by arranging for the cue to be handled by the theory in one way and the response in a different way. We should expect that the data learning problem will become apparent once the theories are reformulated with sufficient computational explicitness.

The rest of the present paper begins to explore the generality of the data learning problem. It first shows that the problem arises also in ACT-R, and then goes on to argue that it occurs in connectionist architectures too. The paper concludes by considering the implications for cognitive theory.

The Data Learning Problem in ACT-R

ACT-R (Anderson & Lebiere, 1998; Anderson, Bothell, Byrne, Douglass, Lebiere & Qin, in press) stores long-term knowledge in two different forms. Declarative Memory (DM) holds information in the form of simple relational units called *chunks,* while Production Memory holds knowledge in procedural form as production rules. By default, facts are stored as chunks in DM. However, retrieval of chunks from DM is a potentially slow and error-prone process. Consequently, when a task which includes the retrieval of information from DM has been much practised, the information becomes incorporated into production rules which then no longer have to access DM during fluent performance. This process is referred to as *knowledge compilation.*

New chunks in DM are acquired by, among other means, perceptual input from the environment. The acquisition of new production rules is a little more complicated (Taatgen & Lee, 2003). Basically, ACT-R examines each pair of production rules that fire consecutively, and tries to compose a new production by combining them. If the first production is of the form *condition1* ∪ *action1* and the second one is *condition2* ∪ *action2,* then the new composed production is roughly

{condition1 ∪ condition2} – action1 ∪ action2

where the '∪' indicates the union of the two conditions. In the case where *action1* requests an item to be retrieved from DM and *condition2* makes use of that item, the actual item retrieved is compiled into the new production rule, so that in future it avoids the need for the retrieval.

Interestingly, this mechanism implies that ACT-R is unable directly to acquire a production rule encoding information presented externally. In other words, it exhibits the data learning problem. Consider an ACT-R model given the information that Paris is the capital of France, and trying to acquire a rule like R1 encoding that fact. It cannot do so directly, because the composition mechanism just described gives a newly created production an action (*action2* as written above) which comes from an existing production, and initially the model will have no production whose action side is *Paris.*

The solution is for the model first to learn the information as a fact in DM. Then, on a later occasion, it can retrieve the fact and compile it into a production rule. The whole process of acquisition consists of three stages:

In the first stage, the ACT-R model is given the question *capital of France?* and the answer *Paris.* It attends to them and acquires a chunk in DM encoding the question-answer relation between them.

In the second stage, the model is given the question again (but not the answer), and it attempts to answer the question itself by using a pair of very general production rules representing a method along the lines of

IF you want to answer a question (Pa)
THEN try to retrieve from DM a fact about that question

IF you want to answer a question (Pb)
and you have retrieved from DM a fact relating that question to an answer
THEN give that answer

In this case the question is *capital of France?* and the answer retrieved is *Paris.* The consequence of firing production Pa followed by Pb is that the model acquires a new production rule corresponding to R1.

In the third stage, on later occasions when the model is presented with the question, the new production rule R1 fires and yields the answer directly.

This process has some noteworthy characteristics:

- Acquiring the recall rule R1 requires the agent to generate the answer from its own internal, cognitive resources, in this case by recall of the relevant chunk from DM.
- The sequence of learning emerges as a by-product of the agent's attempts to answer the question. The agent's behaviour is rational, in that the firing of rule R1 is a faster, less effortful, and more reliable method than continued reliance on the recall route. At no point is there a deliberate attempt to learn, and there is no appeal to any mechanism for "self-programming", i.e. deliberate learning.
- The learning process is inherently multi-pass, because of the implicit informational dependencies in the process. The first pass is needed to acquire the declarative chunk, which is used in the second pass to acquire the direct production. Only in the third and later passes can the production be used.

These characteristics can fruitfully be compared with the corresponding properties of the process for Soar. (The astute reader will notice that the data learning problem relates mainly to the *condition* or *stimulus* side of the learned association in Soar, but to the *action* or *response* side in ACT-R.)

The Data Learning Problem in Connectionist Architectures

At first sight it seems implausible that connectionist models would suffer from the data learning problem, since such models are generally regarded as epitomising the mechanism needed for associating a given input with a specific output. However, closer examination raises doubts as to whether the usual story about associative learning and connectionism is indeed sufficient to enable networks to learn associations directly without encountering the data learning problem.

We consider first the connectionist architectures for supervised learning, such as simple pattern associators or multi-layer feedforward networks (see, e.g., McLeod, Plunkett & Rolls, 1998), trained by back-propagation or similar regimes. Such networks cannot be considered as providing counter-examples to the claimed generality of the data learning problem, for two reasons. First, as conventionally formulated, they require the cue and the response to be treated by the mechanism in quite different ways, with the cue being used as the input to a network and the response being used as the implicit target for training. We have already seen, as in the discussion of stimulus sampling theory above, that such different treatment of cue and response indicates that the theory is not fully computationally explicit and that the theorist is implicitly playing a homunculus-like role in the application of the theory. The second reason is the difficulty of mapping the supervised learning paradigm onto the cognitive reality of the kind of learning situations we are considering. If the supervisor is regarded as being external to the agent, then it simply is not the case that a source outside the agent is providing the information and feedback the numerous times needed by this kind of connectionist learning. If, on the other hand, the supervisor is regarded as being somehow internal to the agent, then it is even more apparent that the theory is postulating a homunculus inside the agent able to configure the input and output of the network appropriately and provide a suitable regime of reward and punishment.

More relevant to the concerns of the present paper are the various kinds of auto-associative and recurrent networks. Since the overt task for these networks is to learn to reproduce (or anticipate) their input, a knowledgeable supervisor is not needed. The natural way to apply such networks to the task of associating a response with a given cue is to treat the cue and the response together as a compound input and learn to map that input onto itself. Doing this forces the network to develop attractors, one for each cue-response pair, and we can then rely on the ability of such networks to "complete" partial inputs in order to reconstruct the pair given just the cue. (It should be noted that the ability of the network to auto-complete is closely analogous to ACT-R's ability to retrieve a complete chunk of information from declarative memory, given just one of its components.)

Whether such auto-completion is feasible and effective for a particular set of cues and responses will depend in complicated ways on the encoding of the input, on the pattern of similarities among the cues and responses, and on the regularity or otherwise of the mapping between them. If we assume that the completion can be learned, then the network will have learned to retrieve the pair (e.g., {*capital of France?, Paris*}) given just the cue (*capital of France?*). If we really want a connectionist network that, given the cue, produces just the response, then the auto-associative network could be used as an internal supervisory mechanism to train a separate feedforward network — as is done, for example, in network models of how a fast-learning hippocampus may function to internally train a slower-learning neocortical system (McClelland, McNaughton & O'Reilly, 1995).

If we try to summarise the situation for connectionist architectures in the same style as we have for Soar and Act--R, we get:

- A computationally explicit connectionist model is unable to learn directly the mapping from cue to response. (This is the data learning problem again.)
- An appropriate kind of auto-associative network can learn the pairs {cue, response}.
- The ability of the network to auto-complete enables it to retrieve a pair given just the cue.
- If necessary, the network with the learned pairs can be employed as the basis for an internal process of learning the direct mapping from cue to response.

The parallels with the case of the production rule architectures will be apparent to the reader.

Concluding Discussion

This paper has defined and explored the data learning problem and the way the problem manifests itself in a symbolic cognitive architecture, Soar; in a hybrid architecture, ACT-R; and in connectionist architectures. Despite their marked differences at the implementation level, the problem arises in a strikingly similar way in all three families of architecture. The paper further argues that the problem should in principle occur for all cognitive architectures.

The core of the phenomenon is that, when an agent has to learn the answer to a question — or in a laboratory setting, has to learn the *response* to a *cue* — then because both the question and the answer have to be presented to the agent in order for it to learn the pairing, the association that is acquired links both the question and the answer to what gets retrieved. In order for the agent to learn a 'direct' association, so that given (just) the question it can recall (just) the answer, the agent has to use the acquired pairing in order to generate the answer for itself as a basis for learning the direct association.

Although there has not been space in this paper to pursue the matter, the data learning problem exerts a deep-seated influence over the nature of human memory (Young & Lewis, 1999). It bears directly on the generative aspects of memory (e.g., Bahrick, 1970); it likely underlies the need for the agent's active involvement in learning; and it perhaps extends to effects such as the importance of self-explanation in conceptual understanding (e.g., Chi, Deleeuw, Chiu & Lavancher, 1994).

If the phenomenon is as ubiquitous as is being claimed here, some account is needed of why it has not been noticed before. The explanation was traced back to the observation that in less computationally explicit theories, such as the many versions of associative learning theory on offer in psychology, the theorist implicitly plays the role of homunculus for the theory, deciding what aspects of the perceived environment are to be regarded as cue and what as response, and the different ways in which they are handled. This unintentional involvement of the theorist in the working of the theory has served to mask the data learning problem until now. It is only when cognitive architectures are formulated explicitly in computational terms that the problem comes to light.

One of the advantages always claimed for computational cognitive modelling is the increased explicitness it forces on the theorist, and the way that it consequently banishes the homunculus from cognitive theorising. The present discussion suggests that the emergence of the data learning problem is a case in point.

References

Anderson, J. R., Bothell, D., Byrne, M. D., Douglass, S., Lebiere, C. & Qin, Y. (in press). An integrated theory of the mind. [Available from http://ACT-R.psy.cmu.edu/papers/403/IntegratedTheory.pdf]

Anderson, J. R. & Lebiere, C. (1998) *The Atomic Components of Thought.* Erlbaum.

Atkinson, R. C. & Estes, W. K. (1963) Stimulus sampling theory. In R. D. Luce, R. R. Bush & E. Galanter (Eds.), *Handbook of Mathematical Psychology,* vol 2, 121-268. New York: Wiley.

Bahrick, H. P. (1970) Two-phase model for prompted recall. *Psychological Review, 77,* 215-222.

Chi, M., Deleeuw, N., Chiu, M. & Lavancher, C. (1994) Eliciting self-explanations improves understanding. *Cognitive Science, 18,* 439-477.

Estes, W. K. (1994) *Classification and Cognition.* Oxford University Press.

McClelland, J. L., McNaughton, B. L. & O'Reilly, R. C. (1995) Why there are complementary learning systems in the hippocampus and neocortex: Insights from the successes and failures of connectionist models of learning and memory. *Psychological Review, 102* (3), 419-437.

McLeod, P., Plunkett, K. & Rolls, E. T. (1998) *Introduction to Connectionist Modelling of Cognitive Processes.* Oxford University Press.

Newell, A. (1990) *Unified Theories of Cognition.* Harvard University Press.

Pavlov, I. P. (1927) *Conditioned Reflexes.* London: Oxford University Press.

Rosenbloom, P. S., Laird, J. E. & Newell, A. (1987) Knowledge-level learning in Soar. *Proceedings of AAAI'87,* 499-504. Los Altos, CA: Morgan Kaufmann. [Reprinted in P. S. Rosenbloom, J. E. Laird & A. Newell (Eds), *The Soar Papers: Research on Integrated Intelligence,* vol 1, 527-532. MIT Press, 1993.]

Taatgen, N. A. & Lee, F. J. (2003) Production compilation: A simple mechanism to model complex skill acquisition. *Human Factors, 45* (1), 61-76.

Vera, A. H., Lewis, R. L., Lerch, F. J. (1993) Situated decision-making and recognition-based learning: Applying symbolic theories to interactive tasks. *Proceedings of the 15th Annual Conference of the Cognitive Science Society,* 84-95. Boulder, Colorado.

Young, R. M. & Lewis, R. L. (1999) The Soar cognitive architecture and human working memory. In A. Miyake & P. Shah (Eds), *Models of Working Memory: Mechanisms of Active Maintenance and Executive Control,* 224-256. Cambridge University Press.

Posters

ADAPT: A Cognitive Architecture for Robotics

D. Paul Benjamin (benjamin@pace.edu)
Computer Science Department, Pace University
1 Pace Plaza, New York, NY 10038 USA

Damian Lyons (dlyons@fordham.edu)
Computer Science Department, Fordham University
340 JMH Fordham University, 441 E. Fordham Rd., Bronx, NY 10458

Deryle Lonsdale (lonz@byu.edu)
Linguistics Department, Brigham Young University, Provo, UT 84602

Introduction

ADAPT (Adaptive Dynamics and Active Perception for Thought) is a cognitive architecture specifically designed for robotics researchers. ADAPT is in the initial state of development. It manipulates a network of schemas that implement a formal model of concurrent, distributed real-time control. Perception is active, which means that all perceptual processing is goal-directed and context-sensitive, even down to the raw sensory data.

The development of ADAPT is motivated by robotics researchers who want their robots to exhibit sophisticated behaviors including use of natural language, speech recognition, visual understanding, problem solving and learning, in complex analog environments and in real time. Many robotics researchers have realized that programming robots one task at a time is not likely to lead to a robot with such general capabilities, so interest has turned to cognitive robotic architectures to try to achieve this goal. ADAPT is a collaborative effort combining Pace University's Robotics Lab – csis.pace.edu/robotlab, Fordham University's Robot Lab – www.cis.fordham.edu/~lyons/rcvlab and the linguistics department of Brigham Young University – linguistics.byu.edu/nlsoar.

Robotics researchers are faced with a major obstacle when attempting to use cognitive architectures to control robots: the architectures do not easily support certain paradigms of perception and control that are mainstream in robotics. This paper focuses on two such paradigms: adaptive dynamics and active perception.

One common assumption that is made by many robotics researchers is that perception is an *active* process, rather than a passive one. This means that perception is goal-directed and context-sensitive at every stage, including the initial processing of input sensory data. Active perception, and in particular active vision (Blake & Yuille, 1992), is a major research area in robotics. For example, if a robot is crossing the street, it will not process all of its visual frame equally, but will filter the data for large blobs that are moving towards the robot, processing in a very coarse way the rest of the visual field. Such a robot will perceive cars very differently depending on the goal and the situation. Existing cognitive architectures can be altered to perceive in this way (Byrne, 2001) but this is the exception rather than the rule. EPIC in particular does not perceive this way, and this is a major limitation in its application to robotics.

A second assumption made by many robotics researchers is that a robot's actions must be modeled by concurrent, distributed, real-time processes. A robot typically possesses a large number of moving components, e.g. gripper joints, wheels or legs, pan-tilt-zoom camera units, microphones, sonars, etc. and these components may all be active simultaneously. In addition, the robot must model processes occurring simultaneously in the environment.

As a result, many robot programs are the composition of concurrent, communicating real-time behaviors that are not typically organized in a hierarchy. Cognitive architectures such as Soar (Rosenbloom, Laird, & Newell, 1993), EPIC (Kieras, Wood, & Meyer, 1997) and ACT-R (Anderson, 1996) are relatively weak in this area. These models focus primarily on sequential search and selection, and their learning mechanisms focus on composing sequential rather than concurrent actions. Furthermore, these architectures tend to model actions hierarchically (with a stack), not in a truly distributed fashion. It is not that these architectures forbid the use or learning of concurrent distributed programs, but rather that this is not a central feature of them, and it is of primary importance in robotics.

Our Approach to Perception and Planning

ADAPT resembles existing architectures in many ways. It is a production-based architecture with a working memory, and matches productions against working memory during each cycle in the same manner as these architectures. All the matching productions fire, as in Soar, and place their results in working memory. ADAPT possesses a long-term declarative memory, as ACT-R does, in which it stores sensory-motor schemas that control its perception and action. All perceptual processors fire in parallel, as in EPIC, but place their low-level sensory data into working memory, where it is processed by the cognitive mechanism.

We draw a distinction between the goals that are task

goals, e.g. "find the blue block", and those that are goals of the architecture, e.g. "start the schema that scans the environment for a segment of a given color". Similarly, we distinguish between task actions, e.g. "pick up the block", and architectural actions, e.g. "start the gripper-closing schema". Our goal is to reason about concurrent goals and actions *in the task*, so we partition the actions and goals into an architectural part and a task part. We restrict the architectural part to one active goal (it has a goal stack) and one architectural action at a time. The architectural part is procedurally represented, i.e. the system can execute the actions but cannot examine their internal representation. We represent the task goals and actions declaratively in working memory as well as procedurally. There can be as many active task goals and actions in working memory as the system wants. We call these goals and actions "schemas".

An Implementation of ADAPT

An initial implementation of ADAPT has been completed. Testing has just begun using a Pioneer P2 robot that is equipped with stereo color vision, microphone and speakers, sonars and touch sensors. The implementation is in Soar, because of the similarity between ADAPT and Soar. A declarative memory has been added to Soar for the general schemas, together with operators that instantiate a general schema, transform it into a set of chunks, and control its execution. Schemas are executed by a runtime system that implements the RS language (Lyons & Arbib, 1989) in the Colbert language, which is a behavior-based language provided with Pioneer robots.

Currently, ADAPT has access only to sonar and bump-sensor readings. This permits simple navigation and pushing behaviors, but nothing more. The existing version of ADAPT has a cycle time of 50ms and is very successful at guiding the robot at basic navigation tasks such as moving from one room to another and avoiding obstacles.

The language component will be provided by NL-Soar (Lonsdale, 2001), which is being ported to Soar8 as part of our collaboration with Brigham Young University. It will be available in July and will be integrated into the system.

The vision component consists of two pieces: a bottom-up component that is always on and is goal-independent, and a top-down active component. Both components exist but the full integration within ADAPT is not complete. We anticipate a full integration by September.

Testing and Evaluating ADAPT

We have selected an important and flexible class of mobile robot applications as our example domain: the "shepherding" class. In this application, one or more mobile robots have the objective of moving one or more objects so that they are grouped according to a given constraint.

This class of tasks is attractive for two reasons. The first is that it includes the full range of problems for the robot, from abstract task planning to real-time scheduling of motions, and including perception, navigation and grasping of objects. This is ideal, because it creates a situation in which complex hierarchies of features and constraints arise.

The second reason is that we can embed lots of other problems in it, especially those that have been examined by cognitive psychology. For example, we can create an isomorph of the Towers of Hanoi task by having three narrow enclosures, with three boxes in the leftmost enclosure. The robot must move them to the rightmost enclosure one at a time. We add the constraint that no object can be in front of a shorter object (so that all objects are always visible by the observer). The three enclosures act as the three pegs in the Towers of Hanoi, and the constraint is isomorphic to the constraint that no disk can be on a smaller disk in the Towers of Hanoi. In this way, the robot can be presented with a Towers of Hanoi problem in a robotic setting with perceptual and navigational difficulties, rather than just as an abstract task. This permits us to evaluate the robot's problem-solving and learning capabilities in a way that may permit comparison with human data.

Summary

Cognitive psychology and robotics have much to offer each other. The development of robot cognitive architectures is an attempt to apply the results of cognitive modeling to the difficult problems faced by robotics research. ADAPT is a cognitive architecture specifically designed to permit robotics researchers to utilize well-known robotics research in areas such as active perception and adaptive dynamics within a cognitive framework. This architecture is still in its infancy, and in particular has not yet been integrated with vision or language. The goals of this research are to expand the capabilities of robots and simultaneously to expand and generalize the capabilities of existing cognitive models.

Acknowledgements

We would like to thank Frank Ritter for many helpful comments and pointers to relevant work.

References

Anderson, J. R. (1996). ACT: A simple theory of complex cognition. *American Psychologist*, 51, 355-365.

Blake, A., and Yuille, A., (Eds,) (1992). *Active Vision*, MIT Press, Cambridge, MA.

Kieras, D.E., Wood, S.D., and Meyer, D.E. (1997). Predictive engineering models based on the EPIC architecture for a multimodal high-performance human-computer interaction task, *ACM Transactions on Computer-Human Interaction* 4, 230-275.

Lonsdale, Deryle, (2001). An Operator-based Integration of Comprehension and Production, *LACUS Forum XXVII*, pages 123–132, Linguistic Association of Canada and the United States.

Lyons, D.M. and Arbib, M.A., (1989). A Formal Model of Computation for Sensory-based Robotics, *IEEE Transactions on Robotics and Automation* 5(3), June.

P. S. Rosenbloom, J. E. Laird, and A. Newell, (1993). *The SOAR Papers*. MIT Press.

Transition of Aesthetic Emotions in Interactive Environments

Ana Constantinescu, Naoko Matsumoto, Daisuke Moriyasu, Hajime Murai, Akifumi Tokosumi
{ana; matsun; dmoray; h_murai; akt}@valdes.titech.ac.jp
Department of Value and Decision Science, Tokyo Institute of Technology
2-12-1 Ookayama, Meguro-ku, Tokyo 152-8552 JAPAN

Introduction

With the rapid expansion in interactive environments, the need for more sophisticated cognitive models of aesthetic emotion transition is even more pressing for both academic and industrial applications. As part of our research to develop a model of aesthetic emotions and their patterns of transition, we have conducted two free report experiments concerning an interactive environment (a driving simulation), and have extracted and analyzed emotion words from collected verbal protocol data. In this research, we considered aesthetic emotions to be an overall configuration of cognitive components, and this configuration is a subjective experience of emotions, and thus of a state of mind. The results offer some insights into the mechanisms and structure of aesthetic emotions.

Experimental Method

We conducted two experiments relating to the experience of driving. In Experiment 1, 13 participants (20-50 years old) took part in a driving simulation, while in Experiment 2, conducted for comparative purposes, two participants (who also participated in Experiment 1) actually drove around the course simulated in Experiment 1. The driving simulation in Experiment 1 consisted of driving simulation software, Touge 3 (ATLUS), running on a Playstation 2 (Sony) connected with a driving simulation kit, including steering wheel, driving seat and pedals (Sparco). In Experiments 1, the participants went around the same course four times. The participants were encouraged to report out aloud anything that might come into their minds as they drove around the course. The verbal protocol was video taped, and was subsequently analyzed according to a classification system developed in our laboratory (Tokosumi, A. et al 2000-2003). The classification table included a total of 51 categories. A part of the category table including emotion related categories in shown in Table 1. All protocol data was classified into the defined categories by the same analyzer to preserve consistency of the analysis.

e.g.: aesthetic emotions "This is great", "Cool";
negative emotions "I' m scared", "Dangerous;

The verbal protocol was:

1. manually transcribed preserving intonation and pausing information.
2. divided into appropriate chunks of text (in this study, we chose the clause as our unit of analysis). Missing semantic elements were added to incomplete clauses by the analyzer, such as making pronoun references explicit.
   e.g.: "This is beautiful!" → "This [view] is beautiful"
3. categorized according to a predefined category definition table, which includes categories related to the experience of driving, such as the route, the background, and meta-information.

Table 1: A Part of the Category Definition System

?* >*>-		?1 >W ?> ?HF	?>?8 > ?> ?HF

Results and Discussion

From Experiment 1, we collected 6,805 utterances from the participants, with a total of 1,451 utterances (21.32% of all utterances) containing emotion words, and within these 370 aesthetic emotions (25.49% of emotion words). We conducted a sequential analysis on the utterances to investigate the transition in emotion during the simulation. The main results are as follows:

1. Emotion utterances tended to be concentrated together.
2. Utterances containing aesthetic emotions are more likely to follow utterances containing other negative emotions (38), course descriptions (29), description of driving techniques (27), and background descriptions (18).
3. Utterances following utterances containing aesthetic emotions are more likely to be positive evaluations of the background (32), and positive evaluations of driving technique (28).

In order to examine the influence of external factors related to the driving course itself, such as its level of difficulty and other characteristics, on the elicitation of aesthetic emotions, we divided the driving circuit into sections, and evaluated these according to various characteristics. The frequency and type of emotion utterances were then examined in terms of these course characteristics. The main results are:

1. Aesthetic emotions are more likely to occur on less difficult parts of circuit (10.42%).
2. Aesthetic emotions are elicited by parts of the course that have changes in terms of background characteristics (7.42%);
3. Aesthetic emotions tend to increase two times during later drives through difficult parts of the course.
4. Although negative emotion utterances tend to be preponderant for difficult parts of the course during the first simulation drive, they tend to decrease to one third during later runs.
5. Negative emotion utterances tend to decrease with respect to course characteristics over the four simulation runs.

From Experiment 2, we collected 1,156 utterances from the two participants, as they actually drove around the course. The data was processed according to the same procedures as for Experiment 1. Analysis of the protocol data from Experiment 2 showed that participant utterances were similar to that obtained in Experiment 1, indicates both accuracy and consistency of our category system.

A cognitive component model of emotions

The proposed computational model is based on our model of literary text appreciation reported in Tokosumi and Matsumoto (2003). It assumes a set of cognitive units particularly related to emotions, including aesthetic emotions. Although the data obtained from the driving experiments doesn't directly support each of those units, we adopt the set as a possible candidate for the complete set of aesthetically oriented cognitive units. Using the framework of cognitive theories of emotion, we can identify cognitive appraisal components underlying the aesthetic emotions:

(a) recognition of quality (completeness: the quality of the expression meets driver's completeness standards?, novelty: the quality meets driver's newness standards?)
(b) meta-awareness of memory / consciousness of explicit memory (first experience: experience the scene for the first time?)
(c) meta-awareness of ability (driving: able to handle with the road / car?)

Similarly, we can also identify action-readiness components:

(d) possession: (want to posses the car?)
(e) re-experience: (want to drive the route again?)
(f) creation: (want to find a similar route / car?)
(g) evangelic (want to tell somebody about the driving?)

Transition of aesthetic experiences

Each cognitive component was implemented as an object (CLOS) in a standard common lisp environment (CLISP). With given set of variables describing the states of outer world (road conditions, visual scenery, car conditions, outcomes of driving maneuvers), the computational model determines corresponding states of every cognitive component, which are to be interpreted as states of mind including aesthetic emotions. The model behaves as a sequence of such discrete states, thus should be understood as a static model.

In the model, the transition of emotions described in previous sections is represented as a transition of activation level of each cognitive component. For instance, drivers tend to concentrate into driving maneuvers in technically difficult part of the circuit. Ordinary problem solving components such as 'action planning' and 'self monitor' which are basically have goal-plan structures dominate major activities of the model and components related to aesthetic emotions remain in relatively lower activation level. Repeated experience however, causes the activation level of problem solving components to decrease and total experience of aesthetics gains relative importance. In the current stage of the development of the model, the activation levels are hand-coded so that the overall states of the model match the experimental findings.

Conclusions

In this project, we formulated a model about the mechanism and structure of aesthetic emotions in an interactive environment. Although closely related to driving simulation, the results offer a significant insight to the general elicitation mechanism of aesthetic emotions and the cognitive appraisal components to be further studied in future work. Future directions include modeling of dynamic transient of emotions, precise mapping to aesthetic experiences.

Acknowledgements

This project has been conducted on behalf of the Toyota Motor Corporation. We thank the members of Toyota Motor Corporation for their support during this project.

References

Frijda, N. H. (1986) *The emotions,* Cambridge: Cambridge Univ. Press.

Ortony, A. Clore, G..L. and Collins, A. (1988) *The cognitive structure of emotions,* Cambridge: Cambridge Univ. Press.

Matsumoto, N. and Tokosumi, A. (to appear) Affection for cohabitant toy dolls. *The Computer Animation and Virtual Worlds Journal.*

Tokosumi, A. (2001) The brain/mind machine: Toward modeling its wish generation processes. In T. Kitamura (Ed), *What should be computed to understand and model brain function?,* 43-51. Singapore: World Scientific Publishing.

Tokosumi, A. and Matsumoto, N. (2003). Appraisal components for aesthetic emotions. *Abstracts of the Fourth International Conference on Cognitive Science,* p.181, Sydney.

Tokosumi, A. (1998) *Cognitive Components of Affective Words,* Abstracts for the 1st International Conference on the Mental Lexicon.

LSDNet: A Neural Network for Multisensory Perception

Matthew C. Costello (costellomc@upmc.edu)
Psychology Department, Duquesne University
514 College Hall, Pittsburgh, PA 15260 USA

Eric D. Reichle (reichle@pitt.edu)
Learning Research and Development Center, University of Pittsburgh
3939 O'Hara St., Pittsburgh, PA 15260, USA

Infant Multisensory Perception

Research on infant multisensory perception has detailed the increasing capacity of infants to differentiate stimuli when it is present to multiple sensory modalities (Bahrick (2002), Lewkowicz & Lickliter (1994)). Bahrick and Lickliter's (2002) *intersensory redundancy theory* posits a two-fold thesis to explain infant and neonate multisensory perception. In the first thesis, infant attention is directed towards amodal features when redundant sensory information is present to two or more sensory modalities. Secondly, when only unimodal input is available, attention is directed towards modality-specific features. Modality-specific features are particular for a given sensory modality (color, scent, pitch, etc), and amodal features are evident in multiple sensory modalities (rhythm, rate, tempo, intensity, etc).
Our model attempts to replicate the overall computational structure to Bahrick and Lickliter's theory and to offer validation to the two-fold thesis. Though their thesis is highly compatible to a computational approach, to date there have been no attempts to create such a model, and as such the LSDNet promises a new direction for multisensory research.

Model Specifics

Our model is a three-layered backpropogated neural network (Fig. 1). Input units consist of four auditory and four visual nodes that feed into three hidden units – two of which represent 'modality-specific' units, and one of which is an amodal representation which receives input from all eight input units. Inhibitory connections exist between all three hidden units. Output units consist of arbitrary A-B response nodes. The network is trained with two stimulus sets, one corresponding to multimodal pickup and one to unimodal pickup.

Preliminary Results

Initial results seem promising. When the network is tested with matching modality nodes, output results reflect the increased capacity of the model to pickup amodal features when receiving multimodal input. Currently we are investigating the effect of restructuring the input units so that given units will feed into the dedicated amodal hidden unit. By doing so we hope to improve upon the network's performance to more cleanly match empirical results.

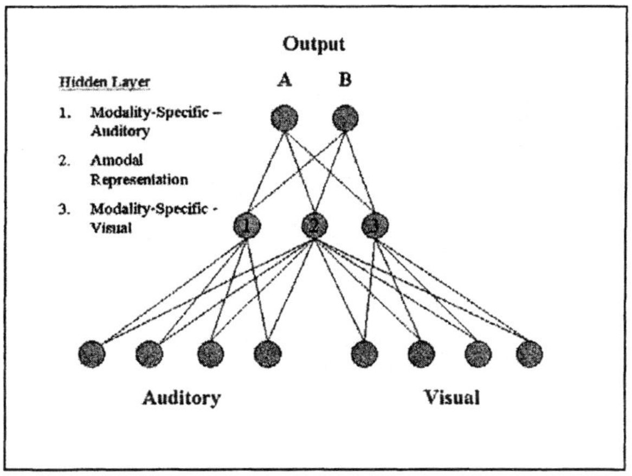

Figure 1: LSDNet

References

Bahrick, L. (2002). Generalization of learning in three-and-a-half month-old infants on the basis of amodal relations. *Child Development, 73*, 3, 667-681.

Bahrick, L., & Lickliter, R. (2002). Intersensory redundancy guides early perceptual and cognitive development. *Advances in Child Development, 30*, 153-187.

Lewkowicz, D.J., &Lickliter, R. (1994). *The development of intersensory perception: Comparative perspectives*. Hillsdale, NJ: Erlbaum.

Constraint-Based Human Causal Learning

David Danks (ddanks@cmu.edu)
Department of Philosophy, Carnegie Mellon University; & Institute for Human & Machine Cognition
135 Baker Hall, Pittsburgh, PA 15213 USA

Theories of Human Causal Learning

Much of human cognition and activity depends on causal beliefs and reasoning. In psychological research on human causal learning and inference, we usually suppose that we have a set of binary potential causes, C_1, ..., C_n, and a known binary effect, E, all typically *present-absent* values of a property or event. The differentiation into potential causes and effect is made on the basis of external factors, including prior knowledge or temporal information.

Given these variables, people are then asked to infer the existence and strength of causal relationships between the C_i's and E from observed data in one of several formats (serially, as a list, or in a summary). The standard measure of people's causal beliefs is a rating of some proxy for causal influence, where a zero rating indicates no causal relationship. The exact probe question varies between experiments, and has been found to significantly impact participants' ratings (e.g., Collins & Shanks, under review).

A variety of theories have been proposed to explain people's causal inferences in this type of highly limited scenario (see Danks, forthcoming, for a theoretical overview and synthesis). One general view for which there is a growing body of evidence is that people's causal beliefs and learning are well-modeled as though they are learning a causal Bayesian network (CBN, henceforth).

Causal Learning with Bayesian Networks

CBNs have proven to be a powerful framework for representing and learning causal structure from observational, experimental, and mixed data (e.g., Pearl, 2000; Spirtes, Glymour, & Scheines, 2000). At a general level, a CBN contains two distinct, related components: a directed acyclic graph (DAG) that represents qualitative causal relationships ($X \rightarrow Y$ means that X is a direct cause of Y); and quantitative information about the strengths of the various causal connections (e.g., a joint probability distribution; a set of a linear equations; and so on). These components are connected through the Markov and Faithfulness/Stability assumptions, which constrain the ways in which causal relationships manifest themselves in observational and experimental data. These assumptions are domain-general, and themselves testable.

There are essentially two different strategies for learning a CBN from data: (i) score-based or Bayesian approaches; or (ii) constraint-based (C-B) approaches. In the former approach, we search either heuristically or exhaustively for the CBN that maximizes P(CBN | observed data). We focus here on the latter approach, in which we determine the set of DAGs that could possibly have produced the observed independencies and associations (given Markov and Faithfulness). C-B algorithms thus take a set of independence and association judgments as input, and output an equivalence class of DAGs, all of which make identical predictions about the observed data. For small numbers of variables, the equivalence class will frequently *not* be a singleton, and so we will have a set of possibilities that cannot be distinguished given the data.

As an example of a C-B algorithm, suppose that we have data on X, Y, and Z. There are six different independencies (conditional and unconditional) that may or may not hold for these three variables. Suppose that some process yields the following statistical judgments about the variables: the only independence (of the six possibilities) is that X and Z are unconditionally independent. If we further suppose that there are no unobserved common causes (latents) of these variables, then there is exactly one DAG that could have produced these data: $X \rightarrow Y \leftarrow Z$. (If we drop the "no latents" assumption, then there might be unobserved common cause(s) of X and Y, or Z and Y, in addition to or in place of the $X \rightarrow Y$, $Z \rightarrow Y$, edge.)

Both types of approaches have been used for rational analyses of causal learning and categorization. Examples using Bayesian approaches include Griffiths, Baraff, & Tenenbaum (2004); and Tenenbaum & Griffiths (2001), (2003). Rational analyses with C-B algorithms include Gopnik, Glymour, Sobel, Schulz, Kushnir, & Danks (2004); and Kushnir, Gopnik, Schulz, & Danks (2003).

Constraint-Based Human Causal Learning

A range of evidence suggests that human causal learning is best-modeled by CBN structure learning, including: learning from manipulations (as opposed to just observations); learning when the variables are not differentiated into causes and effect; and differences in predictive and diagnostic learning. All of these phenomena can be explained both by C-B and Bayesian approaches to CBN structure learning, and have been modeled elsewhere.

Another important piece of evidence for the CBN theory is that people are seemingly able to use domain-specific knowledge to draw causal conclusions based on small sample sizes, and CBN learning algorithms are the only ones currently on offer that have the flexibility to model this adaptive behavior (Griffiths, et al., 2004; Tenenbaum & Griffiths, 2003).

Griffiths, et al. (2004) have drawn a further conclusion based on inferences from small samples: C-B algorithms cannot model this phenomenon—and so are incorrect—because they do not incorporate domain-specific knowledge about the manner or functional form for the operation of possible causal mechanisms. This latter claim about C-B algorithms is based on standard machine learning implementations that use general statistical tests (e.g., χ^2 or correlation) for independence and association judgments.

Griffiths, et al. (2004) do not consider a psychological C-B learning algorithm—as opposed to a rational analysis—because one has not previously been proposed. The most straightforward such algorithm would be to estimate all of the possible independencies (conditional and unconditional) using some method, and then derive the equivalence class of DAGs from these estimates using a "standard" C-B learning algorithm (e.g., the PC algorithm of Spirtes, et al., 2000).

C-B algorithms are not computationally intensive once the independence judgments are provided, and many of their steps are intuitively sensible (e.g., if X and Y are un/conditionally independent, then conclude that there is no direct causal connection). One possible concern with the psychological plausibility of this theory is that the number of possible independencies grows exponentially with the number of variables. However, there is little evidence that people actually can learn causal structures for large numbers of variables without substantial prior knowledge constraints, and so this worry is irrelevant for psychological modeling.

This proposed theory has left relatively open the question of the source of the independence and association judgments. The C-B learning framework simply requires association/independence judgments; machine learning implementations have used general statistical tests, but the framework itself does not require that choice. So a C-B algorithm could, for example, estimate the probability of association using Bayesian statistics, perhaps even using a similar method as that advocated by Griffiths, et al. (2004).

In fact, given information about the particular (type of) causal mechanism, we can use even simpler theories than full Bayesian statistics. For example, if we know or believe that the causal influences operate as though they have causal powers, as proposed by Cheng (1997), then we can use the dynamical theory of Danks, Griffiths, and Tenenbaum (2003) to estimate conditional association in an online manner. A simple transformation of that equation leads to dynamical computation of *un*conditional association, assuming causal influences combine as causal powers. And there are similar dynamical equations for a variety of other underlying "mechanisms," at least when those mechanisms are equated with functional form (see Danks, forthcoming, for an overview of some of the dynamical equations). C-B algorithms can thus provide a psychological theory—and not just rational analysis—of domain-specific causal learning phenomena by incorporating the domain knowledge into the particular statistical tests used to make estimates of independence and association.

Moreover, other proposed psychological theories of causal learning can be modeled as special cases of this psychologically plausible theory of C-B structure learning: the other theories correspond to cases in which people estimate only a subset of the possible independencies. For example, suppose prior knowledge leads us to believe that only the independencies of the C_i's and E conditional on the other C_i's are potentially relevant. (There are a variety of plausible situations in which we might believe this.) If we also believe that causal influences operate like causal powers, we will use the dynamical method of Danks, et al. (2003). The resulting theory is exactly the same as Cheng's (1997) power PC theory. That is, we can model the power PC theory as C-B learning of CBN structure, but where people only estimate a subset of the possibly relevant independencies. A similar story can be told for the other non-CBN psychological theories. Hence, this version of C-B learning explains not only the data supporting CBN structure learning, but also a range of the data that seemingly support non-CBN theories.

Acknowledgments

This research was supported in part by NASA grants NCC2-1399 and NCC2-1377.

References

Cheng, P.W. (1997). From covariation to causation: a causal power theory. *Psychological Review, 104,* 367-405.

Collins, D. J. & Shanks, D. R. (Under review). Conformity to the power PC theory of causal induction depends on type of probe question. *Memory and Cognition.*

Danks, D. (Forthcoming). Causal learning from observations and manipulations. In M. Lovett & P. Shah (Eds.), *Thinking with data.* Erlbaum.

Danks, D., Griffiths, T. L., & Tenenbaum, J. B. (2003). Dynamical causal learning. *Advances in Neural Information Processing Systems 15* (pp. 67-74). Cambridge, MA.: The MIT Press.

Griffiths, T. L., Baraff, E. R., & Tenenbaum, J. B. (2004). Using physical theories to infer hidden causal structure. *Proceedings of the 26th Annual Conference of the Cognitive Science Society.*

Gopnik, A., Glymour, C., Sobel, D. M., Schulz, L. E., Kushnir, T., & Danks, D. (2004). A theory of causal learning in children: causal maps and Bayes nets. *Psychological Review, 111,* 3-32.

Kushnir, T., Gopnik, A., Schulz, L., & Danks, D. (2003). Inferring hidden causes. *Proceedings of the 25th Annual Meeting of the Cognitive Science Society* (pp. 699-703). Boston: Cognitive Science Society.

Pearl, J. (2000). *Causality: models, reasoning, and inference.* Cambridge: Cambridge University Press.

Spirtes, P., Glymour, C. & Scheines, R. (2000). *Causation, prediction, and search.* 2nd edition. Cambridge, MA: The MIT Press.

Tenenbaum, J. B., & Griffiths, T. L. (2001). Structure learning in human causal induction. *Advances in Neural Information Processing 13* (pp. 59-65). Cambridge, MA: The MIT Press.

Tenenbaum, J. B., & Griffiths, T. L. (2003). Theory-based causal inference. *Advances in Neural Information Processing Systems 15* (pp. 35-42). Cambridge, MA: The MIT Press.

Should successful agents have Emotions? The role of emotions in problem solving

Dietrich Dörner (dietrich.doerner@ppp.uni-bamberg.de)
Ulrike Starker (ulrike.starker@ppp.uni-bamberg.de)
Otto-Friedrich-Universität Bamberg
Markusplatz 3
96047 Bamberg
Germany

Emotion and Cognition

In this article we examine whether emotions are helpful or even necessary for problem solving or not. We investigate the behaviour of a Psi-model with and without emotions. Psi is a theory of human "action regulation", a theory about how cognition, motivation and emotion are integrated in controlling human behaviour in complex and dynamic domains of reality (see Dörner et al. 2002). The Psi-theory is partially implemented as a computer program. In Psi cognition, motivation and emotion are conceptualized as information processes, generally as "calculation". How is it possible to generate "irrational" emotional processes as "rational" processes of "calculation"?

According to concepts of Belavkin & Ritter (2003) what is called "emotion" in our eyes is a controlling system which takes into account two relations of an organism to reality, namely the **uncertainty** (unpredictability) of the environment and the (estimated) degree of efficiency of an organism to tackle problems (**competence**). – When uncertainty is high an organism should be prepared for action, especially for flight or aggresssion, should exhibit a tendency to explore its environment, should exhibit a high degree of safeguarding behaviour. When competence is low an organism should try to avoid contact with too difficult problems, should try to enlarge its competence by looking for "efficiency" signals by trying to master problems which hitherto it has not been able to master. Emotions are based on such modes of being prepared for different forms of actions; additionally however internal parameters of cognitive processes are modified according to the degree of competence and certainty. One of these parameter is **arousal** (= general preparedness for action), another one is the **resolution level** of cognitive processes. A low resolution level means rough planning, superficial ("overinclusive") perception and shallow, conservative processes of recall.

A sudden decrease of competence for instance would mean a sudden increase of arousal, decrease of resolution level (to guarantee quick action), incomplete recall of possible modes of action for the situation at hand and rough planning of actions, hence swift and risky (undeliberated) action. This mode of action could be described as "anger".

Prediction of human behaviour

In a series of experiments we examined the relation of the Psi-model to human behaviour. We used different forms of a complex, dynamic, maze like environment (the "island" Detje, 1998; Gerdes 2001) where human subjects had to play a kind of adventure game. They had to take care of the "survival" of a robot, which needed water and fuel (for instance in the form of sunflower seeds or hazelnuts) for its internal steam engine. The subjects had to explore the geographical form of the island, had to plan actions to find water and fuel, to preserve the robot from damage and they had to find as many "nuggets" as possible. The Psi version of an experimental subject was the same robot without the control of a human subject, but with the Psi-theory implementation serving as the robots "soul".

In several experiments it was possible to predict human behaviour to a satisfactory degree.

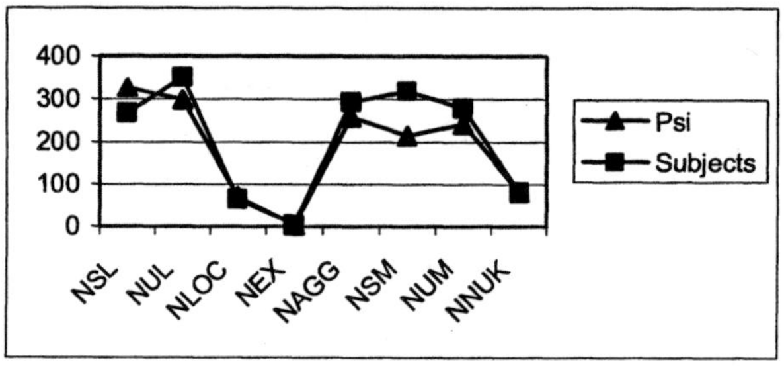

Figure 1: Behavior on the "Island".

In figure 1 the results of an experiment of ElKady & Starker are shown. The averaged profiles of 20 different human subjects can be seen, compared with the averaged profiles of 20 different Psis. The profiles concern 8 different parameters of the behaviour for a 2-hour time period for the human subjects and an equivalent time period for the Psis. The names of the parameters mean:

NSL: number of successful locomotions on the island,
NuL: number of unsuccessful trials to move,
Nloc: number of places visited on the island (exploration activity),
Nex: number of breakdowns of the robot because of missing "food" or water or too much damage,
NAgg: approaches to an object (hazelnutbush, rock, etc.),

NsM: number of objects successfully manipulated,
NuM: number of unsuccessfull trials to manipulate an object,
NNuk: number of nuggets found.

In this experiment we used Psis with different "personalities". The personality of a Psi-subject is the profile of parameters of the emotional regulations; such parameters are for instance the gradient of the increase of arousal (in the case of an increase of a need) or the gradient of the decrease of competence when inefficiency is noticed.

It is obvious from figure 1 that human behaviour and the "theoretical" Psi-behaviour are in good accordance; similar to earlier experiments (Dörner et al. 2002) the canonical correlation between the Psi-Simulations and the human subjects is r=.88 and is highly significant.

Psi with and without emotions

To examine the role of emotions we compared the 20 Psi-subjects with different "personalities" of the above mentioned experiment with the same Psi-subjects for which we fixed the parameters for "arousal" and "resolution level" on a medium level. (The same manipulation was used by Hille 1997 with a forerunner of the actual Psi-system. Hille found out that generally "emotional" systems were more successful than unemotional ones.) Table 1 shows the results of the T-Tests for the two groups of artificial subjects.

Table 1: T-test comparing psis with and without emotions.

T-Test	T	Df
NsL	2,92**	38
NuL	1,34	38
Nloc	-3,05**	20,25
Nex	-7,96**	36,21
Nagg	1,76	38
NsM	20,33**	20,20
NuM	-2,71**	38
NNuk	14,62**	19,20

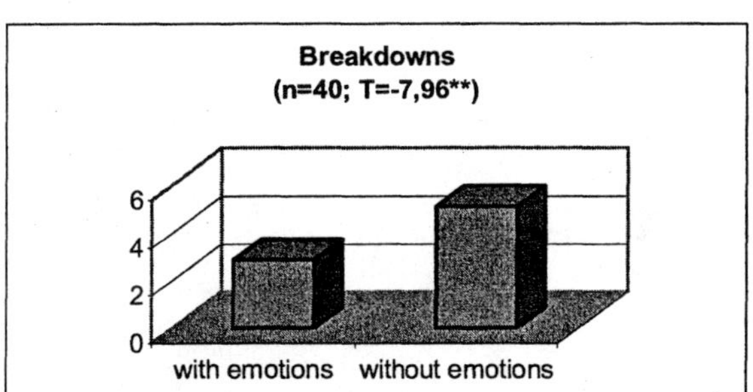

Figure 2: Mean number of breakdowns.

It is obvious that there are highly significant differences with respect to many of the behavioural parameters of the two groups. The emotional Psis especially were much more successfull in collecting nuggets and in preserving themselves from damage (see figure 2 and 3).

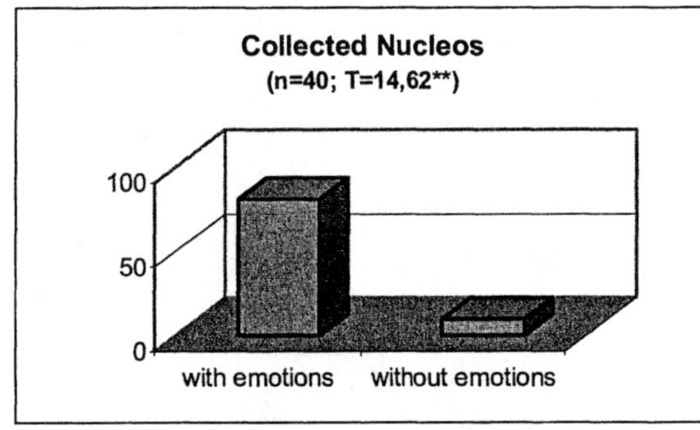

Figure 3: Mean number of collected nucleos.

Discussion

Obviously the emotional modulations of parameters like preparedness for flight or aggression, the modulation of arousal and resolutionlevel increase the strategic flexibility by adapting the behaviour to the respective state of uncertainty and competence. An "emotional" system is more able to adjust its behaviour to the requirements of the situation at hand with respect to uncertainty and competence.

Acknowledgements

The research reported here was conducted in the "Kognitive Plastizität" (Cognitive Plasticity), DFG Do 200/19.

References

Belavkin, R.V., & Ritter, F.E. (2002). The Use of Entropy for Analysis and Control of Cognitive Models. *Proceedings of the Fifth International Conference on Cognitive Modeling* (pp. 75-80). Bamberg: Universitäts-Verlag.

Detje, F. (1998). Das Inselspiel (The Island Game). Computerprogram. Bamberg. Institut für Theoretische Psychologie.

Dörner, D. (2003). The Mathematics of Emotions. *Proceedings of the Fifth International Conference on Cognitive Modeling* (pp. 75-80). Bamberg: Universitäts-Verlag.

Dörner, D., Bartl, Ch., Detje, F. Gerdes, J., Halcour, D., & Schaub, H. (2002). Die Mechanik des Seelenwagens. Bern: Huber.

ElKady, A., & Starker, U. (in preparation). Personality and Problem Solving. A Simulation Study.

Gerdes, J., Dörner, D. (2003). Psi2D Reality-Simulation System. Computerprogram. Bamberg. Institut für Theoretische Psychologie.

Hille, K. (1988). Die „künstliche Seele". (The Artificial Soul) Doctoral Dissertation, Otto-Friedrich-Universität Bamberg.

A New Framework for Cognitive/Perceptual Problems

James L Eilbert (jeilbert@chisystems.com), Tom Santarelli (tsantarelli@chisystems.com)
CHI Systems Inc. , 1035 Virginia Drive, Suite 300
Fort Washington, PA 19034 USA

David M. Carmody
Sytex, Inc,

Introduction

There are a variety of real-world domains where "analysis" of incrementally arriving perceptual input requires the simultaneous solution of a set of hard problems. We claim that basically the same set of interacting problems appears in each of these domains. While people can solve the problems in each of these domains, no automated system is close to solving the full set of interacting problems in any of these domains. The purpose of this paper is to state the set of interacting problems that needs to be solved, and present a cognitively motivated framework in which a solution is possible. Our interest is in a solution to the full set of interacting problems. We are not interested in modeling a particular statistical correlation in a data set that appears under certain conditions. An automated solution would be best, but a man-in-the-loop solution that greatly increases a person's efficiency an/or accuracy would also be acceptable.

The primary domain discussed here is understanding the situation in an image or a sequence of images. The problems and their interactions are shown in Table 1. Solving the situation understanding problem involves much of the brain, so a simple neural network is not a viable approach. What is required is a framework that specifies the process by which a person combines a priori knowledge with the observed evidence. Most of the separate functional capabilities called for in our proposed framework exist already, which give us some confidence that an implementation of the framework can be constructed.

TABLE 1 -- Problem Space Characteristics

Hard Problems	Cause of Problem	Result of Problem	Example	Problem Interactions
Segmenting regions in an image	Distinct objects may have similar statistics in many dimensions	Boundaries between objects may be missed or false boundaries	Walking stick is hard to separate from wood unless it moves.	Object ID & predicate correlation
Distinguishing between situations given only intensity distribution	A large number of objects and arrangements may be consistent with a top-level description of a situation	Knowing a situation is not enough to predict intensity distributions	Knowing the situation is a baseball game does not mean that players are on the field	All but predicting the future
Predicting the future	Situation awareness must be built up through a search sequence	Missed threats and false alerts	Not predicting an aircraft out of commercial lanes could be a weapon.	All
Object identification	Partial info is all that is available in the real world	Occlusion, lighting, or pose can make objects indistinguishable	Misidentify car as a truck	
Predicate Correlation	Situations may be described in a variety of ways	May not recognize different descriptions of the same image	Person over the water vs. person in a boat	
Finding signal in noise and clutter	Many regions in natural images. Pose and lighting make different objects look similar	Finding a particular object is hard even if objects are already separated	Where is Waldo problem	Segmentation (at leastfor clutter)

A Framework for Visual Situation Assessment

The following set of assumptions underlies our proposed framework:

1. Analysis emerges incrementally from cycles of situation understanding
2. Situation understanding has 3 components:
   - A internal map specifying the relative positions of entities potentially important in the situation
   - A consistent story about how the situation reached the current state
   - A subset of an ontology relevant to the situation
3. A situation understanding cycle consists of the following steps:
   - Situation-sensitive segmentation or search for groupings of evidence that hang together
   - Matching groupings to local patterns (activities based on plans without contingencies)
   - Update/refinement of situation story and map based on global pattern (e.g., MPT) or case matching
4. The a priori knowledge needed to update or refine situation understanding
5. People use a similar approach to understanding situations in all domain with these problems

While the neurological substrate and techniques that people employ in doing object recognition, have been the subject of extensive research,. situation understanding has not. Our choices in assembling a situation understanding framework must therefore rely on anecdotal evidence.

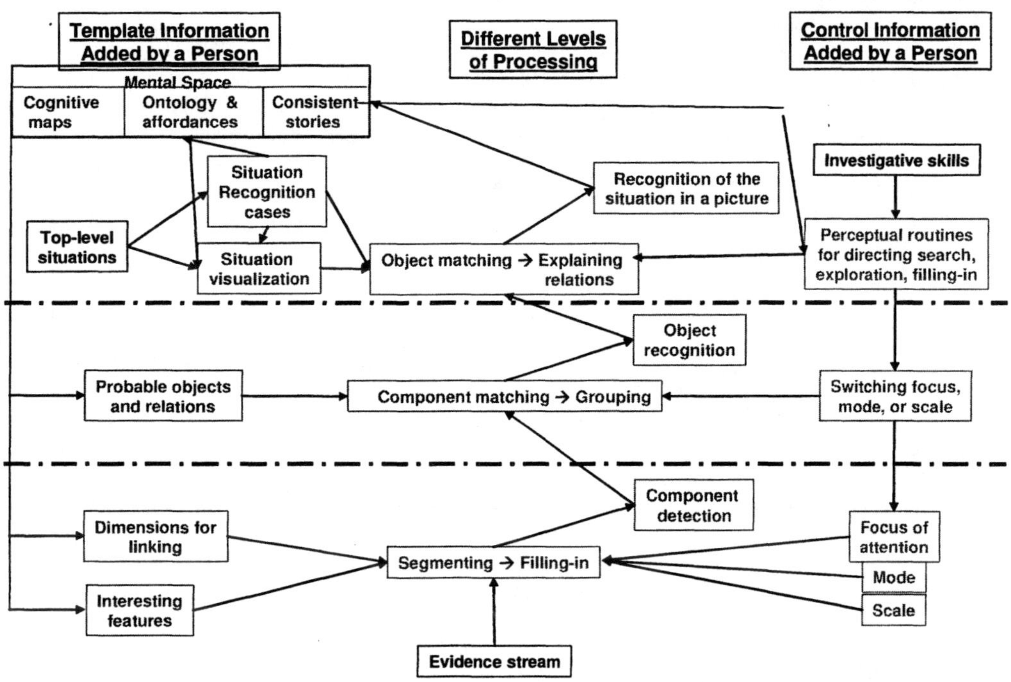

Figure 2: A Model of Cognitive Process Leading to Recognition of the Situation in a Picture

Support for various assumptions in the framework comes from multiple sources. Evidence for a primary role for situation understanding in human vision is that people are not particularly good at recognizing even familiar objects in unfamiliar contexts. For example, *Puzzle* magazine features a section on identifying common objects in real images taken at unusual scales. Looking at the routines that a person employs to visually find and utilize things shows that they have knowledge about what they are likely to see, about what these objects look like in the current context, and what characteristics of the imagery are important when segmenting objects. These types of knowledge must be learned to operate in a new context.

Recently, researchers Upada, Saha, and Lotufo (2002) have found the context-sensitive segmentation techniques work in hard real-world domains that have resisted generic techniques for many years.

Acknowledgments

This research was sponsored and managed by Air Force Research Laboratory Information Directorate, Rome Research Site (AFRL/IF/RRS) under contract F30602-01-C-0200. The views and conclusions contained in this document are those of the authors and should not be interpreted as necessarily representing the official policies, either expressed or implied of Air Force Research Laboratory, or the United States Government.

References

Udupa, J.K., Punam K. Saha, P.K., and Roberto A. Lotufo, R.A. (2002) Relative Fuzzy Connectedness and Object Definition: Theory, Algorithms, and Applications in Image Segmentation. IEEE Trans. PAMI 24(11):1485-1500.

A Recursive Attention – Perception Chaotic Attractor Model of Cognitive Multistability

Norbert Fürstenau (norbert.fuerstenau@dlr.de)
German Aerospace Center (DLR), Institute of Flight Guidance, Lilienthalplatz 7
D-38108 Braunschweig, Germany

Introduction

Attneave (1971) and Lehky (1988) pointed out the similarity of cognitive bistability and electronic multivibrator circuits and suggested analogous neural structures with locking into alternative schemata and exhibiting fatigue. In contrast to the microscopic neural approach I propose to model the dynamics of the macroscopic behavioral variables perception and attention by a phase feedback equivalent circuit which is related to the mean field theory of temporal binding (Schuster & Wagner 1990). Based on a previously outlined nonlinear dynamics model (Fürstenau, 2003), I show that the spontaneous reversals between two perception states with an ambiguous stimulus, e.g. the perspective switching of the Necker-cube, can be explained as self - oscillation between chaotic attractors in attention - perception phase space. The perception state variable is represented by the phase v of a recursive cosinuidal mapping function with feedback delay time T and attention control parameter G. G is proportional to feedback gain g of a corresponding equivalent circuit representing the dynamics of behavioral variables v, G. According to Hillyard, Vogel & Luck (1999) a difference between bias and gain control mechanisms of attention is observed. Like in the multistability model of Ditzinger & Haken (e.g. Haken (1996)) the perception variable is treated as order parameter within the formal framework of Synergetics, with the slowly time varying attention parameter G(t) exhibiting satuation or adaptation. The important aspect of the present approach is the neurophysiologically motivated delay T, giving rise to the chaotic attractors in agreement with Freeman et.al. (e.g. Freeman 2000). The statistical analysis of simulated time series predicts gamma distributions of the perceptual reversal times with mean values and variance in reasonable agreement with experimental results of Borsellino et.al. (1972).

The recursive attention - perception model

The present approach is closely related to the mean field phase oscillator theory of coupled neural groups in the visual cortex (Schuster & Wagner 1990) which was used for modeling the synchronization of neural oscillations as the physiological basis of dynamic temporal binding (Engel et. al.1999). As a kind of minimum architecture allowing for multistability via coupling of the attention and perception dynamics I suggest a nonlinar delayed phase feedback model. An example of a corresponding equivalent circuit from the optics domain was described e.g. in (Watts & Fürstenau, 1989). A simplified block diagram is shown in Figure 1, depicting interference between coherent fields which are associated with the different interpretations of the stimulus. The coupled perception (v_t) – attention (G_t) dynamics for small damping τ is approximated by the recursive equations (time steps $t_j = j\,T$, j = 1,2,...):

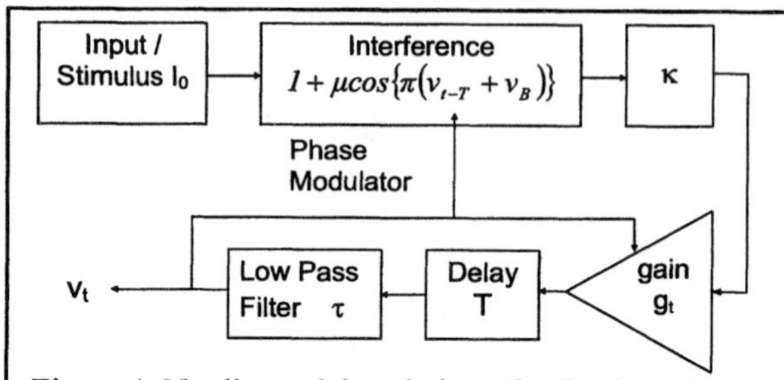

Figure 1: Nonlinear delayed phase feedback equivalent circuit. Interference = superposition of separate coherent phase shifted fields with difference of meaning (contrast) μ, excited by stimulus I_0. v_t = perception state (phase variable), g_t ~ attention (control parameter), κ = conversion factor, τ = damping time

$$v_{t+T} = \frac{G}{1+\tau}\left[1+\mu cos(\pi(v_t+v_B))\right]+\frac{\tau}{1+\tau}v_t \qquad (1a)$$

$$G_{t+T} = G_0(v_b - v_t)/\gamma + G_t(1-1/\tau_G) + L(t) \qquad (1b)$$

Here G_t is a discrete version of a similar attention equation used in the Ditzinger&Haken model. The relative duration of the dominant and suppressed phase of a percept is determined by the bias parameter v_b. In a first approach to model the random disturbances due to dissipative processes a δ-correlated stochastic (Langevin) force L(t) with random amplitude r s_j , $-1 \le s_j \le 1$ is included in (1b), similar to (Haken 1996) and (Lehky 1988). It adds pushes of random amplitude to G_t .

Simulated Perception - Attention Dynamics

If $\mu >$ 0.18 the stationary solution of (1a) becomes multivalued (hysteresis curve $v_{t+T} = v_t = v^*$, (Fürstenau 2003)) and the delayed feedback system allows for spontaneous transitions between percept P1 corresponding to stationary state $v^* < 1$, and P2, corresponding to $2 < v^* < 3$. Figure 2 shows the numerical evaluation of equations (1) for $\mu = 0.7$, $\tau/T = 0.03$, $v_b = 1.5$, $\gamma = 170$, $\tau_G = 1400$, a random noise amplitude r = 0.025, $G_0 = 1$ and normalized time with T :=1. The time series of the perception state shows the low frequency self oscillations between P1 ($v(t) \approx 1$) and P2 ($v(t) \approx 2$), with superimposed high frequency oscillations (f $\ge$ 1/2T). These low amplitude rapid variations of the perception state are due to a combination of limit cycle oscillations, chaotic trajectories in v_t – G_t - phase space and the

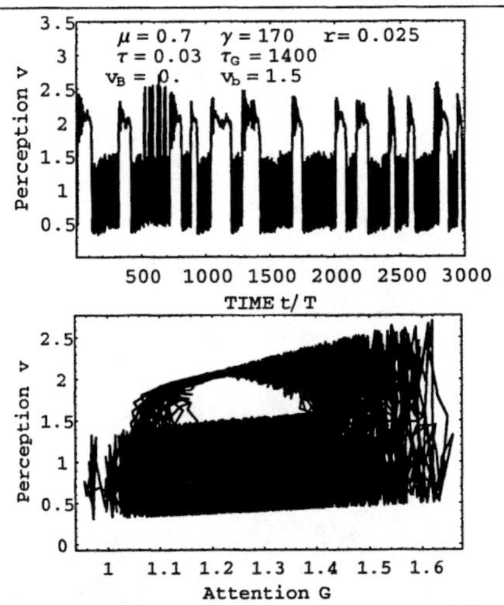

Figure 2: Numerical solution of equ.(1) with 15000 iterations. Top: perception state v_t time series in time units T. Bottom: trajectories in attention – perception phase space.

small random noise (amplitude r = 0.025) of the attention (gain) control parameter. A transition time between P1 and P2 of the order of 5 T is measured. This corresponds to 150 - 200 ms if the recurrence time T is identified with the temporally segmented processing time units of typically 30 - 40 ms as proposed by Pöppel (1990). It is in reasonable agreement with the time interval between (visual) stimulus onset and the beginning of conscious perception (Lamme 2003). The lower graph depicts the trajectories in attention (G) – perception (v) phase space. Besides the limit cycle and chaotic oscillations it exhibits the counterclockwise global movement of v(t) with transitions between P1, P2 due to the coupling with the slowly adapting G(t) (satuation: $v_b - v(t) < 0$ for P2 and $v_b - v(t) > 0$ for P1). The perception speed is determined by $1/\gamma$.

Reversal Time Statistics

Figure 3 depicts the relative frequencies of the perceptual duration times as obtained by averaging 11 time series con-

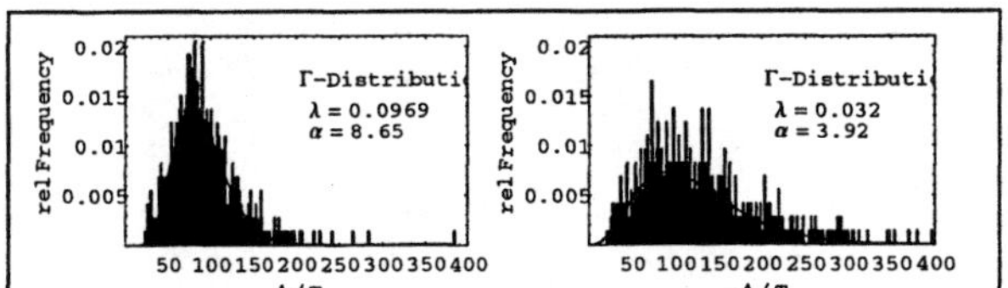

Figure 3 : Relative frequencies of perceptual duration time Δ. Fit with Γ-distribution (solid line). Right: Percept P1 ($v \approx 1$), left: Percept P2 ($v \approx 2$).

sisting of N = 15000 iterations each. Plotted are the two distributions of the perceptual durations Δ(P1), Δ(P2). According to Borsellino et.al. (1972) who evaluated experiments with the Necker cube, the relative frequencies are fitted by a Γ – distribution as probability density with shape parameter α and scale parameter λ, with mean and variance $\Delta_m = \alpha/\lambda$ and $\sigma^2 = \alpha / \lambda^2$ respectively. The X^2 – test suggests acceptance of the Γ-distribution hypothesis for percepts P1, P2 at a significance level of 1%: X^2(P1,18) = 22.1, X^2(P2,19) = 30.3. For percept P1 mean Δ_m = 123 T, standard deviation σ = 62 T; for P2 Δ_m = 89 T, σ = 30 T. In contrast to (Fürstenau, 2003) with purely deterministic time series, the addition of the small random attention noise L(t) in (1b) leads to a significant increase of the variance, whereas the mean values remain roughly the same, indicating the dominant influence of the deterministic (chaotic) dynamics. The contributions of chaotic dynamics as well as random noise agrees qualitatively with experimental results of Richards, Wilson & Sommer (1994) who separated the chaotic and random contributions of nonlinear perception state time series of quite different visual perception phenomena including multistability. With T = 30 ms e.g. Δ_m (P1) = 3.6 s with σ = 1.8 s, in reasonable agreement with the experimental results of (Borsellino et.al.1972). The model parameter values μ, v_b, τ, γ, τ_G may be tuned to match the inter – subject variations of the experimental results.

References

Attneave, F. (1971). Multistability in perception, *Scientific American 225*, 63 - 71

Borsellino A, de Marco A, Allazetta A., Rinesi S, Bartolini B. (1972): Reversal time distribution in the perception of visual ambiguous stimuli, *Kybernetik 10,* 139 - 144

Engel, A.K., Fries, P., König, P., Brecht, M., Singer, W. (1999). Temporal binding, binocular rivalry, and consciousness, *Consciousness and Cognition 8,* 128 - 151

Freeman, W.J. *Neurodynamics*, Springer, London 2000.

Fürstenau N. (2003). Nonlinear dynamics model of cognitive multistability and binocular rivalry, *Proceedings of the IEEE 2003 Int. Conf. on Systems, Man and Cybernetics* . IEEE cat. no. 03CH37483C, pp. 1081-1088

Haken, H. (1996). *Principles of brain functioning*, Berlin: Springer

Hillyard, S.A., Vogel, E.K. Luck, S.J.(1999). Sensory gain control (amplification) as a mechanism of selective attention: electrophysiological and neuroimaging evidence. In G.W. Humphreys, J. Duncan, A. Treisman (Eds.), A*ttention, Space, and Action*, Oxford University press

Lamme, V.A.F. (2003). Why visual attention and awareness are different. *Trends in cognitive Sciences, 7,* 12 – 18

Lehky, S. R. (1988). An astable multivibrator model of binocular rivalry. *Perception vol.17 215-228*

Pöppel E., Ruhnau E., Schill K., v. Steinbüchel N. (1990). A Hypothesis concerning timing in the brain. In: H. Haken, E. Stadler (Eds.*), Synergetics of Cognition,* Springer Series in Synergetics vol. 45, Springer, Berlin

Richards W., Wilson H.R.,. Sommer M.A. (1994). Chaos in percepts. *Biol. Cybern. 70,* 345-349

Schuster H.G., Wagner P. (1990*).* A model for neural oscillations in the visual cortex: 1. Mean field theory and derivation of the phase equations, *Biological Cybernetics, 64,* 77 - 82

Watts, C., Fürstenau, N. (1989). Multistable fiber-optic Michelson Interferometer exhibiting 95 stable states, *IEEE J. Quantum Electron. 25,* 1 – 5

Ontological Aspects of Computing Analogies

Helmar Gust (helmar.gust@uos.de), Kai-Uwe Kühnberger (kkuehnbe@uos.de)
Institute of Cognitive Science; University of Osnabrück, Katharinenstr. 24
49069 Osnabrück, Germany

Ute Schmid (schmid@informatik.uni-osnabrueck.de)
Department of Mathematics and Computer Science; University of Osnabrück, Albrechtstr. 28
49069 Osnabrück, Germany

Introduction

In AI, there is an increasing interest in examining ontologies for a variety of applications. Classical ontologies can have different forms ranging from lattice-like structures (Ganter & Wille, 1996) to less restricted semantic networks (Peters & Shrobe, 2003).

Analogical reasoning has a long tradition in cognitive science and AI. The monograph Gentner, Holyoak & Kokinov (2001) is a good summary of recent theories for analogies. An important tool for modeling analogies is anti-unification (AU), introduced in Plotkin (1970). AU is a framework to compute generalizations of source and target which in turn can be used to establish an analogical relation (Schmid, Gust, Kühnberger & Burghardt, 2003). We will extend AU to so-called heuristic-driven theory projection (HDTP) to model analogical reasoning processes.

The Algorithm HDTP-A

We will present the algorithm HDTP-A computing generalizations together with their corresponding substitutions given a source and a target domain (Table 1). HDTP-A is based on an implementation in Gust, Kühnberger & Schmid (2003).

Table 1: The algorithm HDTP-A

Input: A set of axioms S of the source domain in a language L_S inducing a theory Th_S and a set of axioms T of the target domain in a language L_T inducing a theory Th_T.
Output: A generalized set of axioms G in a language $L^+_{S\oplus T}$ together with corresponding substitutions inducing a theory Th_G.

```
T = axioms of the target domain sorted by a heuristics h
S = axioms of the source domain
G = empty list of axioms of generalized theory
Θ1 = Θ2 = empty substitution
Th_T^{A_h} = Th_T
FOR φ ∈ T
    φ = normal_form(φ)
    SELECT ψ ∈ S
        ψ = normal_form(ψ)
        IF not same_structure(φ, ψ) REJECT
        SELECT (ξ, Θ1, Θ2) ∈ anti_instances(φ, ψ, Θ1, Θ2)
    WITH ξ best according to a heuristics h'
    IF h'(ξ) > a given threshold
        ADD ξ to G
        ADD ξΘ2 to T_T^{A_h}
        REMOVE ψ from S
    ELSE FAIL
END FOR
FOR ψ ∈ S
    φ = transfer(ψ, Θ1, Θ2)
    IF T_T^{A_h} ⊢ ¬φ CONTINUE
    IF oracle(φ) = FALSE CONTINUE
    ADD φ to Th_T^{A_h}
    ADD generalize(ψ, Θ1) to G
END FOR
```

The input is given by S (source) and T (target) inducing corresponding theories Th_S and Th_T. The output of the algorithm is a set of axioms (facts and laws) G inducing a theory Th_G generalizing source and target. The algorithm chooses an axiom from T governed by a heuristics h and searches an axiom from S to generalize both. Possible heuristics h that can be used for choosing axioms are *"Select simple axioms first"*, or *"Select axioms with a maximal number of shared terms w.r.t. already generalized terms"*. Additionally a heuristics h' is needed to select an anti-instance (AI) from all computed generalizations. Examples for such heuristics are *"Select an AI with minimal length of substitutions"* or *"Select an AI with a minimal number of second-order objects"*. After a successful generalization the resulting axiom is added to Th_G. This processes is recursively applied to all axioms in T. Finally remaining axioms in S can be transferred to T governed by the already computed substitutions as long as consistency of the extended theory and consistency w.r.t. to oberservables (checked by experiments) is guaranteed. Examples of such transfers are discussed in Schmid, Gust, Kühnberger & Burghardt (2003).

An Example: The Rutherford Analogy

In Table 2, representations of the source (left side) and the target (right side) of the Rutherford analogy are given. We consider *planet* and *sun* to be objects. Observable properties are *mass*, *distance*, and *force*. We assume certain laws that govern the behavior of objects in the solar system. Concerning the conceptualization of the atom (right side) we assume that objects *electron* and *nucleus* are given and observable properties are *electric charge*, *mass*, and *Coloumb force*. We presuppose that the *electron* and the *nucleus* have a *mass* and an *electric charge*. Finally we can perform an experiment (an abstract representation of the Rutherford experiment) to test whether analogical transfers yield valid results. Here are the computed generalizations:

Source Theory S	Target Theory T	Generalized Theory G
$mass(s) > mass(p)$	$mass(n) > mass(e)$	$mass(Y) > mass(X)$
$rev\_around(p,s)$	$rev\_around(e,n)$	$rev\_around(X,Y)$
$gravity(p,s,t) > 0$	$coloumb(e,n,t) > 0$	$F(X,Y,t) > 0$
$dist(p,s,t) > 0$	$dist(e,n,t) > 0$	$dist(X,Y,t) > 0$

Integrating Ontologies in HDTP-A

Ontologies typically order concepts hierarchically using a subsumption relation. Based on Stumme & Maedche

Table 2: Modeling the physics of the solar system and the atom model

Solar system	Atom model
types *real, object, time*	*types* *real, object, time*
entities *planet : object ; sun : object*	*entities* *electron : object ; nucleus : object*
functions *observable mass:* $object \times time \rightarrow real \times \{kg\}$ *observable dist:* $object \times object \times time \rightarrow real \times \{m\}$ *observable gravity:* $object \times object \times time \rightarrow real \times \{N\}$ *observable centrifugal:* $object \times object \times time \rightarrow real \times \{N\}$	*functions* *observable mass:* $object \times time \rightarrow real \times \{kg\}$ *observable dist:* $object \times object \times time \rightarrow real \times \{m\}$ *observable electric_charge:* $object \rightarrow real \times \{eV\}$ *observable coloumb:* $object \times object \times time \rightarrow real \times \{N\}$
facts *revolves_around(planet, sun)* $mass(sun, t) > mass(planet, t)$ $\forall t : time : gravity(planet, sun, t) > 0$ $\forall t : time : dist(planet, sun, t) > 0$	*facts* $mass(nucleus, t) > mass(electron, t)$ $electric\_charge(electron) < 0$ $electric\_charge(nucleus) > 0$ $\forall t : time : coloumb(electron, nucleus, t) > 0$
laws $\forall t : time, o_1 : object, o_2 : object :$ $dist(o_1, o_2, t) > 0 \wedge gravity(o_1, o_2, t) > 0 \rightarrow$ $\exists force : force(o_1, o_2, t) < 0 \wedge$ $force(o_1, o_2, t) = centrifugal(o_1, o_2, t)$ $\forall t : time, o_1 : object, o_2 : object :$ $dist(o_1, o_2, t) > 0 \wedge \mathrm{centrifugal}(o_1, o_2, t) < 0 \rightarrow$ $revolves\_around(o_1, o_2)$	*experiment* $\forall t : time : dist(electron, nucleus, t) > 0$

(2001), we define ontologies as follows.

Definition 1 *An ontology is a tuple* $\langle C, \texttt{is\_a}, R, \sigma \rangle$ *s.t.* C *is a set of concepts,* $\texttt{is\_a} \subseteq C \times C$ *is a partial order on* C*,* R *is a set of relations, and* σ *is an arity function.*[1]

The crucial feature in this definition is the subsumption relations is_a: Concepts specialize to subconcepts and generalize to superconcepts. In Figure 1, a corresponding hierarchical structure for the solar system of the example above is depicted. The idea is to use such ontologies to block generalizations that do not have a common superconcept. If the ontology has a tree-like structure like in Figure 1, then we need to specify subtrees that allow generalizations. In Figure 1 we would like to allow a generalization of *gravitation* and *centrifugal* and (under certain circumstances) the generalization of *force* and *distance*.

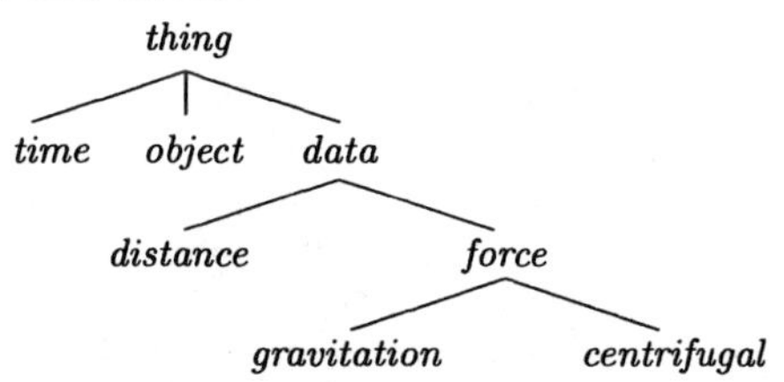

Figure 1: A possible ontology for for the model $\mathfrak{M}_1$.

Definition 2 *Assume an ontology* $O = \langle C, \texttt{is\_a}, R, \sigma \rangle$ *and a set* $C_{gen} \subseteq C$ *are given.*[2] *The generalizable expressions* O' *relative to a function Sort mapping terms to sorts are defined as follows:*

If t_1 *and* t_2 *are terms, then* $\langle t_1, t_2 \rangle \in O'$ *iff there exists* $c \in C_{gen}$ *with* $Sort(t_1)$ is_a c *and* $Sort(t_2)$ is_a c.

If ϕ *and* ψ *are complex, then* $\langle \phi, \psi \rangle \in O'$ *iff all corresponding subterms are generalizable.*

Using the definition of generalizable expressions the modified FOR-loop of the algorithm HDTP-A is depicted in Table 3 (given T and S, and an ontology O).

[1] In this paper, R and σ are not relevant.

[2] The intended interpretation of C_{gen} is a set of leaves of an upper structure, in our example the set {*time, object, date*}.

Table 3: The modified algorithm HDTP-A

```
...
FOR φ ∈ T
    φ = normal_form(φ)
    SELECT ψ ∈ S
        ψ = normal_form(ψ)
        IF not same_structure(φ, ψ) REJECT
        IF not ⟨φ, ψ⟩ ∈ O' REJECT
        SELECT (ξ, Θ1, Θ2) ∈ anti_instances(φ, ψ, Θ1, Θ2)
    WITH ξ best according to a heuristics h'
    IF h'(ξ) > a given threshold
        ADD ξ to G
        ADD ξΘ2 to T_T^{A_h}
        REMOVE ψ from S
    ELSE FAIL
END FOR
...
```

A test relative to a given ontology O is integrated in the algorithm HDTP-A in order to reduce the search space for possible generalizations. The corresponding ontological commitments are clearly domain dependent.

References

Ganter, B. & Wille, R. (1996). *Formale Begriffsanalyse. Mathematische Grundlagen.* Berlin, Heidelberg: Springer.

Gentner, D., Holyoak, K. & Kokinov, B. (2001). *The Analogical Mind. Perspectives from Cognitive Science.* Cambridge, MA.

Gust, H., Kühnberger, K.-U. & Schmid, U. (2003). Anti-unification of axiomatic systems, available online: http://www.cogsci.uni-osnabrueck.de/~helmar/analogy1.ps/

Peters, S. & Shrobe, H. (2003). Using Semantic Networks for Knowledge Representation in an Intelligent Environment. *First IEEE International Conference on Pervasive Computing and Communications*, Forth Worth, Texas, pp. 323-337.

Plotkin, G. (1970). A note on inductive generalization. *Machine Intelligence, 5*, pp. 153-163.

Schmid, U., Gust, H., Kühnberger, K.-U. & Burghardt, J. (2003). An Algebraic Framework for Solving Proportional and Predictive Analogies. In F. Schmalhofer, et.al. (Eds.), *Proceedings of the European Conference on Cognitive Science* (pp. 295-300), Lawrence Erlbaum.

Stumme, G. & Maedche, A. (2001). FCA-MERGE: Bottom-Up Merging of Ontologies. *Proceedings of the 7th Internatioanl Joint Conference on Artificial Intelligence* (pp. 225-230), Seattle, WA.

Applying Fuzzy Logic to Neural Modeling

Sébastien Hélie (helie.sebastien@courrier.uqam.ca)
Department of Computer Science, C.P. 8888 Succ. Centre-Ville
Montréal, PQ H3C 3P8 CANADA

Sylvain Chartier (chartier.sylvain@courrier.uqam.ca)
Department of Psychology, C.P. 8888 Succ. Centre-Ville
Montréal, PQ H3C 3P8 CANADA

Robert Proulx (proulx.robert@uqam.ca)
Department of Psychology, C.P. 8888 Succ. Centre-Ville
Montréal, PQ H3C 3P8 CANADA

What is fuzzy logic?

Fuzzy logic is an extension of Boolean logic which allows intermediate values between True and False. As in Boolean logic, a true statement is expressed by the value "1" and a false statement by the value "0". However, unlike in probability theory, the value must not be interpreted as a confidence level but rather as a Membership Function (MF). Therefore, every statement is "True" to a certain degree and "False" to another.

An interesting property of these MFs is that, because they vary between zero and one, they can be manipulated like probabilities, even though they are interpreted differently.

Fuzzy logic, categorization and neural modeling

Early results in psychology suggest that humans are using fuzzy decision rules. For example, Rosch (1973) has argued in favor of fuzzy boundaries between natural categories. A strong evidence in favor of fuzzy decision rules is seen in an experiment by Labov (1973) which showed that the boundary between the category "bowl" and the category "cup" is a function of size and context and that these categories overlap greatly in medium-size objects (big cups vs. small bowls).

However, connectionist networks, which are increasingly popular in quantitative modeling, are not able to account for these results. For example, multilayer feedforward networks (Rumelhart & McClelland, 1986) learn to fix crisp boundaries and, thereafter, a given item will always be a "bowl" or a "cup". Moreover, it is difficult to understand what processes are actually involved in the classification task. Some methods have been proposed to analyze activation patterns in the hidden layer but the results are unclear.

ANFIS: A neural implementation of fuzzy decision rules

An interesting alternative to standard feedforward connectionist networks would be one which would: 1) implement fuzzy decision rules and, 2) explicitly reveal its decision criterion.

In the field of electrical engineering, Roger Jang (1993) has recently proposed one such model. It is called Adaptive-Network-based Fuzzy Inference System (ANFIS). ANFIS is a multilayer feedforward network which searches for fuzzy decision rules that perform well on any given task. The fuzzy decision rules are implemented as MFs and the model learns the best fitting parameters of the MFs. The architecture of ANFIS is shown in Figure 1.

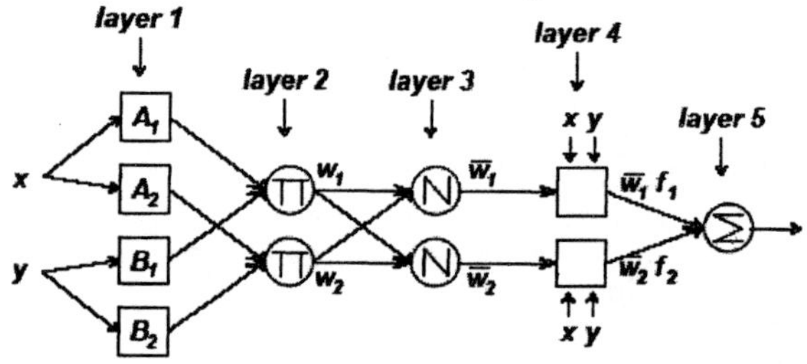

Figure 1: The general architecture of ANFIS (from Roger Jang, 1993).

Even thought ANFIS is a five layer neural network, only two of these layers have adjustable weights (here represented by squares). The first layer is composed of n MFs, each implementing a fuzzy decision rule. Any type of distributions can be modeled by MFs and the set of parameters to minimize is determined accordingly[1]. The second layer computes every possible conjunction of the n decisions rules[2]. The third layer normalizes the conjunctives MFs in order to rescale the inputs[3]. The fourth layer is a standard Perceptron (Rosenblatt, 1958) and associates every normalized MF with an output (weights are called

[1] More precisely, a MF is a distribution which was not scaled to have an area of one but rather a maximum at one.

[2] By assuming that the rules are totally independent, the MF of each conjunction is simply the product of the MFs of the conjuncts.

[3] The normalization procedure is: $x_i' = \frac{x_i}{\sum x_j}$.

consequent parameters). Finally, the fifth layer sums the evidences. The output is a real number. The consequent parameters and the MF's parameters are learned by standard backpropagation.

Categorizing Gabor patches using the XOR rule

The learning capabilities of ANFIS (Roger Jang, 1993) were tested in a categorization task. The stimuli used were Gabor patches and the categorization rule was a standard XOR. The sixteen Gabor patches were varying according to their frequency and orientation (Hélie & Cousineau, 2003)[4]. The network was fed directly with the dimension values used to create the stimuli. Because the Cartesian product of our dimension values is of size sixteen, we used sixteen MFs, modeled by Gaussian distributions, evenly dispersed in the stimulus space. The starting MFs are shown in Figure 2.

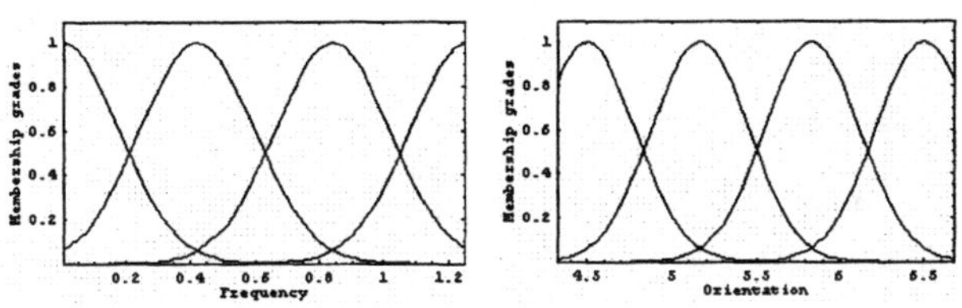

(*a*) (*b*)

Figure 2: Starting MF in the XOR categorization task. *(a)* MF of the frequency. *(b)* MF of the orientation.

The learning curve and rules learned by ANFIS after 10 epochs of training are shown in Figure 3. As seen, ANFIS has no problem learning the XOR categorization task (a). Moreover, ANFIS learns to discard useless MFs and to use only necessary decision rules (c, d). This result is consistent with results in perceptual learning (Livingston, Andrews & Harnad, 1998) as well as the information reduction theory (Haider & Frensch, 1996). Finally, panel (b) of Figure 3 shows that the two categories are well distinguished by ANFIS.

Conclusion

Fuzzy decision-rules are used by humans in everyday life. Therefore, psychological models should be able to manipulate fuzzy rules. We proposed one such model developed by Roger Jang (1993). ANFIS has the form of a neural network and uses fuzzy decision rules that are transparent (contrary to standard backpropagation models). ANFIS is thus an interesting alternative for neural modeling in having the same capabilities as standard backpropagation networks but in being explicit about it decision rules.

Acknowledgments

This research is supported by scholarships from *Le fonds québecois de la recherche sur la nature et les technologies* and the *Natural Sciences and Engineering Research Council of Canada*.

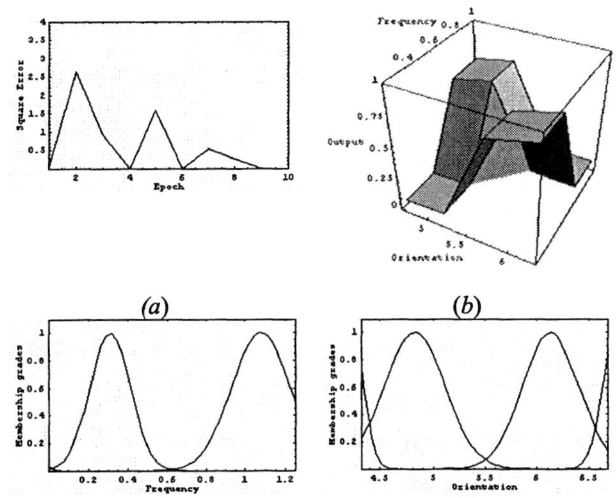

(c) (d)

Figure 3: Performance of ANFIS on a XOR categorization task. (a) Learning curve of ANFIS. (b) Fuzzy rules learned by ANFIS. (c) Resulting MFs of frequency. (d) Resulting MFs of orientation.

References

Haider, H., Frensch, P. A. (1996). The role of information reduction in skill acquisition. *Cognitive Psychology, 30*, 304-337.

Hélie, S., Cousineau, D. (June 2003). Latent interference of task-related knowledge on learning transfer. *13th Annual Meeting of the Canadian Society for Brain, Behaviour, and Cognitive Science*, Hamilton, Ontario.

Labov, W. (1973). The boundaries of words and their meanings. In. Fishman, J. (Eds.), *New Ways of Analysing Variation in English.* Washington DC: Georgetown University Press.

Livingston, K. R., Andrews, J. K., Harnad, S. (1998). Categorical perception effects induced by category learning. *Journal of Experimental Psychology: Learning, Memory, and Cognition, 24*, 732-753.

Roger Jang, J.-S. (1993). ANFIS: Adaptive Network-based Fuzzy Inference System. *IEEE Transactions on Systems, Man, and Cybernetics, 23*, 665-685.

Rosch, E. (1973). On the internal structure of perceptual and semantic categories. In. Moore, T. M. (Eds.), *Cognitive Development and the Acquisition of Language.* New York: Academic Press.

Rosenblatt, F. (1958). The perceptron: a probabilistic model for information storage and organization in the brain. *Psychological Review, 65*, 386-408.

Rumelhart, D. E., McClelland, J. L. (1986). *Parallel Distributed Processing: Explorations in the Microstructure of Cognition.* Cambridge, MA: MIT Press.

[4]The frequencies were: {0.25, 0.5, 0.75, 1} and the orientations were: $\{3\pi/2, 133\pi/80, 73\pi/40, 2\pi\}$.

A Cognitive Model for Spatial Perspective Taking

Laura M. Hiatt (lahiatt@stanfordalumni.org)
CSLI, Stanford University
220 Panama Street, Stanford, CA 94305 USA

J. Gregory Trafton (trafton@itd.nrl.navy.mil)
Naval Research Laboratory
4555 Overlook Avenue, Washington, DC 20375 USA

Anthony Harrison (anh23@pitt.edu)
Learning Research and Development Center, University of Pittsburgh
3939 O'Hara St., Pittsburgh, PA 15260 USA

Alan C. Schultz (schultz@aic.nrl.navy.mil)
Naval Research Laboratory
4555 Overlook Avenue, Washington, DC 20375 USA

Introduction

When communicating with other people, one of the basic things that people must do is take others' perspectives. Most of the experimental work on spatial language and perspective taking has focused on four frames of reference: exocentric (world-based, such as "Go north"), egocentric (self-based, "Turn to my left"), addressee-centered (other-based, "Turn to your left") and object-centric (object-based, "The fork is to the left of the plate") (Carson-Radvansky & Logan, 1997; Carson-Radvansky & Radvansky, 1996; Levelt, 1984). Any time egocentric or addressee-centered frames of references are used, spatial perspective taking is needed: egocentric utterances require the listener to take the speaker's perspective, and addressee-centered utterances require the speaker to have already taken the listener's perspective. As part of a project to make intelligent agents and robots more useful to people, we have been developing cognitive models of spatial cognition and perspective taking (Trafton, Schultz, Cassimatis et al., under review; Trafton, Schultz, Perzanowski et al., under review).

Unfortunately, there are relatively few computational cognitive models of spatial cognition available in order to implement spatial perspective-taking models. One recent entrée to spatial cognition research has been ACT-R/S (Harrison & Schunn, 2003).

ACT-R/S extends ACT-R (Anderson & Lebiere, 1998) to implement a theory about spatial reasoning. It posits that spatial representations of objects are egocentric and dynamically updated (Wang & Spelke, 2002). ACT-R/S represents objects using vectors to the visible sides of the object. It has the ability to track these objects through a configural buffer, a data structure analogous to the other buffers of ACT-R that stores each object once it has been identified. The coordinate vectors of the objects in the buffer are then dynamically updated as the agent moves throughout the spatial domain. The configural buffer, unlike the visual and retrieval buffers of ACT-R, can hold more than one object to account for the fact that animals have been shown to track more than one landmark at once while moving through the world. The other spatial buffers within ACT-R/S, the visual and manipulative buffers, are not a central part of the work described here and are described more fully elsewhere (Harrison & Schunn, 2003).

Perspective Taking Model

In order to demonstrate the results of perspective taking using ACT-R/S, a simple 'fetch' task was designed for the model to perform in a simulated world. In this world, two agents (hereafter referred to as the 'speaker' and the 'robot') are in a room with two wrenches and a screen. In many cases, an utterance is ambiguous given the listener's knowledge, but unambiguous given the speaker's knowledge. Figure 1 is an example. The figure shows a robot and a person facing each other. The robot can see that there are two wrenches in the room, wrench1 and wrench2, but the person only knows about wrench2 because wrench1 is hidden from her. If a person said, "Robot, give me the wrench" (which is understood by our robotic system), the phrase "the wrench" is potentially ambiguous to the robot because there are two wrenches, though unambiguous to the person because she only knows of the existence of one wrench. Intuitively, if the robot could take the perspective of the person in this task, it would see that, from that perspective, wrench2 is the only wrench and therefore "the wrench" must refer to wrench2. Our model of spatial perspective taking uses ACT-R/S to accomplish this task. There are several components to perspective taking that the model goes through in order to successfully accomplish its goals.

Perspective taking process

The production rules involved in the perspective-taking process are the most important part of the model, as they implement the heart of its theory of spatial perspective taking. Taking the perspective of someone at position and orientation B, from position and orientation A, the over all procedure is to: (1) Turn to face position B; (2) Walk to position B; (3) Face orientation B; (4) Extract the desired information from the visual knowledge at this position and orientation; (5) Face position A; (6) Walk back to position A; and (7) Return to orientation A.

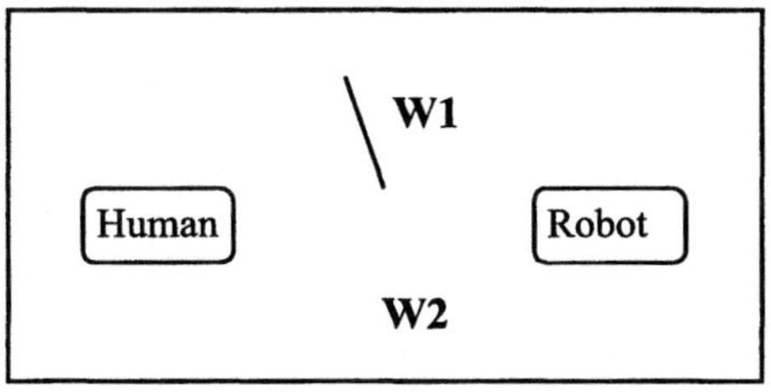

Figure 1: The human can only see one wrench (shown as W2), while the robot can see two.

The key to this process is that all of these movements – i.e. turning and walking – are *mentally* done by only transforming the configural buffer contents by the appropriate vector, leaving everything else the same. Thus the physical location of the robot does not change; it is only its mental location and perspective that changes.

Initial scan for objects The model first uses perspective taking to deduce where it should begin looking for the wrench. If the speaker gives a references such as "in front of me", or "to my left", the robot can use that information to constrain its search. It takes the speaker's perspective and then looks in front of it, or to its left (as indicated by the speaker's initial instructions), and keeps track of that location as it returns to its own perspective. This is where it begins its search for the wrench.

Deciding which wrench to give to the speaker The model also uses perspective taking once a wrench has been found. When it has located a wrench in the desired location, it looks around for obstacles that could possibly block the speaker's view of the wrench. If it finds any such obstacles, it takes the speaker's perspective again in order to judge whether or not the speaker can see that particular wrench from her perspective.

This time, however, once the robot has taken the speaker's perspective, instead of turning to match the speaker's orientation, it turns to face the located wrench. Determining whether or not the wrench is visible by the speaker is then done by comparing the transformed location vectors of the target object with the location vectors of the possible obstacles, making sure that the obstacle's vectors do not completely surround the target object's vectors. This ensures that the speaker has the ability to see at least part of the wrench.

If the speaker can in fact see the wrench, the robot hands that wrench to the speaker. If the speaker cannot, the robot continues to look for a wrench that the speaker can see.

Discussion

Using a person's (hypothesized) representation allows the cognitive agent to undergo perspective taking by imagining movement throughout the world by simply altering the representation of the objects in the configural buffer. This ultimately results in true perspective taking in the sense that the agent's representation of objects after taking the other's perspective roughly matches the second agent's own representation of these objects, allowing the agent to truly see the world as the other does. In the end, this provides a more natural and human-like interaction with the second agent, since the cognitive agent responds as a human plausibly would instead of introducing into the conversation an item (here, a wrench), that the second agent might not even knows exists.

Acknowledgments

This work was partially supported by the Office of Naval Research to Greg Trafton. and Alan Schultz under work order number N0001402WX20374. Additional funding was supplied by DARPA IPTO under the MARS program.

References

Anderson, J. R., & Lebiere, C. (1998). *Atomic components of thought*. Mahwah, NJ: Erlbaum.

Carson-Radvansky, L. A., & Logan, G. D. (1997). The influence of functional relations on spatial template construction. *Journal of Memory & Language, 37*, 411-437.

Carson-Radvansky, L. A., & Radvansky, G. A. (1996). The influence of functional relations on spatial term selection. *Psychological Science, 7*, 56-60.

Harrison, A. M., & Schunn, C. D. (2003). ACT-R/S: Look Ma, No "Cognitive-map"! In *International Conference on Cognitive Modeling*.

Levelt, W. J. M. (1984). Some perceptual limitations on talking about space. In A. J. van Doorn, W. A. van der Grind & J. J. Koenderink (Eds.), *Limits in perception* (pp. 323-358). Utrecht: VNU Science Press.

Trafton, J. G., Schultz, A. C., Cassimatis, N. L., Hiatt, L. M., Perzanowski, D., Brock, D. P., et al. (under review). Using similar representations to improve human-robot interaction.

Trafton, J. G., Schultz, A. C., Perzanowski, D., Adams, W., Bugajska, M. D., Cassimatis, N. L., et al. (under review). Children and robots learning to play hide and seek.

Wang, R. F., & Spelke, E. S. (2002). Human spatial representation: Insights from animals. *Trends in Cognitive Sciences, 6*(9), 376-382.

Changes in Youth's Authority Concept After a Group Accommodative Event

Shinya Iida (wbs22616@mail.wbs.ne.jp)
Fukude Elementary School, Shizuoka, Japan
Department of Computer Science, Shizuoka University
3-5-1 Johoku, Hamamatsu, Shizuoka 432-8011 Japan

Hiroyuki Iida (iida@cs.inf.shizuoka.ac.jp)
Department of Computer Science, Shizuoka University,
3-5-1 Johoku, Hamamatsu, Shizuoka 432-8011 Japan

Abstract

This paper concerns moral development in the framework of cognitive modeling. Traditionally, moral development researches have been conducted, mostly discussing different level of development among subjects of different ages and their developmental tendencies and orders. This methodology works well in finding the ratio of subjects in each level at the time of research, as well as subjects' tendencies. However, it fails to address the paths taken to move from one level to next, as well as from the low-level to high-level.

In order to get accurate information on the process of changes and order, it is necessary to conduct follow-up researches with same subjects and predict the determining factors. This paper discusses the moral development changes taken place during a school group accommodative event. The objective for the event set by school was to "provide activities which would allow students to learn how to function in a group, as well as allow them to have a desirable morale -learning experience. The purpose was to nurture collectiveness and to learn social abilities in a small group environment by interaction, being in the situation where each other's role and relationship are clearly defined. We can say that this activity was relative in providing the necessary information concerning the authority concept development where problems usually exist between the leaders and the followers. Our results indicate that there was a notable difference between males and females in terms of authority concept development.

Method

The study was conducted among 72, public elementary school fifth-grade students (33 males and 39 females) in the Shizuoka Prefecture, Japan. Subjects are divided into groups of six with mixed genders. Then students decided on different roles and responsibilities for each member, including the group leader. The collective accommodative event over two nights and three days took place in the form of a camping school. The questionnaires were prepared based on the materials used in Watanabe's study (1989) on authority and were given both within a week prior to and after the event.

To check the longevity of the changes in the subjects, the third questionnaire was given three months after the event.

Results and Discussions

1. Damon's developmental stage theory:

Concerning authority concept development there are two studies conducted: one by Damon (1980) and the other one by Watanabe (1989) whose study tested whether Damon's findings would hold true with Japanese subjects. However, neither one of those studies did not illustrate how one would advance from one level to the next, nor did it show the relationship between morale education and morale development.

Our understanding of Damon's developmental theory, as indicated in Fig. 1, shows developmental stages in a structure (authority and merit) and sub-stages (absolute-relative). With this interpretation, the grading accuracy had improved to 86%.

2. Changes in authority concept development stages and observation

The authority concept development stage distribution before and after the group accommodating event is shown in Fig. 2. Of those, 25 (34.7%) have seen changes in their development steps. Fig. 3 shows the changes that had taken place: changes made from both level 1-A to 1-B and 0-B to 1-B, had most with 8 people each. It is noteworthy to see a significant number of students had moved up from 0-B to 1-B, skipping 1-A altogether. When observing the changes from 0-B to 1-B, the axis indicating relativity which is found in both does not see any changes. However, the change does take place on the "authority and merit" axis. Also, it is believed that the reason for no movement recorded from 1-B to 2-B is because of the holding of the axis of relativity on 1-B leads to the absoluteness of the "authority and merit".

3.Gender differences

There was no apparent difference between genders recorded when the overall distribution was looked at initially (before the trip: D=0.0219; after the trip=0.207). However, when individual stages were closely analyzed significant difference was revealed: before the event 0-B (male: 15.2%, female: 43.6%), 1-A (male: 39.4%, female: 20.5%); after the event 1-A (male: 36.4%, female: 5.1%). (Prior to the event: 0-B (χ^2=6.55,df=1,p<.05), 1-A (χ^2=4.94,df=1,p<.05). After the event: 1-A (χ^2=7.14,df=1,p<.01)).

To examine this finding further, each individual's developmental changes and answers were analyzed. After the analysis, it was clear that males changed from "0-A→0-B→1-A→1-B", exactly following Damon's developmental stages (Damon, 1980). However, in case of females 44% of them moved up to 1-B without going through 1-A. In case of women, being more "considerate" than men is a part of their nature therefore it is quite possible that the reason for their thoughtfulness has changed from "affection/desire" → "focusing on the authority figure", at least in case of Japanese elementary school female children. Discussion concerning this finding may suggest problems related to Damon's development stages process for females.

4.Educational effectiveness of the group accommodative event. When the third and the final questionnaire was conducted two individuals developmental changes had gone down, rather than up, making the retention rate 92%. When analyzing the differences between the questionnaire given immediately after the event and three months after the event, we can say that with 34.7% of the participants experiencing the advancement in their developmental steps in an event which lasted for less than a week, the impact the group accommodative event had on the authority principle concept was very profound. We feel that this was a study that shed some light on what kind of experience affects children's developmental stages while revealing its educational significance, as well.

FIG. 1: Early Authority Levels.

2	Merit - Respect	2-A	2-B
1	Focuses on the authority figure	1-A	1-B
0	Affection-Desire	0-A	0-B
		Absolute	Relative
		A	B

FIG 2: Subject distribution on developmental stages after the group accommodative event

	0-A	0-B	1-A	1-B	2-A	2-B
Before	4	22	21	15	10	0
After	1	14	14	28	15	0

FIG 3: Developmental change charts

		After	
Before	M	F	T
0-A		0-B	
	1	1	2
		1-A	
	1	0	1

		After	
Before	M	F	T
0-B		1-A	
	1	0	1
		1-B	
	1	7	8
		2-A	
	0	1	1

		After	
Before	M	F	T
1-A		1-B	
	3	5	8
		2-A	
	0	1	1

		After	
Before	M	F	T
1-B		2-A	
	2	1	3

References

Damon, W. (1980) Patterns of change in children's social reasoning: a two-year longitudinal study. Child Development, 51, 1010-1017.

Kurtines, W. M. & Gewirtz, J. (1984) Morality, Moral Behavior and Moral Development, New York: Wiley.

Ministry of Education (1990) Elementary School Suggested Instruction Method. Ministry of Finance Printing

Watanabe, Y. (1989) The relationship between the development of the concept of fairness and authority concept development in adolescence Educational Psychology, 37, 163-171

A Memory-Based Learning Model of Dutch Plural Inflection

Emmanuel Keuleers (Emmanuel.Keuleers@ua.ac.be)
Center for Psycholinguistics, University of Antwerp
Antwerp, Belgium

Dominiek Sandra (Dominiek.Sandra@ua.ac.be)
Center for Psycholinguistics, University of Antwerp
Antwerp, Belgium

A common observation in studies of inflectional morphology is that words which are atypical members of their syntactic class often take a regular inflection, independent of their phonology. For example, in English we would say that the plural of the surname MANN is MANNS although it has the same phonology as the noun MAN which has the irregular plural MEN. The same goes for the borrowed word TALISMAN (plural TALISMANS) as opposed to the noun FIREMAN (plural FIREMEN).

Marcus et al. (1995), refer to these atypical words as *non-canonical roots*, and suggest that their inflection can best be explained by a dual-mechanism model. Irregular words are inflected through an associative memory system, while regular words are inflected by a symbolic rule (e.g. add *-s)*. For non-canonical roots however, access to the associative memory sytem is blocked because they are not marked as nouns or verbs. Consequently, they are inflected by the symbolic rule by default.

The inflection of non-canonical roots presents a problem for single-mechanism models such as the connectionist model for the English past tense proposed by Rumelhart and McClelland (1986). As words in these models are coded in terms of their phonological characteristics, atypical words cannot be distinguished from other words. Adding a feature that marks words as non-canonical entries could resolve this, but as Pinker and Ullman (2002) point out, this would not be very different from the implementation of a symbolic rule. To present a complete alternative to the dual-mechanism model, a single-mechanism associative model should be able to resolve the inflection of non-canonical roots based on characteristics that are not reducable to a symbolic rule.

Using simulations on corpus data, we will try to show that a single-mechanism model can inflect a particular type of non-canonical roots, namely unassimilated borrowings in Dutch, by using orthographic information. The choice for this type of information is grounded in the observation that the spelling of borrowings often deviates from typical Dutch spelling. If the plural of existing borrowings is stored along with their spelling pattern, this orthographic information could drive an analogical process with these stored representations and thus guide the selection of the plural.

Simulations

The dual-mechanism model takes a somewhat particular position with respect to the Dutch plural, in that both its productive plural suffixes (*-en* and *-s*) can be the default, dependent on the phonology of the root form (Pinker, 1999). This double default makes the Dutch plural a highly regular system: about 85 % of all words can be inflected by the default rule and little storage of exceptions is necessary. As non-canonical roots are inflected by the default rule, they should more or less reflect the distribution of the plural suffixes, taking *-en* more often than *-s*. This appears to be the case for some non-canonical roots (e.g. surnames), but as is commonly noted in descriptions of the Dutch plural system, unassimilated borrowings in Dutch tend to take the less frequent *-s* plural, which makes them an interesting test case. The goal for these simulations is to show that an associative model is able to correctly predict the plural of most words in a Dutch corpus, but that it is also able to predict the plural for unassimilated borrowings at least as well as the default part of a dual-mechanism model. This would be evidence that non-canonical roots can be stored and that their inflection is guided by analogical principles rather than by a default-rule.

To test our hypothesis, we compared the plural prediction accuracy of a phonological rule model (equivalent to the default part of the dual-mechanism model) to the accuracy of three single-mechanism models using diffferent information representations. The data-set on which the models were tested consisted of 3145 non-compound singular-plural noun pairs found in the CELEX lexical database for Dutch. The single-mechanism models were implemented using the Tilburg Memory Based Learner (TiMBL) (Daelemans et al., 2003), which can be seen as an extension to classical nearest neighbor models that allows for the specification of a number of parameters, such as number of neighbours, overlap and distance metric and feature weigthing. We made two changes to TiMBL's default parameter values: the modified value difference metric was used instead of the overlap metric and the number of neighbours participating in the class prediction was set to 10. All simulations used the leave-one-out method (for each word the plural was predicted once, using the rest of the data as a training set). Words were represented using the onset, nucleus and coda of the two final syllables. A first model (MBL-phon) used a purely phonological representation, while

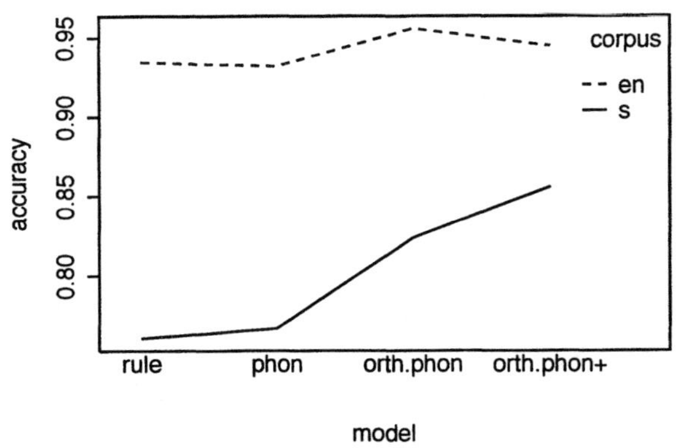

Figure 1: Accuracy of four models on predicting the plural of nouns from the CELEX lexical database.

a second model (MBL-orth-phon) used an additional orthographic representation. Finally, a third model (MBL-orth-phon+) was implemented using a representation where additional features for the distinctiveness of each of the orthographic features were derived using a sequential memory-based learning approach. The distinctiveness measures correspond to the number of phonological neighbors that do not share the orthographic feature. For example, in Dutch, the spelling *fr* is used in almost all words with the sound */fr/*, so its spelling is not distinctive. On the other hand, the spelling *ea* is almost never used in words with the sound */i:/*, so its spelling is highly distinctive. These additional features should give the MBL-orth-phon+ model the potential of finding neighbours that are not very similar in spelling or sound but that are similar in the distinctiveness of their spelling pattern.

Results and Discussion Figure 1 illustrates the results of the simulations. We performed a logit-analysis on prediction accuracy taking the rule model as the baseline. For words which had an *-en* plural in the corpus, neither the MBL-phon nor the MBL-orth-phon+ model were more successful than the Rule model. The MBL-orth-phon model on the other hand, was significantly better at predicting these plurals, $z = 2.970, p < .01$. For words which had an *-s* plural in the corpus, both the MBL-orth-phon ($z = 3.696, p < .001$) and MBL-orth-phon+ model ($z = 5.7335, p < .001$) performed significantly better than the Rule model.

These results show that an associative model can be successfully used to predict the suffix of Dutch plurals and that the addition of orthography as an information source constitutes a significant improvement . For *-s* plurals, where the Rule-model is successful in about 76 % of the cases, the MBL-orth-phon and MBL-orth-phon+ model offer a 6% and 10 % improvement respectively. The fact that the MBL-phon model does not do any better than the Rule model, indicates that this improvement is due to the additional information source, orthography.

Table 1: Words for which the *-s* plural was predicted correctly by the MBL-orth-phon and MBL-orth-phon+ models but for which the Rule model and MBL-phon predicted an *-en* plural.

BACK BARBECUE BIDON BOARD BOY CAKE CAPE CAPUCHON CLAN CLOWN CLUB COAT COUPE CRACK CREPE CROSS DRIVE E ECHELON ESKADRON FILE FRAME FRITE G GAME GOAL HALL HOLE HOME INCH JACK JOULE KAMELEON KARTEL KICK L MISS MOVE P PASS PIECE PLAQUE POLL R RACE RAID RELIEF RIFF SHERIFF SHOP SHOW SNACK SPIKE SPRAY STICK STOCK TAKE TAPE TIC TOUR TRACK TRICK TRUC TRUCK VUE WAGON Y YANK

The question at this point is whether orthographic information is also used in the inflection of unassimilated borrowings, which under the dual-mechanism model are inflected by the default rule. As the Rule model is an implementation of the default it should always inflect these words correctly. Words that are inflected correctly by the MBL-orth-phon(+) model but incorrectly by the Rule model should therefore not include unassimilated borrowings or other non-canonical roots. Moreover, if it is to be shown unequivocally that orthography is the information source which allows for the correct inflection of these words, they should also be incorrectly predicted by the MBL-phon model.

As the appreciation of what constitutes an unassimilated borrowings is mostly qualitative, table 1 lists all the words for which the MBL models using orthography predicted an *-s* plural but for which both the Rule model and the MBL-phon model incorrectly predicted an *-en* plural. These constraints do not make for a very long list, but as can be observed, it contains almost exclusively English words, i.e. non-canonical roots in Dutch. The fact that these words are captured by an associative model using information that is not reducable to a symbolic rule, suggests that associative memory is not blocked for these words and that a default mechanism is not needed to inflect them. The failure of the default rule to inflect these words, poses an additional problem to the dual-mechanism model.

References

Daelemans, W., Zavrel, J., van der Sloot, K., and van den Bosch, A. (2003). Timbl: Tilburg memory based learner, version 5.0, reference guide. Technical report, University of Tilburg.

Marcus, G. F., Brinkmann, U., Clahsen, H., Wiese, R., and Pinker, S. (1995). German inflection: The exception that proves the rule. *Cognitive Psychology*, 29:189–256.

Pinker, S. (1999). *Words and Rules*. London: Phoenix.

Pinker, S. and Ullman, M. T. (2002). The past-tense debate: The past and future of past tense. *Trends in Cognitive Sciences*, 6(11):456–463.

Rumelhart, D. and McClelland, J. (1986). On learning the past tenses of English verbs. In *Parallel Distributed Processing, vol 2.* MIT Press.

A Cognitive/Affective Model of Strategic Behavior - 2-Person Repeated Prisoner's Dilemma Game

Sung-youn Kim (sung-youn.kim@stonybrook.edu)
Department of Political Science, Stony Brook University
Stony Brook, NY 11790-4392 USA

Charles Taber (ctaber@notes.cc.sunysb.edu)
Department of Political Science, Stony Brook University
Stony Brook, NY 11790-4392 USA

John Q. Public, a computational model of political cognition that incorporates both cognitive and affective mechanisms and both on-line and memory-based processing, is employed to study strategic behaviors in a social dilemma situation. Specifically, two John Q. Publics were employed as players in a 2-person repeated prisoner's dilemma game to simulate the experimental data from Rapoport et al. (1976). Compared to previous studies that simulated the same data, this study has two advantages: 1) it is arguably more psychologically realistic in that it incorporates an affective mechanism and 2) the players are modeled as completely independent agents.

The Model

Affective Mechanism

The model is built using ACT-R. Together with the cognitive mechanisms embedded in ACT-R, an affective, attitudinal mechanism was incorporated into the model based on the following axioms;

- (Hot Cognition) Most social concepts in memory are affectively charged. The affective evaluation linked to a concept in long term memory can be positive or negative or close to zero, indicating either a non-attitude or ambivalence. (Abelson, 1963)
- (Attitude Priming) The information in memory that is affectively congruent to the information being processed is more accessible. (Fazio, 1990)
- (Primacy of Affect) Affect can not only be triggered automatically without conscious appraisal of an attitude object, but also is primary in the sense that it comes into working memory before other conscious thoughts and appraisals enter into the judgment process. (Zajonc, 1980, 2001)
- (On-line and Memory-based Processing)
  - (Memory-based Processing) Different, often conflicting considerations and feelings that come to mind on the spot influence the evaluations of objects. The accessibility of situational factors, together with the content and structure of prior beliefs determine what considerations and feelings come to mind on the spot.
  - (Attitude Construction and Colorization) An attitude toward an object is constructed and/or updated continuously in real time. That is, it is colored by those thoughts and feelings that come to mind at the time of information processing.
  - (On-line Processing) An affective summary evaluation (valence) is linked to every object in memory that has been evaluated in the past and is updated continuously on every exposure to new information, thereby reflecting the weighted influence from all momentarily accessible information. That is, the valence of the new information at the time of updating is colored by the respective valence of those thoughts and feelings elicited at the time of exposure.

(For formal presentations and a check of their internal validity in the context of political candidate evaluation, see, Kim, Lodge, & Taber, MPSA 2004 Conference Proceeding.)

Simulation Framework

A player's behavior in each round of the PD game was modeled as follows:

- At each round of play, the model chooses a strategy that it believes or feels is better at the moment of decision-making.
- Upon receiving the outcome, it adjusts its beliefs about strategies and the opponent based on (its perception of) the realized outcome.
- It goes to the next round of play with the updated beliefs.

Most importantly, given the above cognitive/affective mechanisms, the model chooses a strategy that it believes or feels is better at the moment of decision-making. That is, it chooses strategies based on its attitudes toward the strategies at the moment that are 1) continuously updated reflecting past experiences and 2) constructed on the spot reflecting elicited thoughts and feelings.

Multiple agents were incorporated by running them on multiple, separate lisp processes and allowing them to communicate with one another via the 'socket'.

Simulation Results

In the experiment conducted by Rapoport et al (1976), each of 10 pairs of subjects (players) repeatedly played 300 rounds of PD game. The payoffs were (-1, -1), (10, -10), (-10, 10), and (1, 1) for the outcomes (Defect, Defect), (Defect, Cooperate), (Cooperate, Defect), and (Cooperate, Cooperate), respectively.

100 simulations were implemented, meaning that each of 100 pairs of agents played 300 rounds of the game. Important parameters specific to the simulation were two. First, the subjective evaluations of the game payoffs (1, 10, -1, and -10) used in the simulation were 0.1, 0.29, -0.1, and -0.29, respectively, on the conventional -1 to 1 attitude scale. Note that objective game payoffs are different from subjective attitudes toward them. Second, the anchoring and adjusting parameter in on-line updating mechanism was set to be 0.93. Other parameters are left to their 'common' values (e.g., $d = 0.5$).

10 simulations that closely resemble the human data were selected for a direct comparison. The Table 1 shows the correlation in mean frequencies of outcomes between human data and model with both 10 and 100 simulations.

Table 1: General Fit : Correlation of Mean Frequencies

	DD	DC	CD	CC	r	Mean-dev
Human	30	7	8	55		
10 run	41	2	1	56	0.970	0.070
100 run	44	1	1	53	0.944	0.084

The Figure 1 and 2 show the changes in frequencies of mutual cooperation (CC) and mutual defection (DD) over time in human data and the model runs with 10 simulations, respectively.

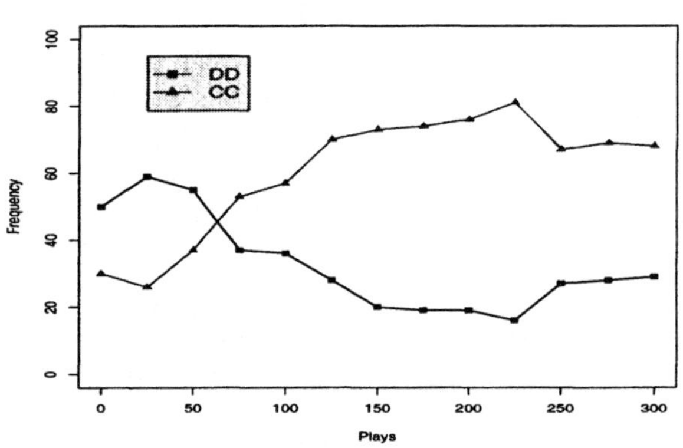

Figure 1: Strategy Shift : Human Data

In all, the simulation results seem to be reasonable in terms of reproducing main observations in the experimental data, namely, the strategy shift, bimodality in outcomes, and a sort of stabilization of outcome frequencies over time. The simulation also provides a coherent

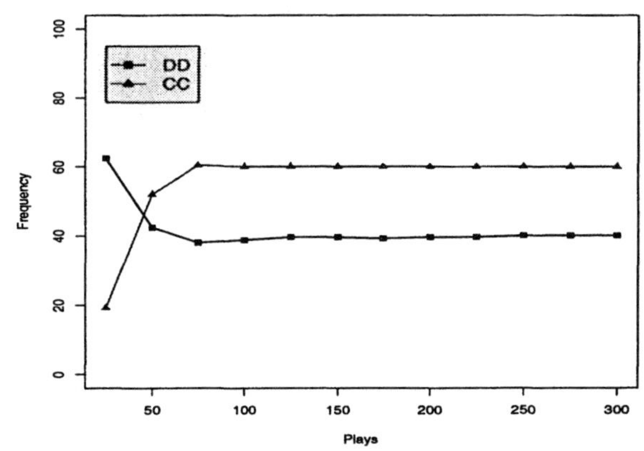

Figure 2: Strategy Shift : Model with 10 simulations

explanation for the main observations based on, especially, the role of affective, attitudinal judgements. (For detail, see, Kim & Taber, MPSA 2004 Conference Proceeding.)

However, the model was not successful in generating enough variance in strategy outcomes; that is, the simulated strategy outcomes were overly bimodal. It seems to be mainly due to the assumed zero variance in players' subjective evaluations of the payoffs. For better simulation, it appears to be necessary to have data where players' subjective evaluations of game payoffs are available.

The model and the simulation framework can be applied to a wide range of different games such as games with more than 2 players or different observational mechanisms.

Acknowledgments

The computational model is drawn from the collaborative work by Kim, Lodge, & Taber.

References

Lebiere, Wallach, & West (2000). A Memory-based Account of the Prisoner's Dilemma and Other 2x2 Games. *Proceedings of International Conference on Cognitive Modeling 2000.* (pp. 185–193).

Kim, Lodge, & Taber (2004). A Computational Model of Political Cognition: The Dynamics of Candidate Evaluation. *Proceedings of Midwest Political Science Association Annual Conference 2004.*

Kim & Taber (2004). A Cognitive Model of Strategic Behavior: The Dynamics of Cooperation in Repeated Social Dilemma Games. *Proceedings of Midwest Political Science Association Annual Conference 2004.*

Lodge & Steenbergen (1995). The Responsive Voter: Campaign Information and the Dynamics of Candidate Evaluation. *APSR 89.* (pp. 309–326).

Zaller & Feldman (1992). A Simple Theory of the Survey Response: Answering Questions versus Revealing Preferences. *APSR 36.* (pp. 579–616).

A Computational Model of Voter - The Dynamics of Political Candidate Evaluation

Sung-youn Kim (sung-youn.kim@stonybrook.edu)
Department of Political Science, Stony Brook University
Stony Brook, NY 11790-4392 USA

Milton Lodge (mlodge@notes.cc.sunysb.edu)
Department of Political Science, Stony Brook University
Stony Brook, NY 11790-4392 USA

Charles Taber (ctaber@notes.cc.sunysb.edu)
Department of Political Science, Stony Brook University
Stony Brook, NY 11790-4392 USA

John Q. Public, a computational model of political cognition which incorporates both cognitive and affective mechanisms, is employed as a voter facing political campaign information. A series of hypothetical, computational experiments show that the model successfully reproduces a set of well-known empirical phenomena found in electoral research and research on political cognition. Specifically, in response to issue and candidate information, the model reproduces 1) practice, recency, and spreading activation effects on recall, 2) cognitive and attitude priming effects, 3) question order and wording effects in survey research, and 4) both on-line and memory-based processing.

Given the space limit, only some of axiomatic characterizations of the model and some of the simulation results were presented here. (For formal presentation and full discussion of the model and simulation results, see Kim, Lodge, & Taber, MPSA 2004 Conference Proceeding.)

The Model

The model is built using ACT-R. Together with the cognitive mechanisms embedded in ACT-R, an affective, attitudinal mechanism was incorporated into the model based on a set of axioms, some of which are the following.

- (Hot Cognition) Most social concepts in memory are affectively charged. The affective evaluation linked to a concept in long term memory can be positive or negative or close to zero, indicating either a non-attitude or ambivalence. (Abelson, 1963)
- (Attitude Priming) The information in memory that is affectively congruent to the information being processed is more accessible. (Fazio, 1990)
- (Primacy of Affect) Affect can not only be triggered automatically without conscious appraisal of an attitude object, but also is primary in the sense that it comes into working memory before other conscious thoughts and appraisals enter into the judgment process. (Zajonc, 1980, 2001)
- (On-line and Memory-based Processing)
  - (Memory-based Processing) Different, often conflicting considerations and feelings that come to mind on the spot influence the evaluations of objects. The accessibility of situational factors, together with the content and structure of prior beliefs determine what considerations and feelings come to mind on the spot.
  - (Attitude Construction and Colorization) An attitude toward an object is constructed and/or updated continuously in real time. That is, it is colored by those thoughts and feelings that come to mind at the time of information processing.
  - (On-line Processing) An affective summary evaluation (valence) is linked to every object in memory that has been evaluated in the past and is updated continuously on every exposure to new information, thereby reflecting the weighted influence from all momentarily accessible information. That is, the valence of the new information at the time of updating is colored by the respective valence of those thoughts and feelings elicited at the time of exposure.

Computational Experiments and Results

The computational experiments consisted of three experiments that were identical except for the model's initial ideological beliefs - typical conservative, moderate, and liberal beliefs. The initial ideological beliefs were derived largely from the NES (National Election Survey) 2000. Within each experiment, an identical set of 24 pieces of campaign information were presented to the model, with varying order and wording and with three intervals. Important parameter values used for the experiments were: 1) anchoring and adjusting ($\rho = 0.7$); 2) the parameter governing the effect of affective strength of chunks on their decay rates ($\alpha = 0.1$); 3) affective congruence effect parameters. All other parameters were left to their 'common' values (e.g. $d = 0.5$). In the below, some of the results were briefly presented.

The Figure 1 shows the changes in attitudes towards two fictitious candidates, James and Edwards, and two political parties, Republican and Democratic Parties, as the set of campaign information is processed (with moderate initial belief). As expected, the valence of attitudes towards and the preference over the objects change over time as the campaign information is integrated into the attitudes towards the objects. In the end, the model was slightly positive toward Democratic Party, slightly negative toward Republican Party, but positive towards

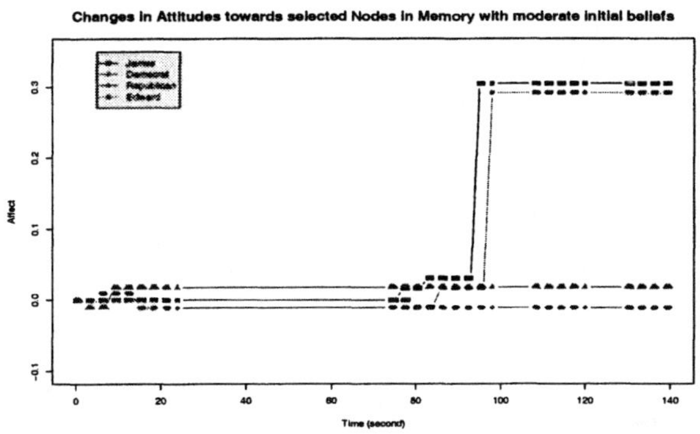

Figure 1: Changes in Candidate Evaluations Over Time

both candidates, though slightly preferring the Democratic candidate (James).

The Figure 2 shows the changes in full activation levels of two chunks, "James" and "Democrat", over time, which is the sum of base-level activation, spreading of activation, and the attitude priming effect.

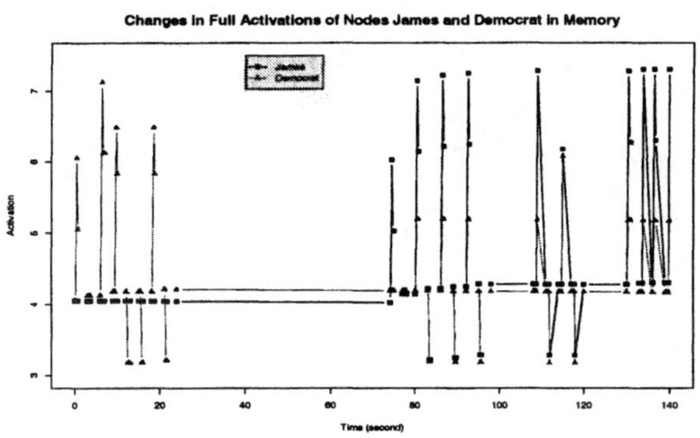

Figure 2: Changes in Full Activations over Time

The Figure 3 shows cognitive and attitude priming effects with liberal initial belief. It shows the full activation levels of four chunks in memory when the model processes the word "Republican" (i.e., "Republican" is a prime). At the moment of processing, "rich" was both semantically associated with and attitudinally congruent to the prime. "dumb" was only attitudinally congruent to the prime. "Edward" was not related to the prime in any way (a base case). "intelligent" was only attitudinally incongruent with the prime. Consistent with cognitive and attitude priming effects, the order of chunks in terms of their activation levels, from the highest to the lowest, were "rich", "dumb", "Edward", and "intelligent".

The Figure 4 shows a question order effect in survey response. It compares the expressions of attitudes toward the fictitious candidates with those when the information about candidates' party affiliations were presented

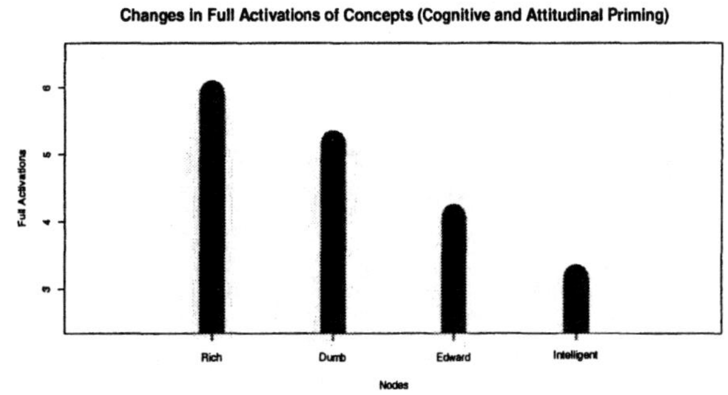

Figure 3: Activation Levels of Chunks with "Republican" as a Prime

later. When the information were presented later, the model's attitudes toward both James and Edward increased mainly due to the changes in influences of the information on its attitudes towards candidates.

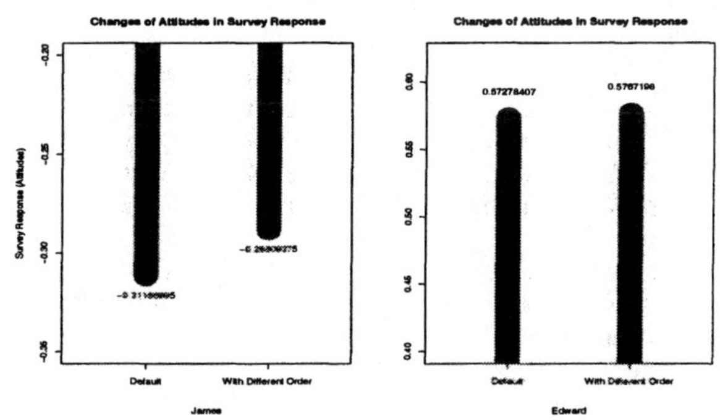

Figure 4: Changes of Attitudes due to Order Effect

Acknowledgments

This paper is drawn from Kim, Lodge, & Taber (MPSA 2004 Conference Proceeding).

References

Kim, Lodge, & Taber (2004). A Computational Model of Political Cognition: The Dynamics of Candidate Evaluation. *Proceedings of Midwest Political Science Association Annual Conference 2004.*

Lodge & Steenbergen (1995). The Responsive Voter: Campaign Information and the Dynamics of Candidate Evaluation. *APSR 89.* (pp. 309–326).

Lodge & Taber (2002). The Primacy of Affect for Political Candidates, Groups, and Issues: An Experimental Test of the Hot Cognition Hypothesis. Manuscript submitted for publication.

Zaller & Feldman (1992). A Simple Theory of the Survey Response: Answering Questions versus Revealing Preferences. *APSR 36.* (pp. 579–616).

New Contribution To Adaptive Temporal Radial Basis Function Applied on TIMIT Corpus

Mesbahi Larbi (**mesbahi_99@yahoo.com**)
SIMPA laboratory, BP 1505, Department of Computer Science,
USTO-MB, Oran, 31000, Algeria.

Benyettou Abdelkader (**aek.benyettou@univ-usto.dz**)
SIMPA laboratory, BP 1505, Department of Computer Science,
USTO-MB, Oran, 31000, Algeria.

Introduction

A successful speech recognition system has to determine features not only present in the input pattern at one point in time, but also features of input pattern changing over time (e.g., Berthold, 1994; Benyettou, 1995). In network design, great importance must be attributed to correct choice of the number of hidden neurons, which helps avoiding problems of overfitting and contributes to reduce the time required for the training without significantly affecting the network performances (e.g., Colla & Reyneri & Sgarbi, 1999), but never looking to architecture adapting effect according to input.

The goal to combine the approach of the RBF with the shift invariance features of the TDNN, can be get a new robust model, this is named temporal radial basis function "TRBF" (e.g., Mesbahi & Benyettou, 2003), but to be more efficient, we have adapt these networks so that they come more dynamic according to their behaviour and features of the object has study. It can be goes more clearly in continuous speech.

Therefore in object to obtain an Adaptive TRBF, we must adapt the TRBF networks, consequently it was necessary to develop an algorithm that permits to solve this type of problem, this algorithm is called "DOLS" which means Dynamic Orthogonal Least Square, that will be presented in this paper.

Dynamic OLS Algorithm

The OLS with its classic version can not adapt our ATRBF network, therefore the original idea of the Dynamic OLS method resides at each iteration, the creation of a hidden centre block and not only one centre, as the size of every block is expressed like suit :

If we consider that the input vector is composed of n characteristic on a temporal input window of Nfe length and if we take the value of input time delay equal to Nde, such as : Nfe ≥Nde. Then the number of neurons composing every block is equal to : Nfe-Nde+1.

The size of every hidden centre is equal to : n x Nde, (Figure 1).

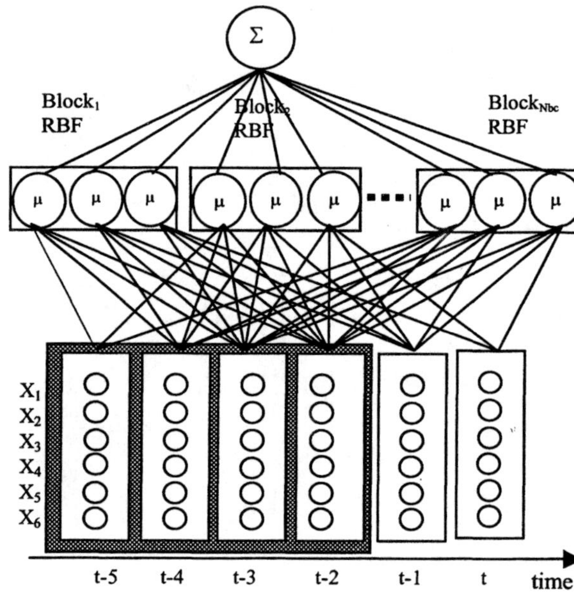

Figure 1: Parameters of this ATRBF network are : n : number of features equal to dim(X)=6, Nde : Time delay equal to 4, Nfe : Size of the input window equal to 6, Nnc : number of hidden neurons by block equal to 3, Nbc: represents the number of centres in hidden bloc, μ : represents a centre of size equal to Nde x n.

With this representation, we can follow the following steps:

$$d = (P\ \theta + E) \qquad (1)$$

The Orthogonalisation of the columns Pi can be gotten by the decomposition of the P matrix in two matrices W and A as:

$$P = W A \qquad (2)$$

Where W : of size Nex M, is the orthogonal image of the P matrix. A : of size M x M, is a superior triangular matrix containing orthogonal coefficients. The A matrix is defined:

Table 1: Representation of A Matrix containing the orthogonal coefficients.

1	$\alpha_{1,2}$				α_{1M}
0	1				$\alpha_{2,M}$
0	0	1			
				$\alpha_{M-2,M-1}$	$\alpha_{M-2,M}$
0			0	1	$\alpha_{M-1,M}$
0			0	0	1

The space begotten by the vectors P_i is the same space begotten by the vector W, then:

$$d = W*G + E \quad (3)$$

Where G=A. θ is the search solution.

$$[err]_i = G_i^2 \, . \, w_i^T \, .w_i \, /(d^T. \, d) \; ; 1 \leq i \leq M \quad (4)$$

At each iteration we calculate the elements of A and W :

$$\alpha_{jk}^{i} = W^i_j * p_i / W^i_j * W_j \quad (5)$$

$$W^i_k = p_i - \Sigma^{k-1} \alpha_{jk}^{i} * W_j \quad (6)$$

The criterion of iteration stop here is not based merely on the Akaike criterion:

$$1 - \sum_{i=1}^{i=M} err_i \leq \varepsilon \quad (7)$$

Modulo M on Nnc=0, where M≠0 (8)

In end of iterations, we calculate the synaptic weights according to the system:

$$G = A * \theta \quad (9)$$

Application

The developed approach has been achieved on a subset of the TIMIT corpus (e.g., Benyettou, 1995; Yak et al, 1995), which is organized of 6 vowels, 6 fricatives and 6 plosives. For our application we reduced the space of study for the case of plosives. Signals have been sampled to 16 KHZS with an analysis Cepstral under the Mel ladder, takes all 20ms in Hamming windows of 25ms giving each 12 MFCCS coefficients and the corresponding residual energy.

This work was achieved on a PIV microcomputer 1.7 GHz with 256 Mo of RAM, developed by the C++ builder and Matlab 6.5 languages. Concerning the training data basis, it has been quantified with a SOM-Kohonen of size 15x15 , generating a training basis of 225 phoneme examples, therefore the size of training basis is about 675 examples, every example is normalized on an input length window equal to 4. The size of the test basis is 450 example at average 150 examples by phoneme. The obtained results can be viewed on the Figure 2

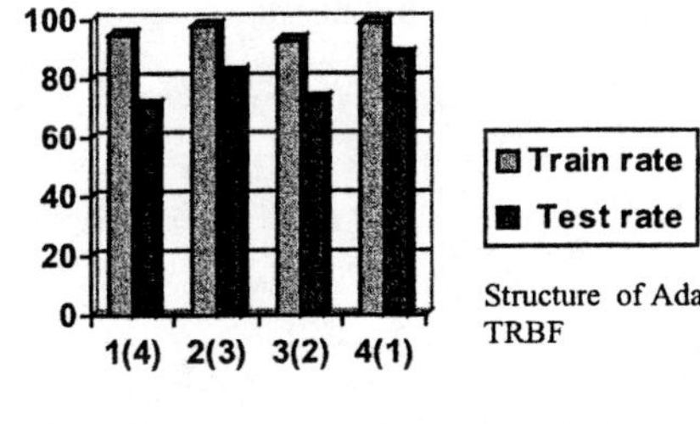

Structure of Adaptive TRBF

Figure 2: Performance according to size of windows x(y), x :Nde, y :Nnc.

The global rate accuracy is about 98% in learning and 83% in test, the Figure 2 shows us the influence of the time delay change on the performance of the network. If the number of the time delay is narrow the performance degrades seen that it has less space time to browse all characteristic.

On the other hand if the time delay is large, there is a risk that the system becomes not shift invariant over time. The best case is choosing the time delay in the median

Conclusion

In this paper we have presented a new approach based on dynamic orthogonal least square developed for adaptive temporal radial basis function , applied to speech recognition. The main advantage with regard to other neural architectures, is the time won in the training, in addition we have few parameters to adjust. The Adaptive TRBF combines advantages of the TDNN that are shift invariant in the time and their capacity of temporal feature recognition and advantages of the RBF in their speed.

We have obtained a 98% in learning rate and 83% in test rate. This shown that our algorithm has good rate of training and test. We hope that this work can be generalized and tested in different domains like pattern recognition and more especially in applications covering the Spatio-Temporal approaches, like mobile robotics, detection of targets, economic fluctuations etc.

Acknowledgments

All my thanks to my supervisor Professor A. Benyettou without missing my Professor N. E. Berrached in master studies director of LARESI Laboratory, Electronic Department, USTO, Oran, Algeria.

References

Berthold, M. R. (1994). A Time Delay Radial Basis Function for Phoneme Recognition. Proceeding of the International Conference on Neural Network, Orlando.

Benyettou, A. (1995). Acoustic Phonetic Recognition in the Arabex System. Int. Work Shop on Robot and Human Communication, ATIP95.44, Japan.

Colla, V., & Reyneri, L.M., & Sgarbi, M. (1999). Orthogonal Least Square Algorithm Applied to the Initialisation of Multi-Layer Perceptron. ESSAN'1999, (pp. 363-369).

Mesbahi, L., & Benyettou, A. (2003). A New Contribution Towards a Temporal Radial Basis Function Applied To Mackey-Glass Time Series. 3 rd ICNNAI Nov 2003, (pp. 72-78). Minsk, Belarus.

Yak, D., & Lin, Q., & che, C., & Jin, L., & Flanagan, J. (1995). Environment-Independent Continuous Speech Recognition. IEEE Automatic Speech Recognition Workshops, Dec 10-13, 1995, (pp.151-152). Utah.

Bilingual Language Learning in Connectionist Networks

Jing Liang (jliang@andrew.cmu.edu)
Department of Modern Languages, Carnegie Mellon University
5000 Forbes Ave., Pittsburgh, PA 15213 USA

Introduction

Connectionist modeling opens up a venue to research on language learning from the statistical point of view. Network input construction offers a way to examine what kind of knowledge could be statistically learned and internal representations from network hidden units help to clarify how the knowledge thereafter could be represented in human minds (Elman, 1991; St. John & McClelland, 1990). Results, thus far, demonstrate the ability of general cognitive learning in various aspects of human language, morphology, syntax, etc., without recourse to innate linguistic rules. This study aims to extend the work of modeling monolingual language learning into the area of bilingual language learning in hopes of drawing further evidence to support the argument of general cognitive capacity in language learning by requiring the network to learn different syntactic structures of two languages using just one learning algorithm and with no innate linguistic structures built in.

Grammar and Corpora

Grammar in the current simulation is a stochastic context-free grammar (SCFG hereafter), including simple transitive and intransitive sentences, sentences with subject- and object-extracted relative clauses, and passive sentences. 10 nouns and 14 verbs in English used in Rhode & Plaut (1999) were borrowed because of good matching in light of meanings between Chinese and English. Actual verbs, however, drop to 7 when translated into Chinese since there is no morphosyntactic change on verbs in terms of tense and plurality. A program was written to generate 11,600 sentences in total with 5,788 sentences in English and 5,812 in Chinese. Additionally, some semantic and pragmatic constraints were incorporated into sentences to make this pseudo-language grammar less artificial. For example, only human beings can walk dogs; only animals can bite human beings, not vice versa; the subject in transitive sentences cannot be the same as the object.

Network Architecture

The network used in the simulation is a simple recurrent network (SRN hereafter), adapted from the one in Elman (1991). The SRN is characterized of the ability of human memory by copying internal representations in the hidden layer at one time step to the context layer and feeding back the input at the next time step together with the memory in the context to the hidden layer for more effective weight updating.

The input layer employs a localist form with each unit representing one word either in English or Chinese. In English, the auxiliary verb is bound with the main verb as one unit. There are no explicit grammatical rules encoded in the network input. Two groups of output units perform the tasks of word prediction and sentence comprehension in terms of agent identification. The SRN is trained to maintain information only from the network's current activation in order to produce appropriate output for the next step. Therefore, the word prediction task forces the network to better retain previous word information such that the network could extract agent information more accurately, particularly in the case of Chinese because it is more difficult to identify the agent in Chinese due to the fact that pre-posed Chinese relative clauses can cause the position of the agent keep changing as the sentence length varies. One small hidden layer is added between the input layer and the main hidden layer as transducers (Elman, 1991) in order to cope with the dramatic changes of sentence structures in bilingual language learning.

Network Training and Testing

The network was trained 400 epochs with the mixed corpus of Chinese and English to simulate the native bilingual learning situation. Different combinations of learning parameters were tried to find the optimal one. A learning rate 0.01, a batch size of 10 words per weight update, initial random weights between $\pm$ 1.0, and a momentum of 0 were used for non-incremental training with back-propagation. The overall performance was measured using mean squared errors.

100 testing sentences, 50 in Chinese and 50 in English, were excluded from the training corpus and later used to examine the generalization of the network performance. For the word prediction task, the test corpus was constructed such that it tested not on the actual next word, but the probabilistic distributions of the next word. Rather than activate the exact next word, the network should activate a likelihood of many words at the same time. Since the grammar is SCFG, theoretically the correct distributions for the next word can be computed based on the corpus. Optimally, the theoretically generated distributions could be matched with the actual output distributions.

Results and Discussion

The overall performance of the network was examined by computing mean squared errors weighted by the percentage of different sentence types in the training corpus. The error rate 0.021 for word prediction and 0.048 for agent identification show that the network successfully learned Chinese and English simultaneously. The network activated the correct output units for agent identification more than what it should not have 99% of the time for English and 97% of the time for Chinese. The relatively low accuracy for Chinese is attributed to the fact that agent identification in Chinese bears the characteristic of unpredictability to some extent because potentially the agent could occur at any position in contrast with agent identification in English where the agent occurs only at the first word position unless in passive sentences with regard to this grammar.

The better evaluation of the network performance entails the scrutiny on the activations of individual sentences. The activation of the word prediction task for one Chinese sentence is shown in Figure 1, in which the units listed for every word indicate the probable units that could be activated after this word. It demonstrates that the network learned co-occurrence of words, especially the collocation between nouns and verbs. In this grammar, all nouns can be "bitten" by at least one other noun; therefore after the verb *yao (bite)* and the constructive particle *de*, all nouns were activated almost evenly. However, after the agent *goumen (dogs)* occurred, the network activated several frequently co-occurrent verbs, e.g. *fei (bark)*, *yao (bite)*, *zhui (chase)*, much more than others. The activation level for the agent identification task also reached 0.954 with the target at 1.

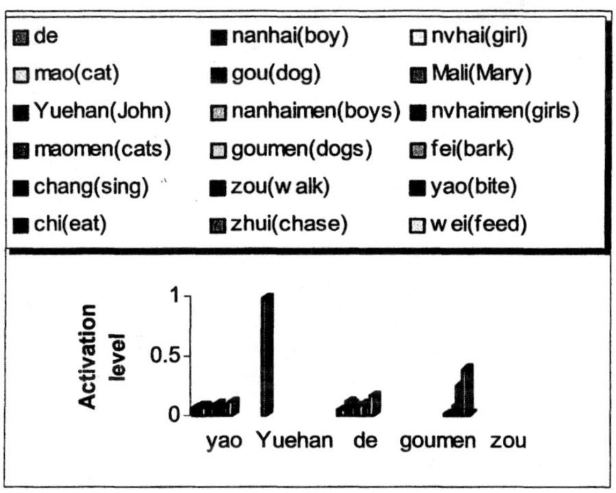

Figure 1: Activation levels for the word prediction task.

As the network successfully learned probabilistic distributions of language corpora, it also constructed internal representations of language structures in the hidden units. After the network was fully trained, weight matrices were extracted and entered into principal component analysis (PCA thereafter). PCA is useful for decreasing the dimensionalities in a multivariate space and characterizing the variation of each dimension in the hidden units. Sentence (1) and (2) were analyzed along the dimension of component 1 and 2 as indicated in Figure 2. Except for two nouns *gou (dog)* vs. *goumen (dogs)*, the principal component loading values for all other nouns are the same for two sentences. This indicates that the network learned to differentiate the structure of singularity vs. plurality in the state space.

(1) *Yao Yuehan de gou (dog) zou. (Singular)*
(2) *Yao Yuehan de goumen (dogs) zou. (Plural)*

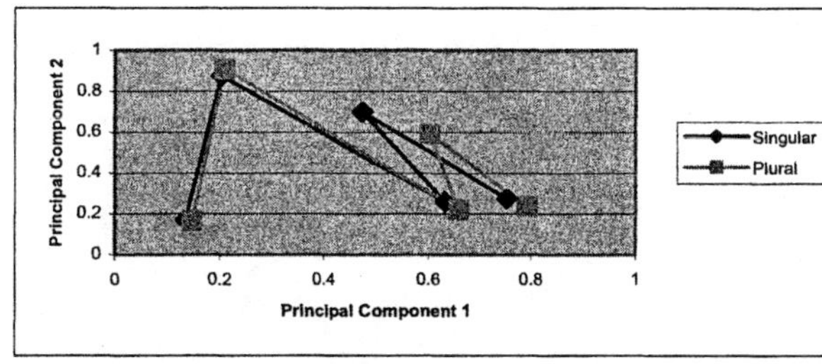

Figure 2: PCA for sentence (1) and (2).

Hierarchical clustering analysis was also used to evaluate whether the network constructed systematic structures for nouns and verbs. Verb similarity analysis was drawn from Chinese as an illustration, where similar verbs were connected earlier along the axis. As shown in Figure 3, the network captured the differences between transitive and intransitive verbs, even including slight ones. For example, zou/liu (*walk*) can be used as a transitive verb occasionally and chi (*eat*) in the corpus is mainly used as an intransitive verb. In conclusion, this simulation successfully testifies that the connectionist network has the ability to learn two different language structures implicitly as human beings do in natural language acquisition using statistical information available in language corpora and forming systematic structures of language knowledge in the end.

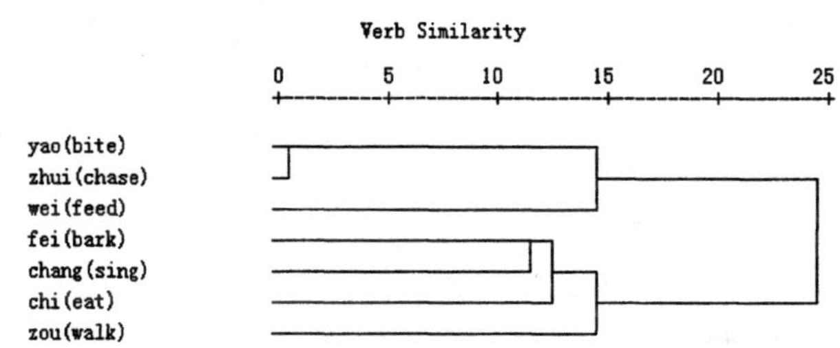

Figure 3: Hierarchical clustering analysis for Chinese verbs.

Acknowledgments

I want to thank Dr. David Plaut and Dr. Brian MacWhinney for some suggestions on network configuration and implementation. Thanks also go to Dr. Douglas Rohde for some tips related to LENS usage.

References

Elman, J. L. (1991). Distributed representations, simple recurrent networks, and grammatical structure. *Machine Learning, 7,* 195-225.

Rhode, D. L. T. & Plaut, D. C. (1999). Language acquisition in the absence of explicit negative evidence: How important is starting small? *Cognition, 72,* 67-109.

St. John, M., & McClelland, J. L. (1990). Learning and applying contextual constraints in sentence comprehension. *Artificial Intelligence, 46,* 217-257.

Emulating a Visuospatial Memory Field Using ACT-R

Don R. Lyon (don.lyon@mesa.afmc.af.mil)
L3 Communications at Air Force Research Laboratory
6030 S. Kent St., Mesa, AZ 85212-6061 USA

Glenn Gunzelmann (glenn.gunzelmann@mesa.afmc.af.mil)
Kevin A. Gluck (kevin.gluck@mesa.afmc.af.mil)
Air Force Research Laboratory, Mesa Research Site
6030 S. Kent St., Mesa, AZ 85212-6061 USA

Introduction

We define visuospatial working memory (VSWM) as the set of cognitive processes used to visualize the locations of things. VSWM is not permanent visual memory; it is a temporary visual workspace used to solve spatial problems. VSWM is ubiquitous in everyday reasoning (for example, visualizing different ways of arranging furniture in a room), and is thought to be particularly important for certain occupations (for example, engineers, architects, and pilots).

Both subjective experience and empirical studies suggest that the capacity of VSWM is limited (c.f. Logie, 1994). Our goal is to model the underlying sources of this capacity limitation. To measure VSWM capacity, we use a new method called *Path Visualization* (PV, Lyon, 2004). PV is similar to some existing methods (e.g. Brooks, 1968; Attneave & Curlee, 1983; Kerr, 1987, 1993; Diwadkar, Carpenter & Just, 2000; Barshi & Healy, 2002) in which participants must visualize relative positions or movements (up, right, down...) in a two-dimensional or three-dimensional array of locations. However PV also requires participants to make continuous timed spatial judgments about intersections in a path defined by such movements. This yields accuracy and response time data that help define the VSWM representation of an imaginary path.

Method

Four paid participants (2 female, 2 male) were each given five 30-trial sessions doing the PV task. During each trial, the participant saw a sequence of 15 text phrases presented on a CRT. Each phrase described the direction and distance (e.g. 'Left 1') of a segment of a path. There were six possible directions (Right, Left, Forward, Back, Up, Down); all distances were one. Each phrase was presented for 2000 msec, followed by a blank screen for 133 msec, then the presentation of the next phrase. As each new phrase was presented, the participant decided whether or not the endpoint of the new path segment intersected with any previously presented part of the path, and pressed a key with the right index finger to indicate 'yes', or the left index finger to indicate 'no'. In the rare event that no key was pressed during the presentation of a text phrase, the response was scored as incorrect, and the presentation of the next phrase proceeded normally. Participants were instructed to respond as accurately and quickly as possible. Small bonuses were paid for maintaining high overall accuracy and low response time.

All paths started at the center of an imaginary 5 x 5 x 5 cube (Figure 1). Segment directions were relative to a fixed frame of reference, so, for example, 'Left 1' was always toward the left side of the cube as depicted in the figure, regardless of the direction of the previous segment. No two successive segments could be on the same axis, so the path always turned to a new axis with each new text phrase. Paths were randomly generated with the restriction that a near-balance of intersection and no-intersection segments exist in the entire corpus of paths. Although a picture similar to Figure 1 was shown in the instructions, no picture was available during performance of the task. The path had to be visualized from the sequence of text phrases.

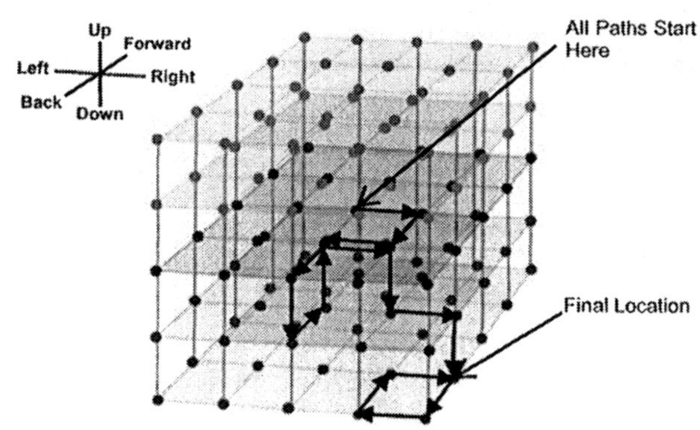

Figure 1: Depiction of a path in an imaginary space.

Model

On the surface, this task may resemble standard verbal memory tasks in which the stimulus is a sequence of words or phrases. However since participants must make continuous spatial intersection judgments, standard models and strategies for verbal memory are insufficient. For example, a model based on serial verbal rehearsal of the text phrases would not necessarily represent the spatial information required for correct responses. Therefore we developed a model that embodies the idea of a *spatial field*. The model assumes that participants attempt to visualize the paths using some isomorphic representation in which spatial relationships between path segments are preserved.

Currently, no spatial-field-like representation exists in ACT-R. Therefore we chose to emulate a spatial field using similarity values in declarative memory. Our model performs the path visualization task as follows: First, it reads a segment descriptor ('Right 1') and generates a declarative representation of the new location. Then, using the new location as a cue, it attempts to retrieve a previous instance in which this location was generated in response to any descriptor. If anything is retrieved, the model responds 'Yes', otherwise it responds 'No'. The model predicts the following pattern of memory errors. (1) Failures to detect an intersection will increase with the number of segments presented between the initial visit to a location and the re-visit. This occurs because activation from the initial visit decays over time. (2) Incorrect 'Yes' responses (false alarms) will increase with the number of visits in the prior path to nearby locations. This is because the similarity of two locations in the model decreases as the Euclidean distance between them increases. Thus, prior visits to locations very close to the current one have a chance to be erroneously retrieved through partial matching. The more such 'near visits' there are, the greater the chance of this kind of error.

Results and Conclusion

The two predictions of the model were tested by doing two separate partitions of the data. First we examined the segments that resulted in an intersection. Participants' responses to intersection segments were partitioned into bins based on the number of segments that intervened between visits to the location. As predicted, accuracy declined with number of intervening segments for all participants (mean χ^2 = 45.8, p's < 0.05). The model captures this effect well (RMSD = 0.045).

The second partition used the segments that did not result in an intersection. These were sorted by number of near visits, that is, the number of times that prior segments visited adjacent locations, defined as one step away from the current decision location on any axis or diagonal. For example, the final location in the path depicted in Figure 1 has five near visits. Accuracy declined sharply with number of near visits for all participants (mean χ^2 = 77.8, p's < 0.001). Further analysis showed that this pattern is not due to speed-accuracy tradeoff, criterion shift or aggregation artifacts. Rather, it appears to reflect interference based on proximity in imaginary space. The spatial field emulation model exhibits this interference (Figure 2) with default values for most ACT-R parameters (RMSD=0.064). We varied retrieval threshold and a scaling factor relating distance to similarity.

These results suggest that the capacity of VSWM is limited by both decay of activation and a location-based interference process that can be emulated using a spatial field. It appears that imaginary proximity has real consequences for visuospatial memory. In this respect, visualization space acts like real space.

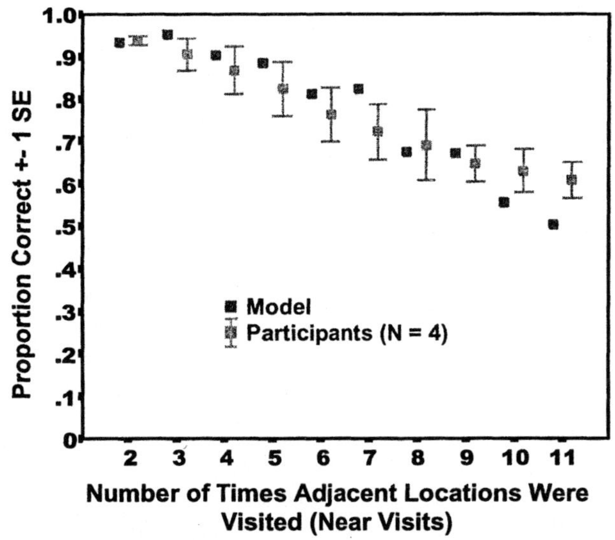

Figure 2. Effect of near visits on accuracy – Human data and model predictions.

Acknowledgments

We thank Ben Sperry and Jerry Ball for software development, Lisa Park for research assistance, and the U. S. Air Force Office of Scientific Research for support (Grant # 02HE01COR). The second author participated in this research as a National Research Council Research Associate at the Air Force Research Laboratory, Mesa, AZ.

References

Attneave, F., & Curlee, T. E. (1983). Locational representation in imagery: A moving spot task. *Journal of Experimental Psychology: Human Perception and Performance, 9*, 20-30.

Barshi, I & Healy, A. F. (2002). The effects of mental representation on performance in a navigation task. *Memory and Cognition, 30*, 1189-1203.

Brooks, L. R. (1968). Spatial and verbal components in the act of recall. *Canadian Journal of Psychology, 22*, 349-368.

Diwadkar, V. A., Carpenter, P. A., & Just, M. A. (2000). Collaborative activity between parietal and dorso-lateral prefrontal cortex in dynamic spatial working memory revealed by fMRI. *NeuroImage, 12*, 85-99.

Kerr, N. H. (1987). Locational representation in imagery: The third dimension. *Memory and Cognition, 15*, 521-530.

Kerr, N. H. (1993). Rate of imagery processing in two versus three dimensions. *Memory and Cognition, 21*, 467-476.

Lyon, D. R. (2004). *Measuring visuospatial working memory using path visualization.* Technical Memorandum 23-1, Air Force Research Laboratory, Mesa, Arizona.

Logie, R. H. (1994). *Visuospatial Working Memory.* Hove: Lawrence Erlbaum Associates.

Irresistibly Attractive Fruitless Feature Dimensions

Toshihiko Matsuka (matsuka@psychology.rutgers.edu)
Psychology Department, Rutgers University – Newark

James E. Corter (corter@tc.exchange.columbia.edu)
Department of Human Development, Teachers College, Columbia University

Stephen Jose Hanson (jose@psychology.rutgers.edu)
Psychology Department, Rutgers University - Newark

Although selective attention allocation has been suggested to be one of the most important processes implemented in the recent computational models of category learning (e.g., Kruschke, 1992), the models' predictions on attention allocation have been virtually ignored by cognitive modeling researchers. Rather almost all modeling studies had focused solely on the models' capabilities in reproducing observed learning curves or classification response patterns, and thus the models' descriptive validities on their selective attention allocation processes have remained still mostly untested. Though they did not directly evaluate the descriptive validities of attention processes in computational models of category learning, Matsuka, Corter, and Markman (2004) conducted a set of simulation studies indicating that information on predicted patterns of attention allocation can be informative in differentiating models. In that study, three recent models of category learning, including ALCOVE (Kruschke, 1992) and SUSATIN (Love & Medin, 1998, Love, Medin, Gureckis, 2004), were compared in multiple aspects, including predicted attention allocation patterns. Matsuka et al. (2004) found that the models perform comparably in reproducing the observed classification response profiles but gave markedly different predictions on selective attention allocation patterns. The results of the simulation studies also indicated the models have different tendencies in their attention allocation patterns. ALCOVE has a tendency to be attracted to a feature dimension that consists of unique elements which help differentiate each exemplar. SUSTAIN on the other hand, has a tendency to be attracted to a feature dimension whose elements are less diverse or more homogeneous.

If these tendencies are generalizable, then their descriptive validities may become questionable in some cases. For example, in order to categorize humans into either male or female, ALCOVE would pay good amount of attention to each person's finger print (i.e., feature dimension consists of unique elements), and SUSTAIN would pay good amount of attention to the fact the they are humans (i.e., feature dimension consist of a constant element). These tendencies, however, in less extreme cases, might be realistic tendencies in humans' "irrational" cognitive processes. In any case, we first should examine the models' mechanisms that are associated with such tendencies in order to evaluate the models' descriptive validities. Specifically, in the present study, we extend Matsuka et al. (2004) to investigate those tendencies in attention allocation associated with two prominent models of category learning, namely ALOCE (Kruschke, 1992) and SUSTAIN (Love et al, 1998, 2004) with a simulation study using an artificial stimulus set.

Table 1: Artificial stimulus set used in the present study.

	Category A				Category B			
Dim 1	1	1	1	1	0	0	0	0
Dim 2	1	1	1	1	1	1	1	1
Dim 3	1	2	3	4	5	6	7	8

Simulations

We investigated aforementioned two computational models of category learning in their predicted attention allocations in a simulated classification task. In particular, we created an artificial stimulus set consisting of two different types of ineffective feature dimensions that are suggested to attract the models' attention in different ways. Table 1 shows the stimulus set used in present study. Note that Dimension 1 is the most informative dimension, perfectly correlated with the category membership. Dimensions 2 is constant and thus non-diagnostic at all, but SUSTAIN is expected to pay a good amount of attention to this dimension. Dimension 3 on the other hand can be informative, but does not allow people to have abstract concepts about the categories (because all elements are unique). Thus, in real word categorization task, Dim3 is not expected to be attended or at least is expected to be less attended than Dim 1. However, ALCOVE is expected to pay a good amount of attention to Dim3, if the claim by Matsuka et al (2004) is valid. Note that all feature values in Dim3 are treated as nominal value differentiating from each element within the dimension, and thus their numeric differences do not have any meaning.

In order to investigate general tendencies in their attention allocation patterns for two different types of ineffective feature dimension, 10,000 simulated subjects with randomly assigned parameter values were trained to classify the artificial stimulus set. The ranges of parameters were [0.1 10] for c and ϕ, [0.001 1] for the two learning rates for ALCOVE. For SUSTAIN the ranges were set at [0.1 10] for β, d, and r, and [0.001 1] for the learning. Note that we tested the original version of SUSTAIN (Love & Medin, 1998). Readers are advised to refer to Kruschke (1992) and Love et al. (1998 & 2004) for the details of ALCOVE and SUSTAIN.

The two models were run in a simulated training procedure to learn the correct classification responses for the stimulus set. The models were run for 20 blocks of training, where each block consisted of a complete set of the training instances.

Results and Discussion

The predicted *relative* attention allocation by ALOCVE and SUSTAIN are summarized in Table 2. As expected ALCOVE, on average, paid the greatest amount of attention to Dimension 3 (i.e., one with all unique elements). In contrast, SUSTAIN allocated the same amount of attention to the diagnostic dimension (i.e., Dim1) and the constant dimension (Dim2). Surprisingly, although there were 10,000 randomly chosen parameter configurations, there was virtually no variability in SUSTAIN's attention distribution pattern (i.e., the standard deviations were less than 0.001 for all dimensions). In other words, regardless of its parameters, SUSTAIN always allocated its attention in the same manner. In contrast, ALCOVE shows greater variability, and ALCOVE did not always pay more attention to Dim 3 than Dim 1 (see Figure 1). Approximately 60% of simulated ALCOVE subjects paid more attention to Dim3.

But what causes ALCOVE to pay more attention to Dimension 3 than Dimension 1? The bivariate scatter plots (or correlation coefficients) suggest when the learning rate for association weights (LRW) is high, ALCOVE is more likely to pay greater amount attention to Dim 3 (Figure 1, top panels), and when it is low, ALCOVE is more likely to pay lesser amount of attention to the dimension. This is because when the LRW is high, ALCOVE learns to form a strong association between each exemplar to the correct category node very quickly, which in turn, during learning phase, sends a stronger signal to the exemplar node that has identical features to the current input stimulus. Then, the feature dimension that distinguishes the exemplar (with strong feedback signal) from other exemplars receives more amount of attention weight update, resulting in distributing the most attention to the unique dimension (Dim3)

This result shows that ALCOVE can model at least two types of learners; domain experts, who are very familiar with many exemplars in a given domain and learn to form associations between exemplars and categories quickly; and ordinary people, those who have more abstract concept on the domain and do not learn about the associations as quickly as the domain experts. Thus, this ALCOVE's tendency seems to be descriptively valid in some extent and/or cases. It is rather interesting, however, that the learning rate for attention is less sensitive to this ALCOVE's tendency in attention allocation than LRW.

The next issue to be discussed is why SUSTAIN pays equal amount of attention to Dim1 and Dim2. This can be attributed to the way SUSTAIN utilizes its reference points (i.e., prototypes and exceptions) in learning. SUSTAIN utilizes only the single most activated reference point (RP) for learning (and classification). In SUSTAIN, the update in attention strength for each dimension is determined as a function of the distance from the most activated RP and the current input stimulus in that dimension (i.e., smaller the dimensional distance, more attention to be allocated to that dimension). SUSTAIN's winner-take-all type of RPs-utilization (or more precisely, receiving no information on each dimension's category-diagnosticity from other RPs during learning) can be problematic, because as we have seen it does not effectively reduce attention distribution to a non-diagnostic constant dimension. This is because regardless of stimulus and/or category structure, if there is a constant dimension, then the constant dimension always has zero distance for all RPs in the corresponding dimension. Since only one most activated RP is used for learning, SUSTAIN, regardless of particular RPs to be most activated, always finds the smallest distance (i.e., zero) in the constant dimension and thus always misperceives the comparative advantage of the constant dimension (as compared with other dimension) resulting in paying a great amount of attention to the dimension.

The most apparent limitation of the present study is that we do not have empirical evidence about how people would utilize those dimensions. However, we, in some extent, showed the mechanisms of ALCOVE and SUSTAIN's tendencies in attention allocations. Empirical data on attention allocation and simulation studies with several different stimulus structures can be very fruitful for this line of work.

Table 2: Predicted relative attention allocations

	ALCOVE		SUSTAIN	
	Mean	Stddev	Mean	Stddev
D1	0.335	0.199	0.477	0.000
D2	0.190	0.125	0.477	0.000
D3	0.475	0.294	0.046	0.001

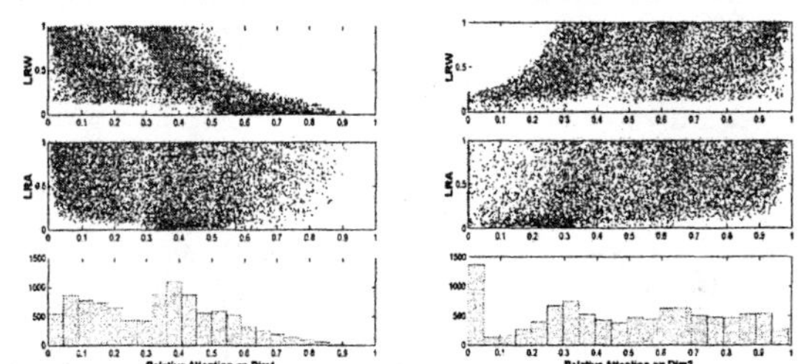

Figure 1: Predicted attention allocation to Dimensions 1 (left bottom panel) & 3 (right bottom panel), and their relationships to the learning rate (LR) for association weights (top panels) and LR for attention (middle panels).

References

Kruschke, J. E. (1992). ALCOVE: An exemplar-based connectionist model of category learning, *Psychological Review, 99.* 22-44.

Love, B.C. & Medin, D.L. (1998). SUSTAIN: A model of human category learning. *Proceeding of the Fifteenth National Conference on AI (AAAI-98),* 671-676.

Love, B.C., Medin, D.L., & Gureckis, T.M. (2004). SUSTAIN: A network model of human category learning. *Psychological Review.*

Matsuka, T., Corter, J. E. & Markman, A. B. (2004). Allocation of attention in neural network models of categorization. Under review..

The simBorg Approach to Modeling a Dynamic Decision-making Task

Christopher W. Myers, Hansjörg Neth, Michael J. Schoelles, & Wayne D. Gray
Cognitive Science Department
Rensselaer Polytechnic Institute
[myersc, nethh, schoem, grayw] @rpi.edu

Motivation

Understanding decision-making processes within dynamic task environments via embodied computational cognitive models proves to be a challenge for the modeling community (see Gonzalez, Lerch, & Lebiere, 2003 for an example). Decisions made by an agent may be the result of explicit, strategic moves, or result from implicit, cost-benefit tradeoffs occurring at the level of 1/3 of a second. Understanding and properly modeling decisions occurring at these different levels is a challenge for the modeler. This paper proposes a novel incremental modeling approach (see Byrne, 2001 and Gonzalez et al., 2003 for examples of other incremental approaches) that promises to inform decision theory as well as limitations within the chosen modeling architecture.

The *simulated cyborg* (or, *simBorg*) approach blends computational embodied-cognitive models of interactive behavior with artificial intelligence based components in a simulated task environment (Gray, Schoelles, & Veksler, 2004). simBorgs combine human and machine components. This combination of high fidelity cognitive modeling (human) and AI (machine) facilitates the development of families of models that allow the modeler to hold components (memory, vision, etc) at different levels of expertise without concern for cognitive plausibility. For example, rather than modeling human problem solving, the modeler can rely on various *black-box* techniques (i.e., cognitively implausible AI), thereby focusing on predicting how subtle differences in costs and benefits in interactive methods affect performance and errors.

The current modeling endeavor adopts the simBorg approach in order to build a family of interactive decision-making agents. The following section will present the task of interest followed by a fraction of the family of simBorgs under development. Finally, we conclude with future directions we plan to lead the family of simBorgs.

Task Description

The dynamic decision-making task is Decision-Making Argus Prime (D-MAP). D-MAP is a scenario–driven simulated radar environment. The display contains a radar display on the left half and a decision-making task (DMT) on the right half. The users' goal was to choose the alternative with the highest threat value (TV) from the DMT. See Figure 1 for an example of D-MAP.

Each DMT contained a minimum of 2 and a maximum of 6 alternatives. The alternatives were presented in a tabular format where each table entry corresponded to a radar target. There was only one correct answer for each DMT. Each DMT was limited in time (60 s), and had a 1 s interval between the offset of one DMT and the onset of another.

A correct decision resulted in the elimination of the chosen target from the radar. Threat values were accessed by moving the cursor to a radar target and clicking it. It is important to note it was not necessary, nor were participants instructed, to access the TVs for all targets in a given DMT. A scenario continued for 720 s or until all remaining targets had a threat value of 1.

The amount of time for the TV to appear after clicking on a radar target varied as a between-subjects independent variable. One group (n = 12) had no time between clicking the target and the appearance of the TV (0-Lock). The second group (n = 12) had a 2 second delay between clicking a target and TV appearance (2-Lock).

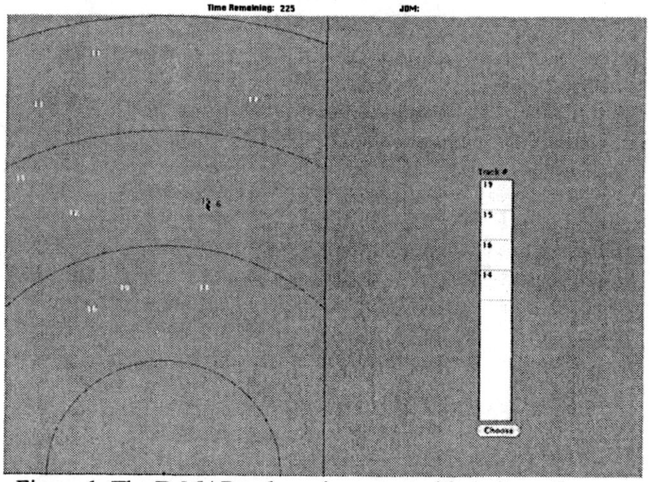

Figure 1. The D-MAP task environment with 4 alternatives occurring in the DMT.

The *Borgs: A Family of simBorgs

This modeling endeavor is based on the simBorg approach of holding particular cognitive processes of decision-making at an expert level and varying others. The models will range from a cognitively *implausible* optimal agent (allBorg) to a cognitively more plausible simulated human user. We title this process the **Borg family of simBorgs*, where the * represents a wildcard for the type of simBorg under development. For example, if visual search is of interest and memory is held optimal, the *seeBorg* is born. Such models allow optimization of one aspect of human performance while allowing other aspects to be realistically simulated (Gray, Schoelles, & Myers, 2002) informing both theory and application. An early goal of the *Borg family of models is to predict when and explain why participants *satisfice* (Simon, 1956) in the D-MAP task environment. It is extremely likely that participants satisfice as a result of the costs embedded in the task environment, therefore it is important to pinpoint which costs (visual search, memory, TV delay, etc) are the greatest contributor.

It is likely that the simBorg approach will uncover adaptively optimal behavior that might be construed as satisficing. For example, suppose *all* threat values are *correctly* recalled for each alternative in a DMT and a decision is made without checking one radar target. This is actually optimal behavior, however it could easily be misinterpreted as satisficing.

All simBorgs are modeled using the ACT-R architecture, however it is important to point out that this method may be employed using any cognitive architecture. Only two of the *Borg models will be discussed at a general level in this abstract due to space limitation.

Normative (allBorg)

The allBorg will be cognitively implausible in nearly all aspects of the decision-making process (memory, visual search, etc) to provide the highest level of performance possible in D-MAP. Only movement times for eyes and hands will be the agent's constraints. Figure 2 provides a process flow of the allBorg agent. The allBorg agent will provide a ceiling of performance that we propose all human participants strive to achieve given the environmental constraints.

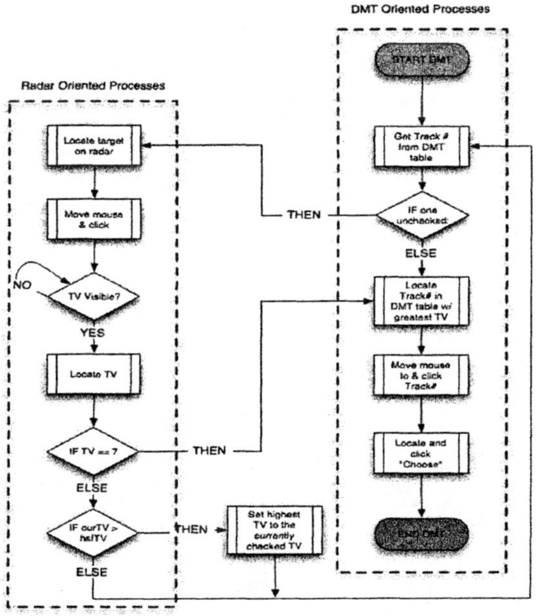

Figure 2. Process flow of allBorg. curTV stands for Current TV being checked and hsfTV stands for Highest So Far TV.

Refining Visual Search (seeBorg)

The seeBorg agent's memory will use ACT-R's declarative memory repository without decay. Thus, seeBorg's memory will always be precise, holding the cost associated with memory to a minimum. In contrast, seeBorg's visual search processes and strategies will be the modeling focus and will provide insight to the benefits of different search strategies and the effect of costs involved.

The primary goal of the seeBorg will be to explore the effectiveness of different search processes. For example, storing radar target positions in memory to be used in later DMTs (Byrne, 2001; Ehret, 2002). However, because the seeBorg's memory is absolute, we will gradually impose the same limitations on seeBorg's memory which humans face, namely decay which leads to errors of omission and commission.

Future Directions & Conclusions

Although all of the models have not been completely constructed (nor discussed in this abstract), we believe that the *Borg family of simBorgs will inform both decision-making theory as well as model development. We believe that what appears to be satisficing on the surface and that may actually be adaptively optimal behavior will naturally fall out of the incremental simBorg approach to modeling. Finally, after fitting the cognitively most plausible agent to the human data, we plan on porting the agent to a version of D-MAP that includes Argus Prime's target classification task (Schoelles & Gray, 2001), allowing us to determine if the same visual search and memory strategies transfer to the new task environment.

Acknowledgments

The work reported here was supported by a grant from the Air Force Office of Scientific Research AFOSR #F49620-03-1-0143, to Wayne D. Gray.

References

Byrne, M. D. (2001). ACT-R/PM and menu selection: applying a cognitive architecture to HCI. *Int. J. Human-Computer Studies* 55, 41-84.

Ehret, B. D. (2002). Learning where to look: Location learning in graphical user interfaces. *CHI Letters, 4*(1), 211-218.

Gonzalez, C., Lerch, J. F., Lebiere, C. (2003). Instance-based learning in dynamic decision making. *Cognitive Science 27,* 591-635).

Gray, W. D., Schoelles, M. J., & Myers, C. W. (2002). *Computational cognitive models ISO ecologically optimal strategies.* In the 46th Annual Conference of the Human Factors & Ergonomics Society. Santa Monica, CA: Human Factors & Ergonomics Society.

Gray, W. D., Schoelles, M. J., & Veksler, V. D. (2004). *Simborgs: Towards the building of simulated human users for interactive systems design.* 48th Annual Conference of the Human Factors and Ergonomics Society, Santa Monica, CA.

Schoelles, M. J., & Gray, W. D. (2001). Argus: A suite of tools for research in complex cognition. *Behavior Research Methods, Instruments, & Computers, 33*(2), 130-140.

Simon, H. A. (1956). Rational choice and the structure of the environment. *Psychological Review 63*(2), 129-138.

'Artificial brain methodology' and an application of StarLogo

Kohei Noda (knoda@valdes.titech.ac.jp)
Department of Value and Decision Science, Tokyo Institute of Technology
2-12-1 Ookayama, Meguro-ku, Tokyo 152-8552 JAPAN

Akifumi Tokosumi (akt@valdes.titech.ac.jp)
Department of Value and Decision Science, Tokyo Institute of Technology
2-12-1 Ookayama, Meguro-ku, Tokyo 152-8552 JAPAN

Introduction

This study aims to construct a comprehensive computational model of depression. In order to achieve an accurate model of depression, it is necessary not only to include subjective emotion experience, but to also include changes in physiological and behavioral states, because all three of these aspects are integral to definitions of major depressive disorders.

If one were only seeking to model subjective emotion experience, it would be possible to develop a symbolic computational model of depressive feeling and thinking. However, as depression is a more comprehensive phenomenon, traditional symbolic computational model are inappropriate, so it is important to construct a model that represents all aspects of depression.

Artificial Brain Methodology

We have proposed a construction methodology in order to develop a complex cognitive model of depression that encompasses all aspects of depression. Stressing the important role that the model of brain has in connecting subjective mental process with behavioral and physiological processes, we refer to this methodology as 'artificial brain methodology'. This methodology consists of three approaches; a 'bottom-up approach', an 'embodied approach', and a 'brain-mind approach.'

The bottom-up approach

If one tries to develop a computational model of human emotions, such as depression, by adopting a top-down approach to the processes underlying human depressive thinking, then it is extremely difficult to attain a comprehensive understanding of depression, which also includes lower-level functioning (i.e., neural mechanisms, behavior, and more primitive processes). This kind of dilemma is also encountered with other types of emotions. To avoid this dilemma, it is important to build a model from a basic level of internal mechanisms, and to link behavior to more complex levels, that represent human-like thinking processes and neural mechanisms. With this approach to modeling it is easier to incorporate important evolutionary aspects of the brain.

The embodied approach

Definitions of emotions include not only mental notions, such as happy and sad, but also physiological changes and behavioral expressions. In seeking to develop a comprehensive model of emotion, it is important to also incorporate these aspects. Accordingly, it is necessary to develop an embodied model, which would encompass both functional aspects of mind and internal agent mechanisms for the subjective body and the environment.

The mind-brain approach

In addition to the bottom-up and embodied approaches, we introduce a new strategy; a brain-inspired modeling methodology which comes from an evolutionary bottom-up approach. In order to take into account emotion-physiological interactions, the brain must also be modeled. In line with the bottom-up approach's notion of modeling from simple structures to more complex brain, we introduce Mclean's three-layered evolutional model of the brain (MacLean, 1982). This model has already been adopted within Damasio's concept of emotions and Sloman's evolvable human-like systems (Damasio, 1994; Sloman & Logan, 2000). In this three-layered brain model, we implement network mechanisms, representations and a goal-based architecture as the principles of the human mind system. This combination of a goal-based architecture and a brain-inspired neural network system unites the bottom-up embodied approaches within a computational symbolic model of the human mind.

A model using the life teaching toolkit 'Starlogo'

In this study, we have sought to simulate depression by utilizing the artificial brain methodology. Although there are various ways of defining depression (i.e., behavioral, subjective-feeling, and neural functional definitions), we have adopted definitions from learned helplessness (a behavioral definition) and Drevet's brain functional definition following the bottom-up approach (Seligman, 1975; APA, 1994; Drevets, 1992).

Based on these definitions, a prototype model of depression was developed. Firstly, to realize the embodied approach, we employed an artificial life simulation tool, 'StarLogo'. 'StarLogo' was developed by the MIT media lab, originally to model decentralized systems, and has been used to model artificial life simulations. As this toolkit has been used to teach artificial life programming at the Department of Computer Science at the University of Zurich, the author realized that it could also be employed to develop a cognitive model according to the artificial brain methodology, because the model requires both a body and an environment, which interact in an artificial life simulation.

With this simulation tool, it is possible to create a simple agent and its environment, including energy sources (fig. 1.). The goal of the agent is to access an energy source before its energy levels fall to zero. The environment and the agent are described in logo with StarLogo. In line with the embodied approach, the agent has been designed to have components for a brain, a mouth, direction-sensors, a stomach, a body, and feet.

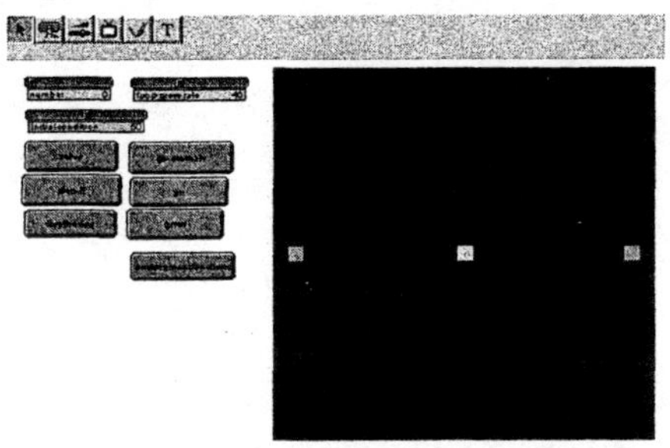

Figure 1: The environment of the model.

Conforming to the mind-brain approach, the brain component of the agent has been developed with a type of neural-network circuit to represent an evolutional three-layered brain. Consistent with the bottom-up approach, the agent brain was constructed from the first layer to the third layer. The first layer consists of modules that respond to hunger signals from the stomach. The second layer consists of a module that memorizes the probabilities of successfully obtaining food in a given direction and reacts according to these probabilities. The third layer consists of modules that represent memories and determine actions based on the contents of these memories. From simulation data for the model, we suggest that depression occurs when third-level decision modules, acting like bias-neurons, restrain the action of the agent (fig. 2).

Conclusion and further research

Based on modeling considerations, we have succeeded in defining depression according to four criteria; (a) negative changes in the environment which exceed agent limitations, (b) repeated failure to attain a goal, (c) inhibition of basic instincts and actions, and (d) devaluation of goals. We have also identified functional aspects of depression when inappropriate values of the system are reset to zero due to both the inhibition of actions and the natural decay of old memories. This mechanism has the function of increasing the survival probability of the agent, because following this kind of inhibition old maladaptive values in the system are lost. In our model, we have partially simulated the 'depressive mood' and 'loss of interest' included in the list of DSM-IV criteria of depressions (APA, 1994), a 'cortex-limbic system-thalamus circuit' (Drevets et. al., 1992) and a 'neuro-modulator model' (Nomura, 1996) from brain science models, as well as 'constraints on action' and 'decreases in appetite' as part of passive stress reactions (Hori, 1991). However, the representations of subjective feeling and thinking, brain mechanisms, and agent structures are still rather simplistic. In future stages of modeling, we shall implement a greater variety of goals, agent instincts and emotions, interactions between multiple agents, and more complex brain and physical structures, as well as complex mental representations. A computer language which can represent human mental process easily more than logo, such as lisp, prolog, or other common language (C++, VB, Java, etc.) will be used. Then, a complete computational model to simulate depression will be developed.

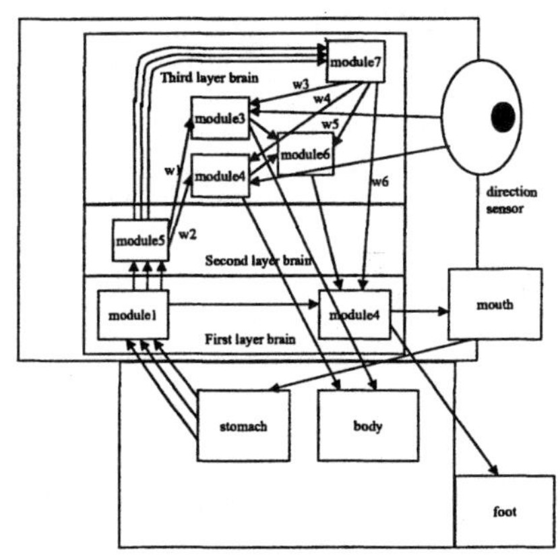

Figure2: The model of an agent.

References

MacLean, P. D. (1982) *Primate Brain Evolution, Method and Concepts.* Plenum Press.

Damasio, A. R. (1994) *Descartes' Error: Emotion, Reason, and the Human Brain* Grosset/Putnum.

Sloman, A. and Logan, B. (2000). Evolvable architectures for human-like minds. In G. Hatano, N. Okada & H. Tanabe (Eds.), *Affective minds.* Amsterdam: Elsevier.

Seligman, M. (1975) *Helplessness.* Freeman & Co.

American Psychiatric Association. (1994) *Diagnostic and Statistical Manuals of Mental Disorders: DSM-IV, fourth edition.* American Psychiatric Association.

Drevets, W. C., Videen, T. O., Price, J. L., Preskon, S. T. & Raichle, M. E. (1992) A functional anatomy of unipolar depression. *Journal of Neuroscience*, 12, 3628-3642,

Nomura, S. (1996) Biological depression research and cognitive therapy. In Y. Ohno & T. Oyaizu (Eds.), *The handbook of cognitive therapy vol. 1.* Tokyo: Seiwa Shoten.

Hori, T. (1991) *The brain and emotion: Mechanism of emotion.* Tokyo: Kyoritsu Shuppan.

An ACT-R Model of Memory Applied to Finding the Optimal Schedule of Practice

Philip I. Pavlik (ppavlik@andrew.cmu.edu)
Department of Psychology, Carnegie Mellon University
Pittsburgh, PA 15213 USA

John R. Anderson (ja+@cmu.edu)
Department of Psychology, Carnegie Mellon University
Pittsburgh, PA 15213 USA

When a person confronts the task of memorizing a collection of facts some questions have to be answered about how to optimize their learning. Perhaps the most important of these questions is how the practices for each item should be scheduled. The most basic suggestion from psychology is that the practices should be spaced as widely as possible. Since Bahrick (1979) it has been clear that wide spacing intervals result in better performance after a long delay. However, while wide spacing is likely to help, it may also be important to consider item differences when making practice scheduling decisions.

Memory items (such as the Japanese-English paired-associates we have used in this research) are not equally difficult for each subject to learn. In postexperimental interviews subjects often mention that some items were easier to remember because they could find some mediator to connect the cue to the response. Further, the data shows that some items are on average easier for all subjects. Perhaps the cue (the Japanese word) for these items is fairly close to some common English word that can be used to mediate between the cue and response.

It is useful to consider how we might optimize learning by attending to this variability in item difficulty. For instance, if we knew an item was easier we could schedule it less often, while if it was more difficult we could schedule it more often. This may have two benefits. First, if we did not use this procedure and instead gave the items equal practice we might expect that some items would be overlearned (the easier items) and some items would be learned poorly (the harder items). This uneven learning might be undesirable. Second, by saving practices on easy items and then distributing them to hard items it might be possible to achieve gains in performance.

To attend to this variability requires a model of memory that does two things. First, it needs to describe accurately how practice in general leads to recall performance by characterizing memory effects due to the frequency, recency, and spacing of practices. Second, to optimize performance for individual items, the model needs to be able to characterize both initial average difficulty for each item (estimated by fitting data from prior experiments, from now on referred to as item difficulty) and the difficulty particular subjects have with each item (which will need to be estimated based on performance with the item over the course of each subject's learning, from now on referred to as subject/item difficulty).

Model

The model we used, an extension of the Adaptive Character of Thought – Rational (ACT-R -- Anderson & Lebiere, 1998) theory, captures the three major effects in memory. The original ACT-R model captures the recency and frequency effects. Our extension of ACT-R (Pavlik and Anderson, 2003) captures the spacing effect. Importantly, it also models the spacing by practice interaction (that more spaced practice causes a larger spacing effect) and the spacing by retention interval interaction (that a longer retention interval causes a larger spacing effect).

In the model, memory strength for an item is represented by a value called activation which is calculated from the frequency, recency and spacing of practices. To capture the item difficulty for any item (both the item and subject/item components) we added a parameter to the model, β, which is added to activation for each item to represent whether the item is easy (positive β) or difficult (negative β). Overall β for any item is the sum of the item and subject/item components. A sigmoidal output function is used to calculate probability of recall as a function of activation.

Optimization

Our optimization routine supposes that training the lowest activation item leads to maximum learning (in terms of activation at a delay). This assumption follows from the model which describes how the long-run activation contribution of each practice depends on the activation for the item at the time the practice occurs.

The optimization procedure for a set of paired-associate items begins by presenting both the cue and response for each pair in a sequence of study trials. Following this, each trial involves the following steps:

1. Calculate the activation for all items.
2. Practice the item with the most to gain at recall (lowest activation item).
3. Update estimate of subject/item β for the practiced item.
4. If all trials completed end, else go to step 1.

The initial introduction of the items is in order from the items with lowest βs to the items with the highest βs. After each item has been introduced it has a history and the routine can calculate its activation, which includes the current β estimate for the item (the sum of item and subject/item components). After computing the activations for all items, the lowest activation item is selected and practiced. Each practice is a test of the item, which is followed by the correct response if the subject supplies the wrong answer. Defining a practice in this way allows us to both give practice and assess performance with each trial.

The routine uses the performance from each trial to update the estimate of subject/item β for the item. The routine starts out by assuming a prior distribution $f(x)$ for subject/item βs determined in the prior experiment (Pavlik and Anderson, 2003). Each time the item is tested the mean and variance of this distribution are updated. If S is the history of success and failure for an item the posterior distribution is calculated according to the Bayesian formula:

$$f(x|S) = \frac{f(x)P(S|x)}{\int f(x)P(S|x)\,dx} \quad (1)$$

where $P(S|x)$ is the probability of S given a value x.

This procedure allows us to estimate the current subject/item β estimate for an item as a function of the history of performance with the item. Essentially, what this involves is successive shifting and narrowing of our subject/item estimate of β as we gather data for an item.

Experiment

As a test of this optimization algorithm, we compared it with 2 control conditions in a within-subjects design. The first control condition scheduled practice using the widest possible spacing, while in the second control condition the schedule was yoked to the schedule produced for the previous subject's optimized words (the schedules did not correspond to the same words for the previous subject.)

The experiment also had 2 between-subjects conditions. In addition to the Japanese-English pairs from the prior experiment, we also used a set of Spanish-English pairs. While the Japanese stimuli had been designed to prevent easy associations, the Spanish language set contained more variability. While some words were as difficult as Japanese, roughly half of the Spanish set was rather easy with close cognates in English.

Subjects learned 60 pairs in the Japanese condition and 90 pairs in the Spanish condition. The word pairs for each subject were randomized into the three conditions so one-third of the pairs was designated in each condition. Since we had estimates of item βs for Japanese pairs, the optimization used these as estimates of the initial item βs. We had no initial item β estimates for the Spanish pairs so assumed 0 values for all these items.

The experiment took place in two sessions separated by 2 days. On session 1, subjects went through 360 trials in which the 3 within-subject conditions were interleaved. On the second session all items were tested twice to determine the effects of conditions.

Because there was little difference between the between-subjects conditions they were aggregated for analysis. As expected, both the wide spacing and optimized conditions (*M*s = 0.798 and 0.818) performed significantly better than the yoked condition (*M* = 0.726). The difference between the optimized and wide conditions was not significant.

To explore the results of the optimization, for all subjects within each within-subjects condition, we calculated the correlation of first session percent correct for each word with second session percent correct for each word. This gave three correlations for each subject, one for each condition. These correlations provided a measure of how likely a word that was responded to incorrectly on session 1 would be responded to incorrectly on session 2. With even or yoked spacing one might expect the correlation would be substantially positive because hard items would still be on average harder on session 2. In the optimized condition however, to the extent that hard items were given more practices and easy items fewer, one should expect that session 1 and 2 performance should be uncorrelated.

Since 10 subjects got 100% correct on session 2 for at least one of the conditions it was impossible to calculate one or more correlations. Statistical tests were performed on the 30 subjects with complete data. After showing the overall difference was significant using a within subjects ANOVA, $F(2, 58) = 52.6$, $p < .001$, we ran several t-tests to compare mean correlations (*M*s 0.54, 0.13 and 0.64 respectively for even, optimized and yoked conditions). Both even spacing and the yoked condition correlations were significantly greater than optimized, $ts > 6$, $ps < 0.001$.

Conclusion

The results of the experiment showed that the optimization routine behaved as expected. While the improvement was not significant compared with even spacing, the correlations showed that the optimization as it currently stands has the desirable property of automatically identifying difficult items and giving them more practice at the expense of easier items. This property might be especially useful in longer term training protocols during which extensive overtraining of well learned items would result in considerable inefficiency.

Acknowledgments

Preparation of this paper was supported by grant N00014-96-01-1491 from the Office of Naval Research.

References

Anderson, J. R. & Lebiere, C. (1998). *The atomic components of thought*. Mahwah. NJ. US: Lawrence Erlbaum Associates. Publishers.

Bahrick, H. P. (1979). Maintenance of knowledge: Questions about memory we forgot to ask. *Journal of Experimental Psychology: General, 108,* 296-308.

Pavlik Jr., P. I., & Anderson, J. R. (2003). An ACT-R model of the spacing effect. In F. Detje, D. Dorner & H. Schaub (Eds.), *Proceedings of the Fifth International Conference of Cognitive Modeling (pp. 177-182).* Germany: Universitats-Verlag Bamberg.

Navigating within a web site: the WebStep model

Sara Rigutti (rigutti@psico.units.it)
Walter Gerbino (gerbino@units.it)

Department of Psychology, University of Trieste
Via Sant' Anastasio 12, 34134 Trieste, Italy

Introduction

Web navigation is a complex activity in which users interact with a system to achieve their goals. Current web-navigation models (Kitajima, Blackmon, & Polson, 2000; Miller & Remington, 2000; Pirolli & Fu, 2003) are based on the label-following strategy, which takes the semantic aspects of navigation elements as fundamental. We propose that goal-directed actions of users are supported by *semantic* and *pragmatic* cues embodied in all navigation elements. Semantic and pragmatic cues are related to context-independent and context-dependent aspects of element meaning, respectively. The intrinsic contents of icons and verbal labels provide semantic cues. Spatial features (position and distribution in the page), as well as structural features of verbal labels (shape and color), provide pragmatic cues.

Here, we focus on the pragmatic cues that differentiate embedded links from navigation-bar options. As regards spatial features, embedded links are placed in the central area of the page while the navigation bar is placed at the top. As regards structural features, embedded links are intermixed with the embedding text and can be perceptually segregated because of shape dissimilarity (underline vs. normal) and color dissimilarity (blue vs. black); while the navigation bar includes only homologous elements, grouped by shape and color similarity, good continuation of button contours, and proximity.

WebStep Model

We developed a new model of web navigation, called WebStep. The model evaluates spatial and structural aspects of embedded links and navigation-bar options to generate predictions about the probability and speed of element selection. WebStep refers to paradigmatic navigation cases in which selection cannot be explained by semantic cues only. In a simple scenario including embedded links and navigation-bar options, navigation elements are choice items with variable utilities. Element utility $U_E(d)$ depends on the distance from the target page, estimated by the user within a hierarchical representation of the web site (Farris, Jones, & Elgin, 2002). In such a representation, generic information connected to thematic macro-categories is at the top of the hierarchy while specific information is at the bottom.

The utility of navigation-bar options is constant over distance (squares in Fig. 1a), because the bar leads to the highest information level consistent with the target, independent of user's position. Bar utility depends on its position in the web page, becoming lower as the distance from the conventional top position increases. This is achieved by modulating bar utility by *R*, an attenuation factor proportional to the ratio between actual and conventional positions.

The utility of embedded links is an inverse function of distance (diamonds in Fig. 1a). Since embedded links lead to specific information, their utility increases as the estimated distance from the target becomes smaller. The utility function of embedded links has been chosen as the positive half of a Gaussian distribution centered in d= 0.

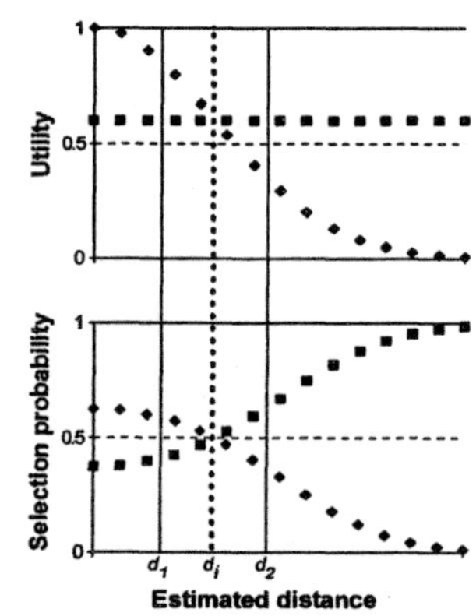

Figure 1: From utility (a) to selection probability (b) of navigation-bar options (squares) and embedded links (diamonds). Both function pairs intersect at d_i, where $U_B(d)= U_L(d)$ and $P_B(d)= P_L(d)= 0.5$.

The probability of element selection (Fig. 1b) results from the following equation:

$$P_E(d) = \frac{U_{En}(d)}{\sum_{i=1}^{N} U_{Ei}(d)}$$

where $P_E(d)$ is the selection probability of element E and N is the number of navigation elements.

Three predictions follow from WebStep assumptions.

1. The greater the estimated distance, the greater is the selection probability of a bar option with respect to the selection probability of a semantically equivalent embedded link. In particular, at the ordinal level $[d> d_i] \chi [P_B(d)> P_L(d)]$ and $[d< d_i] \chi [P_B(d)< P_L(d)]$.
2. At small estimated distances, as the bar approaches the conventional top position the difference between selection probabilities of the embedded link and bar options

decreases (and vice versa at large estimated distances). This formalizes the idea that the bar at the top is more easily identified than the bar at the bottom.

3. The difference between selection probabilities of the two navigation elements increases when the page heading is present. Such a regularity is achieved by implementing a distributed extraction of the estimated distance by means of a MonteCarlo method formalizing the non deterministic nature of the selection behavior.

Simulation

WebStep selection probabilities were computed by a Matlab program using three input parameters:

bar position, from a maximum for the bar on top to a minimum for the bar on bottom;

d_{ideal}, ranging from zero (user in target position) to ∞;

heading: presence= 1, absence= 0.

The program extracted 1000 distance values from a Gaussian distribution of distances centered in d_{ideal}, with heading-dependent width $\chi_{presence} > \chi_{absence}$. For every distance, element utilities were computed and the final utility value was defined by taking the average. Selection probabilities were computed on the basis of final utility values. This process was iterated 50 times to reach higher precision in the selection probability estimation. Final probabilities were computed as averages from 50 iterations. Selection time was computed by means of the linear combination of a component inversely proportional to selection probability and a processing constant for each element: $T(E)= 1/P_E(d) + K_E$. By manipulating free parameters values the intersection between the two utility functions was set at (4.0, 0.6). As a consequence $U_B(d)$ was equal to 0.6 while $U_L(d)$ was the left half of a Gaussian centered in d= 0 with χ= 31.3.

Experiments

To test the empirical validity of WebStep we run two experiments, in which the semantic aspects of bar options and embedded links were equivalent. Every label appeared on both elements, which could be differentiated only on the basis of their spatial and structural features. To test the first prediction we manipulated the starting navigation page. In Exp. 1 the starting navigation page was in the no-target position, inducing a high estimated distance from target (content-target in-congruency). In Exp. 2 the starting-navigation page was in target position, inducing a low estimated distance from target (content-target congruency). In both experiments we tested prediction 2 by manipulating the bar position (top vs. bottom) and prediction 3 by manipulating the the heading (presence vs. absence). The selection rate and time of elements in the starting navigation page were recorded and used as dependent variables.

Results

Obtained results were consistent with WebStep predictions. In Exp. 1 we found a preference for the selection of the bar with respect to the embedded link, χ_1^2= 6.37; $p < 0.05$: *users use the bar to move towards the macro-category congruent with the target.* Furthermore, consistently with second prediction, the bar selection rate increased when the bar was in the conventional top position, [χ_1^2 = 5.31; $p < 0.05$]. Consistently with the third prediction the difference between the selection probability of the two elements increased in presence of the heading [χ_3^2 = 14.20; $p < 0.01$]

The inverse relation between selection time and selection rate (slope= -1.9; r^2= 0.8) confirmed that the choice of an element with low selection probability requests a higher time than the choice of a navigation element with high selection probability.

In Exp. 2, consistently with the first prediction we found a preference for the embedded-link selection when the heading was present, [χ_1^2 = 4.263; $p < 0.05$]: *users use the embedded link to move towards the sub-category congruent with the target.* Furthermore, consistently with second prediction, the difference between selection-rate of the two elements was lower when the bar was at the bottom than at the top. We also found a higher difference between elements selection probability when the heading was present [χ_3^2 = 14.94; $p < 0.01$].

As a final analysis we compared selection times and selection probabilities derived from simulations with values obtained in Exps. 1 and 2. The fit was significant for both probability (r^2= 0.7, $F_{1,15}$= 29.7, $p < 0.001$) and selection time (r^2= 0.5; $F_{1,15}$= 22, $p < 0.001$).

Conclusions

Results are consistent with the idea that navigation depends also on pragmatic aspects of elements, as well as on contextual aspects like the page heading. Our WebStep model fits obtained effects.

References

Farris J.S., Jones K.S. & Elgin P.D. (2002). Users' schemata of hypermedia: what is so spatial about a web site? *Interacting with Computers*, 14, (5), 487-502.

Kitajima M., Blackmon M.H. & Polson P.G. (2000).. A Comprehension-based Model of Web Navigation and Its Application to Web Usability Analysis. In S. McDonald, Y. Waern & G. Cockton (eds.), *People and Computers XIV - Usability or Else! (Proceedings of HCI 2000)*, Springer, pp.357-373.

Miller C. S. & Remington R. W. (2000). A computational model of Web navigation: Exploring interactions between hierarchical depth and link ambiguity. In *The proceedings of The 6th Conference on Human Factors and the Web.*

Pirolli P. & Fu W. T. (2003). SNIFT-ACT: A Model of Information Foraging on the World Wide Web. *Ninth International Conference on User Modeling*, Johnstown, PA.

What goals do students have when choosing the actions they perform?

Ido Roll, Ryan Shaun Baker, Vincent Aleven, Kenneth Koedinger
(idoroll@cmu.edu, rsbaker@cmu.edu, aleven@cs.cmu.edu, koedinger@cmu.edu)
Human Computer Interaction Institute, Carnegie Mellon University
5000 Forbes Ave., Pittsburgh, PA 15217 USA

Introduction

Several models were built recently in the metacognitive level of the students' interaction with Cognitive Tutors, an intelligent tutoring system based on ACT-R theory. After finding suboptimal help-seeking behavior, we built a metacognitive model of desired help-seeking behavior (Aleven et al. in press). In a different Cognitive Tutor, Baker et al. (2004) built a model that identifies misuse of the tutor.

Here we take another step and describe a model of students' goals and strategies, which rely in the basis of their metacognitive actions. By comparing the model's predictions to students' log-files we find the correlation between having the goals and learning gains.

Goals and actions

According to the model, each student has tendencies towards four different local-goals. Each goal is related to a certain strategy, which leads to specific action/s with the tutor (table 1). The unique personal pattern of tendencies categorizes the individual learning process with the tutor.

Table 1: goals, strategies and actions.

Goal	Strategy	Action/s
Learning Orientation (I want to learn as much as possible)	First I try. If I cannot, I ask for help.	Slow attempts, slow help requests.
Performance Orientation (I want to work quickly)	I ask for hints until I get the answer	Fast repeated help requests
Least Effort (I want to get done with minimal mental effort)	I guess repeatedly without thinking much about it. Eventually I get it right	Fast solution attempts, most of them are wrong.
Solve by myself (I am determined to be independent)	First I try. If I cannot, I do as much as I can and guess something, since I don't want to use help.	Slow solution attempts. Some of which are correct and some are educated guesses.

At each point, student can perform one of the following actions: Solve the question, guess an answer, or ask for a hint. Students can perform these actions quickly or slowly. Each attempt to answer the question (whether by solving it or by guessing) has a certain probability of being correct. The correct-rate of solution-attempts is much higher then that of guesses.

The model

In the model, at each action, the student chooses (whether consciously or not) one local goal and acts according to it. The model assumes that the student's likelihood to choose a certain local goal is determined by the following considerations:

1. Individual differences in the tendencies to choose the different goals.
2. Skill level: The student's estimated ability to solve the problem. (We predict this ability using the skill-level that the tutor assigns to the student for each individual problem.)
3. Context: the student's earlier actions and the system's feedback to them (e.g., error messages).

To evaluate the model, we first applied an ACT-R version that identifies local-goals of students' actions within the same context and estimated level (Roll et al. in press).
The model describes the following process (figure 1):

- The student evaluates her ability to solve the question immediately (1), and does so if she can (2)
- If the student needs to spend more time thinking (3) she chooses a local goal (4) and acts upon it (5).

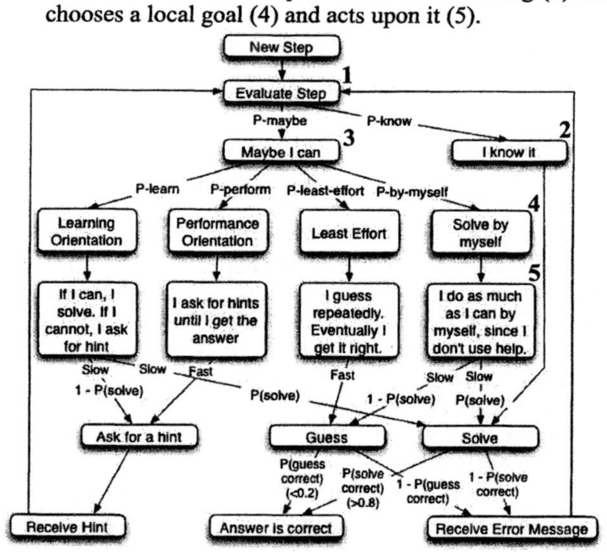

Figure 1: The goals metacognitive model.

Based on the ACT-R model we build an equivalent mathematical model. Though the ACT-R framework was appropriate in modeling the process, switching to a mathematical model enabled us to fit the data and conduct statistical analysis relatively easily.

By comparing the actual performed actions to the model's predictions, we could find the individual students' tendencies towards the different goals.
A total of 57,000 actions, performed by 21 students, were fitted by the model (Aleven et al., in press). The average square deviation between the model results and the original data was 0.003.
We correlated between the tendencies to act upon each goal and between an independent measure of learning outcomes (as defined by the improvement from pre- to post-test divided by maximum possible improvement for the same student), in order to identify the relation between the goals and learning. We found a significant negative correlation between learning outcomes and the tendency to be Performance Oriented ($F(1,19)=6.35$, $p=0.02$, $r=-0.50$). We also found some significant positive correlation between Learning Orientation and learning outcomes in several contexts – for example, across all skills, being Learning Oriented after a hint was significantly correlated with learning gains ($F(1,19)=6.00$, $p<0.03$, $r=0.49$). Having the goal of Solve-by-Myself had a positive correlation with learning outcomes across most of the skills and contexts, but wasn't always significant.
One main flaw is that the model is too local and has 54 parameters per student (for 72 data points per student across all skills and contexts), a number that should be reduced dramatically. The many parameters make it harder to generalize regarding students' general tendencies.

Currently we build a new version of the model with fewer parameters. In this version, besides the set of tendencies, each student is assigned parameters that represent the adaptation to different skills and adaptation to different contexts.
Just like we defined several strategies to choose what action to perform, we now define several strategies to adapt the single set of tendencies to the various skills and contexts.
We relate to the basic set of tendencies of each student as the *basic vector*. At each point, the sum of all tendencies is 1.00. When a certain tendency is getting strengthened by the adaptation vector, it has higher chances of being selected by the model while others maintain the ratio between them.
For example, let us take the following set of tendencies:

Basic vector = (0.26, 0.41, 0.13, 0.20).

This vector describes the tendencies to choose (learning orientation, performance orientation, least effort, solve by myself). When acting after an error, let us suppose that the student increases the tendency to have a least-effort goal while decreases the tendency to solve-by-myself. This student has two adaptation vectors:

Increase least-effort vector = (1, 1, x, 1) $x>1$

Decrease solve-by-myself vector = (1, 1, 1, y) $y<1$

In this case, the chances of having any goal (for example, least-effort) will be calculated in the following way:

$$\frac{(basic\sum value)_j \sum(adaptation\sum value)_j}{\sum_{i=1}^{number\sum of\sum goals}(basic\sum value)_i \sum(adaptation\sum value)_i} = \frac{0.13\sum X}{0.26\sum 1 + 0.41\sum 1 + 0.13\sum X + 0.20\sum Y}$$

We find x and y by fitting the prediction to the data.
Since there are only few strategies that make sense at each context or skill, we expect to find very limited number of adaptation vectors. This will enable us to generalize conclusions as to the efficiency of each vector.

Conclusions and future work

The current model gives a good description of students' specific goals when working with cognitive tutors. In addition, we found a link between the tendency to have various goals and learning outcomes. However, due to its many parameters, it is too difficult to generalize from it about students' overall goals and strategies.
Through identifying *adaptation strategies*, we categorize the interaction of each student using three types of vectors: (i) the basic vector – the student's overall tendencies towards the different goals, (ii) skill adaptation-vectors, and (iii) context adaptation-vectors (first action, after a hint or after an error).
Once completed, these strategies will be correlated with learning outcomes in order to identify which patterns lead to better learning. In addition, the model should be fitted to data from different tutors and its predictions should be compared to other metacognitive models.
We would like to thank John R. Anderson for his helpful advice.

References

Aleven, V., McLaren, B., Roll, I., Koedinger, K. Toward Tutoring Help Seeking: Applying Cognitive Modeling to Meta-Cognitive Skills. To appear at Intelligent Tutoring Systems Conference (2004)

Anderson, J. R. & Lebiere, C.J. (1998). The Atomic Components of Thought, Lawrence Erlbaum Associates inc., Mahwah, NJ.

Baker, R. S., Corbett, A. T., Wagner, A. Z. & Koedinger, K. R., Off-Task Behavior in the Cognitive Tutor Classroom: When Students "Game the System", Proceedings of the SIGCHI conference on human factors in computing systems (2004), p. 383-390, Vol. 6 no. 1.

Corbett, A.T. & Anderson, J.R. (1995), Knowledge tracing: Modeling the acquisition of procedural knowledge. User modeling and user-adapted interaction, 4, 253-278.

Roll, I., Aleven, V., Koedinger, K. (in press) A Metacognitive ACT-R Model of Students' Learning Strategies in Intelligent Tutoring Systems. To appear at Intelligent Tutoring Systems Conference (2004)

White, B.Y. & Federistan, J.R. (1998), Inquiry, Modeling, and Metacognition: Making Science Accessible to All Students. Cognition and Instruction, 16(1), 3-118

A General-Purpose Computational Model of the Conscious Mind

Alexei Samsonovich (asamsono@gmu.edu)
Krasnow Institute for Advanced Study, George Mason University
4400 University Dr. MS 2A1, Fairfax, VA 22030 USA

Kenneth DeJong (kdejong@gmu.edu)
Department of Computer Science and Krasnow Institute for Advanced Study
George Mason University, 4400 University Dr., Fairfax, VA 22030 USA

Elements of the Proposed Approach

Building artificial cognitive systems that possess a concept of "self", as well as cognitive and behavioral abilities of a conscious being is a difficult task, but one with tremendous practical significance and potential. In this work we outline general principles of a new approach to this task and illustrate them with detailed computational analysis based on a simple paradigm. Our approach to this task is grounded on three conceptual components – three key ideas (schemas, charts and the self) that underlie three levels of organization of our cognitive system. They are explained below.

Calculus of Schemas and Mental States

Schemas are the basic elementary building blocks of our cognitive system (Samsonovich & DeJong, 2003). The term "schema" (plural "schemata" or "schemas") was introduced by Kant (1781/1929) and is currently used in a variety of senses in computational and cognitive sciences. Here this term refers to an abstract model or a template that is used to instantiate and to process a particular cognitive category. Thus, schemas are units of semantic knowledge, primitives of action and reasoning, rules, concepts, sensory qualia, etc.

We generally say that a schema has a state, when its instance is bound to some given content. Any connected complex of states (e.g., A-D-C, Figure 1) is also regarded as a state (and is potentially convertible into a schema). For example, a state of seeing a red circle includes states of the schemas of red and a circle. A state generating a proof can be described in terms of states of schemas of logical inference, etc. Here a state is considered "mental", if it is symbolically attributed to a subject of experience (a self). Schemas and states are dynamical objects: they can be created and modified "online".

We implement schemas and their states as data structures, using one universal format, regardless of the category. Each schema has a "head" (specifying its binding terminals, rules and conditions of binding, as well as the expected effects of execution) and a "body" (specifying how, if at all, the schema is executed). Simply speaking, if a schema can be viewed as a class, then its state is an instance of this class.

Dynamical Multichart Architecture

The second key idea of our approach is the dynamical multichart architecture that serves as an infrastructure for the process of cognition. We use a simulationist approach based on an abstract notion of a chart. Each chart represents a simulated mental perspective, providing a room for a mental simulation (i.e., dynamics of mutually bound states) of what one may be expected to experience when placed in that perspective. Therefore, a chart is associated with a particular instance of the subject to whom the mental states are attributed: this is reflected in the chart label (e.g., I-Now, I-Next, I-Previous, I-Imagined, I-Goal, I-Past, I-Meta, He-Now, She-Now, etc.). Typically, this instance of the subject is the imaginary "self" of the cognitive system itself, taken in a particular context (e.g., at a particular moment of time).

Thus outlined system of charts and associated with them mental simulations can be viewed as a simplified, practical implementation of a generalized modal logic framework that may include epistemic, doxastic, deontic, conditional, temporal, alethic and other logics (with mental states regarded as "doxons", "deons", etc.). While each simulation is confined to its chart, charts interact with each other according to the rules of a corresponding logic. E.g., a state A (Figure 1) that may represent a voluntary action initiated in I-Now is copied into I-Next to produce expected effects.

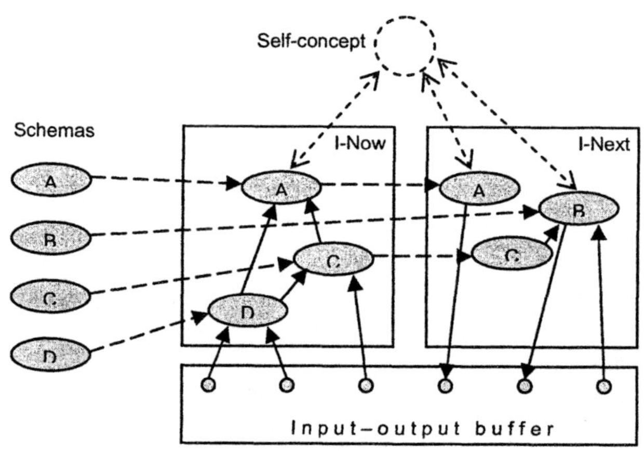

Figure 1: A snapshot of the cognitive system dynamics.

The Self Concept

The third, metacognitive component of our conceptual framework is the self concept. From the above, the reader might expect that now we shall introduce a gizmo operating on states and charts, saying that it represents the self of the system. On the contrary, we postulate that there is no such gizmo, or virtual engine, or homunculus, etc., in any truly conscious cognitive system (we do have "drivers" operating on states and charts that follow simple, mechanistic rules).

In our approach, the self is implemented via a set of self-axioms (Aleksander & Dunmall, 2003; Samsonovich & Nadel, 2005) that globally constrain dynamics of states and charts at a higher level. In particular, this implies a linear organization of the main sequence of charts into a consistent scenario (stream of consciousness) underlying perception, thoughts, attention focus, decisions and "voluntary" actions performed by the system. This scenario (which extends in both directions in time and is continuously updated), when expressed at a verbal level (based on introspective reports that are automatically generated by states), can be controlled by linguistic rules and constraints (instantiating self-axioms) using methods of discourse analysis and the like.

Demonstration Based on a Learning Paradigm

In order to demonstrate the outlined above general principles in action, we consider a paradigm that can be called "leveraging self-learning with the self-concept". In this paradigm, a set of specially designed virtual worlds is used as a "training facility" to help a virtual robot controlled by the cognitive system to develop useful and powerful schemas. Innate schemas may include elementary moves and senses, as well as relevant reasoning primitives. The system repeats the following procedure:

1. Select an action schema and mutate its head to produce an idea of an action that is not available yet as a schema.
2. In each of several encountered situations, take the new header as a challenge and solve it (e.g., by imagined trial-and-error procedure), then execute the solution.
3. Reinterpret own behavior (not imagery) based on the schema of a voluntary action: find an apparent common motivation in the performed intermediate steps in all cases.
4. Based on the above, write the body of the new schema.

As the robot learns essential "tricks" at one level (and compiles them into schemas), it is taken to the next level. At each new stage, previously developed schemas are used for solving new challenges.

The scheme outlined above will be demonstrated in the poster by computer simulations based on a push-push puzzle setup. A minimal set of innate schemas includes a one-step move and some useful cognitive primitives, e.g., the notion of Euclidean distance. At the first stage the robot learns to move in an open space. Then it learns to navigate a maze, to push blocks, to avoid irreversible moves, etc. After that, when given a goal, it is capable of solving simple puzzle configurations and learns to deal with more complex ones.

Related Works

During recent years a tremendous progress has been made in several fields related to intelligent agents possessing higher and meta-cognitive abilities, including logical foundations (e.g., Panzarasa et al., 2002), agent architecture (e.g., works originated from the BDI framework: Bratman, 1987) and practical computational tools (e.g., Soar: Laird et al., 1987, and ACT-R: Anderson & Lebiere, 1998). The presented framework stands closer to the latter, offering more universality and a higher capacity for development and generalization than an apparatus based on chunks and/or productions. Next, our multichart architecture facilitates implementation of modal logics in a practically useful manner. Finally, our third component (the self concept) is completely missing and probably not feasible in Soar, in ACT-R, and in most other computational cognitive models.

Our system of charts can also be related to the framework of possible world boxes proposed by Nichols and Stich (2003), although each our "box" (chart) represents a possible state of agent's awareness rather than a world.

Concluding Remarks

Our present ambition is to create a virtual entity emulating human mind in its most essential abilities, an entity that by itself might be of a great scientific and practical interest for us as a general-purpose, new-generation intelligent artifact. Of a major interest for us are its abilities to learn, to exhibit rational initiative, and to be able to work interactively with a human. Among near-future applications of the proposed model or its equivalents we expect to find tasks that involve:

(a) learning to understand and to develop communication languages (e.g., learning natural language from a human),
(b) learning to do lower-level tasks based on their higher-level description (e.g., automated programming),
(c) learning to assist a human in a given cognitive domain (e.g., rapid decision making in unstable environments),
(d) self-adaptation and survival in complex environments.

References

Aleksander, I., & Dunmall, B. (2003). Necessary first-person axioms of neuroconsciousness. *Lecture Notes in Computer Science,* 2686: 630-637.

Anderson, J.R., & Lebiere, C. (1998). *The Atomic Components of Thought.* Mahwah, NJ: Erlbaum.

Bratman, M.E. (1987) *Intentions, Plans, and Practical Reason.* Harvard University Press: Cambridge, MA.

Kant, I. (1781/1929) *Critique of pure reason.* Translated by N. K. Smith. New York: St. Martin's Press.

Laird, J.E., Newell, A., & Rosenbloom, P. (1987) Soar: an architecture for general intelligence. *Artificial Intelligence* 33: 1-64.

Nichols, S., & Stich, S.P. (2003). *Mindreading: An Integrated Account of Pretense, Self-Awareness, and Understanding Other Minds.* Oxford: Clarendon.

Panzarasa, P., Jennings, N.R., & Norman, T.J. (2002) Formalizing collaborative decision-making and practical reasoning in multi-agent systems. *Journal of Logic and Computation* 12 (1): 55-117.

Samsonovich, A. V., & De Jong, K. A. (2003). Meta-cognitive architecture for team agents. *Proceedings of the 25th Annual Meeting of the Cognitive Science Society* (pp. 1029-1034). Boston, MA: Cognitive Science Society.

Samsonovich A, Nadel L (2005) Fundamental principles and mechanisms of the conscious self. *Cortex,* in press.

How Forgetting Fosters Heuristic Inference

Lael J. Schooler (schooler@mpib-berlin.mpg.de)
Max Planck Institute for Human Development
Lentzeallee 94, 14195 Berlin Germany

Ralph Hertwig (Ralph.Hertwig@unibas.ch)
Institut für Psychologie
Universität Basel, Missionsstraße 60/62
CH-4055 Basel Switzerland

Functions of Forgetting

The notion that forgetting serves an adaptive function has repeatedly been put forth in the history of the analysis of human memory. Bjork and Bjork (1996), for instance, have argued that it prevents obsolete information from interfering with the recall of more current information. Here we explore functions that forgetting may play in memory-based inference strategies. To this end, we bring together two research programs—the program on fast and frugal heuristics (Gigerenzer, Todd, & the ABC research group, 1999) and the ACT-R research program (Anderson & Lebiere, 1998). Specifically, we implement in ACT-R the *recognition heuristic* (RH) (Goldstein & Gigerenzer, 2002), and the *fluency heuristic* (FH), each of which exploit fundamental memory retrieval processes.

The Recognition Heuristic

Consider one of the simplest decisions that can be made: selecting one option from two possibilities, according to some criterion on which the two can be compared. How this decision is made depends on the available information. If the only information at hand is whether or not the decision maker recognizes one of the alternatives, and she suspects that recognition is positively correlated with the criterion, then she can do little better than rely on her own partial ignorance, choosing recognized options over unrecognized ones. This kind of "ignorance-based reasoning" is embodied in the *recognition heuristic* (RH) (Goldstein & Gigerenzer, 2002), which for two–alternative choice tasks can be stated as follows:

***If** one of two objects is recognized and the other is not,*
***then** infer that the recognized object has the higher value with respect to the criterion.*

This minimal strategy may not sound like much for a decision maker to go on, but there is often information implicit in the failure to recognize something, and this failure can be exploited by the heuristic.

Goldstein & Gigerenzer (2002) conducted studies to find out whether people actually use the recognition heuristic. For instance, they presented U.S. students with pairs of U.S. cities and with pairs of German cities. The task was to infer which city in each pair had the most inhabitants. The students performed about equally well on both types of cities. This result is counterintuitive: The students had accumulated a lifetime of facts about U.S. cities that could be useful for inferring population, but they knew little or nothing about the German cities beyond merely recognizing about half of them. According to Goldstein and Gigerenzer (2002), the latter fact is just what allowed them to employ the recognition heuristic to infer that the German cities that they recognized were larger than those they did not. The students could not use this heuristic when comparing US cities, because they recognized all of them and thus had to rely on other methods for making their decisions. In short, the RH works because our lack of recognition knowledge is often not random, but systematic and exploitable.

Recognition Heuristic Exploits Correlations

According to Goldstein and Gigerenzer (2002) the recognition heuristic works, because there is a chain of correlations, linking the criteria (e.g., city population), environmental frequencies (e.g., how often a city is mentioned) and recognition. ACT-R's activation tracks just such environmental regularities. Therefore activation differences reflect, in part, these frequency differences. Thus, it appears that inferences could be based on activation values read off the chunks. However, it has been a long-standing policy in ACT-R that sub-symbolic quantities, such as activation, cannot be accessed directly. We propose, nevertheless, that the system could still capitalize on activation differences associated with various objects by gauging how it responds to them. Two such responses that correlate with activation are (1) whether a chunk associated to a specific object can be retrieved and (2) how quickly it can be retrieved. The first response is key to our implementation of the RH and the second to the FH.

Fluency Heuristic

The use of fluency of reprocessing as a cue in inferential judgment has been termed the *fluency heuristic* (e.g., Jacoby & Dallas, 1981). We adopt this name but define the heuristic in the context of a two-alternative choice task in which the objective is to select one option from two possibilities, according to some criterion on which the two can be compared. According to the fluency heuristic:

***If** one of two objects is more fluently reprocessed,*
***then** infer that this object has the higher value with respect to the criterion.*

To see how this might work, let us assume that American students are sensitive to differences in recognition times. That is, they are attuned to, for instance, being able to recognize instantaneously "Berlin", but taking a moment to recognize "Stuttgart". We suggest that these differences in recognition time reflect, in part, retrieval time differences, which, in turn, reflect the base level activations of the corresponding memory chunks, which correlates with environmental frequency, and finally with city size. Further, rather than assuming that the system can discriminate between minute differences in any two retrieval times, we allow for limits on the system's ability to do this. Thus, if the retrieval times of the two alternatives are within a j.n.d. of, say, 100 ms., then the system must guess.

Simulations

Method

First, like Goldstein and Gigerenzer (2002) we assume that the frequency with which a city is mentioned in the newspaper mirrors its overall environmental frequency. Based on this assumption, we constructed environments such that the probability of encountering a city name on any given day was proportional to its relative frequency in the Chicago Tribune. Second, the model learned about the environment by strengthening memory chunks associated with each city according to ACT-R's base level activation equation. Third, the model's recognition rates were determined by fitting the activation strengths of the city records to the recognition rates of Goldstein and Gigerenzer's (2002) students. These recognition rates, in turn, determine how well the recognition and fluency heuristics perform on the city comparison task.

Results

Figure 1 shows how the performance of the heuristics depends on how quickly chunk activation decays (i.e., ACT-R's parameter *d*); both heuristics peak at intermediate decay rates. In the case of the recognition heuristic, intermediate levels of forgetting maintain a distribution of recognition rates that are highly correlated with the criterion. In the case of the fluency heuristic, intermediate amounts of forgetting increase the chances that differences in the retrieval times of two chunks will be detected. To see why, consider Figure 2 that shows the exponential function that relates a chunk's activation to its retrieval time. Forgetting lowers the range of activations to levels that correspond to retrieval times that can be more easily discriminated. In other words, a given difference in activation at a lower range results in a larger, more easily detected, difference in retrieval time than the same difference at a higher range

Conclusion

The recognition and fluency heuristics can be understood as means to indirectly tap the environmental frequency information locked in the activations. These heuristics will be effective to the extent that the chain of correlations, linking the criteria, environmental frequencies, activations and responses, is strong. Forgetting functions to foster the performance of these heuristics by strengthening this chain.

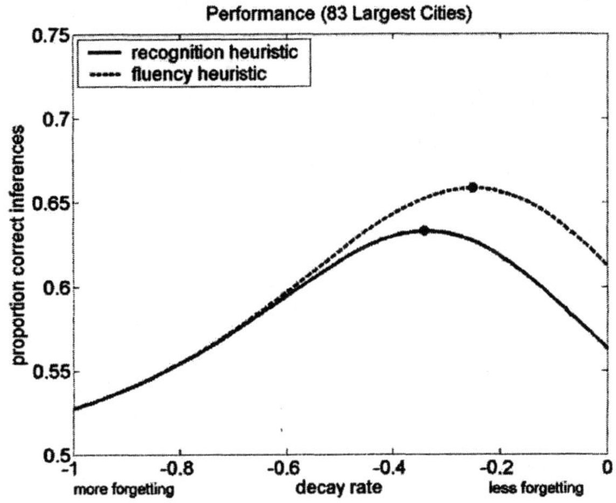

Figure 1. Performance varies with decay rate.

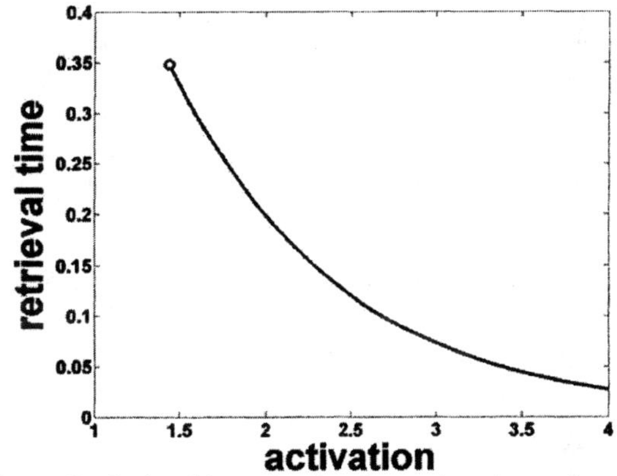

Figure 2. A chunk's activation determines its retrieval time.

References

Anderson, J. R., Lebiere, C. (1998). *The atomic components of thought*. Mahwah, NJ: Lawrence Erlbaum Associates.

Bjork, E. L., & Bjork, R. A. (1988) On the adaptive aspects of retrieval failure in autobiographical memory. In M. M. Gruneberg, P. E. Morris, & R. N. Sykes (Eds.). *Practical Aspects of Memory II*, London: Wiley, 1988.

Gigerenzer, G., Todd, P. M. & the ABC Research Group (1999). *Simple heuristics that make us smart*. New York: Oxford University Press.

Goldstein, D. G. & Gigerenzer, G. (2002). Models of ecological rationality: The recognition heuristic. *Psychological Review*, *109*, 75-90.

Jacoby, L. L., & Dallas, M. (1981). On the relationship between autobiographical memory and perceptual learning. *Journal of Experimental Psychology: General*, *110*, 306-340.

Teaching Computational Modeling to Non-Computer Scientists

Terrence C Stewart (tcstewar@connect.carleton.ca)
Institute of Cognitive Science, Carleton University
1125 Colonel By Drive, Ottawa, ON, K1S 5B6, Canada

Abstract

Most courses on computational modeling involve having students skilled in the computer science implement various computational models. Here, a course is presented which is aimed at students with minimal programming background. The focus is on using a large number of different models on various problems. Students learn how well various algorithms work, rather than how to implement them. They are provided with simple implementations and sample code that makes use of these implementations, and are then asked to adapt the given system to deal with other problems. This material is presented as part of the ongoing development of this course, and it is hoped that others may find it useful for their own courses and research.

Introduction

Courses in computational modeling tend to presume a significant amount of programming skill on the part of the student. Typical projects include implementing neural networks from scratch, or writing one's own genetic algorithm to solve a particular problem. These sorts of projects mean that students who are not specialists in computer science spend a disproportionate amount of time debugging. Furthermore, since the focus in these courses is on implementation, less time is spent discussing what sorts of problems they tend to work well or work poorly on.

In this paper, a course is presented which addresses these problems. The course ran at Carleton University in the winter term of 2004 as Cognitive Science 5001: Cognition and Artificial Systems. The only prerequisite was either graduate standing or a single previous programming course of any sort, and it was open to both graduate and fourth-year undergraduate students within the Institute of Cognitive Science.

Course Organization

Since the goal of the course was to give students practical experience using these various algorithms, the major content (worth 70%) of the course was weekly assignments. After each week's lecture, students were given a working implementation of a model (such as Kohonen Maps or Evolutionary Strategies). They were also provided with working code that used this implementation to perform a typical task. The students were then asked to modify this software to perform different tasks, and to modify the various parameter values of the models. The performance of these models in these conditions was then examined.

To allow the students to demonstrate a more in-depth mastery of one particular algorithm, the final paper for the course involved taking a published paper and replicating those results therein. For this paper, students were also expected to discuss the effects of varying various parameters of the model, and look at the effect of varying the problem representations used. In this way, the students perform a useful role within the cognitive modeling community, as this detailed replication is relatively rare.

The Computational Models

For all of these models, the students worked exclusively with the Python programming language, which was chosen for its ease of use and clear syntax. None of the students had any prior exposure to it. This standardization allowed for direct comparisons between models, and made it easy to combine models together. This author has also found Python to be highly suitable for modeling research outside of the classroom.

The programs were carefully designed to be as easy to read, understand, and modify as possible. Students commented that this significantly increased their confidence that their changes caused the programs to do what they wanted them to do. Furthermore, the success of this clarity was embarrassingly shown when one student (who had no previous programming experience) uncovered bugs in the implementations of two different models. These were quickly fixed, and demonstrate the code's readability.

Cellular Automata

To begin the course, students examined systems both with agents in the world (e.g. Langton's Vants) and without (e.g. Conway's Game of Life and Langton's Loops). Variations of different rules and different types of Vants were investigated. A collective sorting simulation (Deneubourg et al, 1991) was also replicated. This module was used throughout in the course as a grid-world environment.

Genetic Algorithms and Evolutionary Strategies

A wide variety of evolutionary options were investigated. This included various implementations of mutation and crossover, elitism, tournament and roulette selection, rank-based selection, steady-state models, and the recent development of neutral network theory, including extrema selection (Stewart, 2001). These techniques were applied to the traveling salesperson problem, genetic programming, and the evolution of agent behavior on a grid-world.

In a somewhat uncommon aspect of the course, Genetic Algorithms were directly compared to Evolutionary Strategies. Students were exposed to the significant advantage that ES algorithms have on parameter-optimizing problems (in this case, an ant pheromone model). They were also encouraged to investigate the effect of using ES-($\mu/\rho+\lambda$) selection within more typical GA problems.

Perceptrons and Multi-Layer Perceptrons

Here students used back-propagation learning to investigate the abilities of this common neural network architecture. Of interest was the effect of learning rate, momentum, varying numbers of hidden nodes, different approaches to representations, and the significant effect of normalizing input and output data. These systems were also compared to a memorization-based approach to supervised learning to indicate the practical speed advantages of these models.

Simple Recurrent Neural Networks

In this section, the core findings of Elman's (1990) paper introducing this model were replicated. The capabilities of the networks for maintaining context over longer and longer periods of time were investigated, as was the effect of using a non-obvious representation scheme in his syllable-identification problem.

Kohonen Self-Organizing Feature Maps

Kohonen Maps were compared to standard clustering algorithms, and demonstrated in terms of their ability to change the dimensionality of input data. Special attention was given to recent results (such as Touzet, 1997) showing that Kohonen maps can be used as an automatic way of adjusting representations before presenting them to standard supervised learning systems. To show this, the Kohonen map was used as a pre-processor to change the representation of the input data into a form that proved to be more suitable for learning by the backprop networks.

Fuzzy ART and Fuzzy ARTMAP

Both the unsupervised and the supervised versions of Adaptive Resonance Theory were investigated. The advantages in terms of learning rate were discussed, as was their ability to indicate that a particular input is unknown to the network. The models were also dissected to reveal the learned rules. The effect of the vigilance parameter and the order of presentation of the input was also studied.

Q-Learning and Sarsa

Both of these Temporal Difference algorithms were compared, highlighting the effects of the very slight difference between the two. Their behavior on the Cliff grid-world was discussed, and students were able to develop their own worlds to run the algorithms in. Students adjusted the various parameters of the learning algorithm, and saw the effects of changing the reinforcement regime.

ACT-R

In a potentially surprising move, a stripped-down implementation of ACT-R was also included as part of the course. This version only supported the rule-matching and chunk-activation learning aspects of ACT-R. This was sufficient for demonstrating the successful Rock-Paper-Scissors ACT-R model (Lebiere & West, 1999), allowing students to see how the model reacts to different patterns in its opponent's play. By watching the model adjust to different opponent strategies, students gained insight into how the ACT-R learning rule dynamically affects its behavior. Also, it provided some evidence that it is possible to develop an ACT-R model that is separate from the current LISP-based implementation.

Future Additions

In preparation for teaching this course again, a few small changes are being made to the course. The interfaces to the unsupervised and supervised learning algorithms are being standardized so that they can be easily swapped, allowing direct comparisons to be made. Some of the slower code (such as the back-propagation system) is being modified to interface with existing (significantly faster) C implementations, while still maintaining the same simple interface as used in the course. Also, the ACT-R library is being expanded with the missing learning mechanisms.

Conclusions

The course successfully exposed the students to a variety of widely used models, and allowed them more time to work with these models than would have been possible if they were to perform implementations themselves. The implementations given to the students were in a clear and readable language, allowing them to inspect the algorithms themselves to see exactly how they worked. These modules were also flexible enough to be easily combined and used by the students to replicate published research. Indeed, in a number of cases, the students' results revealed mistaken assumptions in the original research. The course was also found to be accessible to students with no previous programming experience.

Everyone is encouraged to make use of the materials developed for this course. The complete lecture notes and all software modules are available online at:

http://chat.carleton.ca/~tcstewar/cgsc5001/

References

Deneubourg, J.L., Goss, S., Franks, N., Sendova-Franks, A., Detrain, C., & Chretien, L. (1991). The Dynamics of Collective Sorting. Robot-like Ants and Ant-like Robots, *From Animals to Animats*, MIT Press.

Elman, J.L. (1990). Finding structure in time. *Cognitive Science, 14*, 179-211.

Langton, C.G. (1986). Studying Artificial Life with Cellular Automata, *Physica 22D*, North-Holland, 120-149.

Lebiere, C., & West, R. L. (1999). Using ACT-R to model the dynamic properties of simple games. *The Proceedings of Cognitive Science.*

Stewart, T.C. (2001). Extrema Selection: Accelerated Evolution on Neutral Networks. *IEEE Congress on Evolutionary Computation*, 25-29.

Touzet, C.F. (1997). Neural Reinforcement Learning for Behaviour Synthesis. *Robotics and Autonomous Systems, Special issue on Learning Robot: the New Wave.*

Comparing Teamwork Modeling in an Empirical Approach

Shuang Sun (ssun@ist.psu.edu) Isaac Councill (icg2@psu.edu)
Xiaocong Fan (zfan@ist.psu.edu) Frank E. Ritter (frank.ritter@psu.edu)
John Yen (jyen@ist.psu.edu)
School of Information Sciences and Technology, the Pennsylvania State University
Information Sciences and Technology Building, University Park, PA 16802 - 6823

Introduction

Modeling teams is becoming an increasingly active topic for the research communities of multi-agent systems, cognitive modeling, and decision making. Many research efforts have tried to use software-agents to model teamwork. However, we found little work has been done comparing different methods with common experiments. The Agent-based Modeling and Behavior Representation (AMBR) project has compared different cognitive process modeling in a military simulation environment (Gluck & Pew, 2001), but it has only examined single agent behavior so far. In teamwork modeling, different architectures or models are often evaluated with different domains, which are not comparable. For example, STEAM (a Shell for Teamwork) used a helicopter combat scenario as a simulator (Tambe, 1997), while CAST (Collaborative Agents for Simulating Teamwork) used a Wumpus grid-world as its test-bed (Yen, Yin, Ioerger, Miller, Xu, & Volz, 2001). Consequently, it is difficult to evaluate teamwork performance between two architectures.

As a first attempt to address the issue, we conducted an empirical study to compare team performance. First, we built agent teams with two different architectures: CAST and Soar. Then, we tested them within the dTank simulation (Councill, Morgan, & Ritter, 2004). dTank is a tank game simulator for agents in a distributed environment. In the experiments, we compared the agents' knowledge and behavior.

Teamwork Modeling

We take a Unified Theory of Cognition (UTC) view of an agent as being composed of two parts: architecture and knowledge. We use the term architecture to refer to mechanisms and structures that process content (knowledge) and generate behaviors.

CAST. CAST has been designed to study teamwork related issues emerging from teams with well-defined structure and process, distributed expertise, and limited communication in time-stress domains. It implements a teamwork model that enables agents to anticipate potential information needs among teammates and to exchange information proactively. For example, the precondition of the "attack-enemy" plan includes the location of enemy. By sharing the condition knowledge, an agent will proactively deliver the location of detected enemy to other agents. Messages from CAST agents are realized through JAVA RMI with a Knowledge Query Manipulation Language or KQML format (Finin, Fritzson, McKay, & McEntire, 1994) that defines both the content of the messages and performatives of the communication.

Soar. Soar is a general cognitive modeling architecture that is designed to model human cognition (Laird, Newell, & Rosenbloom, 1987). Soar does not include any particular functions for teamwork. However, there have been several models of teamwork created in Soar. For instance, Tambe (1997) and his research group developed STEAM—a teamwork model that is based on Soar and joint intention theory (Cohen & Levesque, 1991). Implementing teamwork models using Soar, such as STEAM and Team-Soar, require writing Soar rules to implement collaboration and communication. For example, STEAM includes about 300 domain-independent Soar rules. For the purpose of this study, we developed a Soar team, which contains 22 productions encoded as communication knowledge.

dTank Teams. For comparison, agents in the CAST and Soar teams have similar procedural and declarative domain knowledge. The procedural knowledge contains a two-phase plan: search and attack. A team of agents wander around and search for enemy tanks. Once an enemy is found, the team-members communicate to inform each other of the enemy location and to coordinate their attack. Next, team members attack the target together and destroy it. This process iterates until all the enemies are destroyed. Declarative domain knowledge includes moving directions and obstacle locations.

Procedure. Both the CAST and Soar teams compete against a group of simple agents. We tested each team with the same scenario for 20 times. Each scenario is created with a randomly generated map and a set number of enemies. Agents' actions are recorded for further analysis. The simple agents are also implemented in Soar, except that they have no teamwork but a simpler knowledge representation.

Results

In our study, team agents' behavior is compared with their sequence of actions. In Figure 1, example agents are compared side by side. We find similar patterns for team operations across both teams. For example, initially both teams coordinate with a "hello-team" message, followed by a search plan that starts with a "turn" operator. This similarity results from equivalent collaborative knowledge included in both teams. We also noticed some consistency among attack behavior. For example, the circle in Figure 1 indicates that agents from both teams have a similar aim-and-shoot behavior pattern.

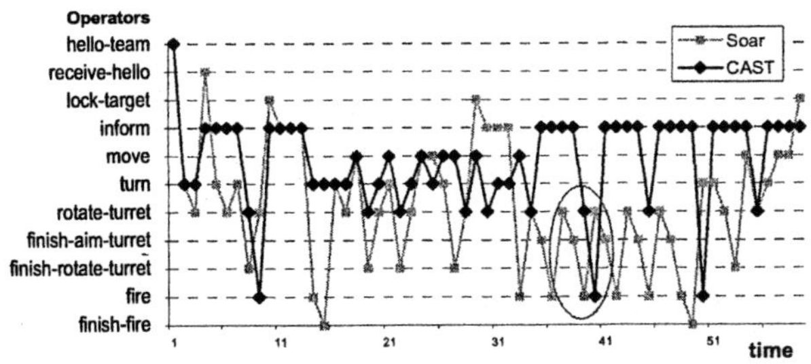

Figure 1: Behavior comparison by process model.

By comparing the frequency of each operator (Figure 2), we found that the most frequent domain non-communicative operator of both teams is rotate-turret. Other operators have comparable counts except the operators that are specially designed for Soar agents such as finish-fire. The most important and surprising difference is the communication operator, inform: a CAST agent communicates much more frequently than a Soar agent does. The CAST architecture is designed to communicate efficiently and a CAST agent should communicate less frequently. Why in this experiment do we see a contradictory result?

Actions	F2_Soar	F2_CAST
finish-aim-turret	22	0
finish-fire	5	0
finish-rottate-turret	12	0
fire	33	14
hello-team	1	1
inform	21	104
lock-target	5	0
move	11	14
receive-hello	1	0
rotate-turret	40	26
trace-target	6	0
turn	18	16

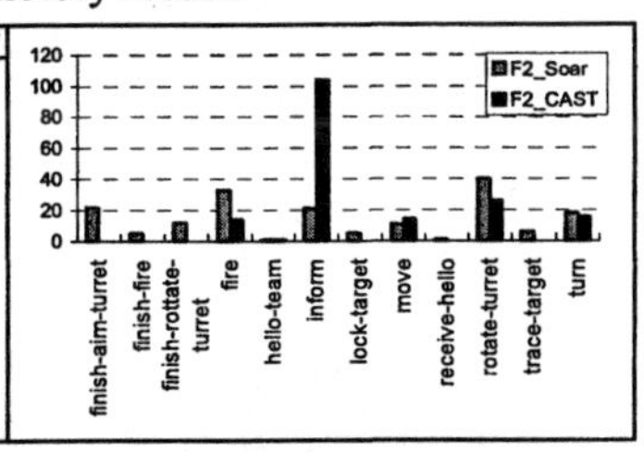

Figure 2: Frequency comparison of operators.

By examining the traces, we found the reason. A Soar agent needs to define the communication explicitly in productions. In this case, we defined the productions for communication after an agent locked onto a target. In contrast, the communication policies in CAST are embedded: whenever an agent observes a new piece of information, it sends the information to others who may need this information. Therefore, although implementing team communication for a CAST agent is easy, it sends more messages and it can be less efficient than ad-hoc designs that are less verbose. More importantly, the difference suggests that some team behavioral differences are difficult to resolve either with knowledge or with architecture alone.

Conclusions

As individual models become aggregated into teams of models, comparing architectures for teamwork modeling will become more important for cognitive engineers making decisions on choosing tools and implementing models. Through a set of experiments, we compared two architectures and their behaviors. Compared with Soar, CAST has more features that are designed specifically for modeling teams, and performed well where increased communication was useful.

Furthermore, we grouped the knowledge into domain-dependent knowledge and domain-independent knowledge. The domain independent knowledge is needed to compliment certain features from the architecture. In the Soar team, for example, knowledge about how to aim a gun at an enemy is domain-dependent; knowledge on how to compose messages is domain-independent. In CAST, communication is a part of the architecture. Therefore, the boundary between domain-independent knowledge and architecture can be blurred. Different architectures may be able to capture or use different domain-dependent knowledge. When we compare the knowledge coded for CAST and Soar, we find that Soar can incorporate more productions for making choice decisions.

Although the above findings are not conclusive, we have learned lessons on the relations between architecture and knowledge. (a) By including equivalent knowledge, a Soar team can perform collaborative behavior that is partially similar to a CAST team. (b) Some team behavioral differences are difficult to resolve with knowledge alone. (c) The experiments suggest that capturing teamwork behaviors as a part of the architecture or as a part of the agent's knowledge is important and perhaps equivalent decision choice for team modeling. (d) Human teams may vary on the teamwork behavior. Is it also affected by differences in knowledge? The question will motivate us to collect human data and compare with the models.

Acknowledgments

This research was supported by ONR (contract N000140210021), an AFOSR MURI grant (No F49620-00-1-0326), and a grant from Lockheed Martin Inc.

References

Cohen, P. R., & Levesque, H. J. (1991). Teamwork. *Nous, 25*(4), 487-512.

Councill, I. G., Morgan, G. P., & Ritter, F. E. (2004). *dTank: A Competitive Enviorment for Distributed Agents* (Technical Report ACS2004-1). University Park: The Pennsylvania State University. acs.ist.psu.edu/dTank.

Finin, T., Fritzson, R., McKay, D., & McEntire, R. (1994). KQML as an agent communication language. *In Proceedings of the Third International Conference on Information and Knowledge Management (CIKM'94).* 456-463. Gaithersburg, MD. ACM Press

Gluck, K. A., & Pew, R. W. (2001). Overview of the Agent-based Modeling and Behavior Representation (AMBR) Model Comparision Project. *In Proceedings of the 10th Computer Generated Forces and Behavioral Representation Conference.* 3-6. Orlando, FL.

Laird, J., Newell, A., & Rosenbloom, P. (1987). Soar: An architecture for general intelligence. *Artificial Intelligence, 33*(1), 1-64.

Tambe, M. (1997). Towards flexible teamwork. *Journal of Artificial Intelligence Research, 7*, 83-124.

Yen, J., Yin, J., Ioerger, T. R., Miller, M. S., Xu, D., & Volz, R. A. (2001). CAST: Collaborative Agents for Simulating Teamwork. *In Proceedings of the Seventeenth International Joint Conference on Artificial Intelligence (IJCAI-01).* 1135-1142. Seattle, WA.

Tekkotsu: Cognitive Robotics on the Sony AIBO

Ethan J. Tira-Thompson, Neil S. Halelamien,
Jordan J. Wales, and David S. Touretzky
{ejt,nsh,jjw,dst}@cs.cmu.edu
Computer Science Department, Carnegie Mellon University
Pittsburgh, PA 15213

Introduction

Tekkotsu (the name means "framework" in Japanese) is an application development framework for the Sony AIBO robot that endeavors to provide an intuitive set of primitives for perception, manipulation, and control. The framework also addresses some of the human-robot interaction problems faced by mobile robot application developers. See `www.Tekkotsu.org`. The contributions of this work lie in three areas:

1. Providing a set of intuitive, high level primitives for robotics that draw on insights from cognitive science.
2. Making robot programming accessible to a wider audience by teaching students to use these primitives to compose complex behaviors.
3. Offering a robotic platform that cognitive modeling systems could use to interface with the physical world.

Figure 1: AIBO surveying the game.

Perception

Tekkotsu provides a "dual coding" (iconic and symbolic) representation of images. The term comes from Paivio's "dual coding theory" of mental representations, which posits parallel verbal and non-verbal (i.e., imagistic) systems with extensive referential connections between them (Paivio, 1986). On this view, cognitive problem solving invokes transformations on items within these representational systems; more demanding tasks require translation between the two. In Tekkotsu, visual routines (Ullman, 1984) are provided for operating on pixelized iconic representations, extraction operators produce symbolic descriptions (lines and ellipses), and rendering operators convert from the symbolic form back to the iconic.

Figure 2: AIBO's view of the board.

Applications can use these primitives to parse the visual world. Our current touchstone task is getting AIBO to play tic-tac-toe on a real board, as shown in Figures 1-3. We use color image segmentation to locate the board and game pieces, and symbolic operations on these lines and ellipses to determine which board positions are occupied.

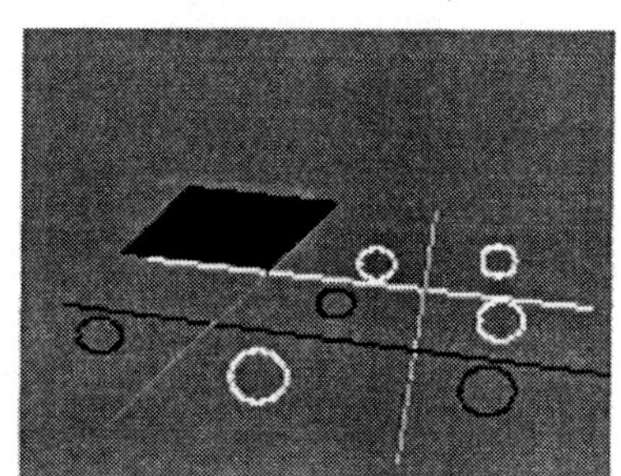

Figure 3: Visual parse of the board, with next move indicated by shading.

"Sketches" computed by visual routines, and symbolic objects (lines and ellipses) extracted from these sketches, are automatically organized into a derivation tree that can be viewed with a graphical browser running on a PC, allowing the programmer to directly examine the robot's "mental imagery". In addition, the robot can

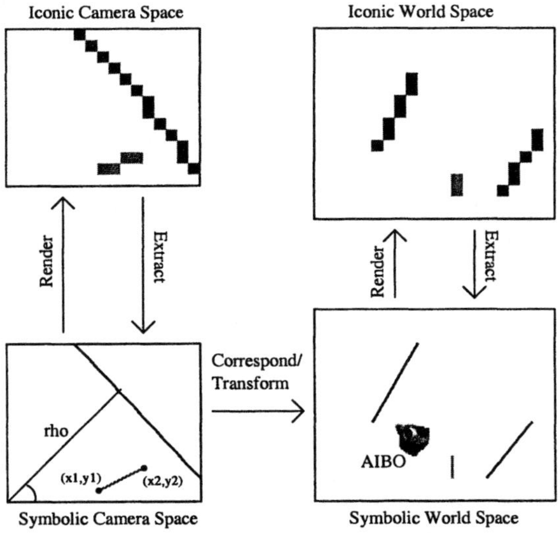

Figure 4: Updating the world map by matching objects in camera space against existing world map structures.

construct sketches specifically for display to the user, as in Figure 3(c) where it indicates its intended move.

Tekkotsu maintains a world map using the same dual coding strategy, as shown in Figure 4. The world map is a persistent allocentric representation built from successive egocentric camera images. Symbolic objects in camera space are matched against their counterparts in the world map, which is updated with each new image frame. The world map is also rendered iconically, allowing visual routines to operate on the map for path planning.

Manipulation

We are currently designing closed-loop manipulation primitives for visually guided pushing or dragging of small objects. Different types of motion are required to push an object along a path, drag or swipe an object away from a barrier, or apply pressure to an object such as a key on a computer keyboard.

A manipulation schema comprises the following elements: (i) point of contact on the object; (ii) type of contact (ballistic, steady pressure, rolling swipe); (iii) intended motion (along a path, perpendicular to surface, etc.); (iv) type of visual monitoring employed; and (v) motion constraints (okay or not okay to hit other objects, keep away from the wall, etc.) We anticipate this approach will make it intuitive for users to develop new behaviors for manipulating novel objects.

Control

Tekkotsu provides a state machine formalism for describing robot behaviors. State nodes and transition arcs are both implemented as computational processes. Communication takes place via an event-based mechanism. For example, a transition may subscibe to a particular stream of sensor events, and when such an event occurs, the transition will deactivate one state and activate another.

The state machine is hierarchical: a state may contain another state machine inside it. And transitions can "fork", allowing multiple states to become active at once. Tekkotsu does not presently incorporate any long term or working memory model, other than its world map representation. Such functionality could be provided by interfacing with widely used cognitive modeling systems such as SOAR or ACT-R. (Tekkotsu can communicate over wireless with processes running on a PC.)

Attention

A longer range goal for Tekkotsu is to integrate perception with manipulation and navigation. For example, it should be possible to construct a path by "visualizing" it on the world map, and then have the robot traverse that path. Similarly, when manipulating an object, parameters such as "point of contact" or "direction to move along" might be expressed as markers in a sketch produced by a visual routine.

A major issue that remains to be addressed is control of attention. The robot's camera has a limited field of view and the head has a limited range of motion. In order to monitor the object it is manipulating, verify its own location, check for nearby obstacles, and keep track of any humans it's interacting with, the robot must divide its attention and interleave a variety of head and body motions with the other actions it is performing. Intelligent scheduling of these motions will require an explicit representation of the robot's short term planned behavior, so that, for example, it does not stop and look backwards just before reaching a goal location where the entire body will need to turn around anyway. An attentional model that managed the robot's limited resources and arbitrated among competing demands would relieve application developers of a significant burden, and perhaps suggest approaches animals may be using in similar circumstances.

Acknowledgments

This work was funded in part by a grant from the Sony Corporation, and by an REU supplement award to NSF grant IIS-9978403.

References

Paivio, A (1986). *Mental representations: a dual-coding approach.* New York: Oxford University Press.

Ullman, S. (1984). Visual routines. *Cognition, 18*, 97–159.

www.tekkotsu.org

A Computational Model for Acquisition of Counterintuitive Concepts

M. Afzal Upal
University of Toledo,
Toledo, OH, 43606
afzal.upal@eng.utoledo.edu

Introduction and Background

In a seminal study, Bartlett (1932) asked British university students to read a culturally unfamiliar Native North American folk tale called "the war of the ghosts" and retell it to others in writing who in turn passed it on to others. Over several generations of retellings culturally unfamiliar concepts were distorted and replaced by more familiar concepts; for instance, a canoe was replaced by a rowboat. Bartlett reasoned that unfamiliar concepts such as canoe are more difficult to represent and thus less likely to be represented and transmitted than familiar concepts. However, recent studies following up on Bartlett's work (Boyer and Ramble 2001, Barrett and Nyhof 2001, Atran 2003) have drawn the seemingly opposite conclusion, namely that under some conditions, counterintuitive concepts (i.e., concepts that violate intuitive expectations) are actually better recalled than relatively more intuitive concepts. Barrett and Nyhof (2001) asked their subjects to remember and retell three of six different Native American folktales containing an equal number of intuitive and counterintuitive concepts. A content analysis of what they remembered showed that subjects remembered significantly larger number of counterintuitive concepts than intuitive concepts. In subsequent experiments, Barrett and Nyhof (2001) and Boyer and Ramble (2001) constructed a number of stories containing an equal number of three different types of concepts: the minimally counterintuitive concepts (i.e., concepts that violate intuitive expectations regarding one or two feature values) such as "furniture that flies in the air", ordinary concepts such as "furniture that stays where you put it", and maximally counterintuitive concepts (i.e., concepts that violate intuitive expectations regarding multiple feature values) such as "furniture that flies in the air, melts when try to grab it, and is made of uranium." When subjects from a variety of cultural backgrounds were asked to recall the stories containing an equal number of three types of concepts, they recalled a significantly larger number of minimally counterintuitive concepts than both ordinary and maximally counterintuitive concepts. The results regarding better recall for minimally counterintuitive concepts visa-vi ordinary concepts are in accordance with a host of studies showing that unusual or distinctive stimuli are generally better remembered than stimuli that are not (see Waddill and McDaniel 1998 for a review). A number of factors have been proposed to account for these effects including:

- more attention being paid to the unusual stimuli resulting in richer spontaneous elaboration and encoding for them and less attention and time being spent on processing the usual stimuli,
- unusual stimuli being encoded in their own category, different from the one into which the common stimuli are clustered,
- unusual stimuli being distinctive in the retrieval set that is formed.
- Boyer and Barrett have argued that it is the economy of representation as simple negation of a feature value and the ability to invoke the counterintuitive concept to make predictions because of its set of non-violated feature values that makes minimally counterintuitive concepts easier to recall than ordinary concepts. They argue that maximally counterintuitive concepts are harder to recall because they have little "inferential potential" as they violate too many intuitive expectations.

There are two major problems with the previous explanations. First, the proposed mechanisms are little more than descriptive accounts that do not really explain as to why people use these concept acquisition and memory mechanisms as opposed to using different concept acquisition and memory mechanisms. Second, they do not explain the reasons for the contradictions among findings of various studies including Bartlett's original study and more recent work that seems to confirm findings regarding better recall for ordinary concepts (Atran 2003). Instead of embedding concepts in the structure of a story, Norenzayan and Atran (2003) simply asked their subjects to remember and recall an equal number of intuitive, minimally counterintuitive, and maximally counterintuitive concepts. Unlike Barrett *et al.* and Boyer *et al.*'s studies, Norenzayan *et al.* observed that intuitive concepts were better recalled than minimally counterintuitive concepts which were better recalled than maximally counterintuitive concepts.

This paper proposes an opportunistic (Francis 1995) concept learning framework that can account for the seemingly contradictory findings. We are currently working to implement the proposed technique in a machine learning system called CICL (CounterIntuitive Concept Learner).

Knowledge-driven Concept Learning

The idea that concept learning is strongly influenced by theoretical and causal knowledge that people possess is not new (Murphy & Medin 1985, Kunda, Miller, & Claire 1990, Rehder 1999). When exposed to a new object belonging to a novel concept, people use their prior knowledge about similar concepts to generate expectations about the new concept. If the new concept is grossly incompatible with their expectations, it is relatively more difficult to acquire than a concept that is in accordance with expectations. Murphy and Wisniewski (1994) reported that when features of a concept are consistent with prior knowledge people found categories easier to learn compared to situations in which features do not fit in with prior knowledge. Pazzani (1991) and Wattenmaker (1995) found that people learned a concept in fewer trials if its definition was congruent with prior knowledge. Heit (1998) found that when subjects were allowed to self pace during a training phase in which they were presented with intuitive concepts (such as "shy person who rarely attends parties") and expectation violating concepts (such as "shy person who attends parties often"), they spent significantly more time processing expectation violating concepts than the time spent processing expectation compliant concepts.

There is also evidence that part of the extra time taken to process expectation violating concepts is spent on trying to find answers to the questions raised by the expectation violations. When Kunda *et al.* (1990) presented subjects with expectation violating concepts such as a blind marathon runner, Harvard educated carpenter, and feminist bank teller they attempted to answer questions such as how could a Harvard educated person become a carpenter? using their prior knowledge about Harvard educated people, about carpenters, and about people in general. Kunda *et al* (1990) report that subjects "had no trouble combining sometimes incongruous social categories. There were no cases in which subjects reported difficulty or failed to complete the task... some subjects spontaneously provided causal narratives that seemed designed to address the question of how a member of one category acquired membership in the other."

An Opportunistic Model of Concept Forgetting

An agent with finite memory resources cannot possibly remember every object it perceives (Altman & Gray 2002). An opportunistic agent should prefer to forget those concepts that require less cognitive processing to acquire. For a rational agent whose goal is to increase its knowledge of the environment to improve its problem solving performance, it makes sense to forget objects that are expectation compliant at a higher rate because they require minimal effort to predict. Similarly, for an agent living in a world full of uncertainty, it makes sense to preferably forget bizarre objects whose occurrence it cannot explain.

We hypothesize that human memory forgets concepts at different rates depending on how much cognitive effort was required to acquire those concepts. We believe that this differential rate of forgetting accounts for the differences between recall rates of ordinary concepts, recall rates of minimally counterintuitive concepts, and recall rates of maximally counterintuitive. The concepts that do not violate any expectations require minimal cognitive effort and hence are most easily forgotten. When presented with minimally counterintuitive concepts people attempt to answer questions such as "why would someone design a piece of furniture that can fly?" What question people ask depends on the concept. For instance, in case of artifact concepts the crucial question is the functionality of the artifact leading to the question why would someone design an artifact with these properties. People learn new concepts as a result of being able to answer such questions. These newly learned concepts are ranked lower for forgetting because of the cognitive effort spent in learning them. The maximally counterintuitive concepts require more cognitive processing than ordinary concepts because they violate expectations triggering the explanation seeking mechanism but the search is quickly abandoned when no explanation can be formed.

Our hypothesis, if true, would also explain the results obtained by Norenzayan and Atran (2003) that show that ordinary concepts are recalled better than counterintuitive concepts when presented without the structure of a story. The prior knowledge that a person possesses impacts the creative explanation formation process. For instance, when Heit (2001) provided subjects with verbal arguments such as "shy people may also go to parties in an attempt to overcome their shyness and to meet new people" the impact of expectation violation disappeared. We hypothesize that stories in conjunction with people's common knowledge about stories act as verbal arguments by providing a context that explains the occurrence of counterintuitive concepts such as "a sobbing oak." When stripped of this context, as in the experiments performed by Norenzayan and Atran (2003) the minimally counterintuitive concepts become maximally counterintuitive concepts and can no longer be justified. However, such concepts can be justified in the context of stories which leads to better recall for them in that context.

Conclusion

Previous models of memory and concept learning are unable to explain differences between recall rates of moderately expectation violating concepts, the recall rates of ordinary concepts, and the recall rates of maximally expectation violating concepts in various situations. This paper has presented a model based on Bayesian belief networks that explains the seemingly contradictory results of the free recall experiments conducted with human subjects. Our theory also gives rise to a number of testable predictions. Plans are in the works to conduct experiments with human subjects to confirm implications of our theory. We are also readying CICL, a computer system designed using the concept learning and memory system, for public release.

Acknowledgement: Thanks to Ryan Tweney, Jason Slone, Nicholas Jutte, and David Mullen for guidance and suggestions.

References

Atran, S. (2003), *In Gods We Trust*, Oxford Univ Press.

Barrett, J. & Nyhof, M. (2001). Spreading non-natural concepts. *Journal of Cognition and Culture*, 1, 69-100.

Boyer, P. & Ramble, C.. (2001). Cognitive templates for religious concepts. *Cognitive Science*, 25, 535-564.

Murphy, G. (2002) The Big Book of Concepts. MIT Press.

Waddill, P. J. & McDaniel, M. A. (1998) Distinctiveness effects in recall, *Memory and Cognition*, 26, 108-120.

State Definition in the Tetris Task: Designing a Hybrid Model of Cognition

V. Daniel Veksler (vekslv@rpi.edu)
Wayne D. Gray (grayw@rpi.edu)
Cognitive Science Department
Rensselaer Polytechnic Institute

In defining state/action pairs for reinforcement learning of the Tetris task, we seek to recognize known game states, as well as to learn new ones according to relevant game features. We propose a model of cognition that uses categorization as a mechanism to sort such features into appropriate state-types, and an attention mechanism based on predicted values of each state as a method for deciding which states/features are relevant.

Reinforcement Learning, Categorization, and Spreading Activation

Models of reinforcement learning (RL) have proven to be successful in learning and predicting values of actions given a particular state of the world (Sutton & Barto, 1998). Models of categorization, on the other hand, have been shown to learn and recognize world states with great precision (Medin & Schaffer, 1978). Finally, models of spreading activation, such as ACT-R, have had the most success in matching human memory recall time and accuracy, as well as explaining human priming capabilities (Anderson & Lebiere, 1998).

We have developed a theory for a general model of cognition based on these three mechanisms – CB-Mineral. CB-Mineral (CB) is a Categorization-Based Memory Network with Reinforcement Learning. It is a specialized neural network in which the number of hidden-layer nodes and their interconnections are not predefined, nor are connection weights trained using supervised learning. Rather, memory nodes, connections, and connection weights are incrementally added, consistent with categorization and reinforcement-learning rules.

The State Recognition Problem

Complex dynamic environments, such as Tetris, pose problems for computational modeling. In pairing action sets with system states, the difficulty for production systems, as well as for stand-alone RL models, is in learning new system states (in production systems like ACT-R, the left-hand side, "If X", of any production, "If X, Do Y", is the qualitative equivalent of State in a State/Action pair). It would be both awkward and cognitively implausible for a modeler to hardcode every possible qualitatively unique world-state for a dynamic environment, even one as simple as Tetris.

Categorization models and various connectionist approaches do very well at mapping abstract features from the environment to distinct categories. However, even our hybrid model that uses such methods to establish the state of the world prior to using RL still requires some sort of a heuristic mechanism to decide when categories are created.

CB Implementation

The CB memory network starts its lifecycle with just the perception and the action nodes. Perception nodes are assigned to represent individual contours and locations in the world, as well as pleasure and pain. Action nodes are assigned to various keystrokes and internal commands.

At every tic, the perception module sends signals to the perception nodes, which then activate their subclass nodes. The activation spreads according to connection weights down to the action nodes, which act directly upon the world.

Categorization

Currently, all highly activated objects are considered to be in Working Memory. CB creates a new memory node for every distinct set of objects in WM. This memory node becomes a subclass to each of the objects in WM (Ex: {—} + {|} = {—,|}). If two or more objects in WM share common features, a superclass object is formed for these objects (Ex: {■, location 1, moving down} ∧ {■, location 2, moving down} = {■, moving down}). Each object can have superclasses and subclasses. For example, each Tetris piece (tetronimo) is a subclass to its contours, and is a superclass to its rotation states and falling patterns.

Action Values, State Values, and Connection Strengths

Unlike the traditional RL algorithms that only have a single reward value regardless of whether it is negative or positive, CB records both pain and pleasure values, similar to the way ACT-R records successes and failures.

The value of an action depends on the values of the states that follow that action and the strengths of connections to those states. A given state's value (love/fear), in turn, is based on its connection strength to the pleasure and pain nodes.

The strength of connection between any two memory nodes is determined by the strength of activation of the two nodes at the time when both are in WM at the same time.

Current Work

It has become apparent that not all highly activated memory nodes should be placed in WM. We are currently looking to construct a separate WM buffer that would serve as the blackboard of the mind. The memory elements placed in this buffer would be only the most relevant (according to some heuristic other than activation level) of the highly activated objects.

We refer to Attention as the mechanism responsible for deciding which of the activated memory objects are sent to WM. CB would attend only the most loved and the most feared of the activated memory elements. This would drive the model to reflect upon and learn about only the most relevant task features.

Learning Tetris States

At the start of the game, CB would only "see" a blank board, the contours of the border, and the first falling tetronimo (see Figure 1). Each piece of the contour activates a perception node pre-assigned to detect that particular shape at that particular location, as well as a perception node for that location, and one for that shape.

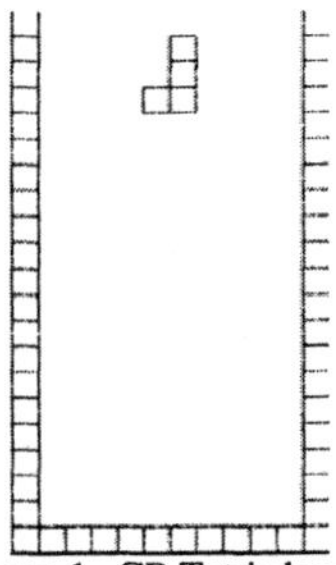

Figure 1. CB Tetris board.

The subclass created out of the co-activated contours makes up what we can semantically translate as that tetronimo (in that rotation state). As the tetronimo moves down one notch, the new combination of active perception nodes is recorded as another object, and so forth. The multiple instances of the tetronimo become the superclasses of an object that may be semantically translated as the "tetronimo falling pattern".

As this pattern is observed multiple times, all the superclass-subclass connections in the memory network are strengthened. With enough connection strength, each "tetronimo" memory object will be activated with just a few distinctive features, and that, in turn, will activate the "tetronimo falling pattern" object. With the creation of the WM buffer, each such object will be pushed into WM for reflection, and all its superclasses – the instances of the falling pattern – will be activated, as well.

Imagination and Planning

The ability to "imagine" future positions of a pattern, as described above, may be the key to planning and decision-making. The model could then keep the bottom of the Tetris board in WM, while mentally going through the various possible states of the current tetronimo in search of a combined state with highly valued responses (as per the rules of RL). The opposite planning pattern may occur as well – the model could also keep the tetronimo in WM, while going through possible states of the board. These two search patterns would be the decision searches for CB in the Tetris task.

Data

Both of the above CB planning patterns were observed in an informal analysis of human eye data during the Tetris task. Collected eye-gaze data suggests that after every newly observed state, subjects quickly search for a complementary object.

After gazing at a new pattern at the bottom of the board, subjects quickly gaze at the top piece to see if it fit into the last observed "hole" in that pattern. Subjects also searched through the bottom of the board for an appropriate "hole" immediately after observing the new tetronimo at the top, with quicker response times when the "hole" matched the tetronimo. We are still looking to do more rigorous analysis of the collected data prior to forming any conclusions.

Summary

In combining the state recognition capabilities of categorization with state prediction of spreading activation and trial-and-error learning of RL, we design the CB-Mineral memory network. The Working Memory buffer and the Attention mechanism are currently being implemented to complete the architecture. Upon completion, CB will be capable of exploring and mastering Tetris as a sample complex dynamic environment. A preliminary look at human eye data suggests a qualitative match to the predicted cognitive flow of the model, leaving the doors open for future research.

References

Anderson, J., & Lebiere, C. (1998). *The atomic components of thought.* Hillsdale, NJ: Lawrence Erlbaum Associates.

Medin, D. L., & Schaffer, M. M. (1978). Context theory of classification learning. *Psychological Review, 85,* 207-238.

Sutton, R. S., & Barto, A. G. (1998). *Reinforcement Learning: An Introduction.* Cambridge, Massachusetts: The MIT Press.

Modeling Visual Behavior of Pilots in the Human Performance Model – Air MIDAS

Savita Verma (saverma@mail.arc.nasa.gov)
NASA Ames Research Center, Moffett Field CA 94035

Kevin Corker (Kevin.Corker@sjsu.edu)
San Jose State University, San Jose, CA 95192

Amit Jadhav (amit_jadhav3@hotmail.com)
San Jose State University, San Jose, CA 95192

Introduction

Increases in automation change the operator's information seeking behavior. New technologies involve important issues of information selection, information integration requirements, and design of operations and interfaces. Prediction of a pilot's control behavior under varying visibility conditions and with new automation, like synthetic vision systems (SVS), can be improved if their visual behavior can be modeled. The role of a pilot in the glass cockpit is that of a supervisor with the main task of monitoring the health of the avionics in an aircraft.

The visual scan pattern is intrinsic to the task of monitoring the instruments in the glass cockpit. Various researchers (Bellenkes et al., 1997) have studied the visual scan pattern of pilots, and concluded that there are differences in the information seeking pattern of novices and experts, and there are also differences is the pattern based on phase of flight (Mumaw et. al, 2000). Pilots have a mental model they use for seeking information in the glass cockpit, that involves T-Scan pattern and cross checking different instruments.

Human Performance Model

The visual behavior of pilots was modeled in the human performance model – Air MIDAS (Man-Machine Integrated Design and Analysis System). MIDAS has agent based architecture and is a closed loop model. The model has been briefly described here, for a detailed description see Corker & Smith (1993). The main components of the model are comprised of the virtual operator's world representation, and a symbolic operator model (SOM) that represents perceptual and cognitive activities of an operator. An important element of the SOM is the Updateable World Representation (UWR). The world representation information (environment, crew-station, vehicle, physical constraints, and the terrain database) is passed through the perceptual and attention processes of the SOM to the UWR. The world information is a complex environmental representation that is created by the researcher or programmer and serves to trigger activities in the virtual operator. The UWR represents the operator's cognitive constraints on procedural completion – it contains the Working Memory (WM), domain knowledge, and required procedural activity structure. The UWR passes information to a scheduler within the SOM that determines the resources available for the completion of the activity. The scheduler views WM and the measures contained within it as a capacity-limited resource. A four-channel activity loading mechanism (Visual, Auditory, Cognitive, and Psychomotor) is representative of the measures contained within WM and these activity load factors are used as constraints on the scheduling process. The scheduler controls the flow of UWR into and out of WM based on its knowledge of activities to be performed, ensuring that the number of nodes in WM at any given time does not exceed the WM node capacity. This cognitive structure interacts with the physical constraints on a virtual operator's performance.

Empirical research data was used to create an augmented visual model within the perceptual process of the SOM. The visual scan pattern data (Mumaw et al., 2000) was used to calibrate the simulation's input data. The researchers (Mumaw et al., 2000) studied the scan patterns of pilots during different flight phases and collected data on dwell durations and dwell percentages.

The Scan Pattern Model

Two forms of information seeking were modeled, the first one was a fixed sequential fixation pattern that scans the instruments and Out The Window (OTW) that pilots are trained to do. The other form of information seeking was directed-fixation, which is coupled with a specific goal or activity, e.g. look at gears when changing them. In both the forms of information seeking; the instruments, their display values, dwell durations, and percentages were a part of the database that resides in the equipment agent. The dwell percentages were used to select the instrument that the agent fixated on. The duration of dwell was calculated in Monte Carlo fashion on the mean and SD dwell durations specified by Mumaw et al. (2000) for that instrument. It was also assumed that every fixation did not result in updation of the UWR. The theoretical basis of modeling the update of UWR was the speed-accuracy function i.e. longer dwell durations resulted in an update of the UWR. If the estimated dwell duration was shorter than the mean duration + _ SD (high speed), the information from that display was not imbibed (low accuracy), and the agent did not update its information. For sake of simplicity, "all or none" information intake was assumed that formed the basis for modeling the speed-accuracy function. If the changing values on the display in the flight deck were correctly perceived by the agent, it triggered activities that were based on those values such as callouts, flap extensions and the like. See Figure 1 for the augmented visual model in Air MIDAS.

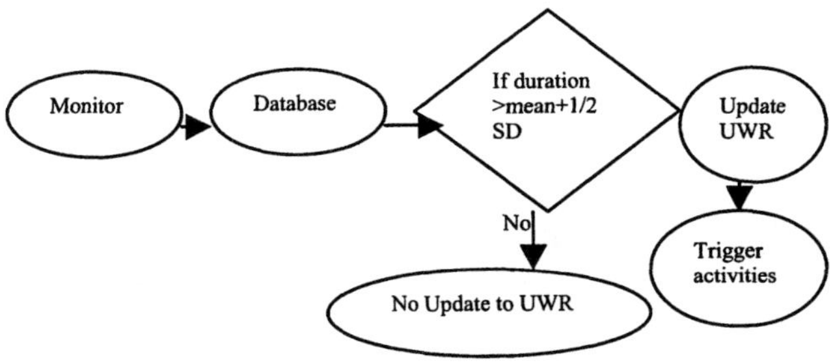

Figure 1: Pictorial representation of the augmented visual model

The model was adapted and data collected for three conditions- nominal approach without SVS, Nominal approach with SVS and Sidestep maneuver with SVS. NASA Human Performance Modeling (HPM) Element (2000) also collected data for the same scenarios using part-task simulation, where the flight deck had SVS superimposed on Primary Flight Display (PFD), and was simulated on a PC. Eye tracking data was collected for the pilots flying in the simulator. Since SVS as an instrument was not a part of the Mumaw (2000) data, the dwell durations and dwell percentages on OTW were used for SVS scan data. Some of the results have been detailed in the following section.

Validation of model

The correlation between the NASA HPM (2000) part-task simulation and the Air MIDAS for percent of fixations data were generally high- normal approach without SVS ($\underline{r}$ = 0.7608), nominal with SVS ($\underline{r}$ = 0. 8782), and sidestep with SVS ($\underline{r}$ = 0. 5538). Only normal approach with SVS is discussed here, for a detailed description of all data see Corker et al. (2002). In the normal approach with SVS, the Air MIDAS model predicted lower dwells on the Navigation Display, OTW and the SVS

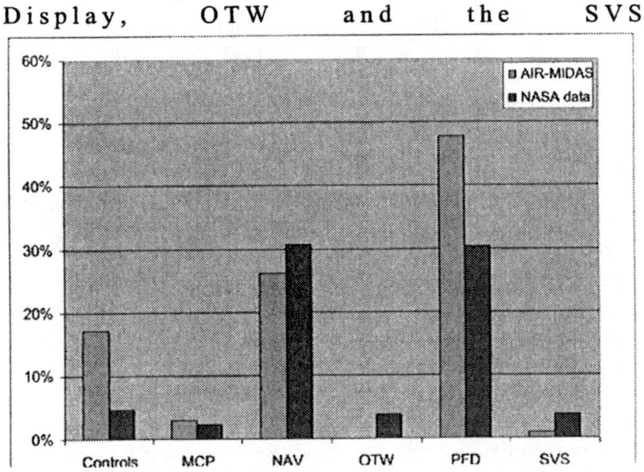

Figure 2: Comparison of fixation percentage between NASA HPM data and model (Air-MIDAS) data for Normal Approach with SVS

displays than did the NASA HPM (2002) simulation. This suggests that when flying with the SVS display, the NASA HPM (2002) flight crews looked at the SVS information to a greater extent than did the model. In short summary, the human flight crew received PFD information from overlays in the SVS, whereas the Air MIDAS model required looking at the PFD (not the SVS) for the same information. It is clear from Figure 2 that while the MIDAS agent looked at PFD about 50% of times, the human operator looked at the PFD and SVS about 35% of time. Thus one can presume that the human pilot was looking for the same information -attitude, altitude, airspeed on the SVS display that was superimposed on PFD, reducing the need to look at PFD separately.

Conclusion

Empirical results on visual scan pattern for pilots in the approach phase of flight were used to model the visual scan pattern in Air MIDAS. The simulation runs were used to collect data in three different scenarios. The results of the simulation were compared against a separate part task simulation done at NASA Ames Research Center. The data correlated well for most of the scenarios, it was relatively low for the SVS scenarios because the SVS technology was not completely modeled. In general, the correlations proved the validity of the model.

References

Bellenkes, A.H., Wickens, C.D., & Kramer, A.F. (1997). Visual Scanning and pilot expertise: The role of attentional flexibility and mental model development. *Aviation, Space, and Environmental Medicine,* 68, 569-579.

Corker, K. & Smith, B. (1993). An architecture model for cognitive engineering simulation analysis: Application to advanced aviation analysis. *AIAA conference on Computing in Aerospace*, San Diego, CA.

Corker, K., Gore, B.F., Guneratne, E., Jadhav, A. & Verma, S. (2002). *Integration of Air MIDAS Human Visual Model Requirement and Validation of Human Performance Model for Assessment of Safety Risk Reduction through the implementation of SVS technologies.* NASA Contract Task Order #: NCC2-1307.

Mumaw, R., Sarter, N., Wickens, C., Kimball, S., Nikolic, M., Marsh, R., Xu, W., & Xu, X. (2000). *Analysis of pilot monitoring and performance on highly automated flight decks* (NASA Final Project Report: NAS2-99074). Moffett Field, CA: NASA Ames Research Center.

NASA Human Performance Modeling Element (2002). *HPM-SVS part-task simulation documentation: Characterizing pilot performance during approach and landing with and without visual aiding.* Moffett Field, CA: NASA Ames Research Center.

Acknowledgements

This research was funded by the NASA Aviation Safety Program and the authors are grateful for the opportunity.

The Environment as Theory: An Example Using the ACT-R/SOS Environment

Robert L. West (robert_west@carleton.ca)
Depts. of Cognitive Science and Psychology,
Carleton University, Ottawa, Ontario Canada

Bruno Emond (bruno.emond@nrc-cnrc.gc.ca)
Institute for Information technology
National Research Council Canada, Ottawa, Canada

One of the most useful and clarifying ideas in cognitive modeling is Anderson's (1993) schema for understanding cognitive models in terms of *frameworks*, *theories* and *models*. In this paper we propose that this schema is equally valuable for describing the environment that a cognitive model acts within. Physics provides us with the correct models for representing the physical environment. However, the interaction between cognition and the physical environment is mediated by transduction processes, which are not fully understood. Thus you cannot simply slap a physics model onto a cognitive model. One way of dealing with this is to assume a transduction model that delivers what the cognitive model needs. Another approach is to drop physics and transduction completely and model the environment as it appears to the cognitive system after transduction.

This is the approach we took in creating SOS version 1.0 (Emond, West, 2003; West, Emond 2002; http://actr-sos.sourceforge.net/). SOS, which stands for Simple Object System, is a modeling system for creating low fidelity environments for ACT-R models to act on. It is meant to compliment ACT-R PM (Byrne & Anderson, 1998), which is a higher fidelity environment system for ACT-R, containing a much more detailed model of how we interact with the environment. The advantage of SOS is that it is easy to learn, quick to use, and very flexible (assuming you understand ACT-R).

SOS v 1.0 allows the modeler to build a chunked representation of the environment that functions according to its own rules, built in by the modeler. Changes occurring in the environment are represented the same way they would be in the ACT-R declarative memory system. That is, by making new chunks available or changing the slot values of existing chunks (new chunks are not actually created but toggled from visible to not visible). Both the perceptual and semantic properties of objects are encoded as slot values. SOS objects also contain a special slot for actions. The action slot specifies a target (i.e., another chunk representing some part of the environment) and an action (i.e., how to change the chunk structure of the target). It is also possible to insert lisp functions into the action sequence to create more complex effects. Thus SOS models how the environment works, but represents the environment as how we see it rather than how it is, physically.

SOS fits well into Anderson's (1993) schema of frameworks, theories, and models. SOS is based on the framework assumption that the end function of transduction is to create a mental model of the physical world in a format acceptable to declarative memory. The SOS code represents a computational theory about how the physical world is coded and how, once coded, it interacts with cognition. SOS assumes that cognition is described by ACT-R and also assumes that the coding of the physical environment results in a chunked representation that is highly similar to the way ACT-R chunks information in declarative memory. Consistent with the current ACT-R architecture, SOS interacts with ACT-R through the ACT-R buffer system. Also, although SOS does not embody a specific model of attention, because SOS focuses on objects it lends itself somewhat more naturally to creating models that use some form of object-orientated attention.

However, just as ACT-R does not specify how specific types of declarative memory content should be organized, SOS does not specify how specific environments should be organized. This must be specified as an SOS model. The important point in creating an SOS model is that chunking can occur based on abstract properties, as well as physical properties. This can be illustrated with a simple SOS model we built called SUE (simulated user environment). SUE is meant to model knowledge driven navigation in graphical computer interfaces such as Windows or OS X. To do this it was necessary to have chunks representing graphical elements such as icons, menu bars, dialogue boxes, etc. However, at a higher level, all objects in the environment are considered as either objects or containers. For example, an icon would be an object and an icon bar would be a container. This abstract structure is central to the process of navigating, modeled using ACT-R, but does not occur in the ACT-R model. Instead it is *delivered* from the environment through the transduction processes.

The SOS architecture makes this distinction very clear. All chunking that occurs in SOS occurs outside of cognition, in the perceptual systems. Note that this implies that the perceptual systems are capable of a certain level of semantic coding (e.g., seeing a bucket as a container). Being clear about this distinction, whatever modeling system you are using, is important because meaningful structure exists in the environment and is also added by the transduction process. What a cognitive model needs to do in order to complete a task is heavily dependent on the type of structure it is presented with. For example, the abstract structure presented by SUE allows the cognitive model to navigate much faster than it could if only low-level visual information were available. Problem solving can also be facilitated by structure from the environment and transduction (e.g., see Chalmers, French, & Hofstadter, 1995). Thus it is important to be clear about what is being coded where. We suggest that treating environmental simulations as models and using the frameworks, theories, models schema is a good way to do this.

References

Anderson, J. R. (1993). *Rules of the Mind.* Hillsdale, NJ: Lawrence Erlbaum Associates.

Byrne, M. D. & Anderson, J. R. (1998). Perception and action. In Anderson, J. R. & Lebiere, C. (Eds.), *The atomic components of thought.* Mahwah NJ: Lawrence Erlbaum.

Chalmers, D., French, R., & Hofstadter, D. (1995). High level perception, representation, and analogy. In Hofstadter, D. (Ed.) *Fluid concepts and creative analogies: computer models of the fundamental mechanisms of thought.*

Emond, B., West, R.L. (2003). Exploring the usability of adaptive menus with a simple object system. Eight Annual ACT-R Workshop, Pittsburg, PA : Department of Psychology, Carnegie Mellon University.

West, R.L, Emond B. (2002). SOS: A simple operating system for modeling HCI with ACT-R. Seventh Annual ACT-R Workshop). Pittsburg, PA : Department of Psychology, Carnegie Mellon University.

The Attentional Blink at 20 items/sec, Model Prediction and Empirical Validation of Lag-2 Sparing

Bradley P. Wyble (bw5@kent.ac.uk)
Centre for Cognitive Neuroscience and Cognitive Systems, University of Kent
Canterbury, UK CT2 7NF

Howard Bowman (h.bowman@kent.ac.uk)
Centre for Cognitive Neuroscience and Cognitive Systems, University of Kent
Canterbury, UK CT2 7NF

Introduction

The Attentional Blink (AB) paradigm uses a Rapid Serial Visual Presentation (RSVP) stream, with two targets (denoted T1 and T2). There is a period of approximately 500 msec during which processing of T1 seems to impair the ability to detect and report T2 (Raymond, Shapiro and Arnell 1992). This suggests that the deployment of attention to processing T1 has a temporal window of a little over half a second. This interpretation is complicated by *lag 1 sparing*, which is the robust finding of almost unimpaired performance on T2 when it immediately follows T1.

Early theories of the AB have posited that the blink is the result of interference between the T1, T1+1, T2 and T2+1 items (Raymond, Shapiro and Arnell 1992). In this formulation, the existence of distractors is key to producing the particular U shape of the AB curve. Specifically, it is the T1+1 distractor that reduces the accuracy of a lag-2 T2 by entering the visual buffer along with the T1.

Another theoretical account posits the AB as the result of the interaction of two stages of processing (Chun and Potter 1995). Consolidation of the T1 in the second stage prevents the processing of T2. T1+1 items are important in causing the blink in that they make T1 processing more difficult

We have implemented a model of the AB, which combines this two-stage architecture with a token-based account of working memory (Kanwisher, 1987). This model makes explicit the prediction that it is the temporal characteristics of the token binding process that determines the shape and time course of the blink, and not the sequential arrangement of targets. Specifically, our model predicts that the shape of the blink curve will be constant with respect to time, such that at a presentation rate of 20 items/sec, the maximal depth of the blink will be obtained at lags 4-6, recovery will occur by lags 10-12, and there will be strong lag-2 sparing. This paper will then describe an experiment that confirms these results empirically. These results are generally incompatible with the intereference theory of Raymond, Shapiro, and Arnell (1992).

Modelling the AB

A complete description of this model is beyond the scope of this submission format. Accordingly, this text will focus on the general architecture of the model. A full specification can be found online at http://www.cs.kent.ac.uk/projects/cncs/online/bw5/iccm2004/.

This model uses a continuous firing rate representation of neural units, arranged in a series of layers representing a series of processing steps from early visual traces, through semantic processing and eventually leading to consolidation in working memory through recurrent sustained activation of working memory tokens. All units are connected by fixed inhibitory and excitatory connections.

Our implementation is strongly influence by the two-stage model of Chun (1997) in which the first stage performs low level visual and semantic processing of the RSVP stream. This stage can hold multiple items simultaneously, each in a localist fashion. However, lateral and feedforward inhibition effects cause weak activation traces for masked items, which is the case for all items in an RSVP stream. This feature sets the stage for T1+1 blanks attenuating the blink (Chun and Potter 1995), which is not described here. Representations in this first stage are akin to *Types* in the *Types-Tokens* distinction of Kanwisher (1987)

The second stage implements a *token* system that binds working memory *tokens* to the *type* information in the first stage. This system is capacity limited in that it can only bind one *token* at a time. In order to protect the integrity of this binding process, a transient attentional mechanism is deactivated during token binding. The unavailability of this resource makes it difficult for a T2 to reach an activation level sufficient to undergo tokenization at a relative lag of 200-500 msec following the T1.

Critically, in this model, it is the temporal dynamics of the *token* binding process that determine the shape of the blink. If the T2 follows closely enough after the T1, it receives a strong benefit from the transient attentional episode triggered by the T1, and is capable of being included in the T1's *token* binding process. The time course of this attentional transient is relatively unaffected by either the strength of the targets, or the number of distractors, that intervene between T1 and T2. Consequently, this model predicts that at a much faster presentation rate, lag 2 sparing should be strongly evident.

The duration of the blink is also independent of the number of items, being instead a function of the time between T1 and T2. The presentation of input to the model assumes that some letters are more visible through the stream of distractors than others, and this is modeled by presenting inputs at varying strengths to the input layer. The result of this variance is such that weaker traces caused by the faster presentation rate affect the depth of the blink and the baseline performance, rather than the duration.

In this work, this model was presented with streams at 10 and 20 items/sec (100 and 50 msec SOA's). The task is similar to that of Chun and Potter (1995), in which targets are categorically distinct from distractors (e.g. letters in a digit stream). For the 20 item/sec presentation rate, it was necessary to increase the strength of bottom-up connections so that higher layers were excited above threshold by the 50 msec stimuli. This modification represents the presumed adaptation of low level feature detectors to the more rapid presentation rate.

Figure 1 illustrates the output of the model in the 10 and 20 item/sec rates. Note that for the faster rate, the leftmost position in the graph represents lag-2, the next lag-4, etc up to lag 16. For the slow rate, data points represent lags 1-8. Criticially, in this model, distractors serve primarily as low level feature masks, as supported by recent data from Maki et al (2003). Distractors are capable of evoking traces in the semantic layer of the network, but these traces are suppressed by a task demand system prior to entry into the token binding system.

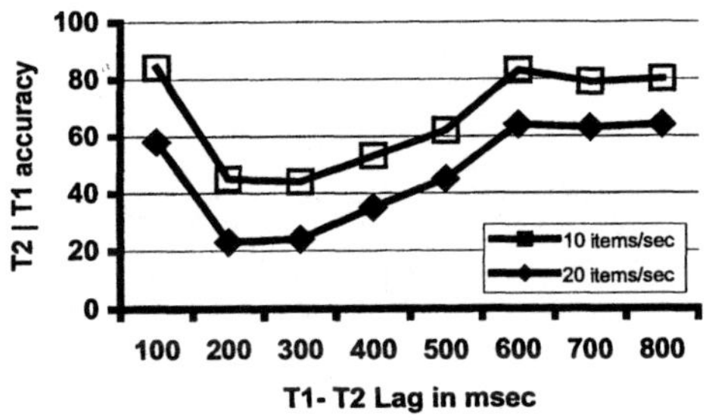

Figure 1: Model data for fast and slow rates.

Empirical Results

Empirical studies of human volunteers were undertaken to test the results of the model in comparing the AB from 10 and 20 item/sec presentation rates in two experiments. In the first, 14 volunteers were presented with two blocks of a task similar to that of Chun & Potter (1995) as described above. Items were 93 msec in duration with no intervening blank space and presented in black on a white background. RSVP streams were 17-21 items in length with the T1 appearing from positions 5-9 and the T2 appearing 1-8 positions later. For the faster presentation rate, items were presented for 53 msec and all RSVP parameters were doubled, presenting a similar temporal profile between T1 and T2. Results are depicted in Figure 2. Results closely match that of the model, with impaired performance for the faster rate, but a blink curve of a nearly identical shape.

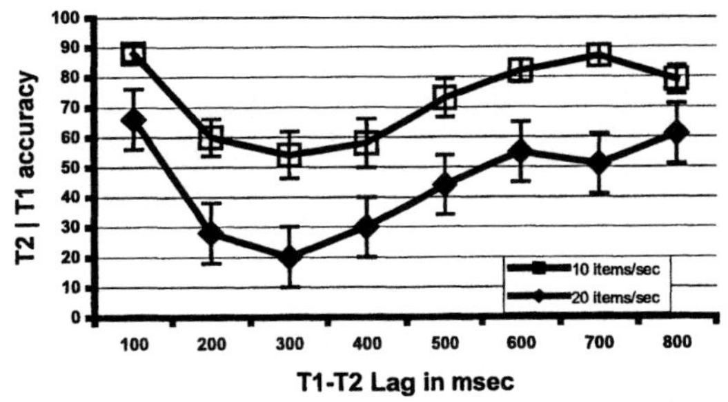

Figure 2: Empirical data for fast and slow rates. Bars represent standard error.

Conclusions

In conclusion, this work presents computational evidence and empirical validation that the AB shape is dictated by the temporal lag between T1 and T2, rather than the sequential arrangement of the targets in between. Consequently, clear Lag-2 sparing exists for a presentation rate of 20 items/sec. This result stands against the predictions of existing formulations of the interference model but is well explained by the two-stage model of Chun & Potter (1995) and Chun (1997). In our formulation, a token binding system serves as the second stage, and the temporal requirements of binding the T1 to a token determine the shape of the blink. If T2 arrives within 100 msec of the T1, it is spared, regardless of intervening distractors, which primarily serve as visual masks in initial stages of processing.

Acknowledgements

This work was performed with the funding of EPSRC grant GR/S 15075/01.

References

Chun, M.M. (1997) Types and tokens in visual processing: a double dissociation between the attentional blink and repetition blindness. *J Exp Psychol, HPP*, 23: 738-755.

Chun, M. M. and M. C. Potter (1995). A Two-Stage Model for Multiple Target Detection in Rapid Serial Visual Presentation. *J Exp Psychol: HPP* 21(1): 109-127

Kanwisher, N. G. (1987) Repetition blindness: Type recognition without token individuation. *Cognition*, 27:117-143.

Maki, W.S., G. Bussard, Lopez, K, and Digby, B (2003) Sources of Interference in the Attentional Blink: Target-distractor Similarity Revisited. *Perception & Psychophysics*, 65 (2), 188-201

Raymond, J.E., K.L. Shapiro et al (1992) Temporary Suppression of Visual Processing in an RSVP Task: An Attentional Blink. *J Exp Psychol, HPP*, 18(3):849-860.

Doctoral Consortium Posters

A Model of Single-page Web Search: The Effect of Interdependence on Link Assessment

Duncan P. Brumby (BrumbyDP@Cardiff.ac.uk)
School of Psychology, Cardiff University,
Cardiff CF10 3YG, Wales, UK

Abstract

When people search a Web page for links that are relevant to their information goal, they attend to the labels and estimate the likelihood that the link will lead to the goal. We have previously found in a simplified single-page menu search task that people sometimes, but not always, assess only a subset of the links available. Importantly, the presence of lower relevance distracters resulted in fewer items being fixated; suggesting people may adjust an independent assessment of the relevance of a link, in order to derive an estimate that is interdependent with the quality of the other links in the choice set. Recently, we have presented an ACT-R model that was inspired by Young's (1998) rational account of exploratory choice. The model makes use of ACT-R's architectural assumptions in order to produce behavior that provides both qualitative and quantitative fits across a range of performance measures including eye-movement data. The behavior of the model differs substantially from previous ACT-R models of Web navigation.

Keywords: Web navigation, single-page menu search, ACT-R cognitive modeling, eye-tracking, decision making

Thesis Research Summary

Consider searching a newly encountered Web page for links that are relevant to the achievement of some information goal. During this activity, people focus on the labeled links in order to derive a subjective assessment of the likelihood that the selection of a given link will lead to the achievement of the current information goal. Although, people tend to select items that are highly relevant to their goal (Pirolli & Fu, 2003), eye movement protocols from a simplified single-page menu search task (Brumby & Howes, 2003) suggest that people do more than assess the *scent* of a label.

SNIF-ACT (Pirolli & Fu, 2003) makes an important contribution to the development of cognitive models that can simulate users searching the World Wide Web for information relevant to an unfamiliar information goal. The model provides a characterization of Web search behavior. Importantly, the model uses ACT-R's declarative memory module in order to derive assessments of a labels *scent* (or the relevance of the label to the information goal). The model accurately predicts that users will select Web labels that have high information scent. However, Brumby and Howes' (2003) data suggest that SNIF-ACT lacks a plausible model of how people search an individual page. This simplification is non-trivial, if people do not always assess all items on a menu page prior to selection.

In a simplified single-page Web search task, Brumby and Howes (2003) asked participants to search single-Web pages in pursuit of an information goal. Each page contained a single goal, or *target*, link and the rest of the links were *distracters*. An eye tracker recorded participant's eye movements while they searched the labeled links. The relevance of the distracter links to the goal was varied while the quality of the target link was held constant.

Brumby and Howes (2003) observed that people do not always assess all of the items available in the choice set prior to the selection of an item, and that they re-fixate a smaller and smaller subset of these items prior to selection. These signature behaviors are consistent with the search of early keyboard-driven menus in which only a single choice can be seen at a time (MacGregor, Lee, & Lam, 1986).

While some previous models do not capture the menu search behavior observed by Brumby and Howes (2003), Young (1998) has presented a rational analysis of exploratory choice. Young's model is particularly interesting because it is sensitive to the implications of the structure of the task environment in many menu search tasks. When searching a menu that contains a given choice set of menu items ($item_1 \ldots item_n$), typically only a single item will lead to the achievement of the information goal. Given some probability estimate of this likelihood it can be assumed that the sum of estimates across all items in the choice set must be equal to one. This *normalization assumption* "reflects real cross-relationships between the judgments about choices made by a person, and cannot be avoided ... the reality is that people are often forced to make rapid and radical revisions of their estimates of the correctness of particular options as they work their way through [the options available]" (Young, 1998, p. 474). A novel prediction to emerge from the normalization assumption is that the relevance of both the target *and* the distracters will affect the decision of whether to select or continue assessment.

Importantly, Brumby and Howes (2003) found that the presence of lower relevance distracters resulted in fewer items being fixated. This finding supports the idea that people may adjust an independent assessment of the relevance of a link, in order to derive an estimate that is interdependent with the quality of the other links in the choice set. Consequently, people may make implicit assumptions about the value of items that they *have not assessed* on the basis of generalization from those that they have assessed. More recent empirical studies have replicated this main finding.

We have presented an ACT-R model of our previous menu search data (Brumby & Howes, submitted). The model is partially constrained by Young's (1998) rational analysis of exploratory choice, and in addition is consistent with the memory constraints imposed by ACT-R (Anderson & Lebiere, 1998). More specifically, the model used ACT-R's declarative memory retrieval mechanism to model how people choose between selection of an item and further assessment of items. This approach is consistent with that employed in previous ACT-R models of Web navigation (SNIF-ACT, Pirolli & Fu, 2003) but makes a novel use of the mechanism by which source-activation models the focus of attention.

We present a brief overview of the model (but see, Brumby & Howes, submitted for more details). In the model the ACT-R goal chunk included *n* slots, one for each label link and each of which, initially, had a value of *unassessed.* We call these *assessment* slots. Assessment of label links was achieved by repeated attempts to retrieve chunks from declarative memory. The successful retrieval of a chunk was assumed to indicate that there was positive information linking the label and the goal, and resulted in the replacement of 'unassessed' values on the goal with the retrieved value. An unsuccessful retrieval resulted in the replacement of a slot value with "not relevant", i.e. equivalent to setting the slot value to nil. Whether or not a chunk was retrieved depended in part on its activation. Activation of a chunk was determined partly by the strength of association S between the goal and the chunk, and more crucially through the amount of *source activation* W (representing the attentional focus). Consequently, the more label links that were in the choice set (as represented by the assessment slots) the lower the amount of source activation W received by a chunk. Conversely, the fewer the number of items in the choice set the greater the amount of source activation W received by a chunk.

In other words, the model was *more likely* to make a positive assessment of an item, when the assessed distracter items were not relevant to the current goal. The model then chose to select an item if it had been judged to be highly relevant to the current information goal (indicated by retrieval of all chunk-types related to the item). Importantly, the model could sometimes retrieve some of the assessment chunks for an item, while other assessment chunks for the item would fail to be retrieved. In this case the item was judged *partially relevant* to the information goal, and given some change in the value of source activation W future assessment of the item may warrant selection.

The aim in evaluating the validity of the model was to match the models performance across a range of dependent variables used in the menu search experiment, including eye-movement data.

The model provided a good fit with the data across most of the dependent variables (the number of items fixated; percentage of trials correct; percentage of self-terminating searches; and time to selection). The model also demonstrated the observed signature behaviors: (1) the model rarely attended all of the items available in the choice set prior to the selection of an item, and (2) reattended a smaller and smaller subset of these items prior to the selection of an item.

The model does not yet capture all of our previous data (Brumby & Howes, 2003). In particular, we have found that the history of information search moderates the local search strategy. After participants had completed trials in which they were more likely to select an incorrect item (because the distracters had been made more attractive), they were more cautious about selection. That is, they assessed more of the items in the choice set and were less likely to select an item immediately following an initial fixation of that item. Extending the current model to account for these findings should be relatively straight forward, because ACT-R's production rule learning mechanism is well suited to modeling the influence of history of successes on operator selection (Anderson & Lebiere, 1998, Chap. 8).

The idea that Web-page search is attentional focusing may seem counter-intuitive. In our model the goal of assessing unassessed items reduces the probability of retrieving information about the currently fixated item. Although, this mechanism predicts the observed behavior it seems counter-intuitive because, given that the goal is presumably under strategic control, an implication is that participants deliberately reduced the probability of retrieval of information associating an item with the goal (at least initially) in order to achieve the desired overall search strategy. Further data is required.

In summary, the current research program has used both empirical studies and cognitive modeling methods, in order to delineate how people might search a newly encountered Web page for links that are relevant to the achievement of some information goal.

References

Anderson, J.R., & Lebiere, C. (1998). *The atomic components of thought.* Mahwah, NJ: Erlbaum

Brumby, D.P., & Howes, A. (submitted). Good enough but I'll just check: Web-page search as attentional refocusing. *6th Internal Conference on Cognitive Modeling,* Pittsburgh, PA

Brumby, D.P., & Howes, A. (2003). Interdependence and past experience in menu choice assessment. In the *Proceedings of the 25th Annual Conference of the Cognitive Science Society,* Boston, MA

MacGregor, J., Lee, E., & Lam, N. (1986). Optimizing the structure of database menu indexes: a decision model of menu search. *Human Factors, 28(4),* 387–399

Pirolli, P., & Fu, W-T.F. (2003). SNIF-ACT: a model of information foraging on the World Wide Web. In *Proceedings of the Ninth International Conference on User Modeling,* 2003

Young, R.M. (1998). Rational Analysis of exploratory choice. In M.Oaksford & N.Chater (Eds.). *Rational Models of Cognition.* Oxford: Oxford University Press

An Analysis of Two (or Three) Models of Visual Attention Allocation

Michael D. Fleetwood (fleet@rice.edu)
Thesis Advisor: Michael D. Byrne (byrne@rice.edu)
Department of Psychology, Rice University
6100 Main St., Houston, TX 77005 USA

The Issue

This work is concerned with examining in a formal quantitative manner what human observers look at and what the objects of their gaze tell them. Researchers have developed a number of analytical models designed to describe and predict the allocation of human attention. The proposed research aims to compare and evaluate the predictions made by three such models and further refine the models where possible.

Many models have been designed to predict the allocation of human attention in the visual domain. One class of such models has focused on visual sampling or monitoring behavior in supervisory control tasks. These models use sampling or scanning as a dependent variable (e.g., Moray, 1986; Senders, 1964, 1983; Sheridan, 1970). In these models, the observer is not looking for a static target, but is rather supervising a series of dynamic processes, such as temperature gauges or aircraft movements, and the key dependent variable is the proportion of visual attention distributed to various "areas of interest" (AOIs) as a function of the quantitative properties of those AOIs.

Psychologists have also developed of a number of models designed to describe and predict the general allocation of human attention. These general attention models do not deal with vision per se, but rather consider attention to be a resource and are concerned with how people allocate this resource. This is usually measured in terms of the amount of time they devote to a task or source of information. Each model brings to the table its own set of advantages and disadvantages as a predictor of human performance. In the research proposed herein, we will examine the ability of a model of general attention to predict attention allocation in the visual sampling domain.

The Current Research

We propose to examine how the attention allocation predictions made by different types of models fare in a common task. There are two primary comparisons that will be made. One, how do the model predictions differ from each other? More specifically, we will examine how models designed to predict the allocation of attention on a coarser grain of analysis fare on a perceptual level.

The second objective of the current research is to make some refinements to the models where possible. One of the overarching goals of the line of research dedicated to modeling visual attention is to determine the factors that influence when and where a person will look next. Moray (1986) identified a number of factors, which under the right conditions, will influence the monitoring strategy of observers. The research proposed here will explicitly investigate four of those factors: the rate at which exogenous uncertainty is generated by the monitored process; the probability that while viewing one source another may show a critical value; the payoff matrix associated with missing or detecting critical values; and the cost of making an observation. Under the scenario that we will be using, we would like to determine the extent to which the above factors contribute to the attention allocation patterns of observers and incorporate the factors into the models accordingly.

Because of the number and complexity of the computational models in this domain, only a subset deemed to be representative are examined here. From the visual sampling domain, the seminal model of visual sampling, developed by Senders (1964, 1983) is examined because of its influence in the domain and because of its power in predicting human performance. The primary parameter of Senders model is bandwidth, or the information generation rate, of a signal. A more recent model, the SEEV model developed by Wickens and colleagues (2001) is a relatively simple model that has been shown to be an accurate predictor of visual attention by airline pilots. Each letter in the models acronym represents a parameter of the model (Salience of the information source; Expectancy, or information rate of the source; Effort to make an observation; Value of the source relative to others). Regarding general models of attention, the Information Foraging model (Pirolli & Card, 1999) has been evaluated as a model of general attention in a wide variety of contexts, and its potential as a model of visual sampling will be evaluated here. Its primary parameters are the amount of information gained from a source, the effort required to examine a source, and the effort required to switch attention to another information source.

The Paradigm

The basic paradigm to be employed in our evaluation of the three models is similar to that developed by Senders and colleagues (Senders, 1964, 1983). Observers will be placed in front of a computer screen displaying four ammeters located in the four corners of the screen. The task of the observer will be to monitor the array of ammeters and to press a button whenever the pointer of any ammeter enters a "danger zone." The eye movements of participants will be the principal dependent measure recorded.

In Sender's study, the meters were driven by quasi-random forcing functions such that their movements appeared random to observers. Senders was primarily concerned with the bandwidth of the instrument as a determinant of sampling behavior, and thus only manipulated the bandwidth of the four instruments. As a result, the bandwidth of the instrument and the alarm frequency (when the pointer enters the "danger zone') were inextricably linked, i.e., the higher the bandwidth the greater the alarm frequency. The system we will be using, developed by Alex Kirlik and Alex Kosorukoff at the University of Illinois, dissociates bandwidth and alarm frequency. Thus we will be able to determine the extent to which each variable drives sampling behavior.

Other than dissociating bandwidth and alarm frequency, two additional manipulations will be performed. The second manipulation we will undertake will be to associate a value with the detection of an alarm on each of the different dials. A different number of points will be associated with each dial, and the observer will earn points by correctly detecting an alarm on each dial. The task of the observer will be to earn a maximum number of points. The purpose of the manipulation is to determine the influence that the relative importance of an instrument (as determined by the number of points received for detecting an alarm) plays in the observer's attention allocation strategies. Although not explicitly addressed in Senders's model, the SEEV model and the Information Foraging model both predict that the relative value of the information source will influence sampling behavior.

A final manipulation will involve manipulating the effort required by observers to make an observation. This will simply involve altering the distance between the ammeters, such that more or less effort will be required of observers to shift their attention from one dial to another. Again, this is not explicitly addressed in Senders's model, but the SEEV model and the Information Foraging model both predict that the effort required to observe an information source will influence sampling behavior.

Expected Results

The aggregate aspects of the eye tracking data that will be evaluated for comparison with the models are: the proportion of time spent observing each instrument, the average dwell time per instrument, the frequency with which each instrument is observed, and the transition matrix of glances from instrument to instrument.

With respect to the first stated goal of the research, comparing the predictions made by the three models, we expect the Senders model to make relatively less accurate predictions as the task becomes more complex. In first proposed experiment, there are only two factors determining the accuracy of the model to predict human behavior, bandwidth and alarm frequency. The Senders model has been shown to predict sampling behavior well given at least one of these constraints, bandwidth. However, Senders's model does not account for the additional experimental manipulations of the relative value of the instruments and the effort of observation. We expect the Information Foraging model to show consistent and possibly even improved performance in predicting human behavior as the additional experimental manipulations are performed. Each of these considerations is accounted for in the model, and the power of the model to account for complex data may be limited in the earlier manipulations. Regarding the SEEV model, we expect it to maintain a consistent level of behavior prediction. Each of the manipulations is accounted for in the model; however, the model has traditionally be evaluated on a relatively coarse grain of analysis (the proportion of time spent observing different AOI), and its predictions may suffer as the predictions are extrapolated to make more specific predictions, such as the average dwell time.

Regarding possible refinements to the models, there is a clear opportunity to refine the original Senders model by incorporating the parameters of alarm frequency, relative value, and effort of observation. Indeed, Senders anticipated these concerns with his original model and has proposed several additional models that may account for them (Senders, 1983). To our knowledge, these newer Senders models have not been evaluated experimentally, and the proposed research will provide an opportunity to do so. The SEEV model incorporates each of the proposed manipulations as parameters, and the proposed research may involve supplementing the basic algorithm in order to account for finer-grained inputs. Regarding the Information Foraging model, it has not been evaluated as a model of perception, and the proposed research will evaluate its adequacy in this domain.

References

Moray, N. (1986). Monitoring behavior and supervisory control. In K. R. Boff, L. Kaufman, & J. P. Thomas (Eds.), *Handbook of perception and performance*, Vol II (pp 40-1-40-51). New York: Wiley & Sons.

Senders, J. (1964). The human operator as a monitor and controller of multidegree of freedom systems. *IEE Transactions on Human Factors in Electronics, HFE-5*, 2-6.

Senders, J. W. (1983). Visual scanning processes. Netherlands: University of Tilburg Press.

Sheridan, T. B. (1970), *On how often the supervisor should sample*. IEE Transactions on Systems Science and Cybernetics, SSC-6(2), 140-145.

Wickens, C. D., Helleberg, J., Goh, J., Xu, X., & Horrey, W. J. (2001) Pilot task management: Testing an attentional expected value model of visual scanning. (ARL-01-14/NASA-01-7). Savoy,IL: University of Illinois, Aviation Research Lab.

Pirolli, P., Card, S. (1999). Information foraging. *Psychological Review*, 106, 643-675.

Learning and Strategy Selection in Probabilistic Environments

Wolfgang Gaissmaier (gaissmaier@mpib-berlin.mpg.de)
Center for Adaptive Behavior and Cognition, Max Planck Institute for Human Development
Lentzeallee 94, 14195 Berlin, Germany

Abstract

Many decisions have to be made on the basis of knowledge about correlational structures in the environment. It has been found that people with a low working memory capacity perform better in a covariation detection task (Kareev, Lieberman, & Lev, 1997). This has been attributed to the assumption that they can only consider smaller samples which are more likely to bear a correlation parameter that exceeds the population parameter. Our data and results from a reinforcement learning model on an extended version of their task do not clearly support this account. As alternative explanations differences in reinforcement learning, hypothesis generation and strategy selection are considered. A very simple strategy (payoff maximization) is most successful in this environment. Differences in capacity and strategy selection are to be further studied in a cue based social categorization task, namely to predict political party preferences. Here, people have to apply knowledge about correlations acquired in the real world. It is hypothesized that people with a lower capacity use simpler strategies that could be, again, even more successful than more complex strategies because they exploit the structure of the environment. This could be reflected in behavioral data and in differential model fit to a variety of models like exemplar based models or Categorization by Elimination.

Keywords: Working Memory Capacity, Covariation Detection, Reinforcement Learning Model, Simple Strategies, Social Categorization

Introduction

We are concerned with the question how people learn correlational structures of the environment and how they make decisions on the basis of this knowledge. A major focus is on interindividual differences with regard to cognitive capacity. We are studying as well covariation detection in an artificial environment as a social categorization task in which knowledge about correlations has to be applied that participants acquired not in the lab, but in the real world.

Covariation Detection

The starting point for this project was the argument that cognitive limits are useful with regard to covariation detection (e.g., Kareev, 1995b). The observation is that it is more likely to find a correlation parameter that exceeds the population parameter in small samples. Therefore, people with a lower working memory capacity and hence a smaller sample size to consider should detect correlations earlier and perform better in a covariation task, which has been demonstrated empirically (Kareev et al., 1997).

Experiments

We conducted two experiments to test and model the original finding that low capacity people perform better in a covariation task. An additional hypothesis deduced from the small sample account is that they are also better in detecting a change in the correlational structure of the environment. As in the original experiment, working memory capacity was assessed with a digit span test. The original finding was replicated in the first but not in the second experiment, thus it seems to be a weak and unstable effect. It is worth noting that the probability of replicating a result at the same level of significance (and in the same direction) is only 50% (Goodman, 1992). Contrary to the predictions by the small sample account there was a high capacity advantage after a change in the first experiment. In the second experiment we did not find any differences between low and high capacity people, neither before nor after a change. Therefore, we focus on the first experiment with regard to modeling.

Modeling

Every model was fitted to each individual separately since we want to relate capacity to model parameters. A naïve window model that tries to translate the small sample idea directly could not capture the low capacity advantage. But we were able to model it with a reinforcement learning model (Camerer & Ho, 1999) with a decay, a sensitivity and an initial attraction parameter, where we forced the variance in each of the parameters separately by fixing the other two to their means. All three versions were able to capture the low capacity advantage on covariation detection, but only the initial attraction version was related to capacity and could predict behavior after a change.

Conclusions

The small sample account is not clearly supported by our data. First, the deduced hypothesis of a low capacity advantage after a change does not hold, we find either no effect or the opposite. Second, the model version with the decay parameter which has the strongest connection to memory has to be rejected. Instead, an initial attraction parameter model is successful, indicating a faster learning process of low capacity people in the beginning, but not later on. Still, faster learning can be interpreted as relying on smaller samples. But it is also congruent with the finding of Weir (1964) that children use the simple but most successful payoff maximization strategy (i.e. always choose the more frequent option given a color) earlier in a similar task because they are simply reinforcement driven. [illegible], in contrast, develop complex hypothesis and apply [illegible]x strategies because they believe that there exists a [illegible] solution, but they end up worse. As capacity differs between children and adults (Kail, 1984) and plays an important role in hypothesis generation (Dougherty and Hunter, 2002) this could be an explanation for the low capacity advantage. Consistent with this explanation, a simple strategy model (win stay, lose shift) fits better to the behavior of low capacity people.

Predicting Political Preferences

Appropriate behavior in social situations depends strongly on what other people think, e.g. what they like or dislike. As it is often impossible really to know what they think we have to rely on inferences about their opinions based on cues. Imagine that you are having dinner with an important person that you have never met before. If you share similar political opinions, this may be a good topic of conversation, if not, it may be wiser to avoid it. Asking directly is risky, so you infer what the other person thinks based on, e.g., statements regarding other topics he or she made. This inference process can be seen as a social categorization process.

To be able to socially categorize others, e.g. according to party preference, one has to know correlated structures in the environment (e.g. Rosch, 1978). There is a broad literature reporting relations between preferences for particular political parties, the social category that interests us here, and a variety of variables like, e.g., self interest and economic beliefs (Allen & Ng, 2000).

We are interested in the accuracy of people's knowledge about these correlations and want to study which strategies they use when they apply this knowledge to a social categorization task. The hypothesis is that people with a lower cognitive capacity use simpler strategies that could even be more successful than more complex ones as most people whose party preferences have to be predicted do not have complex patterns of opinions but rather one dimensional ones (Gigerenzer, 1982). Furthermore, we want to asses each participant's own opinions and party preference, as it is known that people often overestimate homogeneity in outgroups (i.e. supporters of other parties), but not in their ingroup (Tajfel & Turner, 1986) which could have an influence on the accuracy of different strategies.

Task

The task will be to infer which political party people prefer. Target people are selected from the European Values Survey, a dataset which documents individual data on opinions and party preferences. They are successively presented to our participants who are allowed to ask questions like "Do you have confidence in the armed forces?" which serve as cues. Possible parties are those that are currently represented in the German parliament.

Modeling

We want to test which model best describes the actual behavior of people. In most studies concerning social categorization the categories are the independent rather than the dependent variable. But there are models of categorization that can be transferred to social categorization, e.g. exemplar based models (Nosofsky, 1988) or Categorization by Elimination (CBE, Berretty, Todd & Blythe, 1997). These models also reflect different strategies. Thus, a differential fitting to people's behavior as well as their behavioral data allows us to see which strategies individuals were most likely to use. It is probable that there are interindividual differences, and maybe they relate to memory capacity.

Acknowledgements

I want to thank my advisors Lael Schooler and Jörg Rieskamp for their great help and support.

References

Allen, M.W. & Ng, S.H. (2000). Self interest, economic beliefs and political party preference in New Zealand. *Political Psychology, 21*, 323-345.

Berretty, P.M., Todd, P.M., & Blythe, P.W. (1997). Categorization by Elimination: A fast and frugal approach to categorization. In M.G. Shafto and P. Langley (Eds.), *Proceedings of the Nineteenth Annual Conference of the Cognitive Science Society* (pp. 43-48). Mahwah, NJ: Lawrence Erlbaum Associates.

Camerer, C.F., & Ho, T. (1999). Experience-weighted attraction learning in normal-form games. *Econometrica, 67*, 827-874.

Dougherty, M. R. P., & Hunter, J. E. (2002). Hypothesis generation, probability judgment and individual differences in working memory capacity. *Acta Psychologica, 113*, 263-282.

Gigerenzer, G. (1982). Der eindimensionale Wähler. *Zeitschrift für Sozialpsychologie, 13*, 217-236.

Goodman, S.N. (1992). A comment on p-values, replication and evidence, *Statistics in Medicine, 11*, 875-879.

Kail, R.V. (1984). *The development of memory in children* (2nd ed.). New York: Freeman.

Kareev, Y. (1995b). Through a narrow window: Working memory capacity and the detection of covariation. *Cognition, 56*, 263-269.

Kareev, Y., Lieberman, I., & Lev, M. (1997). Through a narrow window: Sample size and the perception of correlation. *Journal of Experimental Psychology: General, 126*, 278-287.

Nosofsky, R. M. (1988). Similarity, frequency, and category representations. *Journal of Experimental Psychology: Learning, Memory, & Cognition, 14*, 54-65.

Rosch, E. (1978). Principles of categorization. In E. Rosch & B. Lloyd (Eds.), *Cognition and Categorization.* Erlbaum: Hillsdale, NJ.

Tajfel, H. & Turner, J.C. (1986). The social identity theory of intergroup behaviour. In S. Worschel & W.G. Austin (Eds.), *Psychology of Intergroup Relations* (2nd ed.). Chicago: Nelson-Hall.

Weir, M. (1964). Developmental Changes in Problem-Solving Strategies. *Psychological Review, 71*, 473-490.

Emergence of Bayesian Structures from Recurrent Networks

Sébastien Hélie (helie.sebastien@courrier.uqam.ca)
Department of Computer Science, C.P. 8888 Succ. Centre-Ville
Montréal, PQ H3C 3P8 CANADA

Abstract

The problem of representational form has always limited the applicability of cognitive models: where symbolic representations have succeeded, distributed representations have failed, and vice-versa. Hybrid modeling is thus a promising venue, which however brings its share of new problems. For instance, it doubles the number of necessary assumptions. To counter this problem, we believe that one network should generate the other. This would require specific assumptions for only one network. In the present project, we plan to use a recurrent network to generate a Bayesian network. The former will be used to model low-level cognition while the latter will represent higher-level cognition. Moreover, both models will be active in every task and will need to communicate in order to generate a unique answer.

General Problem

In cognitive science, the problem of representational form is crucial. During the cognitive revolution, the computer metaphor was used to model human intelligence, which was thus seen as a set of symbol-manipulating syntactic processes (Turing, 1936). These processes were modeled as a series of conjunctive conditions and consequential actions (known as "IF-THEN" rules). This modeling approach is referred to as the classical view (Russel & Norvig, 1995).

In the late seventies, another metaphor became increasingly popular for modeling cognitive processes, namely: the brain. The connectionist (or "neural") networks proposed during this period were mostly unsupervised networks, either competitive (Grossberg, 1976; Kohonen, 1982) or recurrent (Anderson, Silverstein, Ritz & Jones, 1977). Another class of neural networks, first proposed by Rosenblatt (1958) and further developed by Rumelhart and McClelland (1986), was based on error backpropagation. These supervised networks have since been proven to succeed in fitting complex data but are often criticized on the grounds of their biological implausibility.

Connectionist networks, in general, differ from the classical view in that they are based on a conception of intelligence as a set of parallel processes acting on distributed representations. Moreover, while the distinction between processes and representations is quite straightforward in classical models, here this difference is a bit fuzzier.

Finally, in the early nineties, pioneers of the second wave of artificial intelligence started using symbol-manipulating probabilistic models which were either operating in a serial (e.g. Hidden Markov Models, see Rabiner, 1989) or parallel manner (e.g. Bayesian Networks, see Pearl, 1988).

Each of the previous approaches has important shortcomings. For example, while symbolic models can easily be used to model tasks involving language and reasoning (e.g. the use of recursive rules, Marcus, 2001), they are considered less useful for modeling low-level similarity-based perceptual processes. This has been shown to be better achieved through the use of connectionist models (Anderson & al., 1977; Grossberg, 1977).

Two important points are of interest here. First, approaches can be classified using their representational format and this classification is informative about what kind of task it will do best. Second, and more importantly, we see that where one approach succeeds, the other falls short of explanation.

Hypothesis

Since humans are able to use both recursive rules and similarity-based judgments, they must be able to use both kinds of representations. First, we will argue that low-level implicit knowledge is best modeled by an unsupervised neural network (using distributed representations), while higher-level cognitive operations are best modeled by symbolic networks. Second, those two kinds of representational format must be able to coexist and work together, sharing the same goals.

Proposed Thesis

In the proposed project, we will use an unsupervised neural network (either competitive or recurrent) to model low-level cognition. For modeling higher-order processes, because we believe that all levels of cognition share the same parallel architecture, we will use a parallel probabilistic architecture, namely a Bayesian network (Pearl, 1988). Moreover, the use of Bayesian networks is consistent with Anderson's rational analysis of cognition (Anderson & Milson, 1989; Oaksford & Chater, 1998). The main problem is: How can an unsupervised neural network build, from scratch, a fully functional Bayesian network?

First, the unsupervised neural network will have to generate nodes (symbols) and connections for a Bayesian network. Second, each network will eventually be individually functional, and they will each be capable of learning independently. Finally, both networks will have to communicate in an interactive way in order to produce a single answer from separate processes.

What has already been done

The first part of this thesis consists in identifying which properties are needed by a connectionist network to construct a Bayesian network. First, the network must be able to identify what is important in the environment and generate symbols according to those findings. Second, it must be able to feed the Bayesian network with the probability distribution of it environment.

This second property was used to compare a hard competitive neural network[1] with the Brain-State-in-a-Box (BSB, Anderson et al., 1977), which was chosen because it is the simplest form of recurrent networks (our findings will thus generalize to other, more recent recurrent networks). We tested these networks with four environmental distributions: a bimodal distribution consisting of two Gaussian distributions, a step distribution, the exponential distribution, and the uniform distribution (Hélie & Proulx, 2003). The choice of these distributions was motivated by the fact that they could either vary the frequency of the exemplars, the frequency of the categories, both, or neither. Results show that the competitive network is not sensitive to the frequency of the exemplars. Moreover, this network was unable to learn categories that were relatively rare. The recurrent network, on the other hand, was sensitive to both frequency of exemplars and frequency of categories. The former affected the position of the eigenvectors of the weights matrix, while the latter affected the magnitude of associated eigenvalues.

The results of this research thus suggest the use of recurrent networks to generate a Bayesian network.

What is left to do

Once the lower-level network has been chosen, the next step will be to find out how this network can generate useful symbols from the environment. Recurrent networks are particularly interesting because of their dynamic properties: these dynamic properties may thus be used in symbol generation. For example, we can see activation from the environment as a trajectory in a multidimensional space and try to find useful trajectories to create symbols.

Another important unresolved issue is the nature of the interaction between both models (once they will be fully functional). Should this interaction be just a crosstalk or should it involve weights that can be trained? In the later case, should these weights affect the learning of the networks (e.g.. BAM, see Kosko, 1988)? Should networks compete to give the answer (Logan, 1988) or, work together, adding evidence until a criterion is reached (Cousineau, Lacroix & Hélie, 2003)? Clearly, a lot is left to explore.

[1] Hard competitive networks update the weights of only one winning unit. They must be distinguished from soft competitive networks were the weights of several winning units are updated simultaneously.

Acknowledgments

This research is supported by scholarships from *Le fonds québecois de la recherche sur la nature et les technologies* and the *Natural Sciences and Engineering Research Council of Canada*.

References

Anderson, J. A., Silverstein, J. W., Ritz, S. A., Jones, R. S. (1977). Distinctive features, categorical perception, and probability learning: Applications of a neural model. *Psychological Review, 84*, 413-451.

Anderson, J. R., Milson, R. (1989). Human memory: An adaptive perspective. *Psychological Review, 96*, 703-719.

Cousineau, D., Lacroix, G. L., Hélie, S. (2003). Redefining the rules: Providing race models with a connectionist learning rule. *Connection Science, 15*, 27-43.

Grossberg, S. (1976). Adaptive pattern classification and universal recoding: I. Parallel development and coding of neural feature detectors. *Biological Cybernetics, 23*, 121-134.

Hélie, S., Proulx, R. (2003). Effet de la distribution environnementale sur l'apprentissage dans un réseau autoassociatif. *26ième Congrès Annuel de la Société Québécoise pour la Recherche en Psychologie*. Montréal, Qc.

Kohonen, T. (1982). Self-organized formation of topologically correct feature maps. *Biological Cybernetics, 43*, 59-69.

Kosko, B. (1988). Bidirectional Associative Memories. *IEEE Transactions on Systems, Man, and Cybernetics, 18*, 49-60.

Logan, G. D. (1988). Toward an instance theory of automatization. *Psychological Review, 95*, 492-527.

Marcus, G. F. (2001). *The Algebraic Mind: Integrating Connectionism and Cognitive Science*. The MIT Press, Cambridge, MA.

Oaksford, M., Chater, N. (1998). *Rational Models of Cognition*. Oxford University Press, Oxford.

Pearl, J. (1988). *Probabilistic Reasoning in Intelligent Systems: Networks of Plausible Inference*. Morgan Kaufmann, San Mateo, California.

Rabiner, L. R. (1989). A tutorial on hidden markov models and selected applications in speech recognition. *Proceedings of the IEEE, 77*, 257-285.

Rosenblatt, F. (1958). The perceptron: a probabilistic model for information storage and organization in the brain. *Psychological Review, 65*, 386-408.

Rumelhart, D. E., McClelland, J. L. (1986). *Parallel Distributed Processing: Explorations in the Microstructure of Cognition*. Cambridge, MA: MIT Press.

Russell, S., Norvig, P. (1995). *Artificial Intelligence : A Modern Approach*. Prentice-Hall, Upper Saddle River, NJ.

Turing, A. M. (1936). On computable numbers with an application to the Entscheidugsproblem. *Proceedings of the Mathematical Society, série 2, 42*, 230-265.

Methodology for Comparing Agent-based Models of Land-use Decisions

Tei Laine (telaine@cs.indiana.edu)
Indiana University
Computer Science Department and the Cognitive Science Program
Bloomington, IN 47405-7104 USA

Abstract

The focus of the research is mainly methodological. The goal is to develop a framework for comparing computational, agent-based models for land-use decision making. The framework will allow studying spatial patterns emerging from different distributions of agent characteristics, learning and communications schemes, initial spatial configurations, and varying spatial suitabilities. The framework will also facilitate the assessment of model complexity, thus enabling a rigorous practice of model comparison and selection.

Keywords Model selection, agent-based modeling, learning, decision making, land-use

Introduction

A common problem in computational modeling of cognitive, natural or social phenomena is *validation* of the model; how can one tell the model is the right one, i.e., it captures the essential mechanisms behind the modeled empirical phenomenon? The solution is to compare model predictions to the empirical data and measure the fit between them. One can then use this measure draw conclusions about the plausibility of model assumptions.

The problem of this approach is that the fit does not tell us anything about the model's goodness with respect to unseen data or alternative models designed to explain the same phenomenon. Once the validation issue is settled, the next question is the problem of model *selection*; how can one tell the model is the best one, i.e., one that explains the target phenomenon most accurately? One way is to compare several models with each other and report their relative goodnesses. However, this approach does not necessarily treat the models fairly, since some of them may achieve their superior performance due to their inherent complexity or number of ad hoc assumptions.

This research project's goal is to design and implement a framework for comparing complex real world agent-based decision making models, which accounts for the trade-off between the quality of models' performance and the complexity with which they achieve the performance. In the next chapter an overview on study's task domain, land-use decision making, is given. Then some issues regarding model selection and model complexity are addressed. The extended abstract concludes by briefly describing the proposed modeling framework.

Land-use decision making

This research focuses on a specific type of agent-based models (Janssen, 2004); models that are designed to explain how human land-owners make land-use decisions, and to explore land-cover patterns emerging from these decisions. The decision maker's goal is to find an effective way of using his/her assets — size and quality of land, technology, education and experience — in allocation of available resources — labour and land — to different uses. There are a number of factors to be considered ranging from the suitability of the land, dictated by various bio-physical variables, to the expected monetary or non-pecuniary returns from the uses. The optimal or good decision does not solely depend on careful consideration of the above factors, but also on the decisions of neighboring owners and the use of their land. Modeling land-use decision making is challenging, because the domain is inherently multidisciplinary. It involves human factors, environmental dynamics, and a number of ecological, economic, social and institutional variables and several kinds of interactions between them. Finally, the models are used to understand human cognitive processes, but seldom in this domain does any empirical data exist about them; yet inferences need to be made from land-cover data.

Model comparison and complexity

Computational modeling is a relatively established method of studying human cognition. The philosophical and theoretical issues of designing and implementing models have been discussed (see for instance (Cooper, Fox, & Farringdon, 1996)), but the question of what should be done once the model exists is practically unaddressed. The discipline lacks efficient methods of comparing model performance and assessing their complexity.

Good scientific practice prefers simple models, which explain the target phenomenon with few well-founded theoretical principles. The goal of the modeling endeavor is to increase understanding of the underlying mechanisms of the phenomenon, not to fit data. A model that fits the data perfectly does not necessarily capture the essential mechanisms behind the modeled phenomenon, but due to excess complexity is flexible enough to fit random noise. Model complexity is defined as a characteristic of the model that makes it fit varying patterns of data (Myung & Pitt, 1997; Pitt, Myung, & Zhang,

2002).

In general, model selection methods proposed for computational modeling of cognition are rare. Pitt *et al.* (Pitt et al., 2002) propose a selection method that emphasizes the generalizability of the model, and is based on *minimum description length (MDL)* (Grünwald, 2000). The MDL measure combines the maximum likelihood factor and penalty terms that account for the number of free parameters and the functional form. Busemeyer and Wang (Busemeyer & Wang, 2000) propose a *generalization criterion* method for model selection. The criterion is related to *cross-validation*, but instead of using different data sets in calibration and validation stages, it uses different designs. Myung and Pitt (Myung & Pitt, 1997) propose a *Bayesian method* that is based on maximizing the marginal likelihood, which can be obtained by normalizing goodness of fit with model complexity term that accounts for number of parameters, their functional form and the range of parameter values.

Proposed modeling framework

Most of the techniques discussed above apply to mathematical models only, but the models of interest in this research are not necessarily easily formulated in equations. They may also introduce other potential sources of complexity that enhance the fit but not necessarily the explanatory power. Thus, the goal of this research is to develop a methodology for comparing complex agent-based models of land-use, which allows the assessment of model complexity and quality of performance along various dimensions. Within the framework the performance can be evaluated both qualitatively and quantitatively (e.g., focus on amount of changes in land-cover or the patterns of changes), and either on global or individual level (i.e., assess prediction accuracy at the level of individual decision maker or the whole modeled area). The complexity assessment can take into account the number of parameters (i.e., degrees of freedom) and the algorithmic complexity (i.e., the requirements for time and space).

The main components of the general framework are *agents*, that represent land-owners, and the *landscape*, an abstract rectangular area divided into equal-size cells. A cell is the basic unit in which land-related information is encoded and it also defines the resolution at which land-use decisions are made.

In the general modeling framework a number of characteristics can be defined for these components. Agent characteristics are the size and location of the land they own and make decisions about, size of the household, initial wealth, number of neighbors and the relationship to them, learning and decision strategy they use, the activities (land-uses) they may take part in, and finally a list of individual parameters, that give a preference for each of these activities. Landscape characteristics are the area of land, number of possible land-covers and a list of suitability parameters associated to each land-use. Exogenous parameters to the model include monetary returns from different land-uses and costs associated to changes in uses.

The specific models implemented in this framework define possible land-covers (e.g., primary forest, farmed land, residential areas) together with the initial cover, the actual available land-uses (e.g., farming, development), relationship between uses and resulting covers, agent learning and decision strategies (e.g., expected utility maximizer, reinforcement learner), and the model specific parameters related to each of these (e.g., learning rate, number of hidden units).

Summary

Within the proposed framework several models, assuming different agent and landscape characteristics, will be implemented and compared. The framework will allow assessment of model performance by measuring prediction accuracy in various resolutions of decisions and output metrics. The framework will also enable the quantification of model complexity, by taking into account both architectural and computational factors, such as degrees of freedom, and time and space requirements. Finally, the framework will facilitate understanding of the models by making explicit their intuitive and hidden assumptions.

Acknowledgments

The research is partially funded by NSF grant SES-0083511.

References

Busemeyer, J. R., & Wang, Y.-M. (2000). Model comparisons and model selections based on generalization criterion methodology. *Journal of Mathematical Psychology, 44*, 171-189.

Cooper, R., Fox, J., & Farringdon, J. (1996). A systematic methodology for cognitive modeling. *Artificial Intelligence, 85*(1), 3-44.

Grünwald, P. (2000). Model selection based on minimum description length. *Journal of Mathematical Psychology, 44*, 133-152.

Janssen, M. A. (2004). Agent-based modelling. In *Modelling in ecological economics.* Edward Elgar.

Myung, I. J., & Pitt, M. A. (1997). Applying occam's razor in modeling cognition: A bayesian approach. *Psychonomic Bulletin & Review, 4*(1), 79-95.

Pitt, M. A., Myung, I. J., & Zhang, S. (2002). Toward a method of selecting among computational models of cognition. *Psyhcological Review, 109*(3), 472-491.

Generalization and Discrimination in a Semantic Network Trained with Semi-Supervised Learning

Rebecca J. Robare (RobareOwl@att.net)
Psychology/Experimental Cognition, Graduate Center, CUNY
Department of Psychology, City College of New York
138th St. at Convent Ave
New York, NY 10031

Keywords: semi-supervised learning, semantics, language learning

Abstract

The goal of the proposed research is to explore semantic learning in an artificial neural network trained with a semi-supervised learning paradigm. Semi-supervised learning permits both labeled and unlabeled data to be used in training the network, giving the training procedure greater ecological validity than seen in fully supervised learning, and improving the generalization ability of the network beyond that seen in networks trained with full supervision. As there is evidence that both supervised and unsupervised learning are necessary for learning word meaning, semi-supervised learning can provide a balanced approach to investigating semantic learning.

Semi-supervised learning can be defined as the use of both labeled and unlabeled data in the training of an artificial neural network. Such learning takes the form of expectation-maximization (EM) algorithms (i.e., Nigam, McCallum, Thrun, & Mitchell, 2000), support vector machines (SVM) (i.e., Chen, Wang, & Dong, 2003), and a variety of other forms. In the computer sciences, these algorithms have been used successfully in applications for various practical purposes, especially text classification in documents or Web sites. Semi-supervised learning has the following advantages over more traditional, fully supervised learning: (1) it is less costly in terms of time and labor to train from unlabeled examples than to manually label many items of information (Nigam et al., 2000); (2) the trained classifier or network shows better generalization, that is, it is more successful at appropriately classifying novel items of information (Anagnostopoulos et al., 2003); and (3) overfitting of the data is thereby avoided (Nigam et al., 2000). Improved generalization is, perhaps, the most important of these advantages. In fully supervised learning, when performance on a set of patterns is perfect, there will be no generalization; the network will not correctly classify novel patterns. Using unlabeled input patterns can improve generalization performance at test because the network is not overtrained on the specific labeled examples included in the training set (Anagnostopoulos et al., 2003; Chen, Wang, & Dong, 2003). Semi-supervised learning has the following advantages over learning mechanisms that are entirely unsupervised: (1) class labels can be learned; and (2) training exemplars that are similar on a physical dimension but should appropriately be classified differently, can be (Nigam et al., 2000).

Semi-supervised learning has proven successful in computer science applications. However, it has not been applied to any of the problems in cognition toward which artificial neural networks are often used. In connectionist psychology, learning is either fully supervised or unsupervised; the "environment" in the form of the "teacher" or labeled training examples provides either perfect feedback or none (O'Reilly & Munakata, 2000). This absence of semi-supervised learning is a gap in psychology's simulation of and understanding of human cognition. I offer the following two justifications for researching semi-supervised learning in psychology.

First, environments are inconstant and may not provide feedback consistently or perfectly. One example of this is language. It is often claimed by linguists that the feedback present in a language learner's environment is not sufficient to prevent learners from making linguistic mistakes that they do not, in fact, make (Pinker, 1999). And although some lexical items, such as some nouns and verbs, may be learned without supervision (Roy, 2000), it seems unlikely that generative syntax can be learned without some sort of supervision mechanism (Bloom, 2000). Semi-supervised learning is an elegant possible solution to this problem.

Second, semi-supervised learning is at least as biologically plausible as supervised or unsupervised learning. Mechanisms for both have been found within the brain, and these are generally assumed to operate over all synapses (O'Reilly & Munakata, 2000). In other words, both kinds of learning can operate over the same synapse simultaneously. If this is the case, then observing that the synapse is plastic as a result of both kinds of learning, under various environmental conditions, this could constitute semi-supervised learning at the synaptic level.

Preliminary research on the utility of semi-supervision has been encouraging. For the purposes of constructing neural networks, semi-supervised learning can be used with learning algorithms already applied in psychology. I have used the Leabra algorithm (O'Reilly & Munakata, 2000; O'Reilly, 2001) sucessfully with semi-supervised learning. Leabra is ideal for this purpose because it combines error-driven and Hebbian learning, thus providing a framework in which semi-supervised learning can operate. Furthermore, Leabra was designed with the intent that it be biologically plausible. Applying semi-supervised learning in this structure, therefore, can be seen to not contradict current knowledge about synaptic learning.

Using the PDP++ software (O'Reilly et al., 1995) it is easy to construct a data set for training the network that includes both labeled and unlabeled data. It is the use of unlabeled data in this context that is novel to psychology; earlier networks have used only labeled or unlabeled data, never both. Even Leabra has been used only to apply Hebbian learning to data that is otherwise learned in a fully supervised fashion. Simple Leabra networks trained by semi-supervised learning so far have shown task learning as good as that of fully supervised networks and generalization that is better in a task requiring a non-linearly separable discrimination.

My dissertation project will be to explore the potential of semi-supervised learning as a substrate for language learning. Specifically, I will be exploring semantic learning occurring in a supervised or unsupervised versus a semi-supervised fashion. There is reason to believe that semantic learning requires both types of learning processes (Bloom, 2000), so semi-supervised learning very naturally assumes this role. The overarching intention is to make this learning as realistic as possible, and to this end it is anticipated that word meanings will be trained using inputs that represent sentences, "images" or some sort of corresponding visual input (see Roy, 2000), and other words, thus giving the training set a context that approximates in a general way the context of the language learner. It is not a good approximation; but it is a better approximation than a network that merely trains associative word-to-word mappings and attempts to make inferences about semantic organization.

According to Bloom (2000), sometimes the words a language learner hears are presented in a context in which the meaning can be extracted, such as when the mother says "cookie" and gives the child a cookie, and sometimes when it cannot be, as when the caregiver says, "Do you want a cookie?" when no cookies are present. Note that the first of these scenarios can be interpreted as a correlation-detection situation (unsupervised learning) and the second as a situation of reinforcing the expectation of a cookie with the eventual presence of a cookie (supervised learning). Neither type of learning on its own will suffice. Two distinct but interdependent learning abilities are required to extract the correct information from each scenario and put that information together to determine the meaning of "cookie." Semi-supervised learning is ideal in this situation; it allows the accurate simulation of both learning mechanisms and environmental conditions, which are necessary for the development of precise models of language acquisition and other cognitive processes.

References

Anagnostopoulos, G. C., Bharadwaj, M., Georgiopoulos, M., Verzi, S. J., & Heileman, G. L. (2003). Exemplar-based pattern recognition via semi-supervised learning. *Proceedings of the IEEE-INNS-ENNS International Joint Conference on Neural Networks (ISCNN '03)*. Portland, Oregon. Retrieved 28 March 2003 from pegasus.cc.ucf.edu/~mbharadw/thesis/gca-ssl-icjnn-2003.pdf.

Bloom, P. (2000). *How children learn the meanings of words*. Cambridge, MA: MIT Press.

Chen, Y., Wang, G., & Dong, S. (2003). Learning with progressive transductive support vector machine. *Pattern Recognition Letters 24*, 1845-1856.

Nigam, K., McCallum, A. K., Thrun, S., & Mitchell, T. M. (2000). Text classification from labeled and unlabeled documents using EM. *Machine Learning 39*, 103-134.

O'Reilly, R. C. (2001). Generalization in interactive networks: The benefits of inhibitory competition and Hebbian learning. *Neural Computation 13*, 1199-1241.

O'Reilly, R. C., Dawson, C. K., McClelland, J. L., & Carnegie Mellon University. PDP++.

O'Reilly, R. C., & Munakata, Y. (2000). *Computational explanations in cognitive neuroscience.* Cambridge, MA: MIT Press.

Roy, D. (2000). Learning from multimodal observations. *IEEE International Conference on Multimedia and Expo (I)*, 579-582. Retrieved 7 May 2003 from dkroy.www.media.mit.edu/people/dkroy/.

World Models, Action Selection, Embodied Concept Formation, and Conditioning

Terrence C Stewart (tcstewar@connect.carleton.ca)
Institute of Cognitive Science, Carleton University
1125 Colonel By Drive, Ottawa, ON, K1S 5B6, Canada

Abstract

There are many computational models whose broad purpose is to allow an agent to learn via experience to perform effectively in a given environment. However, it is uncommon to see these models directly compared to each other, or to empirical data of real creatures adapting to their environments. Here, a comparison methodology is proposed involving various known results in classical and operant conditioning and concept formation. The project involves examining a broad selection of computational models in various environments, and also mixing and matching components from these different models.

Keywords

Hybrid modeling, reinforcement learning, action selection, conditioning, concept formation, world models, sensory pre-processing, model comparisons, supervised learning, unsupervised learning

Introduction

The goal of many computational modelers is to develop agents that can learn to perform well within their environment. To this end, we have seen a proliferation of diverse algorithms and approaches for solving this task. Furthermore, many of these models draw upon other research to define the components of their model. We see back-propagation systems being used within TD(λ) learning (Tesauro, 1995), Genetic Algorithms used to support Action Selection (Farritor & Dubowsky, 2002), Kohonen SOMs combined with Q-Learning (Smith, 2002), and so on.

This variety of approaches is matched with a variety of test problems. Agents learn to control elevators, play board games, navigate grid-worlds, control physical robots, track visual objects, to name just a few. For each situation, results change depending on the sensory representations chosen, the particular implementations of the model's components, and the method whereby the results are judged.

When reading the published results of these models, one is often led to wonder how well a particular model would fare in a completely different domain. Could Tani's Hierarchical RNNs (Tani & Nolfi, 1999) play backgammon successfully? How well would Sarsa or Q-Learning deal with this or that particular robot navigation task? It is rare to find a wide variety of computational models that have been tested on exactly the same task in exactly the same manner (see Gershenson, 2003 for an exception).

This lack makes it difficult to accurately judge the development of the field. When new models (or new variants of old models) are presented, they are often accompanied by a completely new task. Without a comprehensive system of comparison, it is difficult to know the actual strengths and weaknesses of the models. Furthermore, the capabilities of these models are seldom compared with those of real living creatures.

A Unified Approach

Figure 1 shows the framework to which this research restricts itself. A variety of computational models can be seen as particular implementations of this system. The World Model can be implemented by any Supervised Learning scheme. The Sensory and Action States can either be manually set, or modified via any Unsupervised Learning scheme. There are numerous ways of implementing Action Selection, and the various Reinforcement Learning methods can be used to organize the system as a whole.

An advantage of presenting a common framework is that we can start to directly mix and match aspects of these algorithms. Touzet (1997) presents results of using self-organizing maps to automatically reorganize sensor data within a Q-Learning system. This same technique could clearly be applied other cognitive models. Tani's suggestion of using a self-prediction system to help suggest potential actions (Tani & Yamamoto, 2002), or implementing a bias towards novel situations by rewarding incorrect predictions (ibid.), could also be applied to other computational models than his Hierarchical RNNs.

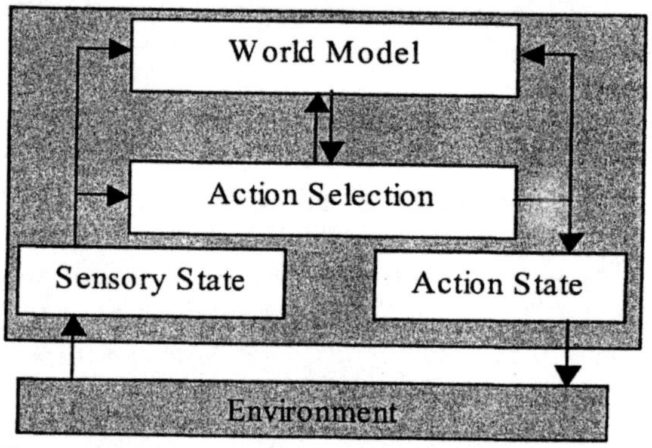

Figure 1: The Generic Agent Model. All models to be investigated fit within this framework.

Most models do not require much coaxing to fit within this framework. They may differ in terms of what exactly is learned by the World Model (does it predict future states or future rewards?). They may have more or less tightly coupled modules (Action Selection systems can make use of neural network world model weights to help find promising actions to perform). But even models as seemingly different as Distributed Adaptive Control or even ACT-R can be seen as falling within this generic framework.

Empirical Data

The existing research literature contains a tremendous variety of tasks and environments that are used to study model performance. From these, we can choose a set of problems to test these algorithms on. Many of the robotic systems cannot be used due to equipment limitations, but any of the simulated worlds (including games and multi-agent systems) are useful domains for comparison.

In addition to these traditional test systems, there is also the potential to compare the performance of these algorithms to the behavior of real living organisms. In (Stewart & Wood, 2001) and (Stewart, 2000), we presented some initial work comparing one model's learning capabilities with the known capabilities of classical and operant conditioning. All reinforcement learning systems could be investigated in these terms. Experiments mirroring those done on living creatures can be performed, measuring rates of acquisition, extinction, specialization, and generalization.

Furthermore, these sorts of comparative studies can also shed light on the ability of the various computational models to form concepts. In the afore-mentioned papers, we identified particular high-level concepts (i.e. regularities of its environment that were not directly represented in the sensory state) that the model was able to identify, and others that it was not able to. This is an embodied approach to concept formation, where an agent is deemed to have recognized an environmental pattern based on its ability to react appropriately to its presence. By combining this with the conditioning research, it is possible to examine those aspects of conditioning which require the modification of concepts (such as generalization and specialization). Existing models of conditioning (such as Kakade & Dayan, 2000) do not discuss such situations.

The Process

The research is divided into three stages. First, the set of supervised systems for developing world models will be compared on a set of test environments. Each model (e.g. ARTMAP, back-prop multi-layer perceptrons, or recurrent neural networks) will be trained using identical input data. The domains will include various grid-worlds, games, and robot navigation environments. The training will be done by giving the model a set of state-action-reward data over time of a particular agent interacting with a world. These various models would then be compared in terms of their ability to predict the future state, reward, or discounted value. Importantly, each model would be trained using exactly the same set of training data. A range of test situations will be used, specifically including ones that would be comparable to real conditioning tasks. For example, Jakobi (1998) discusses a T-maze robotic learning task that is comparable to T-maze tasks given to rodents.

In the second stage, a comparison of full reinforcement learning systems will be performed. Since most of these models make use of supervised learning components to learn the regularities of their environment, we can make use of the results of the previous stage to select promising world model implementations. The environments from the first stage would continue to be used here. The models can be compared to each other based on their resulting total reward. Furthermore, the models' performance can be compared to that of real world creatures in conditioning experiments. For example, if an animal was trained to associate a particular stimulus with a reward, a characteristic acquisition curve would appear in its responses. If this association was extinguished by presenting the stimulus without the reward, a characteristic exponential decay in response should occur. After this, if time passes and then the stimuli is presented again without the reward, the animal will still 'spontaneously recover' the association to a certain degree. This sort of experiment can be performed on any of the reinforcement learning models, allowing us to compare them to the behavior of real living creatures.

In the final stage, various ways of extending this learning model can be investigated. One of these is the use of unsupervised learning to automatically adjust the sensory and motor representations. ARTMAP is an early example of this, and Touzet (1997) gives promising results on using Kohonen SOMs to improve Q-Learning. To perform this comparison, we repeat stage two, with various approaches to automatically pre-processing the sensory data.

References

Farritor, S., and Dubowsky, S. (2002). A Genetic Planning Method and its Application to Planetary Exploration. *ASME Journal of Dynamic Systems, Measurement and Control, 124:4*, 698-701.

Gershenson, C. (2004). Cognitive paradigms: which one is the best? *Cognitive Systems Research, 5:2*, 135-156

Jakobi, N. (1998). Minimal Simulations for Evolutionary Robotics. PhD thesis, University of Sussex.

Kakade, S. & Dayan, P. (2000). Acquisition in Autoshaping. *Adv. in Neural Information Processing Systems, 12.*

Smith, A. J. (2002). Applications of the Self-Organising Map to Reinforcement Learning. *Neural Networks, 15*, 1107-1124.

Stewart, T.C. (2000). Learning in Artificial Life: Conditioning, Concept Formation, and Sensorimotor Loops. MPhil Thesis, University of Sussex.

Stewart, T.C. & Wood, S. (2001). Conditioning and Concept Formation in Embodied Agents. *AAAI Spring Symposium.*

Tani, J. & Nolfi, S. (1999) Learning to perceive the world as articulated: an approach for hierarchical learning in sensory-motor systems, *Neural Networks, 12*, 1131-1141.

Tani, J. & Yamamoto, J. (2002). On the dynamics of robot exploration learning, *Cognitive Systems Research, 3:3*, 459-470.

Tesauro, G. J. (1995). Temporal difference learning and TD-Gammon. *Communications of the ACM, 38*, 58-68.

Touzet, Claude F. (1997). Neural Reinforcement Learning for Behaviour Synthesis. *Robotics and Autonomous Systems, Special issue on Learning Robot: the New Wave.*

Student Modeling for Collaborative Medical Problem-Based Leaning

Siriwan Suebnukarn (ssiriwan@cs.ait.ac.th)
Computer Science and Information Management Program, Asian Institute of Technology
PO Box 4, Pathumthani, 12120 Thailand

Background and Objectives

Today a great many medical schools have turned to a problem-based learning (PBL) approach to teaching. PBL instructional models vary but the general approach is student-centered, small group, collaborative problem-based learning activities. While PBL has many strengths, effective PBL requires the tutor to provide a high degree of personal attention to the students. In the current academic environment where resources are becoming increasingly scarce and costs must be reduced, providing such attention becomes increasingly difficult. This is exacerbated by the fact that medical school faculty, in particular, often have limited time to devote to teaching. As a consequence, medical students often do not get as much facilitated PBL training as they might need or want.

Our proposed work combines concepts from Intelligent Tutoring System with those from Computer Supported Collaborative Learning to develop an intelligent group-based medical PBL system. Our proposed work departs from previous efforts to incorporate user modeling into computer supported collaborative environments by focusing on modeling individual and group problem solving behavior. Since problem solving in group PBL is a collaborative process, modeling individuals and the group is necessary if we wish to develop tutoring algorithms that can do things like focus the group discussion, promote collaboration, and suggest peer helpers.

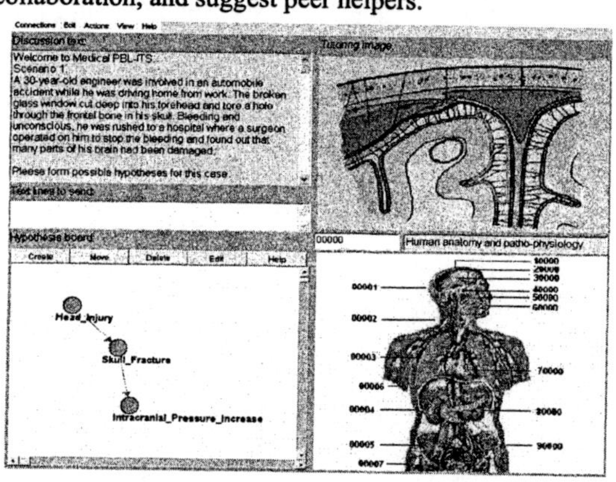

Figure 1: COMET Student interface

We have developed an initial prototype system called COMET (Suebnukarn & Haddawy, 2004) that is designed to provide an experience that emulates that of live human-tutored medical PBL sessions as much as possible while at the same time permitting the students to participate from disparate locations. The system uses Bayesian networks to model individual student knowledge and activity, as well as that of the group. It uses the models to generate tutorial hints to guide group problem solving activity (Fig. 1).

Student Clinical Reasoning Model

Generating appropriate tutorial actions requires a model of the students' understanding of the problem domain. This modeling task is necessarily wrought with uncertainty since we have only a limited number of observations from which to infer each student's level of understanding. Thus we have chosen to use Bayesian networks as our modeling technique. We model each student with an instance of our general Bayesian network student model. The group is reasoned about by combining information from the models of the individual students, as described below. Figure 2 shows a portion of the student model corresponding to the head injury scenario in Figure 1. The model contains two types of information; (1) the hypothesis structure based on the differential diagnosis of the scenario (the right group of nodes in Fig. 2); and (2) the application of medical concepts in terms of anatomy and patho-physiology (the left group of nodes in Fig. 2) to derive the hypotheses.

We represent the hypothesis structure following the model of Feltovich and Barrows (1984), which defines three categories of illness features: enabling conditions, faults, and consequences. Enabling conditions are illness features associated with the acquisition of illness, which can make an individual more susceptible than usual to illness in general or to particular illnesses. Faults are the major real malfunctions in illness. Consequences are the secondary consequences of faults within the organism, and generally comprise different types of signs and symptoms. Each hypothesis node has parent nodes, which have a direct casual impact on it. All hypothesis nodes have two states, indicating whether or not the student knows that the hypothesis is a valid hypothesis for the case.

The application of medical concepts is represented in terms of three kinds of nodes: goals, general medical knowledge, and apply actions. Every hypothesis node (except the root, which represents the scenario itself) has a unique *Apply* node as parent. The *Apply* node represents the application of a medical concept to a goal in order to derive the hypothesis. For example the *Apply3* node indicates that the student is able to use knowledge of the *Blood_Flow_Decrease* medical concept to infer that *Brain_Damage* is a consequence of *Brain_Infection*. Each hypothesis node thus has a conditional probability table specifying the probability of the hypothesis being known conditioned on whether the parent hypotheses are known and whether the student is able to apply the appropriate piece of knowledge to determine the cause-effect relationship. The conditional probability tables for the *Apply* nodes are simple AND gates.

To construct the model which took about one person-month, we started from an initial Bayesian network built based on information extracted from medical textbooks and from interviews with a neurosurgeon and a physiologist from Thammasat University Medical School. The initial

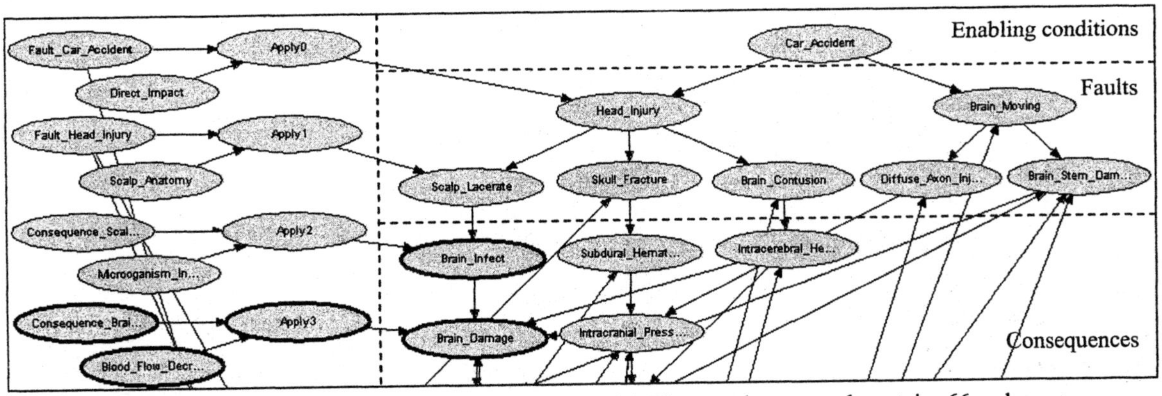

Figure 2: Part of the Bayesian network student model. The complete network contains 66 nodes.

model was refined by learning structure and parameters using data collected from a medical problem-based learning tutorial.

The student model is used to reason about the state of knowledge of each student, as well as about problem solving behavior of each student and of the group. The general student model is instantiated for each student by entering that student's medical background knowledge as evidence. We make the assumption that once a hypothesis in the domain model is mentioned by one student in the group, every student knows that hypothesis. So as hypotheses are mentioned, they are instantiated in each student model. Following commonly accepted practice in medical PBL, we assume that students should and generally do enumerate the possible hypotheses by focusing sequentially on the various causal paths in the domain, linking enabling conditions with faults and consequences. So for each student, we must determine what causal path he is reasoning along, which we do by identifying the path of highest probability. This is computed as the joint probability of the nodes along the path. It is also necessary to identify a causal path that can be used to focus group discussion, particularly when the discussion seems to be diverging in different directions. This is done as follows. We take the most likely path for each student and compute the sum of the probabilities of that path over all the students. We then take the path with the highest sum. The idea is that we want to identify the path that has much of the attention of much of the group and has at least one member whose attention is focused on that path.

Our automated tutor takes on the role of guiding the tutorial group to construct possible hypotheses for the case by the use of specific open-ended questions. From our study of the tutoring session transcripts, we identified eight hint strategies commonly used by experienced human tutors: 1) focus group discussion using general hint, 2) focus group discussion using specific hint, 3) promote open discussion, 4) deflect uneducated guessing, 5) avoid jumping critical steps, 6) address incomplete information, 7) refer to experts in the group, and 8) promote collaborative discussion. We developed algorithms to generate each of these types of hints, using as input the interaction log and the Bayesian network student models.

Evaluation

In order to evaluate the appropriateness and quality of the hints generated by our system, for each of the eight hinting strategies, we created three scenarios under which COMET would generate that hint. Ten experienced medical tutors were asked to analyze each situation and indicate whether they would provide a hint and what hint they would provide. Our results show that COMET's hints agree with the hints of the majority of the human tutors with a high degree of statistical agreement (McNemar test, p = 0.652, Kappa = 0.773).

Conclusions and Future Work

We have described a Bayesian network clinical reasoning model that integrates the hypothesis structure based on the differential diagnosis of the case and the application of the corresponding medical concepts in the problem solving process. Results from our initial evaluation show a high degree of agreement of tutoring dialogues between our system and human tutors.

We are currently working on evaluating the accuracy of a number of alternative Bayesian network student models that capture student's actions of creating hypotheses as well as the causal links between them. The models are being evaluated in terms of accuracy in predicting individual student actions and in terms of accuracy of identifying group reasoning paths. The ultimate test of the effectiveness of our work is how it impacts student learning. So we intend to compare the effectiveness of student learning with COMET versus student learning with human tutors.

Acknowledgments

We thank Hugin Expert for providing us the use of the Hugin Researcher software. Thanks to Thammasat University Medical School for their participation in the data collection and system evaluation.

References

Suebnukarn, S., & Haddawy, P. (2004). A collaborative intelligent tutoring system for medical problem-based learning. *Proceedings of the 9th International Conference on Intelligent User Interface* (pp. 14-21). Madeira, Portugal.

Feltovich, P. J., & Barrows, H. S. (1984) Issues of generality in medical problem solving. *Tutorials in problem-based learning: A new direction in teaching the health professions*. The Netherlands: Van Gorcum.

Modeling the ability to reason about oneself and others in educational contexts

Karin Zondervan (karinz@ai.rug.nl)
Rineke Verbrugge (rineke@ai.rug.nl)
Niels A. Taatgen (niels@ai.rug.nl)
Department of Artificial Intelligence, University of Groningen
Grote Kruisstraat 2/1, 9712 TS Groningen, Netherlands

Introduction

There is a huge gap between educational science on the one hand and cognitive science on the other. Cognitive scientists investigate learning and cognitive skill acquisition in a fundamental way. The result of their research almost never reaches real-world educational practice. Educational scientists, on the other hand examine realistic learning situations and improve instructions and educational settings to fit the learner's needs. They cannot do so without resorting to results from cognitive science. On certain topics, there is little fundamental work they can rely on. However, the educational field is rapidly changing and many new forms of learning and approaches to learning have emerged in the last few years. Educational science is in the forefront of investigating and evaluating new forms of learning like discovery learning, self-regulated learning and cooperative learning (see e.g., Hoek, van der Eeden & Terwel, 1999). Especially the self-regulation and social interaction aspects of learning receive much attention these days. To be able to provide fundamental cognitive results that can be of service for educational science and instruction design, the inclusion of social processes into cognitive modeling seems timely.

Cognitive load theory

Educational research gratefully embraces cognitive load theory (CLT; see Paas, Renkl & Sweller for an overview) as a framework for investigating the interaction between information structures and human cognition. CLT is based on several assumptions regarding cognition: a virtually unlimited capacity of long-term memory, schema theory of mental representations of knowledge, and limited processing capacity for working memory. Based on CLT, educational researchers have been able to generate a variety of new and sometimes counterintuitive instructional designs and procedures. However, cognitive load cannot be measured directly. In this respect cognitive modeling had a lot to offer, because cognitive models can predict in a very precise way the form, interaction and size of the different kinds of cognitive load involved in a certain learning task.

Reflection on oneself and others

Most of the time, cognitive load is investigated in the context of self-regulated learning. However, in our opinion, cooperative learning might just as well profit from cognitive load theory. Seen from a cognitive point of view the two forms of learning call on quite similar cognitive mechanisms. We propose that in both learning situations reflection is an important process. Figure 1 depicts how we perceive the cognitive connection between self-regulated learning and cooperative learning. Central to our approach is the hypothesis that performing a task and simultaneously reflecting upon this task can be seen as a form of dual tasking. In some learning situations this reflection upon the task concerns mainly reflecting upon the learner's own knowledge and reflecting upon the process. In cooperative learning situations reflection upon what *others* know or believe is crucial. Regardless of what the nature of the reflection might be, we assume that a reflection process cannot start until cognitive resources are freed, due to automation of the task at hand (see Figure 2). For novices, reflection is not yet possible, because they place a heavy burden upon declarative memory for performing the task at hand. As declarative memory is also involved in reflection, performing a task and reflecting on it can be thought of as a dual-task. We think of reflection as the complement of automation. Whereas automation can be described as a process from declarative to procedural knowledge,

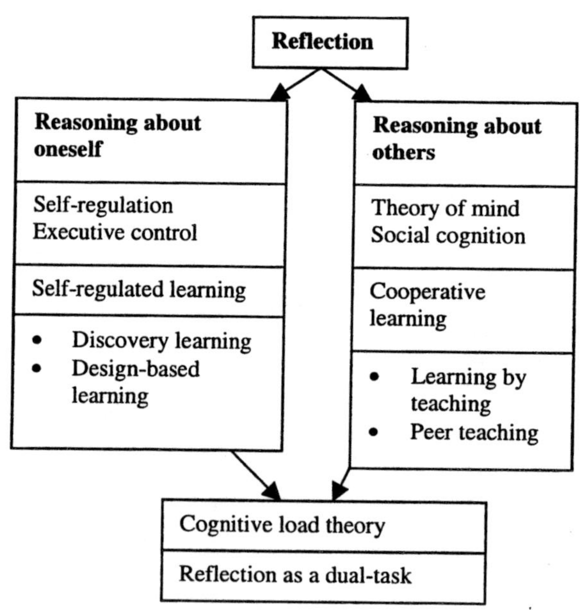

⁻igure 1: Associating cognition and education

reflection can be thought of as a process from procedural knowledge to new, more abstract declarative knowledge.

Developing new declarative knowledge

Karmiloff-Smith (1992) describes a developmental process in which children acquire explicit knowledge through a process of *representational redescription*. She conducted experiments with children who already had mastered a

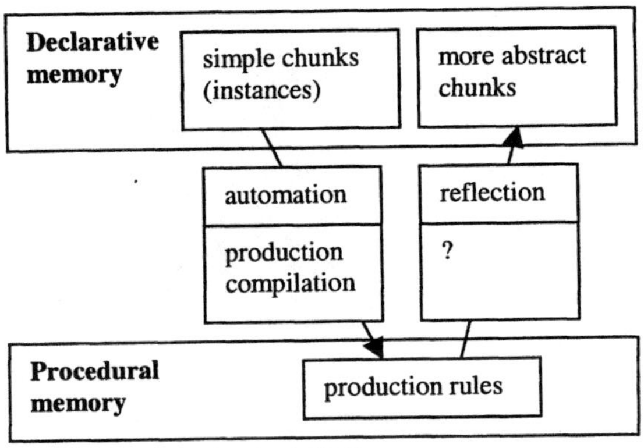

Figure 2: Reflection in ACT-R

certain task. On a micro time-scale of a few hours as well as on a macro time-scale as long as several years she demonstrated a development in which the internal representations of the children changed. Due to the development of more abstract and explicit representations, the children became able to give verbal justifications and descriptions of their behavior in the learning task. A simulation model in ACT-R of this representational redescription process (Zondervan & Taatgen, 2003) explicates some of the representations and redescriptions involved. However, it is unclear how verbalization is reached. In order to answer this important question, a reflection mechanism is needed in ACT-R. Since production rules are not directly accessible, we assume the existence of a reflective process that generates new declarative knowledge in order to reach possible verbalization of what has been learned. Computationally modeling how learners reach verbalization of the learned content can be seen as a first step in investigating social cognition and interactive learning.

Theory of mind and productive helping

Cooperative learning settings can be seen as situations in which two or more agents collaborate, changing roles of "teacher" and "student". Sometimes also other forms of cooperative learning, e.g. collaborative discovery take place. We are investigating forms of collaborative learning in which *peer teaching* and *learning by teaching* are central themes. Our focus is on productive helping in cooperative groups, as experimentally assessed by Webb, Farivar and Mastergeorge (2002). Intuitively, people must rely on their so-called "theory of mind" in order to effectively provide help. A huge amount of developmental experimental research concluded children aged six already have a sophisticated adult-like theory of mind, enabling them to understand the actions of others in terms of underlying mental states and to distinguish between their mental states and those of others. However, Keysar, Lin and Barr (2003) argue that even adults sometimes forget to account for deviations in others' mental states while performing actions, for instance in speech interactions. It is unclear what causes these egocentric mistakes. We assume they are related to cognitive load and the dual-task of executing a task and reflecting on oneself and others. The more the task at hand is automated, the more cognitive resources are available for reflection on what others might or might not know. Whether people can learn to prevent this type of failures in general remains an open question.

Social interaction with models

To investigate how helping can be productive for learning we are going to develop cognitive models of help-givers (teachers or peer teachers) and help-seekers (students) in learning situations. We aim at direct interaction between humans and cognitive models. Playing against a cognitive model in a game setting has proved to be feasible already. In a similar vein we hope to develop cognitive models with which humans can actually interact in cooperative learning settings, dynamically changing the roles of teachers and learners. These models will be a valuable extension of current research in the field of intelligent tutoring systems, in which cognitive models are used as learner models but not as teacher models. Besides that, this type of research will have implications for social cognition research and research on embodied communication. Above all that, it can enrich research endeavors to display and explain what our cognitive models are actually doing, by letting the models explain it themselves!

References

Hoek, D., van den Eeden, P., & Terwel, J. (1999). The effects of integrated social and cognitive strategy instruction on the mathematics achievements in secondary education. *Learning and Instruction, 9*, 427-448.

Karmiloff-Smith, A. (1992). *Beyond modularity: A developmental perspective on cognitive science*. Cambridge, MA: MIT / Bradford Books.

Keysar, B., Lin, S., & Barr, D.J. (2003). Limits on theory of mind use in adults. *Cognition, 89*, 25-41.

Paas, F., Renkl, A., & Sweller, J. (1998). Cognitive load theory and instructional design: Recent developments. *Educational Psychologist, 38*(1), 1-4.

Webb, N.M., Farivar, S.H., & Mastergeorge, A.M. (2002). Productive helping in cooperative groups. *Theory into Practice, wntr.*

Zondervan, K., & Taatgen, N.A. (2003). The deteminers model: A cognitive model of macro development and U-shaped learning in a micro domain. *Proceedings of the 5th international conference on cognitive modelling* (pp. 230). Bamberg: Universitätsverlag Bamberg.

Author Index